CLINICAL MICROBIOLOGY AND INFECTION CONTROL

Clinical Microbiology and Infection Control

ELAINE LARSON, R.N., PH.D., F.A.A.N.
Robert Wood Johnson Clinical Nurse Scholar
University of Pennsylvania
Philadelphia, Pennsylvania

BLACKWELL SCIENTIFIC PUBLICATIONS, INC.
Boston Oxford London Edinburgh Melbourne

Library of Congress Cataloging in Publication Data

Larson, Elaine.
Clinical microbiology and infection control.

Includes index.
1. Medical microbiology. I. Title: [DNLM:
1. Microbiology. 2. Bacterial infections—Microbiology.
3. Microbiological technics. QW 4 L334c]
QR46.L27 1984 616'.01 83–15894
ISBN 0–86542–011–4

88 87 86 85 5 4 3 2 1

Blackwell Scientific Publications

Editorial Offices:
52 Beacon Street, Boston, Massachusetts 02108, USA
Osney Mead, Oxford OX2 0EL, England
8 John Street, London, WC1N 2ES, England
9 Forrest Road, Edinburgh, EH 1 2QH, Scotland
99 Barry Street, Carlton, Victoria 3053, Australia

Distributors:
USA
Blackwell Mosby Book Distributors
11830 Westline Industrial Drive
St. Louis, Missouri 63146

Canada
Blackwell Mosby Book Distributors
120 Melford Drive
Scarborough, Ontario, M1B 2X4

Australia
Blackwell Scientific Book Distributors, Pty., Ltd.
31 Advantage Road
Highett, Victoria 3190

Outside North America and Australia
Blackwell Scientific Publications Ltd.
Osney Mead
Oxford OX2 OEL
England

CONTENTS

II. Specific Microorganisms, *Continued*

III. Clinical Considerations, *Continued*

PREFACE

In clinical practice, one encounters daily the results of interactions between the patient and the microbial world. These interactions, however, are often unanticipated and unrecognized. Even when the patient exhibits classical signs and symptoms that we associate with infection and disease (fever, inflammatory response, elevated white blood count), we have only a superficial knowledge of the forces at work in the patient's body.

This textbook is designed for health care practitioners, such as nurses and infection control practitioners, as a basic general reference with clinical applications from the field of microbiology. It is not a laboratory manual; diagnostic tests are described from the viewpoint of the clinician who will be obtaining specimens and using results as a part of a problem-solving approach to anticipate patient needs and plan appropriate interventions. Neither is this text designed for the microbial physiologist. Detailed information regarding microbial genetics and physiology are not included. Because parasites are relatively rare in developed countries, parasitology is discussed only superficially, and information on that subject should be sought in more definitive texts.

As you use this text, our goals for you are that you will be able to

1. Describe life cycles, including cell structure, genetics, metabolic and nutritional needs of microorganisms important in human disease;
2. Identify the complex relationship between host and microorganism in the infection process;
3. Differentiate characteristics and situations in which specific microorganisms become part of the body's normal inhabitants or become human pathogens;
4. Enhance in your clinical practice the patient's host defenses against microinvaders;
5. Recognize clinical signs and symptoms of infection, and the patient's physiologic responses to disease;
6. Participate in the appropriate diagnosis and treatment of infectious diseases.

Because we wish the text to be as inclusive as possible, bacteria and viruses that may be uncommon or obscure are included when they are known to have caused human disease. To assist the reader in sorting out the more important and prevalent microorganisms, we have listed them in the *Key Points* of each chapter.

As a rule, characteristics common to a group of organisms (e.g. the gram-positive bacilli) are discussed in a general section. The more important members of each group are discussed individually, using the following format:

Biologic characteristics
Epidemiology
Pathogenesis
Host responses
Control and treatment

Other organisms of lesser importance are discussed in paragraph form. Each chapter ends with "Clinical Implications," information that is not included in the discussion of individual organisms.

CONTRIBUTORS

Katherine Talmage Alkire, R.N., M.N.
Instructor, Intercollegiate Center for Nursing Education
Spokane, Washington

Terry F. Cicero, R.N., M.N., C.C.R.N.
Lecturer, Department of Physiologic Nursing
University of Washington
Seattle, Washington

Anna Lee DeHaven, M.S., Ph.D.
Professor, Advanced Nursing Science
College of Nursing
University of Delaware
Newark, Delaware

Annette J. De Vito, R.N., M.N.
Pulmonary Clinical Nurse Specialist
Department of Nursing
Northwest Hospital
Seattle, Washington

John Steven Miller, R.N., M.N., C.E.N.
Nursing Instructor
Yakima Valley Community College
Yakima, Washington

Gertrud A. Schmidt, B.S., M.T. (A.S.C.P.)
Laboratory Supervisor, Clinical Microbiology
University Hospital
Seattle, Washington

Ortwin W. Schmidt, M.S., M.T. (A.S.C.P.), Ph.D.
Assistant Laboratory Director
School of Medicine
University of Washington
Seattle, Washington

Carol Y. Sun, R.N., B.S.N., M.N.
Oncology Clinical Nurse Specialist
Oncology Service
Riverton Hospital
Seattle, Washington

Flora Mae Wellings, R.N., M.S., Sc.D.
Director, Epidemiology Research Center
State of Florida
Department of Health and Rehabilitative Services
Tampa, Florida

I BASIC PRINCIPLES

1 CLASSIFICATION OF MICROORGANISMS

KEY POINTS

- There are three major biologic kingdoms: animals, plants, and protists (microorganisms).
- Because of their unique characteristics, viruses are not classified in any kingdom.
- Microorganisms, like plants and animals, are classified according to genetic relatedness and distinguishing characteristics into phyla, classes, orders, families, tribes, genera, and species.
- Taxonomic divisions are essential to organize knowledge about related organisms.
- The classification of microorganisms is constantly changing as new organisms are discovered or as additional distinguishing biochemical or morphologic characteristics of organisms are recognized.

This book is concerned with the study of one of the three biologic kingdoms of the world: the Protista. The other two kingdoms, plants and animals, share characteristics with the microbial world and until the late nineteenth century microorganisms were classified either as plants or as animals. Algae, for example, photosynthesize; fungal and bacterial cells divide by transverse fission, as do members of the plant kingdom. Protozoa, on the other hand, because they move actively and are heterotrophic, might be considered unicellular animals. Protists, however, can be distinguished by their one-celled structure, as compared with plants and animals that consist of multiple highly differentiated cells. The Protista are considered to be descendants of the primitive ancestors that gave rise to plants and animals.

Protists, because of their small size, are the most recently discovered biologic beings. In the nineteenth century, fungi and protozoa were the first microorganisms recognized as agents of disease. In the mid 1800s, Koch discovered the tubercle bacillus; in 1876, he produced unequivocal proof that a bacterium was the etiologic agent of a disease (anthrax). The first virus known to infect plants, tobacco mosaic virus, was discovered in 1892. Protists are classified into two groups, based upon the complexity of their cell structure. The more advanced higher protists are eukaryotic, containing a true nucleus bounded by a membrane, whereas the lower protists are prokaryotic, containing a single chromosome without a nuclear membrane. Viruses are classified separately because of their unique structure and mode of replication.

TAXONOMY

The taxonomic classification of all biologic beings makes it possible for scientists to compare the differences and similarities of the diverse organisms they study, and to organize and communicate data in a meaningful way. The systematic taxonomic scheme is largely attributable to the eighteenth century Swedish naturalist, Linnaeus. The classification divisions of all living things are summarized in Table 1-1.

The classification of microorganisms is not always easy, however. A species is usually defined as a group of organisms whose members can interbreed. However, because most

TABLE 1-1. Taxonomic Divisions of Living Organisms (examples in parentheses)

Kingdom (Protista)
Phylum or division (Protophyta)
Class (Schizomycetes)
Order (Eubacterioles)
Family (Enterobacteriaceae)
Tribe (Escherichieae)
Genus *(Klebsiella)*
Species *(pneumoniae)*

microorganisms can reproduce asexually, it is not possible to divide them into species in the classic manner. Rather, microorganisms are classified either on the basis of specific common characteristics (the "key" method of taxonomy) or on the basis of genetic similarity and common ancestry. Examples of "keys" might be capsules that surround certain organisms, the ability of an organism to accept a chemical stain, the reliance of an organism on oxygen for respiration, and the toxins released by an organism (see Table 1-2). The obvious problem with this classification approach, despite its practical usefulness, is that it is not mutually exclusive; one organism can be "keyed" multiple times.

The attempt to group organisms that share evolutionary ties is hindered by a lack of data. The most successful phylogenetic attempts, and the most extensive reference for the classification of bacteria, is *Bergey's Manual of Determinative Bacteriology,* first published in 1923 and now in its eighth edition (1974). This manual is used throughout the world and contains a complete list of known bacterial species. The latest edition has diverged from the classical orders and has rearranged bacteria into 19 groups according to morphologic appearance and other

TABLE 1-2. Common Keys by Which Microorganisms are Classified

Key	*Classifications*	*Organisms classified*
Nutritional requirements	Autotrophs Heterotrophs a. Parasites b. Saprophytes	All (higher protists, lower protists, viruses)
Gram stain	Gram-positive Gram-negative	Most bacteria (excluding spirochetes, mycobacteria, *Mycoplasma*)
Shape	Varied, though specific to genus Rod, Coccus, Spiral — Bacteria Helical, Icosohedral — Virus	All
Need for oxygen	Aerobic Anaerobic (facultative, obligate)	Bacteria
Method of replication	Sexual Asexual a. Binary fission b. Replication and reassembly in host cell	All
Motility	Motile Nonmotile	All
Relative size		All

descriptive variables such as Gram stain reaction, spore formation, and oxygen requirements.

Several other classification schemes are also in current use. Computers can now match organisms in terms of numerous characteristics. DNA hybridization is another classification technique with much promise. The technique compares the DNA of one organism with that of others. Five criteria are used to classify organisms into species: genome size, guanine plus cytosine (G + C) content, thermal stability of related DNA sequences, and relatedness of DNA under conditions optimal for DNA reassociation and supraoptimal for DNA reassociation. The technique can be applied to any organism, though it is not in widespread use at this time.

The names and classification of many organisms have changed and will continue to change as new species or additional subgroupings of characteristics are found; for example, a single species may become a genus comprised of several different species, or organisms may be reclassified in different families or genera. To cite two such occurrences, *Enterobacter agglomerans,* the bacteria responsible for a nationwide epidemic related to contaminated intravenous fluid in 1973, has been recently reclassified from the Erwineae to the Klebsiellae tribe. *Acinetobacter anitratus,* a normal flora of skin, has been known by a variety of different names including *Acinetobacter calcoaceticus, Achromobacter anitratus, Bacterium anitratum, Herellea vaginicola, Moraxella glucidolytica,* and *Pseudomonas calcoacetica.* (The entire genus, *Acinetobacter,* includes organisms formerly classified as *Achromobacter, Moraxella, Herellea,* and *Mimae.*)

Even within species, organisms may vary greatly, with numerous strains and biotypes identifiable. Despite confusion about and continual revision of taxonomy, the names by which we identify microorganisms in clinical practice are the genus (e.g., *Legionella*) and the species (e.g., *pneumophila*), both of which are italicized or underlined. The first letter of the genus (genera is plural) is capitalized, and that of the species is in lower case letters. In summary, organisms are classified based on overall morphologic and biochemical patterns. These patterns are determined using tests on large numbers of strains.

HIGHER PROTISTS (EUKARYOTES)

The divisions of the Kingdom Protista are summarized in Table 1-3. The cells of the higher protists are 5 to 20 times larger than bacterial cells (Fig. 1-1). Protist cells are complex and resemble those of plants and animals in that each cell contains a distinct nucleus. This nucleus has many chromosomes, which are visible only during cell division, and a mitotic apparatus to assure that the two daughter cells receive equal chromosomes during division. The nucleus is surrounded by a nuclear membrane (Fig. 1-2). The cytoplasm contains structures such as vacuoles, endoplasmic reticulum, and mitochondria. Algae and other eukaryotic photosynthetic microorganisms contain chloroplasts. The cell membrane that encloses the cell is composed of lipoprotein. Algae and fungi have thick cell walls composed of polysaccharide; protozoa do not have cell walls. Protozoa also possess cilia; one cell may have thousands. In all eukaryotes the cilia and flagella (which are a longer version of cilia) are of similar composition: two central fibers surrounded by nine outer fibers. These motility organelles differ significantly from those of prokaryotes. The

TABLE 1-3. Divisions of the Kingdom Protista

Higher protists (eukaryotes)	*Lower protists (prokaryotes)*	*Viruses*
A. **Algae**	A. **Schizophyceae**	***DNA**
Chlorophyta	Cyanrobacteria	
Euglenophyta	B. **Schizomycetes** (bacteria)	***RNA**
Chrysophyta	Myxobacteria	
Pyrrhophyta	*Spirochetes	
Phalophyta	*Eubacteria	
Rhodophyta		
Cyanophyta		
Cryptophyta		
Chloromonadophyta		
B. **Protozoa**		
*Mastigophora		
*Sarcodina		
*Sporozoa		
*Ciliophora		
C. **Fungi**		
*Phycomycetes (primitive fungi)	Yeasts may be classified in any of the three higher classes of fungi.	
*Ascomycetes		
Basidiomycetes		
Deuteromycetes		
D. **Slime molds**		

*Contains species significant in causing human infectious diseases.

5-10 micrometers*
Plant and Animal Cells

0.5-2.0 micrometers
Bacteria

0.02-0.3 micrometers (20-300 nanometers**)
Viruses

* 1 micrometer (μm, also called a micron) = 1/1000 millimeter (mm) = 1/1,000,000 meter (39 inches)
** 1 nanometer (nm) = 1/1000 micrometer

FIGURE 1-1. Relative diameter of plant and animal (eukaryotic) cells, bacterial (prokaryotic) cells, and viruses.

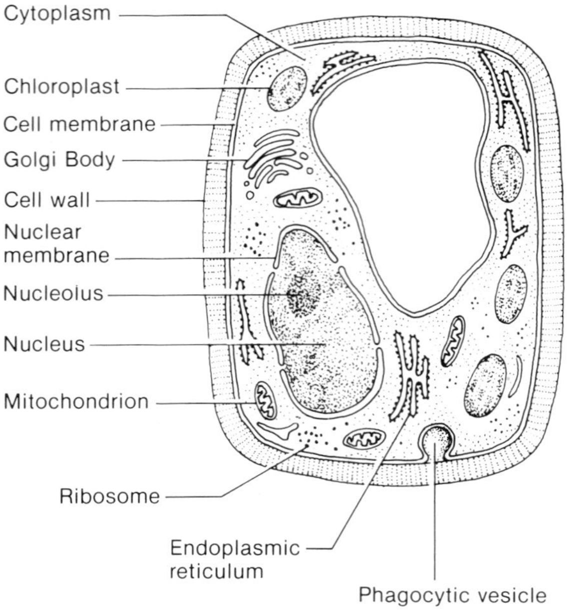

FIGURE 1-2. Eukaryotic plant cell.

flagella of all eukaryotes are monotrichous (attached to one end of the cell) and produce a whiplike movement. Prokaryotic and eukaryotic cells are compared in Table 1-4.

Algae and Slime Molds

Algae contain chlorophyll, are free living, self-nourishing (autotrophic), and are of little significance in human disease. Slime molds are unique in that they possess a loose mass of creeping cytoplasm called a plasmodium. The plasmodium can flow in all directions, may sometimes even be seen with the naked eye, and produces swarm spores or nonflagellated "mysamoebae." Slime molds are also unimportant in human disease.

Protozoa

Protozoa are nonphotosynthetic higher protists, many being motile by means of flagella or pseudopods (false feet). The four classes are Mastigophora, Sarcodina, Sporozoa, and Ciliophora, all of which contain species potentially harmful to humans. Protozoa are discussed in more detail in Chapter 23.

Mastigophora. These protozoa are motile, owing to one or more polar whiplike flagella or an undulating membrane (as in trypanosomes). They are usually elongated or oval. They may possess a mouth section or form pseudopods with which to engulf food. Some types are ingested with food and reside in the intestinal tract, others if inoculated (such as with the bite of an infected insect) may cause diseases such as sleeping sickness. Reproduction is usually asexual. Pathogens in this class include *Giardia, Trypanosoma, Trichomonas,* and *Leishmania.*

Sarcodina. These include amoeba and amoebalike organisms that move by temporary extension of parts of the cell to form pseudopods, into which cytoplasm streams. Most species are free living in water and soil, and are harmless to humans. Many species may inhabit the intestinal tract; only a few, however, cause disease (the classic example is *Entamoeba histolytica* which is a cause of dysentery).

Sporozoa. These are usually nonmotile throughout their life cycle and may reproduce in two ways: asexually in humans or other vertebrate hosts and sexually in an insect vector. Their life cycle is complex, and they can survive only in the tissues of higher

TABLE 1-4. A Comparison of Eukaryotic and Prokarytic Cells

Feature	*Eukaryotic*	*Prokaryotic*
Structure	Complex, more advanced	Simple, more primitive (except for cell wall)
Nucleus	True nucleus with nuclear membrane, many chromosomes, mitotic apparatus	"Nucleus" is a single chromosome with no nuclear membrane or mitotic apparatus
Cytoplasmic structures	Endoplasmic reticulum, digestive vacuoles, mitochondria, chloroplasts (in algae); larger ribosomes	No mitochondria, chloroplasts, or vacuoles; smaller ribosomes
Cell membrane	Lipoprotein, contain sterols	Phospholipids and proteins, only *Mycoplasma* contain sterols
Cell wall	Not present in animal cells; composed of polysaccharide in microorganisms or silica in diatoms; much thicker than in prokaryotes when present	Complex cell wall consisting of peptidoglycan, which provides high tensile strength; gram-positive organisms contain teichoic acids and polysaccharides, gram-negative bacteria contain lipoprotein and lipopolysaccharide
Structures of motility	Some may have flagella (long) or cilia (short) consisting of 2 central fibrils surrounded by 9 fibrils	Flagella may be monotrichous (single polar), lophotrichous (tuft of polar flagella), or peritrichous (over the entire cells); gram-negative bacteria have finer organelles called pili or fimbriae
Metabolic mechanisms	Glycolysis	A wide variety, including the ability to fix nitrogen

organisms. The class gets its name from the sporozoites, the cells it forms in sexual division. The agents of malaria and toxoplasmosis are included in the sporozoa.

Ciliophora. Named, of course, from the cilia that cover their outer membranes, Ciliophora are complex protozoa possessing a mouth and primitive intestine. In the vegetative phase they contain two nuclei. Reproduction may be sexual by conjugation or asexual. This class contains only one human parasite, *Balantidium coli,* which causes dysentery in rare instances.

Fungi

Fungi reproduce both sexually and asexually, and are classified by their reproductive methods. Those fungi for which a sexual cycle has not been observed are termed "imperfect." (Fungi in class Deuteromycetes are imperfect.) If both a sexual and an asexual cycle have been observed, they are called "perfect." These include the other three classes of fungi. The four classes of fungi and the characteristics of each are summarized in Table 1-5.

Some fungi grow in the form of yeasts, being unicellular with colonies morphologically similar to bacteria, whereas other fungi grow as molds, whose distinctive appearance is due to the formation of tubular branching filaments called hyphae (Fig. 1-3). The multicellular hyphae multiply, branch, and become colonies called mycelia. The aerial or reproductive mycelium is that portion of the mold colony that grows above the surface of the substrate. The portion that grows into the substrate is called the vegetative mycelium. Many fungi may grow in either the yeast or mold form, depending upon environmental conditions.

Asexual reproduction occurs by means of spores, either sporangiospores in the phycomycetes or conidia in the other three classes. These spores must not be confused

TABLE 1-5. Classification of Fungi

Class	*Characteristics*	*Examples*	*Role in human disease*
Phycomycetes (Zygomycetes)	Primitive; sexual fusion of cells produces a spore called a zygote; asexual spores (sporangiospores) are released from a sac called a sporangium	*Rhizopus nigricans* (bread mold), *Mucor, Basidiobolus*	Only in the debilitated; fairly rare opportunistic pathogens
Ascomycetes	Largest class of fungi, including yeast used in baking and brewing; asexual reproduction by budding cells called blastospores; sexual ascospores are contained in a sac (ascus)	*Trichophyton, Blastomyces, Histoplasma, Aspergillus,* some *Candida*	Several disease-producing agents
Basidiomycetes	Sexual spore (basidiospore) is contained in a club-shaped basidium; asexual spores are on tips of hyphae	Common mushroom, *Cryptococcus*	*Cryptococcus* is only human pathogen; some mushrooms are edible, others are poisonous
Deuteromycetes (Fungi imperfecti)	Sexual cycle has not been observed ("imperfect fungi")	Imperfect forms of *Candida, Cryptococcus, Coccidioides, Sporothrix, Epidermophyton*	Localized skin infections (ring worm, "athlete's foot") and serious systemic infections

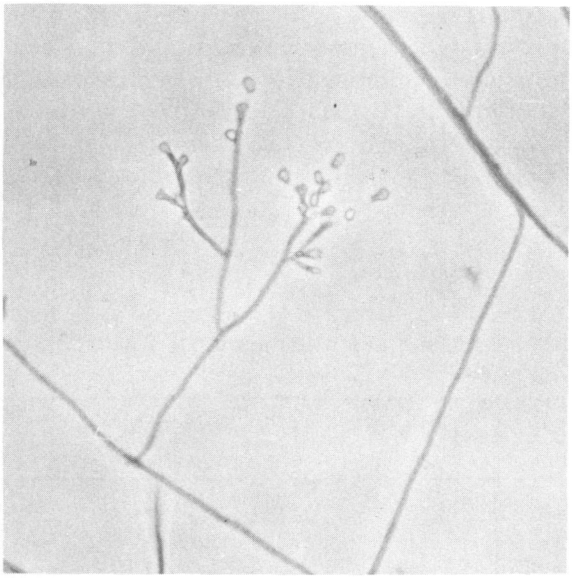

FIGURE 1-3. Filamentous form (hyphae) of fungus (*Oidiodendren,* magnification 40X). (Courtesy of Dr. James Plorde.)

with sexual spores which are also present in some fungi. Some fungi may produce several types of asexual spores.

LOWER PROTISTS (PROKARYOTES)

Lower protists have a different cell structure than the higher protists (Fig. 1-4). Bacteria also have a particular system of genetic transfer, which is discussed in the next chapter. In general, prokaryotic cells are smaller and more primitive than eukaryotic cells. They range in size from 0.3 micrometers (μm) (*Rickettsia* and *Chlamydia*) to 2.0 μm—about 2 to 100 times larger than viruses. The prokaryotic cell does not contain a true nucleus, but rather a single coiled chromosome. The nuclear area lacks a mitotic apparatus and has no nuclear membrane. The cytoplasm is also much simpler

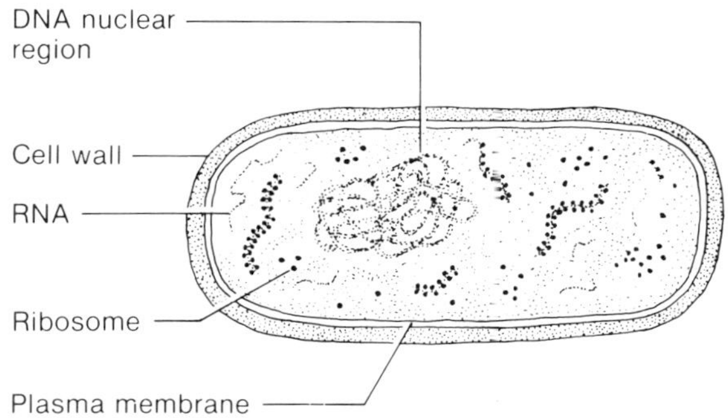

FIGURE 1-4. Prokaryotic cell *(Escherichia coli).*

than in the eukaryotic cell, containing no mitochondria or chloroplasts. In the prokaryotic cell, the cell membrane serves the same functions as those in the eukaryotic cell and is similar except for its absence of sterols. Only *Mycoplasma* among the lower protists contain steroids in the cell membranes. The cell wall is the only structure of the prokaryotic cell that is more complex and highly organized than that of the eukaryotic cell. The construction of bacterial cell walls is discussed below

Bacteria (Schizomycetes)

The cell wall of bacteria is unique and serves several purposes: it prevents the cell from rupturing by providing osmotic protection, it is necessary for cell division, and in gram-negative bacteria it is the source of endotoxin. A peptidoglycan layer in all bacterial cell walls provides strength, so that the walls of most bacteria are rigid.* A bacterium's

*There are exceptions to this. A small and medically unimportant group of bacteria known as Archaebacteria possess a cell wall that does not contain peptidoglycan. Their cell membrane composition and nucleic acids are also very different from other bacteria.

response to the Gram stain (Fig. 1-5) is determined by components of its cell wall, which are listed for gram-positive and gram-negative bacteria in Table 1-6. The classification of bacteria of medical importance according to Bergey's eighth edition is shown in Table 1-7.

Cyanobacteria (Schizophyceae)

Cyanobacteria (formerly called blue-green algae) are intermediate in size between eukaryotic algae and prokaryotic bacteria. They are classified as lower protists because of their prokaryotic cell construction, such as the absence of chloroplasts as cytoplasmic inclusions. These organisms contain their photosynthetic apparatus under the cell membrane. Cyanobacteria have characteristic amoeboid movement and are unimportant in human disease.

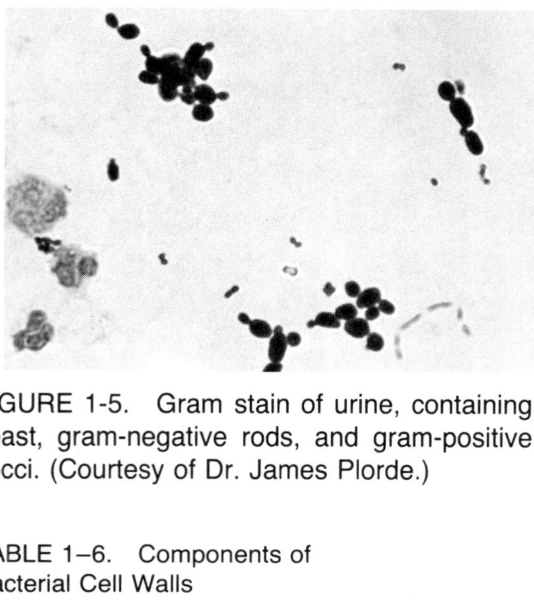

FIGURE 1-5. Gram stain of urine, containing yeast, gram-negative rods, and gram-positive cocci. (Courtesy of Dr. James Plorde.)

TABLE 1–6. Components of Bacterial Cell Walls

Gram-positive	*Gram-negative*
Peptidoglycan	Peptidoglycan
Teichoic acids	Lipoprotein
Polysaccharides	Outer membrane (phospholipid bilayer)
	Lipopolysaccharide

TABLE 1–7. Selected Major Groups of Prokaryotes of Medical Importance (Based on Bergey's Eighth Edition)

Group	*Sample families or genera*
Spirochetes	*Treponema, Borrelia, Leptospira*
Spiral and curved bacteria	*Campylobacter*
Gram-negative aerobic rods and cocci	*Pseudomonas, Brucella, Bordetella, Francisella*
Gram-negative facultatively anaerobic rods	Enterobacteriaceae, *Vibrio*
Gram-negative anaerobic bacteria	*Bacteroides*
Gram-negative cocci and coccobacilli	*Neisseria*
Gram-negative anaerobic cocci	*Veillonella*
Gram-positive cocci	*Staphylococcus, Streptococcus, Peptococcus*
Endospore-forming rods and cocci	*Bacillus, Clostridium*
Gram-positive asporogenous rod-shaped bacteria	*Lactobacillus, Listeria, Erysipelothrix*
Actinomycetes and related bacteria	Coryneform types, *Propionibacterium, Mycobacterium, Nocardia, Actinomyces, Streptomyces*
Rickettsia	*Rickettsia, Chlamydia*
Mycoplasma	*Mycoplasma*

TABLE 1-8. Comparison of Bacteria and Viruses

Feature	*Bacteria*	*Viruses*
Size	0.3–2 micrometers	0.02–0.3 micrometers
Mode of replication	Cellular enlargement and division	Synthesis of viral components by host cell, reassembly of new viruses in host cell
Habitat	May live in or outside of other cells	Obligate intracellular parasites
Major components	Nucleus, cytoplasmic structures, cell membrane, cell wall	Nucleic acid core, capsid, capsomere, envelope
Host range	Plants and animals	Plants, animals, unicellular organisms such as bacteria or algae
Energy production	Oxidation-reduction reactions, fermentation	Host cell provides energy for replication, extreme metabolic simplicity

VIRUSES

Viruses, because of their entirely different structure and reproductive cycle, are not classified as either prokaryotes or eukaryotes. Bacteria and viruses are compared in Table 1-8. Viruses are divided into families, genera, and species on the basis of morphology, mode of replication, size, and type of nucleic acid (see Table 1-9). As expected, virus taxonomy has undergone several major revisions in the past decade. Davis et al (1980), have suggested a clinical classification scheme in which viruses are grouped by their pathogenic characteristics in humans (see Table 1-10). This clinical classification scheme has been used to discuss viruses in Chapters 14–17 of this text.

In the next two chapters we will describe the characteristics of bacteria and viruses as background for understanding their role in human disease.

CLINICAL IMPLICATIONS

For many of us, our first experience with the microbial world was seeing the "little swimming bugs" under a low-power light microscope in school or at a museum. (Leeuwenhoek, the "Father of Modern Microscopy," in the seventeenth century called these organisms "animalcules") For the first time we became aware that we shared the world with millions of unseen creatures living their own parallel existence. We have come quite a distance from our understanding of microorganisms as "little swimming bugs" to our present classification systems, and we have a great deal more to learn.

TABLE 1–9. Criteria Used to Classify Viruses

Type of nucleic acid present	DNA or RNA
Form of nucleic acid present	Double-stranded or single-stranded
Size	Range: 18–300 nm
Shape	Icosahedral, helical, complex, or uncertain
Site of capsid assembly	Cytoplasm or nucleus
Virion cover	Naked or enveloped
Reaction to lipid solvents	Resistant or sensitive

TABLE 1-10. Grouping of Viruses by Pathogenic Characteristics in Humans

Classification by major target organs	*Specific virus*	*Portal of entry*	*Other affected organs*
Respiratory viruses	Influenza A, B, and C	Respiratory tract	
	Parainfluenza	Respiratory tract	
	Respiratory syncytial	Respiratory tract	
	Measles	Respiratory tract	Brain, skin, lung
	Mumps	Respiratory tract	CNS, testes, ovaries, pancreas
	Adenoviruses	Respiratory tract	
	Rhinoviruses	Respiratory tract	
	Coxsackieviruses (some)	Respiratory tract	CNS*
	Echoviruses (some)	Respiratory tract	CNS,* skin
	Reoviruses	Respiratory tract	?
	Lymphocytic choriomeningitis	Respiratory tract	CNS
	Coronavirus	Respiratory tract	
Enteric viruses	Polioviruses	Gastrointestinal tract	Muscles, CNS*
	Coxsackieviruses	Gastrointestinal tract	CNS,* skin
	Echoviruses	Gastrointestinal tract	CNS,* skin
	Rotavirus (reovirus)	Gastrointestinal tract	
	Hepatitis A and B; C (?)	Gastrointestinal tract and blood	Liver*
Neurotropic viruses	Polioviruses	Gastrointestinal tract	Upper respiratory tract
	Coxsackieviruses	Gastrointestinal tract	Upper respiratory tract
	Echoviruses	Gastrointestinal tract	Upper respiratory tract
	Rabies	Skin and blood	
	Mumps	Respiratory tract	Testes, ovaries, pancreas
	Arboviruses	Blood	
	Herpes simplex	Respiratory tract; genitalia	Skin
	Virus B	Respiratory tract; blood	Respiratory tract
	Varicella-zoster	Respiratory tract	Skin, cornea
	Lymphocytic choriomeningitis	Respiratory tract	Respiratory tract
Dermotropic viruses	Poxviruses	Respiratory tract	Respiratory tract, viscera, CNS
	Measles	Respiratory tract	Lung, brain
	Varicella-zoster	Respiratory tract	CNS, cornea
	Coxsackieviruses	Gastrointestinal tract	Upper respiratory tract
	Echoviruses	Gastrointestinal tract	CNS, gastrointestinal tract
	Herpes simplex	Skin	Peripheral ganglia, cornea, genitalia

TABLE 1-10. *(continued)*

Classification by major target organs	*Specific virus*	*Portal of entry*	*Other affected organs*
	Rubella	Respiratory tract	Respiratory tract
	Human wart (papillomaviruses)	Skin	
	Molluscum contagiosum	Skin	

*Major involvement in clinical disease.

Reprinted by permission of the publisher from Davis BD, et al. Microbiology, 3rd edition. Hagerstown, Md: Harper and Row, 1980.

Because of changes and improvements in our methods of classifying microorganisms, the clinician must be aware of the fact that names and methods of identifying species of clinical importance may change over time. In addition, one cannot assume that several isolates of the same species of organism, even if from the same patient, are indeed the same strain. Such considerations are important during epidemiologic investigations of cross-infection or outbreaks.

The newer methods of identification such as DNA hybridization will provide means to make more sophisticated differentiation of organisms. These techniques are not in general use at this time, however. Our understanding of the classes and types of viruses is still in its early stages, primarily because viruses are the most recent microorganisms to be discovered and because most are very difficult to cultivate.

SUGGESTED READINGS

Alexopoulous CJ, Mims CW. Introductory Mycology, 3rd edition. New York: John Wiley and Sons, 1979.

Brenner DJ. Characterization and clinical identification of *Enterobacteriaceae* by DNA hybridization. Prog Clin Pathol 1978;7:71-117.

Buchanan RE, Gibbons NE, eds Bergey's Manual of Determinative Bacteriology, 8th edition. Baltimore: The Williams and Wilkins Co, 1974.

Davis BD, et al. Microbiology, 3rd edition. Hagerstown, Maryland: Harper and Row, 1980.

Edwards PR, Ewing WH. Identification of *Enterobacteriaceae,* 3rd edition. Minneapolis: Burgess Publishing Co, 1972.

Mathews REF. Classification and nomenclature of viruses: third report of the International Committee on Taxonomy of Viruses. Intervirology 1979;12:129-296.

Prescott GW. The Algae: A review. Boston: Houghton Mifflin Company, 1968.

2 BACTERIA: STRUCTURE AND GENETICS

KEY POINTS

- All cells—animal, plant, and those of the smallest known microorganisms—are derived from the same basic genetic material (nucleic acids), and use similar mechanisms for the production of energy.
- Bacterial cells are structurally different than animal cells.
- Bacteria can transmit genetic information in several ways.
- Plasmid-mediated resistance to antimicrobial agents is a major concern in infectious disease.

Since pre-Biblical times, the concept of contagion has been recognized. Even though it was thought that pestilence and epidemics were punishments from superior beings for evil doing and causative disease agents were not recognized, people did use various means to prevent the spread of infection from one individual to another. These included the segregation of lepers and prohibitions against eating pork and certain kinds of fish.

Despite the fact that Antony van Leeuwenhoek first observed microorganisms under the microscope in 1673, it was not until the nineteenth century that the light microscope was developed to the point where some of the internal structures of microorganisms could be distinguished (Fig. 2-1). With that development, the study of bacteria and the science of microbiology was born.

BACTERIAL CELL STRUCTURE

Morphology

Bacteria range in diameter from 0.3 to 2.0 μm. They are therefore many times smaller than the smallest human cell. Bacteria are often classified according to morphology (shape) (Table 2-1). Species are also distinguished by their colonial morphology on

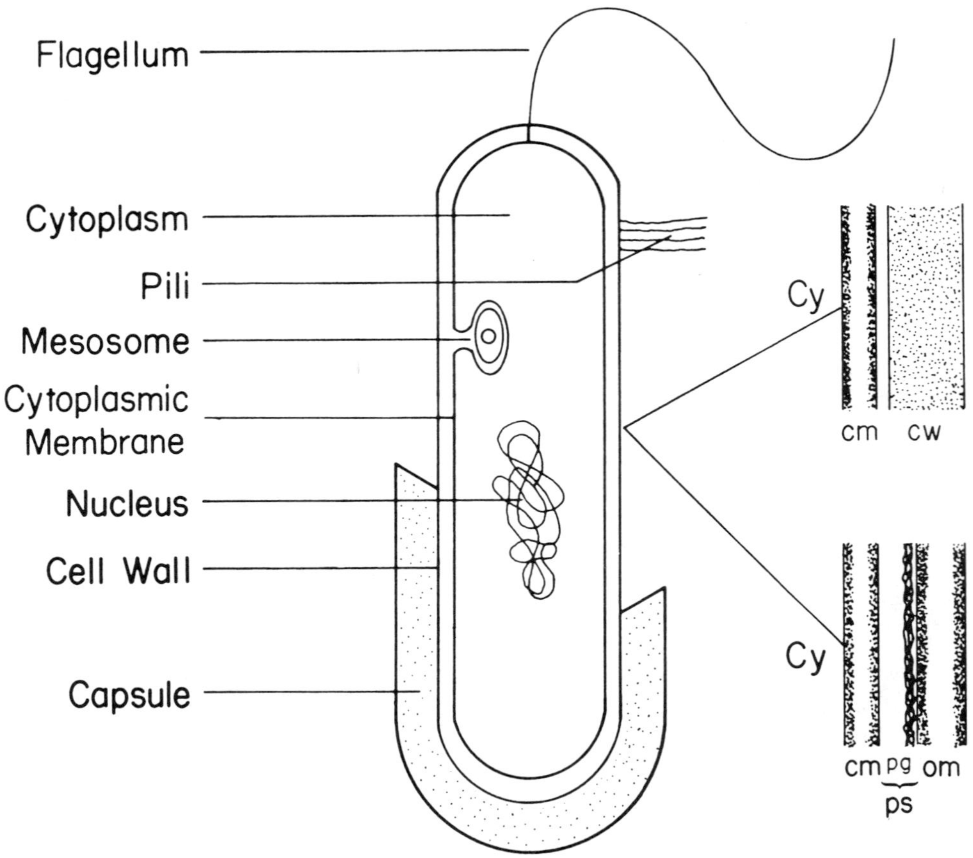

FIGURE 2-1. Diagrammatic representation of a typical bacterial cell. (Not all structures are encountered in all cells.) The insert at the upper right shows structural details of cell envelope of gram-positive cells, lower insert shows that of gram-negative cells. (Cy = cytoplasm; CM = cell membrane; CW = cell wall of gram-positive bacteria; OM = outer membrane of gram-negative cell envelope; pg = peptidoglycan; ps = periplasmic space.) (Reprinted with permission of the publisher from Freeman BA. Burrow's Textbook of Microbiology, 21st edition. Philadelphia: WB Saunders, 1979.)

solid media such as agar (Fig. 2-2). Colonies range in size from pinpoint to a swarming mass covering the agar's surface. Their form may be smooth, irregular, flat, conical, convex or, a combination of these. They may be mucoid or dry, glistening, striated, or varying in color and pigmentation.

Outer Layers

The cell envelope of a typical bacterium consists of the cell wall and the cytoplasmic membrane. In addition, some bacteria may possess flagella, if they are motile; pili (also called fimbriae), which resemble flagella

TABLE 2-1. Shapes of Bacteria

Shape	*Description*	*Example*
Bacilli	Rod-shaped, cylindrical	
Cocci	Round	
Diplococci	Round, in pairs	
Streptococci	Round, in chains	
Staphylococci	Round, in clusters like grapes	
Sarcinae	Round, in bundles	
Coccobacilli	Short bacilli, rods	
Spirilla	Curved	

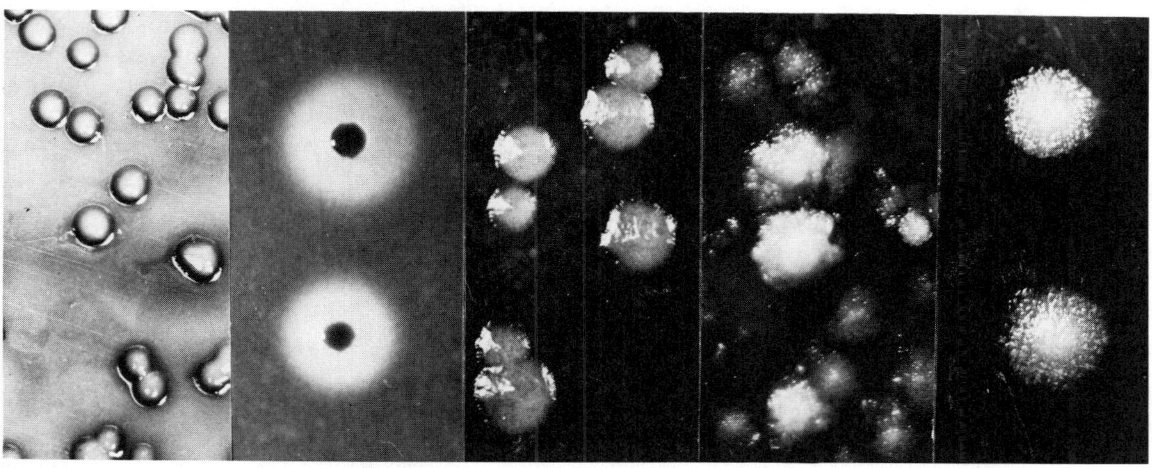

FIGURE 2-2. Morphologic types of bacterial colonies.
(*left* to *right*) Raised smooth viscous colonies of gonococci on chocolate agar; beta-hemolytic streptococci colonies on blood agar showing clear zone of hemolysis and slightly matte, slightly irregular edge of typical colony; colonies of typhoid bacillus on nutrient agar showing typical irregular edges (maple-leaf appearance); colonies of tubercle bacillus on Lowenstein's medium showing characteristic roughened appearance; colonies of anthrax bacillus on nutrient agar in typically rough, virulent form. (Reprinted with permission of the publisher from Freeman BA. Burrow's Textbook of Microbiology, 21st edition. Philadelphia: WB Saunders, 1979.)

except that they are shorter, thinner, and do not function as motility organelles; or capsules. In contrast to the internal structures of a bacterium, flagella, pili, and capsules have no biochemical function and the cell can maintain its physiologic processes without them. The simplest bacterial cell envelope is that of the *Mycoplasma*. It is merely a membrane without a peptidoglycan (mucopeptide, murein) layer. Hence, it is a much less rigid and stable cell than other bacterial types. Gram-positive bacteria, as discussed briefly in Chapter 1, have both a cell membrane and cell wall. Gram-negative organisms possess a cell membrane and a more complex cell wall which includes some unique components.

Cell Wall

The strength of the cell wall is due to its outer peptidoglycan layer. Glycan strands, which are similar in most species, are crosshatched with peptide chains to form a rigid covering that surrounds the cytoplasmic membrane. The composition of the peptide chains varies between species. Bacteria are under high osmotic pressure; this peptidoglycan lattice prevents their rupture.

Bacterial responses to the Gram stain are due, in a way not completely understood, to their cell wall components. Gram-positive cell walls constitute about 30 percent of the dry weight of the cell. They contain primarily peptidoglycan, along with various polysaccharides, proteins, and teichoic acids. Though they do not help to serve as an osmotic barrier, teichoic acids do play two important roles: they serve as major surface antigenic determinants for gram-positive bacteria, and they help to maintain proper ionic conditions in the cell by binding magnesium ions.

The gram-negative cell wall, though it is thinner than gram-positive bacterial walls, is more complex. In gram-negative bacteria, peptidoglycans form the innermost of four polymeric layers. From the outside, the layers are in the following order: lipopolysaccharide, phospholipid layer, lipoprotein, and peptidoglycan. The lipopolysaccharide (LPS) is unique to gram-negative bacteria. It is synonymous with *endotoxin,* and is highly toxic to humans. LPS is released only when the cell lyses. LPS consists of three regions: lipid A, core polysaccharide, and O-specific polysaccharides, the characteristics of which are summarized in Table 2-2. LPS is essential to the development of the gram-negative cell wall.

The outer phospholipid layer of gram-negative bacteria does not carry out active transport, but it does act as a hydrophobic barrier. This prevents large molecules

TABLE 2-2. Layers of Gram-Negative Bacterial Cell Wall Lipopolysaccharide (Endotoxin)

Layer	*Characteristics*
Lipid A	Responsible for toxicity; located in hydrophobic zone of outer membrane; linked to peptidoglycan
Core polysaccharide	Constant in all species; may result in some nonspecific resistance to infection
O-specific polysaccharide	Called O antigen, major surface antigens which induce specific immunity

like antimicrobials from entering the cell, and hydrolytic enzymes from leaving the cell. This outer membrane is partially responsible for the relative resistance of gram-negative bacteria to antimicrobial agents. The lipoprotein links the outer membrane and the peptidoglycan layer.

When the peptidoglycan layer is digested, bacterial cells can survive only in a hypertonic medium as protoplasts from gram-positive cells, which retain their outer membrane, or spheroplasts, from gram-negative cells, which retain their outer membrane along with fragments of their cell wall.

Like protoplasts and spheroplasts, *L-phase variants* or *L forms* lack a cell wall. L forms which are induced may be quite fragile and revert to their normal form if the inducing agent (often lysozyme or, in the case where clinical infection is being treated, penicillin) is removed from the environment. Unlike protoplasts and spheroplasts, however, they may occasionally grow slowly and divide in serum. Spontaneous L forms are also produced by some bacterial species and resemble *Mycoplasma* in their growth requirements. If L forms persist in clinical disease, they can cause difficult therapeutic problems because of their resistance to antimicrobial agents that act on the cell wall (see Chapter 12).

Cytoplasmic Membrane

The cytoplasmic membrane is composed of protein and phospholipids and is similar in all bacteria of clinical importance, except that only the mycoplasmal cytoplasmic membrane contains sterols. The membrane serves primarily as an osmotic barrier to exclude external molecules and ions and retain internal metabolites. The cytoplasmic membrane is an extremely important structure in bacterial cells because the cytochrome enzymes are present in the membrane rather than in cytoplasmic inclusion bodies as is the case with eukaryotic cells. These enzymes are important to many of the functions listed on Table 2-3. Despite its essential functions, the cytoplasmic membrane does not help to maintain the shape of the cell, but it is bonded to the cell wall. Small, internal organelles called *mesosomes* occur in invaginations in the cytoplasmic membrane. They probably function during cell division to initiate septal closing and may also help regulate the division of DNA between the two daughter cells (septal mesosomes). Lateral mesosomes have secretory functions.

Flagella

Flagella are very long, thin motility organelles that are difficult to see with normal staining techniques (Fig. 2-3*a–d*). By means of oxidative phosphorylation, flagella are able to move bacteria by helical rotation either clockwise (which creates a tumbling motion) or counterclockwise (for a smooth swimming motion). The flagella also serve a function in chemotaxis. Bacteria can detect concentration gradients that decrease as the cell moves toward them. A cell that is moving toward an attractant such as an

TABLE 2-3. Functions of the Bacterial Cell Membrane

1. Provides an osmotic barrier
2. Binds ribosomes in the cell
3. Participates in
   a. synthesis of proteins, lipids, cell wall components
   b. oxidative phosphorylation
   c. electron transport
   d. DNA replication
4. Facilitates cell motility and chemotaxis
5. Provides means by which a membrane potential can build up

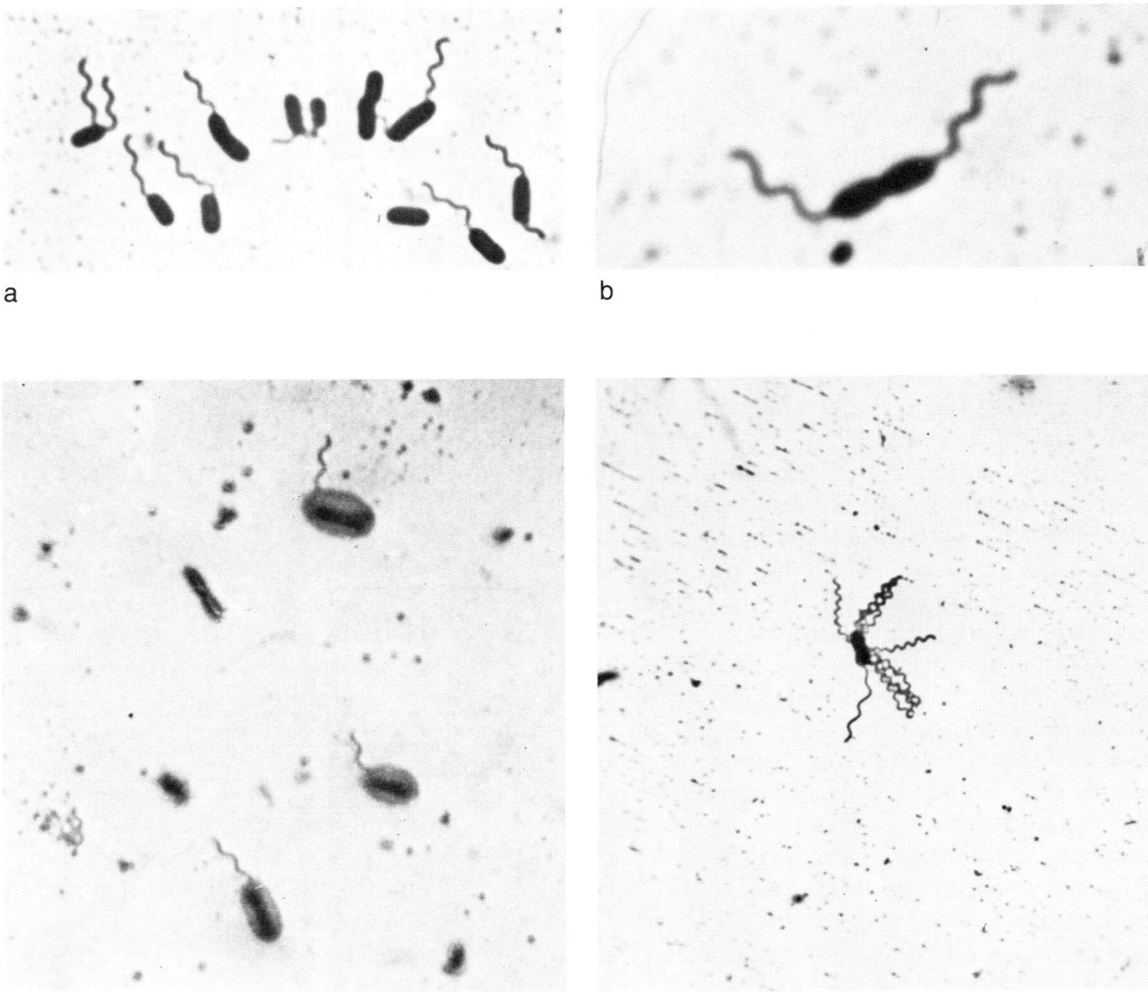

a b c d

FIGURE 2-3. *Pseudomonas aeruginosa.*
(a) Typical flagellation of a young culture.
(b) A dividing organism.
(c) Direct stain from peritoneal fluid of infected guinea pig.
(d) Curly peritrichous flagella.
(All stained with Leifson Flagella Stain, magnification 2000×.) (*a, b,* and *d* courtesy of Dr. E. Leifson. Used with permission of the publisher from Leifson E. Atlas of Bacterial Flagella. New York: Academic Press, 1960. *c* courtesy of Dr. E. Leifson.)

amino acid or a sugar moves in a more directed, less random manner.

Flagella are most common in rod-shaped bacteria. They may be unipolar (monotrichous), tufted at one end of the cell (lophotrichous), or surrounding the cell completely (peritrichous). They are of variable length, usually several times the length of the cell, and are not tapered at the end. *Proteus* species, which swarm on the surface of an agar plate and cover other organisms present, possess huge numbers of flagella. Spirochetes move by contraction of an axial filament and do not possess flagella. The structure of the flagella in various bacterial species differs. Flagella are attached to the organism by a set of rings (S and M in gram-positive bacteria, S,M,L, and P in gram-negative bacteria) and a hook. When antibody, primarily IgG, specific to the flagellar ("H") antigens is produced, cells lose their motility.

Pili

Pili are like flagella, but they are shorter, thinner, straighter, and not involved in motility (Fig. 2-4). Again, they are more common in gram-negative bacteria, and are composed predominantly of protein. They are probably essential for bacterial adherence to host surfaces, and are also necessary for conjugation. The M protein of the streptococci (discussed in Chapter 7) is an outer layer of pili and serves as the major surface antigen of streptococci.

Capsules

Some bacteria, especially in certain environments, produce a layer of gelatinous material (usually a polysaccharide) called a *slime layer* or *capsule.* This material is excreted by the cell and does not interfere with permeability. Most species that produce capsules do so only when it is advantageous; for example, in the presence of serum. Mutations are also common, with smooth, mucoid colonies being formed by encapsulated cells and rough colonies by nonencapsulated cells. Capsules are important in pathogenicity of an organism because they enable the organism to escape phagocytosis. The capsule varies from a thin layer to a layer that is thicker than the diameter of the cell itself. The capsule is antigenic ("Vi" antigens), and the Quellung reaction or capsule swelling test is a precipitation reaction between antibody and the "Vi" antigen (Fig. 2-5).

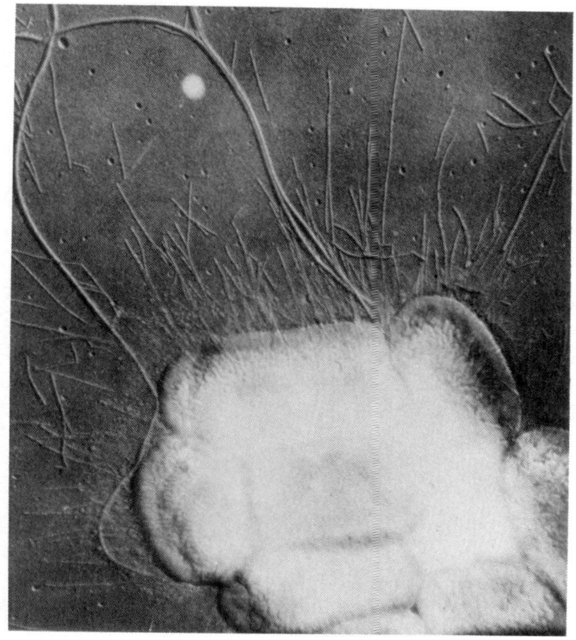

FIGURE 2-4. Pili. Piliated strain of *Escherichia coli* grown in liquid medium without aeration. Each cell possesses hundreds of pili (diameter 7 nm), and their presence promotes aggregation. Many isolated, broken pili are also seen. A few flagella, much longer and of larger diameter (14 nm), extend from the cells to the edge of the photograph. (Platinum shadowed, > 45,000×, reduced.) (Photograph by Charles C. Brinton, Jr. Reprinted with permission of the publisher from Davis BD, et al. Microbiology, 3rd edition. Hagerstown, Md: Harper and Row, 1980.)

Inner Structures

Nuclear Region

The prokaryotic "nucleus" is simply a single DNA chromosome. There is no nuclear membrane or mitotic apparatus. The chro-

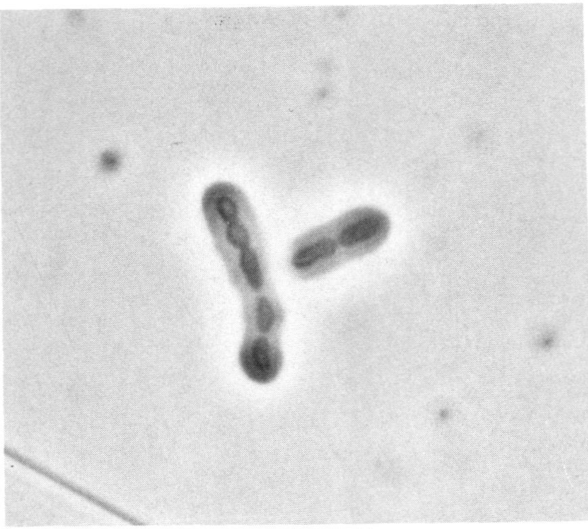

FIGURE 2-5. Encapsulated pneumococci showing a positive capsular swelling (Quellung) test. (Courtesy of Dr. James Plorde.)

mosome is about 1000 times as long as the nuclear region, and is folded and compacted into a supercoiled shape.

Cytoplasm

The cytoplasm of bacterial cells is less complex than that of eukaryotic cells. It contains spherical granules composed of polyphosphate in which are stored phosphate, polysaccharides, and sulfur. Other, more complex, membrane-bound granules contain polymers of poly-β-hydroxybutyric acid and granulose, which can be used as sources of carbon and energy. The metachromatic granules of *Corynebacterium* are stores of inorganic phosphate called volutin. Ribosomes are submicroscopic cytoplasmic inclusions that are responsible for protein synthesis. Ribosomal RNA makes up about 80 percent of the total cytoplasmic RNA.

Spores

When there are limited supplies of essential nutrients such as nitrogen, carbon, or phosphorous, certain bacteria will form resting cells called *spores*. Spores are highly resistant to destruction by heat, freezing, radiation, drying, or chemicals. The spore allows the cell to survive until once again favorable environmental conditions prevail. Organisms that form spores include members of the genera *Clostridium, Bacillus, Actinomyces,* and *Sarcina*. Only three organisms of importance in human disease are spore formers: *Clostridium tetani* and *perfringens,* and *Bacillus anthracis*. The producing mother cell is called a sporangium, and the process of spore formation is called sporulation. The spore is a spheroid body, which is released from the cell after lysis. The "drumstick" appearance of *C. tetani* is due to a spore located at the end of the cell (terminal spore), though spores in other species may be central or subterminal. Spores are metabolically inactive.

Sporulation is activated by a state of cell starvation; only one spore per cell is formed. Components of the spore are listed in Table 2-4. Large amounts of calcium dipicolinate in the spore account in part for their resistance to dessication. Germination, the process by which a spore again becomes a vegetative cell, occurs when the environment is no longer depleted of necessary cell nutrients. Before the cell can germinate in a favorable environment, dormancy must be broken by an activator such as heat, abrasion, treatment at acid pH or with oxidizing agents, or aging. Apparently, the activator works by causing damage to the spore's coat. Once activated, germination of the spore is rapid and irreversible and a new vegetative cell is produced.

MICROBIAL GENETICS

The mechanism by which the genetic code is maintained and transferred from generation to generation has only recently been

TABLE 2-4. Components of Bacterial Spores

Component	*Description*
Core or spore membrane	Spore protoplast including nucleus, protein-synthesizing apparatus
Spore wall	Contains peptidoglycan; becomes cell wall when the spore germinates
Cortex	A thick layer containing cross-linked peptidoglycan
Coat	An impervious protein coat or covering which endows resistance to chemical agents
Exosporium	A lipoprotein and carbohydrate membrane of unknown function

known. In 1944, several American scientists discovered deoxyribonucleic acid (DNA), and in 1954 the chemical structure of DNA was described by Watson and Crick. The nucleic acid DNA and also ribonucleic acid (RNA) contain the genetic information of all cells. In all prokaryotes, eukaryotes, and many viruses the genetic code is contained in DNA; in some viruses it is in RNA.

Both DNA and RNA are comprised of smaller units called *nucleotides*. Nucleotides have three components: a phosphate (PO_4) group, a sugar (deoxyribose in DNA and ribose in RNA) and one of four bases: adenine, cytosine, guanine, and in DNA, thymine, or in RNA, uracil. DNA in all species (human and bacteria, for example) is identical, but nucleotide sequences differ. Bacterial chromosomes contain about 5×10^6 nucleotide pairs. Human cells contain about 1000 times more DNA per cell than bacterial cells.

The actual information in the genetic code is determined by the sequence in which the four base units are ordered and coiled. Each nucleic acid is extremely long and thin, containing thousands of sequences. DNA is usually composed of complementary double-stranded helices (coils) with inverse polarities. The complementary strands are sometimes called "sense" and "antisense" or "nonsense" strands. The sense strand contains the information necessary for the production of amino acids. RNA usually occurs as a single strand. DNA and RNA are compared in Table 2-5.

DNA and RNA Synthesis

DNA synthesis occurs only at specific times called synthetic or S phases. Replication moves, usually in both directions, from a point of origin called a *replicating fork* or *replicator* to a terminal point. Replication is directed by a number of enzymes and begins at certain sites that are the same in related bacterial strains. In double-stranded DNA, the two strands begin to unwind and each acts as a template on which a new strand is assembled.

In double-stranded DNA, adenine (A) is always opposite thymine (T) and cytosine (C) is always opposite guanine (G). The amount of G + C as compared to the amount of A + T can be determined and is called the *G + C ratio* or *G + C content*. It is possible to use the G + C ratio to identify organisms, since all organisms of the same species have the same G + C ratio. Likewise, all organisms in the same genus have G + C ratios that are very similar.

TABLE 2-5. Differences Between DNA and RNA

	DNA	*RNA*
Sugar	2′-deoxyribose	Ribose
Bases	Adenine, guanine, cytosine, thymine	Adenine, guanine, cytosine, uracil
Strandedness	Usually double	Usually single
Relative content of nucleotides	Guanine = cytosine, adenine = thymine	Not necessarily equal content
Hydrolyzed by alkali?	No	Yes
Basic nucleic acid in	All eukaryotic and prokaryotic cells, many viruses	A few viruses

Hydrogen bonds form between adenine and thymine and between guanine and cytosine so that the strands reproduced are not identical but rather complementary. Where the template carries guanine, the new strand will carry cytosine; where the template carries adenine the new strand will carry thymine. This is called the *semiconservative mechanism of replication* since the daughter molecules have information only semiconserved (not identical) from the parent. There may be several independent units of replication, *replicons,* present and active in a bacterial cell at any given time. These may include, in addition to the chromosome, plasmids and bacteriophage particles (discussed in Chapter 3). Mechanisms of synthesis of DNA and RNA are summarized in Table 2-6.

The sequence of DNA bases directs protein synthesis, which occurs within the cell in several steps and involves a number of enzymes that perform specific functions listed in Table 2-7. Through *transcription* the genetic code of the DNA in the cell's nuclear chromosome is passed to messenger RNA (mRNA), which moves into the cell's cytoplasm to serve as the "working tape" to control and direct protein synthesis. The "master tape" of DNA never leaves the nucleus.

Specific transfer RNA (tRNA) molecules in the cell cytoplasm translate the nucleotide sequences by recognizing codons (groups of three bases) that specify amino acids. There are 64 possible combinations of three bases or codons. These code for 20 different amino acids. Codons called "nonsense" or "termination" codons, rather than specifying an amino acid, serve to start or stop protein synthesis. Ribosomal RNA (rRNA) is about one-third protein and two-thirds RNA, and serves as the protein manufacturing apparatus. The process by which the ribosome moves along the RNA forming a growing chain of polypeptides is called *translation.* (Fig. 2-6 illustrates these processes.)

Changes in inherited characteristics of bacteria due to modification in DNA nucleotide sequencing can occur naturally through mutation or through mechanisms discussed under "Inheritance in Bacteria"; such recombinations can also be induced.

Mutation

A mutation, or change in the inherited sequence of nucleotides, can only be detected if it results in an observable or measurable change in a cell's properties (a phenotypic change). Such changes might affect physiology, protein synthesis, or morphology.

TABLE 2-6. Synthesis of Nucleic Acids

DNA	RNA
1. Formation of short link of RNA (initiator) 2. DNA fragments (Okazaki pieces) attach to RNA initiator 3. Replication occurs in both strands simultaneously, directed by enzymes (polymerases) 4. Molecule unwinds (at about 400,000 turns/second) to allow rewinding of the semiconserved replica 5. Some viruses replicate DNA molecules with information from RNA genomes (retroviruses)	1. Holo RNA polymerase (core polymerase and sigma factor) molecule is bound to template of promoter site 2. Specific protein factor (sigma factor) is released and initiates formation of RNA molecule antiparallel to the template 3. A termination protein (rho factor) recognizes a signal from a specific sequence of sense strand of DNA molecule 4. After termination of synthesis, core enzyme separates from DNA template

Though mutations are rare, occurring in only 1 in every 100,000 to 10,000,000,000 cells, their prevalence may be increased by selection. For example, only methicillin-resistant mutants of *Staphylococcus aureus* will grow in a culture medium containing methicillin. This same phenomenon occurs, of course, when antimicrobial therapy is employed in clinical practice.

"Point mutations" occur if there is a change in only one base pair from the original nucleotide sequence. These "accidents" probably result from a shift of electrons in a pyrimidine or purine base, and are often reversible. DNA mutations may be deletions (loss mutations), insertions (gain mutations), or replacement of nucleotides (these are called frame shifts and account for the majority of spontaneous mutations). Nonsense mutations also occur if a nonsense codon terminates growth of a polypeptide chain before protein synthesis is complete. These, of course, are lethal if proteins required for replication or metabolism are not manufactured. Mutations can be induced by radiation, certain alkylating agents, acridine dyes, and nitrous acid.

Elements That Transfer Genetic Information

Insertion sequences (IS) are elements pres-

TABLE 2-7. Activities in Nucleic Acid Metabolism Mediated by Enzymes

Activity	Function	Examples
Polymerization	To copy a template strand of nucleic acid	DNA or RNA polymerases
Modification	To effect changes in nucleotides or in transcribed chain length	DNA ligases, transferases, methylases, restriction endonucleases
Degradation	To degrade nucleic acids	DNAases or RNAases, endonucleases

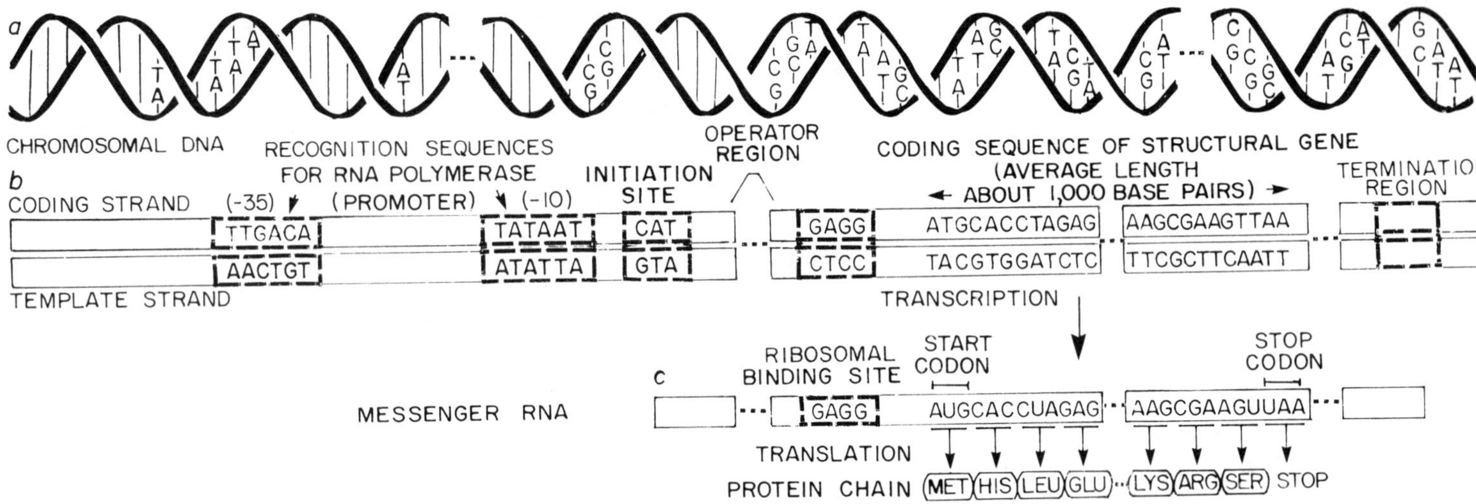

FIGURE 2-6. Genetic information is stored in the double helix of DNA. *(a)* Each strand of the helix is a chain of nucleotides, each comprising a deoxyribose sugar and a phosphate group, which form the strand's backbone, as well as one of four bases: adenine (A), guanine (G), thymine (T), and cytosine (C). The information is encoded in the sequence of the bases along a strand. The complementarity of the bases (A always pairs with T, and G with C) is the basis of the replication of DNA from generation to generation and of its expression (shown here for bacterial DNA) as protein. Expression begins with the transcription of the DNA base sequence *(b)* into a strand of messenger RNA *(c)*, which corresponds to the coding strand of the DNA except for the fact that uracil (U), replaces thymine. Transcription into RNA and translation into protein are regulated by special sequences in the DNA and RNA respectively. The transcribing enxyme, RNA polymerase, binds to a promoter region that (in the bacterium *Escherichia coli*) has the specific sequences shown (or minor variations of them) about 10 base pairs and 35 base pairs before a transcription-initiation site. Beyond the end of the structural gene a termination region causes the polymerase to cease transcription. In some genes an operator sequence, which can bind a repressor molecule, provides an extra control. Messenger RNA is translated in cellular organelles called ribosomes. Each triplet of bases (codon) encodes a particular amino acid and specifies its incorporation into the growing protein chain. A ribosomal binding site on the RNA allows translation to begin at a "start" codon, which is always AUG for the amino acid methionine *(MET)*. Translation proceeds until a "stop" codon is reached (UAA is one of three possibilities) that signals the end of translation and detachment of completed protein chain from ribosome. (Adapted, and reprinted with permission of the publisher from Hopwood DA. The genetic programming of industrial microorganisms. Scien Am 1981;245:90-102.)

TABLE 2-8. Methods of DNA Repair

Method	*Description*
Photoreactivation	Occurs in cells damaged by ultraviolet light; an enzyme activated by light combines with pyrimidine dimers, reactivating damaged cells
Excision repair (dark reactivation	Four steps mediated by enzymes: incision into damaged portion, excision from the DNA, filling in the gap, and rejoining of the free ends
Postreplication repair	Crossover exchange of undamaged segments of DNA between reconstituted daughter molecules
Error-prone repair (SOS system)	Induced by ultraviolet light, thymine starvation, some chemicals such as nitrosoguanidine; activated by proteolytic cleavage of gene repressors; can cause letal mutations in some cells and results in repair of others; a "last resort" to save cells

ent in bacterial or plasmid DNA that consist of 700 to 1400 base pairs. They can result in deletions and translocations of DNA segments as well as the integration of plasmid replicons. Certain DNA sequences called *transposons* are also capable of transporting genetic information between the cell chromosome and replicons such as plasmids.

DNA Recombination

Genetic recombination occurs naturally and serves several useful functions. Damaged or defective DNA sequences are repaired by several mechanisms, at least one of which involves recombination of undamaged segments. (See Table 2-8 for a summary of various methods of DNA repair.) Generalized recombination can occur at any site on the DNA chain. It involves a number of enzymes, particularly endonucleases and exonucleases, and usually results from the need to repair damaged DNA. Site-specific recombination occurs when genes are integrated into or exchanged in the bacterial chromosome (Fig. 2-7).

In the early 1970s a technique was developed that made it possible to artifically

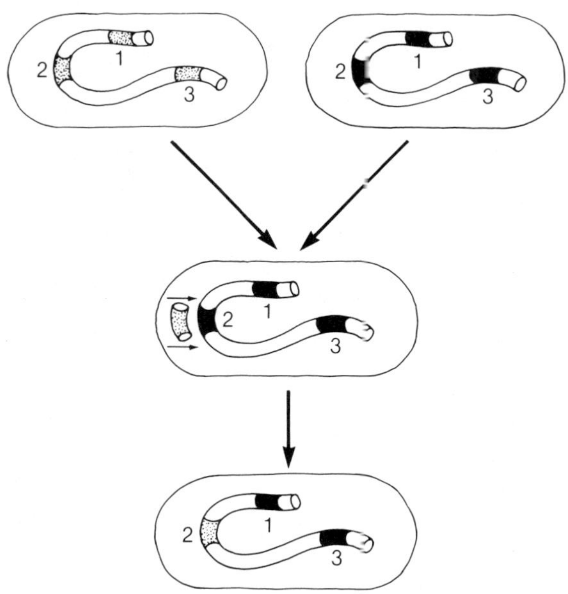

FIGURE 2-7. DNA recombination can occur by a process called crossing over in which the DNA from one bacterium pairs with a matching region of another bacterium.

splice together DNA segments and reimplant these hybrid genes in bacteria or other hosts. This ability to recombine DNA has clinical promise for genetic analysis, for evaluation of hereditary diseases, and for the use of bacteria to manufacture substances like interferons or hormones. However, the obvious potential dangers of producing organisms with toxic, unpredictable, or unknown characteristics must be considered. The federal government has made some limited attempts at regulation of such recombination experiments; some state governments have developed mandatory regulations.

Inheritance in Bacteria

The amount of time required for chromosomal replication varies; in *Escherichia coli*, for example, it takes about 45 minutes. Cell division takes place when replication of the chromosome is completed. Bacteria replicate by binary fission in which two daughter cells, both with identical genes, are produced (Fig. 2-8). They are generally of equal size, though the budding of yeasts results in one smaller daughter cell. Division starts with the cytoplasmic membrane, which invaginates and bisects the cell with a thick transverse wall. This septum eventually separates from the outside of the cell and works inward.

Bacteria have three other unique mechanisms for transferring or recombining genetic material. These are called transformation, transduction, and conjugation (see Table 2-9).

Transformation

Some gram-positive and gram-negative bacteria, called *competent* cells, are able to incorporate high molecular weight DNA from their environment. This process of transformation involves the direct uptake of soluble DNA by such cells from closely related donor cells. Transformation is highly specific because the nucleotide sequences of donor and recipient DNA must be very similar. The phenomenon has been observed in *Neisseria, Bacillus, Staphylococcus, Streptococcus, Acinetobacter,* and *Hemophilus* species. Characteristics required for competence are summarized in Table 2-10. Transformable DNA may be released spontaneously from cells or may be chemically extracted. This process can be artificially induced in some gram-negative bacteria by increasing the permeability of the cell envelope. This artificial process of transformation is called *transfection.*

Transduction

In transduction, genetic material is transferred from one bacterium to another by a virus called a temperate bacteriophage (see Chapter 3). When such a virus is replicating in its bacterial host, some bacterial DNA may be incorporated with the viral DNA. When these new virus particles are released from their bacterial host they infect new bacteria, transferring the DNA to their new host. Transduction is useful as a tool for genetic mapping of the bacterial chromosome. It is a more common phenomenon than transformation, and occurs in numerous species of organisms.

Frequently the fragment of DNA that is introduced into a recipient cell (called the *exogenote*) is neither combined nor destroyed. It persists but cannot be replicated and is transmitted from the recipient cell to only one of its two daughter cells. It is transmitted without replication in further cell divisions. This common occurrence is called *abortive transduction.*

There are two major types of successful

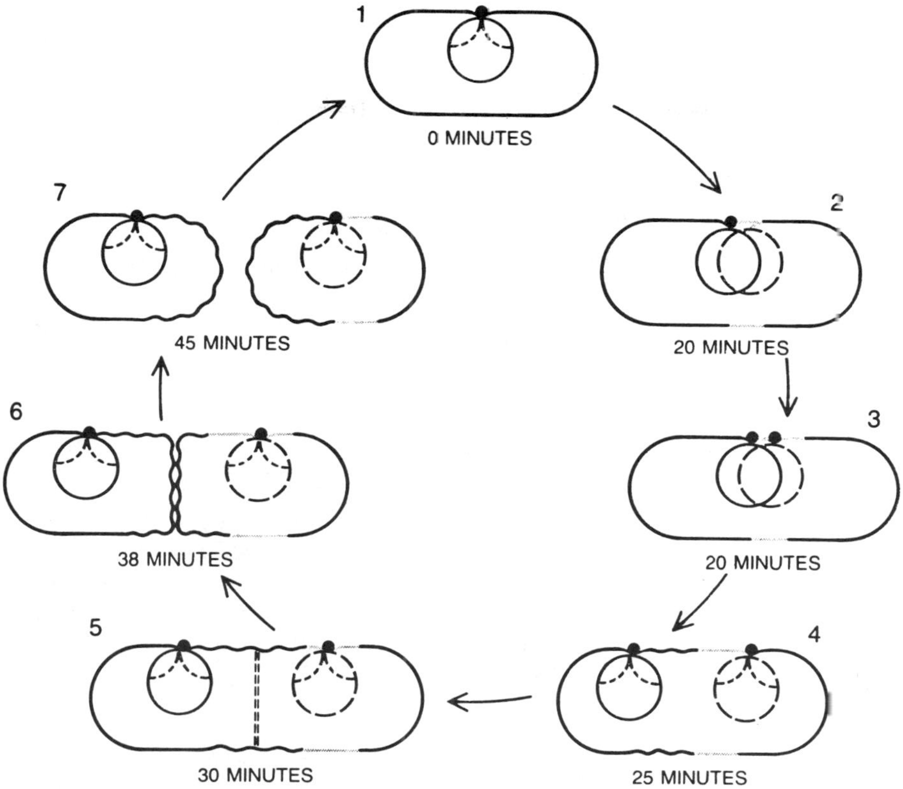

FIGURE 2-8. Reproduction of bacteria is accomplished asexually by cell division. The diagram shows seven stages in the life cycle of the colon bacterium *Escherichia coli,* which has a doubling time of 45 minutes. In the newborn cell *(1)* the single circular chromosome of DNA is already being replicated *(broken lines).* After 20 minutes the new chromosome is complete and is affixed to an attachment site within the cell *(2,3).* By 25 minutes the two chromosomes have begun to replicate *(4)* and a septum, or dividing membrane, appears in the middle of the cell *(5).* By 38 minutes the septum is a wall *(6).* Seven minutes later division is complete *(7).* (Reprinted with permission of the publisher from Phaff HF. Industrial microorganisms. Scien Am 1981;245:76-89.)

transductions: generalized and restricted. In *generalized transduction,* any portion of the chromosome may be transmitted. In *restricted transduction,* genes can be transferred from only a local region of the donor chromosome near the site of phage attachment. Penicillinase plasmids and other resistance markers in *S. aureus* have been shown to be transferred by generalized transduction.

Specialized transduction occurs when certain temperate phage strains integrate into a site on the host bacterium's chromosome. When the host bacterium dies or undergoes

TABLE 2-9. Methods of Bacterial Genetic Transfer

Method	*Description*
Binary fission	Cell divides from cytoplasmic membrane inward, producing two cloned daughter cells
Conjugation	Cells mate after contact via a conjugation bridge (pilus); genetic material is transmitted from donor (male) to recipient (female) cells
Transduction	DNA is transferred from one cell to another by a nonlethal virus (bacteriophage) that infects the cell
Transformation	Cell takes up naked DNA of closely related species, which is then replicated

lysis, some bacterial genes next to the phage chromosome may be carried along in the replicating process as in restricted transduction. Under certain circumstances, when replication is complete the resulting phage is defective and cannot multiply because a portion of its chromosome has been replaced by the bacterial gene. If a simultaneous infection of the host cell with a healthy phage (lambda, λ) occurs, it can serve as a helper, providing the missing functions of the defective phage by complementation. Thus, maturation of both phages is possible.

In a normal culture containing temperate phage, only a few phage are transducing.

TABLE 2-10. Attributes Necessary for Competence (Ability to Take Up Free DNA) Among Bacterial Cells

1. Molecular weight of DNA molecules is greater than 4×10^5
2. Takes up double-stranded DNA more readily than single-stranded
3. Donor and recipient must be closely related organisms
4. Only certain species are competent (*Neisseria, Bacillus, Streptococcus, Staphylococcus, Acinetobacter* are examples)
5. In any given bacterial population, only a small portion of cells are competent at any time

This lysate results in low-frequency transduction (LFT), at a rate of about 1 in 100,000. Phage that result from induction of transduced cells, however, yield high-frequency transduction (HFT) lysates, a rate of 1 in 10 to 1 in 100. Much genetic material is transferred via transduction by viruses that infect bacterial cells.

Conjugation

Conjugation (Fig. 2-9*a,b*) is a method of genetic transfer that requires direct contact between two bacterial strains. It is a primitive form of sexual intercourse during which the bacterial chromosome or extrachromosomal genetic elements called *plasmids* or *episomes* are transferred between a donor (male) and a recipient (female) cell. Chromosomal DNA from the donor can replace DNA in a like segment of the strand of the recipient DNA. Plasmids remain in the cell cytoplasm as separate pieces of DNA. Episomes may become integrated into the chromosome or can remain separate as do plasmids.

Conjugation is promoted by certain plasmids called sex factors. The donor cell contains a sex plasmid called the F (fertility) factor, which is a circular piece of DNA coding for 40 to 60 proteins. The F factor func-

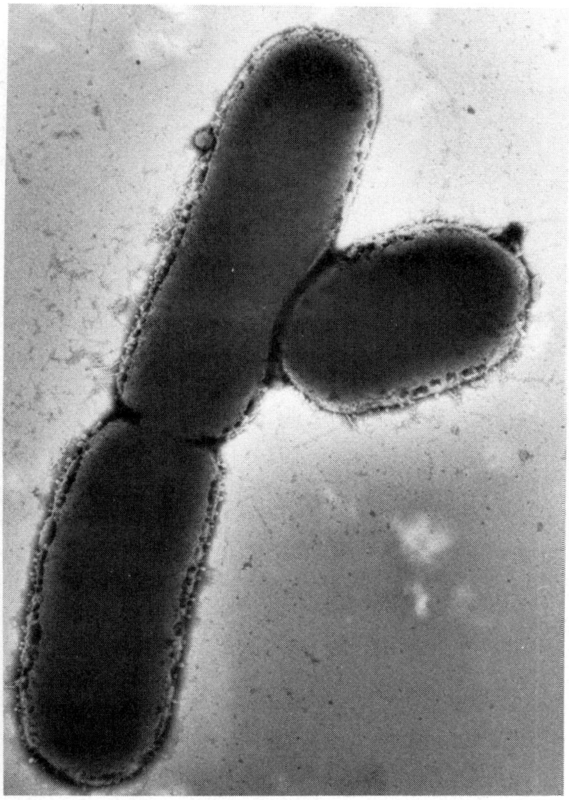

FIGURE 2-9a. Side-by-side pairing of *E. coli* in conjugation. The donor (Hfr) and recipient (F –) cells are attached by a threadlike F pilus *(upper right).* (Reprinted with permission of the publisher from Achtman M. Cell-cell interactions in conjugating *Escherichia coli:* role of F pili and fate of mating aggregates. J Bacteriol 1978;135:1053-61.)

tions and replicates independently of the bacterial chromosome. Donor cells contain the F factor and are called F^+; recipient cells are F^-. However, when conjugation occurs, the F factor may be transferred so that the F^- cell becomes F^+. Since F^- cells are more common in nature than F^+, spontaneous loss of the F factor must occur frequently.

Conjugation is made possible by an appendage called an F pilus (sometimes called the sex pilus), possessed by male strains the synthesis of which is controlled by genes on the F factor. Each male may contain several (up to five) pili. These pili unite the male and female cells and serve as bridges through which DNA (either from a plasmid or from the bacterial chromosome) is transferred. Conjugation has been demonstrated in most gram-negative species as well as in many gram-positive species. It occurs both within species and between related species such as between *E. coli* and *Shigella* species, *S. aureus* and *S. epidermidis, Strep. pneumoniae* and *Strep. faecalis.*

Certain strains of F^+ cells, called high frequency of recombination (Hfr) strains, possess the ability to transfer chromosomal genes more efficiently. In these cells, the F factor becomes inserted into the bacterial chromosome. With Hfr cells, the F factor is rarely transferred during conjugation but genetic recombination is frequent since the F factor is an episome in the bacterial chromosome itself. This differs from F^+ cells in which transfer of F^+ is high, but the frequency of recombination is low. The stages of conjugation between a Hfr and F^- cell are summarized in Table 2-11.

The F and other sex factors that initiate conjugation are only some of several types of plasmids. Plasmids (Fig. 2-10) are, under normal conditions, nonessential to the cell's survival, but they can endow their hosts with characteristics that are advantageous in certain situations. Plasmids are DNA molecules coding for about 5 to 150 polypeptides. Numerous genera of bacteria harbor plasmids; most commonly gram-negative bacteria, but also *Staphylococcus, Streptococcus, Bacillus,* and *Mycoplasma.* Plasmids transfer at different frequencies. F factors are transferred at high frequency; R (resistance) factors at low frequency. In a recent

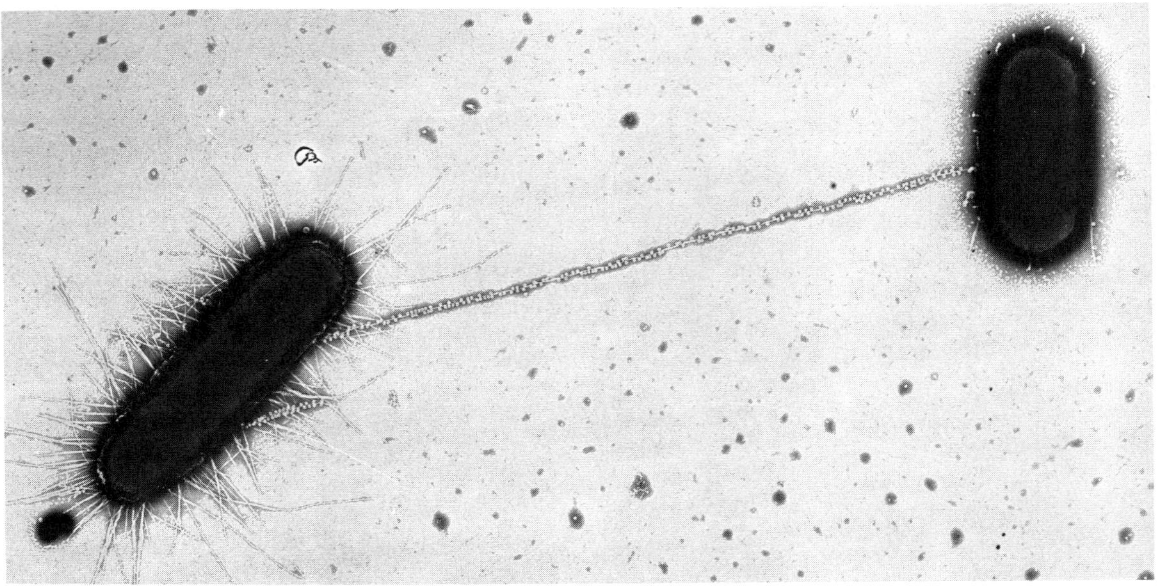

FIGURE 2-9b. The bacterial cell covered with numerous appendages is a genetic donor connected to a recipient cell (without appendages) by a structure called an F pilus. The F pilus is necessary for the genetic donor to transfer bacterial genes to the recipient. The F pilus itself is controlled by genes contained in a separate infectious genetic element, the F plasmid. This class of bacterial pili are called sex pili.

The F pilus is labeled along its length by special virus particles which infect donor cells through the F pilus. The numerous appendages on the donor are called type 1 pili and have no role in conjugation. They are required for *Escherichia coli* to colonize the intestine (its normal habitat) and for the *E. coli* to cause diarrhea. Vaccines for human and animal diarrhea have been prepared from these pili. This class of bacterial pili are called somatic pili. (Photograph by Charles C. Brinton, Jr. and Judith Carnahan. Reprinted with permission of the publisher from Freeman BA: Barrow's Textbook of Microbiology, 21st edition. Philadelphia: WB Saunders, 1979.)

TABLE 2-11. Stages of Conjugation Between an Hfr and an F^- Cell*

Stage	*Description*
Specific pair formation	Formation of a stable union between donor F^- pilus and recipient
Effective pair formation	Bridge formation requiring donor cell energy
Chromosome mobilization	Opening of chromosome at site of F^- integration
Chromosome transfer	DNA polymerase synthesizes a complementary strand
Recombination	May occur once the segments of chromosome pair

*Adapted from Sokatch and Ferretti 1976.

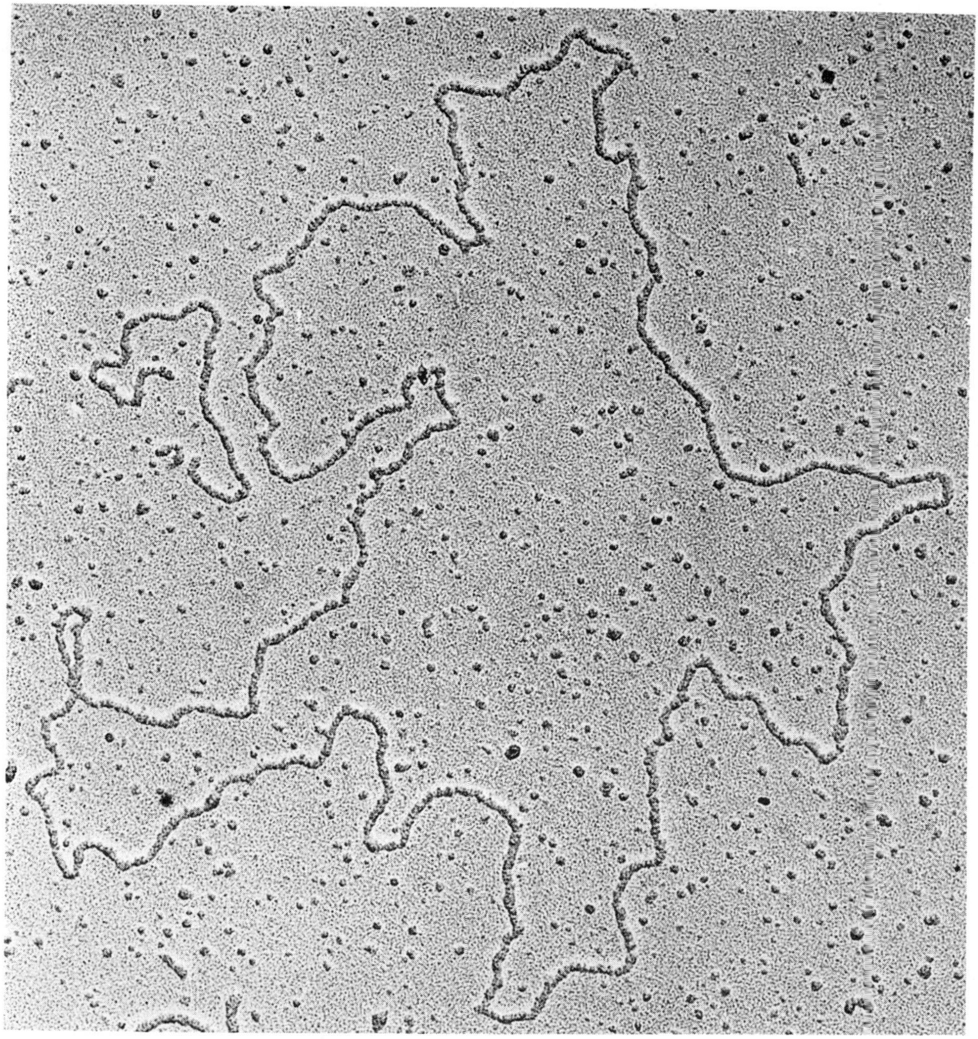

FIGURE 2-10. Plasmid isolated from a bacterium is enlarged 115,000 × in the electron micrograph on the opposite page. Plasmids are small, circular (closed-loop) molecules of the genetic material DNA that exist outside the bacterial chromosome. They play a major role in the genetic programming of microorganisms because they can be transferred from one strain to another (even to a different species) and can serve as vectors for introducing completely new genetic information, by means of recombinant DNA techniques, into bacteria. This plasmid was isolated from *Streptomyces coelicolor* by Mervyn Bibb and was shadowed with platinum, exaggerating its thickness. The double helix of DNA is about 10 micrometers in circumference, comprising about 30,000 base pairs of DNA—enough for some 30 genes. (Reprinted with permission of the publisher from Hopwood DN. The genetic programming of industrial microorganisms. Scien Am 1981;245:90-102.)

TABLE 2-13. Types of Fermentation Used by Bacteria for Energy Production

Pathway	*Type*	*Examples of organisms*
Glycolytic (Embden-Meyerhof)	Lactic fermentation	*Lactobacillus casei* and *Streptococcus cremoris* (used in first step of cheese making), pathogenic streptococci
	Alcoholic fermentation	Yeast, propionibacteria
	Mixed acid (formic) fermentation	Enterobacteriaceae
	Butanediol (acetoin) fermentation	*Enterobacter; Bacillus*
	Methane fermentation	Mostly in GI tracts of ruminants and some adult humans
	Butyric-butylic fermentation	Some strict anaerobes (e.g., *Bacteroides*)
	Mixed amino acid fermentation	Occurs in putrefaction and gangrene
Phosphogluconate	Heterolactic fermentation	*Leuconostoc mesenteroides*
	Ketodeoxygluconate (Entner-Doudoroff pathway)	*E. coli* (to metabolize gluconate); pseudomonads

as carbonate, nitrate, or sulfate, are used to produce ATP. Facultative anaerobes can use either oxygen or nitrate as final electron acceptors, or they may switch to fermentation in the absence of air (Table 2-14). Aerobic respiration is a very efficient process, yielding about 20 times as much ATP per mole of glucose metabolized as fermentation. Major energy-producing pathways that use oxygen (respiratory pathways) are pyruvate oxidation and the tricarboxylic acid (TCA or Krebs) cycle (Figs. 2-11, 2-12).

Microorganisms can be classified according to the means by which they derive their energy. *Phototrophs* are organisms that use light as their energy source. *Chemolithotrophic autotrophs* derive energy from the oxidation of inorganic compounds and use CO_2 as their carbon source. *Autotrophs* can grow without organic carbon or nitrogen. This group includes *Nitrobacter, Clostridium aceticum,* algae, purple sulfur bacteria, and nonsulfur purple bacteria, none of which are of importance in human disease. Most bacteria and animals are *chemoheterotrophs,* using organic molecules for metabolic energy.

Metabolic Regulation

Cells differ in their ability to utilize substrates in their environment. Substrates are nutrients in the growth environment that the bacteria convert to essential biologic constituents. All bacteria require water, an energy source, carbon and nitrogen sources (about 10 percent of the dry weight of bacterial cells is nitrogen), and growth factors. *Growth factors* are substances that the organism cannot produce but that are necessary for its growth. Hence, they must be obtained from the environment. Particular microenvironments (mouth, GI tract, soil) contain different growth factors and are inhabited selectively by organisms because of their specific metabolic needs. Substrate levels in the environment serve as one regulatory mechanism for biosynthesis. If substrates are present in the medium, the cells

TABLE 2-14. Methods of Energy Production Used by Bacteria

Type of bacteria	*Examples*	*Method of energy production*
Obligate aerobes	Tubercle and some spore-forming bacilli	Respiration
Facultative organisms	Many yeasts, enterobacteria, staphylococci	Fermentation or respiration
Aerotolerant anaerobes	Lactobacilli, streptococci, peptococci	Fermentation (though they may grow in the presence or absence of air)
Obligate anaerobes	Clostridia, propionibacteria	Fermentation (some may tolerate oxygen at low tension, called microaerophilic)

will not synthesize them even if they are capable of doing so. However, if the supply of certain substrates in the medium or environment is exhausted, internal preferential synthesis of that substrate can occur in some organisms present and they will continue to grow.

Bacteria actively transport nutrients from their environments across their cell membranes. This requires energy expenditure. Specific transport systems exist for sugars, amino acids, and inorganic ions. At least 30 specific chemoreceptors, the binding proteins of active transport, function as the mediators of bacterial chemotaxis. Bacteria are attracted to sugars and amino acids and are repelled by substances such as alkali and acid (Fig. 2-13). Negative and positive feedback loops intrinsic to the bacterial cell provide genetic regulation of metabolism via regulatory proteins called activators or repressors.

Growth Patterns

Bacterial generation times range from about 10 minutes to several days, though most species undergo binary fission about every hour. Within an 8-hour period, one bacterial cell can undergo enough divisions to produce greater than one million cells. Growth is measured in terms of increases either in numbers or in the density (dry weight) of cells. Number of organisms may be directly counted by appropriate dilution and plating techniques. Cell density is often measured indirectly by measuring the amount of certain constant cell components like nucleic acid or specific enzymes.

Five phases of bacterial growth can be delineated. The *lag phase* occurs while organisms generate sufficient metabolites, or while mutants able to survive in a new medium begin to multiply. The *acceleration phase* begins when cell numbers begin to increase. The *exponential phase* begins when the cells are multiplying at a constant rate, according to the following formula:

$$\text{generations/hour} = \frac{\text{growth constant}}{\ln 2} \quad \text{where}$$

$$\text{growth constant} = \frac{\ln(\text{no. cells at a given time/no. cells at time 0})}{\text{time}}$$

(text continues on page 41)

The Krebs Tricarboxylic Acid Cycle

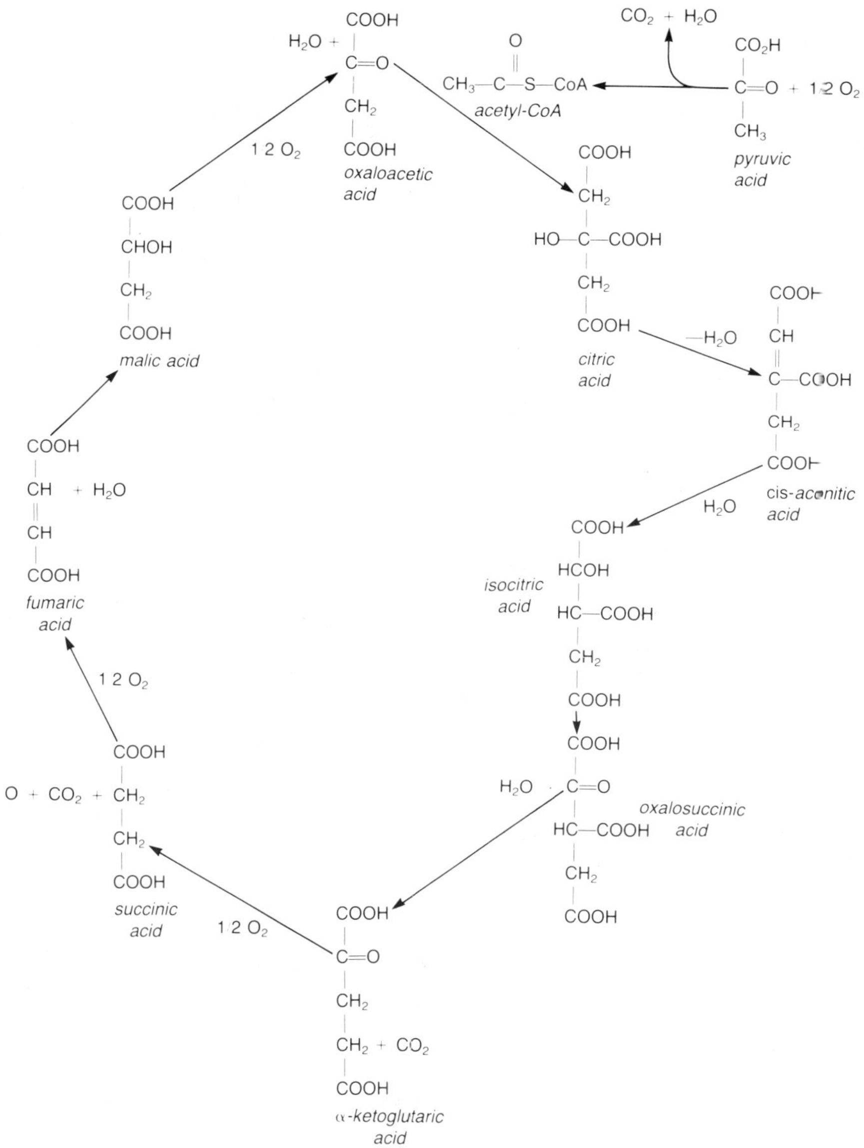

Net reaction:

$C_3H_4O_3 + 2\ 1/2\ O_2 \rightarrow 3\ CO_2 + 2\ H_2O$

pyruvic acid

FIGURE 2-11. The Krebs tricarboxylic acid cycle. (Reprinted with permission of the publisher from Freeman BA. Burrow's Textbook of Microbiology, 21st edition. Philadelphia, WB Saunders, 1979.)

Lipoic Acid and Pyruvate Oxidation

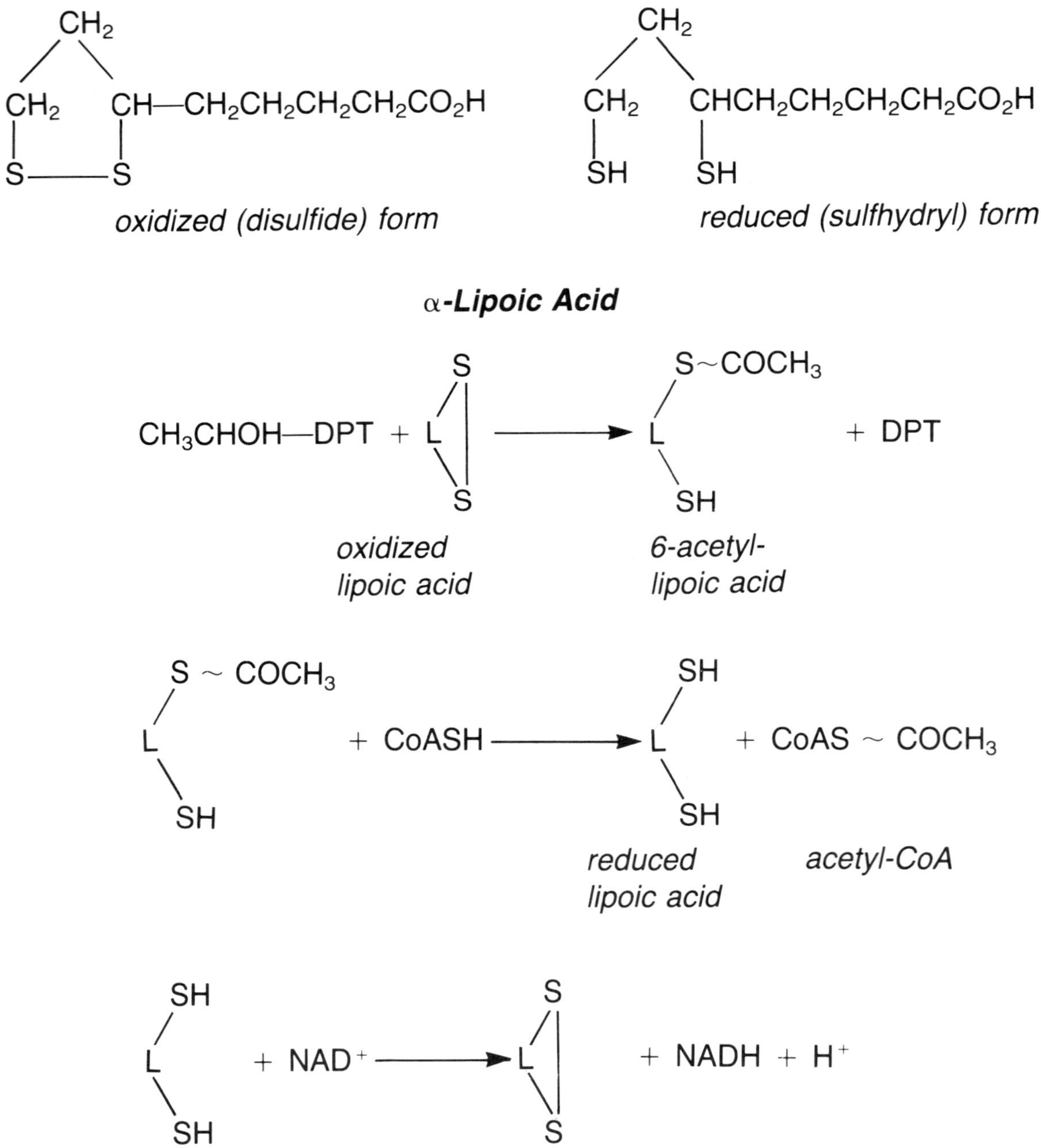

FIGURE 2-12. Lipoic acid and pyruvate oxidation. (Reprinted with permission of the publisher from Freeman BA. Burrow's Textbook of Microbiology, 21st edition. Philadelphia: WB Saunders, 1979.)

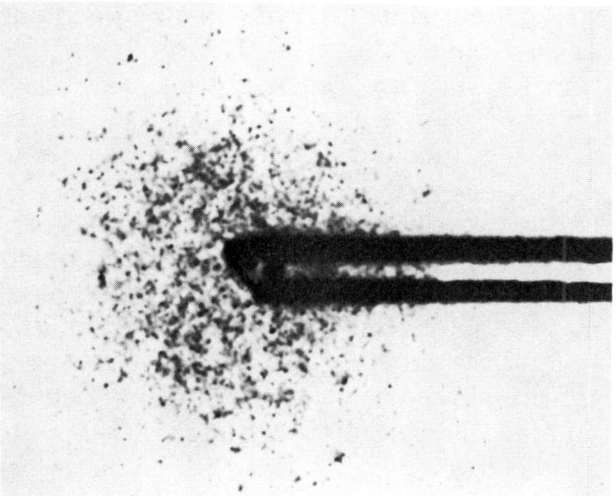

FIGURE 2-13. Positive chemotaxis of *E. coli* cells toward the tip of a capillary containing a concentrated solution of glucose. (Photo by J. Adler. Reprinted with permission of the publisher from Davis BD, et al. Microbiology, 3rd edition. Hagerstown, Md: Harper and Row, 1980.)

Cells continue to grow exponentially until nutrients in the medium are exhausted, or until toxic metabolic products accumulate. When bacteria are growing maximally, they enlarge, have few granules, and stain more deeply. They can remain in this stage if fresh medium is supplied before the cells stop exponential growth. The *retardation phase* then begins, and growth rate decreases. During the maximum stationary phase growth nearly ceases so that the cell count is balanced between deaths and new synthesis. The *decline phase,* when increasing numbers of organisms die, then ensues. Death of bacterial populations also occurs exponentially. Old and dying cultures contain numerous unusual forms such as balloonlike shapes and filaments. These involutional forms result from breakdown in cell permeability.

Environmental conditions such as temperature, nutrient concentrations, and the presence of antimicrobial agents affect the rate and form of bacterial growth. It is possible for certain bacteria to grow at temperatures ranging from below freezing to above boiling, though most organisms of clinical concern grow optimally around body temperature (about 37° C). These are called mesophilic organisms, as opposed to psychrophiles, which grow optimally in cooler temperatures, 15° C and below, and thermophiles, which thrive in temperatures over 45° C.

CLINICAL IMPLICATIONS

The increasing incidence of infections due to plasmid-mediated multiresistant organisms is of major concern to health care personnel. Though organisms that commonly carry such R factors are not necessarily "virulent," they do cause significant, sometimes lethal disease in the hospitalized patient or in other compromised hosts. Once such an organism colonizes a patient, it may become necessary to treat the infection with highly toxic antimicrobial agents. The decision regarding whether the disease or the treatment is worse then becomes a difficult one to make. And, of course, even if the colonized patient is not endangered by the plasmid-carrying bacterium, the risk of cross-infection to others must be considered when any decision to treat the patient is made.

Organisms commonly seen in the hospital that carry R factors are primarily gram-negative bacteria, especially *Serratia marcescens, E. coli,* and *Klebsiella pneumoniae.* The major reservoir for the multiresistant bacteria (as well as for L forms that can result from antimicrobial therapy) is the urinary tract, and the most common source of cross-infection is the urinary catheter. Respiratory tract infections, bacteremias, and wound infections have also been re-study, eight different enteric strains of six different species were found to contain a single epidemic-causing R factor (Tompkins, Plorde, and Falkow 1980). Since complementary strands of plasmid DNA are made by both donor and recipient cells during conjugation, both cells retain the plasmid information.

Plasmids are classified either on the basis of the traits that they incur on the host (e.g., "sex factor," "resistance factor," "virulence plasmid"), or on the basis of their genetic relatedness. Plasmids that are closely related are incompatible in the same host cell. Bacteria can harbor a number of different, nonrelated plasmids. Based on their genetic relatedness, there are at least 40 different incompatibility groups.

The major group of plasmids of clinical concern are resistance (R) factors, which code for bacterial resistance to many antimicrobial agents. Such plasmids have been found to be the cause of nosocomial epidemics. Plasmids coding for both antimicrobial resistance and enterotoxin production have been found in enteropathic *E. coli*. Toxigenicity in *C. tetani* may also be plasmid associated. Functions mediated by plasmids are listed in Table 2-12.

TABLE 2-12. Factors Transferred by Plasmids

Transfer (sex) factors
Production factors
Toxin
Bacteriocins
Virulence enhancing factors
Cell surface antigens
Hemolysin, streptolysin, H_2S
Resistance (R) factors
Antimicrobial agents
Heavy metals
Specific phages
Ultraviolet light
Modification factors
DNA restriction
Specialized recombination systems
Fermentation of lactose or sucrose
Degradation of octane or naphthalene

MICROBIAL GROWTH

Pathways of Energy Production

All living organisms require energy for growth and functioning, and thus have pathways for energy breakdown and release (catabolism) and for energy storage and the production of new cellular materials (anabolism). All pathways for energy production result in the formation of energy-rich bonds carried by adenosine triphosphate (ATP).

ATP is formed by the processes of *oxidation* (the removal of electrons from a substance; in this case, hydrogen [H^-] from glyceraldehyde phosphate) and *reduction* (the addition of electrons from a substance; in this case, adding hydrogen [H^-] to nicotinamide-adenine dinucleotide, NAD, or nicotinamide-adenine dinucleotide phosphate, NADP).

All energy utilization pathways lead from ATP. There are four major processes by which the high energy phosphate bonds in ATP are formed and energy is produced: photosynthesis, fermentation, aerobic respiration, and anaerobic respiration. In *fermentation,* organic compounds such as glucose are used for energy production. Several types of fermentation used in bacterial energy production are listed in Table 2-13. Since this process occurs in the absence of air, organisms using fermentation are anaerobes. *Aerobic respiration* is a metabolic process in which oxygen is necessary for energy production. Organisms that require molecular oxygen for energy production are strict aerobes. In *anaerobic respiration,* inorganic compounds other than oxygen such ported. Epidemic transmission of species containing R factors has been traced to contaminated respiratory therapy equipment, faulty nursing technique, and inadequate handwashing by personnel. The emergence of these strains is clearly related to overuse of broad-spectrum antimicrobial agents.

The need for control measures, then, is evident. In terms of medical therapy, liberal unrestrained use of antimicrobial agents must be prevented. Barrier nursing techniques must be used to prevent cross-infections. Some hospitals have developed special isolation procedures as an added precaution if patients become colonized (not necessarily diseased) with such organisms. A sample of such an isolation procedure is shown in Figure 2-14.

We have come full cycle in our infectious disease therapy. At the beginning of this century we were dependent on prevention and hygiene. In the "antibiotic era" many diseases that had devastated children and adults (streptococcal sore throat with resultant scarlet fever or rheumatic fever, tuberculosis, diphtheria, pneumococcal pneumonia) were controlled by antimicrobial agents. We became comfortable depending upon these same agents to treat nosocomial infections. Now, however, resistant bacteria emerge faster than new therapeutic agents can be developed. Hence, we must again depend upon prevention and hygiene as primary modes of defense against these highly varied and versatile organisms, the bacteria.

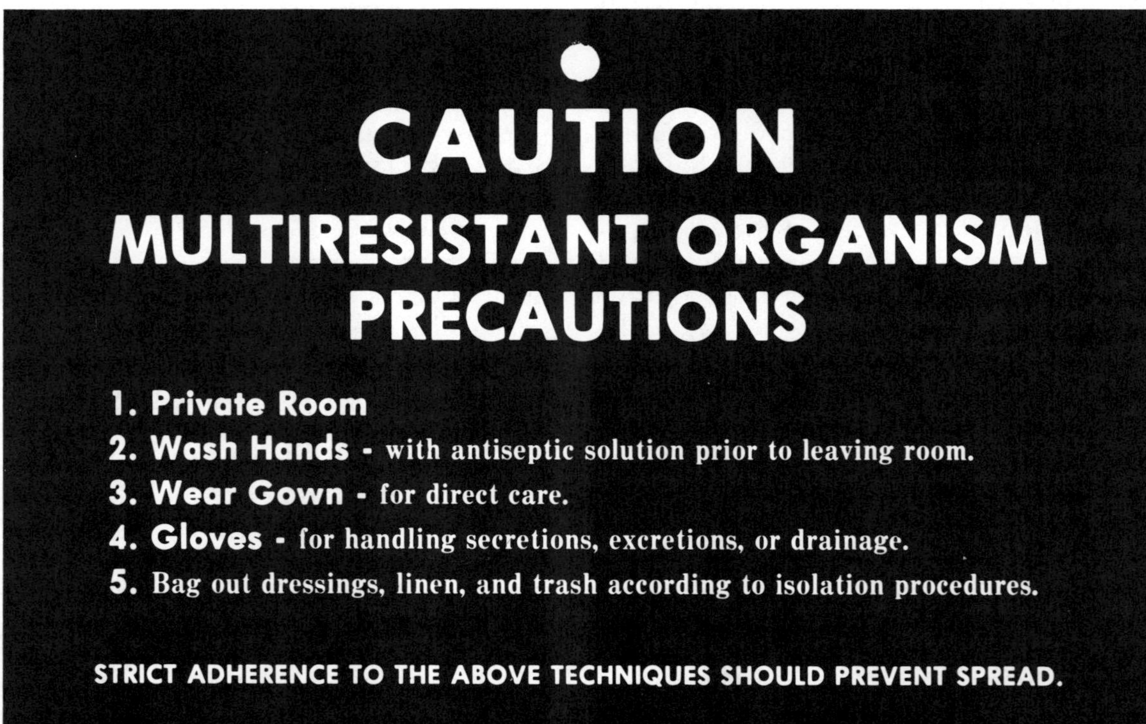

FIGURE 2-14. Isolation sign used for patients with infections due to multiresistant organisms.

SUGGESTED READINGS

Callahan WS. The applicability of genetic transformation to bacterial identification. J Am Med Technol 1980;42:13-16.

Castle M, et al. Outbreak of a multiply resistant *Acinetobacter* in a surgical intensive care unit: epidemiology and control. Heart Lung 1978;7:640-644.

Chambon P. Split genes. Scien Am 1981;244:60-71.

Demain AL, Solomon NA. Industrial microbiology. Scien Am 1981;245:66-75.

Echeverria P, Murphy JR. Enterotoxigenic *Escherichia coli* carrying plasmids coding for antibiotic resistance and enterotoxin production. J Infec Dis 1980;142:273-286.

Gray A. DNA—the amazing helix. Midwife Health Vis Commun Nurs 1980;16:295-296.

Hadden CT. Postreplication repair of ultraviolet-irradiated transforming deoxyribonucleic acid in *Bacillus subtilis*. J Bacteriol 1981;145:434-441.

Harper HA, Rodwell VW, Mayes PA. Review of Physiological Chemistry, 17th edition. Los Altos, Calif: Lange Medical Publishers, 1979.

Hopwood DA. The genetic programming of industrial microorganisms. Scien Am 1981; 145:90-102.

Krimsky S, Ozonoff D. Recombinant DNA research: the scope and limits of regulation. Am J Publ Health 1971;69:1252-1259.

Laird WJ, et al. Plasmid-associated toxigenicity in *Clostridium tetani*. J Infec Dis 1980;142:623.

Markowitz SM, et al. Sequential outbreaks of infection due to *Klebsiella pneumoniae* in a neonatal intensive care unit: implication of a conjugative R plasmid. J Infect Dis 1980; 142:106-112.

Meijers JA, Winkler KC, Stobberingh EE. Resistance transfer in mixed cultures of *Staphylococcus aureus*. J Med Microbiol 1981;14:21-39.

Phaff HJ. Industrial microorganisms. Scien Am 1981;245:76-89.

Schaberg DR, et al. An outbreak of nosocomial infection due to multiply resistant *Serratia marcescens:* evidence of interhospital spread. J Infec Dis 1976;134:181-187.

Smith MD, et al. Transfer of plasmids by conjugation in *Streptococcus pneumoniae*. Plasmid 1980;3:70-79.

Sokatch JR, Ferretti JJ. Basic Bacteriology and Genetics. Chicago: Year Book Medical Publishers, 1976.

Tompkins LS, Plorde JJ, Falkow S. Molecular analysis of R-factors from multiresistant nosocomial isolates. J Infec Dis 1980;141:625-636.

3 VIRUSES: STRUCTURE AND GENETICS

KEY POINTS

- Viruses are classified separately from other microorganisms based on their small size, simplicity of structure, and dependence on host cells for metabolic needs and replication.
- Viruses consist of a core of nucleic acid surrounded by a capsid (cover) and, in some cases, a lipoprotein envelope.
- Unlike other living cells, the genetic code of viruses may be contained in *either* DNA or RNA.
- Viruses are obligate intracellular parasites.
- Viruses can infect animal, plant, or bacterial cells. A virus that infects bacteria is called a bacteriophage (which is often shortened to phage).
- Viral diseases are not as amenable to antimicrobial chemotherapy as are other infectious diseases, though vaccines to some viral diseases are available.

In 1892 a Russian named Ivanovski discovered that extracts from diseased tobacco leaves could pass infection to other such plants, even after all the bacteria were filtered out. The agent that caused the plant disease was clearly smaller than bacteria. These "filterable agents" he named viruses, which means "living fluids."

Viruses are particles of nucleic acid, either DNA or RNA (but not both), surrounded by a covering. Though a few small bacteria such as the *Rickettsia* and *Chlamydia* overlap in size with the viruses, in general the viruses are the smallest known life forms (Fig. 3-1). Only poxviruses are large enough to be seen under a light microscope as small dots (0.3 μm or 300 nm); other viruses must be visualized by electron microscopy. Viruses are distinctive in their simplicity of structure, their minimal metabolic systems for energy production and biosynthesis, and their parasitic method of replication.

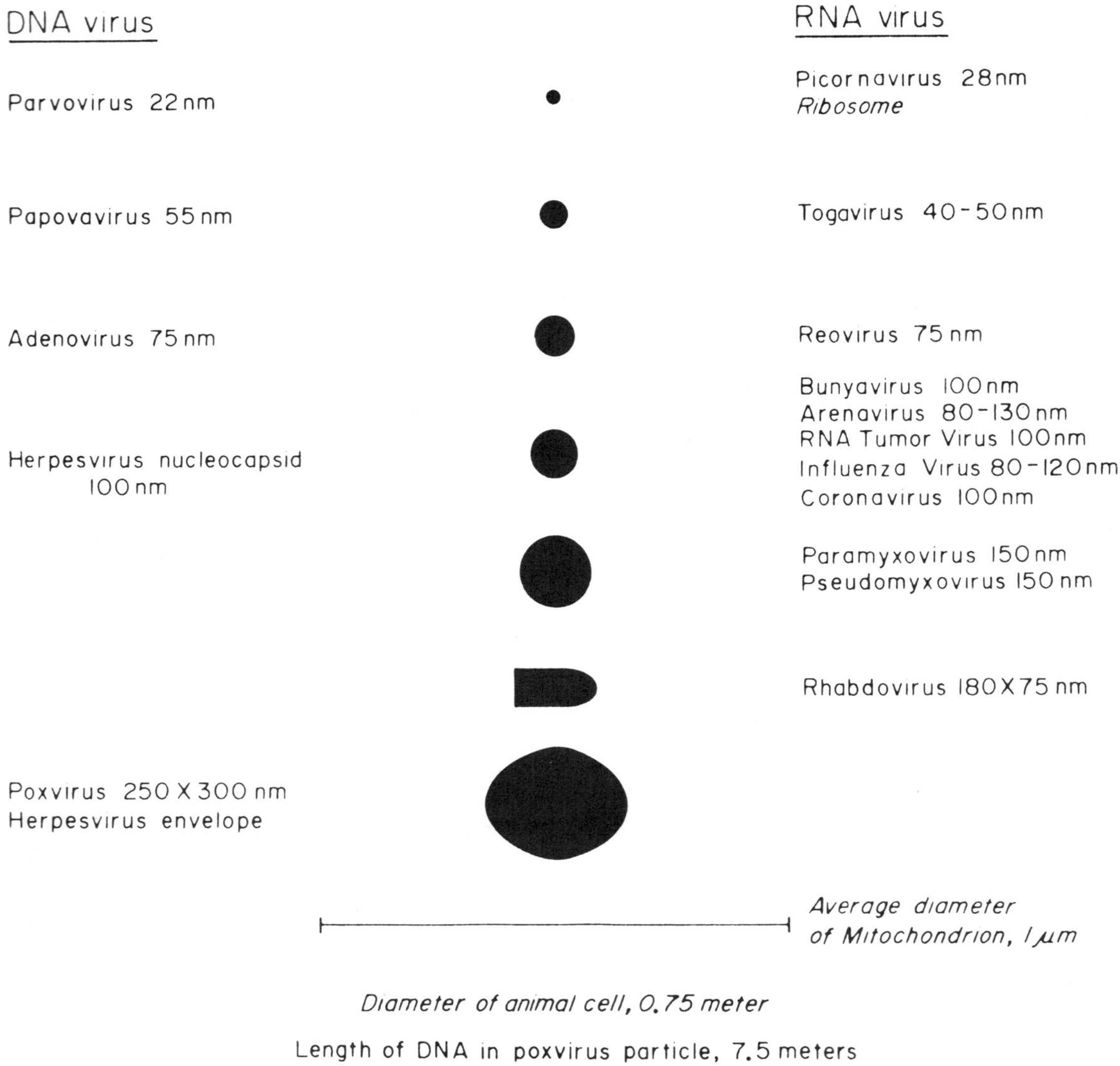

FIGURE 3-1. The relative sizes of the principal families of animal viruses. Unless otherwise indicated, the scale is the same for all. (Reprinted with permission of the publisher from Joklik WK. Principles of Animal Virology. New York: Appleton-Century-Crofts, 1980.)

VIRAL STRUCTURE

A virus particle, also called a *virion,* is always composed of the nucleic acid genome and a capsid (covering), which is often enclosed by an envelope.

Nucleic Acid Genome

Unlike bacteria and other living cells, viral nucleic acid may be either DNA or RNA, linear or circular, double or single stranded, or single stranded in short sections of otherwise double-stranded particles. A major division of viruses is based on whether they contain RNA or DNA.

Capsids

Capsids serve to protect viral genomes and to help introduce them into host cells. They are proteins composed of many identical subunits called capsomers.

Each polypeptide chain of the capsomers is coded for by the viral nucleic acid. A minimal amount of genetic information is required to manufacture capsomers since they are all identical. Most virions are helical or icosahedral in shape; a few have more complex appearances. Helical capsids are spiral with a single rotational axis. The nucleic acid is usually linear, as an extended filament. These spiral-shaped viruses are generally encased in a lipoprotein envelope. Icosahedral capsids resemble soccer balls. They have 20 sides, each an equilateral triangle, in which the nucleic acid is condensed or coiled. The patterns are arranged in 5:3:2 rotational symmetries as shown in Figure 3-2. This is a stable and efficient form for the structure of identical base units. Smaller capsids (picornavirus) contain 60 capsomers; larger capsids (adenovirus) contain 252.

Some complex capsids include spikelike glycoprotein appendages that help initiate infection. The complex poxvirus is shaped like a biconcave disc and contains lipoprotein layers. Bacteriophages are composed of a head and a tail (see p. 55). The combination of the nucleic acid genome and the viral capsid is called the *nucleocapsid.*

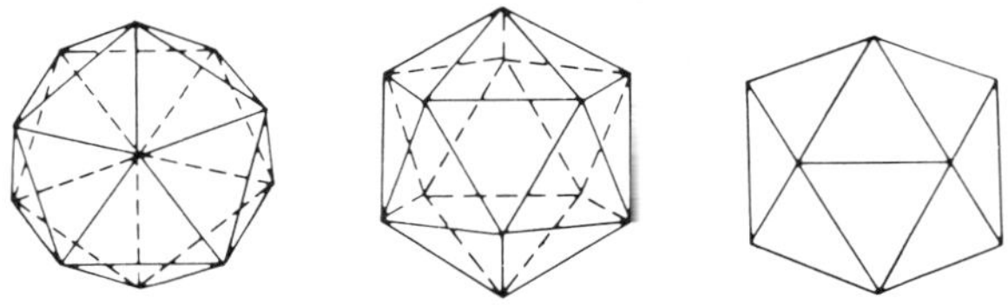

FIGURE 3-2. The icosahedron viewed normal to 5-, 3-, and 2-fold rotational axes. Edges of the upper and lower surfaces are drawn in solid and broken lines, respectively. The 5-fold rotational axes pass through the vertices *(left);* the 3-fold rotational axes pass through the centers of the triangular focus *(center)* and the 2-fold rotational axes pass through the edges *(right).* In this view, the edges on the upper and lower surfaces coincide. (Reprinted with permission of the publisher from Joklik WK. Principles of Animal Virology. New York: Appleton-Century-Crofts, 1980.)

Envelopes

All but six families of animal viruses are enclosed by a membrane composed of two lipid layers and glycoproteins. These viral envelopes resemble the selectively permeable membranes of other cells. They are easy to study because mutants can be produced whose progeny have specific changes in envelope polypeptides. This allows the researcher to determine the influence of protein composition on membrane properties.

The lipid component of the envelope is derived from the host cell. The lipids, phospholipid, neutral lipid (cholesterol), and glycolipids, are in a bilayer and may differ when virions transfer from one host to another. Viral envelopes generally do not include host-cell glycoproteins. Viruses contain only one or two surface glycoproteins as compared to hundreds in mammalian cells. Most enveloped viruses are spherical and flexible. Exceptions are the bullet-shaped rhabdovirus and some elongated influenza viruses. Virions may contain as many as 30 to 40 different polypeptide chains during envelope formation.

VIRAL GENETICS

Viruses contain no ribosomes and few enzymes that make them capable of producing energy. Hence, they are totally dependent on their host cells for their replication. Once replicated, they do not grow. Because viruses are obligatory intracellular parasites, they cannot be grown on artificial laboratory media, but require living cells in the form of tissue or cell cultures, live animals, or embryonated eggs.

All viruses except retroviruses are haploid, containing only a single copy of nucleic acid. The nucleic acid content of viral particles ranges from 1 to 50 percent, making it possible to code for the equivalent of about 4 to 500 average-sized protein molecules. Some viruses have repetitious nucleotide sequences on their terminal ends. This redundancy is related to the mode of replication in viruses such as RNA tumor viruses and the DNA-containing adenoviruses, parvoviruses, and poxviruses. Some viruses have proteins linked to their nucleic acids by covalent bonds, which may also function in replication.

Most viral nucleic acids are linear; papovavirus has double-stranded DNA (dsDNA) in supercoiled circles, and bacteriophage DNA is frequently circular. The nucleic acid usually occurs in long unbroken strands, but may in certain viruses (influenza, arenavirus, rotavirus) consist of 2 to 11 segments. This segmentation increases the efficiency of recombination, which occurs by reassortment of these segments. For example, the reovirus has ten segments of dsRNA. Each is transcribed to serve as mRNA and a template for synthesis of a complementary strand by the host cell. The progeny are dsRNA, but it is unknown how they get a complete set of 10 RNA segments.

Host-cell biosynthesis is inhibited during viral replication; the host cells are so preoccupied producing viral components that their own metabolic needs are neglected. This is why viral infection of a cell very often results in lysis or death of the cell.

The study of viruses has revealed the fact that RNA as well as DNA can store genetic material. As in bacteria, viral DNA is transcribed into mRNA and is then translated into proteins. However, some viruses (see class V below) do not contain the "sense" strand of their RNA. They do, however, contain an enzyme that produces positive strand, genetically effective RNA. The general properties of DNA and RNA viruses are compared in Table 3-1.

Viruses can be classified according to the type of nucleic acid they contain. The nucleic acid of double-stranded DNA (dsDNA) viruses, *class I viruses,* is replicated in a symmetric and semiconservative manner similar to bacterial DNA. Replication takes place in two well-defined phases. In the early stage a portion of the genome is transcribed into early mRNA, the remainder being transcribed after DNA synthesis, the late mRNA. In vaccinia virus, for example, during the early phase about 40 to 50 percent of the viral genome is transcribed in mRNA. In the late phase, after further uncoating that is mediated by the core mRNA, the remainder of the genome is also transcribed. Late mRNA is translated into structural viral components.

The single-stranded DNA (ssDNA) viruses, or *class II viruses,* replicate in a semiconservative and asymmetric manner. The ssDNA, after uncoating, is converted to an intermediate dsDNA replicating form (RF). Hence, replication of the strand is initiated and the new portion displaces the unreplicated single strand. Double-stranded RNA (dsRNA) viruses, or *class III viruses,* replicate by mediation of an RNA polymerase

TABLE 3-1. General Properties of DNA and RNA Viruses

Property	DNA	RNA
Strands	Double (except parvovirus)	Single (except reovirus)
Molecular weight	1.5×10^6-160×10^6	1×10^6-15×10^6
Configuration	Mostly linear or circular, often hypercoiled	Several forms, often complex
Color reaction to acridine orange	Yellow (except parvovirus)	Red (except reovirus)
"Cap structure"	None	Present

(called transcriptase), which transcribes the nucleic acid segments into mRNA. The mRNA of these viruses, like that of higher cells, has a methylated terminal oligonucleotide called a "cap structure," which enhances initiation of translation of mRNA into protein.

There are two types of single-stranded RNA (ssRNA)-containing viruses, called positive-strand and negative-strand RNA. Positive-strand viruses (+RNA) have an RNA strand that can serve as mRNA so that genetic information is translated directly from it. Negative-strand RNA viruses (−RNA), on the other hand, must first be transcribed into strands of the opposite polarity before the RNA can combine with ribosomes and serve as mRNA. All parvoviruses, except adeno-associated satellite virus (AAV), are −RNA. The AAV, however, is unique in that both plus and minus strands are produced in different virus particles.

The +RNA viruses are *class IV*. They produce a polymerase to synthesize viral protein after which the genome RNA directs +RNA strand replication. *Class V*, −RNA viruses, require transcriptase for infectivity since their RNA cannot function as mRNA to transcribe genetic information. They differ from +RNA because they use a complementary copy of virion RNA as mRNA and they synthesize proteins from individual mRNAs rather than making polyproteins, which are then split.

Class VI viruses, also called retroviruses (*re*verse *tr*anscriptase *viruses*), have a +RNA genome, but they are able to convert into an integrated DNA genome by a process called reverse transcription, which reverses the usual flow of genetic information. That is, the enzyme reverse transcriptase is used to direct the formation of DNA to serve as a template for new mRNA. The nucleic acid of these retroviruses exists inside the host cell as integrated DNA. These six classes of viruses are listed in Table 3-2.

Integration into the host DNA is one mechanism of viral persistence and occurs in many species of viruses with varying degrees of efficiency. Viral genomes can be joined to host DNA at many sites in the viral DNA sequence; there is no site specificity. Such integration, however, is not necessary and generally does not occur during the lytic cycle (when the host cells are killed). The integrated genome can induce host transformation by changing the expression of critical host genes or by encoding proteins that can change the phenotypic expression of the cell (Fig. 3-3). Genetic interactions that occur between viruses and hosts are summarized in Table 3-3.

Viral proteins (which determine biologic properties) can undergo modifications in-

TABLE 3-2. Classification of Viruses by Nucleic Acid Structure

Class	Description	Examples
I	dsDNA*	Poxviruses, herpesviruses, papovaviruses, adenoviruses
II	ssDNA†	Parvoviruses
III	dsRNA	Reoviruses
IV	Positive-strand RNA (+RNA); contains RNA which serves as mRNA	Togaviruses, picornaviruses
V	Negative-strand RNA (−RNA); contains RNA opposite in polarity to mRNA	Orthomyxoviruses, paramyxoviruses, rhabdoviruses
VI	Retroviruses; +RNA genome that directs formation of DNA to serve as a template for new mRNA	Retroviruses

*ds = double stranded
†ss = single stranded

TABLE 3-3. Viral Genetic Interactions In Host Cells

Interaction	Description
Complementation	When two viruses with defective genes or mutations cannot multiply alone, they may complement each other to produce progeny comprising both mutants. Permits survival of viruses with nonfunctioning or defective genes.
Interference	In mixed infections, there is usually reduced replication of one virus.
Multiplicity reactivation	After ultraviolet irradiation, damaged viruses cooperate (probably by recombination) to produce intact genomes.
Phenotypic mixing and masking	When two closely related viruses infect the same cell, the resultant capsid may be hybrid (mixed) or may be of the other virus (masked or transcapsidated).
Recombination	Usually occurs in double-stranded DNA viruses; involves reassortment of nucleic acid segments so that some progeny have a unique combination of genes derived from both parents.

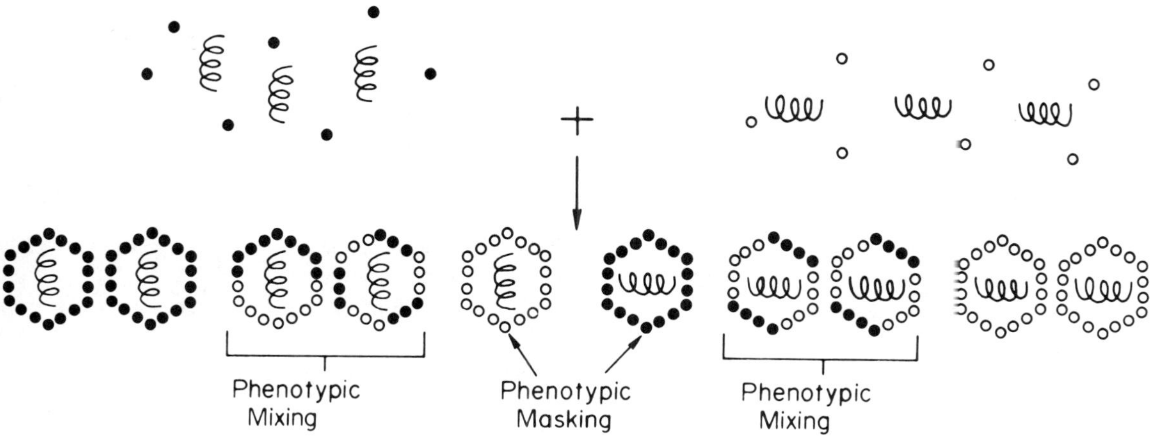

FIGURE 3-3. Phenotypic masking and phenotypic mixing. Simultaneous infection with two related viruses is illustrated. Either genome can be encapsidated in capsids that are composed exclusively of homologous capsomers, or mixed capsomers (phenotypic mixing), or exclusively heterologous capsomers (phenotypic masking). (Reproduced with permission of the publisher from Joklik WK. Principles of Animal Virology. New York. Appleton-Century-Crofts, 1980.)

side or outside infected cells. These modifications include fusion with other proteins or nucleic acids, phosphorylation, addition or deletion of sugars or lipids, or proteolysis. Viruses are subject to spontaneous mutation at a rate similar to that of bacterial and other cells (about 1 per 1,000,000). Mutagenesis in the laboratory can result from treatment of the virus with various chemicals or ultraviolet radiation. As in bacteria, point and deletion mutations occur. Lethal mutations are often nonsense mutants in which the codon for some amino acids is prematurely changed to a termination codon, or temperature sensitive (ts) mutations in which an enzyme is formed that cannot function in the usual temperature range of the virus. Other common mutations include enzyme deficiencies, drug resistance, changes in plaque size, or the ability to grow at higher than normal temperature. Interference between viruses occurs; one virus can inhibit adsorption of another or can prevent the translation of its mRNA.

MEASUREMENT OF VIRAL GROWTH

Titration is the measurement of numbers of particles per unit volume. Vital growth is measured by several methods: plaque, pock or focus formation, or serial dilution end point. Susceptible cells growing on a culture plate, when inoculated with serial dilutions of virus will lyse. Cells infected by the virus die in 2 to 14 days following inoculation. Areas where cells die can be observed as empty circles or plaques on the surface of the agar plates. A small amount of virus is added to a vast excess of cells (100–300 viral particles to 10^6–10^7 cells) so that the plaques are countable.

In a sense, this can be considered the inverse of bacterial colony counting when the numbers of bacteria are quantitated by counting colony-forming units (CFU). However, the plaque count enumerates the areas in which a virus particle has multiplied and caused lysis of surrounding cells, and is counted in plaque-forming units (PFU).

Within one plaque, as many as 10^6 PFU of viral particles are present. Since the dose response curve is linear and each plaque derives from a single virion, it is possible to calculate the number of viruses present.

Viruses that are not cytotoxic cannot be measured by counting plaques, but they may cause formation of lesions (pocks) on chorioallantoic membrane of chick embryos so that pock-forming units can be counted. Unfortunately this technique is less reliable. Other viruses cause increased growth rate or morphologic changes, which can be visibly identified as foci. Then foci-forming units (FFU) are counted. Viruses can also be counted by serial dilution. The dilution that first indicates cell destruction is the end point, the assumption being that it contains at least one viral particle.

Some viruses do not cause any detectable cytopathic changes in cells, but they do agglutinate red blood cells. For such viruses, hemagglutination tests are diagnostically helpful. Complement fixation and precipitation with antiserum for viruses, however, are tests of low sensitivity and are not routinely used.

The ability of certain viruses, the bacteriophages or phages, to infect and lyse specific strains of bacteria is used for epidemiologic purposes to identify various species of organisms such as *Staphylococcus aureus* and *Escherichia coli*. The strains of bacteria to be identified are inoculated with a series of phages. Susceptibility patterns to phages are unique to each particular strain. Such tests take only a few days.

VIRAL REPLICATION IN A HOST CELL

As previously mentioned, viruses can only multiply in host cells. Hosts may be bacteria (for viruses called bacteriophages), other protists, or plant or animal cells. Replication occurs only when viral infection of a cell is initiated. The phases of viral infection of a cell are listed in Table 3-4. Scanning electron microscopy (SEM) or surface replication (SR) using transmission electron

TABLE 3-4. Phases of Viral Infection of a Cell

Phase	*Occurrence*
Adsorption (attachment)	Adherence of the virus to host cell surface; mediated by specific host receptors
Penetration	Entry of virion or subviral units through the cytoplasmic membrane of the host; may occur by fusion of viral envelope with cell membrane or by phagocytosis
Uncoating (eclipse)	Loss of virion integrity within host cell; components are disassembled and nucleic acid replication occurs
Synthesis and assembly	Viral structural components are made, some coded for by host cell and some by viral genes; parts are assembled
Release	Virions either bud from host cell membrane or are released by cell lysis

microscopy (TEM) of platinum replicas of surfaces of dried or frozen cells are techniques used to study stages of viral replication on the host-cell surfaces. Identification is easy when particles have a characteristic shape, but pleomorphic virions may be confused with normal protrusions of the uninfected cell.

Adsorption (Attachment)

For attachment of the virion to a host cell to occur, the cell must possess receptors. Receptor sites and numbers vary depending on host species, tissue from which the cell originated, and the cell's physiologic state. Viruses are not active during cell entry, rather, adsorption occurs as a result of complementarity between the viral surface and receptor molecules on the host-cell surface. One in every 1000 to 10,000 random collisions of a virus with a cell surface results in attachment. The site of union is possibly a spike, fiber, or structural or capsid protein on the virus, and is usually a glycoprotein on the host cell. The receptor sites on the cell are very specific and can be counted. The rate of adsorption is independent of temperature but is directly proportional to the cell concentration; the more viral particles present, the more likely adsorption will occur. Many viral particles may become adsorbed to a single host cell; some animal cells contain as many as 10^5 viral receptor sites.

Penetration

There are two primary mechanisms by which virions enter the host: phagocytosis and direct crossing of the cytoplasmic membrane. Naked virions usually enter the cell by phagocytosis, whereas enveloped viruses most frequently use the second mechanism. Some viruses such as the influenza virus possess enzymes that destroy host-cell membranes, thereby facilitating their penetration.

In naked particles, penetration and uncoating are simultaneous events, whereas the two stages are separate for enveloped virions. Enveloped viruses, after fusing with the host-cell membrane, penetrate the plasma membrane directly and liberate their nucleocapsid into host cytoplasm. The envelope is thus lost. They may also be phagocytized with the free nucleocapsid entering the cytoplasm via the phagocytic vacuole. Phage capsids do not enter host cells; the empty viral capsid remains outside the cell after injection of the genome. Once virions enter the host cell they are no longer accessible to antibody.

Uncoating (Eclipse)

Uncoating is called the eclipse stage because infectious viral particles have "disappeared"; they are not recoverable from the cell for a period of time during which the viral nucleic acid is freed from the capsid so that replication and reassembly can occur. It is during this stage that new viral components are being manufactured. In nonenveloped viruses, uncoating is probably initiated by interaction with host-cell components. Uncoating for these naked virions occurs in the host cell nucleus of cytoplasm after the particles have been released from the phagocytic vacuole.

Uncoating of enveloped viruses occurs in two stages: loss of envelope at penetration, and lateration of the capsid for uncoating. In some viruses (for example, poxviruses and reoviruses) a transcriptase is required to complete the uncoating process. These first three stages—adsorption, penetration, and uncoating—are inefficient. Only a few of the viral particles that contact host cells actually complete these phases and get to the point of replication.

Synthesis and Assembly

When uncoating is complete, synthesis of viral components can begin. Synthesis occurs in three stages: replication of viral nucleic acid, synthesis of viral proteins, and assembly of the various components. Viral nucleic acid may be replicated in the host-cell nucleus or in the cytoplasm, but viral proteins are synthesized in the cytoplasm. The site of origin of replication of nucleic acids in the chain is usually unique to each type of virus. Once primed (replication always requires an initiator or primer), chain elongation always occurs in the same direction. A variety of proteins are required. Smaller viruses depend on host components; larger viruses (such as the lytic phages) code for their own proteins and enzymes that are required for their replication.

The assembly of genomes and capsids is a process called *morphogenesis*. Assembly is directed by formation of bonds between proteins and nucleic acids, and proteins and proteins. In naked icosahedral viruses, after nucleic acid synthesis, capsid polypeptides are synthesized and capsomers associate to form the shell, which still lacks the viral genome. After nucleic acid enters the capsid, proteins are cleaved to form the mature virion. Following assembly, icosahedral virions may concentrate at the maturation site and form intracellular crystals that are observable in infected cells.

In enveloped viruses, the nucleocapsids are formed as with naked virions. The proteins necessary for capsid information are synthesized in cytoplasmic ribosomes. Virus-specific envelope proteins replace host proteins during viral replication. Adherence of the nucleocapsid with the host-cell membrane causes the membrane to curve and bud. Synthesis in some viruses is more complex. In the poxvirus, for example, before the viral DNA enters the capsid, membranes are formed that contain virus-specific lipids. If simultaneous infection with two closely related viruses occurs, phenotypic mixing or masking can occur (Fig. 3-3).

Release

Emergence of viral structures in diffuse or patchy areas on infected cell surfaces occurs within 3 to 20 hours after infection. Cytopathic changes in the host cell may occur at this time. These changes include cell retraction, acquisition of unusual configurations, or giant cell formation. Release is often by cell lysis, especially with naked virions. Most viral progeny reside in and are released from the host cytoplasm. Adenoviruses and papovaviruses are integrated into the cell nucleus; when the cell dies naturally, they are released. Polioviruses have a unique release mechanism, a type of "reverse phagocytosis." Virus-filled vacuoles release their contents by erupting through the cell membrane.

Some enveloped viruses release from host cells by budding at the host's outer plasma membrane. The budding site is characteristic of specific viruses. The bud containing the completed virion detaches by infolding of the host-cell membrane, which has been altered to contain lipoprotein specific to the infecting virus. The virion incorporates this lipoprotein envelope and buds from the host cell without causing lysis. Many such particles (thousands per hour) may be released from the host.

BACTERIOPHAGES

In addition to viruses that inhabit human and other animal cells, viral particles are also known to parasitize almost every group of bacteria. The bacterial viruses are called bacteriophages or phages. It is important that we know something about this group of complex viruses for several reasons. Much of what is known about viral structure and functioning has been studied in phages; phages cause morphologic and biochemical changes in bacteria, which are sometimes important in identifying bacteria or which affect their ability to cause disease in humans; phage typing is an important epidemiologic tool for the identification of certain strains of bacteria; and transduction, the transfer of genetic material from one bacterial cell to another through a phage, is an important method of genetic recombination.

Phages require a complex capsid structure that enables them to penetrate the rigid bacterial cell wall. Most phages have a head that contains the viral genome and some internal protein, and a tail covered by a contractible sheath. Tails vary greatly in complexity, but all are used for attachment to the host bacterium. They have a central tube through which the viral nucleic acid can pass and be "injected" into the host, in a manner similar to a syringe (Fig. 3-4). RNA phages lack tails, so a protein molecule serves as the adsorption structure. Terminal tail fibers or filaments are often present and act as attachment structures.

The most extensively studied phages are the coliphages T1 through T7, which attach to *E. coli* strain B. These T phages are shaped in two icosahedral halves connected by a hexagon. Each phage is highly specific for certain hosts. Components of the bacterial cell wall determine which phages will be adsorbed.

Most phages contain dsDNA, though some contain ssDNA or dsRNA. They adsorb through the bacterial cytoplasmic membrane after attaching to specific host receptor sites. There are small RNA phages, however, that attach only to the sex pili and use them as channels for injecting their nucleic acid into the cells. Their structure is relatively simple: they contain only three genes. Another group of phages attaches to flagella (Fig. 3-5), which propel the phage to the cell's surface (only motile bacteria can serve as hosts for these phages). Some phages can even inhabit encapsulated bacteria by producing a hydrolyzing enzyme that provides access through the capsule to the cell wall.

Tail attachment to host cell surfaces (Fig. 3-6), even before the viral genome is injected, causes the host's cytoplasmic membrane to become disorganized, increasing permeability, and causing cell metabolic changes with resultant interference with protein synthesis. This sometimes results in cell lysis and is called "lysis from without" since no viral replication has yet occurred in the cell.

Under ideal conditions (adequate concentrations of both phage and host bacteria, and physiologic stability), most contacts between a phage and a susceptible host will result in adsorption; hundreds of phage particles, covering the entire bacterial cell surface, can adsorb into one host. Unrelated phages are mutually exclusive and cannot infect the same cell; genetically related phages can infect a single bacterium. The viral progeny of such infection will include all types of phage present, plus new types that result from genetic recombination. More

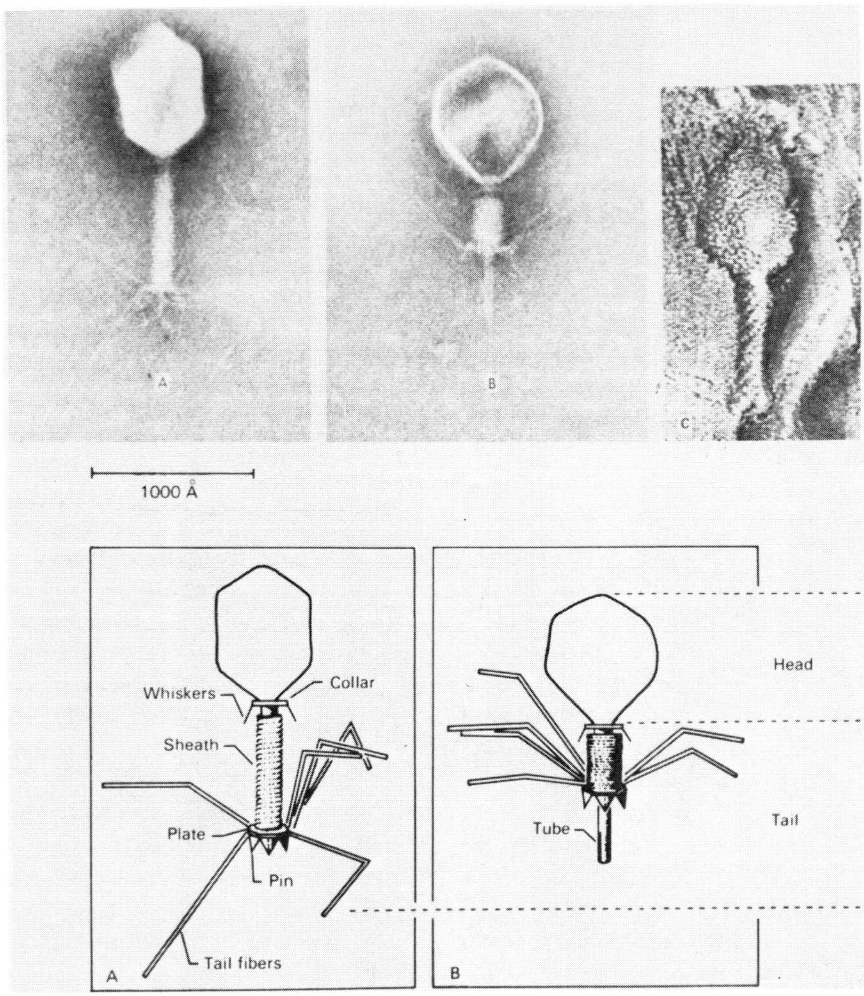

FIGURE 3-4. Electron micrographs of bacteriophage T2. *(a)* Phage before injection, with full head of extended sheath. *(b)* Phage after injection, with empty head and contracted sheath. *(c)* Helical structure of the sheath. (*a* and *b,* negative staining. *c* obtained by freeze-etching. The phage was embedded in ice that was then fractured; the place of fracture was covered with a thin layer of evaporated metal, which was then photographed. (Reprinted with permission of the publisher from Davis BD, et al. Microbiology, 3rd edition. Hagerstown, Md: Harper and Row, 1980.)

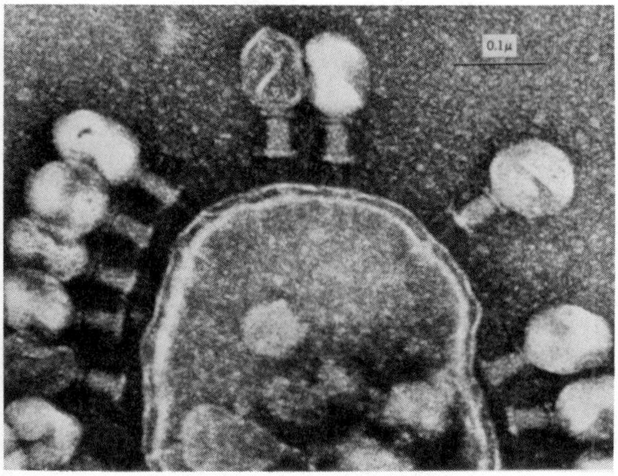

FIGURE 3-5. A partially lysed cell of *E. coli* B with attached virions of phage T4. Note an empty capsid, the contracted tail sheaths, and the interval between tail plate and cell wall. The "collar" is also well evident. (Courtesy of Dr. LD Simon. Reprinted with permission of the publisher from Luria SE, et al. General Virology, 3rd edition. New York: John Wiley and Sons, 1978.)

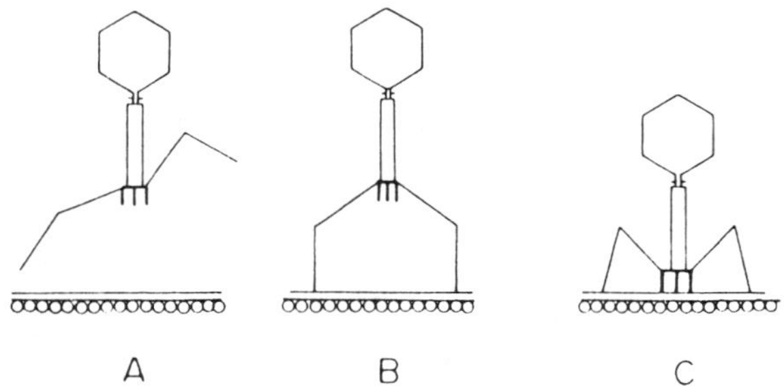

FIGURE 3-6. Assembly of bacteriophage T4. (Reprinted with permission of the publisher from Primrose SB, Dimmock NJ. Introduction to Modern Virology. Oxford: Blackwell Scientific Publications Ltd, 1980.)

than 25 genetically related phages can contribute to the phage progeny from one host bacterium.

Once inside the host cell, replication of phage nucleic acid is similar to that in animal DNA or RNA viruses. Phages undergo an eclipse period of several minutes. The results of infection, however, may take two different courses: lysis or lysogeny. Whether the infection is lytic or lysogenic depends on the type of phage (T-even phages, T2, T4, T6, are usually virulent, residing primarily in the human GI tract); condition of the host (lysogeny is favored if cells are thymine starved); and environmental conditions (temperature, number of phage adsorbed per cell). *Temperate* phages establish lysogeny; *virulent* or *productive* phages are lytic. Some phage such as lambda may become either lysogenic or lytic depending on which genes are expressed as well as the physiologic conditions of the host.

Virulent (lytic) phages cause alterations of host cytoplasmic membranes after which phage lysozyme causes the formation of holes in the cell wall with resultant host-cell death and release of phage progeny.

In lysogenesis, phage nucleic acid becomes integrated into the host bacterium's chromosome. Usually as many as 90 percent of cells infected with a temperate phage become lysogenic. The phage thus integrated is called a *prophage*. The presence of prophage renders the host immune to superinfection with the same or related phages. Note the difference between host *resistance* to phage infection due to lack of surface receptors, and host *immunity* to a specific phage due to the presence of prophage in the host chromosome. Cell phenotypes may be altered by prophage (a process called *lysogenic conversion),* but in most cases the host cell appears normal and cannot be differentiated from other cells. Bacteria containing a prophage may spontaneously convert to a lytic state, or the prophage may leave the host viable and uninfected. The causes of this are unknown. The prophage

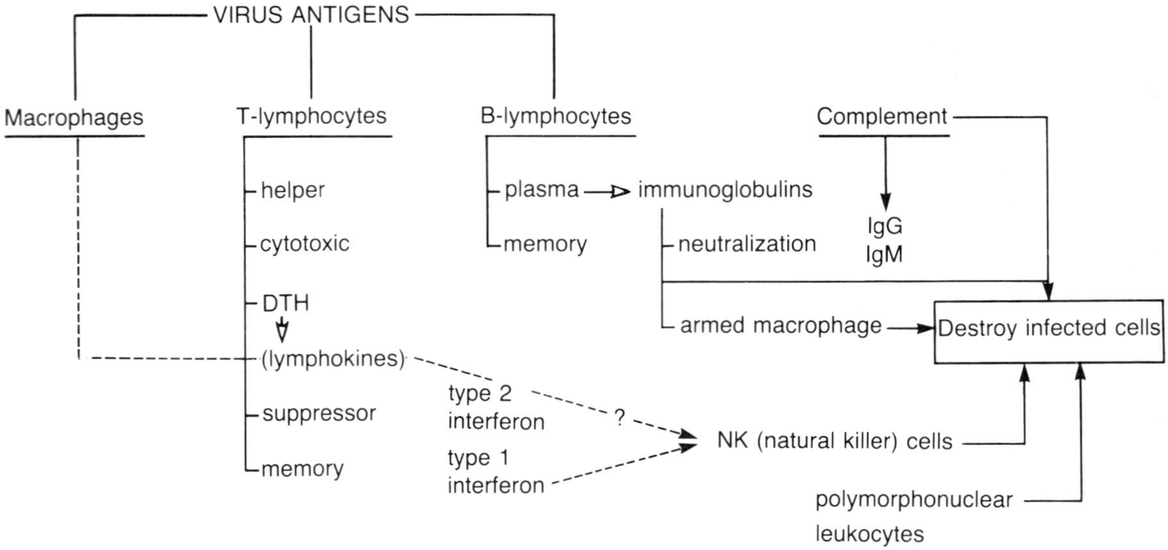

FIGURE 3-7. Summary of responses of the immune system and associated cells to viruses and virus-infected cells. (Reprinted with permission of the publisher from Primrose SB, Dimmock NJ. Introduction to Modern Virology. Oxford: Blackwell Scientific Publications Ltd, 1980.)

state is an inheritable entity; it persists in the host-cell chromosome but does not itself replicate. Turbid plaques indicate that some of the phage present have resulted in lysogeny rather than cell lysis.

Prophages will eventually produce vegetative phages again, but replication is controlled by bacterial rather than viral genes. Induction, change from prophage to vegetative phage, can occur spontaneously, or artificially by ultraviolet light, irradiation, thymine starvation, carcinogens, and alkylating agents. Induction requires that genetic repressors in the host cell be inactivated or bypassed. Some latent prophages can be transmitted with the bacterial chromosome or as extrachromosomal elements, plasmids.

A temperate coliphage (a phage that infects coliforms), phage Mu (mutation), induces mutations. Its DNA has heterogeneous ends that hybridize to *E. coli* DNA, integrating anywhere in the bacterial chromosome. Recombination of bacterial and phage DNA causes loss of some host DNA and formation of a hybrid. It is possible with phage Mu to transpose genetic material from the bacterial chromosome to other phages or to plasmids.

Prophages can cause their lysogenic hosts to exhibit new genetically acquired features. Though the expression of phage genes by bacterial hosts is regulated primarily by transcription, phages also cause transduction of bacterial genes, during which host DNA is incorporated into the phage head and transferred to another host cell by that means. The portion of host DNA is called the *exogenote*.

After replication, viral DNA pools in the host cell. Molecules are randomly drawn from this pool, which can be the equivalent of DNA for about 50 virions. Final assembly occurs in a particular sequence. Each protein component has binding site specific for the preceding component produced. The DNA enters the capsid (head) after it has been formed. When the capsid is full, an endonuclease severs the extraneous DNA. Briefly, the phases of phage assembly are as follows:

Baseplate of phage tail with about 15 gene products is formed.
Tail tube is assembled.
Shell of phage head is synthesized.
Tail and head combine.
Tail fibers (necessary for attachment) are attached to the tail.

CLINICAL IMPLICATIONS

Viruses may have originated by regressive (backward) evolution, or more likely from cellular genetic material that has become able to exist independently. They are the agents about which we know the least and against which we can provide the least protection once infection has begun.

Unfortunately, viruses also cause the majority of infectious diseases. Some viral diseases like the common cold, influenza, chickenpox, cytomegalovirus, and herpes simplex infect practically the entire population of the world.

The fact that there are as yet no simple, inexpensive methods for cultivation of viruses complicates both the problems of diagnosis and treatment. Though we do have vaccines available against some viral diseases (for example, polio, rubella, and hepatitis B), there are many for which specific control or treatment modalities are unavailable. The antimicrobial drugs used to treat bacterial and parasitic infections are ineffective against viral disease. Hence, once a person becomes ill with a viral infection, we are dependent primarily on the body's inborn immunologic capabilities to combat the disease (Fig. 3-7).

It is important for the clinician to appreciate the complex interaction between an infecting agent, in this case the virus, and the host's immune response. Because viruses are obligate intracellular parasites, and antibody cannot access the insides of infected host cells, T-cellular immunity (see Chapter 6) plays an important role in determining the outcome of viral invasion (Fig. 3-8). Even when a virus successfully infects

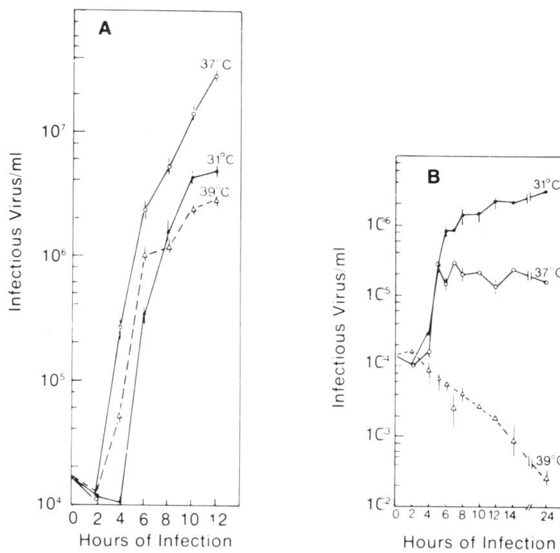

FIGURE 3-8. The effect of incubation temperature on the replication of wild-type and temperature-sensitive vesicular stomatitis viruses in baby hamster cell cultures. *(a)* Wild-type vesicular stomatitis virus. Note that this virus grows well at all three temperatures although 37°C is optimal. *(b)* A temperature-sensitive mutant of vesicular stomatitis virus. Note that this virus grows best at 31°C, virus yields are reduced at 37°C, and no growth occurs at 39°C. (Reprinted with permission of the publisher from Youmans GP, Paterson PY, Sommers HM. The Biologic and Clinical Basis of Infectious Disease, 2nd edition. Philadelphia: WB Saunders, 1980.)

a host, clinical disease may not develop. Depending upon numerous factors related to the host's defenses—as well as the nature and extent of the viral attack and environmental circumstances such as temperature (Fig. 3-8) and pH—the infection may be aborted or may result in replication and transmission to other cells of new virus (Fig. 3-9). Finally, infection may be asymptomatic, chronic, recurring, acute, latent, or slow, possibly taking years to manifest itself (Table 3-5). (Specific families of viruses important in human disease and chapters in which they are discussed are listed in Table 3-6.)

In the succeeding decades it can be anticipated that great strides will be made in our understanding of and ability to treat viral disease. In the meantime, spread of viral infections directly from person to person or via vectors and vehicles is well documented and we are dependent upon hygiene and barrier techniques to limit disease.

(text continues on page 63)

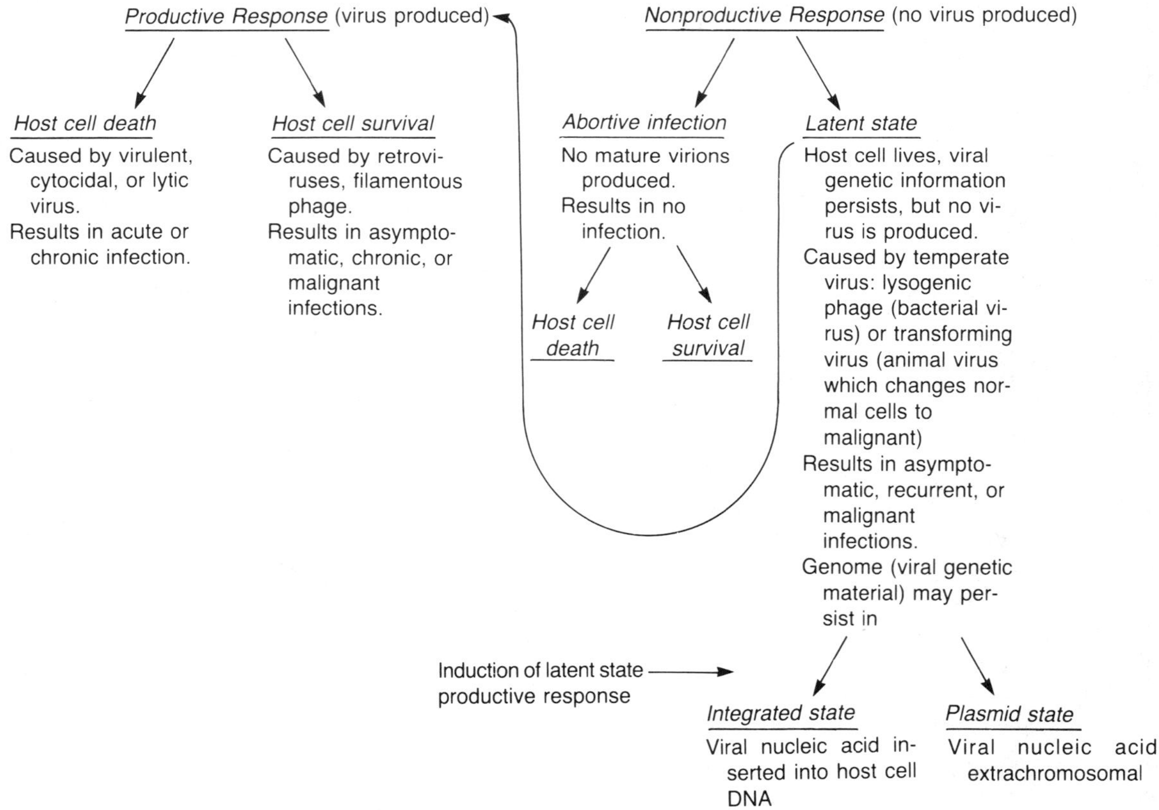

FIGURE 3-9. Possible results of viral infection of host cells.

TABLE 3-5. Different Types of Viral Infections and Characteristics of Their Respective Virus-Host Relationships

Fate of virus in host	*Type of infection*	*Characteristic virus-host relationships*	*Specific examples*
Eliminated; does not persist	Acute	Short incubation periods (2 days to 2 or 3 weeks) Virus recoverable before but not after disease onset; recovery common Recovered host immune to same or closely related viruses	Variola (usually results in disease) Poliomyelitis (rarely results in disease)
Not eliminated; persists in host tissues for months, years, or for life	Latent	Acute primary infection, recovery and subsequent relapses-remissions of disease Virus recoverable during primary and relapsing phases of infection; not recoverable from target tissue during remissions Hose immune response demonstrable but ineffective in preventing relapse	Herpes simplex ("fever blisters") Varicella-zoster ("shingles")
	Chronic	Variable incubation period, outcome, and course Virus persists and regularly recoverable Host immune response demonstrable but does not influence pattern of disease	Rubella syndrome Cytomegalovirus infection Hepatitis B (serum hepatitis) infection
	Slow	Incubation periods of months or years Relentless, progressive, and lethal course of disease Virus recoverable before onset and throughout course of disease Host immune response unpredictable: invariably absent; inconstant and weak; constant and exaggerated	Kuru Progressive multifocal leukoencephalopathy Subacute sclerosing panencephalitis (many would qualify as "chronic" infection)

Reprinted with permission of the publisher from Youmans GP, Paterson PY, Sommers HM. The Biologic and Clinical Basis of Infectious Diseases, 2nd edition. Philadelphia: WB Saunders Co, 1980.

TABLE 3-6. Families of Viruses Important in Human Disease

			Discussed in chapter					
Family	*Representative viruses*	*Type*	*Enteric 14*	*Respiratory 15*	*Neurotropic 16*	*Dermotropic 17*	*Hepatitis 18*	*Oncogenic 19*
Adenoviridae	Adenovirus	dsDNA		X				
Arenaviridae	Lymphocytic choriomeningitis	ssRNA			X			
	Lassa fever				X			
Bunyaviridae	Calif. encephalitis	ssRNA			X			
Coronaviridae	Infectious bronchitis	ssRNA		X				
Herpetoviridae	Herpes, Epstein-Barr, cytomegalovirus	dsDNA				X		X
Orthomyxoviridae	Influenza	ssRNA		X				
Papovaviridae	Papilloma, polyoma	dsDNA				X		X
Paramyxoviridae	Parainfluenza, measles, mumps, Newcastle, respiratory syncytial	ssRNA		X				
Parvoviridae	Adeno-associated	ssDNA		X				
	Norwalk-like agent		X					
Picornaviridae	Enteroviruses	ssRNA	X					
	Rhinoviruses			X				
Poxviridae	Smallpox, vaccinia	dsDNA				X		
Reoviridae	Rotavirus	dsRNA	X	X				
Retroviridae	Avian leukosis	ssRNA						X
Rhabdoviridae	Rabies	ssRNA			X			
Togaviridae	Rubella, yellow fever, encephalitides	ssRNA			X			
Unclassified	Hepatitis	ssRNA					X	

SUGGESTED READINGS

Barnhart ER, Ash RJ. Similarities and differences between viral and cellular membranes. Prog Med Virol 1979;25:89-112.

Dubois-Dalcq M, Rentier B. Structural studies of the surface of virus-infected cells. Prog Med Virol 1980;26:158-213.

Herbert E. Biosynthesis and processing of cellular and viral poly-proteins. Nature 1980; 288:115-116.

Herskowitz I, Hagen D. The lysis-lysogeny decision of phage λ: explicit programming and responsiveness. Ann Rev Genet 1980;14:399-445.

Joklik WK. Principles of Animal Virology. New York: Appleton-Century-Crofts, 1980.

Joklik WK. Virus synthesis and replication: reovirus vs. vaccinia virus. Yale J Biol Med 1980;53:27-39.

Korant BD, Lively MO, Powers JC. Limited proteolysis and viral replication. Biochem Soc Trans 1980;8:417-419.

Lodish HF. Synthesis and assembly of viral membrane proteins. Prog Clin Biol Res 1980;40:259-273.

Luria SE, et al. General Virology, 3rd edition. New York: John Wiley and Sons, 1978.

Makowski L. Structural studies of the assembly of simple viruses. Prog Clin Biol Res 1980;40:233-258.

Mitra S. DNA replication in viruses. Annu Rev Genet 1980;14:347-397.

Weinberg RA. Integrated genomes of animal viruses. Ann Rev Biochem 1980;49:197-226.

4 THE PROCESS OF INFECTION: HOST-MICROORGANISM INTERACTIONS

KEY POINTS

- The clinical response to infection with a microorganism is the result of many characteristics of both the host and the infecting agent.
- Factors that influence host responses often determine if clinical disease follows infection. Hence, there is increased emphasis on altering host responses to infection as well as on altering (attenuating) the infectious agent or preventing its spread in the body.
- Physical and chemical barriers and phagocytosis are the body's most important defenses against infection.
- Characteristics of hosts that influence immunity to infectious diseases include age, ethnicity, hormonal and metabolic factors, nutrition, and environmental influences.
- Characteristics of microorganisms that are important to their ability to become established in a host include the ability to adhere to specific surfaces of host tissue, invasiveness, and the ability to resist phagocytosis.

Even though specific immunologic defense mechanisms such as the ability to produce B and T cells (see Chapters 5 and 6) develop in the first three months of fetal life, newborns are highly susceptible to disease. This susceptibility to infection is due to deficiencies in phagocytic function and in the inflammatory response. Ironically, babies with agranulocytosis (which causes phagocytic dysfunction) can survive only a few days before developing life-threatening infections, whereas infants with severe congenital combined immune deficiencies may not manifest susceptibility to infection until after the first month of life. Clearly, phagocytosis, despite its function as a primitive defense against microbial invasion, is one of the body's most important protective mechanisms.

This chapter discusses factors that constitute the body's primary initial defense against microbial invasion. These include physical and chemical barriers of skin, mucous membranes, and body secretions; phagocytosis; and the inflammatory response. In addition, this chapter includes a description of the characteristics of microorganisms that determine whether or not they may cause infection or disease in the human host.

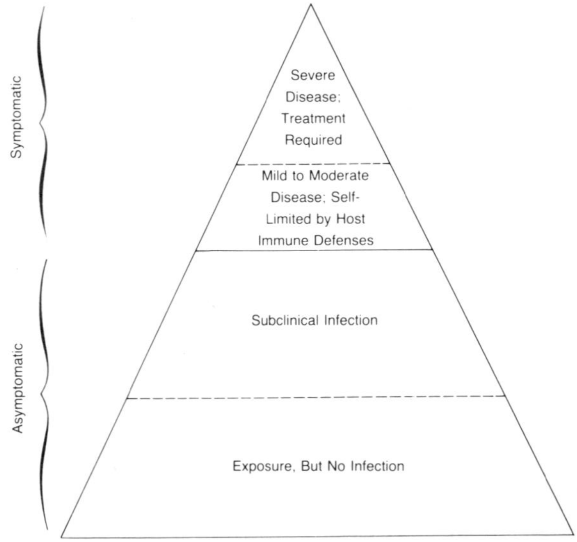

FIGURE 4-1. Host responses to exposure to microorganisms.

THE PROCESS OF INFECTION

In clinical settings, the terms infection and disease are often used interchangeably. We may say that a patient has a bacterial infection. In truth, all humans are infected from birth with many species of microorganisms. These organisms live in symbiosis with their hosts and may never or only rarely result in disease (Fig. 4-1). Thus, infection with a microorganism does not necessarily mean that disease is present. To avoid confusion of these terms, the word *colonization* may be used as a substitute for "infection" to denote the establishment of an organism as part of the body's microflora.

A host may also be a *carrier* of an organism. A carrier harbors an organism for varying periods of time; the organism does not become established as part of the normal flora and does not necessarily cause disease. The organism is usually rid from the body, either before or after causing overt disease in the host. Such carriers can be the source of transmission of disease to other susceptible individuals. For example, on several occasions hospital personnel have been found to be the asymtomatic carriers of strains of beta-hemolytic streptococci, group A, which have resulted in hospital-associated epidemics. These instances occured primarily among male and female operating room staff who were subsequently shown to shed the organisms from the vaginal or rectal areas despite their own lack of symptoms (Berkelman et al. 1982).

During epidemics, carriage rates of certain organisms such as the meningococcus and, without immunization, the polio virus, may reach close to 100 percent, especially under crowded conditions such as those in military barracks or boarding schools. Only a small proportion of those persons infected, however, will manifest clinical disease. Such carriers have developed *inapparent* or subclinical infections, which, in addition to being a potential source of infection to others, can incur to the individual immunity to future disease. Indeed, infection may become apparent only after diagnostic serologic tests are performed. Others may develop *dormant* or *latent* infections. These differ from subclinical infections in that the individual once infected carries the organism for very long periods of time, possibly throughout his or her life. Though the infection may be asymptomatic for long periods of time (even years), under certain circumstances clinical disease recurs. The herpesviruses are common examples of agents that cause latent infection.

When disease does occur, the question of causation may be a difficult one to address. For some time early in this century, for example, influenza (caused by one of several influenza viruses) was attributed to *Hemophilus influenzae,* a normal bacterial inhabitant of the pharynx. *H. influenzae* was isolated from the respiratory tracts of most people with influenza during the great epidemic of 1918. It could have also been isolated from the throats of most people without influenza, but it was not until later that testing for the bacteria among normal individuals was done. In this example, only two of the four Henle-Koch postulates, which have been traditionally used as the rules of causation for disease, were met. These postulates, originally proposed in the nineteenth century are as follows:

1. The organism must be isolated from every case of the disease.
2. The organism should occur in no other disease as fortuitous and nonpathogenic.
3. The organism can be isolated and grown in pure culture.
4. Inoculation of this culture can produce a similar disease in a susceptible species of experimental animal, and the organism can again be recovered from lesions in these animals.

These postulates have worked surprisingly well, although there have always been diseases, even at the time that the postulates were first proposed, that did not meet these criteria. The agents of leprosy, syphilis, and hepatitis, for example, cannot be grown in vitro. Asymptomatic infections with "pathogens" occur, and many diseases may be caused by organisms normally considered harmless and designated "nonpathogens." Other possibilities that can greatly complicate the establishment of causation are (1) the same clinical syndrome may be caused by a variety of agents and (2) the same organism may produce a variety of symptoms. Huebner has proposed a somewhat different set of criteria for establishing that a virus is the cause of a specific disease (Evans, 1976). Huebner's eight criteria are as follows:

1. The virus must be well established by passage in animal or tissue culture.
2. The virus must be isolated from human specimens and not be a contaminant.
3. An increase in antibody should result from active infection.
4. The virus should be fully characterized in terms of hosts, lesions, susceptibility to physical agents, etc.
5. The virus must be constantly associated with a well-defined clinical entity and, if possible, should be isolated from diseased tissue.
6. Humans inoculated with the agent in "double-blind" studies should reproduce the clinical syndrome.
7. Epidemiologic studies identify patterns of infection and disease.
8. The disease can be prevented by a specific vaccine.

These same criteria could be applied to bacteria or other infectious agents.

There is clearly an intimate interaction of factors that determine if a microbial invader will be prevented from entrance into the host, if it will enter as a saprophyte or part of the normal flora, if it will be carried temporarily by the host and eventually shed without harm to either the host or the organism, or if it will result in disease and destruction of host cells. We have increasingly come to appreciate the significant and individual role that the host plays in this interaction.

THE HOST AND THE PROCESS OF INFECTION

Natural Immunity

Natural immunity is immunity that occurs without previous contact with an infecting organism. Its basis is genetic, though the mechanisms involved in natural resistance to certain infectious agents are poorly understood. Natural immunity is specific to species or to individuals within a species. For example, humans are immune to the distemper virus, which is pathogenic to dogs. Humans and chimpanzees, on the other hand, are the only species susceptible to the gonococcal bacterium. Differences in individual response to the same infectious agent, chickenpox in children for example, clearly have some genetic basis.

The human leukocyte antigen (HLA) is the major histocompatibility antigen system in humans. HLA is discussed in Chapter 6. Work with animals, however, indicates that resistance to infections is probably controlled by many genes. Genetic influences may also affect responses to vaccines and other therapeutic drugs. Spruance and Chow (1980) concluded that though they found little variation in susceptibility of epidermal cells to herpesvirus among individuals with frequent recurrences of disease, there may well be genetically controlled host factors that determine clinical response to herpes infection. There are numerous genetic defects of immunologic functions that render individuals unusually susceptible to infections (Table 6-10). Acquired diseases, especially malignancies such as Hodgkin's disease, lymphoma, leukemia, and myeloma, depress the immune system. Patients with such diseases are plagued with many secondary infections caused by common as well as esoteric organisms.

Host factors which influence natural immunity are discussed below.

Age

Some diseases (such as rickettsial diseases, measles, polio, herpes, or mumps), if acquired during early childhood, are less severe than if acquired during adult years. In general, however, the extremes of age are times when susceptibility to infection is highest. Some viruses with mild effects on adults have devastating teratogenic influences on the fetus. Such viruses include cytomegalovirus (low incidence of microcephaly, chorioretinitis, motor disability), rubella (deafness, mental retardation, heart defects), group B coxsackieviruses (congenital heart lesions), and herpes simplex type 2 (CNS malformations). Neonates respond poorly to challenges by most microorganisms and, as mentioned in the introduction to this chapter, have immature phagocytic function. In general, the immune responses tend to peak during childhood and begin to decline after sexual maturity is reached. The thymus and other lymphoid tissues reach their maximum size at the time of puberty. However, after puberty the skin contains more saturated fatty acids, and resistance to fungi such as ringworm is increased.

In mice the total number of B cells either does not change or increases with age, but there is clear evidence from animal and human studies that T cell function declines with age. The primary, but not necessarily the secondary, antibody response also decreases with age. These and other general involutional changes in immune function help to explain the increased incidence of infectious disease and malignancies in the elderly. Autoimmunity, an immunologic reaction against self-antigens, is also increasingly common in individuals over 60 years of age.

In a recent study of disease caused by *Hemophilus influenzae,* it was found that though incidence rates were highest in infants, deaths were primarily in the aged. Black children were also found to be significantly more susceptible to meningitis than were white children, whereas Hispanic children manifested clinical forms of infection other than meningitis. The authors concluded that age and race may have affected susceptibility to different forms of disease caused by the same organism (Granoff and Basden 1980).

Ethnicity

American Indians are known to have increased susceptibility to tuberculosis. Whether there is a biochemical basis for this or whether it may be due to nutritional or other variables is not clear. Persons of dark skin have a much greater risk of developing disseminated coccidioidomycosis than do light-skinned persons. Black individuals with sickle cell anemia, on the other hand, are highly resistant to malaria.

Hormonal and Metabolic Factors

Hormonal influences on viral disease are exemplified by the fact that pregnant women have more complications from influenza and polio, probably because of increased production of steroids and changes in metabolic activity which accompany pregnancy. Persons with diabetes mellitus and adrenal dysfunction are subject to increased numbers of acute and chronic bacterial infections. Indeed, most diseases that alter metabolism or disturb circulation (congestive heart failure, shock, hypovolemic states) increase susceptibility to infection by changing serum pH, depressing phagocytosis, or elevating serum glucose levels (which impairs phagocytosis). Administration of glucocorticoids has profound effects on the immune response in both bacterial and viral infections. These effects are summarized in Table 4-1. Hyperosmolarity and hypophosphatemia also inhibit phagocytosis.

Nutrition

Diet has a profound effect on the composition of the intestinal flora. Breast-fed babies have a specific GI flora composed primarily of lactobacilli and bifidobacteria. Human breast milk is high in carbohydrate and low in protein, resulting in high lactose production. These bacteria, which produce acid from carbohydrate and tolerate a pH of 5.0, thrive in this environment. Since cow's milk contains less carbohydrate, the bowel flora

TABLE 4-1. Effects of Antiinflammatory Agents (Corticosteroids) on the Immune System

Markedly increase susceptibility to infectious diseases because of the following activities:

1. Reduce chemotactic activity
2. Delay attachment of leukocytes of blood vessel epithelium
3. Antiphagocytic
4. Decreased production of interferon
5. Depress cell-mediated immunity
6. In large doses, directly suppress formation of antibodies
7. Suppress inflammatory response
8. Eliminate some thymus-dependent lymphocytes

of bottle-fed babies resembles that of the human adult, with few lactobacilli and a greater variety of other microorganisms. Breast milk is also rich in the antibody, secretory IgA, and provides protection against intestinal infection in breast-fed babies.

Small, thin, undernourished animals who have had restricted total calorie and fat intake have been shown to possess improved cellular immune response, T cell function, and skin graft rejection. This is in contrast to what is known about humans—mild to moderate undernutrition greatly decreases the host's ability to respond effectively to infection. Among undernourished Colombian infants, impaired cell-mediated immunity was noted before age two years. The devastating effects of infectious diarrheas and measles in underdeveloped countries, as well as some of the ethnic differences in disease susceptibility, may well be due to nutritional deficiencies. There is evidence that deficiencies in trace metals (especially zinc) may be the important determinants of nutritional immunologic effects rather than fat and calorie restrictions (McMurray 1981).

Environmental Influences

Overexposure to cold and cigarette smoking adversely affect the body's resistance to respiratory infection by depressing ciliary action. The inflammatory response is depressed in chronic alcoholics. A depressive effect on host resistance has been noted in the first 48 hours after pulmonary infection with the influenza virus. Body temperature also influences susceptibility to viral infections. As temperature increases, replication of most viruses slows. Hence, a rise in body temperature may well contribute to recovery from viral disease. There is also evidence that elevated body temperature increases the uptake of thymidine in lymphocytes, enhances phagocytic activity, increases leukocyte migration, and aids in the localization of infection (Bernheim, Block, and Atkins 1979). Radiation and cytotoxic drugs depress bone marrow activity and production of lymphocytes (see Table 4-2 for list of cytotoxic drugs that depress the immune response). Anesthetic agents (ether, barbiturates, chloroform, halothane, ni-

TABLE 4-2. Cytotoxic Drugs That Depress the Immune System

Alkylating agents*
bisulfan *(Myleran)*
chlorambucil *(Leukeran)*
cyclophosphamide *(Cytoxin)*
melphalan (*Alkeran,* L=PAM)
mechlorethamine hydrochloride (nitrogen mustard, *Mustargen*)
triethylene thiophosphoramide *(Thiotepa)*
Antibiotics
actinomycin-D (dactinomycin, *Cosmegen*)
adriamycin
bleomycin
daunomycin (daunorubicin, *Rubidomycin*)
mitomycin-G *(Mutamycin)*
Antimetabolites
5-azacytidine
azathioprine *(Imuran)*
cytosine arabinoside (*Ara-C,* cytarabine, *Cytosar-U*)
hydroxyurea *(Hydrea)*
fluorouracil (5-FU, *Adrucil*)
mercaptopurine (6-MP)
methotrexate
thioguanine (6-TG)
Nitrosureas
BCNU (carmustine, *BiCNU*)
CCNU (lomustine)
Other agents
decarbazine *(DTIC-dome, DIC)*
L-asparaginase *(Elspar)*
cisplatin *(Platinol)*
procarbazine hydrochloride *(Matulane)*
vinblastine *(Velban)*

*Brand names are in *italics.*

trous oxide) are known to produce leukopenia and depress phagocytic, T cell, and B cell functions. Antimicrobial agents may increase susceptibility to certain infections by inhibiting normal flora and changing the body's ecologic balance. An example is the development of pseudo-membranous colitis caused by intestinal overgrowth of *Clostridium difficile* following antimicrobial therapy. Physiologic and psychologic stress probably lowers host resistance by increasing secretion of adrenal hormones and decreasing blood supply to some tissues. Climate also affects resistance; in warm, moist climates fungal skin infections abound.

Physical Barriers to Infection

With rare exceptions (such as the bacterium of tularemia), organisms cannot enter intact skin. However, with a break in the skin, microbes that normally inhabit the sebaceous glands and hair follicles, as well as others from the environment, can be readily introduced into the body. The skin is the first and most important physical barrier to potential invaders. The mucociliary activity in mucous membranes is another major defense. Bacteria tend to stick to mucus that covers the walls of the mucous membranes. In the respiratory and GI tracts, cilia tend to move these bacteria toward the natural orifices; to the back of the tongue where secretions are swallowed, and to the GI tract to be eliminated as the bulk of the feces. Microorganisms comprise about 50 to 60 percent of the dry weight of the stool. Peristaltic action also serves to move intestinal contents downward. The flushing action of urine in the bladder and tears in the eyes also serve as effective means of ridding the body of invaders. Patients with depressed cough or swallowing reflexes, those receiving mechanical ventilation and those who undergo surgery are at increased risk of respiratory infection resulting from depression of tracheal mucociliary flow.

Chemical Barriers to Infection

The acid pH of the skin and the secretion of saturated fatty acids by the sebaceous glands have an inhibitory effect on bacterial growth. The indigenous microbial flora of the skin and mucous membranes of the body (upper respiratory tract, vagina, mouth) help in several ways to protect the skin from invasion by more pathogenic species. Some strains of staphylococci and streptococci, for example, have been shown to inhibit the growth of *Legionella pneumophila*. This *bacterial interference* may be due to competition for essential nutrients or for attachment sites. The normal flora also help to maintain a low pH on the skin. Some gram-positive bacteria on the skin can convert lipids secreted by the sweat glands into unsaturated fatty acids, which exert antibacterial activity against fungi and gram-negative bacteria.

Lysozyme (muramidase) is an enzyme that dissolves cell walls of certain gram-positive bacteria. This enzyme and perhaps others are present on the skin, in mucus from the nose and other body orifices, and, in large quantities, in tears. Saliva also contains hydrolytic enzymes. The acid pH of the stomach readily kills most bacteria. Only individuals with achlorhydria have an indigenous microbial flora in their stomachs.

In the genitourinary tract, the urine is usually sterile as a result of low pH, phagocytosis in the bladder mucosal cells, and the presence of urinary secretory IgA. Bactericidal substances are present in prostatic fluid. In the vagina, the low pH maintained by lactobacilli, which are a major component of the indigenous flora, inhibits the growth of more pathogenic species.

Interferons

Most cells of the body, especially lymphocytes and macrophages, produce substances called interferons in response to viral invasion, some intracellular bacterial and rickettsial infections, endotoxin, and other inducers (Table 4-3). Interferons are acid-stable (pH 7.2), heat-resistant, nondialyzable glycoproteins that appear within hours in the blood and tissues where virus is present. Interferons appear to play a major role in the body's primary defense against viruses; their appearance in the blood occurs several days before antibodies are formed. The mechanisms for their interference with viral synthesis are not altogether clear. They

TABLE 4-3. Types of Interferon Inducers

- Type I inducers
  - Class A inducers (contain nucleic acid; relatively potent inducers)
    - RNA animal viruses
    - DNA animal viruses
    - Plant viruses
    - Insect viruses
    - Mycophages
    - Bacteriophages
    - Double-strand RNA of normal cells
    - Synthetic double-strand RNA
  - Class B inducers (relatively weak inducers)
    - Intracellular microbes
      - Bacteria
      - Rickettsias
      - Mycoplasma
      - Protozoa (toxoplasma, plasmodia)
      - Chlamydia (TRIC agents)
    - Microbial products
      - Lipopolysaccharides (endotoxin, lipid A)
      - Fungal polysaccharide (mannan)
      - Protein (from *E. coli*)
    - Chemicals
      - Polymers
        - Polycarboxylic polymers (pyran, polyacrylic acid, polymethacrylic acid)
        - Polysulfates (polyvinyl sulfate)
        - Polyphosphates (phosphorylated polysaccharides)
        - Natural and synthetic single-strand RNA
      - Low molecular weight substances
        - Cycloheximide
        - Kanamycin
        - 9-Methylstreptimidone
        - Tilorone [2,7-bis(2-diethylamino)ethoxyfluoren-9-one dihydrochloride] and its derivatives

TABLE 4-3. *(continued)*

Type I inducers
BL20803 (4,3-dimethylaminopropylamino-1,3-dimethyl-1*H*-pyrazalo-3,4-*b*) quinolone dihydrochloride
Basic dyes (toluidine blue, methylene blue, trypaflavine, acridine orange)
N,N-dioctadecyl-*N′,N′*-bis(2-hydroxyethyl)propanediamine
MA 56 (bis-ω-piperidinylacetyldibenzofuran hydrochloride)
Quinacrine; acranil
10-Carboxymethyl-9-acridanone
Actinomycin D
MDMP (2,4-methyl-2,6-dinitroanilino-*N*-methyl propionamide)
DRB (1-β-D-ribofuranosylbenzimidazole)
Type II inducers (in immunocompetent cells)
Antigens (in sensitized immunocytes)
Mitogens (in T and B lymphocytes)
Phytohemagglutinin
Concanavalin A
Pokeweed mitogen
Bacterial lipopolysaccharides
Staphylococcal enterotoxin A

Reprinted with permission of the publisher from Galasso GJ, Merigan TC, Buchanan RA, (eds). Antiviral Agents and Viral Diseases of Man. New York: Raven Press, 1979.

probably work by causing the production of a viral inhibitory protein that interferes with viral messenger RNA.

Because each interferon is species-specific, and perhaps for other reasons as well, their activity is effective only in the species from which the interferon was derived. Thus, chick or mouse interferon is ineffective in human cells and vice versa. On the other hand, interferons are not virus-specific, as are antibodies. They have an inhibitory effect against a host of different viruses. This is probably again due in part to the fact that their protective effects stem from intracellular induction of host metabolism and they do not act on virus particles directly. Interferons have no effect on extracellular virus.

Interferons have been shown to increase phagocytic action of macrophages, to alter plasma membranes, to affect cytotoxicity of lymphocytes, and to inhibit delayed hypersensitivity, allograft rejection, graft-versus-host rejection, and tumor cell rejection. There is currently much enthusiasm about the potential use of interferons both as antiviral and as antitumor agents. A recent breakthrough in development for the clinical use of interferons is the discovery of a method to use bacteria to grow genetic material that will lead to the production of interferon in amounts adequate for therapeutic purposes. Some of the problems and potentials for clinical use of interferons are summarized in Table 4-4. Side effects during trial use to date have been relatively minimal: fever, local tenderness at injection site, nausea, and vomiting. Use of interferons is one of the most promising new treatments for infectious diseases since the advent of chemotherapeutics.

TABLE 4-4. Use of Interferon for Treatment of Viral Diseases

Advantages	*Problems*
Active against many different viruses	Species specific; each interferon is most active in cells of species from which it was produced
Few known toxic or side effects to host	Difficult to get in pure form; lack of homogeneity of batches
Very potent; little is required	Action is reversible; effectiveness of short duration
Does not include selection of resistant mutants	Ineffective after viral synthesis begins in the cells

Phagocytosis

The amazing cellular defense mechanisms involving several types of scavenger cells in the host are activated whenever a foreign substance enters the body or blood stream. The nomenclature of the cells of the immune system can be somewhat confusing and is reviewed here. White blood cells, or leukocytes, are responsible for defending the body against foreign invaders. There are five kinds of circulating leukocytes: monocytes, neutrophils, eosinophils, basophils, and lymphocytes. These cells derive from a common stem cell and differentiate in the bone marrow to perform varying functions. Characteristics of these five types of leukocytes are described in Table 4-5.

Neutrophils (polymorphonulear leukocytes). Neutrophils, PMNs or PMLs, are the major cells responsible for phagocytosis (Fig. 4-2). They are also called granulocytes. They are very short-lived; the mature cell circulates in the blood for six or seven hours, and once they leave the circulatory system to enter the tissues they do not return. Thousands are produced in the bone marrow daily. Maturation of the stem cells takes two to three days in the marrow and most of the mature neutrophils remain there in a reserve pool. Only about 1 percent of the mature neutrophils are in the circulation, and about half of those in the peripheral blood stream adhere to vessel walls.

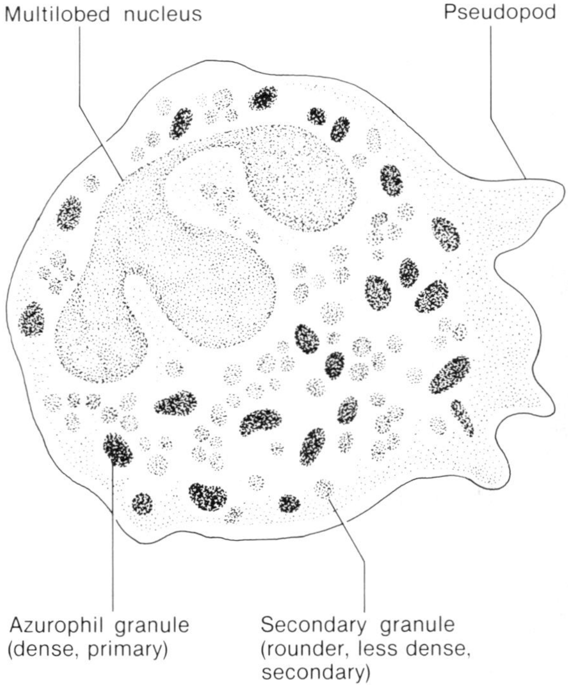

FIGURE 4-2. Schematic of mature neutrophil.

TABLE 4-5. Differentiation of Types of Leukocytes and Their Immune Functions

Cell type	*Alternate names*	*Serum values-cells/mm³ (proportion of total)*	*Functions*	*Active circulating life span*
Granulocytes		3650	Phagocytosis	12-14 hr
Neutrophil	Polymorphonuclear leukocyte; microphage	(0.60)		
Eosinophil		150 (0.02-0.04)	Regulate immediate-type hypersensitivity reaction; control helminthic infections	
Basophil		30 (0.002-0.005)	Involved in acute allergic reactions	
Lymphocytes		2500 (0.25-0.30)	B cells: production of immunoglobulins; T cells: regulation of cellular immunity	B cells: days-weeks; T cells: months-years
Macrophages	Macrophages, fixed or free; tissue histiocytes	430 (0.05-0.07)	Phagocytosis	24-48 hr

Neutrophils derive most of their metabolic energy from glycolysis; therefore, they can function in anaerobic as well as aerobic environments. They cannot divide, but they come equipped with lysosomes—granules containing substances necessary to ingest and degrade microorganisms. The huge stores of neutrophils in the marrow make it possible for the body to mobilize a massive defense very rapidly if microbial invasion occurs. Individual neutrophil counts vary daily with activity and other factors. Several substances in the blood help to regulate neutrophil production and release from the marrow. These include neutrophil-inducing and colony-stimulating factors. However, the mechanisms of their activity are poorly understood.

Eosinophils. Esinophils also develop in the bone marrow and are characterized by red-staining granules that contain the enzyme peroxidase. They play a much lesser role in immune defenses, and are found predominantly in tissues rather than in the circulation. Like other leukocytes, they have surface membrane receptors for immunoglobulins and complement factors, and are selectively attracted by an eosinophilic chemotactic factor excreted by lymphocytes. They are chiefly involved in allergic, immediate hypersensitivity responses and are present in large numbers on the skin when a topical allergen is applied in allergic persons. Eosinophils are also involved in control of helminthic (worm) parasites.

Basophils. The least common type of leukocyte, basophils, contain blue granules that obscure the cell nucleus and are rich in histamine. They play a role in some acute allergic reactions. *Lymphocytes* control cell-mediated and humoral immunity and are discussed in Chapter 5.

Monocytes. The precursors of macrophages, called monocytes, also originate in the bone marrow and emerge into the blood stream with phagocytic and bactericidal properties similar to neutrophils, except that they have fewer lysosomal granules. Most of the lysosomes are formed after a mature macrophage is "activated" by contact with an antigen. These monocytes, in contrast to neutrophils, maintain the ability to differentiate and multiply. They remain in the circulation for several days, after which time they emigrate into tissues and are long-lived, persisting there for weeks or months.

Monocytes are the precursors of several types of macrophages: fixed macrophages and wandering macrophages in the spleen and pulmonary alveoli (also called tissue histiocytes). Macrophages are uniquely suited for each tissue site to which they migrate by their ability to undergo considerable differentiation. They also maintain their capacity to divide in these various tissue sites. This monocyte-macrophage system of cells, also called the *reticuloendothelial system* (RES), thus includes phagocytic cells in blood, liver (called Kupffer cells), spleen, bone marrow, lung, lymphoid tissue, and other tissues. The primary function of the RES is to filter and destroy microorganisms and other particulate matter from the blood stream. During acute inflammation, neutrophils appear in the area first, followed by macrophages that respond to particular macrophage chemotactic factors.

The Phagocytic Process

The following stages in phagocytosis are discussed: chemotaxis, attachment, ingestion, degranulation, killing of the microorganism, and disposal of ingested material. These stages are listed in Table 4-6.

Chemotaxis is the movement of cells or organisms in response to substances in the environment. By this mechanism, phagocytes are biochemically guided from the blood stream across vascular walls (this process is called diapedesis) to the site of microbial invasion. Chemotactic factors include polypeptides produced by bacteria during replication, products made by lymphocytes during antigenic stimulation, fibrin, and collagen degradation products. When the phagocytes (predominantly neutrophils and, later, macrophages) reach the affected area, *attachment* of the foreign particle to the surface of the phagocyte's plasma membrane occurs. Some organisms readily attach to phagocytes and are thus readily phagocytized. Others, such as bacteria containing large polysaccharide capsules *(Streptococcus pneumoniae, Klebsiella pneumoniae)* can be phagocytized only if they are trapped on a rough surface where they cannot escape (called surface phagocytosis) or if opsonization has occurred. Op-

TABLE 4-6. Stages of the Phagocytic Process

1. Chemotaxis
2. Attachment
3. Ingestion
4. Degranulation
5. Killing
6. Disposal of ingested material

sonization, the process of enhancing with antibody the organism's attachment to the phagocytic plasma membrane, is discussed in detail in Chapter 5.

After attachment occurs, the cell membrane of the phagocyte forms pseudopods and ingestion of the invader results. The invader is pinched off and enters into a phagocytic vacuole. This ingestion requires considerable energy expenditure on the part of the phagocyte. In neutrophils and most macrophages, energy derives from a glycolytic pathway; some macrophages use aerobic oxidative mechanisms. After formation of the phagocytic vacuole, a process called *degranulation* occurs. The phagocyte's lysosomal granules disrupt and discharge their contents of hydrolytic enzymes and microbicidal contents into the vacuole. At this point, the structure is called a phagolysosome (digestive vacuole). Intracellular *killing* of the microorganism usually occurs within 1 hour of ingestion, by mechanisms which are not as yet totally explained. Numerous antimicrobial agents are expelled from the lysosomes, including hydrolases, lyzozyme, lactic acid, and lactoferrin. The bactericidal halogenating system uses myeloperoxidase from the lysosomes, hydrogen peroxide, and chloride ions, resulting in death of both bacteria and viruses. Release of these enzymes may lead to injury of local tissue as well as to the ingested material. After killing, *disposal* of the microorganisms must ensue. Some phagocytes remain viable after ingesting foreign material. These will migrate to lymphatics and deposit their debris in the regional lymph nodes. Leukocytes that die will either disintegrate locally or be removed via mucociliary action.

Of course, all organisms are not killed by phagocytosis. Such intracellular parasites as some viruses, *Mycobacterium tuberculosis,* and *Rickettsia* may survive and multiply in phagocytes. In other infections, however (such as with the poxviruses), macrophages are successful and important as inactivators. If the invader survives phagocytosis the next level of immune defense, specific antibody formation, comes into play. Antibody formation is the topic of Chapter 5.

Phagocytic function is essential to the maintenance of health. Phagocytosis is facilitated in tissues with tightly knit structure (lungs) and inhibited in the presence of necrosis and abscesses where circulating antibodies (opsonins) have difficulty reaching the injured area. Clinical signs of phagocytic dysfunction include recurrent infections with common pathogens, infections with organisms of low virulence, infections at unusual sites, and poor response to antimicrobial therapy.

The Inflammatory Response

The gathering of phagocytic cells at the site of infection is the single most important event in the inflammatory response. Characteristics of acute inflammation are summarized in Table 4-7. Of course these classic signs do not appear if the inflammation occurs in deeper body tissues. The inflammatory process serves several important functions in the body's defense against infection, such as

1. It initiates the migration of phagocytes to the area of injury.
2. It causes dilatation of local blood vessels and increased blood vessel permeability, enhancing the transport of complement and antibodies.
3. It promotes the accumulation of antimicrobial substances from the blood to the appropriate area.

TABLE 4-7. Characteristics of Acute Inflammation

Symptom	Cause
Redness	Dilatation of local arterioles and capillaries
Heat, increased temperature	Blood vessel dilatation
Swelling	Accumulation of fluid due to plasma leakage from the dilated blood vessels
Pain	Pressure due to fluid accumulation; release of chemical substances such as beta lysin, lyzozyme
Pus	Pressure of massive numbers of neutrophils

Prostaglandins (long-chain fatty acids) may be mediators of the inflammatory response, since they increase capillary permeability and vasoconstriction (the exact mechanisms for this are unclear).

Acquired Immunity

Immunity may be acquired if an individual has contact with a foreign antigen and produces antibodies, or if the individual receives antibodies from an external source (Table 4-8).

Passive Immunity

Passive immunity is temporary and occurs when antibodies are administered to an individual. Much of the protection of neonates against infection is due to transplacentally acquired antibody. This *natural passive immunity* lasts for 6 to 8 weeks after birth. Breast milk also has a high concentration of secretory immunoglobulin and may serve to protect the child during early months of life.

Artificial passive immunization is used in the following two situations:

1. Immediately following exposure to certain viral infections such as measles or

TABLE 4-8. Types of Acquired Immunity

Type	Method of Acquisition	Example
Passive	Natural	Neonate receives maternal antibodies through placenta or breast colostrum or milk
	Artificial	Parenteral administration of gamma globulin or antitoxin
Active	Natural	Natural infection that results in the formation of antibodies
	Artificial	Immunization with treated infectious agents, whole or parts, resulting in the formation of antibodies

hepatitis B. Specific immune globulins are made from the pooled sera of individuals who have recently recovered from the disease. Therefore, the globulin contains high antibody titers to the particular infectious agent. After administration of such globulin, the disease may not be prevented but the severity of symptoms can be modified. Gamma globulin is of no value after symptoms of disease have begun; it must be administered as soon after exposure as possible.

2. Following exposure to diseases such as botulism, tetanus, and diphtheria, which are caused by organisms that produce toxins. Passive immunization in the form of antitoxin is invaluable since no lag time is required for the onset of the incurred immunity. Its effectiveness lasts 2 to 3 weeks.

Active Immunity

The development of active immunity is a slow process, taking days to weeks and persisting for years. In active immunity the body produces its own antibodies following contact with a foreign antigen. This contact may be in the form of infection, administration of live or killed microorganisms or their products, or absorption of bacterial products such as toxins or toxoids (immunization).

Killed vaccines are prepared from organisms grown in animal cell culture and inactivated by heat, formalin, phenol, or some other chemical agent. Usually a series of injections are required to produce a durable antibody response. Living, attenuated vaccines are prepared by drying, numerous passages through artificial culture or through an animal host to induce mutation, or by use of a closely related strain (BCG and smallpox). The word vaccination originated from the latin word *vacca,* meaning cow, since attenuated cowpox virus was used to immunize humans against smallpox. Treatment of the organisms in these ways serves to maintain their antigenic properties and keep them alive while destroying their virulence and disease-producing ability. A single dose usually incurs lasting immunity.

A *toxoid* is a substance prepared from bacterial exotoxin. It maintains antigenicity but the toxicity is destroyed by heat or formalin. Toxoids differ from the antitoxins that incur passive immunity but do not elicit the formation of antibody by the host. Toxoids are injectable and, as with killed vaccine, require a series of injections and intermittent "boosters" to enhance the anamnestic (secondary) response. Components of microbial cells such as the lipopolysaccharide of gram-negative bacteria or capsular antigens of *Streptococcus pneumoniae* have been used as immunizing agents, but with only minimal success.

Immunization is a general term referring to the induction of antibody formation whether or not a live infectious agent is used. Vaccination has been used in the past as a more specific term in which a live agent is used to induce an immune response; currently these terms are often interchanged, but this is not precise, accurate usage. Research continues in search of new vaccines for diseases such as trachoma and herpes. Mumps and measles vaccines have been widely used only since the 1970s. With the eradication of smallpox, vaccination against it is no longer necessary.

The national "swine flu" immunization program in 1976 taught us several things. First, the predicted swine influenza strain did not appear as anticipated on an epidemic level, therefore the widespread immunizations were unnecessary. Influenza vaccines have a distinct limitation because prevalent strains change from year to year, and even within strains antigenic drift is sufficient that a vaccine that is efficacious one year may be useless in another. A new vaccine must be made based on the best predictions of what strain will be common. These predictions are usually accurate, but always involve an element of educated guessing. Second, among a small minority of persons (about 1 per 100,000), immunization (not just against swine flu virus, but probably against most other organisms as well) can result in a serious sequela, Guillain-Barré syndrome. When this relationship between immunization and Guillain-Barré was recognized, the immunization program was discontinued. For diseases such as influenza, immunization is recommended only for those who are at high risk of serious morbidity such as the aged or the immunosuppressed. Immunization against influenza has also been proposed for pregnant women because there is evidence that fetal protection ensues. Examples of diseases for which there is active and passive immunization are listed in Table 4-9.

TABLE 4-9. Agents Used for Passive and Active Immunization

Agent	*Examples*	*Suggested use*
Passive artificial	Immune human serum globulin	For hypogammaglobulinemia or dysgammaglobulinemia with recurrent bacterial infections, often concurrently with antimicrobial therapy; to modify clinical effects of measles or hepatitis A
	Hepatitis B immune globulin (HBIG)	To prevent or minimize symptoms of hepatitis B after parenteral exposure
	Mumps hyperimmune gamma globulin	To prevent orchitis in nonimmune adult males
	Rabies human immune globulin (HRIG)	Exposure to bite of rabid animal; may suppress antibody response to rabies vaccine; infiltrate half of dose around entry site and give the rest intramuscularly
	Tetanus equine antitoxin	Emergency use in suspected tetanus when adequate immunization has not been maintained. May cause hypersensitivity
	Tetanus immune globulin (TIG)	Superior to above because hypersensitivity is minimal and antibody protection persists longer
	Vaccinia hyperimmune globulin (VIG)	Immunodeficient persons who develop vaccinia gangrenosum after smallpox vaccination
	Varicella-zoster immune globulin (ZIG)	May prevent chickenpox if administered within 72 hours of exposure; recommended primarily for immunocompromised children

TABLE 4-9. *(continued)*

Agent	*Examples*	*Suggested use*
Active artificial immunization		
Killed vaccines	Cholera	Travelers: Middle East, Asia
	Influenza	Aged, immunodeficient
	Plague	Travelers: Middle East, Asia, South America
	Rabies	After exposure give human diploid cell vaccine (preferred) or duck embryo vaccine
	Injectable polio vaccine	Immunodeficient individuals (for normal persons, use live attenuated)
	Typhoid fever	Travelers in endemic areas
	Typhus fever	Travelers: Southeast Europe, Africa, South America
Living, attenuated vaccines	Bacillus Calmette-Guerin (BCG)	Living avirulent strain of tubercle bacilli; only for tuberculin-negative, heavily exposed individuals (family members of cases, medical personnel in high endemic areas); results not totally satisfactory
	Hepatitis B virus	High risk persons like health care personnel, IV drug abusers, male homosexuals
	Measles-mumps-rubella	May be given together or separately; routine immunization of preschool children. Rubella should not be used within 3 months of pregnancy
	Oral polio vaccine (OPV)	Trivalent vaccine given 3 times at 6-8-week intervals, booster at 1 year; routine immunization
	Smallpox	No longer used routinely since smallpox is essentially eradicated; do not use during pregnancy, in immunosuppressed, in presence of eczema; vesicles at injection site must be present to indicate "take"
	Yellow fever	Travelers: Africa, South America
Toxoids	Diphtheria	Routine immunization with pertussis and tetanus in single injection; series of three in first year of life, booster 1 year later, then at 10-year intervals
	Tetanus	See above
Bacterial parts	Pertussis (bacterial antigen)	See above
	Cholera	Experimental

THE MICROORGANISM AND THE PROCESS OF INFECTION

Because of the intimate interaction between host response and a microorganism's ability to produce disease, terms describing attributes of the "strength" of the organism's pathogenic power are not easily differentiated. Some clinicians use the words "pathogenic" and "virulent" as synonyms. Others describe virulence as the degree of pathogenicity that an organism possesses. Clearly the words pathogen and nonpathogen are relative, since microbes previously considered to be harmless are now recognized as quite capable of causing disease under certain circumstances. For example, *Serratia marcescens* was used in the past by students of microbiology to trace the passage of bacteria in the GI tract. It was considered to be such a harmless organism that it was swallowed and feces were subsequently examined for the colonies producing the typical red pigment of some strains of *S. marcescens*. Yet during the 1970s a small epidemic of disease caused by *S. marcescens* occurred in San Francisco. The organism is now a common cause of hospital-associated and community-acquired infectious disease.

Since essentially all microorganisms are capable of producing disease in some host given appropriate conditions, we can use the word *pathogen* to describe an organism that causes disease with some frequency, and the word *virulence* to differentiate the *degree* of pathogenicity that an organism possesses. Thus, a virulent pathogen is one that requires very few organisms in a healthy host to cause disease. A nonvirulent or opportunistic pathogen can cause disease in a compromised host.

We discuss characteristics of microorganisms in the establishment of infection in three steps: First, the microbe must gain entrance into the host. Second, it must become established and multiply in the host. The concepts of invasiveness and adherence are essential at this step. The first two steps are necessary for colonization or infection to take place. The pathogenicity of some microbes may result primarily from their high invasiveness even if they are not toxigenic. Third, if clinical disease is to ensue, the organism must produce some cytotoxic effects in the host. These cytotoxic effects determine the virulence of the organism. They may result from toxigenicity (the ability of an organism to produce potent toxic substances), from direct damage of cell lysosomes and alterations of infected cell membranes, or simply from continued, uninterrupted growth of the infecting organisms.

Entrance of Organisms into Host

Physical and chemical barriers to microbial invasion were discussed as attributes of the host. Despite intact host defenses, organisms do enter the body, most frequently through the respiratory tract, the GI tract, and breaks in the integumentary system. Many microbes are very specific regarding the types of tissues that they can invade. *Neisseria gonorrhoeae,* for example, selectively thrives in the genitourinary tract and generally does not enter the body through other routes. This characteristic of selectivity is called *organotropism* and is due primarily to nutritional determinants and to the organism's ability to adhere to specific tissues. *Staphylococcus aureus* colonizing the anterior nares is quite harmless and is part of the normal flora of about one-fourth of the healthy population. The same organism, however, when introduced into an open wound during surgery can cause fulminant clinical disease. The St. Louis and Japanese B encephalitis viruses are capable of causing disease in mice only if they are injected directly into the brain. These are examples of what happens in situations where organisms are allowed to enter the body by routes that are normally inaccessible. Hospitalized patients with invasive lines (urinary catheters, intravenous therapy, endotracheal intubation) and patients who have suffered traumatic injuries, surgery, and burns are highly susceptible to infection because the ability of potential pathogens to enter the host is enhanced.

Establishment and Multiplication of Organisms in Host

For an organism to become established in a host, it must first become attached or adhere to cells within the host. A microbe's organotropism is in large part explained by preferential adherence factors, especially in the respiratory and intestinal tracts. The flora of various parts of the mouth and GI tract are quite distinct. *Streptococcus sanguinis* and *Strep. salivarius* adhere to epithelial cells of the mouth and cheek but not of the teeth, and *Strep. mutans* primarily adheres to the enamel surface of teeth but not to the tongue or cheek. *E. coli,* on the other hand, attaches well to intestinal and urinary bladder cells, but not to cells of the upper respiratory tract. For viruses, as well, response to infection is determined by viral attachment or release of viral nucleic acid into the cells. Resistance to infection may be caused by failure of viral adsorption. Surface receptors, which enhance adherence, are present on some organisms but not others. They may reside on pili or on fine filamentous fringe.

Organisms that possess the selective ability to adhere strongly to epithelial surfaces may have increased virulence. Streptococcal strains from children with rheumatic fever have been shown to adhere more avidly to pharyngeal cells of patients with acute rheumatic fever than strains that were not associated with rheumatic fever (Reed et al. 1980). In a study of *Neisseria meningitidis* harvested from buccal epithelial cells of patients and carriers, it was concluded that increased adherence could have facilitated host colonization, promoted carriage, and altered virulence (Craven et al. 1980). In studies of *Mycoplasma pneumoniae,* virulence and adherence to tracheal epithelial cells have been correlated. Likewise, a study of bacterial superinfections following viral upper respiratory infection concluded that such infections were in part due to alterations in mucosal cells that led to increased bacterial adherence (Fainstein, Musher, and Cate 1980). In the GI, respiratory, and GU tracts, large doses of infecting microbes are usually necessary to cause disease because the invading cells are competing for attachment sites. Normal flora may protect the body from persistent colonization with more virulent species by binding to attachment sites, making them unavailable to invaders.

When a microbe has established itself by adhering to host cells, it must then spread and multiply. Some organisms remain localized; cell multiplication and infection remain near the site of entry. These organisms are relatively noninvasive and spread primarily by cell contact, colonizing only epithelial surfaces. Such microbes include wart viruses, *Corynebacterium diphtheriae, Vibrio cholerae,* and *Bordetella pertussis* (the virulence of the latter three organisms is due to toxigenicity, but not invasiveness). Other organisms, intermediate on the scale of invasiveness, are able to penetrate and destroy the epithelial cell layer. *Staphylococcus, Streptococcus,* and *Shigella* species can produce disease without penetrating the submucosal layer. Highly invasive organisms like *Bacillus anthracis* and *Yersinia pestis* can readily pass into submucosal layers and spread throughout the body. Some viruses (herpes simplex) may also travel along nerves. Spread of microorganisms throughout the body may be hematogenous, which occurs when organisms enter the blood stream, most commonly via the lymphatic system.

Invasiveness of microorganisms may be enhanced by the elaboration of enzymes such as hyaluronidase, coagulase, streptokinase, and staphylokinase, by certain gram-positive bacteria, and by collagenases of *Clostridium*. Many organisms that are highly invasive may be widely distributed throughout the body without causing disease. Thus, invasiveness and virulence are neither synonymous nor necessarily related. There may be invasive nonpathogens, noninvasive nonpathogens, invasive pathogens, and noninvasive pathogens.

Besides adherence and invasiveness, the organism's response to phagocytosis is a third determinant of whether colonization or disease results. Some organisms possess direct antiphagocytic mechanisms, such as the ability to produce leukocidins. Microbes may be extracellular, facultative intracellular, or obligate intracellular parasites. Extracellular parasites are usually killed readily by phagoctyes. They produce disease in individuals having impaired defenses. Other extracellular parasites, as mentioned previously, may be protected from phagocytosis by possession of a polysaccharide capsule. "Rough strains" of *Strep. pneumoniae,* for example, which do not have such capsules, are readily phagocytized and avirulent. "Smooth strains" with large capsules are capable, in the absence of specific antibody, of producing severe pneumoccocal pneumonia because of their ability to "slip away" from phagocytes.

Some organisms, even after ingestion by phagocytes, are able to resist being killed. These facultative intracellular parasites (such as *Brucella* species, many fungi, and *Mycobacteria tuberculosis*) often produce more chronic diseases involving macrophages and other cells of the reticuloendothelial system. Viruses, *Chlamydia, Rickettsia,* and a few other bacteria are obligate intracellular parasites unable to multiply outside of cells. They may reside in reticuloendothelial cells, in blood vessel endothelial cells (Rickettsiaceae), or even in voluntary muscle cells *(Mycobacterium leprae)*. Table 4-10 lists common disease-producing organisms according to whether they are extracellular or intracellular invaders.

The establishment and multiplication of some viruses are also enhanced because they

(text continues on page 87)

TABLE 4-10. Classification of Disease-Producing Microorganisms by Mechanism of Multiplication

Type	*Organism*	*Diseases produced*	*EC parasite*	*Facultative IC parasite*	*Obligate IC parasite*
B	*Acinetobacter calcoaceticus,* var. *anitratus, lwoffi*	Opportunistic infections	X		
B	*Actinobacillus mallei*	Actinomycosis	X		
B	*Actinomyces israelii*	Subcutaneous, skin granulomatous disease		X	

TABLE 4-10. *(continued)*

Type	Organism	Diseases produced	EC parasite	Facultative IC parasite	Obligate IC parasite
F	*Aspergillus fumigatus*	Opportunistic infections		X	
B	*Bacillus anthracis*	Skin infection, pneumonia ("Wool sorter's Disease)	X		
	B. cereus, B. subtilis	Food poisoning	X		
B	*Bacteroides fragilis*	Opportunistic	X		
F	*Blastomyces dermatitidis*	Blastomycosis		X	
B	*Bordetella pertussis*	Whooping cough	X		
B	*Brucella* (all species)	Undulent fever		X	
F	*Candida albicans*	Candidiasis		X	
B	*Chlamydiae psittaci*	Psittacosis			X
	C. trachomatis	Trachoma			X
B	*Citrobacter freundii, C. diversus*	Opportunistic infections	X		
B	*Clostridium botulinum*	Botulism	X		
	C. perfringens	Gas gangrene	X		
	C. tetani	Tetanus	X		
F	*Coccidioides immitis*	Coccidioidomycosis		X	
B	Coliforms *(E. Coli, Klebsiella, Enterobacter, Enterococcus, Proteus)*	Opportunistic infections	X		
B	*Corynebacterium diphtheriae*	Diphtheria	X		
B	*Coxiella burnetti*	Endocarditis			X
F	*Cryptococcus neoformans*	Cryptococcosis		X	
F	*Epidermophyton* (all species)	Athlete's foot	X		
B	*Francisella tularensis*	Tularemia		X	
B	*Hemophilus* (all species)	UTI, conjunctivitis, meningitis	X		
F	*Histoplasma capsulatum*	Histoplasmosis		X	
B	*Leptospira* (all species)	Leptospirosis	X		
B	*Listeria monocytogenes*	URI, skin lesions, abortion, meningitis, etc.		X	
F	*Microsporum* (all species)	Skin, hair infections	X		

TABLE 4-10. *(continued)*

Type	Organism	Diseases produced	EC parasite	Facultative IC parasite	Obligate IC parasite
B	*Moraxella* (all species)	Opportunistic infections		X	
B	*Mycobacterium leprae*	Leprosy		X	
	Mycobacterium (other species)	Tuberculosis, respiratory infections		X	
B	*Mycoplasma pneumoniae*	Pneumonia	X		
B	*Neisseriae gonorrhoeae*	Gonorrhea	X		
	N. meningitidis	Meningitis	X		
F	*Nocardia* (all species)	Nocardiosis		X	
B	*Pasteurella multocida*	Hemorrhagic fever	X		
B	*Providencia* (all species)	Opportunistic infections	X		
B	*Pseudomonas pseudomallei*	Melioidosis	X		
	Pseudomonas (other species)	Opportunistic infections	X		
B	*Rickettsia* (all species)	Many diseases			X
B	*Salmonella typhi*	Typhoid fever		X	
	Salmonella (other species)	Diarrhea, food poisoning		X	
B	*Serratia* (all species)	Opportunistic infections	X		
B	*Shigella* (all species)	Dysentery	X		
F	*Sporothrix schenckii*	Sporotrichosis	X		
B	Staphylococci, streptococci (all species)	Many infections	X		
B	*Treponema pallidum*	Syphilis		X	
F	*Trichophyton* (all species)	Skin infections	X		
B	*Vibrio cholerae*	Cholera	X		
V	Viruses (all)	Many diseases			X
B	*Yersinia pestis*	Plague		X	

B = bacterium
F = fungus
V = virus

may be able to multiply well at elevated body temperatures, they may be poor inducers of interferon production or resistant to its action, or they may block biosynthesis of host cells. In summary, the major factors that influence the microbe's ability to become established and multiply in a host are adherence to specific epithelial cells, invasiveness, and ability to resist ingestion or killing by phagocytic cells.

Cytopathic Effects on Host

Viruses produce host cell destruction in a number of ways: by depressing cell synthesis, by causing discharge of hydrolytic enzymes from cell lysosomes into the cytoplasm, or by altering host cell membranes so that they are targets for immunologic destruction. They may also have a direct toxic effect, similar to that caused by toxins elaborated by certain virulent bacteria. For example, mumps virus causes lysis of red blood cells, vaccinia virus kills macrophages, and intravenous injection of influenza viruses in mice causes hemorrhage and cell necrosis.

Toxigenicity in bacteria results from the production of exotoxin by a number of species, or from endotoxins by enteric bacteria, *Bordetella pertussis,* and *Yersinia pestis*. The basis for elaboration of exotoxin does not usually rest in the bacterial chromosome, but is the result of infection with temperate bacteriophage in the cases of diphtheria, botulism, and streptococcal food poisoning. The exotoxin of enteropathogenic *E. coli* and some staphylococcal toxins are coded for by plasmids. The endotoxin of gram-negative bacteria, on the other hand, is the lipopolysaccharide component of the bacterial cell wall, which is released when the bacteria die. The exotoxins of *Corynebacterium diphtheriae, Shigella dysenteriae,* and several clostridial species are among the most potent poisons known. Enterotoxins, substances that react in the gut to cause disease, are elaborated by certain bowel flora or other bacteria. A comparison of exotoxins and endotoxins is found in Table 4-11, and

TABLE 4-11. A Comparison of Exotoxins and Endotoxins

Criteria	*Exotoxins*	*Endotoxins*
Produced by	Gram-positive or gram-negative bacteria	Gram-negative bacteria only
Method of production	Coding in temperate bacteriophage or plasmids	Portion of gram-negative bacterial cell wall
Potency	Usually highly toxic	Less potent
Metod of liberation	Excreted by living cells	Liberated when cells lyse
Stability	Inactivated by heat over 60°C; relatively unstable; may be converted into toxoids by heat, acid, formalin	Heat stable; cannot be converted into toxoids
Composition	Polypeptides	Lipopolysaccharide
Antigenicity	Highly antigenic; stimulates formation of antitoxin	Carbohydrate portion antigenic, but no antitoxin formed
Clinical effects	None act the same; have specific effects; do not produce fever	Have common pharmacologic effects including leukopenia, intravascular coagulation, hypotension, fever

specific characteristics of major exotoxins and enterotoxins are described in Table 4-12. Extracellular enzymes produced by some bacteria, though not toxins, increase their disease-producing ability (Table 4-13).

Virulence can be measured by injecting known quantities of organisms into an experimental animal model and recording the numbers of animals that die at each dosage level. Numerous animals are needed because of individual variation even among one species. The dose at which 50 percent of the animals are killed is called the lethal dose 50 (LD_{50}). The virulence of several bacterial strains or the differences in species response to the same strain of bacteria can be compared in this manner. Data for humans are obtained by observing the natural course of disease and by comparison to experimental animals that are known to respond to a particular organism in a manner similar to humans. Factors affecting the development of disease in a susceptible host are summarized in Table 4-14.

CLINICAL IMPLICATIONS

In this chapter we have discussed the important attributes of the host and the microorganism that are determinants in the process of infection. In our clinical practice,

TABLE 4-12. Characteristics of Various Exotoxins and Enterotoxins

Organism	*Disease*	*Characteristics*
Exotoxins		
Clostridium botulinum	Botulism	Excretes a single highly potent toxin that grows in food anaerobically; blocks release of acetylcholine at synapses and neuromuscular junctions; causes paralysis
Clostridium perfringens, septicum, novyi	Gas gangrene	At least 10 exotoxins produced; under anaerobic conditions spores from soil or feces germinate; vegetative cells ferment carbohydrates, produce lactic acid and gas, causing necrosis and hemolysis in host
Clostridium tetani	Tetanus	Requires anaerobic conditions; vegetative forms produce exotoxin, which blocks neural transmission resulting in intense muscle spasm
Corynebacteria diphtheriae	Diphtheria	Only strains lysogenized by beta bacteriophage are toxigenic; arrests protein synthesis, causing necrosis of heart muscle, kidney, nerve tissue, epithelium
Pseudomonas aeruginosa	Unknown	Appears to have action identical to diphtheria exotoxin; antitoxins found in human sera, though role in human disease is unknown

TABLE 4-12. *(continued)*

Organism	*Disease*	*Characteristics*
Exotoxins		
Shigella dysenteriae	Dysentery	Produces protein that affects GI tract and CNS; causes extremely severe infection, lethal in experimental animals
Staphylococcus aureus	Skin necrosis	Contains several hemolysins
Streptococcus pyogenes (beta-hemolytic streptococcus, group A)	Rash of scarlet fever; "strep throat"	Erythrogenic exotoxin; controlled by temperate bacteriophage; mechanism of action unknown
Enterotoxins		
Bacillus cereus	Food poisoning	Stimulates adenylate cylase in small intestine, increasing concentration of cyclic adenosine monophosphate (cAMP)
Clostridium perfringens	Food poisoning with diarrhea	As above *(Bacillus cereus)*
Escherichia coli	Diarrhea	Action similar to *Bacillus cereus;* controlled by a plasmid; causes excess excretion of water and chloride, inhibition of sodium reabsorption
Staphylococcus aureus	Food poisoning	Acts on neural receptors in gut and stimulates vomiting center; produced especially in protein and carbohydrate foods
Vibrio cholerae	Massive diarrhea	As above *(Bacillus cereus)*

TABLE 4-13. Extracellular Enzymes that Increase Disease-Producing Ability of Certain Bacteria

Enzyme	*Produced by*	*Effects*
Coagulase	Staphylococci	Coagulates plasma; forms fibrin walls around lesions, protecting organisms from phagocytosis
Collagenase	*Clostridium perfringens*	Disintegrates collagen, enhancing spread of organisms
Fibrinolysin (streptokinase)	Hemolytic streptococci	Dissolves coagulated plasma; enhances spread of streptococci in tissues
Hemolysins	Many gram-positive bacteria and gram-negative rods	Dissolves red blood cells and tissue cells
Hyaluronidase	Clostridia, staphylococci, streptococci	Hydrolyzes hyaluronic acid, promoting bacteria spread in tissues
Leukocidins	Many gram-positive and gram-negative bacteria	Kills leukocytes

TABLE 4-14. Factors Affecting the Process of Infection

Factors	*High-risk state(s)*
Host	
Age	Infants, aged
Genetics	Defects (see Table 6-10)
Hormones	Diabetes, pregnancy, adrenal dysfunction
Inflammatory response	Conditions that interfere with inflammatory response (chronic alcoholism)
Medical therapy	Radiation, cytotoxic drugs, steroids
Nutritional state	Malnutrition
Phagocytic function	Circulatory disturbances; neutropenia; chemotactic defects
Skin, mucous membranes	a. Breaks in skin b. High moisture: sweating c. Interference with flow of urine, tears, saliva, cough reflex, ciliary action d. Changes in gastric acidity
Microorganism	
Adaptation and selectivity of organisms	Presence of organisms in susceptible tissue
Adherence	High affinity for attachment to epithelial cells
Invasiveness	
Antiphagocytic mechanism	Production of capsules or leukocidins
Intracellular parasitism	Ability to multiply in phagocytic cells
Extracellular enzymes	Elaboration of collagenase, streptokinase, etc.
Microbial antagonism	Alterations in normal flora (e.g., antimicrobial therapy)
Toxigenicity	Ability to produce exotoxins or endotoxins
Ability to resist phagocytosis	Organisms with capsules
Ability to enter intact skin	Infection with tularemia bacterium, some parasites

there are important implications for preventive and protective patient care.

1. Assess the status of the patient's host defenses.
   a. What physical barriers to infection have been interrupted?
   b. Are there chemical barriers to infection that are malfunctioning?
   c. Does the patient exhibit signs of normal phagocytic ability?
2. Recognize patients who are at high risk of infection caused by
   a. Genetic impairment in host defenses
   b. Physiologic status (hormonal imbalance, extremes in age, diabetes mellitus, adrenal malfunction, presence of other infections, circulatory problems, malnutrition)
   c. Therapy (mechanical ventilation,

invasive lines, surgery, radiation, corticosteroids, or cytotoxic drug treatment)
3. Recognize that the inflammatory response may be inadequate or poorly differentiated in certain conditions (when the infection occurs in deep tissues, when host defenses are suppressed).
4. Enhance the natural and nonspecific barriers to infection which are present in the patient. For example, maintain adequate nutrition, hormonal, and acid-base balance. Prevent the entrance of organisms through unusual routes.
5. Be alert to early clinical manifestations of infection (Table 4-15).

(text continues on page 94)

TABLE 4-15. Clinical Evidence of Infection by Body System

Body system	*Signs or symptoms of infection*	*Appropriate specimen*
Skin (cutaneous infections)		
The skin is a covering that protects the body and serves as the body's first line of defense against invasion by organisms. When the skin is scratched, cut, or broken in some other way, this line of defense is weakened and creates an opportunity for organisms to enter the body.	• Heat • Pus • Pain • Redness • Swelling	Aspirate or tissue sample taken directly from infected site Swab specimen from wound site
Skin infections can vary in size from small, contained infections (e.g., minor stitch abscess, surgical wound infection) to large, uncontained infections (e.g., in cases of extensive burns).		
Gastrointestinal system		
Parts of the gastrointestinal tract (e.g., peritoneum) are normally sterile; others (e.g., colon) are not. Sources of infections in the gastrointestinal tract may be exogenous, as in the case of shigellosis; or endogenous, as in the case of leakage into the peritoneum, which can be related to natural occurrences or surgical procedures.	• Cramps • Diarrhea • Vomiting • Fever	Stool Rectal swab
Gastrointestinal fluids leaking into the peritoneum can lead to bacteremia (i.e., pathogenic organisms in the blood) and can result in failure of multiple organ systems of the body.		Fluid from intraabdominal abscess

TABLE 4-15. *(continued)*

Body system	*Signs or symptoms of infection*	*Appropriate specimen*
Urinary system Urinary tract infections are the most common precursors to gram-negative bacteremia. Most occur with endogenous flora. *E. coli* is the most common pathogen. Infections may occur in the upper or lower tract. *Upper.* The upper urinary tract is composed of the kidneys and ureters. Infections of the upper tract are not as common as infections of the lower tract. *Lower.* The lower urinary tract is composed of the urethra and bladder. The bladder is a hollow muscular structure that normally evacuates small amounts of bacteria. Malfunctions or the introduction of large numbers of organisms may result in nosocomial infections in the lower tract. These infections occur most frequently when indwelling catheters are used.	• Back or suprapubic pain • Pyuria • Frequent or painful urination • Fever	Aspirated urine Clean voided midstream *or* catheterized urine (special collecting techniques may be used to differentiate upper from lower urinary tract infections) Clean voided midstream urine
Respiratory system Infections in the respiratory tract are most commonly caused by aspiration of microorganisms (which may be endogenous or exogenous) from the upper airway into the lower respiratory tract. Infections may occur in the upper or lower respiratory tract. Signs and symptoms of infection (e.g., fever, dyspnea, purulent sputum) are also often produced by chronic respiratory diseases themselves (e.g., bronchitis, COPD), thus making it difficult to distinguish between new infections and those that are chronic. *Upper.* The upper respiratory tract is made up of the respiratory organs above the neck (i.e., nose, pharynx, sinuses, ears). *Lower.* The lower respiratory tract consists primarily of the lungs, which are highly vascular. Infections here are	• Pain • Purulent drainage • Swelling • Redness • Pain • Cough • Infiltrates on chest x-ray • Purulent sputum • Fever • Dyspnea (difficult, painful breathing)	Discharge from the site Throat swab Transtracheal or direct aspirate Expectorated sputum

TABLE 4-15. *(continued)*

Body system	*Signs or symptoms of infection*	*Appropriate specimen*
Respiratory system likely to manifest themselves systematically and are more serious than upper respiratory tract infections.	• Rales, ronchi	
Muscoloskeletal system		
Muscles. Muscles protect the body in a manner similar to the skin except that the muscles are deeper inside the body and closer to the organs. Most infections are exogenous and are usually produced by trauma (e.g., severe burns).	• Inflammation • Pain • Heat • Swelling • Purulent drainage	
Bones. The most common bone infection is osteomyelitis. Exogenous wound infections may occur following trauma or surgical procedures (e.g., mediastinotomy, orthopedic prosthetic implant such as a hip). In the case of prosthetic implants, a deep surgical wound infection with no invasion of the bone can occur.	• Pain • Purulent discharge • Loss of bone continuity seen on x-rays	Aspiration or tissue sample
Circulatory system The circulatory system is composed of the blood vessels and the heart. Infections may be localized (e.g., purulent phlebitis) or systemic (e.g., bacteremia, that is, pathogenic microbes present and multiplying in the blood stream). Localized cardiac infections (e.g., mitral valve infection) may become systemic if bacteria get in the bloodstream and attach themselves to various organs as the blood circulates through the body. In general, infections in the heart come from endogenous sources. The most common circulatory infections are • localized infection such as purulent phlebitis (i.e., infection of the blood vessels) • systemic infection such as subacute bacterial endocarditis (i.e., infection in the heart) and the structures immediately adjacent to the heart (e.g., valves), primary bacteremia.	• Redness • Swelling • Purulence • Bacteremia • Chills • Fever • Pain	Aspiration or tissue sample Purulent drainage Semiquantitative culture of IV catheter* Blood

TABLE 4-15. *(continued)*

Body system	*Signs or symptoms of infection*	*Appropriate specimen*
Neurologic system. The neurologic system includes the brain and the spinal cord. Localized peripheral infections in the neurologic system are rare and occur only as a result of trauma. Infections that can occur in the central nervous system include encephalitis and meningitis. Bacterial infections of the neurologic system are usually caused by endogenous flora.	• Purulent CSF • Pain at the site of infection • Interference with neurologic functions at the site of infection • Fever	CSF
Reproductive System *Male.* There is no reason to differentiate the structures associated with reproduction from the internal and external genitalia. Localized skin infections such as herpetic lesions and internal infections such as urethritis or epididymitis are common.	• Purulent discharge • Heat • Pain • Swelling	Discharge from the infected site
Female. External infections include vaginitis and localized skin infections. Internal infections involve the uterus, fallopian tubes, or ovaries.	• Purulent discharge • Heat • Pain • Swelling	Culture directly from the infected site Endocervical swab

*As described in Maki DG, Weise CE, Sarafin HW. A semiquantitative culture method for identifying intravenous catheter-related infection. N Engl J Med 1977;296:1305.

Reproduced from course materials developed by Centers for Disease Control for the Criterion Referenced Course: Management Skills for Infection Control Nurses, June 1982.

Routes of Transmission

Infectious agents are spread between individuals by several different routes. Many agents are spread by *direct contact* with infectious materials or individuals, especially by means of the hands or mouth. *Sexual* contact and *parenteral* injection of agents are two specific forms of direct contact that have become more important as diseases such as gonorrhea, genital herpes, and hepatitis B increase in prevalence.

Insect vectors also spread certain diseases, especially those caused by *Rickettsia* and viruses, by the parenteral route. Other organisms enter the body through minor breaks in the skin when contaminated objects like animal carcasses are handled (tularemia, anthrax), or when the skin is broken as in surgery or trauma (anaerobic infections such as gas gangrene, staphylococcal and streptococcal infections).

Fomites, which are contaminated objects such as clothing, books, or bedding, can serve to indirectly spread infection. Organisms such as the tubercle bacillus and spore forming organisms are resistant to drying and can survive well in the environment for long periods of time. Respiratory infections are spread by contaminated aerosols called *microdroplets* during talking, coughing, etc., or by contamination of the hands with infected secretions and subsequent hand contact with another person. The common cold is most frequently spread by this latter method.

In the *fecal-oral route,* food or water is contaminated with infected fecal material directly or on the inadequately washed hands of food handlers. Enteric infections such as hepatitis A, food intoxications, and the dysenteries and diarrheas are spread in this manner. Flies may also spread enteric infections by mechanically depositing feces on food as they move from one item to another. Ingesting contaminated milk or animal products can spread diseases such as parasitic infestations, tuberculosis, and brucellosis. *Transplacental transmission* of some diseases such as syphilis occurs from mother to fetus.

As each organism in subsequent chapters is discussed, its mode of transmission will be described. It is vital for the health care practitioner to understand how a particular disease is spread in order to plan appropriate preventive strategies.

SUGGESTED READINGS

Adler WH, Jones KH, Brock MA. Aging and immune function. In: Behnke JA, Finch CE, Moment GB. Biology of Aging. New York: Plenum Press, 1978.

Berheim HA, Block LH, Atkins E. Fever: pathogenesis, pathophysiology, and purpose. Ann Intern Med 1979;91:261-270.

Berkelman RL, et al. Streptococcal wound infections caused by a vaginal carrier. JAMA 1982;247:2680-2682.

Craven DE, et al. Adherence of isolates of *Neisseria meningitidis* from patients and carriers to human buccal epithelial cells. J Infec Dis 1980;142:556-568.

Cross AS, et al. Evidence for the role of toxin A in the pathogenesis of infection with *Pseudomonas aeruginosa* in humans. J Infec Dis 1980;142:538-546.

Dahl MV. Chemotaxis: death march of the phagocyte (editorial). Arch Dermatol 1979; 115:1407-1408.

Evans AS. Causation and disease: the Henle-Koch postulates revisited. Yale J Biol Med 1976;49:175-194.

Fainstein V, Musher DM, Cate TR. Bacterial adherence to pharyngeal cells during viral infection. J Infec Dis 1980;141:172-176.

Flesher AR, et al. *Legionella pneumophila:* growth inhibition by human pharyngeal flora. J Infec Dis 1980;142:313-317.

Gardner IO. Effect of influenza virus on susceptibility to bacteria in mice. J Infec Dis 1980;142:704-707.

Golshan N, Barker DF, MacGregor RR. Granulocyte function in recipients of renal transplants from live related donors and from cadavers. J Infec Dis 1980;142:876-883.

Granoff DM, Basden M. *Haemophilus influenzae* infections in Fresno County, California: a prospective study of the effects of age, race and contract with a case on incidence of disease. J Infec Dis 1980;141:40-46.

Groenwald SL. Physiology of the immune system. Heart Lung 1980;9:645-650.

Johnson, RS. Interferon: cloudy but intriguing future (news). JAMA 1981;245:109-110, 115-116.

Makinodan T, Kay MMB. Age influence on the immune system. Adv Immunol 1980;29:287-323.

McKee WM, et al. Anal carriage as the probable source of a streptococcal epidemic. Lancet 1966;2:1007-1009.

McMurray ON, et al. Development of impaired cell-mediated immunity in mild and moderate malnutrition. Am J Clin Nutr 1981;34:68-77.

Overall JC, Yeh TJ, Kern ER. Sensitivity of

herpes simplex virus types 1 and 2 to three preparations of human interferon. J Infec Dis 1980;142:943.
Pichichero ME, et al. Breast milk antibody to capsular polysaccharide of *Haemophilus influenzae* type B. J Infec Dis 1980;142:694-698.
Puck JM, et al. Protection of infants from infection with influenza A virus by transplacentally acquired antibody. J Infec Dis 1980; 142:844-849.
Reed WP, et al. Streptococcal adherence to pharyngeal cells of children with acute rheumatic fever. J Infec Dis 1980;142:803-810.
Repine JE, Clawson CC, Goetz FC. Bactericidal function of neutrophils from patients with acute bacterial infections and from diabetics. J Infec Dis 1980;142:869-875.
Rosenstreich DL. Genetics of resistance to infections. Nature 1980;285:436-437.
Schonberger LB, et al. Guillain-Barré syndrome following vaccination in the national influenza immunization program, United States, 1976–77. Am J Epidemiol 1979;110:105-123.
South MA. The newborn's susceptibility to infection. Cutis 1979;24:23-24.
South MA. Nutritional deficiency and the immune response. Cutis 1980;26:343,346,350.
Spruance SL, Chow FS. Pathogenesis of herpes simplex labialis. I. Replication of herpes simplex virus in cultures of epidermal cells from subjects with frequent recurrences. J Infec Dis 1980;142:671-675.
Walton B. Effects of anaesthesia and surgery on immune status. Br J Anaes 1979;51:37-41.
Weller PF, Goetzl EJ. The human eosinophil roles in host defense and tissue injury. Am J Pathol 1980;100:791-820.

5 HUMORAL IMMUNITY/ ANTIGEN-ANTIBODY REACTIONS

KEY POINTS

- Antibodies are produced by vertebrates *specifically* in response to contact with an agent perceived by the organism's immune system as foreign; only certain cells of the lymphatic system, B cells, are capable of producing antibodies.
- Activation of the complement system and opsonization are the major mechanisms by which antibody mediates killing of microbial invaders.
- There are five major types of antibody: IgG, IgM, IgA, IgD, and IgE.
- IgM is most important during the first encounter with a foreign substance (called an *antigen*). It is abundant in the *primary response*.
- IgG is the most common antibody and is the major antibody found when the same foreign substance is again encountered (during the *secondary response*). The secondary response is much more rapid and results in a greater production of antibody than does the primary response.
- IgA is the major antibody found in body secretions such as tears and saliva. The function of IgD is not well understood; IgE is an important factor in the allergic response.
- Complement is a system of proteins in the blood that can result in the killing of foreign invaders.

The nonspecific host defenses described in Chapter 4 dispose of about 90 percent of microbial invaders. These mechanisms are available immediately upon threat of microbial invasion. As the body's nonspecific defenses respond to microbial invasion, specific immune responses are also mobilized and antibodies to the foreign invader are formed, a process that takes from 4 to 10 days. Specific humoral defenses are mediated by the antibody (Ab) molecule. Specific defenses are often necessary to amplify nonspecific defenses in case of rapidly multiplying or highly virulent pathogens, and to incur lasting immunity to specific antigens.

Immunity due to antibodies is called *humoral immunity* because antibodies are housed in the body fluids such as serum, saliva, colostrum, and nasal secretions. *Cellular immunity,* on the other hand, is focused primarily in one type of blood cell called the T lymphocyte. Cellular immunity is discussed in Chapter 6.

THE NATURE OF SPECIFIC IMMUNITY

Antigens

Specific humoral immunity results from exposure of the host to a foreign substance called an *antigen* (Ag, *anti*body *gen*erating), which elicits the formation of antibodies (Abs), also called immunoglobulins (Igs). Only vertebrates have immune systems that produce Abs. Antigens must have several characteristics: they must be recognized as "foreign" by the host (Abs are not generally formed to self-components) and they must have a molecular weight of at least 10,000 atomic mass units. The more foreign and the larger the Ag, the stronger the response. Some components of the body such as the lens of the eye or brain cells may be recognized as foreign by the body. If trauma results in injury to such areas, an immune response results. These self-antigens are called *occult* antigens.

Antigens are usually proteins or polysaccharides (such as the polysaccharide used to identify blood groups). Nucleic acids are poor Ags, but single-stranded forms are better than double-stranded forms. Lipids in general can only be antigenic if coupled with a protein, polysaccharide, or polypeptide as a hapten. *Haptens* are substances that can elicit Ab formation when coupled in this manner. The "carrier" protein is required for immunity to develop, but after this occurs, the hapten alone can induce Abs. Haptens may be of lower molecular weight than complete Abs.

The parts of the Ag against which Abs are formed are called the *antigenic determinant sites.* In bacteria, for example, portions of the cell itself can serve as Ags (somatic or O Ag and surface Ags), as can other cell components such as the flagella (H Ags), capsule (K Ags), or by-products such as exotoxins. There are multiple antigenic determinant sites on each cell; hence, many types of Abs can be formed against one Ag. Antigenic determinants are quite small; consisting of 4 to 8 amino acids or 3 to 6 glucose molecules. Serum albumin, as an example, has 400 amino acids so that 2×10^4 to 2×10^8 antigenic determinants are possible. An antigenic determinant site fits into a cavity of the specific Ab (Fig. 5-1*a*). Generally the affinity of an Ab is less if the Ag determinant is not identical to the determinant that elicited production of that Ab. Cross reactions, however, do occur between similar Ags. An Ab made to antigen A can react with antigen a if they have identical or very similar Ag determinant sites (Fig. 5-1B).

The antigenicity of substances varies. Insoluble Ag, for example, usually induces Ab formation better than soluble Ag. *Adjuvants* are substances used to improve the effectiveness of antigens. These are helpful in developing maximally effective—yet safe—immunizing agents by reducing the amount of Ag necessary for successful immunization. Some promising adjuvants include muramyl dipeptides (MDP), polynuleotides, and liposomes.

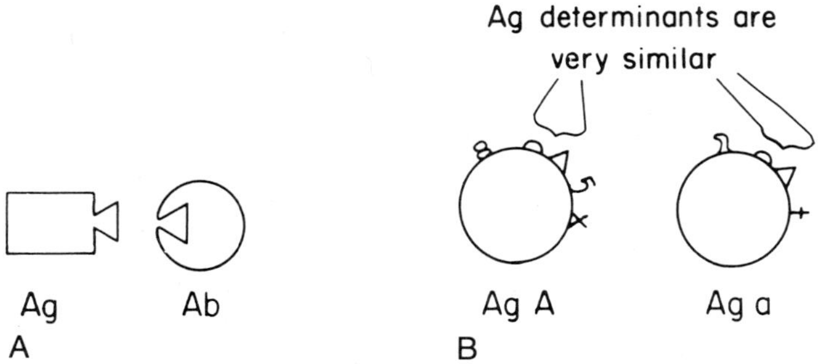

FIGURE 5-1. Antigen/antibody reactions. (A) Antigen and antibody must fit exactly for high affinity. (B) Two different antigens may cross react with the same Ab if antigen determinant sites are very similar, but affinity will be less.

Antigens of Blood Types

The four major blood types, called A, B, O, or AB, are determined by the presence of specific antigens on the red blood cells. The A antigen is present in type A blood, B antigen in type B blood, both A and B antigens in type AB blood, and neither in type O blood. To type blood, serum containing antibodies to type A (anti-A) is added to one drop of the blood to be typed and serum containing antibodies to type B (anti-B) is added to another drop. Antibody specific to the antigen in the blood will cause the red blood cells to clump or agglutinate (Table 5-1).

A second major blood group antigen is the Rh factor, so named because it is also found in the red blood cells of *rhesus* monkeys. Individuals may either have the Rh factor present in their red blood cells (Rh positive) or not (Rh negative). Though there are other blood antigen systems, the ABO and Rh systems are the two that must be crossmatched in blood donors and recipients to prevent transfusion reactions.

TABLE 5-1. Hemagglutination Test for Blood Typing

	Agglutinated by	
Blood type	*Anti-A Serum*	*Anti-B Serum*
A	+	−
B	−	+
AB	+	+
O	−	−

Cell That Mediate Specific Immunity

Specificity of the immune response is conferred by two types of lymphoid cells: the T and B lymphocytes. Both cell types originate in the bone marrow from a stem cell. The stem cell is the progenitor of several cell lines including red blood cells, white blood cells (discussed in Chapter 4), and lymphocytes. Though both T and B cells are produced within the bone marrow, T cells are also dependent for normal differentiation on a functioning thymus while B cells function independently of the thymus.

B cells are the antibody-producing cells. They get their name from the fact that their production was first discovered in the bursa of Fabricius in birds. In mammals, the bone marrow probably serves an analogous function. After leaving the bone marrow, lymphocytes undergo further differentiation in the spleen and lymph nodes. B lymphocytes are concentrated in subcapsular nodules and T lymphocytes accumulate between nodules. The germinal centers in the subcapsular nodules contain rapidly dividing B cells (Fig. 5-2*a, b*). These germinal centers are very large during periods of high Ab production. Other lymphoepithelial structures such as tonsils, Peyer's patches in the intestines, and the appendix are also sites of lymphocytic proliferation.

When Ag is bound on the cell surface of B lymphocytes, they are transformed into large plasma cells that synthesize Abs. Macrophages are also very important in the Ag/Ab reaction because they nonspecifically phagocytose Ag and "present" it on their surface so that the Ag is accessible to Ab-producing lymphocytes. When processed in this manner by macrophages, hundreds of times less Ag is needed to induce Ab production.

T cells, despite the fact that they do not produce Abs, mediate cellular immunity as well as function in several capacities essential to humoral immunity. T cells recognize Ags and regulate the type and intensity of cellular and humoral responses. In the thymus, these cells differentiate for helper or suppressor functions. B cells require "helper" T lymphocytes to form certain types of immunoglobulins. A discrete subset of T cells are balanced in a circuit to determine the outcome of Ag retriggering: the inducer (helper) T cells activate other T cells, B cells, and macrophages to enhance Ab production, whereas the suppressor cells inactivate the helpers when Ab production is adequate. T cells produce a long-lived subpopulation that probably enhances immunologic memory. Basic characteristics of T and B cells are compared in Table 5-2.

Thus, T and B lymphocytes result in the following two types of specific immunity:

1. B cells are responsible for Ab production. *Humoral immunity* results from these Abs circulating in the blood.
2. T cells do not make Ab, but they can react with Ag and are responsible for cell-mediated or *cellular immunity*.

The production of Abs by B lymphocytes will be discussed in the following section. Specific functions of T cells in regard to cell-mediated immunity will be discussed separately in Chapter 6.

ANTIBODIES

Antibody Formation

It is an astounding fact that Abs are extremely heterogeneous. Humans produce about 10^5 to 10^7 separate types of Ab molecules, but the exact mechanisms contributing to this diversity are not understood. Theories run the spectrum—from the postulation that all production of specific antibodies is already programmed in the genes of the fetus, to the hypothesis that Ab formation is solely controlled by the environment or depends upon chance exposure to antigens (Table 5-3). The theory most compatible with existing evidence is called the clonal selection theory, which hypothesizes that each individual carries a host of lymphoid cells, each of which can react with only one or a few closely related Ag determinants. These Ab molecules become activated when Ag is encountered and proliferate to form clones when they contact an Ag with which they can specifically react.

Antibody itself can help regulate the production of more Ab in the two following ways:

1. In the presence of high concentrations of Ab, the Ag determinant sites are

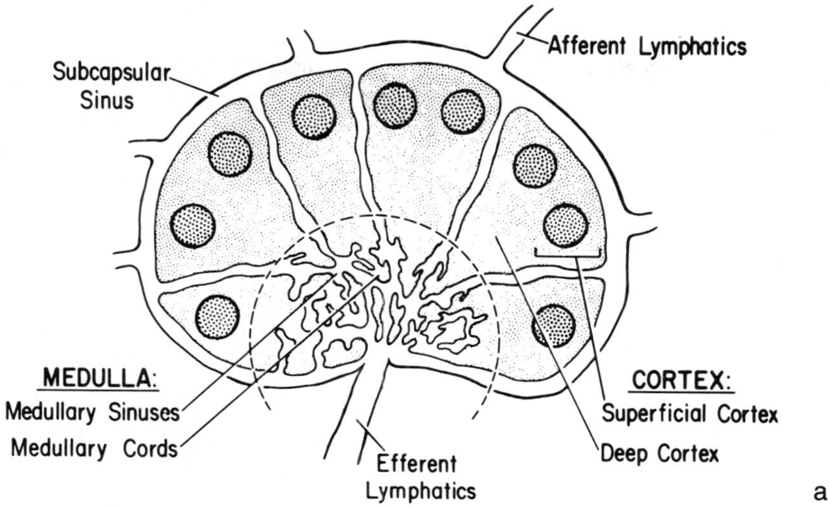

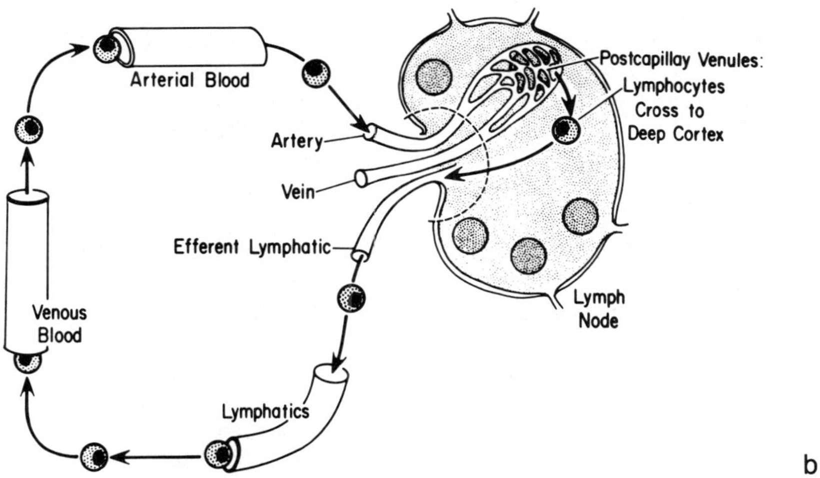

FIGURE 5-2. *(a)* Schematic diagram of lymph node where lymphocytes undergo differentiation after leaving the bone marrow. *(b)* Lymphocytes enter the node through the arteries, cross into the deep cortex, and leave by way of the effluent lymph. The cells reach the venous circulation when the major lymph ducts connect with veins: in most cases, at the neck. (Reprinted with permission of the publisher from Benacerraf B, Unanue ER. Textbook of Immunology. Baltimore: Williams and Wilkins, 1979.)

TABLE 5-2. Comparison of T and B Lymphocytes

B lymphocytes	*T lymphocytes*
Derive from bone marrow	Derive from bone marrow
20 percent of recirculating pool of small lymphocytes	65-80 percent of recirculating pool of small lymphocytes
Do not require functioning thymus	Require functioning thymus
Short life: days to weeks	Long life: months to years
Functions: largely responsible for specific Ab formation in the host; may work alone to produce AB or may need T cell cooperation	Functions: largely responsible for cell-mediated immunity; provide protection against most viruses, slow-acting bacteria such as Tb, fungal infections; responsible for graft rejection
Differentiate into Ab-producing cells	Do *not* differentiate into Ab-producing cells
Each cell carries only one type of Ig, but some carry both IgD and IgM. Some long-lived B cells serve as "memory cells"	Several types: cytotoxic, helper, suppressor
B cell defects cause inadequate Ig formation, but do not affect cell-mediated immunity	T cell defects cause defects in cell-mediated immunity. May also result in impaired Ab synthesis in spite of intact B cell system.
Responsible for hypersensitivity that is immediate (within minutes, lasting less than 1 hr)	Responsible for delayed hypersensitivity (onset many hours, lasting days)
Examples: anaphylactic shock, serum sickness, antibiotic or pollen allergy, asthma reaction	Examples: tuberculin hypersensitivity, skin graft rejection, reaction to poison ivy or oak
Since reaction is associated with circulating Ab, it can be passively transferred in serum	Since reaction is not associated with circulating Ab, it cannot be transferred passively by serum
Requires previous experience with that Ag	Nonspecific response requiring no previous experience with specific Ag
Primary response: 4-10 days after initial exposure to Ag	
Secondary response: 1-2 days after exposure	
Effective primarily against extracellular invaders	Effective against intracellular invaders

masked or blocked, making them inaccessible.

2. In the presence of low Ab concentrations, Abs may interfere with the cooperation of the B cell with helper T cells to inhibit production.

Most Ag/Ab reactions are T cell dependent in that Ab develops an increased affinity for

TABLE 5-3. Theories of Antibody Synthesis

Theory	*Description*
Clonal selection theory	Currently, most acceptable. Ab-producing cells inherit their ability to respond to a particular Ag. When this Ag is encountered, proliferation of the specific Ab-producing cell results
Germline theory	Most Ab genes are separately encoded and change by mutation and evolution
Somatic theories	There is a small set of variable-region (V) genes; somatic processes create variability
Point mutations	Each individual possesses each of a few V genes; mutations occur as needed in lymphoid cells
Recombination	Parts of basic V genes are scrambled and cells recombine to match different regions
Episome	Pool of minigenes to specify many hypervariable regions

Ag and undergoes other changes only under the influence of T lymphocytes. Ab responses to the lipopolysaccharide of gram-negative bacteria and a few other Ags are T cell independent.

Antibody Functions

Antibodies are specialized proteins called immunoglobulins (Igs). The function of Igs is to provide specific and effective mechanisms with which to assure the killing of foreign invaders. Antibodies protect the body from foreign invaders by causing microbial death, by neutralizing toxins, by preventing the attachment of microorganisms to mucous membranes (IgA type antibody), and by opsonization. Antibodies are highly specific and have been demonstrated to discriminate between two proteins that are identical except for a single amino acid residue.

Opsonization enhances the ability of macrophages to phagocytize microorganisms (Fig. 5-3). *Opsonins* are IgG or IgM type Abs, which—often with the help of the C3b

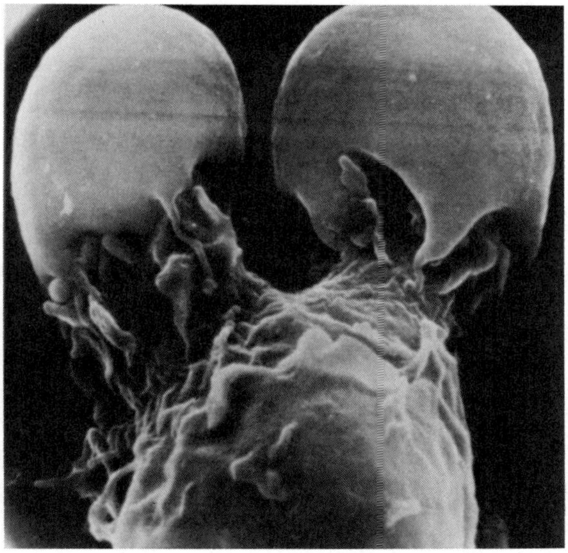

FIGURE 5-3. Beginning of phagocytosis of antibody-coated erythrocytes by a macrophage. The active role of pseudopods in engulfment is similar, but less strikingly seen (because of size), with Ab-coated bacteria. (Scanning electron micrograph courtesy of M.E. Bayer. Reprinted with permission of the publisher from Davis BD, et al. *Microbiology,* 3rd edition. Hagerstown, Md: Harper and Row, 1980.)

and C5b fragments of complement—coat foreign particles to increase their susceptibility to phagocytosis. It has been shown in animal models that opsonization and complement are necessary for clearance of pneumococci from the lung in pneumococcal pneumonia. Antibody formation is a major mechanism for killing viruses as well as bacteria. Serum Abs neutralize free virus in extracellular fluids in cooperation with complement or cytotoxic lymphocytes (Fig. 5-4). Antibodies may be able to interfere with synthesis and expression of viral proteins.

Primary and Secondary Antibody Responses

An individual's first exposure to an Ag results in a rise in serum Ab titer within 4 to 10 days. This does not mean that Ab is not produced earlier, but rather that the quantities are insufficient to be detected by current technology. The importance of Ab production was illustrated dramatically in the preantimicrobial era with pneumococcal pneumonia. Pneumococci, with their mucoid capsules, can resist phagocytosis. By remaining unaffected by the body's nonspecific defenses, they caused fulminant pneumonia with high mortality. The disease was characterized by a "crisis" after several days during which the patient either died or the fever "broke" and recovery began. Those who got well at this crisis point were those who had produced sufficient Ab to opsonize the phagocytosis of the pneumococci.

This first production of Ab is termed the *primary response*. The primary response peaks 1 to 10 weeks after Ag exposure and slowly drops thereafter (Fig. 5-5). After some infections immunity lasts for years and persistent detectable Ab levels exist for a lifetime. In other cases, Ab may fall below detectable levels. Depending upon the particular Ag and the strength of the Ab response, reinfection may be possible after some passage of time.

A *secondary response,* quite different than the primary response, occurs when the individual has a subsequent exposure to the same Ag. This exposure may be in the form of an immunization "booster" or of reinfection, and may occur years after first exposure. On subsequent exposure, Abs previously formed are primed and ready to respond immediately to Ag. This results from the production of long-lived memory cells, primarily by B cells but in some cases by T lymphocytes. Thus the secondary Ab response is much more rapid, lasts longer, and Abs produced have increased affinity for the antigen. Some cross reaction between antigenically similar organisms may occur; an example of this is the basis for influenza vaccination. Because of these characteristics, the secondary response is also called the *anamnestic* (recall) *response.*

Whereas in the primary response IgM is the major Ig formed, IgG is primarily formed in the secondary response. By mechanisms that probably involve recombination of genetic chromosomes on the Ab molecule, the same B cells and their clones that on primary response produce IgM then begin to produce IgG. This IgM-IgG switch is responsible for the increased affinity of Ab molecules in the anamnestic response. See Table 5-4 for a comparison of primary and secondary responses.

Antibody Structure

There are five types of antibodies. Fortunately, however, antibodies have been relatively easy to study since each plasma cell makes only one type of Ig. Cells for study can be obtained from humans and from many animal models, including "nude" mice whose

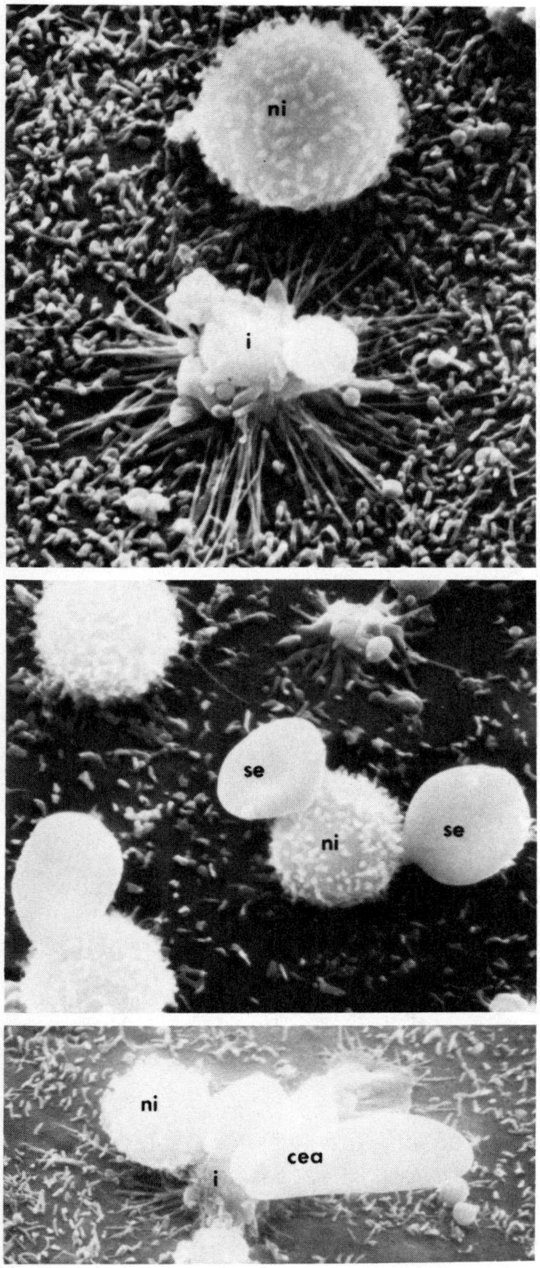

FIGURE 5-4. Antibody-dependent lymphocyte-mediated cytotoxicity (ADLC) on measles virus-infected cells as studied in the scanning electron microscope. *(Top)* Anti-measles virus antibody and human peripheral blood lymphocytes have been incubated for 4 hours at 37°C with measles virus-infected Vero cells. Approximately 50 percent of the absorbed lymphocytes interact strongly with the target cell *(i)*. The rest retain their original shape, that is, are noninteracting *(ni)*. Extreme deformation of the lymphocyte is due to the fact that Vero cells are very resistant to ADLC. Susceptible target cells are usually lysed at this time. (magnification 7000 ×) *(Middle)* Lymphocyte populations interacting and noninteracting with measles virus-infected cells are identical by in situ erythrocyte rosetting. T cells that are absorbing sheep erythrocytes (se) are noninteracting (ni) with the target. (4,400 ×) *(Bottom)* Fc receptor-carrying cells that are absorbing chicken erythrocytes coated with anti-chicken erythrocyte IgG (cea) are the interacting population. (3,500 ×) (Reprinted with permission of the publisher from Dubois-Dalcq M, Rentier B. Structural studies of the surface of virus-infected cells. Prog Med Virol 1980;26:158-213.)

spleens provide almost pure populations of B cells. Even subunits of Igs can be studied. For example, about 50 percent of patients with multiple myeloma (a tumor of plasma cells) excrete in their urine monoclonal Ab light chains called Bence Jones proteins.

Another method to study antibodies is *electrophoresis*. Electrophoresis separates

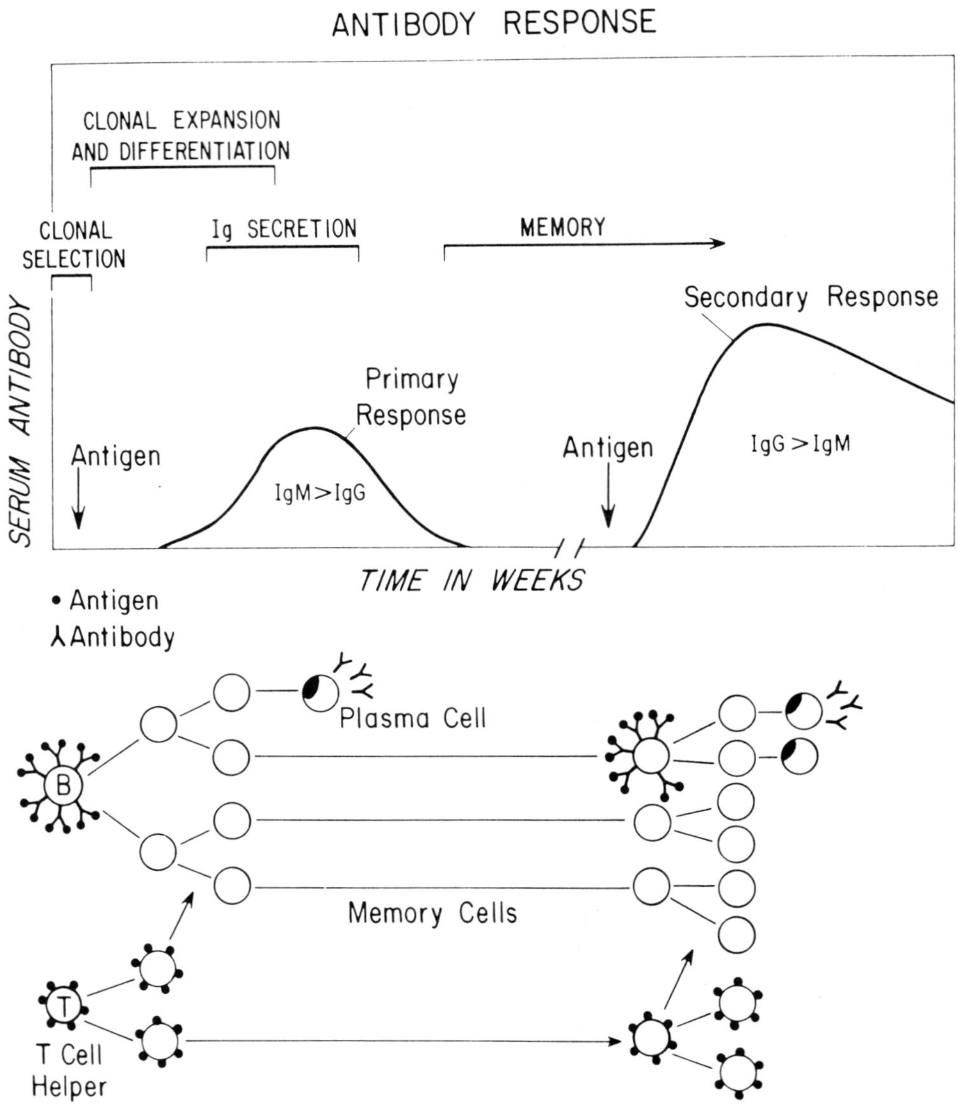

FIGURE 5-5. Antibody response in the serum against a conventional tymus-dependent antigen. The *top panel* depicts the serum levels of antibody after primary and secondary challenges. The *bottom panel* illustrates the cellular events in B cells, T cells and plasma cells. (Reprinted with permission of the publisher from Benacerraf B, Unanue ER. Textbook of Immunology, Baltimore: Williams and Wilkins, 1979.)

TABLE 5-4. A Comparison of the Primary and Secondary Antibody Response

Primary	*Secondary (amamnestic)*
IgM molecules predominate	IgG molecules predominate
Lag time before appearance of Ab about 10 days	Lag time about 2 days
Peak antibody concentration in serum: 0.5 μ/ml	Peak concentration: 80 μ/ml
Abs have moderate affinity for Ag	Abs have very high affinity for Ag
No memory cells available	Many more B cells, including memory cells, capable of responding to Ag

proteins, usually in agar gel, where they migrate in an electric field a distance according to their molecular weight. Albumins migrate the shortest distance, gamma globulins migrate farther than other proteins. In this way, it was determined that Abs are gamma globulins. After separation of serum proteins by this method, specific antisera may be allowed to diffuse via a trough for 18 to 24 hours into the gel along the lines of migration of the proteins. Immune precipitation of the proteins occurs and curved lines showing the areas of precipitation are formed. This is called *immunoelectrophoresis,* a technique invaluable for studying heterogeneity, specificity, and other characteristics of Abs (see Chapter 21).

Synthesis of Abs occurs in the ribosomes of plasma cells. Abs (Igs) are Y-shaped glycoproteins (Fig. 5-6) made up of four polypeptide chains connected by disulfide bridges. There are two identical heavy molecular weight (H) chains, which are flexible and bend at a hinge region. Each chain contains about 430 amino acids. The amino acid sequences of H chains differ and give the five classes of Igs their names. Two identical light molecular weight (L) chains are the same in all classes of Igs, containing 212 amino acids in two equal halves. The valence of an Ab indicates the number of Ag combining sites. Two H and two L chains combine to form a monomeric Ab.

Antibodies can be split by a plant protease (papain) into two large identical fragments with univalent Ag-binding capacity. These are called Fab (fragment antigen-binding). A third fragment, which can be separated by ion-exchange chromatography, is called Fc (fragment crystalline) because it can be crystallized. The Ab molecule is thus composed of mirror-image, identical halves. Antigen-combining regions are made of one L and one H chain.

H chains are connected by disulfide bonds. These bonds also bridge the H and L chains, except in IgA molecules in which the light chains are bridged to each other instead. L chains are of two types, kappa or lambda. On any one Ab molecule both L chains are the same. The ratio of kappa to lambda chains varies by species and type of Ig.

The antibody molecule has two regions. The upper part of the H and L chain is the *constant* (C) region, the lower part is the *variable* (V) region. The variable regions are the antigenic reacting sites. An Ab molecule is able to "fit" an antigenic determinant site because of variations in primary sequences of amino acids on the L and H chains and because of the three-dimensional shape of

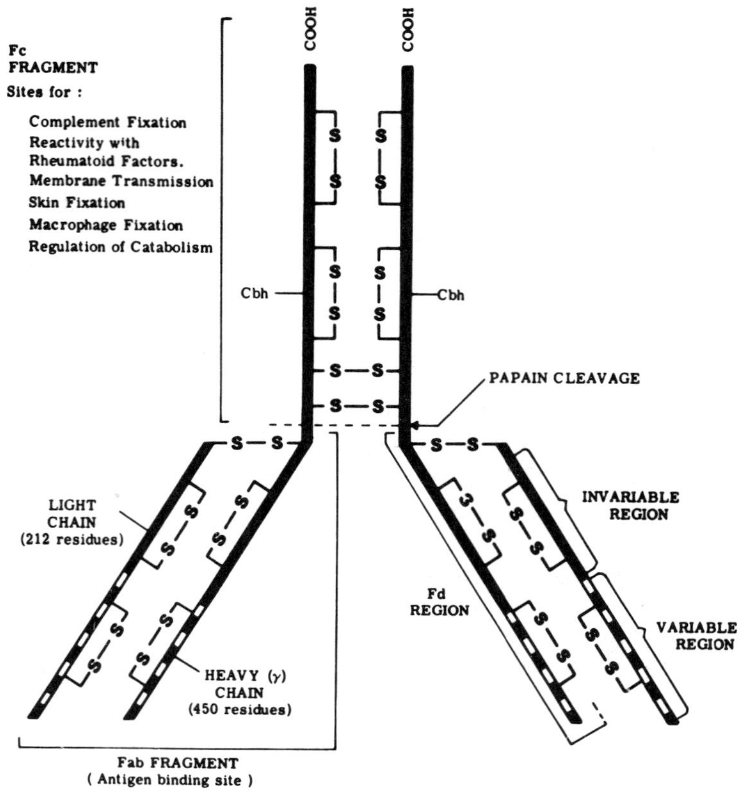

FIGURE 5-6. Schematic representation of the four-chain structure of human IgG showing interchain and intrachain disulfide bridges, papain fragments, and location of effector functions. (Reprinted with permission of the publisher from Glynn LE, Steward MW. Structure and Function of Antibodies. New York: John Wiley and Sons Ltd, 1977.)

the molecule, which makes many combinations of L and H chains possible in the variable region.

Antibodies may be classified by isotype, allotype, or idiotype. An *isotype* is one of the five major classes of Igs. *Allotypes* occur when there are variations in members of the same class or species. They result from minor substitutions on genetic chains that code for immunity, and are useful for identifying serum Igs in a manner analogous to blood typing. An *idiotype* is the name for a homogeneous Ab.

Antibody Types

Five major classes of the structurally related proteins called immunoglobulins (antibodies) have been identified, named for their varying heavy chains: IgG, IgM, IgA, IgD, and IgE (Fig. 5-7).

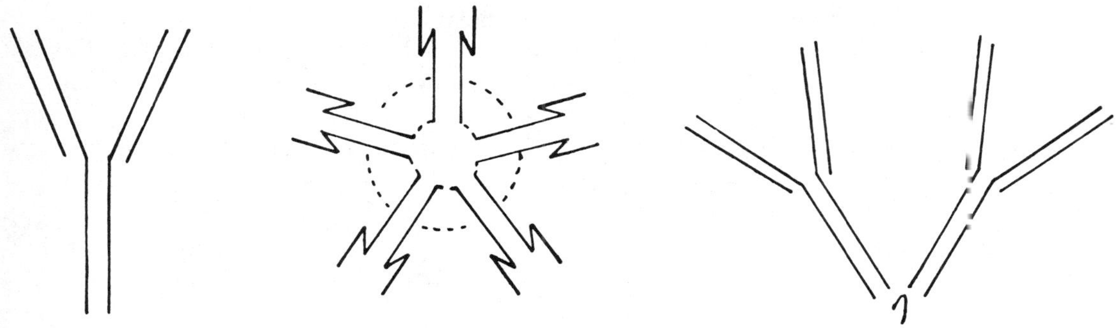

IgG, usually Y or T shaped

IgM, pentameter

sIgA (secretory IgA), usually a dimer connected by a J chain

FIGURE 5-7. Structure of major classes of antibodies.

IgG. IgG is the most important Ig in serum, comprising 70 to 75 percent of circulating Ab. Since it is the most common Ig, it has been well studied. Based on differences in amino acid sequences in their H chains, there are four identified subclasses (IgG1 through IgG4). There are also 25 presently known allotypes or antigenic varieties designated as Gm1, Gm2, etc. Thus it is possible for the serum of a single individual to contain 200 varieties of IgG. IgG passes the placental barrier, binds to neutrophils, and fixes complement.

IgM. IgM is a macromolecule that occurs primarily as a pentameter of five subunits (occasionally as a polymer or monomer), containing 10 L chains and H chains and one J (joining) chain attached to the Fc portion of the H chains by disulfide bonds. IgM is rich in carbohydrate, the function of which may be to increase solubility of the molecule and help determine its shape. It is the first Ig to be produced after an infection, but has a lower binding affinity than IgG. It is the only other Ig besides IgG that fixes complement. IgM is an important Ab against organisms (bacteria and protozoa) that are blood-borne and cause septicemia. It does not cross the placenta. There is a secretory form of IgM, the distribution of which is similar to secretory IgA.

IgA. IgA usually occurs in four-chain monomers, but may have various shapes including dimer, trimer, or multimer. Two subclasses (1 and 2) have been identified. Secretory IgA (sIgA) usually is a dimer with a J chain and a secretory component (SC), which is a polypeptide chain made by epithelial cells. This component helps to protect the molecule from digestion by proteolytic enzymes. The function of the J chain is unknown, but may well be important since it is widely distributed. IgA without the SC is found in internal secretions such as cerebrospinal fluid, synovial, amniotic, and pleural fluids, and the eye. Secretory IgA is found in external secretions such as tears, milk, saliva, and genitourinary and respiratory secretions. There is no evidence that nonsecretory IgA is less effective in secretions than sIgA; both are more effective against local than systemic infection and can activate complement by the properdin (alternative) pathway. IgA binds effectively to neutrophils.

IgD. This uncommon Ig is heat labile and has only one disulfide bridge between H chains. It is rich in carbohydrate and has no known function; it may be involved in immune tolerance.

IgE. IgE (previously called reagin) is also, like IgG, nonsecretory IgA, and IgD, a four-chain monomeric unit. It does not display serologic reactions. It has no known subclasses and is involved in anaphylaxis. Antibodies that bind to membranes of mast cells and basophils (called homocytotropic Abs) may be from several classes, but are primarily IgE. These mediate many allergic reactions and hypersensitivity. Though the effects of IgE are predominantly harmful, IgE may also opsonize macrophages against helminthic infections.

Characteristics of the five classes of Ig are summarized in Table 5-5.

COMPLEMENT

The complement system is a system of about 20 proteins synthesized in the liver and found in the plasma of vertebrates. Its primary function is the mediation of Ag/Ab reactions resulting in lysis of foreign invaders and the initiation of the inflammatory response. Its other important biologic functions are in chemotaxis, kinin formation, opsonization, anaphylatoxin formation, and

TABLE 5-5. Characteristics of Immunoglobulins

Ig	*Molecular weight*	*Ag-combining sites*	*Functions*	*Percent total Ab*	*Half-life (days)*	*Characteristics*
IgG	146,000	2	Most important circulating Ig	70-75	20	Transmitted placentally and via colostrum; fixes complement
IgM	950,000	5	Appears first in primary response	5-10	10	Fixes complement; does not cross placenta
IgA	160,000 or 380,000	4	Important in secretions (tears, milk, mucous, etc.)	15-20	6	Resists digestion by proteolytic enzymes
IgD	184,000	2	Unknown; may be involved in immune tolerance	~1	3	
IgE	188,000	2	Unknown how it functions beneficially; related to allergic reactions	0.05	2	Homocytotropic

immune adherence. The complement reaction can be activated in two ways: either by Ag/Ab combinations with Igs in classes IgM, IgG1, and IgG3 (the classical pathway), or by an alternative pathway that does not require Ag/Ab reactions. Each protein in the complement system must be activated in the appropriate sequence to form a "cascade" of interacting components. The proteins or components of the complement system are termed C1 through C9. A bar over a symbol (C,$\overline{2}$) denotes enzymes formed during the complement activation process.

The Classical Activation Pathway

The classical pathway may be activated by Ag/Ab complexes, by aggregated IgG or IgM, or by certain substances such as DNA, heparin, some cell membranes and enzymes. It is intitated by the binding of C1, the first protein in the complement system, to the Fc region of the Ig. The C1 macromolecule, which consists of three separate proteins, C1q, C1r, and C1s, held together by a calcium-dependent bond, is then combined with C4, creating $C\overline{1,4b}$, which in turn reacts with C2 to produce $C\overline{4b,2a}$, or C3 convertase. C3 has several fragments: C3a, C3b, C3c, and C3d. When C3b attaches to $C\overline{4b,2a}$, the enzyme C5 results as the substrate. The components C5 through C9 then become activated in sequence. As shown in Figure 5-8, these remaining components attach directly onto the antigen (e.g., a microbial cell). Only after all nine components are activated does cell lysis occur as a result of a lesion produced in the Ag cell membrane.

The Alternative Activation Pathway

The alternative pathway is also called the *properdin pathway,* named for one of the normal serum proteins that participate in this pathway. In this pathway a series of six normal serum proteins are converted and lead to the direct activation of C3; C1,C2 and C4 are bypassed. The activators of this pathway include polysaccharides, lipopolysaccharides, or glycoproteins such as capsules of some yeasts and bacteria, and the cell wall lipopolysaccharide of gram-negative bacteria. Classical and alternative complement pathways are compared in Figure 5-9.

Function of Complement

C3b that is attached to Ab is in turn attached to the foreign invader. Many phagocytic cells have receptors to C3b. Adherence of the C3b fragment to this receptor results in phagocytosis and killing of the Ag. This is the mechanism of *opsonization* of organisms by the combination of Ab and complement. C3 and C5 fragments can also generate a factor that increases permeability of mast cells, causing release of histamines and other mediators. This *anaphylatoxin* effect causes increased capillary permeability, bronchospasm, and vasodilation. *Chemotaxins* are produced by several complement fractions. These substances attract leukocytes to the area of foreign invasion. Thus the complement system enhances the phagocytosis of antigens in two distinct ways: (1) by modifying the Ag so that it can be more readily phagocytosed (opsonization) and (2) by attracting more phagocytic cells to the area. Cell *lysis* results from injury to the cell membrane due to the complement reaction. The exact mechanism of this damage is not known. Normal host cells may also be damaged by these same processes, resulting in allergic vasculitis or glomerulitis.

The role of complement in protection against bacterial infection has been well

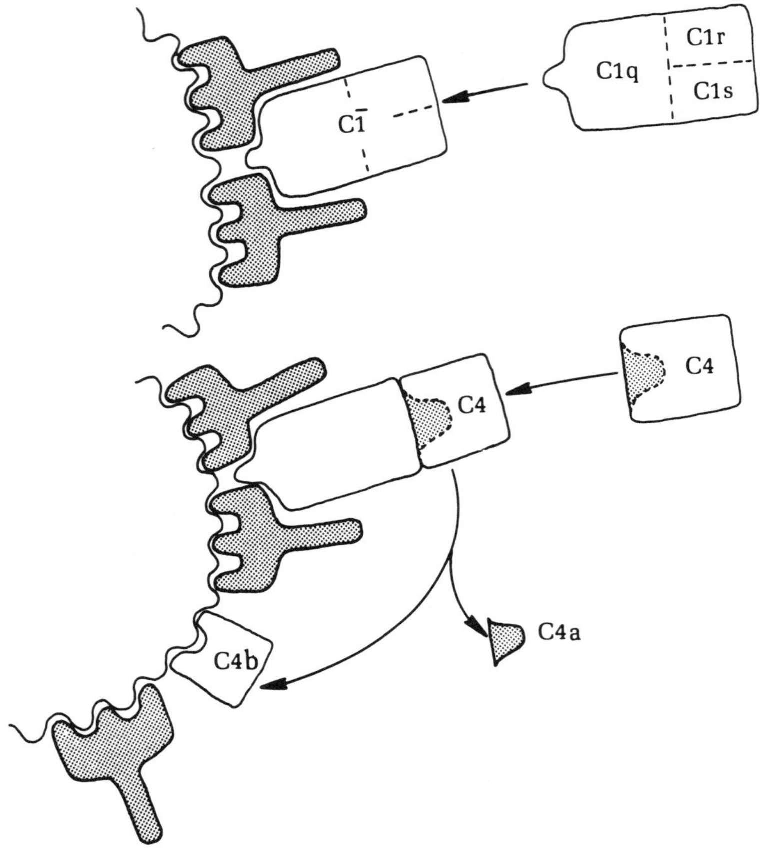

FIGURE 5-8. In the activation of C1, the C1q, C1r, C1s complex is presumed to bridge two antibody molecules. Contact of activated C1 with C4 results in alteration of the latter so that C4b attaches to the antigen. The remaining complement components combine with C4b and do not attach to the antibody molecules. (Reprinted with permission of the publisher from Barrett JT. Basic Immunology and its Medical Application, 2nd edition. St Louis: CV Mosby, 1980.)

documented. It has been shown, for example, that optimal recovery from pneumoncoccal sepsis requires intact classical and alternative pathways. Complement, especially in conjunction with Ab, is also important in neutralizing free virus in the blood. Several mechanisms have been proposed for this. Complement and Ab may coat virions, may aggregate viral particles, may facilitate opsonization, and may directly lyse viruses with lipid envelopes. Complement may also neutralize some viruses even in the absence of Ab. These functions of complement are summarized in Table 5-6.

	Classical	*Alternative*
Activated By	Immune complexes (IgG1, IgG3, IgM) Aggregated Igs Ab-coated erythrocytes (blood transfusion reaction) Heparin Some cell membranes & enzymes	Polysaccharides: Endotoxin IgA, IgE aggregates Anesthesia Trypsinlike enzymes Some virus infected cells

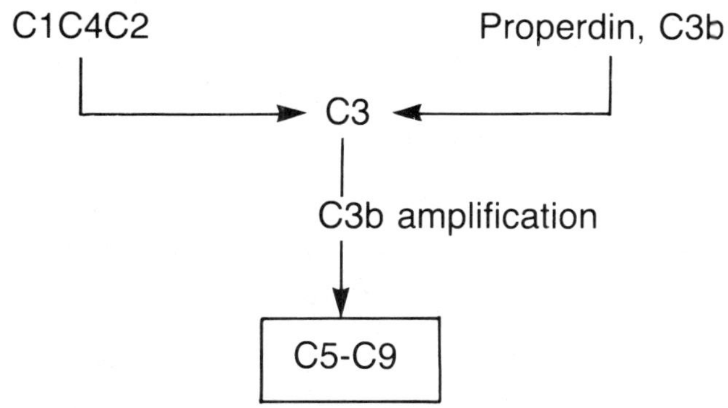

FIGURE 5-9. A comparison of classical and alternative complement pathways.

TABLE 5-6. Functions of Complement

Function	*Fragment(s) primarily responsible*
Adherence and opsonization	C3b, C5
Anaphylatoxin effect	C3, C5
Chemotaxis	C3; C5; C5, C6, C7 complex
Hemolysis and cytolysis	All

CLINICAL IMPLICATIONS

The amounts (titers) as well as types (particularly IgG or IgM) of antibodies produced in response to contact with an antigen can be measured. At times it can be critical to determine whether an individual has a primary or a secondary immune response. This can be done by measuring IgM levels (the major antibody type present in the primary immune response) and levels of IgG (which is produced in much greater quantities during a secondary immune response). This is important, for example, when a woman in the first trimester of pregnancy is exposed to rubella (which can cause severe fetal abnormalities) and her previous immune status is unknown. If, after about 2 weeks, there is a rise in IgM antibody titers, the woman has developed a primary infection with the rubella virus and the fetus is at risk of congenital malformations. If, however, there is an increased titer of IgG, the woman has a reinfection (which occurs in about 4 percent of individuals who have rubella) and the risk to the fetus is minimal.

Antigen/antibody reactions are harmful as well as helpful. Erythroblastosis fetalis, a disease of newborns in which the baby's red blood cells are destroyed, can occur in women who are Rh negative (do not have the Rh factor in their blood) and have babies with Rh-positive blood. Antibodies against the child's Rh factor are produced by the mother so that in subsequent pregnancies, her circulating anti-Rh antibody attacks the fetus's Rh-positive blood. Fortunately, the mother can be given gamma globulin containing anti-Rh antibody at the time of delivery. The gamma globulin combines with the infant's Rh factor, effectively preventing the mother from producing her own anti-Rh antibody.

Other harmful antigen/antibody reactions (hypersensitivities) are discussed in Chapter 6.

SUGGESTED READINGS

Barrett JT. Basic Immunology and Its Medical Application, 2nd edition. St Louis: CV Mosby, 1980.

Edelman R, Hardegree MC, Chedid L. Summary of an international symposium on potentiation of the immune response to vaccines. J Infec Dis 1980;141:103-111.

Fudenberg HH, et al. Basic and Clinical Immunology, 3rd edition. Los Altos, California: Lange Medical Publications, 1980.

Glynn CE, Steward MW, eds. Structure and Function of Antibodies. New York: John Wiley and Sons, 1981.

Guckian JC, Christensen GD, Fine DP. The role of opsonins in recovery from experimental pneumococcal pneumonia. J Infec Dis 1980;142:175-190.

Hoffman MK. Antibody regulates the cooperation of B cells with helper cells. Immunol Rev 1980;49:79-89.

Hood L, et al. Antibody diversity. Develop Immunol 1979;3:173-179.

Hosea SW, Brown EJ, Frank MM. The critical role of complement in experimental pneumococcal sepsis. J Infec Dis 1980;142:903-909.

Reinherz EL, Schlassman SF. Regulation of immune response-inducer and suppressor T-lymphocyte subsets in human beings. New Engl J Med 1980;303:370-373.

Sissons JGP, Oldstone MBA. Killing of virus-infected cells: the role of antiviral antibody and complement in limiting virus infection. J Infec Dis 1980;142:442-448.

Williams BD: The complement system. Br J Anaesth 1979;51:7-11.

6 CELLULAR IMMUNITY/ HYPERSENSITIVITIES

KEY POINTS

- There are four kinds of T cells that are involved in defense against microbial invaders: helper, suppressor, cytotoxic, and delayed hypersensitivity cells.
- The body's immune system can also cause adverse reactions, called hypersensitivities.
- Cellular immunity is most helpful in protection against intracellular microbial invaders such as the tubercle bacillus, *Chlamydia, Rickettsia,* viruses, and fungi.
- Autoimmune disease occurs when the body fails to recognize one of its components as "self" and forms antibodies against itself.

The cell-mediated immune responses, in contrast to the humoral responses that are mediated by Ab, were named because of the belief that T cells had to have direct contact with Ags before they could respond. This is not necessarily the case, however, since T cells, like B cells, excrete substances responsible for their specific reactions. Nevertheless, cellular or cell-mediated immunity is the name given to protection that results from T cells.

FUNCTIONS OF CELLULAR IMMUNITY

While antibody is most effective against extracellular antigens, cellular immunity is especially important to the body's control of intracellular microbial invaders, including many bacteria like the tubercle bacillus, *Chlamydia* and *Rickettsia,* as well as fungi and viruses (humoral and cellular immunity are compared in Table 6-1). Cellular immunity is mediated by lymphokines,

TABLE 6-1. A Comparison of Humoral and Cellular Immunity

	Humoral	*Cellular*
Lymphocytes	B	T
Effector molecules	Ab	Lymphokines, "activated" macrophages
Protection against	Toxin-induced disorders; some viruses; bacteria with capsules	Especially intracellular parasites; fungi; mycobacteria; viruses
Passive transfer	With serum	With T lymphocytes or transfer factor (TF)
Major activities	Fix complement for cell lysis, opsonization to enhance phagocytosis	Activate macrophages to enhance phagocytosis and killing of Ag, direct killing of target cells by cytotoxic lymphocytes

which are large polypeptides excreted by T cells when they are stimulated by Ags or haptens. They are analogous to Igs excreted by B cells, though the lymphokines lack Ag specificity. Lymphokines include the following:

1. Blastogenic factor, which initiates growth of lymphocytes
2. Chemotactic factor, which helps attract macrophages
3. Lymphotoxins, which destroy target cells
4. Interferons (see Chapter 4)
5. Migration inhibitory factor (MIF) and leukocyte inhibitory factor (LIF), which prevent the migration of macrophages and PMNs, respectively
6. Macrophage aggregation factor, which causes agglutination of macrophages
7. Macrophage activating factor, which activates macrophages to become more cytotoxic (called "angry" or "activated" macrophages)

The enhancement of macrophage effectiveness is one of the most important aspects of T cell-mediated immunity.

Macrophages have two functions in cell-mediated immunity. In the "afferent" stage they activate T cells by presenting Ag on their surface. In the "efferent" stage, activated (killer) macrophages kill organisms. These activities are mediated by proteins, called monokines, that are released by macrophages. Monokines include interferons (also, as just mentioned, produced by T cells), lymphocyte activating factor (LAF), T cell activating factor, thymocyte differentiating factor, genetically related factor (which helps convert T cell precursors to helper T cells), and B cell activating factor. Hence there is a cycle of complementary, cooperative activity among B and T lymphocytes, macrophages, and other cells of the immune system that work in unison to effect optimal cellular protection.

T cells contain a polynucleotide component called transfer factor (TF) that is capable of passively incurring delayed hypersensitivity to a sensitized recipient 1 to 7 days after injection of cell extract. TF may have therapeutic usefulness in patients with chronic mucocutaneous candidiasis, lepromatous leprosy, neoplasms, and other conditions in individuals whose cell-mediated immunity is deficient. TF is not really a lymphokine because it is a normal part of T cells rather than being formed in response to contact with an Ag.

TABLE 6-2. Functions of T Lymphocytes

Helpful	*Harmful*
Production of lymphokines to facilitate Ag killing	Contact dermatitis
Resistance to certain infections (primarily intracellular parasites)	Production of autoantigens (autoimmune disease)
Resistance to certain neoplasms	Rejection of allografts
Immunologic memory	Mediation of some drug allergies (penicillin)
Cooperation with B cells to yield an Ab response	

TYPES OF T CELLS

Functions of T cells, summarized in Table 6-2, are performed by at least four subsets of T cells: helper T cells (T_H), suppressor T cells (T_S), cytotoxic T lymphocytes (CTLs, killer T cells, or T_K), and T cells of delayed hypersensitivity (T_{DHY}). The subsets can be differentiated by three distinct surface Ags, Lyt-1,2,3 characteristics of which are described in Table 6-3. The T cells that are lysed only by anti-Lyt-1 (Ly-1 T cells) recognize Ags on macrophages only in conjunction with products of the H-2 I region of the major histocompatibility complex (see "Genetic Influences on the Immune System" in this chapter). The Ly-1 group consists of T_H and T_{DH} cells, the Ly-2,3 group consists of T_S and CTL cells, and the Ly-1,2,3 T cells serve as the main regulators of immune response by controlling the balance of T_H and T_S lymphocytes.

T Helper and T Suppressor Cells

Despite the fact that T cells do not produce Ab, the production of Ab by B cells would

TABLE 6-3. Ly (lymphocyte) Antigens of T Cells

Type	*Location*	*Characteristics*
Ly-1	30 percent of peripheral T cells	Lack Fc receptors; lysed only by anti-Lyt-1; T helper (T_H) and T cells of delayed hypersensitivity (T_{DH}) are in this group; specificity restricted to products of H-2 I region of major histocompatibility complex
Ly-2		Involved in allotypic suppression
Ly-2,3	5-10 percent of peripheral cells	Lysed by anti-Lyt-2 or anti-Lyt-3; T suppressor (T_S) and T killer (T_K) cells are in this group
Ly-1,2,3	All thymocytes and 50 percent of peripheral T cells	Nearly all cells in early postnatal life are Ly-1,2,3, precursors of Ly-1 and Ly-2,3 cells; main regulators of immune response by controlling balance of T_H and T_S

be markedly impaired by T cell deficiency. Activated T_H cells stimulate the production of B cells bearing Ab by releasing a "helping factor." They are also active in facilitating the production of memory cells and possibly of cytotoxic T lymphocytes. The increased affinity of Abs, which occurs over time, requires activated T_H cells. Activated T_S cells hinder Ab formation either by blocking T_H (T_S-2 cells) or by acting on B cells (T_S-1 cells). Both T_H and T_S cells are necessary for balanced Ab production; immune tolerance can result from excessive T_S activity.

Cytotoxic T Lymphocytes

Following acute infection, particularly with viruses, cytotoxic T lymphocytes (CTLs) may be generated. These are specific for the target cell and contact between CTL and target cell is required for lysis to take place. As opposed to other forms of the cellular immune response, macrophages do not participate in these interactions.

There are apparently two mechanisms of action for CTLs: antibody-dependent, cell-mediated cytotoxicity (ADCC), and natural cytotoxicity, which does not depend upon prior contact with Ag. The cells involved in the later mechanism are called natural killer (NK) T cells. There is some controversy regarding whether these two types of cells might in fact be the same, and research in this area continues. The exact mechanism by which these lymphocytes cause cell lysis is not known. The target cells "explode," whereas complement-lysed cells merely leak out their contents and retain their shape. "Killer" T cells and "killer" macrophages are compared in Table 6-4.

HYPERSENSITIVITIES

Harmful immune reactions are called hypersensitivities or allergies. Substances that cause allergic reactions are called *allergens*. Hypersensitivity is the opposite of immunity and like immunity can be Ab mediated or T cell mediated. There are three types of

TABLE 6-4. Differences Between "Killer" T Cells and "Killer" Macrophages

	Killer T cells	*Killer macrophages*
Synonyms	Cytotoxic T lymphocytes (CTLs), T killer (T_K) cells	Activated macrophages, "angry" macrophages
Elicited by	Cells carrying new surface Ag, tissue allografts	Lymphokine, macrophage activating factor (MAF), excreted by T cells
Transmission	Via Ag-specific T cells, not with antiserum	Via Ag-specific T cells, not with antiserum
Activity	Lyse target cells	Augmented ability to phagocytize and lyse Ag
Under genetic (H-2) control	Yes	Yes
Mechanisms of lysis	Killer T binds in a conjugate to target cell, damaged cells explode by some unknown mechanism	Increased lysosomal enzymes and ability to spread; release of superoxide anion (O_2^-) and H_2O_2

Ab-mediated hypersensitivities, called anaphylactic, cytotoxic, and immune complex. Because Ab-mediated reactions usually occur in less than 8 hours, they are called immediate-type hypersensitivities. T cell-mediated hypersensitivity is called delayed hypersensitivity because reactions are not evident for 24 to 48 hours after contact with the allergen. Characteristics of the various hypersensitivities are summarized in Table 6-5.

Antibody-mediated Hypersensitivities

IgE Hypersensitivity (Anaphylaxis)

Anaphylaxis, derived from Greek and meaning "backward protection," can be a life-threatening systemic reaction or a localized reaction. It is usually mediated by IgE antibodies (rarely by IgG1 or IgG2) that, 2 to 3 weeks after sensitization with an Ag, can bind to receptors on mast cells and basophils when the Ag is again introduced. Mast cells

TABLE 6-5. Hypersensitivities

	Ab-mediated (immediate)			*T cell-mediated (delayed)*
	IgE (anaphylactic)	*IgG or IgM (cytotoxic)*	*IgG or IgM (immune complex)*	
Mechanism	IgE on mast cells + Ag	IgG or IgM + Ag + complement	IgG or M + Ag + complement	Ag-specific T cells + Ag + "killer" macrophages
Hypersensitivity type	I	II	III	IV
Cells involved	Mast cells, basophils	Platelets, white and red blood cells	Tissue cells	T lymphocytes
Response time for skin test	30 min	—(No skin test)	3-9 hr	24-48 hr
Transfer via serum?	Yes	Yes	Yes	No
Therapy	Epinephrine, antihistamines, desensitization	Passive administration of antiserum + Ag	Remove source of exposure, steroids	Corticosteroids, avoid contact, antilymphocyte serum (ALS) or antilymphocyte globulin (ALG), antihistamines will *not* work
Diseases	Systemic: anaphylaxis Local: asthma, hay fever, urticaria, food allergies	Rh disease, transfusion reaction, thrombocytopenia	Serum sickness, Arthus reaction, rheumatoid arthritis, autoimmune diseases, nephritis	Two major types: contact dermatitis (poison ivy, etc.) and allergy of infection (tuberculin reaction, etc.); possible tumor immunity

are nonmotile connective tissue cells primarily in lungs, GI and GU tracts, and skin. They have receptors for IgE on their surface membranes. When Ag is introduced into a sensitized host, it combines with the cell-bound IgE and stimulates the mast cells to degranulate, resulting in the release of mediators such as histamines, prostaglandins, serotonin, eosinophil chemotactic factor of anaphylaxis (ECF-A), and heparin from mast cell and basophil granules. Mediators are summarized in Table 6-6.

Fatal systemic anaphylaxis in humans results from cardiovascular collapse or res-

TABLE 6-6. Mediators of Ab-Mediated (Immediate) Hypersensitivities

Mediator	*Characteristics*
Anaphylatoxins	Peptides that induce the release of histamine from mast cells; some originate from C3a and C5a fragments of complement
Eosinophil chemotactic factor of anaphylaxis (ECF-A)	Peptide present in mast cell granules; attracts eosinophils to specific tissues affected by allergen
Heparin	Responsible for subsequent release of histaminase in liver
Histamine	Substance derived from histadine in platelets, basophils, and mast cells; results in vasodilation, contraction of smooth muscle, increased capillary permeability; treated with antihistamines
Kinins	Peptides with activities similar to histamine: patent vasodilation, smooth muscle contraction, edema from increased capillary permeability. Bradykinin, derived from blood proteins by enzymatic activity, causes protracted muscle contraction
Platelet-activating factor	Released by IgE-sensitized basophils; aggregates platelets and causes release of serotonin
Prostaglandins	Exact role uncertain; action is dose-related; may prevent or enhance release of histamine and SRS-A (see below) from sensitized cells
Serotonin	Formed from tryptophan primarily in platelets; sections resemble histamine: increased capillary dilatation and permeability, smooth muscle contraction; of only minor importance in human hypersensitivities
Slow-reacting substance of anaphylaxis	Unsaturated hydroxyl acid synthesized by mast cells during allergic reaction; causes prolonged smooth muscle contraction; unaffected by antihistamines

piratory obstruction due to pronounced effects on smooth muscle. Despite the fact that minute doses of Ag (as little as 0.001 mg of protein) can cause death in a sensitive individual, most anaphylactic events are not fatal. Signs and symptoms include itching, hives and redness of skin, abdominal cramps, vomiting and diarrhea, respiratory distress due to bronchospasm, and shock. The response after intravenous injection of Ag is almost immediate, within 3 to 5 minutes. Highly sensitized individuals may develop anaphylaxis after such exposures as insect bites, eye drops, eating shellfish, and intramuscular or subcutaneous injection of drugs like penicillin or animal sera. Local reactions such as hay fever, asthma, urticaria, or vomiting occur from ingestion or inhalation of allergens such as pollens, grasses, weeds, dust, molds, danders, fish, chocolate, fruits, or nuts.

Treatment of hypersensitivity includes avoidance of the allergen whenever possible, administration of epinephrine or aminophylline, antihistamines (for allergic rhinitis but not useful in asthma), and corticosteroids for long-term effects. Temporary desensitization is possible if it is necessary to contact Ag (if a foreign serum must be given to a sensitive person, for example) by injecting minute doses of Ag at short intervals. Sensitivity, however, recurs after several days or weeks. Unfortunately, contact with Ag serves as a "booster" so that subsequent doses will cause even more serious sequelae. A second treatment approach is to increase levels of IgE in response to antigens other than those causing the allergic reaction, so that receptors on the mast cells will be unable to react to the sensitizing antigen.

Cytotoxic Hypersensitivity

In cytotoxic hypersensitivity, Ab (IgG or IgM) reactions can occur with surface Ags that are natural or created by rearrangements of surface proteins or polysaccharides. Complement is fixed and cell lysis results. Such reactions are responsible for some types of autoimmune disease such as thrombocytopenia or hemolytic anemia, and for disease resulting from foreign Ags such as Rh disease of the newborn and transfusion reactions. Rh incompatibility of the fetus and mother can sometimes be prevented by passive injection of human Ab to Rh antigen into Rh-negative women who have Rh-positive mates and children. When this occurs, the mother's production of Abs against the fetus's Rh Ags is inhibited (see "Clinical Implications," Chapter 5).

Immune Complex Hypersensitivity

Immune complex hypersensitivity reactions involve the deposition of Ab/Ab complexes of IgG or IgM in connective tissue, venules, joints, small arteries, or glomeruli. Complement activation causes precipitants to form and be deposited in vessels and tissues. Tissue injury results from these precipitants. The severity of the reactions depends upon the ratio of Ag to Ab in the complexes; when complexes have an excess of Ag, it is not possible for them to be completely cleared by phagocytic cells. Various types of tissue injury are induced, the prototypes of which are the Arthus reaction and serum sickness.

The *Arthus reaction* occurs when an Ag is injected intradermally into an immunized animal. The precipitation of Ag/Ab complexes in the vascular system results in blockage of local blood vessels, necrosis, and skin sloughing. Healing occurs slowly. *Serum sickness* occurs 8 to 12 days after injection of foreign serum. This is the time required for Abs, primarily IgG, to be formed, to complex with the Ag, and to attach to mast cells. This causes a gradual release of histamine, well below the level seen in anaphylaxis (which is mediated by

IgE). These Ag/Ab complexes deposited in various body sites cause nephritis, urticaria, joint swelling, enlarged lymph nodes, neutropenia, and fever. A unique aspect of serum sickness is that a previous sensitizing dose is not required because of the initially large Ag dose given. The anamnestic response does occur if a second Ag dose is given, in which case the symptoms occur much sooner, within 2 to 5 days. Fortunately the disease is self-limiting, and complete recovery occurs after several days when all Ag has been eliminated. Other immune complex diseases include rheumatic fever and glomerulonephritis.

T Cell-mediated (Delayed) Hypersensitivities

Delayed hypersensitivities are the only types of hypersensitivities that are T cell rather than Ab-dependent. Delayed hypersensitivities are mediated by the lymphokines excreted by T cells of the subset T_{DH}. There are, unfortunately, no therapeutic techniques for treating these hypersensitivities. Antiinflammatory agents are used for symptomatic relief. There are two major classifications of delayed hypersensitivities: allergies of infection (as exemplified by the tuberculin reaction) and contact dermatitis (Fig. 6-1). The tuberculin skin test is described under "Skin Testing for Hypersensitivities" and those viral, bacteria, fungal, and protozoal agents that induce delayed skin reactions are summarized in Table 6-7. Skin hypersensitivity reactions persist throughout life.

Dermal contact hypersensitivities can result from numerous agents (for example, metal, cosmetic components, rubber, wool, poison ivy, and poison oak). Such allergies are under genetic regulation and there is little effective therapy. Contact avoidance, though not always possible, is advised, especially for individuals who have familial tendencies to contact hypersensitivities. Delayed hypersensitivities are transferable to unsensitized individuals with extracts of lymphocytes but not with serum. Viral infections in mice have been shown to interfere with delayed hypersensitivity reactions.

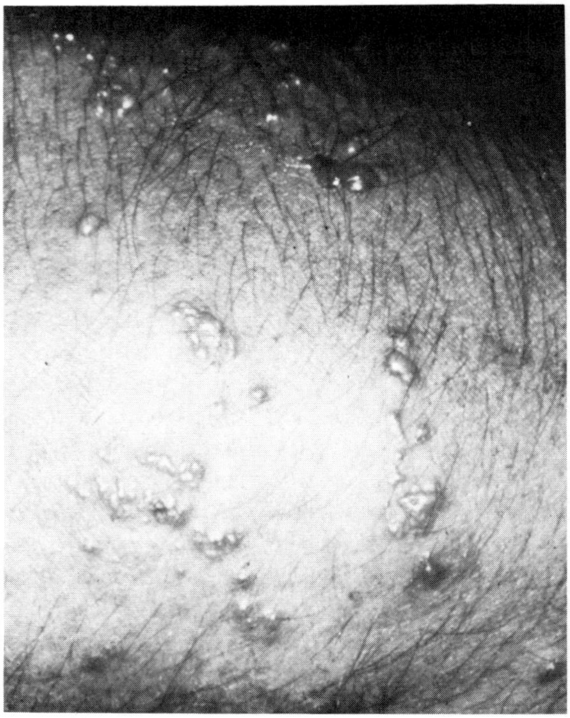

FIGURE 6-1. Poison ivy, a typical example of delayed, T cell-mediated hypersensitivity. (Courtesy of Centers for Disease Control, Atlanta, GA).

Skin Testing for Hypersensitivities

Hypersensitivities can be diagnosed by skin testing. In Ab-mediated hypersensitivities, a small (0.1 ml) intradermal injection of Ag

TABLE 6-7. Infectious Diseases That Have Skin Tests for T Cell (Delayed) Hypersensitivity

Bacterial	*Viral*	*Fungal*	*Protozoal*
Brucellosis	Mumps	Blastomycosis	Leishmaniasis
Diphtheria		Candidiasis	
Leprosy		Coccidioidomycosis	
Lymphogranuloma venereum			
Psittacosis			
Streptococcal infections			
Tuberculosis			
Tularemia			

is given. After several minutes, the characteristic reaction of immediate-type hypersensitivity can be noted in sensitized individuals: itching, edema, and an irregular wheal surrounded by redness ("flare"). This reaction usually peaks within 15 minutes and disappears within several hours. This is in marked contrast to the T cell-mediated skin reaction that develops much more slowly, taking 48 to 72 hours to reach its peak after intradermal injection of Ag. The tuberculin test usually serves as the model for delayed-type skin reactions. No reaction occurs until 1 to 2 days after injection. The lesion that then develops in sensitized persons is red, indurated, and is less edematous than that which occurs in immediate-type reactions. The thickened swelling of the skin remains for days or weeks. There is no correlation between the extent of skin reaction and the level of clinical disease.

One of the difficulties in interpretation of skin tests of the delayed type is that the reaction, once present, persists throughout life. Thus it cannot be used to diagnose acute disease, but merely to indicate past or present infection. A second problem is that skin-testing reagents may themselves be antigenic, causing conversion of serologic tests of a noninfected person to positive. This, of course, can make definitive diagnosis of disease impossible. Fortunately, the purified protein derivative (PPD) used for tuberculosis testing is not antigenic and serologic tests are not used in the diagnosis of tuberculosis so that skin testing for this disease is devoid of these two problems. However, bacille Calmette-Guerin (BCG), the attenuated bovine organism used for immunization against tuberculosis, causes permanent conversion of the tuberculin skin test to positive, and this is one of its disadvantages. Skin tests are available for a number of infectious diseases in which T cell-mediated immunity plays an important role.

AUTOIMMUNITY

Autoimmune disease occurs when an individual fails to recognize a part of his or her own tissue as "self" and antibodies are formed against some component of their own body. It is not totally clear why this process of mounting an immune response to self-antigens occurs, but some possible explanations are listed in Table 6-8. Some of the more common autoimmune diseases include

TABLE 6-8. Possible Mechanisms of Autoimmunity

Mechanism	*Characteristics*
Cross-reactive Ag	Cross reactions between foreign and normal Ags may occur
Loss of self-tolerance	B cells may be changed by mutation or excessive stimulation by T_H cells, causing changes in B cell recognition of Ag
Maturation Ag	Ags formed after fetal life and maturation of the immune system are not recognized as self
Mutation	Lymphocytes may mutate so that self-Ags are no longer recognized
Neoantigen	Self-Ags may be altered in some manner to render them unrecognizable to immune system
Sequestered Ag (occult)	A hidden self-Ag (such as the lens of the eye) that does not normally reach the circulation can engender an immune response
Viruses	Viruses modify host cells so that they are not recognized as self

systemic lupus erythematosus, rheumatoid arthritis, and hemolytic anemia. Certain of these processes are associated with a genetic component, with antibodies, or with cellular immunity (Table 6-9). In general, autoimmune disease becomes more common with aging—when the immune system loses sensitivity.

GENETIC INFLUENCES ON THE IMMUNE SYSTEM

Clearly there is a hereditary influence on many immune reactions. This is most obvious with hypersensitivities such as hay fever, asthma, or other allergies. Most of the work on genetics of the immune system has

TABLE 6-9. Autoimmune Disorders Associated with the Immune Response Systems and the HLA System

HLA-associated (genetic)	*Ig-associated (humoral)*	*T cell-associated (cellular)*
Addison's disease	Graves' disease	Aspermatogenesis
Dermatitis herpetiformis	Hemolytic disease	Hashimoto's thryoiditis
Graves' disease	Myasthenia gravis	Postinfectious encephalomyelitis
Multiple sclerosis	Poststreptococcal glomerulonephritis	Postvaccinial encephalomyelitis
Myasthenia gravis	Rheumatic fever	Sympathetic ophthalmia
Rheumatoid arthritis	Rheumatoid arthritis	
Systemic lupus erythematosus	Skin diseases: pemphigus vulgaris, bulbous pemphigoid, dermatitis herpetiformis	

been done in animal models (mice and guinea pigs), with which it is possible to inbreed species to allow more detailed analyses of functions and locations of immunoregulatory genes. In many cases findings have been corroborated in humans; in other cases, though conclusions seem to be sound and make biologic sense, they are yet to be demonstrated in humans. Genetic disorders with effects on the immune system are summarized in Table 6-10.

Immune responsiveness, as mentioned in Chapter 4, is in part controlled by a group of genes called the major histocompatibility complex (MHC). The H-2 complex genes in the MHC plays a major role in foreign graft rejection. In humans one group of regulatory genes is called the human leukocyte antigens (HLA). HLA are glycoproteins containing five closely related genes, HLA-A, HLA-B, HLA-C, HLA-D, and HLA-DR. They are called HLA proteins because they were first discovered on leukocytes. About 80 different HLA Ags have been discovered. HLAs are located on the sixth chromosome. Immune response (Ir) genes are closely linked to genes expressing HLAs. These Ir genes regulate both Ab and T cell-mediated immunity. In T cell immunity the ability of macrophages to respond to Ag and present

TABLE 6-10. Genetic Syndromes Causing Increased Susceptibility to Infection

Syndrome	*Effects on immune system*	*Mechanisms of action*	*Effects on patient*
Ataxia-telangiectasia	IgE, IgA deficiencies; T cell dysfunction resulting in poor maturation of B cells	Poor development of thymus	Recurrent sinopulmonary infections causing bronchiectasis
Chediak-Higashi syndrome	Normal phagocytosis, but impaired intracellular killing	Myeloperoxidases are not released from lysosomes	May be treated with ascorbate
Chronic granulomatous disease	Certain bacteria (especially *S. aureus* and gram-negative bacteria) are phagocytized but not killed	Granulocytes are deficient in specific enzymes that cause intracellular killing of microorganisms (e.g., NADH dehydrogenase)	Fatal in childhood
DiGeorge's syndrome	T cell deficiency	Maldevelopment of organs derived from third and fourth pharyngeal pouches; lack of thymus	May be treated with fetal thymus grafts with varying success; patient often dies of associated abnormalities (cardiac defects) or infection; a few patients may develop functional T cells with time

TABLE 6-10. *(continued)*

Syndrome	*Effects on immune system*	*Mechanisms of action*	*Effects on patient*
Familial benign neutropenia	Neutrophil count 50-200/mm^3	Largely unknown	Bacteremia rare; may be only slight increase in URIs; may have spontaneous remission
Infantile genetic agranulocytosis	Absence of granulocytes	Unknown	Rapidly fatal
Severe combined immunodeficiency (SCID)	Functional impairment of humoral and cell-mediated immunity	X-linked or autosomal recessive trait; may be enzyme deficiency involved with nucleic acid metabolism, adenosine deaminase (ADA)	Usually fatal before age 1 year
Wiskott-Aldrich syndrome	Serum IgM decreased; inability to make antibodies to polysaccharide antigens; impaired response to protein antigens	May be a primary defect in the B lymphocyte	Occurs only in boys (usually die in childhood of bleeding, infection, or lymphoreticular malignancy)

it for T cell activation is influenced by the Ir gene locus. The extent of formation of T suppressor cells is genetically controlled by immune suppressor (Is) genes.

Histocompatibility testing is done to determine the extent of genetic matching between potential organ donors and recipients. Such testing includes HLA typing and mixed lymphocyte culture (MLC), in which donor and recipient cells are mixed. Because there are so many types, it is impossible, except in the case of identical twins, to get a perfect match of HLA antigens between an organ donor and recipient. At least the major HLA antigens, however, must be matched or rejection of the graft will occur. Evidence of DNA synthesis and cell growth is also evidence of donor/recipient incompatibility.

CLINICAL IMPLICATIONS

An adequate assessment of the status of the patient's immune system is essential for planning patient care. A patient history provides information regarding signs and symptoms of possible familial immune defects. For example, children with humoral immune deficiencies will fail to thrive and may have severe malabsorption and diarrhea as well as recurrent bacterial infections. Children with cell-mediated deficiencies are prone to viral, fungal, and bacterial infections. Immunization, allergy history, and history of present illness should be obtained. Natural and acquired diseases that cause immune deficiency are listed in Tables 6-11 and 6-12.

(text continues on page 131)

TABLE 6-11. Diagnosis and Treatment of the Various Types of Immunologic Host Defects (In Children)

	Cell-mediated immune defects	*Phagocytic immune defects*	*Complement system defects*	*Humoral immune system defect*
Age at clinical onset	≤3 months	Usually ≤3 months, but any age possible	Usually ≤3 months, but any age possible	≥3 months
Clinical clues to diagnosis	Severe or recurrent: pneumonia, acute otitis media, diarrhea, oral or perineal candidiasis, eczema, failure to thrive, autoimmune disease	Severe or recurrent: pneumonia, acute otitis media, furunculosis, draining lymphadenitis, hepatic abscess, osteomyelitis, stomatitis, eczema	Severe or recurrent: pneumonia, acute otitis media, sinusitis	Severe or recurrent: pneumonia, acute otitis media, sinusitis, diarrhea, failure to thrive
Recovered organisms highly indicative of immunodeficiency	From lungs: CMV, varicella-zoster virus, *Pneumocystis carinii* From mouth or perineum: *Candida albicans* persistent or recurrent despite topical therapy, *Cryptococcus, Nocardia*	*Staphylococcus aureus, S. epidermidis, C. albicans, Serratia marcescens*	*Streptococcus pneumoniae, Neisseria* species	Nontypeable *Haemophilus influenzae, Strep. pneumoniae, Giardia lamblia* from gut
Arm of immune system likely involved	T lymphocytes with or without B lymphocytes and antibody involvement	Phagocytes	Complement	B lymphocytes and antibody
Other illnesses to consider in diagnosis	Cystic fibrosis, Down's syndrome, congenital infection with CMV, acrodermatitis enteropathica	Tuberculosis	Cystic fibrosis	Reversible obstructive pulmonary disease, cystic fibrosis

TABLE 6-11. *(continued)*

	Cell-mediated immune defects	*Phagocytic immune defects*	*Complement system defects*	*Humoral immune system defect*
Basic screening tests	Skin tests; lymphocyte transformation to phytohemagglutinin; quantitative IgM, IgA; platelet count	WBC count and differential, WBC nitroblue tetrazolium dye reduction, IgE	Total hemolytic complement, C3, C4 levels	Quantitative immunoglobulins, isohemagglutinins in patients who are not in blood group AB
More specific tests to assess point of abnormality	T cell number; T cell subgroup; T cell response to mitogens, antigens, calcium ionophore; lymphokine production; leukocyte adenosine deaminase (ADA); monocyte chemotaxis and antigen processing; platelet sizing (Wiskott-Aldrich syndrome)	Polymorphonuclear neutrophil and monocyte tests: chemiluminescence, phagocytosis, bactericidal assay, chemotaxis, adherence, monocyte antigen processing, antistaphylococcal IgE	Functional and immunochemical analysis of classical and alternative complement pathway factors	Immunoglobulin response to protein and polysaccharide antigens
Possible immunodeficiency diagnoses	Severe combined immunodeficiency (with and without ADA deficiency), thymic hypoplasia, ataxia-telangiectasia (onset of infection usually after 3 months), varied immunodeficiency (onset of infection usually after 3 months), mucocutaneous can-	Neutropenia primary or secondary to drugs or disease; chronic granulomatous disease, leukocyte glucose-6-phosphate dehydrogenase deficiency; high IgE, low chemotaxis syndrome; Chédiak-	C3 deficiency; C3b inactivator deficiency; factor B deficiency; C6, C7, or C8 deficiency	X-linked (Bruton's) agammaglobulinemia, selective IgA deficiency, immunoglobulin deficiency with increased IgM or with normal IgM and without IgG- and IgA-bearing B lymphocytes, antibody defi-

TABLE 6-11. *(continued)*

	Cell-mediated immune defects	*Phagocytic immune defects*	*Complement system defects*	*Humoral immune system defect*
	didiasis, Wiskott-Aldrich syndrome, selective IgA deficiency if infant not breast-fed	Higashi syndrome		ciency with normal or increased gamma globulin, kappa chain deficiency, varied immunodeficiency
Possible therapy	Bone marrow transplan-qltation, RBC transfusions (ADA deficiency), thymic epithelium, thymosin, appropriate antimicrobials and pulmonary toilet	Appropriate antimicrobials, WBC transfusions for life-threatening illness	Fresh-frozen plasma replacement during acute infection, appropriate antimicrobials	Immune serum globulin, fresh-frozen plasma, appropriate antimicrobials and pulmonary toilet

Reprinted with permission of the publisher from Regelmann WE. Recurrent infection in the child. Postgrad Med 1982; 71:157-178.

TABLE 6-12. Acquired Diseases Causing Immune Deficiency

Disease	*Immune defect*
Acquired immune deficiency syndrome (AIDS)	Depressed cellular immunity
Amyloidosis	May be Ab light-chain disorder
Lymphomas	
Hodgkin's disease	T cell deficiency
Lymphosarcoma	Lymphatic cell tumor
Burkitt's lymphoma	May be B cells made abnormal due to viral infection
Acute and chronic lymphocytic leukemia	May be T- or B cell inadequacy
Multiple myeloma (plasmacytoma)	Tumor of plasma cells
Sarcoidosis	Granulomatous disease of lung, spleen, liver, lymph nodes
Waldenström's macroglobulinemia	IgM myeloma

Since immune functions begin to decline after sexual maturity, and especially after age 65, age is an important variable to note, as is nutritional status. Malnourished individuals have severely depressed cellular and humoral immune responses. Physical assessment should include observing for lymph node enlargement, presence of skin reactions, edema, itching, localized signs of infection or inflammation, evaluating respiratory effort, GI motility, and spleen or liver size. Tests that may be used to assess functioning of T cell-mediated immunity are summarized in Table 6-13.

Cytotoxic drugs, radiation, and immunosuppressive drugs greatly increase patients' risk of infection. Antilymphocyte serum (ALS) produces generalized immunosuppression. Serum sickness may result from administration of ALS since it is produced in heterologous species. After one dose of steroids, a prompt decrease in circulating lymphocytes, especially T cells, occurs due to redistribution (not killing) of cells. This action peaks in 4 to 6 hours and returns to normal after 24 hours. There is a decrease in inflammatory response, in part due to the fact that there are fewer PMNs in the epithelial lining of the blood vessels. Long-term use of steroids results in a decrease in Igs, especially IgG. The inhibitory actions of steroids on the immune system are summarized in Table 4-1.

Cytotoxic drugs (Table 4-2) kill dividing cells that are not yet secreting Ab, especially cells of bone marrow and the GI tract, and germinal cells of gonads. Long-term complications of therapy with these agents include increased risk of infection and cancer. Radiation produces lymphopenia, and surgical removal of spleen, thymus, or lymph nodes affects immune responses by eliminating lymphocyte-producing organs. Patients undergoing these medical therapies,

TABLE 6-13. Tests Evaluating Competence of Cell-Mediated Immune Response

Type of Test	*Interpretation*
Skin test	Apply 1 percent dinitrochlorobenzene (DNCB) in acetone to skin, remove after 24 hr. When 0.03-0.1 percent test dose is reapplied 7-14 days later, a delayed type response indicates healthy cellular immune system
Skin graft rejection	Rejection of graft from unrelated individual indicates competent cellular immunity
In vitro	
Lymphocyte blast transformation	Healthy sensitized lymphocytes transform into large blast cells when exposed to Ag
Macrophage migration inhibitory factor	Healthy sensitized lymphocytes produce MIF in presence of Ag
Rosette formation	Healthy T lymphocytes form rosettes when incubated with sheep RBCs

therefore, must be of particular concern to the clinician in planning infection control measures.

Mechanisms for Enhancing Deficient Host Defenses

Once infectious disease occurs, antimicrobial therapy is the most important therapy available (see Chapter 26). Gamma globulin may be administered to persons with IgG cell deficiencies (not useful with IgM or IgA deficiencies). Routine immunization of children is invaluable. Some therapies that are in use (though some are still experimental) include polyvalent pneumococcal polysaccharide vaccine for children with sickle cell anemia, transfusions of white blood cells for neutropenia, bone marrow transplantation, fetal thymus grafting for thymic defects, administration of transfer factor, antiserum for sepsis caused by gram-negative bacteria, BCG or drugs like levamisole to enhance T cell function in the aged or in those with viral diseases or neoplasms.

Preventive care is the area in which clinicians can most effectively intervene to control infection. Those at highest risk of infection are individuals who are hospitalized or severely ill. Specific measures to enhance and protect the immune system include (1) meticulous aseptic care of indwelling lines such as intravascular devices and urinary catheters; (2) maintain closed system; (3) avoid the use of indwelling lines whenever possible; (4) initiate isolation techniques for protection of extensive burns or traumatic wounds; (5) take particular care with patients who have malignancies or are immunosuppressed; (6) use barrier techniques for patients with clinical infections; (7) and detect signs and symptoms of infection early.

Acquired Immune Deficiency Syndrome

Since 1980 a new disease has been identified in the United States. It is called acquired immune deficiency syndrome (AIDS). Eight hundred and ninety cases of AIDS were reported to the Centers for Disease Control between June 1981 and January 1983. AIDS is characterized by

1. The presence of a disseminated malignancy or opportunisitc infection.
2. Evidence of depressed cellular immunity without obvious cause.

AIDS has been observed primarily in patients less than 60 years of age.

To diagnose AIDS, an analysis of cell-mediated immunity is done. This includes an evaluation of the ratio of helper to suppressor T cells (if it is less than 0.5, the individual is immunocompromised) and the intradermal injection of a battery of antigens. If there is no cutaneous response to three or four common antigens, the individual is anergic (without an immune response). There are two major theories regarding the etiology of AIDS. One proposes that repeated stimulation of the immune system leads to the development of a tumor or infection; the other proposes that there is one causative infectious agent such as the hepatitis B virus. Since individuals with no risk factors have developed the disease after contact with diseased patients, the latter theory seems more likely. Transmission by blood products or sexual contact seems most likely.

About 75 percent of AIDS cases have occurred among homosexual or bisexual men, 13 percent among intravenous drug abusers, 6 percent among Haitians with no history of homosexuality or drug abuse, and 5 percent in other individuals. The most common manifestations of the disease are pneumonia due to *Pneumocystis carinii*, Kaposi's sarcoma (a soft-tissue tumor that tends to occur in the extremities), and other opportunistic infections with agents such as the mycobacteria other than tuberculosis (MOTT) group. Afflicted individuals eventually succumb and there appears to be a rapid rise in incidence. Therefore, there will certainly be intensive study of AIDS until etiologic agents and treatment are better understood. Patients hospitalized with AIDS should be placed on isolation precautions similar to those used for Hepatitis B—isolation of blood, blood products, and other body secretions.

SUGGESTED READINGS

Barrett JT. Basic Immunology and Its Medical Application, 2nd edition. St Louis: CV Mosby, 1980.

Bartlett JG: New developments in infectious diseases for the critical care physician. Crit Care Med 1983;11:563-573.

Centers for Disease Control. Update on acquired immune deficiency syndrome (AIDS)—United States. Morb Mort Week Rep 1982; 31:507-513.

Dolan B. Rapid desensitization. Am J Nurs 1982;82:1532-1534.

Fundenberg HH, et al. Basic and Clinical Immunology, 3rd edition. Los Altos, California: Lange Medical Publications, 1980.

Gallucci BB, Reheis CS. Infection, nutrition and the compromised patient. Top Clin Nurs 1979;1:23-33.

Katsura Y, Minato N, Nishikawa SI. Role of virus-replicating T cells in the suppression of delayed-type hypersensitivity. J Immunol 1981;126:1529-1533.

Lind M. The immunologic assessment: a nursing focus. Heart Lung 1980;9:658-661.

Movat HZ, ed. Inflammation, Immunity and Hypersensitivity. Hagerstown, Md: Harper and Row, 1979.

Rana AN, Luskin A. Immunosuppression, autoimmunity, and hypersensitivity. Heart Lung 1980;9:650-657.

Portaro JK. HLA typing—an exciting discovery. Life Health 1980;95:8-9.

Sissons JGP, Oldstone MBA. Killing of virus-infected cells by cytotoxic lymphocytes. J Infec Dis 1980;142:114-119.

II SPECIFIC MICROORGANISMS

7 GRAM-POSITIVE COCCI

KEY POINTS

- Reaction to the Gram stain (positive or negative) is one of the major tests used to differentiate bacterial species.
- Many gram-positive cocci (staphylococci and streptococci) are normal flora of the human body and are also important causes of infectious diseases among healthy and ill persons of all ages.
- The most important pathogens among the gram-positive cocci are *Staphylococcus aureus, Streptococcus pneumoniae,* and *Streptococcus pyogenes* (synonymous with beta-hemolytic streptococcus, group A).
- *Staphyloccus aureus* is a normal flora of the anterior nares, but causes a variety of infections including boils and impetigo, endocarditis, osteomyelitis, and septicemia.
- *Streptococcus pneumoniae* (the pneumococcus) is a normal flora of the oropharynx, but causes pneumococcal pneumonia, especially in the elderly and debilitated.
- *Streptococcus pyogenes* is not considered to be a normal inhabitant of the human body. The most important diseases it causes include impetigo and "strep throat" with two sequelae of major concern: rheumatic fever and glomerulonephritis.

Gram-positive cocci are classified into three major groups and 12 genera (Table 7-1) of which two, the staphylococci and the streptococci, are of particular concern in human disease (Table 7-2). Along with the gram-negative *Neisseria,* staphylococci and streptococci are termed the *pyogenic cocci* and are the major causes of pyogenic infections as well as being normal inhabitants of the human skin, upper respiratory, and intestinal tracts. They are generally extracellular and invasive. All forms important to humans are nonmotile. The gram-positive cocci may be indistinguishable from each other by Gram stain, though each genus produces a unique morphologic ar-

TABLE 7-1. Classification of Gram-positive Cocci

Aerobic/facultatively anaerobic		*Anaerobic*
Family A:	Micrococcaceae	Genus:
Genus:	*Micrococcus**	*Coprococcus*
	*Staphylococcus**	*Peptococcus**
	Planococcus	*Peptostreptococcus**
Family B:	Streptococcaceae	*Ruminococcus*
Genus:	*Streptococcus**	Species:
	Leuconostoc	*Sarcina ventriculi*
	Pediococcus	*Gaffkya anaerobia*
	*Aerococcus**	*Gemminger formicilis*
	Gemella	

*Contains species causing human disease

rangement (Fig. 7-1). Staphylococcal, meaning "bunch of grapes," cells divide geometrically in three planes, causing the formation of groups of cells in grapelike clusters. Streptococci, on the other hand, divide in such a manner as to form short or long chains. *Streptococcus pneumoniae* (pneumococcus) is usually found in pairs (hence the former genus name *Diplococcus*) with its adjacent sides flattened so that they resemble a double-edged sword or lancet. *Sarcina,* uncommon in humans, tend to grow in tetrads. Only genera important in human disease are discussed in this chapter.

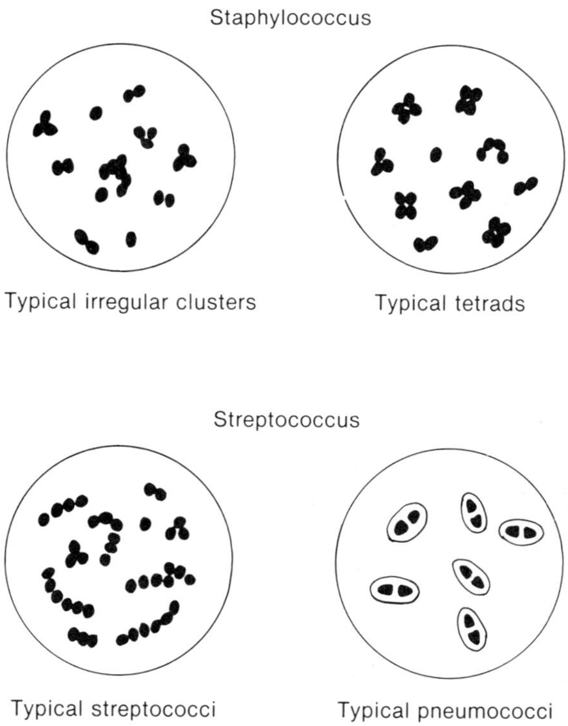

FIGURE 7-1. Morphological arrangements of staphylococci and streptococci.

STAPHYLOCOCCI

Staphylococci, of the family Micrococcaceae, make a large contribution to human's normal flora of protective layers such as skin, and mucosa of the oropharynx and nasopharynx. The genus *Staphylococcus* also accounts for a high proportion of acute and chronic suppurative lesions. Cell size varies but the average diameter is about 1 μm. Staphylococci grow readily on bacteriologic media aerobically and as facultative anaerobes. Temperatures of 37°C and pH of 7.4

TABLE 7-2. Gram-positive Cocci—Pathogenic to Human *(Staphylococcus, Streptococcus)*

Organism	*Disease manifestations*
Staphylococcus	
Staphylococcus aureus	Skin infections: "pimples," boils, hair-follicle infection or abscess; carbuncles; pustules; wound infection; impetigo Secondary localization: meningitis; endocarditis; puerperal sepsis; pneumonia; osteomyelitis; septic arthritis; septicemia Food poisoning: vomiting and diarrhea Toxic shock syndrome: fever; vomiting; diarrhea; erythematous macular, sunburnlike rash progressing to peeling of palms of hands and soles of feet; renal dysfunction; oliguria; ↑ BUN; ↑ Bilirubin; hypotension; peripheral edema; pulmonary syndrome; vaginal discharge (purulent); thrombocytopenia; disseminated intravascular coagulation (DIC)
Staphylococcus epidermidis	Skin infections: pimples, acne, "stitch abscess" Secondary localization: meningitis; ventriculitis; bacterial endocarditis; septicemia
Streptococcus	
α-hemolytic streptococcus *(S. viridans)*	Secondary localization from dental trauma; subacute bacterial endocarditis; arthritis; meningitis; urinary tract infections
S. pneumoniae	Pneumonia; (associated with chronic bronchitis) meningitis; peritonitis
β-hemolytic streptococcus Group A or *S. pyogenes*	Sore throat (tonsillitis, pharyngitis); scarlet fever; cellulitis; puerperal fever; pneumonia, immunologic reaction (acute rheumatic fever and glomerulonephritis), circulatory involvement; lymphangitis; impetigo
Group B	Neonatal septicemia and meningitis; postpartal endometritis; respiratory infections; endocarditis
Groups C and G	(Less serious infections); tonsillitis; sinusitis, bacteremia; endocarditis
Group D enterococcus	Urinary tract infections; wound infections; ulcerative skin lesions; subacute bacterial endocarditis

yield optimal growth. Staphylococci are among the most resistant of nonspore-forming bacteria to drying, heat, and high concentrations of salt. For example, organisms may remain viable after 14 weeks of drying. There are about 30 antigens in staphylococci, but they are not useful for identification or typing. Phage typing, however, is

TABLE 7-3. Species of Staphylococci in Humans

Species	*Coagulase*	*Cause of human disease?*	*Prevalence in humans*
S. aureus	+	Yes	20-60 percent
S. capitis	−	Possibly	Rare
S. cohnii	−	Possibly	Rare
S. epidermidis	−	Yes	Almost 100 percent
S. haemolyticus	−	Probably	5-20 percent
S. hominis	−	Probably	5-20 percent
S. saprophyticus	−	Yes	5-20 percent
S. sciveri	−	Possibly	Rare
S. simulans	−	Probably	5-20 percent
S. warneri	−	Possibly	Rare
S. xylosus	−	Possibly	Rare

used extensively for identifying strains of *S. aureus*. All staphylococci are nonmotile and usually nonencapsulated, though a few strains of *S. aureus* do form capsules. They may produce beta (complete) hemolysis or no hemolysis.

There are at least 11 species of staphylococci (Table 7-3) that have been isolated from humans, but only two, *S. aureus* and *S. epidermidis,* are of particular significance. There are a number of distinguishing features between the two (Table 7-4), the most important of which is the ability to clot plasma (coagulase production). Coagulase production is directly related to virulence, and the coagulase positive *S. aureus* is the most virulent of the staphylococci. The two other genera in the family Micrococcaceae will not be discussed in this chapter. Micrococci are occasionally present in human specimens and on rare occasions cause opportunistic infections; the *Planococcus* is a motile genus that plays no role in human disease.

STAPHYLOCOCCUS AUREUS

Biologic characteristics. *S. aureus* (Fig. 7-2) are hardy, typical staphylococci. It is the only species that produces an enzyme (coagulase), which clots plasma. Most strains are

TABLE 7-4. Comparison of *S. aureus* and *S. epidermidis*

	S. aureus	*S. epidermidis*
Colony color	Usually golden yellow	Usually white
Coagulase	+	−
Hemolysis	Usually beta hemolytic	Usually nonhemolytic
Habitat	Primarily anterior nares	Primarily skin
Mannitol fermentation	+	−
Toxin production	+	−
Virulence	High	Low

beta hemolytic, but some may be nonhemolytic. Many strains of *S. aureus* form carotenoid pigments that produce colonies of a characteristic yellow-golden color; some strains may produce white colonies morphologically identical to those of *S. epidermidis*. Unlike gram-negative bacteria, *S. aureus* are incapable of conjugation, but plasmid transfer of such factors as antimicrobial resistance or toxin production takes place by transduction (transfer of genetic material by a lysogenic bacteriophage).

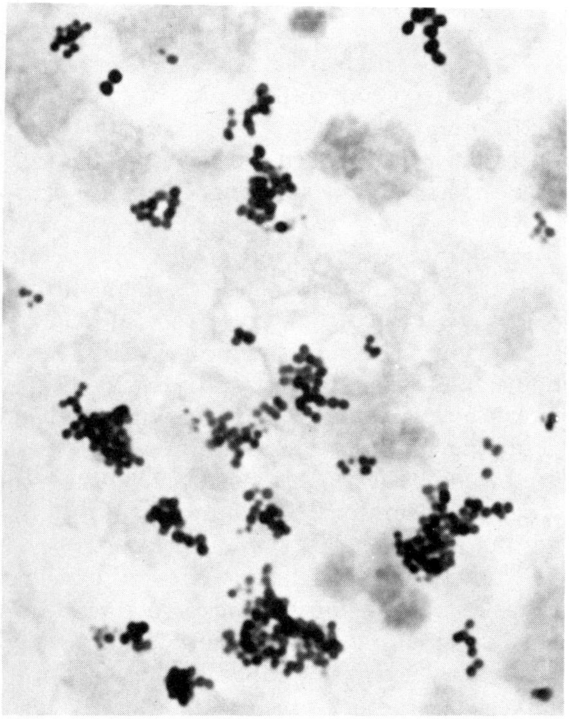

FIGURE 7-2. *S. aureus* in cerebrospinal fluid. Notice the grapelike clusters. (Courtesy of Dr. James Plorde.)

Strains are typed by the phages that lyse them; for example, type 80/81 *S. aureus* is lysed by phages 80 and 81. Phage typing is accomplished by dropping standardized suspensions of 5 to 60 phages in particular positions on a plate on which *S. aureus* are growing. Though phage types do not necessarily correlate with virulence, typing is useful to trace the source of nosocomial infections with *S. aureus*. There are six major groups of phage-typed strains (Table 7-5) that share general characteristics. About 30 percent of strains are nontypable. Generally the antigenic structure of *S. aureus* is of minimal importance in taxonomy or typing. Though certain phage types have been associated in the past with hospital epidemics, these vary over time and in different geographic areas.

Epidemiology. *S. aureus* is widely distributed in air, dust, and fomites, but the human skin and mucous membranes are its major natural habitat. This organism is found on the anterior nasal mucosa of about

TABLE 7-5. Phage Groups of *Staphylococcus aureus*

Group	*Phage numbers*	*Usual characteristics*
I	29, 52, 52A, 79, 80	Penicillinase producers
II	3A, 3B, 3C, 55, 71	Produce epidermolytic toxin (exfoliatin)
III	6, 7, 42E, 47, 53, 54, 75, 77	Produce enterotoxin
IV	42D	Produce enterotoxin
V	Nontypable	Resistant to bacteriophages
VI	94, 95, 96, 81, 187	Miscellaneous

30 to 50 percent of healthy adults, in the throats of many, on the skin of about 20 percent, and in the feces of about 20 percent.

S. aureus can invade the sebaceous glands and is the usual cause of skin infections such as boils, furuncles, and carbuncles. These are especially prevalent during adolescence when the sebaceous glands are hyperactive. Another skin infection, impetigo, causes itching vesicles, primarily in children. Because the blisterlike lesions are often extensive and on open skin areas such as the arms and legs, impetigo can spread rapidly among children, especially in crowded settings like nurseries, day care centers, or large families. Inpetigo may be caused either by *S. aureus* or by virulent streptococci.

S. aureus is of major concern as a cause of nosocomial epidemics, especially in the following high-risk groups:

1. Newborn infants. Newborns do not possess mature immune systems until they are about 6 months of age. In the interim their major defenses are the acquired maternal antibodies. Newborns are readily and rapidly colonized by *S. aureus;* as many as 90 percent of those born in hospitals may carry this organism in their anterior nares or umbilical stump within 2 weeks of birth. Most infants only become colonized, but some will become ill with pneumonia, conjunctivitis, skin lesions, or septicemia.
2. Individuals who have prosthetic heart valves or prosthetic joint surgery. In such patients, staphylococcal endocarditis or osteomyelitis can occur.
3. Compromised hosts, individuals with debilitating conditions such as victims of burns, trauma, cancer, diabetes, and liver dysfunction.

S. aureus may be spread by the respiratory tract, direct contact, or by contaminated fomites. The risk of transmitting *S. aureus* from a nasal carrier is low, unless the hands are contaminated with nasal secretions. Direct person-to-person contact rather than airborne spread seems to be most important.

Food can be contaminated from boils or other infections during handling. If the strain of staphylococci that contaminates food is one which produces an enterotoxin, the organisms will multiply in the food, form toxin, and result in food intoxication when ingested. (This is called an intoxication rather than food poisoning because symptoms are due to ingestion of toxin rather than the organism itself.) *S. aureus* is one of the most common causes of food intoxication. Symptoms of staphylococcal food intoxication occur very suddenly, within a few hours of eating the contaminated food. The victim is very uncomfortable, with nausea, vomiting, diarrhea, and fatigue that lasts up to 24 hours. The disease, though extremely uncomfortable, is not usually dangerous and recovery is spontaneous.

Women who use tampons regularly are at risk for toxic shock syndrome since tampons seem to be an excellent foreign body on which *S. aureus* can proliferate. Toxic shock syndrome, caused by *S. aureus,* was first described in 1978. In late 1979, 12 cases reported to Centers for Disease Control (CDC) were strongly associated with the use of tampons in menstruation (Fig. 7-3). Most cases have been identified in the United States; cases have also been reported in New Zealand and England. In Utah, during 7 months of study in 1980, 52 women had toxic shock syndrome and there was a significant relationship to tampon use. Of cases reported to CDC since 1978, 96 percent occurred in menstruating women. Most cases (94 percent) had vaginal cultures positive for *S. aureus* as compared to the usual prevalence of *S. aureus* in the vagina of 2 to 15

TOXIC-SHOCK SYNDROME

TOXIC-SHOCK SYNDROME – Reported cases by date of onset, United States, January 1970-December 1980

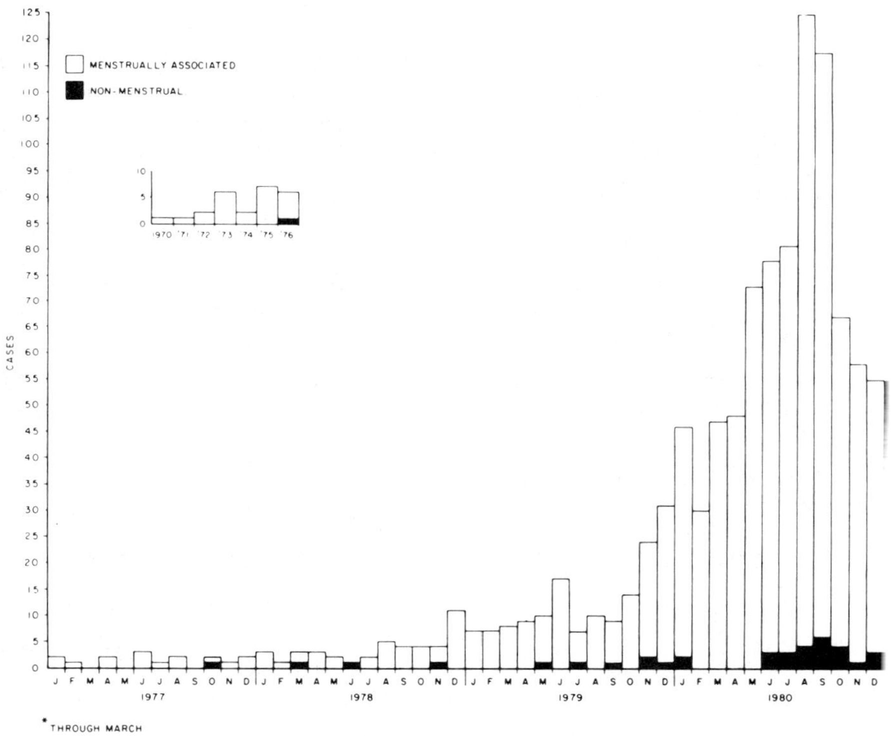

FIGURE 7-3. Reported cases of toxic shock syndrome; by date of onset, United States, January 1970 through December 1980. (Reprinted from Centers for Disease Control. Toxic-shock syndrome—United States, 1970-1980. Morb Mort Week Rep 1981;30:25-28.)

percent of women. The incidence of toxic shock syndrome is about 3 per 100,000 menstruating women per year. Cases in men and nonmenstruating women also occur, though rarely. By 1980, 941 cases had been reported. The increase in numbers of cases is certainly due partly to improved recognition and reporting, as is the case with any newly described syndrome. Since September 1980 there has been a decrease in reported cases, which may be due to less use of tampons. Based on telephone interviews conducted in July 1980, 80 percent of women used tampons. By December 1980, this proportion had decreased to 55 percent.

Pathogenesis. Many characteristics contribute to the virulence of *S. aureus*. A few strains have capsules, all *S. aureus* possess certain surface antigens that are antiphagocytic. Additionally, their cell wall peptidoglycans inhibit leukocyte migration, and they produce a number of enzymes and toxins (Table 7-6). One cell wall component called protein A is especially important for inhibiting phagocytosis, which it does by binding to the Fc receptors of phagocytes. Coagulase activity is associated with virulence and is probably responsible for abscess formation with walling off of infection. A virulent staphylococcus is usually coagulase positive and produces either alpha or gamma hemolysins. Exfoliatin is produced by about 40 percent of group II phage strains. It is a protein transmitted by lysogenic phage which causes scalded skin syndrome. There are two types: type A is heat stable, type B is heat labile. About 50 percent of strains produce one to five types (A to E) of enterotoxin, the majority produced by strains 6/47 and 42D. These strains are responsible for staphylococcal food poisoning.

Host reponses. Common clinical forms of staphylococcal infection as listed in Table 7-2, include localized skin infections (Fig. 7-4) such as furuncles (boils), wound infections, or abscesses; osteomyelitis; pneumonia; bacteremia; food poisoning; and toxic shock and scalded skin syndromes. Local infections (a "pimple," for example) are usually painful with intense inflammatory reaction that subsides quickly when pus is drained. When cell layers around the core of infection are broken due to trauma or manipulation, spread of organisms through

TABLE 7-6. Toxins Produced by *Staphylococcus aureus*

Toxin	*Action(s)*
Staphylolysins	
Alpha toxin (hemolysin)*	Hemolytic; increases permeability of cell membrane; dermonecrotic in rabbits, low activity in humans
Beta hemolysin	More potent than alpha hemolysin, but found primarily in animal strains
Gamma hemolysin*	Hemolytic, leukotoxic
Delta hemolysin	Less dermonecrotic and leukocytic
Epsilon hemolysin	Present in *S. epidermidis*
Leukocidin	Lyses leukocytes
Enterotoxins	Directly act on intestinal wall, causing severe nausea and vomiting
Exfoliative toxins (exfoliatin)	Disrupts desmosomes that bind cells in the granular layer of epidermis, causes skin peeling
Coagulase	Clots plasma
Hyaluronidase	Depolymerizes ground substance of tissues ("spreading factor")
Staphylokinase	Dissolves fibrin clots
Bacteriocin (staphylococcin)	Inhibits or kills other closely related gram-positive cocci

*Greater than 95 percent of *S. aureus* form one of these toxins, 82 percent form both.

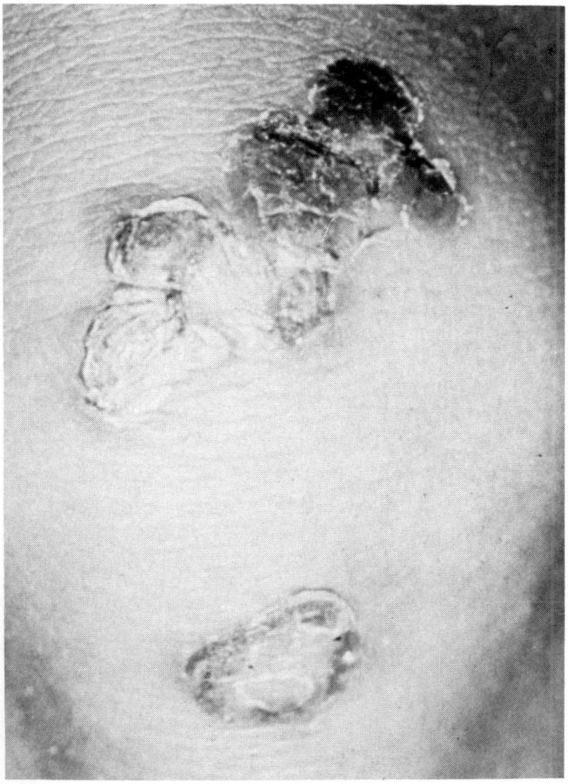

FIGURE 7-4. Staphylococcal impetigo. (Courtesy of Dr. Lawrence Corey.)

the lymphatics or the blood stream results. Common systemic infections include meningitis, endocarditis, and septic arthritis. *S. aureus* is the most common cause of osteomyelitis, which is more prevalent in children but may also occur in adults. Mortality among untreated individuals with osteomyelitis is about 25 percent. However, death now is rare, and chronic sequelae are also less frequent.

About 1 percent of bacterial pneumonias are caused by *S. aureus,* most often following influenza. The onset is abrupt with fever, chills, and pleural pain. Pulmonary abscesses, empyema, pneumothorax with pus, and pneumatocele may occur. Mortality is high, up to 50 percent even with antimicrobial therapy, especially if bacteremia is present. Staphylococcal bacteremia may result from local infections or from invasive procedures, especially surgery or intravenous therapy. Anemia and elevated white blood cell counts are common, as are hypoxia and cyanosis. Endocarditis, especially with prosthetic heart valves in place, is a serious complication that usually requires surgical removal of the infected valve. Mortality with appropriate therapy is about 20 percent. *S. aureus* may also cause mastoiditis, otitis, and mastitis among nursing mothers.

Strains of *S. aureus* that produce enterotoxin cause a large proportion of cases of food poisoning. Symptoms occur within 6 hours of ingesting contaminated food. Most commonly implicated foods are pastries with cream fillings, other milk products, poorly cooked meats (often ham), and salad dressings. The incubation period is short because the toxin has already been formed while the organisms were multiplying in the food. There is an abrupt onset of nausea, vomiting, abdominal cramps, prostration, and diarrhea. There is no fever, since the disease is a toxicosis and not an infection. The disease is usually epidemic in nature since contaminated food has often been prepared in bulk or poorly stored. Diagnosis is confirmed by the presence of *S. aureus* in remaining food. The symptoms subside after about 1 day and there are rarely fatal outcomes. A second type of intestinal disease also caused by *S. aureus* is enterocolitis, which occurs when the organism proliferates in the bowel, usually as a result of alterations in the normal flora following antimicrobial therapy (the epidemiology is similar to pseudomembranous colitis caused by *Clostridium difficile;* see Chapter 8).

Toxic shock syndrome usually has its onset 2 to 3 days after the beginning of menstruation with the development of high fever, diarrhea, hypotension, peripheral edema, disseminated intravascular coagulation, renal failure, erythroderma, and desquamation (peeling) of skin on the palms, soles, and other affected areas of skin. It is usually possible to isolate *S. aureus* from the vagina and cervix but not from the blood. Recovery takes 7 to 10 days, but the recurrence rate is about 30 percent. Mortality is 10 to 15 percent. Scalded skin syndrome, most common in neonates and young children, is caused by strains of *S. aureus* that elaborate erythrogenic toxin. Some forms of scalded skin syndrome are exfoliative dermatitis (Ritter's disease (Fig. 7-5), toxic epidermal necrolysis, bullous impetigo, and staphylococcal scarlatina (similar to streptococcal scarlet fever)). Patients usually recover because deeper layers of skin are spared.

Control and treatment. Scrupulous hygienic practices (e.g., food preparation, aseptic wound dressing and surgical techniques) are necessary to reduce the risk of staphylococcal infection. Immunity to staphylococcal disease is primarily cellular. Coagulase is a poor antigen; leukocidin and many of the toxins are antigenic. Adults, who have low incidence of exfoliative staphylococcal skin disease, probably possess antitoxin to the toxin exfoliatin. Active immunization against staphylococcal disease is in the experimental stages.

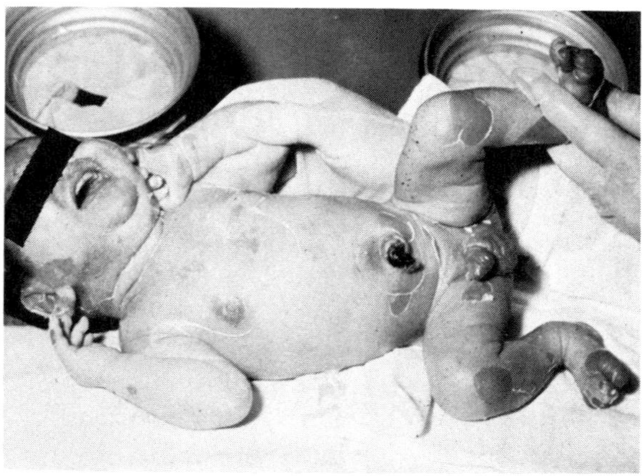

FIGURE 7-5. Ritter's disease in a newborn. (Courtesy of Dr. Lawrence Corey.)

Staphylococcal food poisoning can be prevented by the refrigeration of food before and after cooking to prevent multiplication of the toxin, the adequate cooking of meats, and the segregation of handlers with open staphylococcal lesions from food preparation. Individuals with skin or respiratory infections due to *S. aureus* require isolation in the hospital (see "Clinical Implications"). Hexachlorophene (HCP) bathing of infants has proven to be valuable in minimizing staphylococcal colonization in newborn nursery populations. In 1973, however, its use was restricted because of the discovery that neurologic damage is a rare but serious sequela of HCP use. Nevertheless, the benefits of HCP for infant bathing still outweigh the risks during staphylococcal epidemic periods in nurseries, even though it is not recommended for routine bathing.

When a staphylococcal infection is localized as in an abscess, drainage of pus is the essential first step in treatment since organisms located in the central part of lesions will be inaccessible to antimicrobial agents. Antimicrobial susceptibility testing is essential for all staphylococci because resistance is common. Penicillinase, one of many beta-lactamase enzymes, inactivates penicillin. The ability to produce penicillinase can be acquired by a strain of *S. aureus* through transduction or by mutation. When penicillins were first introduced, most strains of *S. aureus* were sensitive, but about 80 percent are now penicillin resistant as a result of penicillinase production. Sensitive strains of *S. aureus* replace resistant ones without the pressure of antimicrobials, so judicious use of such drugs should decrease resistance.

Methicillin resistance has also gradually increased since the 1960s. In 62 hospitals surveyed by CDC during 1979 to 1980, 23 (37 percent) reported at least one isolate of methicillin-resistant *Staphylococcus aureus* (MRSA), but only six hospitals had more than five cases. MRSA accounted for 1 to 61 percent of *S. aureus* bacteremias, and the incidence of such bacteremias ranged from 7 to 84 per 10,000 hospital admissions. Since that time, however, the prevalence of MRSA in the United States continues to rise. MRSA has been endemic in Europe since 1970, and has increased as a nosocomial isolate from 2.4 percent in 1975 to 4.9 percent in 1980 in the United States. Most isolates to this point have been reported by a small number of hospitals in which MRSA has become endemic. Methicillin resistance is due to some intrinsic factors rather than to penicillinase production, and seems to be naturally present in a small number of strains. Gentamicin resistance, like penicillin resistance, is due to enzymes transmitted by plasmids. Vancomycin is the treatment of choice for MRSA, but since it is very toxic, its use should be limited to severe infections.

Recommendations from CDC to prevent toxic shock syndrome are the following:

1. Women should refrain from use of tampons, use them intermittently, or change them frequently.
2. If fever, vomiting, or diarrhea develop during menstruation, the tampon should be immediately removed and a physician called.

Antimicrobial agents are used to treat toxic shock syndrome. They probably do not affect the disease course, but appear to prevent recurrences.

Hyperbaric oxygen is efficacious in treating osteomyelitis caused by *S. aureus,* which is otherwise often refractory to treatment because the organisms are inaccessible to both phagocytes and chemotherapeutic agents in the bone. Hyperbaric oxygen works primarily by enhancing phagocytosis. Inoculation of newborns or individuals with recurrent staphylococcal infections with an "avirulent" strain of *S. aureus* takes advantage of the phenomenon of bacterial interference to prevent colonization with more pathogenic or epidemic-causing strains. This technique is quite successful at preventing other strains from colonizing. However, the "avirulent" strain itself has occasionally caused clinical disease, so this technique has minimal usefulness.

Other Staphylococci

Biologic characteristics. The coagulase negative staphylococci, especially *S. epidermidis (S. albus* or "white" for the usual color of its colonies), but also occasionally *S. saprophyticus* (sometimes classified as *Micrococcus*) and other species, are usually nonhemolytic, do not ferment mannitol, and can be low-grade pathogens. Strains of *S. epidermidis* as well as those of *S. aureus* (79 and 55 percent, respectively) possess moderately large plasmids (molecular weight 15×10^6 to 29×10^6), whereas other strains of staphylococci contain only small plasmids. Strains of *S. epidermidis* can be identified by antibiogram, by phage or biotyping, and by plasmid profiles.

Epidemiology. *S. epidermidis* and other species are widely distributed in air and in the environment. They are found on skin, and on mucous membranes of the upper respiratory and gastrointestinal tracts of humans and animals. Although usually harmless, these organisms can cause disease when they get beneath the skin into underlying tissues, usually via skin injury. *S. epidermidis* is an important cause of infection related to invasive monitoring, I.V. therapy, and cardiac surgery. One survey of 842 staphylococcal isolates from a hospital over an 8-month time period suggested that the proportions of infections caused by coagulase-negative staphylococci is increasing and that antimicrobial resistance is also more common. Over half (436 out of 842) of the isolates were coagulase-negative staphylococci (Pennington 1981).

Pathogenesis. As with other opportunistic pathogens, the development of clinical disease depends on a combination of factors related to integrity of host defenses and invasiveness of procedures. *S. epidermidis* plays some part in the pathogenesis of minor skin lesions. More recently it has achieved greater clinical attention as an organism with a propensity to colonize the fibrin overlying heart valves and other plastic prostheses. Clinical disease does not appear to link to specific virulence factors, though some strains produce hemolysin.

Host responses. As a rule, infections due to coagulase-negative staphylococci are not virulent, but there have been some cases of fatal septicemia. Meningitis, ventriculitis, and bacterial endocarditis also occur.

Control and treatment. Infected prostheses usually are removed and replaced. Measures to prevent, treat, and control disease due to coagulase-negative staphylococci are similar to measures used for other opportunistic pathogens: aseptic technique to prevent the entrance of the organism into uncolonized body sites, avoidance of invasive procedures whenever possible, and antimicrobial therapy.

STREPTOCOCCI

Streptococci from the family Streptococcaceae contain over 100 types of organisms in one genus. Other members of the same family are the *Aerococcus, Gemella, Leuconostoc,* and *Pediococcus,* which are of little or no importance in human disease. Streptococci comprise a major portion of mouth and pharyngeal flora, as well as inhabiting the bowel and genitourinary tracts of humans. They are also found in water, dust, milk, and vegetation. They are generally commensals, though two species, *S. pyogenes* and *S. pneumoniae,* are often pathogens. Other species, of course, may become opportunistic pathogens.

Streptococcal species share many characteristics. In general they are gram positive (as cells age they become gram negative), 0.5 to 1.0 μm in diameter arranged in pairs or chains. Except for *S. faecium* they are nonmotile. Some strains have capsules that are nonantigenic. The organisms ferment glucose, reduce nitrates, and are usually soluble in bile salts. They are catalase and oxidase negative. Optimal growth takes place in blood or serum media at 30 to 40°C with a pH range of 7.4 to 7.6. Growth and hemolysis are aided by 10 percent carbon dioxide. Streptococci are facultative anaerobes in general and can withstand drying conditions for several days to several weeks in droplet nuclei.

Various methods are used to group species of streptococci (Table 7-7). The most common is by type of hemolysis, which may

TABLE 7-7. Methods of Grouping Species of Streptococci

Grouping	Method
Types of hemolysis	Based on the extent to which strains hemolyze red blood cells. Three types: alpha, beta, or gamma (none)
Lancefield grouping	Precipitin tests that use cell wall antigens tested against specific antisera produced in animals
Subtyping	Based on strain-specific proteins (e.g. M proteins in group A streptococci)

be complete (beta), partial (alpha), or none (gamma). Table 7-8 describes these patterns (see also Fig. 7-6). Streptococci are also grouped according to shared nontoxic cell wall carbohydrate antigens (called C antigen) into about 20 types, A through T. These are called Lancefield groups, named after the person who developed the typing technique. Thus, group A streptococci, for example, share a common A group antigen. Most strains pathogenic to humans are both beta hemolytic and in Lancefield group A. However, the groupings are not mutually exclusive. That is, though most group A streptococci are beta hemolytic, a rare strain can be nonhemolytic. Likewise, group B streptococci are usually beta hemolytic but some may be alpha or nonhemolytic, and so forth. In addition, not all streptococci pos-

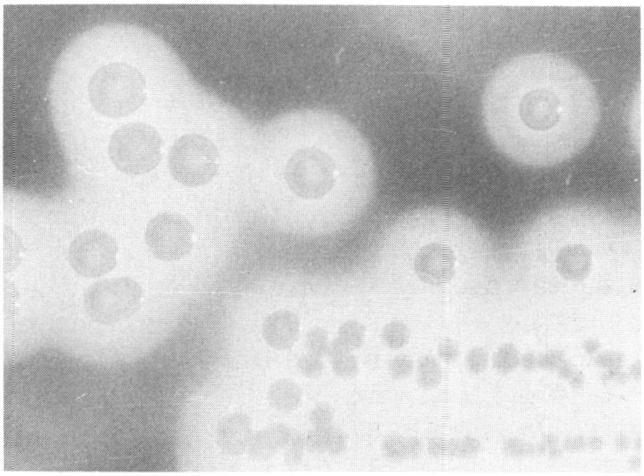

FIGURE 7-6. Colonies of beta-hemolytic streptococcus group A, growing on blood agar. Notice the large zone of complete hemolysis surrounding each colony. (Courtesy of Dr. James Plorde.)

TABLE 7-8. Types of Hemolysis on Blood Agar Plates

Type	Action	Appearance	Example
Alpha (α)	Partial destruction of erythrocytes	Zone of brownish or greenish discoloration around colonies	*S. salivarius*
Beta (β) (see Fig. 7-6)	Complete destruction of erythrocytes	A clear, colorless zone surrounding colonies	*S. pyogenes*
Gamma (γ) or none	No destruction of erythrocytes	No zone of discoloration around colonies	*S. faecalis*

sess Lancefield group antigens (viridans streptococci, *S. pneumoniae*). This complicates the typing of streptococci. Within a particular Lancefield grouping, streptococci can be further differentiated by precipitin or coagglutination tests. For group A streptococci, the M protein, an antigen found in the cell wall, is used for typing. On the basis of types of M proteins, group A streptococci can be divided into about 60 strains. Other subgroups can be identified by counter-immunoelectrophoresis, fluorescent antibody, coagglutination (Figure 7-7), and latex fixation tests.

One or more species of streptococci may be contained in the groups or types identified by the above methods. For example, all group A streptococci are of one species, *S. pyogenes*. Therefore the terms "beta-hemolytic streptococci, group A" and *S. pyogenes* are often used synonymously. On the other hand, there are a number of streptococci in Lancefield group D, and it would be incorrect to use the terms "group D streptococci" and *S. faecalis* or "enterococci" as synonyms (which has been done at times). All alpha-hemolytic streptococci were formerly classified as *S. viridans*, but they are now further speciated. Major groupings of streptococci and their common characteristics are listed in Table 7-9. There is one species each in groups A, B, and F, three in group C, and many (about 20) species in group D streptococci found in humans. Group G streptococci do not have a species name. The other groups are unimportant in human disease. In this chapter the streptococci that will be discussed in detail are those most commonly related to human disease: group A, group B, and the pneumococcus.

Streptococcus pneumoniae

Biologic characteristics. Streptococcus pneumoniae (pneumococcus or formerly *Diplococcus pneumoniae)* is an encapsulated aerobic or faculatively anaerobic gram-positive coccus that frequently occurs in pairs (Fig. 7-8) growing back to back with their nonadjacent ends slightly pointed (lancet shaped). Most strains are alpha hemolytic. There are 84 serotypes, distinguished by

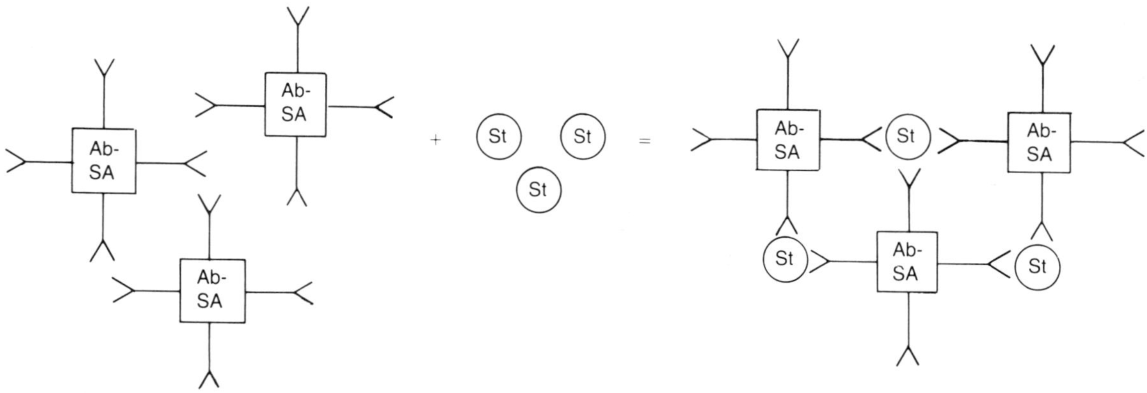

Ab-SA = *S. aureus* coated with group-specific streptococcal antigens
St = A suspension of ungrouped streptococci

FIGURE 7-7. Coagglutination test for grouping streptococci.

TABLE 7-9. Classification of Streptococci Isolated from Humans

Lancefield group	*Species or genus*	*Habitat*	*Disease agent?*
A	*S. pyogenes*	Pharynx, anus, vagina	Commonly
B	*S. agalactiae*	Genital, GI tracts	Commonly
C	*S. equisimilis*	Pharynx of humans and animals	Possibly
	S. equi	Pharynx of humans and animals	Rarely
	S. zooepidemicus	Pharynx of humans and animals	Rarely
D			
Enterococci	*S. avium*	GI tract	Possibly
	S. durans	GI tract	Rarely
	S. faecalis	GI tract	Occasionally
	S. faecium	GI tract	Occasionally
Nonenterococci	*S. bovis*	GI and respiratory tracts, milk	Rarely
E	*S. uberis*	Pharynx, GI tract	Rarely
F	*S. anginosus*	Pharynx, GI tract	Rarely
(microaerophilic)	*S. minutis*	Pharynx, GI tract	Rarely
G	*S. canis*	Pharynx of dogs and humans	Rarely
H	*S. sanguis*	Pharynx	Probably not
K,L,M		Pharynx of humans, dogs, pigs	Probably not
N	*S. lactis*	Milk	No
	S. cremoris	Milk	No
O-T		Pharynx, other sites	Possibly
Viridans	*S. mitis*	Respiratory tract	Occasionally
	S. salivarius	Respiratory tract	Occasionally
	S. mutans	Respiratory tract	Occasionally
	S. viridans	Respiratory tract	Occasionally
Pneumococci	*S. pneumoniae*	Respiratory tract	Commonly
Anaerobic	*Peptostreptococcus* *Peptococcus*	Skin, respiratory or GI tract	Occasionally

complex polysaccharide capsular substances.

The Quellung reaction is used to identify *S. pneumoniae*. A positive test results when the organism's capsular polysaccharide binds with type-specific antiserum, changing the refractive index and causing the capsule to swell and become visible. There are rough (R) strains that do not possess a capsule. Typing of pneumococci is not done routinely because the same antimicrobial therapy is effective regardless of pneumococcal type. Typing was important before the advent of antibiotics because treatment then was the administration of type-specific antiserum. Colonies are often large and mu-

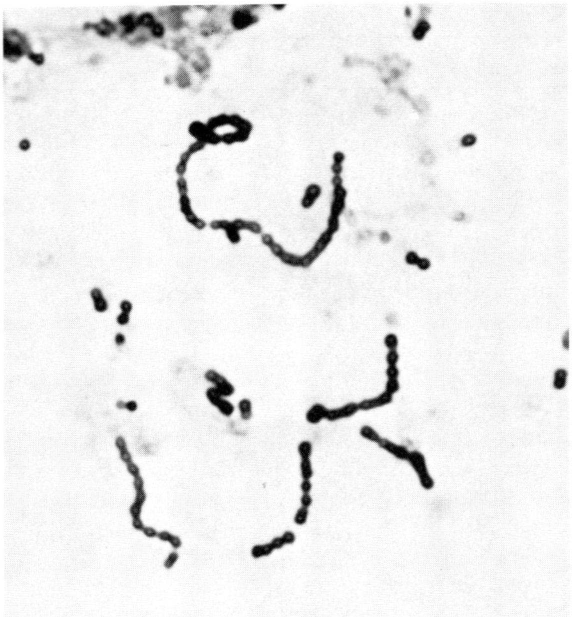

FIGURE 7-8. *Streptococcus pneumoniae* (pneumococcus) in blood. Notice the long chains of cocci. (Courtesy of Dr. James Plorde.)

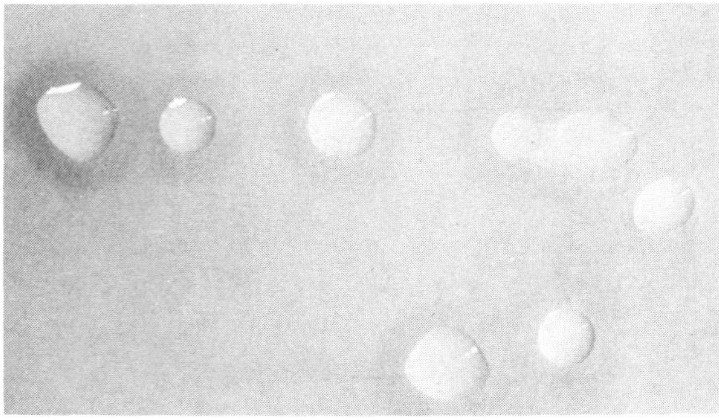

FIGURE 7-9. *Streptococcus pneumoniae* colonies growing on blood agar. The colonies are large and mucoid because of large amounts of capsular material surrounding the pneumococci. (Courtesy of Dr. James Plorde.)

coid due to the formation of large amounts of capsular material (Fig. 7-9).

Pneumococci produce autolytic enzymes that are activated by agents such as bile salts. The bile solubility test (in which pneumococci are lysed) differentiates *S. pneumoniae* from other species of streptococci. The autolytic enzymes eventually result in loss of Gram stain retention (organisms become gram negative) and the death of the pneumococci in culture.

Antigenic components of *S. pneumoniae* are listed on Table 7-10. Some K antigens cross-react with those of other streptococci, *Klebsiella,* and *Salmonella* strains. The somatic R antigen (named because it was originally obtained from rough strains) is a surface protein. The type-specific M antigen is a protein, but antibodies against it are not protective. The species-specific C substance is a carbohydrate precipitated by C-reactive protein (CRP), which can be detected in the blood during some acute illnesses including pneumococcal and other streptococcal diseases such as rheumatic fever. Transformation (the transfer of naked DNA to a competent host; see Chapter 2) was first discovered in *S. pneumoniae* and it occurs in nature at a rate that is low, but is not precisely known.

Epidemiology. *Streptococcus pneumoniae* is a normal upper respiratory inhabitant in 30 to 70 percent of people. In a recent study, 79 of 82 infants carried at least one strain of pneumococci during the first two years of life. There were 31 clinical infections in 24 infants, mostly otitis media, but also bacteremia and meningitis. Fifteen percent of acquisitions of pneumococci resulted in disease. The duration of carriage decreased with successive types carried (Gray, Converse, and Dillon 1980).

Before antimicrobial agents were available, pneumococcal pneumonia was a major

TABLE 7-10. Antigenic Components of *Streptococcus pneumoniae*

Antigen	*Description*
Capsular (K)	About 84 polysaccharide types
Somatic	
R antigen	A protein near the cell's surface, not well defined, not present in all species
C substance	Group-specific cell wall antigen precipitated by C-reactive protein (CRP)
M antigen	Type-specific somatic proteins

cause of death. Pneumococci still cause 90 percent of bacterial pneumonias and are the most frequent cause of purulent meningitis in adults (about 1.7 per 100,000). The mode of transmission is by droplet. Pneumococcal disease probably results from changes in host defenses among those already colonized as much as from new acquisition of the organism. Though any serotype can cause disease, a few strains are responsible for most disease. Infection is most common in winter and disease occurrence is sporadic. The incidence of pneumonia is highest, as is the case with many infectious diseases, at the extremes of age.

The epidemiology of *S. pneumoniae* is similar to that of the meningococcus; that is, it has a high carrier rate with occasional occurrence of disease in closed populations (institutions, military camps), with particularly virulent strains or in high-risk groups. Epidemics of pneumococcal pneumonia, however, are unusual, and the disease tends to be more sporadic or endemic than meningococcal disease.

Pathogenesis. Only encapsulated strains are pathogenic. Virulence is dependent on invasiveness and capsular protection against phagocytosis. Virulence varies by strain (type 3 being the most virulent) and by size of capsule (within the same strain, the larger the capsule, the greater the virulence). Strains that cause disease vary in children and adults. Pneumolysin, a hemolytic toxin produced, and neuraminidase may have some toxic effects, but are not of major importance in pathogenesis. Viral upper respiratory infections predispose to pneumococcal pneumonia. After organisms reach the lungs, there is a large accumulation of edema fluid in which the pneumococci multiply and spread locally. Generally disease only develops in the presence of decreased host defenses. Pneumococcal pneumonia is very common among alcoholics who have increased chance of aspiration of contaminated material into the lungs and depressed phagocytic function.

Host responses. In the majority of cases pneumococcal colonization does not result in disease. Therefore, the presence of the organism in sputum is not diagnostic of disease. When disease occurs it is classically a lobar pneumonia characterized by cough, fever, high WBC count, severe pain on inspiration, and lobar consolidation. Regardless of serotype, the clinical picture is markedly similar. The onset is extremely abrupt with sudden chill, high fever, tachycardia, and the production of rusty colored sputum. Signs and symptoms of pneumonia caused by various bacteria are compared in Table

7-11. Before specific therapy was available, an affected individual recovered if there was production of adequate antibody to opsonize the phagocytosis of the pneumococci. Untreated mortality is about 30 percent. The mortality rate among those treated with antimicrobial agents is about 15 percent, higher at the extremes of age or in debilitated individuals. Complications of pneumococcal pneumonia include lung abscesses and atelectasis, pleural effusions, empyema, and pericarditis.

Clinical syndromes other than pneumonia also occur. Pneumococcal bacteremia

TABLE 7-11. Signs and Symptoms of Various Bacterial Pneumonias

Causative organism	*Signs and symptoms*	*Occurrence*
Beta-hemolytic *streptococcus,* group A *(S. pyogenes)*	Abrupt onset, high fever, empyema, chest pain, vomiting	Rare except after "strep throat," influenza, or measles. Not nosocomial
Enterobacteriaceae (*E. coli, Serratia, Pseudomonas,* etc.)	Not distinguishing, usually accompanied by other infections. High mortality	Usually nosocomial in those receiving antimicrobials; also in aged, debilitated
Francisella tularemia	Mucoid sputum, hemoptysis, dyspnea, cough, cyanosis, pleurisy. Prolonged recovery	Due to contact with infected animals, especially rabbits, squirrels, muskrats
Klebsiella pneumoniae	Very similar to pneumococcal; tenacious "red currant jelly" sputum, diffuse lung involvement, often with cavitation; may require surgical excision; lung abscesses and empyema frequent. Mortality high, even with treatment	About 50 percent in alcoholics, mostly in adults, often nosocomial
Mycobacterium tuberculosis	Symptoms may be mild or lacking, may have nonproductive cough and no organisms present in sputum	Usually debilitated adults
Mycoplasma pneumoniae	Slow onset, cough or respiratory symptoms not necessarily predominant; may cause mild illness for many weeks. Rarely serious or fatal	Children and young adults, usually community acquired
Staphylococcus aureus	Multiple chills; lung abscesses and pneumatoceles formed early in disease, empyema common. Mortality high	Usually in infants or accompanying influenza; often nosocomial
Streptococcus pneumoniae (pneumococcus)	Extremely abrupt onset, shaking chill, high fever, tachycardia, severe pleuritic pain, cough with rusty sputum, dyspnea. Mortality about 15 percent	Usually community acquired, but may be nosocomial, especially in extremes of age, debilitated, alcoholics

occurs occasionally in children less than 3 years old without any other signs or symptoms. Meningitis caused by pneumococci is the second most common type (next to meningococcal) of meningitis in adults, and is also seen with some frequency (though not as often as that caused by *H. influenzae*) in children. Symptoms are those classically seen in meningitis and diagnosis is confirmed by culture of the spinal fluid, which is cloudy with increased protein and decreased glucose. With therapy, mortality is about 25 percent. Endocarditis, arthritis, paralytic ileus, and impaired liver function occasionally occur. In pneumococcal peritonitis, seen in young girls on rare occasion, the presumed place of entry for the organism is the vagina.

Control and treatment. Other pharyngeal normal flora interferes with growth of the pneumococcus so that ecologic balance of the oropharanyx is protected. Type-specific immunity results from the formation of anticapsular antibody, so recurrences are usually due to different strains. The *Francis skin test* can be used to detect circulating antibody and immunity to pneumococci; after intradermal injection of capsular antigen, the site becomes immediately red and swollen. Though type-specific antisera are no longer used for treatment, a multivalent vaccine first licensed in 1977 and active against 14 serotypes of pneumococci is available. It is protective, but the duration of effectiveness is not known. In one year, 573 clinical isolates of *S. pneumoniae* received by CDC from 44 hospitals were typed. About 70 percent of those were among the 14 K strains included in the new vaccine. Use of the vaccine, however, is neither cost effective nor indicated for the general population. Its use should be reserved for those persons at high risk of disease.

The treatment of choice for pneumococcal disease is penicillin. It works especially well on organisms at the outer edges of the edematous zone in the lungs, but less well in the central zone where the organisms are multiplying more slowly. Hence, early initiation of therapy, before organisms spread extensively, is important for maximal effectiveness. Of great concern is the discovery in 1977 in South Africa, Canada, and the United States of pneumococcal strains highly resistant to penicillin. Resistance to tetracycline, chloramphenicol, rifampin, erythromycin, and clindamycin have also been noted recently, both among diseased individuals and healthy carriers. Though person-to-person spread in hospitals is rare, patients with pneumococcal pneumonia should be placed in respiratory isolation, and should be segregated from individuals with high susceptibility to infection (the aged, debilitated, and immunosuppressed).

Streptococcus pyogenes

Biologic characteristics. *S. pyogenes* is synonymous with group A beta-hemolytic streptococcus, although a very rare strain may be nonhemolytic. There are more than 60 types identified by the M protein that is on the cell surface like a covering of tiny hairlike projections (miniature pili). M protein is poorly antigenic, but is related to virulence. Subtyping can also be done on the basis of a protein T antigen (resistant to trypsin) and, in a few strains, an R (named because it was originally discovered in rough strains) antigen. *S. pyogenes* produces a bacteriocin which lyses organisms from groups B, C, E, G, L, and N streptococci. Like all streptococci, these organisms tend to form chains rather than clusters like the staphylococci. *S. pyogenes* requires complex media such as blood agar for growth. About 95 percent are sensitive to bacitracin whereas other beta-hemolytic streptococci are not. This difference in sensitivity is frequently used to differentiate types of streptococci.

Epidemiology. Despite the fact that about 5 percent of the general population are colonized for varying lengths of time with group A streptococci, the organism is not considered to be part of the normal flora since it is such a frequent cause of symptomatic disease. The major danger of streptococcal disease is long-term sequelae: rheumatic fever and glomerulonephritis. Its most common habitat is the oropharynx, and it is especially common in children in the tonsillar tissues. Probably every individual becomes infected at some time during life and, because there are many strains, reinfection is common. It is also occasionally carried in the nose, vagina, and rectum, and such carriers can be important disseminators of the organism.

S. pyogenes is a major cause of clinical pharyngitis. It is responsible for about 25 percent of upper respiratory tract infections, though asymptomatic or mild infections are probably common, constituting as many as 20 to 40 percent of cases. There are an estimated 7 million cases of *S. pyogenes* disease annually in the United States. The organism is normally carried for 2 to 3 weeks following pharyngitis. In general, carriers are less infectious than diseased persons as they less frequently carry strains with large amounts of M protein, though epidemics of surgical wound infections have been traced to asymptomatic anal and vaginal carriers among operating room staff.

Streptococcal skin infections (impetigo) are endemic in the southern United States and in tropical areas. Transmission can be by droplet, direct hand-to-hand or hand-to-lesion contact, airborne, or by contaminated fomites or dust. Intimate direct contact is by far the most important route. Occasional outbreaks due to contaminated food have been reported. *S. pyogenes* is the organism responsible for puerperal sepsis or "childbed fever" classically described by Semmelweiss and Holmes in the mid-nineteenth century. Women hospitalized during childbirth were at high risk of dying from streptococcal infection that was being transmitted from unwashed hands of medical personnel who acquired the organism during autopsy examinations of individuals who had died of infection.

The epidemiology of the two long-term, nonsuppurative sequelae of streptococcal infection—rheumatic fever and glomerulonephritis—are quite different and suggest that different mechanisms are involved in their pathogeneis (Table 7-12). Rheumatic fever can be caused by any strain of *S. pyogenes* and occurs at a low but constant rate after less than 3 percent of untreated cases of streptococcal pharyngitis. The sex distribution is even and recurrent attacks are frequent. Once an individual has had rheumatic fever there is a 25 to 50 percent chance of recurrence following subsequent episodes of streptococcal throat infection.

Glomerulonephritis, on the other hand, is caused by a few nephritogenic strains and may follow either throat or skin infection. The attack rate is quite variable from 0 to 25 percent and there is a 2 to 1 male preponderance of cases. Because type-specific immunity develops and there are comparatively few causative strains, recurrences are rare. Due to the use of penicillin, the incidence of rheumatic fever has decreased by four-fold since World War II in the United States, although it is still endemic in other parts of the world. Scarlet fever, caused by streptococcal strains that produce erythro-

TABLE 7-12. Comparison of Poststreptococcal Sequelae: Rheumatic Fever and Glomerulonephritis

	Rheumatic fever	*Glomerulonephritis*
Affected body parts	Joints, heart valves	Kidneys
Probable mechanism	Some streptococcal antigens cross-react with heart tissue; heart-reactive antibodies produced by host (autoimmunity)	Some strains of streptococci have antigens that resemble (cross react with) human glomerular antigens; or, may be caused by deposition of antigen/antibody complexes
Onset	2-4 weeks after *pharyngeal* infection (not skin)	1-2 weeks after pharyngeal *or* skin infection
Residual effects	Rheumatic heart disease in about 50 percent of cases	Usually none, rarely chronic renal disease
Recurrences	Common	Rare
Attack rate after streptococcal infection	Less than 3 percent, stable	0-25 percent, variable
Streptococcal strains involved	Any	Only a few

genic toxin is almost nonexistent in this country now.

Pathogenesis. There are many factors associated with the virulence of *S. pyogenes* (Table 7-13). Streptolysins O and S are both hemolysins produced not only by *S. pyogenes* but also by group C and G strains. Streptolysin S is nonimmunogenic whereas streptolysin O is both antigenic and is most important in virulence. Antistreptolysin O (ASO) antibody is produced after streptococcal infection. ASO titer is a well standardized test for streptococcal infection. Although a single high titer (such as 1:1200; sometimes expressed as 1200 Todd units) is very suggestive of recent streptococcal infection, obtaining paired sera—one acute and one convalescent—is the most accurate way to use the test. Intermediate titers (125 to 500 Todd units) cannot be interpreted with single tests. This test is useful retrospectively, but treatment must be initiated before final results are available in order to prevent long-term complications.

The amount of M protein in a streptococcal strain is also of prime importance in determining virulence, and seems to function primarily by enhancing adherence and protecting the organism from phagocytosis. Group A streptococci adhere better to pharyngeal cells from persons who have had acute rheumatic fever than from others. The markedly decreased incidence of rheumatic fever in recent years corresponds with the fact that isolates of streptococci now contain much less M protein.

There is evidence that pharyngitis may be partially a result of a delayed hypersensitivity reaction. First, cutaneous reactivity to purified M protein increases with age. Second, children under 2 years of age rarely develop clinical symptoms of pharyngitis even when a throat culture yields large amounts of *S. pyogenes*.

Some types such as type 12, which may

TABLE 7-13. Factors Associated with the Virulence of *Streptococcus pyogenes*

Factor	*Mechanism*
Capsule	Antiphagocytic
Diphosphopyridine nucleotidase (DPNase)	May be leukotoxic
Hyaluronidase	"Spreading factor," may aid in the spread of streptococci in tissues
Leukocidin	Kills leukocytes
M protein	Antiphagocytic, increases adherence
Streptodornase	Depolymerizes DNA
Streptokinase	Lyses fibrin
Streptolysin O	Hemolyses RBCs; causes necrosis of parenchymatous organs; induces arthritis, renal tubular necrosis
Streptolysin S	Hemolyses RBCs, causes release of acetylcholine, cardiotoxic, provokes Shwartzman reaction

have antigens similar to those of human glomeruli, are nephritogenic and only these cause glomerulonephritis. The disease may also be caused by the deposition of antigen/antibody complexes in the renal glomeruli; the exact pathogenic mechanism is not known. Streptococcal glomerulonephritis can occur after either pharyngitis or impetigo. Rheumatic fever occurs only after there has been a pharyngeal infection untreated for at least a week. The exact mechanism by which rheumatic fever develops in a small proportion of individuals with untreated streptococcal pharyngitis is unknown, though autoimmunity is certainly a component. Rashes such as scarlet fever sometimes accompanying streptococcal infection are due to the production of erythrogenic toxin. The toxin can be produced by any strain of group A streptococci. The ability to produce toxin is transmitted by a lysogenic phage.

Host responses. The most common clinical syndrome caused by *S. pyogenes* is acute pharyngitis or tonsillitis, usually occurring in children of school age (Fig. 7-10). After an incubation period of 3 to 5 days, the affected individual experiences headache, sore throat, fever, and tender, swollen cervical lymph nodes. Occasionally otitis media, mastoiditis, and sinusitis also occur. Persons with "strep throat" are usually moderately ill for 1 or 2 days and the pharynx is red, swollen, and there is a grayish-white exudate. However, it is not possible to make a certain diagnosis or to exclude the diagnosis unless a culture is taken (i.e., there are no pathognomonic symptoms). If there is an accompanying skin erythema, the syndrome is called scarlet fever, but this, as mentioned previously, is currently rare.

Impetigo is a skin disorder characterized by reddened vesicles that rupture and become crusted (Fig. 7-11). Streptococci can secondarily infect chickenpox lesions or streptococcal lesions can occur in the absence of other disease. Regional lymph node enlargement usually accompanies the disease and diagnosis is confirmed by isolating

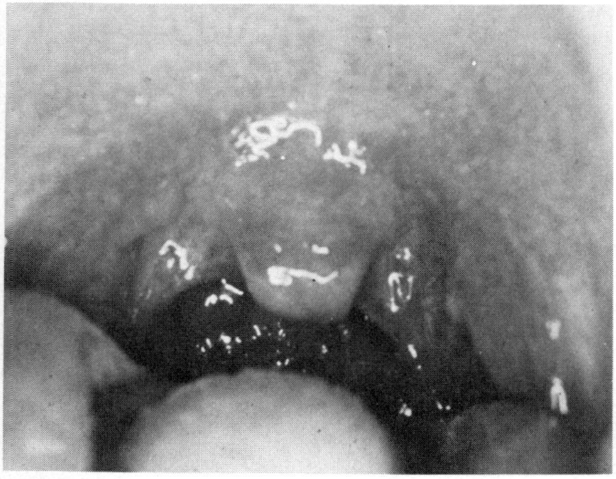

FIGURE 7-10. Streptococcal throat infection with moderate redness and early exudate on one tonsil. (Courtesy of Centers for Disease Control, Atlanta, GA.)

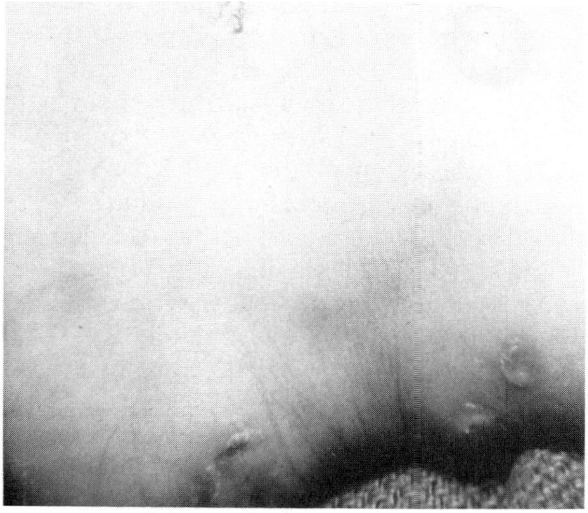

FIGURE 7-11. Impetigo on ankle. This case was first thought to be syphilis. (Courtesy of Centers for Disease Control, Atlanta, GA.)

streptococci group A or C (usually A) from the lesions. The group A streptococcus is the most common cause of cellulitis and lymphangitis in which fine red streaks extend upward from an extremity. *S. pyogenes* is also the usual cause of erysipelas, an acute skin disease most commonly affecting the face. The skin is usually sore and itchy and after 1 to 4 days purplish raised lesions accompanied by edema occur. There is usually fever and may be bacteremia. Recovery is usually within a week; however, mortality among untreated can reach about 15 percent. *S. pyogenes* pneumonia can occur as a secondary event following influenza, but is otherwise rare. Other common clinical manifestations of streptococcal disease are wound infections, bacteremia, meningitis, septic arthritis, pericarditis, and orbital cellulitis (Fig. 7-12).

Rheumatic fever is a poststreptococcal systemic inflammatory disease causing polyarthritis, carditis, and/or neurologic dysfunction (chorea). In most, but not all, cases there is a history of streptococcal infection within the previous 2 to 4 weeks, and the ASO titer is elevated. The most common presentation is low-grade fever and pain and swelling in various joints, which subside and then involve other joints ("migratory ar-

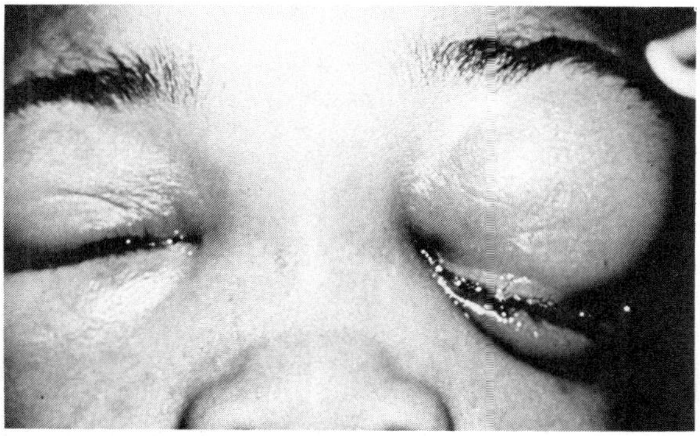

FIGURE 7-12. Orbital cellulitis. (Courtesy of Dr. John Chandler.)

thritis"). Heart murmurs, evidence of carditis, are often but not always present. The major long-term problem is a high incidence of resultant mitral stenosis. Such rheumatic heart disease has been a major cause for heart surgery in young adults, but is declining as the incidence of rheumatic fever decreases. Chorea (irregular and uncontrolled muscle movements) may occur later, after a latent period of several months and is accompanied by emotional lability and speech disorders. Prognosis depends on the major manifestations and the organs involved. Most patients recover, but cardiac sequelae, even congestive heart failure, and death can occur. Once an individual has had rheumatic fever, there is a much greater chance of developing the disease again, and the same clinical manifestations usually recur.

Glomerulonephritis can have a spectrum of manifestations from no symptoms at all to frank renal failure. Symptoms are those of renal dysfunction: hypertension, oliguria, hematuria and proteinuria, fluid retention with edema. Complement levels are decreased for 2 to 6 weeks, indicating a possible immunologic defect. Complete recovery is usual in 90 percent of cases and recurrences are rare. Some individuals may develop chronic renal disease.

Control and treatment. Type-specific immunity to streptococci exists and streptococcal vaccine against M protein is feasible, though not yet developed. Immunity to erythrogenic toxin is demonstrated by the *Dick test*. If circulating antitoxin (antibody) is present and the individual is immune to scarlet fever, intradermally injected toxin will be neutralized and *no* erythema will ensue. Redness at the site indicates lack of immunity.

Since it is not possible to accurately diagnose streptococcal pharyngitis without a culture, it is of great importance that individuals, especially school aged children with sore throats, be microbiologically screened for group A streptococci whenever a sore throat occurs. It is quite safe to delay treatment for 24 to 48 hours until results of the test are available, since there is no danger of rheumatic fever for at least a week.

The serious sequelae of streptococcal infection can be prevented with rapid treatment. Unfortunately, however, some cases of streptococcal infection are asymptomatic or mild and may not be brought to medical attention. In addition, patients and their families may not realize that antimicrobial therapy for "strep throat" is not for the purpose of symptomatic relief (symptoms subside with or without treatment), but to prevent serious complications. Thus, the 10-day course of oral penicillin that is necessary to eradicate the streptococci is often not completed by patients and treatment is inadequate. For this reason, the best treatment is often a single intramuscular dose of long-acting penicillin. All *S. pyogenes*, fortunately, are susceptible to penicillin and no antimicrobial susceptibility testing is necessary. A satisfactory alternative for individuals with penicillin sensitivity is erythromycin, but not tetracycline, since about half of strains are resistant to it.

Treatment for rheumatic fever is symptomatic. Aspirin is used to decrease pain and inflammation; if the patient has cardiac involvement, corticosteroids may be used. Since individuals who have had rheumatic fever are so prone to disease recurrence, they must continue on prophylactic penicillin therapy for extended periods of time, sometimes for life, to prevent streptococcal reinfection. Sulfonamides can be used as a substitute for penicillin prophylaxis but are useless in eradicating streptococci once they are established in the tissues. Penicillin is also given to individuals with glomerulonephritis to eradicate any streptococci remaining, but again, treatment is symptomatic and antimicrobial agents will not change the course of the disease. No prophylaxis is necessary after glomerulonephritis. Rapid passaging of streptococci from person to person, as in epidemics of scarlet fever or "strep throat" seems to increase virulence of strains. Barrier and isolation methods that interrupt the spread of streptococci probably help to minimize virulence.

Group B Streptococci

Biologic characteristics. Group B streptococci *(Streptococcus agalactiae)* are usually beta hemolytic and react with antisera against the group B specific cell wall antigen. There are four types based on various polysaccharide capsular antigens (Ia, Ib, II, III). Type III causes the most severe clinical disease. Type Ic is based on a protein antigen and other strains are nontypable.

Epidemiology. *Streptococcus agalactiae* is a normal inhabitant of the human pharynx, bowel, and vagina. Of healthy women, 5 to 20 percent are cervical carriers; about one-third of women during the third trimester of pregnancy are genitally colonized with group B streptococci. Newborns acquire the organism at birth from their mothers (about 75 percent of newborns whose mothers are carriers will in turn become colonized). Babies also acquire the organism in the newborn nursery by nosocomial transmission. Breast milk and hands of personnel have been shown to be sources. Though less than 1 percent of colonized infants become clinically ill, the group B streptococcus is probably the major cause of neonatal sepsis and meningitis. In a 4-year survey at one institution, group B streptococci caused the majority of streptococcal bacteremias among obstetric patients (a few cases were the result of groups D and A, viridans streptococci, and *S. pneumoniae*). Complications developed in babies of about 20 percent of these bacteremic mothers (Gibbs and Blanco 1981).

Pathogenesis. Group B streptococci are invasive and possess many of the same toxic and virulence mechanisms (except for the M protein) as do group A streptococci.

Host responses. Most colonizations with group B streptococci are asymptomatic. When clinical disease in newborns occurs, the onset is usually about 5 days after birth with fever and few localizing signs. The septic syndrome often involves pneumonia resembling respiratory distress syndrome (RDS) with tachypnea and a hazy chest film. Meningitis often occurs and mortality is high, about 50 percent. Disease of later onset is usually due to streptococci acquired postnatally and has a lower mortality rate. Group B streptococci can also cause osteomyelitis, septic arthritis, otitis media, and orbital cellulitis (Fig. 7-12) in infants and children. It is a common cause of postpartal endometritis and, rarely in adults urinary tract, respiratory tract, or wound infection, meningitis, and peritonitis. Group B streptococci are a common cause of mastitis in cows.

Control and treatment. As with other streptococci, penicillin and ampicillin are used to treat disease, though they are not useful in eradicating the carrier state. All group B streptococci, however, are not necessarily susceptible to penicillin as are group A strains. The organism is generally of no concern except in pregnant women. A capsular antigen vaccine is theoretically possible for pregnant women, but none has been developed. Use of triple dye or silver sulfadiazine on the umbilical stump decreases colonization of babies with both beta-hemolytic streptococci and *S. aureus* and may also be useful in decreasing clinical disease.

Other Aerobic Streptococci

There are three species of group C streptococci. All cause human infections, though they are more commonly found in animals. Most human infections are due to the eight serotypes of *S. equisimilis*. There are also eight serotypes of *S. zooepidemicus*, and one of *S. equi*. They inhabit the oropharynx and vagina, and are beta hemolytic. Pathogenic species produce streptolysin O. Group C streptococci are occasionally isolated from human sputum, pus, or blood.

Group D streptococci may be nonhemolytic, alpha, or rarely, beta hemolytic. The group is sometimes called the enterococci, but in fact, of the five species of group D streptococci that have been isolated from the human gut, four are enterococci, and one species, *S. bovis*, is not an enterococcus. The enterococci can be differentiated by their ability to grow in 6 percent sodium chloride and by their resistance to penicillin. *S. bovis*, the nonenterococcus, is usually sensitive to penicillin. Enterococci comprise the second most common type (after *Escherichia coli*) of aerobic bowel flora. They can grow at a wide temperature range, 10 to 45°C, and at a pH of 9.6. Nonenterococcal group D streptococci are also normal bowel flora, but occur in fewer numbers. Group D streptococci are a frequent cause of nosocomial infections, causing about 20 percent of subacute bacterial endocarditis and 10 percent of urinary tract infections (Fig. 7-13).

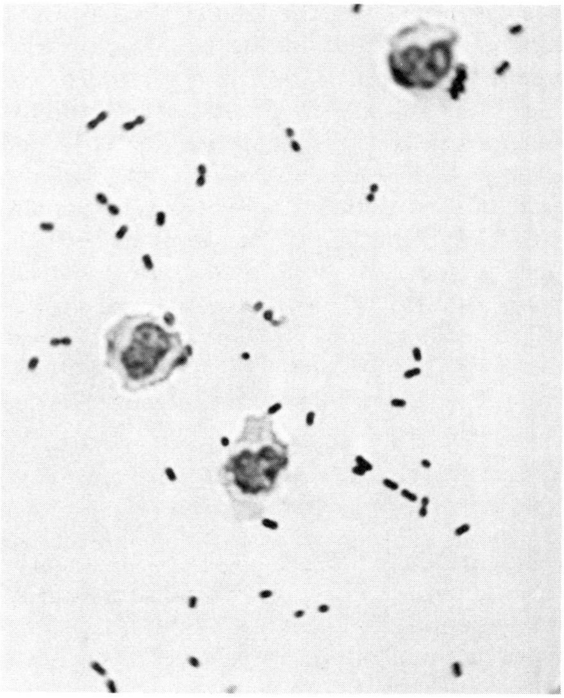

FIGURE 7-13. Urine containing enterococci. (Courtesy of Dr. James Plorde.)

Streptococci have been isolated from the human oral cavity, where they may be found in dental infections. They are susceptible to penicillin, chloramphenicol, and tetracycline. Only rare infections are caused by members of other groups. Aerococci have not been isolated from healthy humans. They are found in the air and occasionally in blood or urine during disease. Their taxonomy is poorly developed.

The viridans group of streptococci are all alpha hemolytic and include about 10 species. They do not share the cell wall group antigens used to type other streptococci. They are predominantly normal flora of the oropharynx, and are a probable cause of dental caries and periodontal disease. Species such as *S. mutans*, cause 50 to 70 percent of cases of bacterial endocarditis, especially following extensive dental manipulation or surgery. *S. mutans* has seven serotypes based on cell wall antigens and produces bacteriocins that lyse members of streptococcal groups A, C, D, G, H, L, and O. The organism can be a rare cause of pneumonia. Viridans streptococci are susceptible to penicillin.

Anaerobic Gram-Positive Cocci

Anaerobic gram-positive cocci are listed in Table 7-1. Only peptococci and peptostreptococci are significant opportunistic pathogens of humans, and to a lesser extent a few strains of viridans streptococci *(S. anginosus, sanguis, morbillorum)*. They are a prominent part of normal mouth, upper respiratory tract, bowel, skin, and female genital flora. Anaerobic streptococci are either alpha or nonhemolytic. About one-third of anaerobic organisms isolated from clinical specimens are streptococci, with *Peptococcus magnus, P. asaccharolyticus, P. prevotii,* and *Peptostreptococcus micros* most common.

Anaerobic gram-positive cocci have caused septic arthritis, endocarditis, empyema, soft-tissue abscesses, respiratory tract and genitourinary tract infections in females. Most of these infections are mixed with other anaerobes, especially *Bacteroides*. There is also synergism with *S. aureus* in some infections, especially in abdominal or chest incisional abscesses. Most of these organisms are sensitive to penicillin and resistant to tetracycline.

CLINICAL IMPLICATIONS

Should infection develop in a patient, physical isolation and environmental control to prevent cross infections are required. Great care should be exercised to prevent contamination of the environment by lesion drainage or pus. Rigid handwashing protocol, careful disposal of dressings, disinfection of bedding, dishes, thermometers, and other fomites, and barrier techniques such as wearing gowns and masks are necessary for preventing dissemination of gram-positive cocci. Staphylococcal pneumonia requires isolation since the organisms are present in sputum and can be disseminated through the air as well as on skin and in wounds. This is of particular concern because staphylococci are so resistant to drying and can survive in the nonliving environment. Localized infections such as an abscess or wound infection can be contained with wound and skin precautions. Methicillin and penicillin resistance of *S. aureus* can be minimized by such isolation as well as thoughtful and careful use of antimicrobial agents.

Attempts to treat nasal carriers of *S. aureus,* especially individuals such as health care personnel working with high-risk patients, have been made over the past few decades with only minimal success. Nasal sprays and antimicrobial ointments will only temporarily reduce the carrier state. Emphasis is better placed on helping the carrier to prevent spread of their colonizing *S. aureus* by careful handwashing techniques (evidence suggests that direct contact rather than droplet or aerosol spread is the major means of transmission). Only rarely has it been necessary to remove an individual who is carrying a highly resistant strain of *S. aureus* from direct patient care.

The emergence of a strain of *S. aureus* resistant to any agent besides penicillin is indication for an epidemiologic investigation to discover its source. Fortunately there seems to be no apparent difference in virulence among strains resistant or susceptible to therapeutic agents, but the treatment problem for a resistant strain is one of concern. Treatment of infections caused by penicillinase-producing strains is with penicillinase-resistant penicillins such as methicillin, nafcillin, or oxacillin. Cephalosporins or aminoglycosides are also used. MRSA can be treated with vancomycin. Antimicrobial resistance of *S. aureus* to any agent can occur. There was a recent report of one extensive outbreak over a period of 16 months in which 120 infections occurred in 82 patients. Infections were due to oxacillin- and aminoglycoside-resistant *S. aureus* and included wound infections, pneumonias, urinary tract infections, I.V. site infections, and bacteremias. The outbreak was associated with one surgical ICU, and 90 percent of *S. aureus* isolates were the same phage type. Three nurses working in ICU were nasal carriers, another, with dermatitis, was colonized on the hands. Removal of these carriers from the unit and strict infection control measures were necessary to stop the epidemic (Craven et al. 1981).

The remarkable reduction in serious sequelae from streptococcal infections has been due certainly in part to changes in virulence of the organism, but also to our understanding of its epidemiology so that preventive measures can be taken. Parents of school aged children must be cognizant of the potential risks of untreated "strep throat" and be aware of the fact that cases may be very mild and cannot be diagnosed by specific symptoms. Many viral causes of sore throat can mimic streptococcal tonsillopharnygitis. Public education is the responsibility of nurses and other health care professionals.

It is also the responsibility of the health care practitioner to obtain an adequate throat specimen for culture. False-negative results can be frequent if the culture is taken improperly. It is essential that the posterior pharynx or tonsils be visualized well and swabbed, even though the gag reflex is briefly stimulated. Swabbing the tongue or uvula will yield primarily normal mouth flora.

SUGGESTED READINGS

Broome CV, et al. Epidemiology of pneumococcal serotypes in the U.S., 1978-1979. J Infec Dis 1980;141:119-123.

Centers for Disease Control. Followup on toxic shock syndrome—U.S. Morb Mort Week Report 1980;29:297-299.

Centers for Disease Control. Methicillin-resistant *Staphylococcus aureus*—U.S. Morb Mort Week Report 1981;30:140.

Centers for Disease Control. Toxic shock syndrome—U.S., 1970-1980. Morb Mort Week Report 1981;30:25-28.

Centers for Disease Control. Toxic shock syndrome—Utah. Morb Mort Week Report 1980;29:495-496.

Craven DE, et al. A large outbreak of infections caused by a strain of *Staphylococcus aureus* resistant to oxacillin and aminoglycosides. Am J Med 1981;71:53-58.

DeLaat ANC. Microbiology for the Allied Health Professionals, 2nd edition. Philadelphia: Lea and Febiger, 1979.

Fine JD, et al. Adult scalded skin syndrome fatally complicated by mixed gram-negative sepsis and cellulitis. Cutis 1981;27:162-167.

Gibbs RS, Blanco JD. Streptococcal infections in pregnancy. Am J Obstetr Gynecol 1981;140: 405-411.

Gray BM, Converse GM, Dillon HC. Epidemiologic studies of *Streptococcus pneumoniae* in infants: acquisition, carriage, and infection during the first 24 months of life. J Infec Dis 1980;142:923-933.

Hargiss C, Larson E. The epidemiology of *Staphylococcus aureus* in a newborn nursery from 1970 through 1976. Pediatr 1978;61:348-353.

Holzman RS, Florman AL, Lyman M. Gentamycin resistant and sensitive strains of

Staphylococcus aureus. Am J Epidemiol 1980;112:352-361.

Kloos WE, Orban BS, Walker DD. Plasmid composition of staphylococcal species. Can J Microbiol 1981;27:271-278.

Mader JT, et al. A mechanism for the amelioration by hyperbaric oxygen of experimental staphylococcal osteomyelitis in rabbits. J Infec Dis 1980;142:915-922.

Moellering RC, et al. A novel mechanism of resistance to penicillin-gentamycin synergism in *Streptococcus faecalis*. J Infec Dis 1980; 141:81-86.

Nurses Action Group. Fighting the hidden hazards. Nurs Mirror 1981;154:22-26.

Parisi JT, Hecht DW. Plasmid profiles in epidemiologic studies on infections by *Staphylococcus epidermidis*. J Infec Dis 1980;141:637-643.

Pennington JE. Pneumococcal vaccine and diabetes (letter). JAMA 1981;245:2026-2027.

Reed WP, et al. Streptococcal adherence to pharyngeal cells of children with acute rheumatic fever. J Infec Dis 1980;142:803-810.

Schumann D. How to help wound healing in your abdominal surgery patient. Nurs 80 1980; 10:34-40.

Staphylococcal infection, toxic shock syndrome and tampons (editorial). New Zeal Med J 1981;93:149.

Veraldo PE, et al. Clinical distribution and antibiotic sensitivities of staphylococcal strains isolated over an 8-month period. J Clin Pathol 1981;34:443-447.

Wroblewski SS. Toxic shock syndrome. Am J Nurs 1981;81:82-85.

8 GRAM-POSITIVE RODS

KEY POINTS

- The gram-positive rods are a diverse group, morphologically ranging from filamentous (actinomycetes) to clubbed and elongated shapes, and clinically ranging from normal flora (propionibacter) to strict pathogens *(Clostridium botulinum)*.
- Most spore-forming bacteria are gram-positive rods.
- The important pathogens of the gram-positive rods include *Actinomyces* species (actinomycosis), *Bacillus anthracis* (anthrax), *Clostridium botulinum* (botulism), *C. perfringens* (food poisoning), *C. tetani* (tetanus), and *Corynebacterium diphtheriae* (diphtheria).

The genera of gram-positive rods that are of interest in human disease include *Actinomyces, Arachnia,* and *Nocardia* (genera of the order Actinomycetales), *Bacillus, Clostridium, Corynebacterium, Erysipelothrix, Lactobacillus, Listeria, Eubacterium, Propionibacterium,* and *Bifidobacterium.* See Table 8-1 for a list of major species.

ACTINOMYCETES

Actinomycetes are not strictly gram-positive rods, but rather a group of branching, filamentous bacteria that are pleomorphic, rod- or club-shaped, and that somewhat resemble fungi. The taxonomy of these organisms is in a state of change. Several species of the aerobic genus *Streptomyces (erythreus, griseus,* and *lincolnensis)* are important as the source of about 90 percent of antimicrobial agents, and cause clinical disease on rare occasion (Fig. 8-1). They normally inhabit soil and only rarely cause actinomycosis. Two genera in the family Actinomycetaceae *(Actinomyces* and *Arachnia)* cause actinomycosis in animals and humans. A third genus in the same family, *Bifidobacterium,* contains species that are normal anaerobic GI flora. These organisms were formerly called *Lactobacillus bifidus. Nocardia,* from the family Nocardiaceae, is an aerobic genus of actinomycetes that causes nocardiosis.

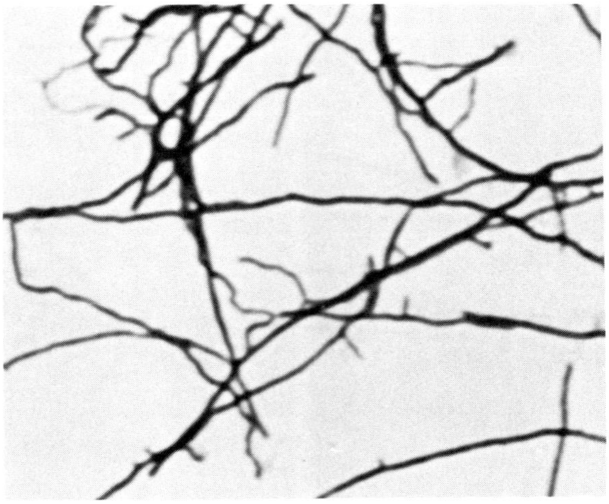

FIGURE 8-1. Disc space aspiration of *Streptomyces* species. (Courtesy of Dr. James Plorde.)

Actinomyces

Biologic characteristics. *Actinomyces,* which means "ray-fungi," are anaerobic free-living gram-positive bacteria with branching filaments that fragment into forms similar to diphtheroids (Fig. 8-2). They require rich culture media for growth. They are pleomorphic, and are usually club shaped. *Arachnia* and *Actinomyces* are the two genera in the family Actinomycetaceae and will be considered together since they cause the same clinical syndrome. Some species are normal flora of the mouth and may be involved in the formation of dental caries (*A. odontolyticus, naeslundii, viscosus, israelii*). *A. bovis* produces actinomycosis, or "lumpy

TABLE 8-1. Gram-Positive Rods Important to Humans

Genus	*Aerobic*	*Anaerobic*	*Species*	*Diseases*
Actinomyces		X	*bovis*	"Lumpy jaw" in cattle
			israelii	Actinomycosis
			viscosus, odontolyticus	Actinomycosis; may cause rare oral disease
			naeslundii	
Arachnia		X	*propionica*	Actinomycosis
Bifidobacterium		X		Normal fecal flora
Bacillus	X		*anthracis*	Anthrax
			cereus	Food poisoning
			subtilis	Rare opportunistic infections
			megaterium, polymyxa	Soil flora
Clostridium		X	*botulinum*	Botulism
			difficile, sordelli, butyricum	Enterocolitis
			perfringens	Gas gangrene, food poisoning
			septicum, novyi, histolyticum, bifermentans, sporogenes	Gas gangrene
			tetani	Tetanus
Corynebacterium	X		*diphtheriae*	Diphtheria
			(diphtheroids)	Rarely, endocarditis, septicemia

TABLE 8-1. *(continued)*

Genus	Aerobic	Anaerobic	Species	Diseases
			vaginale	Vaginitis; rarely, bacteremia
Lactobacillus		X	*acidophilus, casei, bulgaricus, fermenti, thermophilus, leichmannii*	Normal flora of GI tract, vagina, mouth; may be related to dental caries
Listeria	X		*monocytogenes*	Listeriosis; meningoencephalitis or intrauterine infection
Nocardia	X		*asteroides, brasiliensis*	Nocardiosis
Propionibacterium		X	*acnes*	Possible role in acne; rare opportunistic pathogen
			avidum, granulosum, freundenreichii, lymphophilum	
Erysipelothrix	X		*rhusiopathiae (incidiosa)*	Erysipelas
Eubacterium		X	*alactolyticum, combesii,contortum, foedans, lentum, limosum, moniliforme*	Rare opportunist

jaw," in cattle. Human disease caused by *A. bovis* has not been proven. Actinomycosis in humans is most frequently caused by *A. israelii,* but may also be caused by other species including *A. viscosus, odontolyticus, naeslundii,* and *Arachnia propionica.*

Epidemiology. Actinomycosis occurs worldwide but is rare in humans. In one study in India, there were 12 cases proven by culture from 1967 to 1977. This, of course, is greatly underestimated since culturing is often not done. It occurs primarily in adults. Infection seems to be endogenous, with poor oral hygiene being a predisposing factor (Koshi 1981).

Pathogenesis. Actinomycetes may become opportunistic pathogens after oral or other

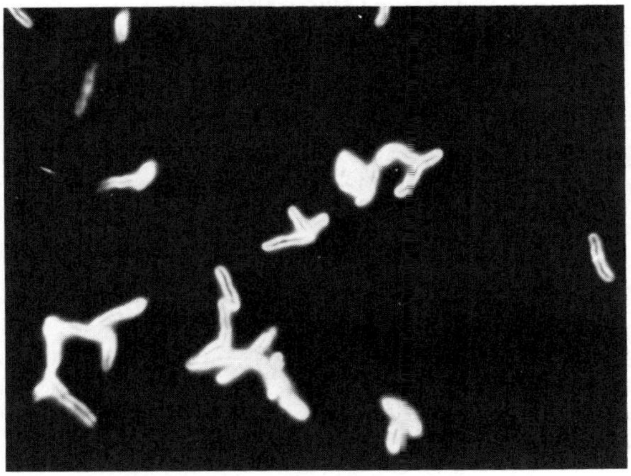

FIGURE 8-2. *Actinomyces israelii* under darkfield microscopy. Short branches similar to diphtheroids. (Courtesy of Dr. James Plorde.)

surgical trauma. Infection is also associated occasionally with intrauterine contraceptive devices and possibly procedures such as laparoscopy. Organisms that result in disease have no special characteristics such as invasiveness or virulence that distinguish them from organisms of the normal oral flora.

Host responses. Actinomycosis is most commonly manifested as a cervical adenitis with connective tissue abscess formation. However, purulent abscesses can occur throughout the body, especially in the abdomen and chest wall. As opposed to the lumpy jaw in cattle, lesions in humans rarely involve the bone. *A. viscosus* as well as *A. israelii* have been shown to cause pulmonary infection and chronic pneumonia in apparently healthy adults. *A. israelii* has also caused spinal abscesses. The presence in pus or infected sputum of clusters of yellow colonies of actinomycetes (called *sulfur granules,* Fig. 8-3, even though they contain no sulfur) is pathognomonic.

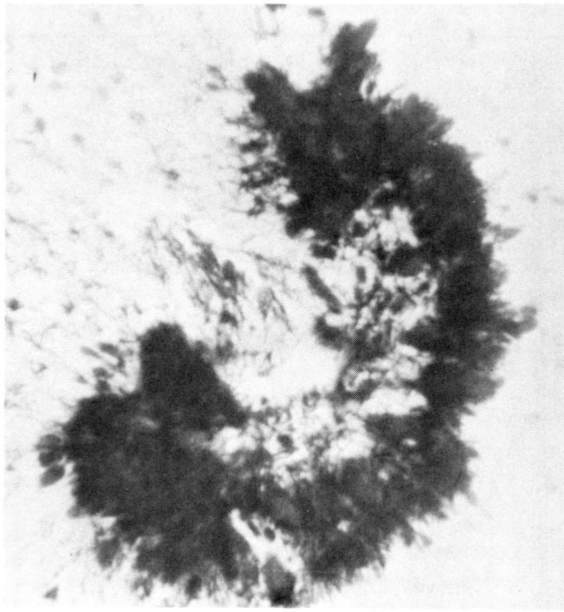

FIGURE 8-3. Typical sulfur granule of actinomycosis. (Courtesy of Dr. James Plorde.)

Control and treatment. Because these species are part of the normal flora, immunity does not develop against them. Thus, once tissue invasion occurs it usually continues until antimicrobial therapy is initiated. Treatment consists of drainage and excision of diseased tissue and administration of penicillin. Chloramphenicol, streptomycin, and tetracycline may also be used. Even with this treatment, about 10 percent of cases, especially abdominal and thoracic actinomycosis, are not cured.

Nocardia

Biologic characteristics. *Nocardia* are aerobic, thin, gram-positive, branching rods (Fig. 8-4). Many species are acid-fast and resemble *Mycobacterium.* In fact, *N. caviae* shares antigenic determinant sites with *M. leprae* though other nocardial species do not. Their growth requirements are simple (Fig. 8-5) and they are inhabitants of soil and plants rather than of animals. Several other species, *N. asteroides, N. brasiliensis,* and *N. somaliensis* have been demonstrated to cause nocardiosis in humans (Fig. 8-6).

Epidemiology. Nocardiosis occurs worldwide but is rare. There has been an increase in reported incidence of nocardiosis since 1975, probably due to increased awareness of the disease. There were 456 cases worldwide reported in the literature between 1950 and 1979, but currently there are estimated to be 500 to 1000 cases per year in the United States. Males are more often affected than females, adults of middle-age more often than children.

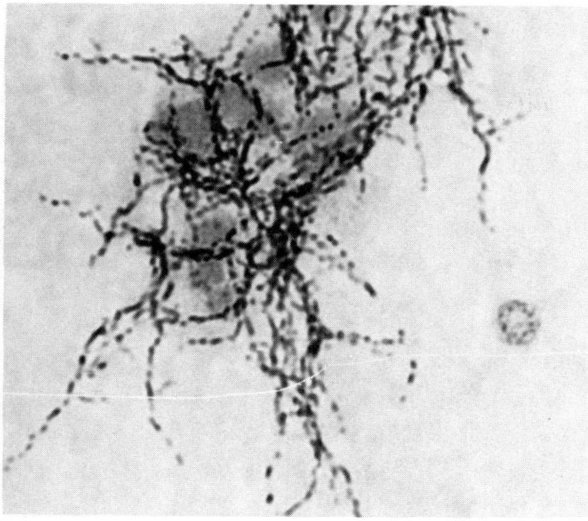

FIGURE 8-4. Sputum containing *Nocardia.* (Courtesy of Dr. James Plorde.)

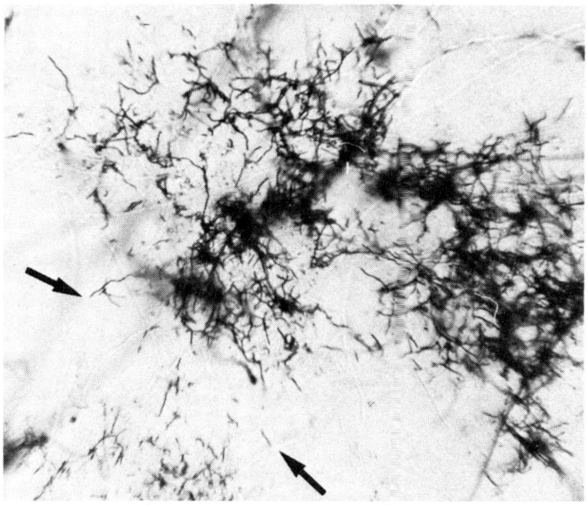

FIGURE 8-5. Colony of *Nocardia asteroides* on tap water. *Arrows* indicate development of aerial hyphae from substrate (vegetative hyphae). Magnification 700 ×. (Courtesy of Dr. Saroj K. Mishra, Robert Koch Institute, Rockville, MD)

Pathogenesis. *Nocardia* may be inhaled, resulting in lung infection, or it may enter through the skin and cause chronic subcutaneous abscesses (mycetomas). Sinus tracts develop by direct extension of lesions. With either mode of entry, other body parts may become affected by hematogenous spread. Brain and kidney abscesses can result.

Host responses. Pulmonary infection can cause fulminant pneumonia, or may be subclinical. In the early stages, pulmonary nocardiosis may be confused with tuberculosis. The disease has no characteristic symptomatology. *Nocardia* is usually an opportunist, but it can also affect healthy individuals. Once the disease begins to spread, it becomes progressive and is fatal in about 50 percent of cases, even with treatment. *N. asteroides* has been shown to cause peritonitis in patients receiving continuous ambulatory peritoneal dialysis (CAPD).

Control and treatment. It is possible but unproven that transmission between compromised hosts can occur. However, since the major reservoir is soil and plants, there are no particular preventive measures. Sulfadiazine and other sulfonamides are the most effective treatment. Penicillins are ineffective because of beta-lactamase production by nocardial species. Important adjuncts to therapy are abscess drainage and correct differential diagnosis so that appropriate chemotherapy is initiated.

PROPIONIBACTERIACEAE

Two genera from the family Propionibacteriaceae, *Propionibacterium* and *Eubacterium,* may rarely cause abscesses, empyema, or may be present in the blood or spinal fluid as opportunistic pathogens. *P. acnes* has been known to cause infection of prosthetic heart valves. They are anaerobic

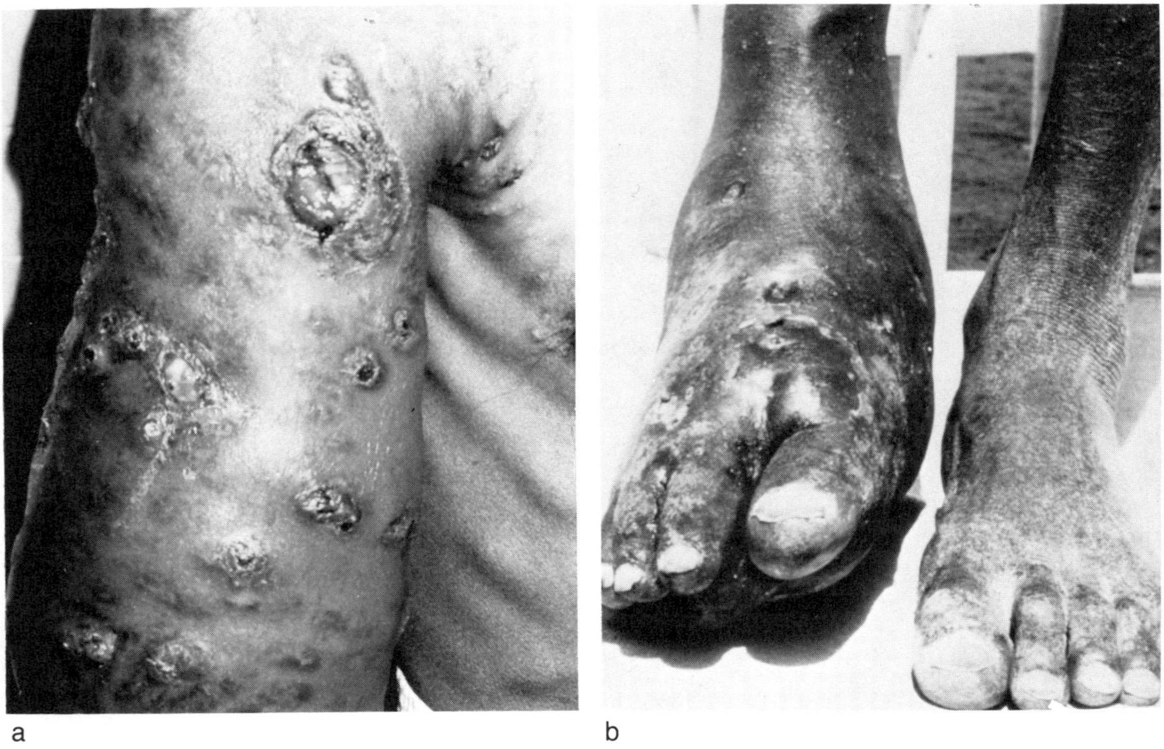

FIGURE 8-6. *(a)* Arm with lesions due to *N. brasiliensis. (b)* Foot with lesions due to *N. somaliensis.* (Courtesy of Centers for Disease Control, Atlanta, GA 30333.)

diphtheroidlike organisms. *P. acnes* is a normal inhabitant of the skin of almost everyone, and also present in conjunctiva, respiratory and GI tracts of humans, and in the soil. It produces lipases which break off free fatty acids from skin lipids, producing tissue inflammation. This activity may contribute to the development of acne. Both *P. acnes* and *P. granulosum* have been isolated from purulent face and chest lesions. Eubacterial species inhabit the GI and GU tracts of about 25 percent of humans.

BACILLUS

Members of the genus *Bacillus* are large gram-positive aerobes (Fig. 8-7) or facultative anaerobes that form spores under environmental conditions unfavorable for their growth. They are nonmotile and ubiquitous in water, soil, and air. Only three species cause human disease: *B. anthracis,* which causes anthrax, *B. cereus,* an occasional and recently recognized cause of food poisoning, and *B. subtilis,* which rarely causes opportunistic disease in compromised hosts.

Bacillus anthracis

Biologic characteristics. *Bacillus anthracis* is very large (1×5 μm), which is probably why it was the first organism demonstrated to cause a disease. This was accomplished by Koch in 1877. It is nonmotile, and has

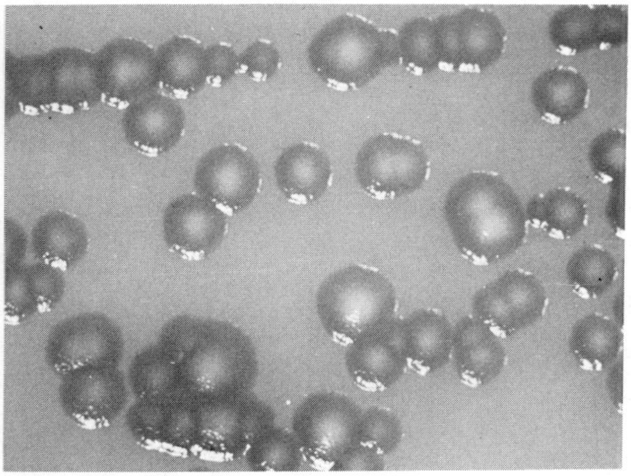

FIGURE 8-7. *Bacillus* species on blood agar. (Courtesy of Dr. James Plorde.)

square ends and center spores. Spores survive for months to years in soil and animal skins; they are not formed in living hosts. Colonies on blood agar (BA) are curly and rough, often called "medusa head." Strains may be encapsulated (in which case they form smooth, S colonies) or unencapsulated (rough, R colonies). Both virulent and avirulent strains form capsules. A phage that attacks only R strains can be used in a medium to select for S strains. There are two groups of antigens: cellular and a part of the exotoxin. The cellular antigens are derived from capsule polypeptide and the cell wall polysaccharide, which is not important for virulence.

Epidemiology. Anthrax is primarily a disease of livestock such as cattle, goats, sheep, horses, pigs, mink, dogs, or birds. After death of a diseased animal, spores are released into the earth where they may survive for years. Other livestock graze and ingest spores, which then germinate in their bodies. Spores can also remain viable in bone for weeks to months, and from there they can be transmitted to new animal hosts. The disease in humans is rare (1 to 2 cases per year in the United States; Fig 8-8), usually occurring in processing plant workers, farmers, veterinarians, or those contacting excreta, hides, or hair of diseased animals, in which spores can also survive for years. *B. anthracis* was sought in 139 samples of sewage in Great Britain and no strains were isolated (Jones et al. 1981).

Pathogenesis. Two components are important for virulence of the organism: an exotoxin and the capsule. The exotoxin, because it is difficult to cultivate in vitro, was not discovered until 1954. Even unencapsulated avirulent strains produce toxin, so it appears that both toxin and capsules are necessary for virulence. There are three components of the exotoxin, factors I, II, and III, as listed in Table 8-2. These factors work together rather than independently. Though the toxin does damage to leukocytes, the site of its action is unknown.

Host responses. The two major forms of human disease are cutaneous and respiratory. If infection occurs through the skin, a primary lesion or *malignant pustule* forms. It consists of a vesicle that forms a shallow

TABLE 8-2. Components of Toxin Produced by *Bacillus anthracis*

Factor	*Name*	*Activity*
I	Edema factor (EF)	Causes edema in host
II	Protective antigen (PA)	Induces antibody formation
III	Lethal factor (LF)	Responsible for lethal effects

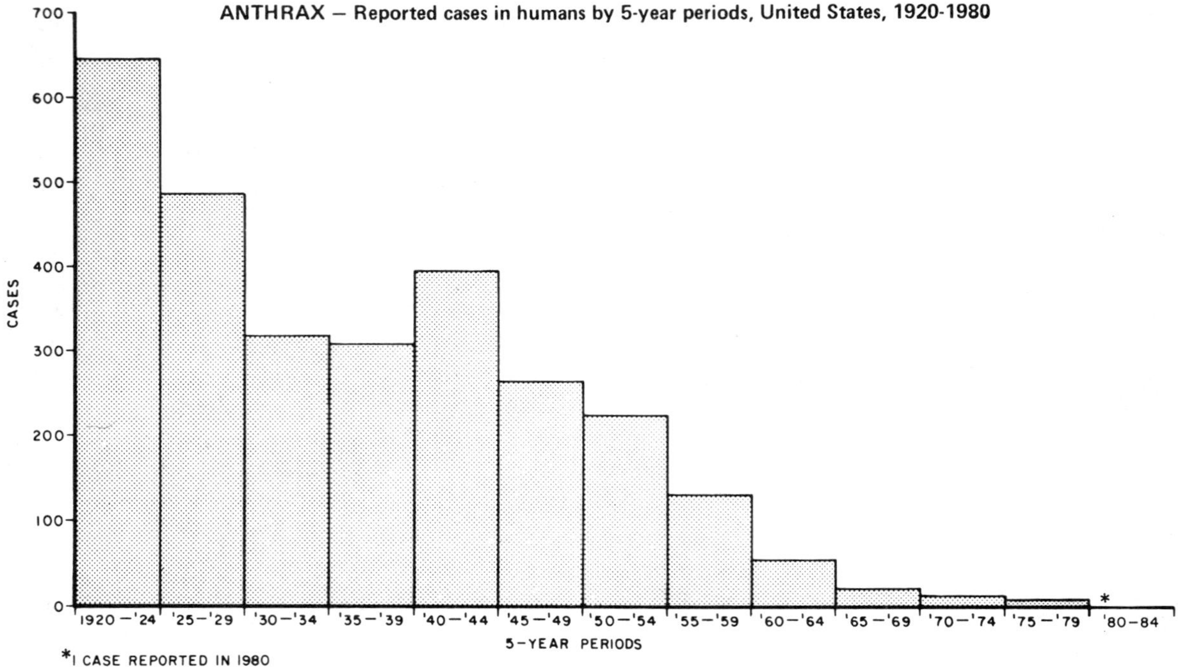

The single reported case of anthrax for 1980 occurred in a 30-year-old rendering-plant worker in Colorado. Anthrax was diagnosed in cattle in the vicinity of the rendering plant, and carcasses from involved premises were processed at the plant.

FIGURE 8-8. Anthrax—Reported cases in humans by 5-year periods, United States, 1920 to 1980. (Reprinted from Centers for Disease Control. Annual Summary 1980. Morb Mort Week Rep 1981;29:21.)

ulcer and black eschar. The disease is usually self-limited but may occasionally spread, causing serious systemic disease, septicemia, or meningitis. Mortality in untreated cases is about 20 percent with death often being sudden due to shock, pulmonary thrombosis, or respiratory failure. "Woolsorter's disease," an acute pneumonia caused by inhaling endospores also has a high mortality rate. It is extremely rare in the United States. Recovery from an episode results in immunity.

Control and treatment. The removal of diseased animals from flocks or herds and the rapid elimination of carcasses has resulted in the near eradication of anthrax in the United States. Most infections are now caused by contact with imported hides or hair (goatskin drums, yarn, saddle pads). An attenuated vaccine can be used in animals and high-risk humans, but its effectiveness is not optimal and frequent boosters are required to maintain immunity. Penicillin is the drug of choice for treatment. Tetracycline and erythromycin are also used. However, treatment is only effective before bacteremia develops. Once bacteremia occurs, mortality is high despite chemotherapy.

Bacillus cereus

In the 1950s *Bacillus cereus* was recognized as a cause of two forms of food poisoning. The first has a short incubation period, less than 8 hours, and is characterized primarily by vomiting. It is caused by a heat-stable enterotoxin. The second syndrome, caused by a heat-labile toxin, has a longer incubation period and is characterized primarily by diarrhea. Of 5300 foodborne outbreaks reported to the Centers for Disease Control, from 1966 to 1979, only 19 were attributed to *Bacillus cereus*. However, the organism could well have been responsible for some of the two-thirds of reported outbreaks in which no pathogen was found. The food most often implicated is rice, occasionally others such as sprouts, macaroni, and cheese. In one series of samples of dried grains and cereals tested for *B. cereus,* 56 percent were contaminated (Blakey and Priest 1980).

Other Species of *Bacillus*

Most other species like *Bacillus megaterium* and *B. polymyxa* are considered to be saprophytes in soil, water, decaying matter, and dust. The antimicrobial agent polymyxin is derived from *B. polymyxa. B. subtilis* is a ubiquitous saprophyte and common laboratory contaminant. On rare occasions it causes bacteremia, wound infection, meningitis, conjunctivitis, or endocarditis in compromised hosts. *Bacillus* species other than *B. anthracis* are resistant to penicillin and cephalosporins because they produce beta lactamases.

CLOSTRIDIA

Clostridia are anaerobic spore-forming bacilli that commonly inhabit animal and human GI tracts and soil. Most species are harmless saprophytes, but at least 12 species can cause several types of human disease, including botulism, food poisoning, gas gangrene, pseudomembranous colitis, and tetanus. Their virulence is due to production of potent exotoxins. All clostridial species except *perfringens* are motile with peritrichous flagella. *C. perfringens* is the only species that possesses a capsule. Clostridial spores (Fig. 8-9). are quite resistant to heat, drying, and disinfectants. Though all species contain somatic and flagellar antigens, they are not useful for identification or serotyping of species.

Clostridium botulinum

Biologic characteristics. *Clostridium botulinum* has seven antigenic types, A through G, based on distinct forms of neurotoxin produced. Only types A, B, and E have been known to produce human disease. These toxins are among the most potent poisons

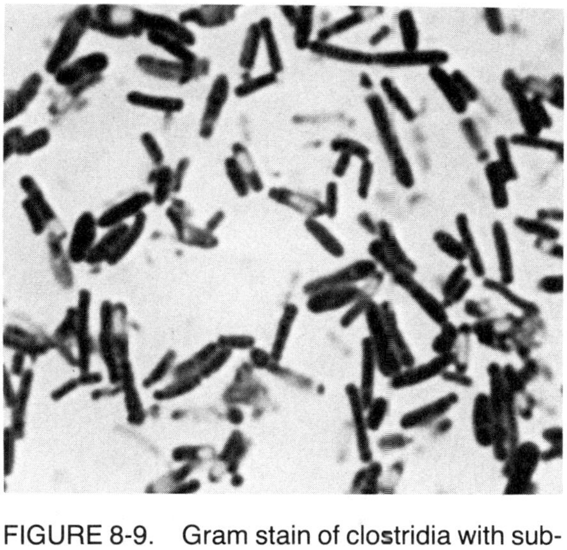

FIGURE 8-9. Gram stain of clostridia with subterminal spores. (Courtesy of Dr. James Plorde.)

known, with lethal human dose probably less than 1 μg. Toxigenicity has been demonstrated to be related to prophage infection and can be transmitted between species by such phages. The spores are quite heat resistant, but the toxin is heat sensitive; it is destroyed after 20 minutes of boiling.

Epidemiology. *Clostridium botulinum* lives in soil worldwide and frequently contaminates foods (Fig. 8-10). The major cause of botulism is inadequately sterilized home-canned foods, especially alkaline foods like olives, corn, string beans, or fish. Forty cases of botulism were reported in the United States in 1979; 24 infant botulism, 3 wound, 5 undesignated, and 8 food-borne (Morris and Hatheway 1980). Wound and infant botulism appear to be increasing (Fig. 8-11). Infant botulism occurs in babies 8 weeks to 6 months old. It may be more likely to occur at that young age because the GI flora has not yet become well established.

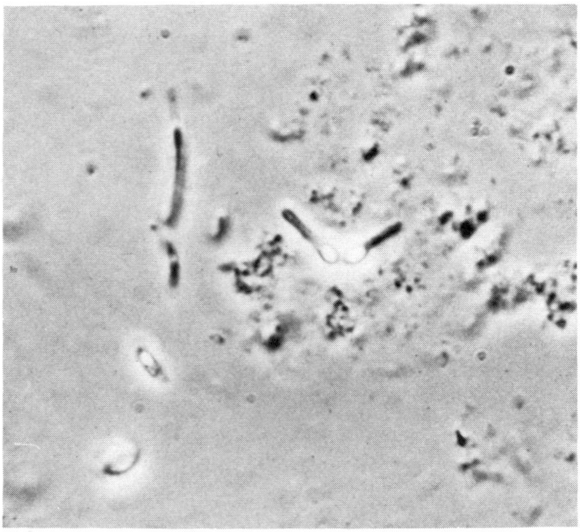

FIGURE 8-10. Phase microscopic isolation of *Clostridium botulinum* from a food sample. (Courtesy of Dr. James Plorde.)

Pathogenesis. Food-borne botulism is an intoxication resulting from ingestion of neurotoxin that has been formed in food in which the spores were not killed. After absorption from the GI tract, the toxin attaches to presynaptic terminals, blocking release of acetylcholine. The characteristics of the organisms causing food-borne disease and infant botulism are indistinguishable. However, infant botulism is due to ingestion of spores with their subsequent germination and toxin formation in the GI tract. Botulism toxin can be isolated from the feces but not serum of affected infants. In wound botulism, toxin is formed in a wound that has been infected with the spores of *C. botulinum*.

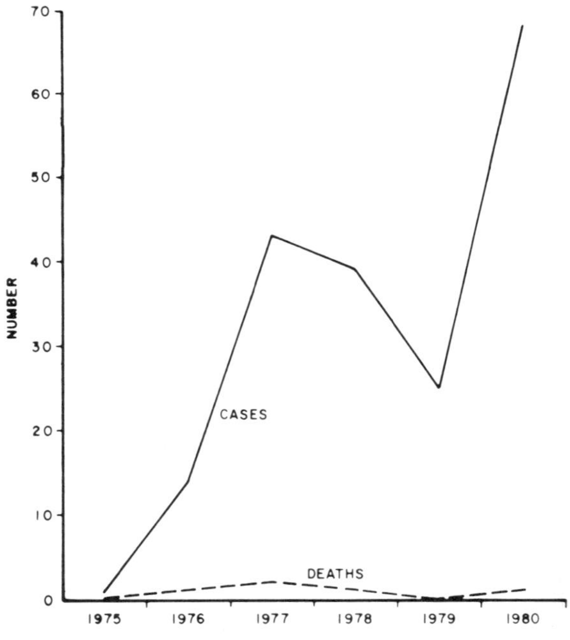

FIGURE 8-11. Botulism (infant)—reported cases and deaths by year, United States, 1975 to 1980. (Reprinted from Centers for Disease Control. Annual Summary 1980. Morb Mort Week Rep 1981;29:26.)

Host responses. There are three forms of disease: food-borne botulism, infant botulism, and wound botulism. Symptoms are similar in both food-borne and infant botulism, with primary involvement of the central nervous system being manifest 1 to 3 days after ingestion of contaminated food. Cranial nerve paralysis (speech and swallowing difficulty, droopy eyes) and muscle flaccidity are early symptoms. Diseased babies are characterized by the term, "floppy infant." There is no fever since the disease is primarily an intoxication rather than an infection. Almost without exception, respiratory support is required for bulbar paralysis, which can be fatal. However, with intensive therapy, overall case fatality rates are as low as 6 percent. *C. botulism* may be a cause of sudden infant death syndrome.

Control and treatment. The most important preventive measures are the proper preparation of home-canned foods as well as strict regulation of commercial canning procedures. Overtly spoiled foods should never be used, but toxin may be present even in foods that appear normal. Hence, all home-canned foods should be boiled for at least 10 minutes before ingestion. Antitoxins to the three types of toxin causing human disease are available, but their efficacy is uncertain. They are usually given together intravenously, preferably before neurologic impairment has begun. Once the nerve endings are affected, antitoxin is of only minimal value. All exposed individuals in a food-borne outbreak should be immediately treated, even if they are free of symptoms. Hypersensitivity to the foreign serum occurs, as is the case with any antiserum. Toxoids that evoke an antibody response have been used to immunize cattle and could be used for humans at high risk (such as laboratory workers). Individuals who have recovered from botulism are not immune to future incidents, however, because the amount of toxin necessary to induce antibody formation would be lethal.

Clostridia of Gas Gangrene

Biologic characteristics. A variety of clostridial species can cause gas gangrene, including primarily *C. perfringens* (90 percent), and also five others listed in Table 8-1. Most infections are mixed, with several species of *Clostridium* as well as aerobic gram-negative bacteria present. *C. perfringens* usually does not form spores in infected wounds.

Epidemiology. Clostridial species are ubiquitous in soil and in the human GI tract. Almost all persons carry *C. perfringens* in the gut in quantities as great as 10^7 per milliliter of feces. Gas gangrene results from wounds contaminated by fecal material or soil that contain clostridial spores (motor vehicle accidents, orthopedic injuries, postpartum uterus, abdominal surgery). In dead tissue where anaerobic conditions exist, clostridial spores can germinate and grow. The toxin spreads to internal organs from the contaminated wound. The incidence of gas gangrene, even in contaminated wounds, is only 2 percent with appropriate preventive treatment.

Pathogenesis. Gas gangrene is caused by clostridial exotoxins that are hemolytic and lethal. *Clostridium perfringens* has five distinct types of toxin, as shown in Table 8-3. Other clostridial species produce distinct exotoxins.

Host responses. The presence of *C. perfringens* in a wound culture as a contami-

TABLE 8-3. Toxins of *Clostridium perfringens*

Toxin	*Effect*
Alpha, lecithinase	Splits lecithin in cell membrane causing necrosis and cell death
Beta	Causes necrosis, death
Epsilon	Causes necrosis, death
Iota	Causes necrosis, death
Theta	Causes hemolysis

nant is common and does not mean that gas gangrene is present or that patient isolation is advisable. However, under anaerobic conditions, gas can accumulate and muscle and connective tissue are destroyed. A less serious form of infection is anaerobic cellulitis in which clostridia form gas and spread locally. Gas gangrene is much more invasive and may be lethal. Its onset is about 1 to 3 days after introduction of spores by surgery or injury. The gas produced in the subcutaneous tissues results in crepitus (crackly sound and feeling when touched). Fever and foul-smelling drainage, the result of the breakdown of proteins, is present. Shock and sometimes death result from rapid destruction of muscle and spread of the toxin.

Control and treatment. The major determinant of the outcome of clostridial contamination of wounds is the care with which the wound is cleansed and surgically debrided. Most cases of gas gangrene can be prevented by meticulous wound cleansing. If clostridial wound infection occurs, debridement is again necessary. Amputation is imperative if gas gangrene of a limb is extensive. Penicillin is the antimicrobial agent of choice. Clindamycin, chloramphenicol, erythromycin, and tetracycline are sometimes used. Antitoxin against several clostridial species is available but not in clinical use. Hyperbaric oxygen also has some efficacy.

Clostridium perfringens Food Poisoning

Some strains of *Clostridium perfringens* type A produce an enterotoxin during sporulation. This enterotoxin causes over-secretion of sodium and water, and reduced sodium absorption in the gut with resultant diarrhea. It, along with *S. aureus,* are the major causes of diarrheal food intoxications in the United States. These intoxications are usually associated with meat or contaminated serving dishes. About 3000 cases are reported in the United States each year. Contaminated meat contains heat-resistant endospores that are activated during the cooking process and germinate during cooling of food. Large numbers, 10^8 to 10^9, of the vegetative organisms must be ingested to cause symptoms. The clostridia again sporulate in the gut and enterotoxin is released after about 12 hours. The resultant diarrhea (sometimes vomiting and abdominal pain as well) lasts several days, is not accompanied by fever, and does not require treatment. A carrier state may last for several weeks after recovery.

Clostridium tetani

Biologic characteristics. *Clostridium tetani,* like all clostridia species, is a spore-forming anaerobe present in soil and the GI tracts of some animals. It is not, however, a normal inhabitant of the human GI tract. Its spores are terminal, resulting in a drumstick-shaped organism. *C. tetani* species share one common O antigen and are typed by ten distinct H, flagellar, antigens. All strains produce the same exotoxin that causes tetanus.

Epidemiology. About 100 cases of tetanus per year are reported in the United States (Fig. 8-12). Tetanus spores are introduced into the body by puncture wounds, cuts, incisions, or even the umbilical stump of neonates. The disease is completely preventable by active immunization with toxoid or use of antitoxin immediately after exposure. Drug abusers are at high risk of infection.

Pathogenesis. If a wound contaminated by tetanus spores is necrotic, infected, or improperly cleansed, anaerobic conditions allow the germination of the spores. The incubation period varies from days to years, depending on the condition of the wound. Spores can be dormant in a wound and germinate after a number of years. A neurotoxin (tetanospasmin), second only to the toxin of botulism in potency, is released by vegetative cells when they lyse. The organism itself is much less invasive than *C. perfringens*. Infection does not spread but remains near the local area of injury. However, the toxin is carried to the central nervous system either through the circulatory system or by retrograde axonal transport. It is fixed in the CNS and there blocks postsynaptic inhibition of spinal motor neurons,

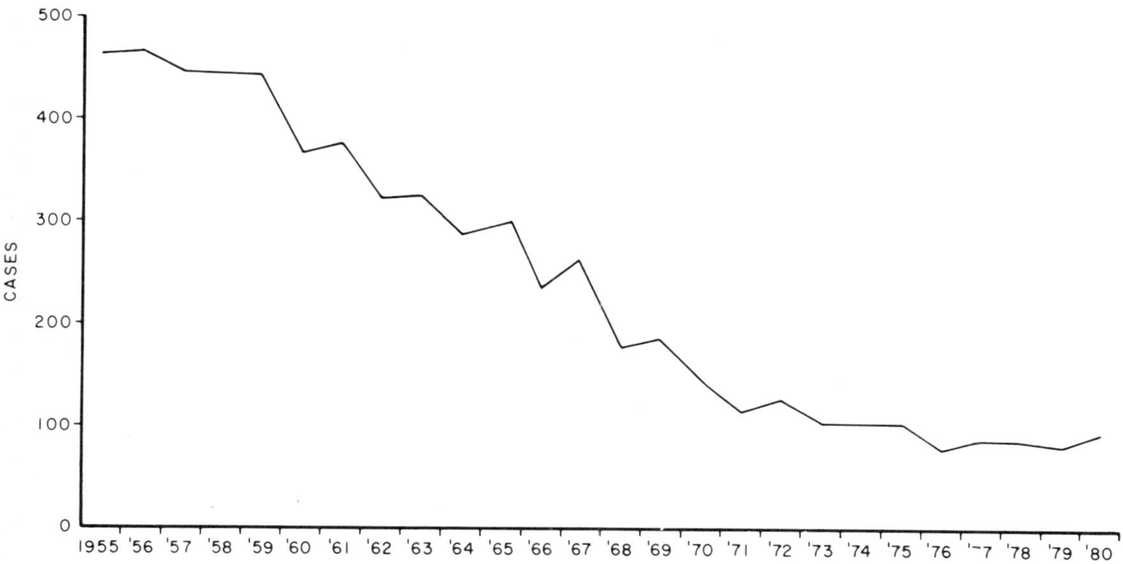

After continuously decreasing from 1955 through 1976, the number of reported cases of tetanus has remained stable. This is most likely related to inadequate vaccination levels in a portion of the population.

FIGURE 8-12. Tetanus—reported cases by year, United States, 1955 to 1980. (Reprinted from Centers for Disease Control. Annual Summary 1980. Morb Mort Week Rep 1981;29:83.)

causing hypersensitivity of excitatory impulses with resultant muscle spasms.

Host responses. Tetanus is usually generalized rather than localized. Local tetanus can occur in persons with partial immunity or with small inocula of *C. tetani*. With local tetanus, only local muscles spasm; this persists for weeks before subsiding. Generalized tetanus is characterized by muscle rigidity, painful spasms, and trismus ("lockjaw"). Convulsions are common though spinal fluid is normal. Mechanical ventilation is usually required to prevent death from respiratory paralysis. Fever and mental confusion are usually absent unless secondary infection, aspiration pneumonia, dehydration, or other causes are also present. Mortality may be as high as 50 percent. Certain diagnosis of the organism is made by production of toxin and its neutralization by special antitoxin.

Control and treatment. Up-to-date immunization with tetanus toxoid (as part of the DPT—diphtheria, pertussis, tetanus—vaccine) is absolutely protective. After the initial course of three doses during infancy, boosters are recommended after 1-year, 4-year, and 10-year intervals subsequently throughout life. Following injury, two courses of action are recommended:

1. For previously immunized persons, administer a booster of toxoid, which will result in an antibody response within 14 days. Also administer antitoxin as below.
2. For nonimmunized individuals, do not give toxoid since the resultant antibody response would be too slow to be effective. Instead, administer tetanus immune globulin (TIG), or, if unavailable, equine antitoxin. The latter, of course, carries a much higher risk of hypersensitivity and serum sickness. A test of hypersensitivity to horse serum must precede injection of the antitoxin.

Thorough wound cleaning and debridement is essential for all traumatic wounds. Penicillin is given to prevent further bacterial growth. Generalized tetanus requires intensive supportive care with administration of sedatives and muscle relaxants, and meticulous assistance with respiratory effort (using care to prevent pulmonary infections or aspiration pneumonia).

Clostridial Enteric Infections

Gastrointestinal disease in which clostridia have been implicated include enteritis necroticans, necrotizing enterocolitis, and pseudomembranous colitis. Enteritis necroticans, not seen in the United States, has a high mortality rate and is probably associated with type C *Clostridium perfringens*. Necrotizing enterocolitis (NEC) occurs in neonates, sometimes in epidemics. It causes bowel necrosis and has a high mortality rate. Though stool colonization with clostridia has been found to significantly correlate with epidemics of NEC, clostridia are probably not related to the initial pathogenesis of the disease. Toxin of *C. difficile* has been isolated in some cases and *C. butyricum* has also been implicated. *C. difficile* primarily, but also *C. sordelli,* have been associated with antibiotic-associated pseudomembranous enterocolitis (PMC). The use of broad-spectrum antimicrobial agents allows the overgrowth of resistant clostridial species with release of large amounts of enterotoxin in the intestine. In one series, 80 percent of cases of PMC that occurred after antimicrobial administration were related to ampicillin, clindamycin, and cephalosporins (Bartless 1981). Both *C. difficile* and *sordelli* respond well to treatment with oral vancomycin.

Other Clostridial Infections

Clostridia, especially *C. perfringens,* may be a major cause of septic abortion. *C. perfringens* may occasionally cause eye infections, brain and liver abscesses, postpartal uterine infections, peritonitis, and bacteremia (especially after surgery of the GI or biliary tract, septic abortion, or with infected decubiti). *C. septicum* has caused gangrene and sepsis in patients with malignancies.

CORYNEBACTERIA

Corynebacteria are nonmotile, nonencapsulated, nonspore-forming gram-positive rods with club-shaped ends that are pleomorphic and make palisade or "Chinese letter" arrangements in culture (Fig. 8-13). Many species are normal flora of skin, conjunctiva, mucous membranes, and the respiratory tract. They contain metachromatic granules and deposits of polyphosphates. Though they grow well on simple media, two selective and differential media are used when diphtheria is suspected because of the characteristic growth of *Corynebacterium diphtheriae* as described in Table 8-4 (Fig. 8-14 *a, b*). All strains that cause disease produce exotoxins.

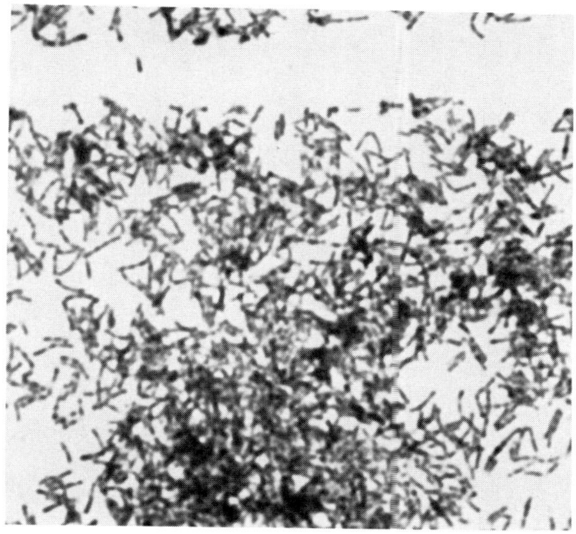

FIGURE 8-13. *Corynebacterium diphtheriae* in "Chinese letter" arrangements. (Courtesy of Dr. James Plorde.)

TABLE 8-4. Appearance of *Corynebacterium diphtheriae* in Differential Media

Medium	*Contents*	*Colonial appearance*
Loeffler's	25 percent broth, 75 percent coagulated serum; high phosphate concentration	Small, gray, granular with irregular edges
McLeod's	Blood agar with potassium tellurite	Mitis: black, small, hemolytic, glossy, convex Gravis: gray, large, nonhemolytic, irregular, striated, dull Intermedius: small, nonhemolytic, intermediate between mitis and gravis

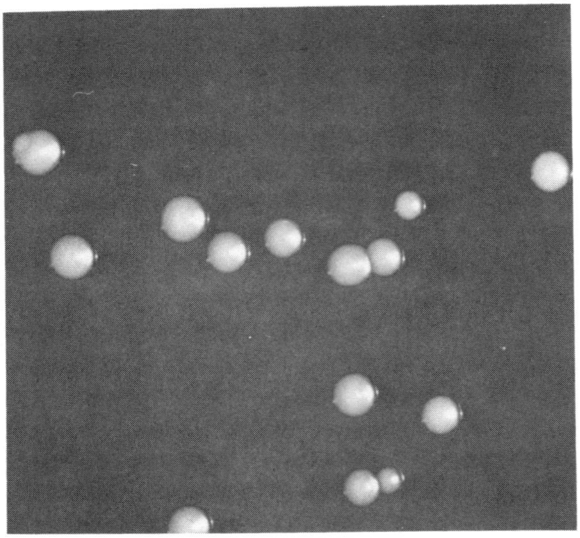

a

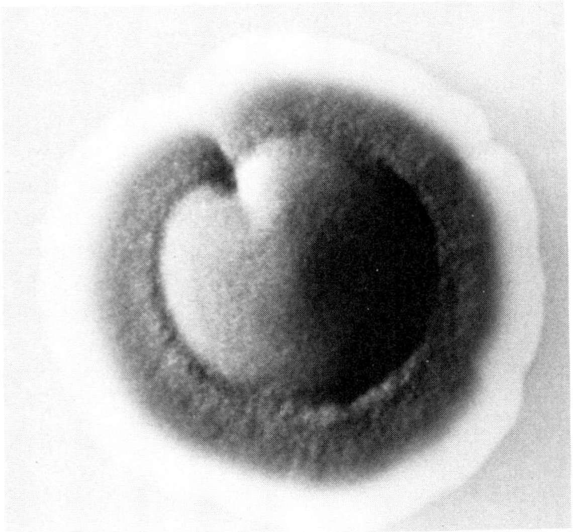

b

FIGURE 8-14. *(a)* Corynebacteria on blood agar. *(b) Corynebacterium diphtheriae* on Tinsdale medium, which contains proteose, L-Cysteine, sodium chloride, and sodium thiosulfate. (Courtesy of Dr. James Plorde.)

Corynebacterium diphtheriae

Biologic characteristics. *Corynebacterium diphtheriae* are obligate aerobes. There are toxigenic and nontoxigenic strains. Toxigenicity is under control of a phage gene. Toxigenic strains are lysogenic, containing a prophage. Acquisition of the phage results in toxigenicity.

Epidemiology. About 200 cases of diphtheria are reported annually in the United States with about 10 deaths (Fig. 8-15). The disease was prevalent before the advent of immunization. Most individuals acquired the disease and subsequent immunity at an early age. Immunity is not life-long, either from natural infection or artificial immunization. Within the community, levels of immunity vary tremendously. Without boosters, levels of antitoxin in the blood of adults are not protective. The highest attack rates continue to be in children 1 to 9 years of age. Minority racial groups have attack rates 5 times greater than Caucasians. American Indians have the highest incidence, and may account for as many as 50 percent of cases in some areas. Diphtheria has a seasonal variation; most cases occur during September through January. Outbreaks occur in certain subpopulations: indigent inner city dwellers, migrant workers, institutionalized adults. Such outbreaks are usually localized within families or geographic areas. Cutaneous diphtheria is increasing, perhaps due to increased recognition. Person-to-person spread results from respiratory dissemination of droplets.

Pathogenesis. The three biotypes of *C. diphtheriae*—gravis, intermedius, and mitis—were originally named because of the relative severity of the resulting disease.

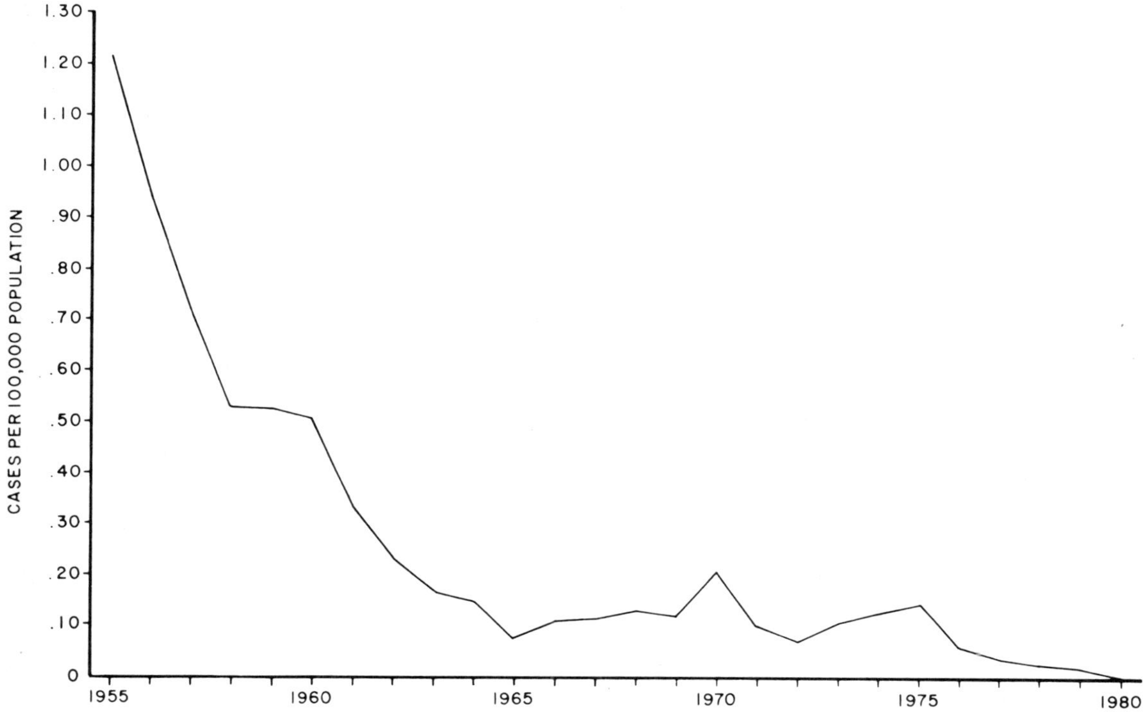

The diphtheria case rate has continued to decrease in recent years; however, the death-to-case ratio has remained stable, suggesting that diphtheria is still difficult to treat. The best method for preventing deaths from diphtheria is vaccination.

FIGURE 8-15. Diphtheria—Reported case rates, United States, 1955 to 1980. (Reprinted from Centers for Disease Control. Annual Summary 1980. Morb Mort Week Rep 1981;29:30.)

However, the designations may be misleading because each type can be toxigenic or nontoxigenic. For example, among strains isolated from skin lesions in persons living in Seattle, the following association between biotype and toxigencity was noted:

	Biotype	*Percent nontoxigenic*
1977	All mitis and intermedius	86
1979	99 percent gravis	96

Hence, virulence and toxigenicity are not well correlated with biotype.

C. diphtheriae is not invasive and rarely enters the circulation; its virulence is primarily a result of local elaboration of toxin. The toxin is a heat-labile protein that inhibits the protein synthesis of surrounding cells. Activated toxin consists of two fragments, one that mediates enzymatic reactions to halt protein synthesis resulting in cell death (fragment A) and another that is essential for the binding and penetration of toxin to the cell (fragment B). Because there are so many corynebacteria that are throat commensals, tests of toxigenicity are required to confirm a diagnosis of diphtheria. The *Schick test* is used to determine an individual's immune response to diphtheria and the possible need for immunization. In nonimmune persons, a local necrotic reaction will occur 4 to 5 days after intradermal injection of a small amount of diphtheria toxin. If there is adequate circulating antitoxin, there will be no reaction. However, individuals may develop hypersensitivity to the toxin; an allergic reaction can be confused with the positive reaction that indicates the need for immunization.

Host responses. The most common syndrome caused by *C. diphtheriae* is membranous pharyngitis called diphtheria. Following infection via the upper respiratory tract, the organisms multiply on pharyngeal mucous membranes. Toxin is elaborated, absorbed into mucous membranes causing inflammation and local destruction of tissue. Necrotic and inflamed epithelium produces a characteristic *pseudomembrane* on the soft palate, usually over the posterior pharynx and tonsils (Fig. 8-16). The pseudomembrane can extend into the nasopharynx or down into the larynx. If the membrane is lifted, bleeding occurs. Membranous laryngitis is often unrecognized until tracheal intubation is performed. It may cause dangerous mechanical obstruction of breathing. Cervical adenitis is common, but infection remains localized. Mild cases of sore throat with membrane formation can result from infection with nontoxigenic diphtherial strains. Absorbed toxin can also result in otitis media; necrosis of heart, liver, kidney, or adrenal tissue; peripheral nerve paralysis; or vaginitis. Despite only a mild fever, the patient is extremely ill.

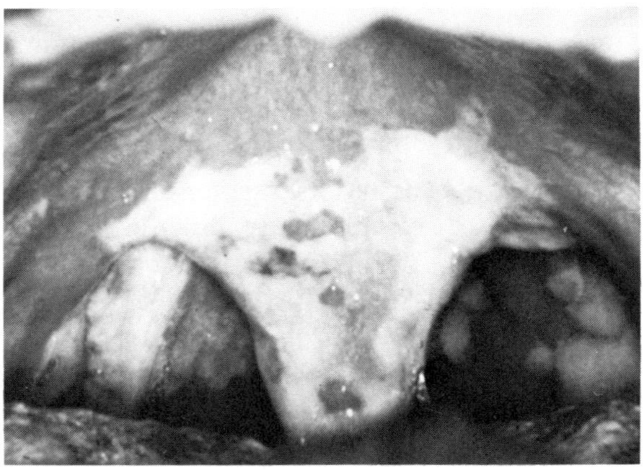

FIGURE 8-16. Soft palate pseudomembrane of diphtheria. (Courtesy of Dr. James Plorde.)

Cutaneous diphtheria can occur, especially in the tropics and in debilitated individuals. It is more common among the indigent in urban slum areas, where resulting lesions are called "wine sores." Most infections are with nontoxigenic strains and are found in shallow crusted ulcers that resemble impetigo. These lesions may serve for long periods of time as a reservoir for *C. diphtheriae*. In Seattle, from 1972 to 1978, about 750 such skin carriers were reported. Nasopharyngeal carriers who have recovered from diphtheria can spread the organism for several weeks after symptoms have been eliminated. Immunity results from antibody formation against the toxin. Temporary passive immunity among newborns is acquired by transplacental passage of antitoxin from resistant mothers.

Control and treatment. The foundation of control is the maintenance of a high level of active immunization in the community and the eradication of carriers in the general population. Toxoid as an alum precipitate or adsorbed on aluminum phosphate gel is given, along with tetanus toxoid and pertussis vaccine. A course of injections begins at age 3 to 4 months with boosters at age 3 to 4 years, age 6 to 8 years, and during the teen years. Since immunity is relatively short-lived (sometimes only a few months or years), boosters are especially important.

It is imperative that early diagnosis be made. The only definitive therapy is antitoxin, which works on fragment B of the toxin by preventing toxin attachment to epithelium. Once the toxin is fixed to the cell, antitoxin is useless. One intravenous injection of 20,000 to 100,000 U should be given as soon as possible, after testing for hypersensitivity to animal serum. Penicillin and erythromycin arrest toxin production and inhibit growth of *C. diphtheriae,* but have no effect on the clinical course of the disease. Antimicrobial agents may also be used to help eradicate the organism in asymptomatic carriers. Patients with diphtheria must be isolated to prevent droplet spread.

Other Corynebacteria Species

Common coryneform commensals (called diphtheroids or coryneform bacteria) are listed in Table 8-5. *Propionibacterium acnes* is the anaerobic diphtheroid of skin and was formerly called *Corynebacterium acnes*. *C. minutissium* causes a skin infection in pubic and axillary regions called erythrasma. *C. hemolyticum* has been shown to mimic diphtheria and may be an occasional cause of severe, prolonged throat infections. Coryneform bacteria are rare causes of endocarditis. A new multiply resistant corynebacterium was isolated from the blood of a septic patient in 1977. The organism was subsequently recovered from 30 other patients, with clinical disease present in 17. Similar organisms (JK organisms) have been isolated among other compromised hosts (Young et al. 1981).

Gardnerella vaginalis (formerly called *Hemophilus vaginalis* and *Corynebacterium vaginale*) is a coryneform bacteria that is a normal inhabitant of the vagina in about 25 percent of women. It may cause "nonspecific" vaginitis and, rarely, bacteremia in postpartal women or in newborns.

TABLE 8-5. Common Species of *Corynebacterium*

Species	*Habitat*	*Associated diseases*
aquaticum	Water sources; confused with *Listeria monocytogenes*	None
diphtheriae	Human respiratory tract, skin, conjunctiva	Diphtheria, pharyngeal or cutaneous
equi	Human female genital tract, respiratory tract	Rarely, pulmonary nfections, endocarditis
haemolyticum	Human skin, throat	Rarely, throat, skin infections, septicemia
hofmannii (pseudodiphtheriticum)	Human upper respiratory tract	Probably none
minutissimum	Human skin	Superficial skin infection, erythrasma
ulcerans	Upper respiratory tract	Possibly tonsillitis
vaginale	Female genital tract	Vaginitis, urethritis puerperal fever, septicemia
J K diphtheroid	Skin	Sepsis in immunocompromised hosts

ERYSIPELOTHRIX RHUSIOPATHIAE

Erysipelothrix rhusiopathiae (insidiosa) is a nonspore-forming aerobic to microaerophilic rod resembling *Listeria*. It occurs worldwide in water, in fish, many other animals, and decaying organic matter. Infection occurs from contact through skin abrasions with infected meat, fish, manure, bones, hides, or poultry. The resultant cutaneous disease, *erysipelas,* is usually painful, red, and spreads locally. (Erysipelas, however, is more commonly caused by *Streptococcus pyogenes*.) Septicemia and endocarditis have also occurred. Treatment is with cephalosporins, clindamycin, erythromycin, or penicillin.

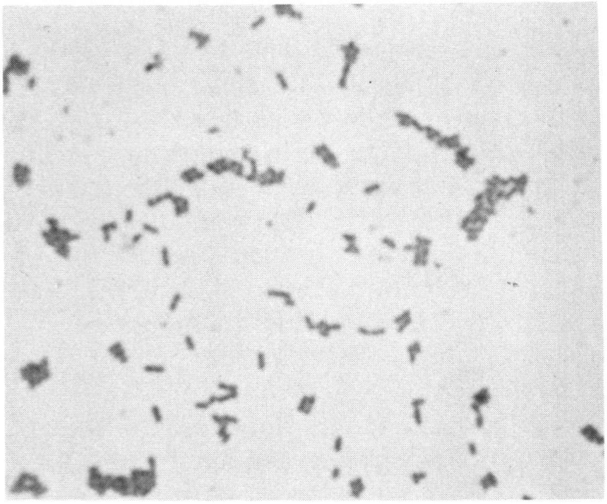

FIGURE 8-17. *Listeria monocytogenes* on Gram stain. (Courtesy of Dr. James Plorde.)

LACTOBACILLI

Lactobacilli are members of the normal flora of the GI tract, the vagina, and the mouth. They are large gram-positive rods, and are microaerophilic or anaerobic. They are important in preventing growth of pathogens in their microenvironments. They are nonpathogenic, but may play a role in the development of dental caries.

LISTERIA MONOCYTOGENES

Listeria monocytogenes is an intracellular parasite of macrophages that resembles diphtheroids and is often mistaken in the laboratory for a contaminant. It is aerobic or microaerophilic and can be pleomorphic, looking like cocci or even gran-negative rods (Fig. 8-17). It is short, nonspore-forming, and has a rough filamentous form. The organism was named for Lister and for its resultant monocytosis. It sometimes causes an infectious mononucleosis-type syndrome. Listeriosis is primarily a disease of fish, birds, and animals. The organism may live in soil and may be carried in the GI tracts of humans and animals for long periods of time. Such asymptomatic infection may be widespread. There are four main serologic groups and 11 serotypes based on somatic (O) and flagellar (H) antigens. Types 1 and 4b are the most common causes of human disease.

There are about 100 cases of listeriosis reported each year in the United States with about 25 percent mortality. Unrecognized disease is probably common. The organism causes two forms of neonatal sepsis. Listerial septicemia (formerly called granulomatosis infantiseptica) is a prenatal infection that may result in spontaneous abortion, stillbirth, prematurity, or disease in the first few days of life. It has a very high mortality rate. Listerial infection occurring after the first few days of life has a better prognosis. Intrauterine infections are caused primarily by types 1a and 1b, whereas the later neonatal sepsis or meningitis is often due to type 4b. The major syndrome in infants and adults is a meningoencephalitis, which has a mortality rate of about 70 percent if untreated. Septicemia also occurs in pregnant women and other adults, especially those with lymphomas or immune deficiencies. Pregnant individuals may be more susceptible to listerial infections because of changes in cellular immunity during pregnancy, since resistance to infection depends on T cell-mediated immune functions and nonspecific activity of the mononuclear phagocyte system. Transmission can be from an infected animal to a person (zoonosis) or from vegetation or soil to a person (sapronosis), or horizontally from mother to fetus via the placenta. Person-to-person spread may also be possible. Treatment is usually with ampicillin; penicillin and tetracycline may also be used. A combination of penicillin and gentamicin seems to be synergistic.

CLINICAL IMPLICATIONS

Although most of the diseases caused by gram-positive rods are relatively uncommon, early diagnosis is often essential to prevent lethal consequences. The clinician's observation and recognition of characteristic and pathognomonic symptoms may be invaluable. For example, in patients suffering from diarrhea who are concurrently receiving broad-spectrum antimicrobial therapy, the possibility of PMC should be considered. One can be cognizant of the possibility of infection with commensals such as coryneform bacteria (or even skin infection with *C. diphtheriae*), actinomyces, or clostridia in high-risk groups such as indigent urban dwellers, individuals who have had dental surgery, women who have experienced spontaneous or septic abortions, or persons with malignancies.

SUGGESTED READINGS

Arfania D, et al. Uncommon causes of peritonitis in patients undergoing peritoneal dialysis. Arch Intern Med 1981;141:61.

Bartless JG. Antimicrobial agents implicated in *Clostridium difficile* toxin-associated diarrhea or colitis. Johns Hopkins Med J 1981; 149:6-9.

Belsey MA, LeBlanc DR. Skin infections and the epidemiology of diphtheria: acquisition and persistence of *Corynebacterium diphtheriae* infections. Am J Epidemiol 1975;102:179-184.

Bittner J. The clinical significance, taxonomy and special methodologic problems of the pathogenic clostridia. Infec 1980; 8 (Suppl 2): S 117-S 122.

Blakey LJ, Priest FG. The occurrence of *Bacillus cereus* in some dried food including pulses and cereals. J Appl Bacteriol 1980 48:297-302.

Brooks GF, Bennet JV, Feldman RA. Diphtheria in the United States, 1959-1970. J Infec Dis 1974;129:172-179.

Cashore WJ, et al. Clostridia colonization and clostridial toxin in neonatal necrotizing enterocolitis. J Pediatr 1981;98:308-311.

Dezfulian M, Dowell VR. Cultural and physiological characteristics and antimicrobial susceptibility of *Clostridium botulinum* isolates from foodborne and infant botulism cases. J Clin Microbiol 1980;11:604-609.

Eng RH, et al. Infections caused by *Actinomyces viscosus*. Am J Clin Pathol 1981;75:113-116.

Green SL, LaPeter KS. Pseudodiphtheric membranous pharyngitis caused by *Corynebacterium hemolyticum*. JAMA 1981;245:2330-2331.

House MJ. Abdominal actinomycosis complication of laparoscopy? Br J Obstet Gynaecol 1981;88:459-460.

Jones PW, et al. The occurrence and significance to animal health of leptospira mycobacteria, *Escherichia coli, Brucella abortus* and *Bacillus anthracis* in sewage and sewage sludges. J Hyg, Camb 1981;86:129-137.

Kahn FW, Gornick CC, Tofte RW. Primary cutaneous *Nocardia asteroides* infection with dissemination. Am J Med 1981;70:859-863.

Kannangara DW, Tanaka, T, Thadepalli H. Spinal epidural abscess due to *A. israelii*. Neurol 1981;31:202-204.

Katlic MR, Derkas WM, Coleman WS. *Clostridium septicum* infections and malignancies. Ann Surg 1981;193:361-364.

Koopman J. The role of cutaneous diphtheria infections in a diphtheria epidemic. J Infec Dis 1975;131:239-244.

Koshi G, et al. Brain abscess and other protean manifestations of actinomycosis. Am J Trop Med Hyg 1981;30:139-144.

Mitchell RG, Etches PC, Day DG. Non-toxigenic clostridia in babies. J Clin Pathol 1981;34:217-220.

Morris JG. *Bacillus cereus* food poisoning. Arch Intern Med 1981;141:771.

Morris JG, Hatheway CL. Botulism in the US, 1979. J Infec Dis 1980;142:302-305.

Schaal KP, Schofield GM, Pulverer G. Taxonomy and clinical significance of Actinomycetaceae and Propionibacteriaceae. Infec 1981;8 (Suppl 2):S 122-S 130.

Sturm R, et al. Neonatal necrotizing enterocolitis associated with penicillin-resistant, toxigenic *Clostridium butyricum*. Pediatr 1980;66:928-931.

Talwar P, Sehgal SC. Pulmonary nocardiosis. Eur J Respir Dis 1980;61:103-107.

Thomas JR, et al. Emetic food poisoning caused by *Bacillus cereus*. Arch Intern Med 1981; 141:766-767.

Tonjum S, et al. Hyperbaric oxygen treatment in gas-producing infections. Acta Chir Scand 1980;146:235-241.

Wilcke BW, Midura TF, Arnon SS. Quantitative evidence of intestinal colonization by *Clostridium botulinum* in four cases of infant botulism. J Infec Dis 1980;141:419-423.

Yoshikai Y, et al. Relationship between non-specific activity of macrophages and immune responses to *Listeria monocytogenes*. Immunol 1980;40:295-301.

Young VB, et al. Emergence of coryneform bacteria as a cause of nosocomial infections in compromised hosts. Am J Med 1981;70:646-670.

9 GRAM-NEGATIVE ENTERIC BACTERIA

KEY POINTS

- Most enteric bacteria are normal flora or saprophytes that do not under normal circumstances cause disease. However, they can become "opportunistic pathogens" and are of major concern as causes of nosocomial (hospital-associated) infection.
- The most common gram-negative enteric bacteria causing opportunistic infections are *Escherchia coli, Pseudomonas aeruginosa, Klebsiella pneumoniae, Enterobacter* species, and *Proteus* species. However, essentially every organism in the coliform group has been known to cause infection.
- The major pathogens in the group are *Salmonella* (food poisoning, typhoid, and serious diarrheas), *Shigella* (bacterial dysentery), and *Vibrio cholerae* (cholera).
- All gram-negative bacteria possess endotoxin as a component of their cell walls. Endotoxin can cause serious symptoms in humans; it can cause circulatory shock and death.
- Gram-negative bacteria have been commonly isolated from the hands of health care personnel; hands appear to be the primary mode of transmission.
- The fermentation of lactose is a simple but important differential test used in the laboratory for gram-negative bacteria. Enteric pathogens *(Salmonella, Shigella,* and *Vibrio)* do not ferment lactose, whereas the most common coliforms *(E. coli, Klebsiella, Enterobacter)* do.

The gram-negative enteric bacteria (sometimes called coliforms) are, for the most part, a helpful and important component of the human body's normal flora. They protect the gastrointestinal tract against invasion by more virulent pathogens as well as synthesizing large amounts of vitamins B and K. Only in unusual circumstances are these bacteria responsible for clinical disease. Pathogenic enteric bacteria include species of *Salmonella, Shigella,* and *Vibrio.*

There are multiple species of heterogeneous, gram-negative, rod-shaped bacteria that are indigenous to humans and other animals. This group, which includes members of the families Enterobacteriaceae, Pseudomonadaceae, Vibrionaceae, and Bacteroidaceae, is often collectively referred to as the enteric bacteria (Table 9-1). These organisms generally range in size from 1 to 2 × 2 to 6 micrometers (μm) and are found in the intestinal tract as normal flora or as pathogens, though some genera are saprophytes found in water and soil.

Families can be differentiated from each other by a number of criteria. Initial division can be made between the fermenters of sugars and the nonfermenters. The Enterobacteriaceae and Vibrionaceae are aerobes or facultative anaerobes that obtain their energy by fermentation or respiration. Pseudomonads are strict aerobes which can oxidize sugars but cannot ferment them. *Bacteroides* are strict anaerobes and together with bifidobacteria (anaerobic gram-positive rods) make up 95 to 99 percent of the intestinal flora. Differentiation can also be made based upon type of flagella. Those members of the Enterobacteriaceae that are motile possess peritrichous flagella, and pseudomonads and vibrios have polar flagella.

A plethora of biochemical tests is used to distinguish the species of gram-negative bacteria. Some of the more common tests are listed in Table 9-2. A general test of differentiation is the ability to ferment lactose. Selective media such as *Salmonella-Shigella* (S-S), eosin-methylene blue, and MacConkey agar are used for ease of identification of lactose fermenters. MacConkey agar contains bile salts that inhibit the growth of gram-positive organisms. Lactose fermenters on MacConkey agar turn bright pink, other organisms are white.

Tests listed in Table 9-2 include the Voges-Proskauer test, which identifies the ability of organisms to produce acetoin, an intermediate in the production of butylene glycol from glucose fermentation; the ability of organisms to use citrate as the sole source of carbon; the production of indole from the metabolism of trytophan; and the production of the enzyme oxidase, which indicates that carbohydrates are used principally by oxidation and only rarely by fermentation. Tests of an organism's ability to ferment a variety of sugars in addition to glucose and lactose are also performed to classify enteric organisms. Four tests are often used together to differentiate *Escherichia coli* (and *Edwardsiella tarda*) from the other coliforms *(Enterobacter, Klebsiella,* and *Serratia)*. The tests collectively are known by the acronym IMViC and are described in Table 9-3.

Coliform is a general term used for facultative gram-negative bacteria that inhabit the intestinal tract, usually without causing disease. Several genera of the family Enterobacteriaceae are included in this group. Coliforms include lactose fermenters *(E. coli, Klebsiella, Enterobacter),* delayed lactose fermenters *(Citrobacter, Arizona),* and those that are lactose-negative *(Edwardsiella, Providencia, Serratia).*

These organisms have complex antigenic structures, composed of three types of antigens (Fig. 9-1): somatic (O from the German words "ohne hauch," which mean without film or cloud), capsular (K for "kapsel" or "Vi" in salmonella), and flagellar (H from the German word "hauch," a name for the

(text continues on page 191)

TABLE 9-1. Enteric Gram-Negative Bacteria Important in Humans

Family	Genus	Species	Normal flora*	Pathogen	Other habitat
Enterobacteriaceae	*Citrobacter*	See Ch. 10			X
	Edwardsiella	*tarda*	X		
	Enterobacter	*aerogenes*	X		
		agglomerans			X
		cloacae	X		
		gergoviae	X		
		sakazakii	X		
	Erwinia				X
	Escherichia	*coli*	X		
	Klebsiella	*oxytoca*	X		X
		ozoenae	X		X
		pneumoniae	X		
		rhinoscleromatis		X	
	Proteus	*mirabilis*	X		
		morganii	X		
		vulgaris	X		
	Providencia	*stuartii*	X		
		rettgeri	X		
		alcalifaciens	X		
	Salmonella	*cholerasuis*		X	
		enteritidis		X	
		typhi		X	
	Serratia	*marcescens*			X
		rubidea			X
		liquefaciens			X
	Shigella	*boydii*		X	
		dysenteriae		X	
		flexneri		X	
		sonnei		X	
	Yersinia	See Ch. 10			
Bacteroidaceae	*Bacteroides*	See Tab. 9-12	X		
	Fusobacterium	See Tab. 9-12	X		
Pseudomonadaceae	*Pseudomonas*	See Tab. 9-9			X
	Aeromonas	See Ch. 10			X
Vibrionaceae	*Vibrio*	See Tab. 9-11		X	
	Campylobacter	See Tab. 9-11		X	

*All normal flora can become opportunistic pathogens.

TABLE 9-2. Some Biochemical Reactions of Common Gram-Negative Bacteria

	Motility	*Lactose fermenter*	*Gas producer*	*Glucose fermenter*	*H_2S producer*	*Voges-Proskauer*	*Urease producer*	*Citrate utilizer*	*Indole producer*	*Oxidase*
Escherichia coli	±	+	+	+	−	−	−	−	+	−
Acinetobacter	−	−	−	V	−	−	±	±	−	−
Citrobacter	+	±	+	+	V	−	+	+	V	−
Enterobacter	+	+	+	+	−	+	V	+	+	−
Flavobacter	−	−	−	−	−	−	V	V	+	+
Klebsiella	−	+	+	+	−	+	+	+	V	−
Proteus	+	−	+	+	+	−	+	V	V	−
Providencia	+	−	V	+	−	−	V	+	+	−
Pseudomonas	+	−	−	±	−	−	−	+	−	+
Salmonella	+	−	+	+	+	−	−	+	−	−
Serratia	+	−	±	+	−	+	−	+	−	−
Shigella	−	−	−	+	−	−	−	−	±	−
Vibrio	+	−	−	+	−	−	−	+	+	+
Yersinia	−	−	−	+	−	−	+	−	−	−

\+ = usually positive
− = usually negative
± = may be either positive or negative
V = reaction variable, depending upon species

TABLE 9-3. The IMViC Test for Differentiation of *Escherichia coli* and Other Coliforms

Part	*Test*	*Reactions*	
		E. coli	*Other coliforms*
I	Formation of indole from tryptophan	+	−
M	Methyl red test to measure pH. Red = high acidity; yellow = low acidity	+	−
Vi	Voges-Proskauer reaction: detects the presence of acetyl-methylcarbinol, an end product of glucose fermentation	−	+
C	Ability to use citrate as a sole source of carbon	−	+

thin film that develops on surface of media on which motile bacteria are growing). The O antigen is one of the three regions of the lipopolysaccharide (endotoxin) component of the cell wall of gram-negative bacteria (see Table 2-12). K antigens are present on organisms that have capsules, and H antigens are present on motile organisms. Organisms, as in Figure 9-2, are serotyped on the basis of these three antigens (O, K, and H). Strains are named and classified; for example, *E. coli* O78:K80:H12 means that the strain has the 78 somatic antigens, 80 capsular antigens, and 12 flagellar antigens. Gram-negative bacteria contain hundreds of such antigens. In addition, all enterobacteria share a common antigen called the Kunin or enterobacteria-common antigen (ECA). ECA's role in pathogenicity is unknown.

Organisms can also be classified on the basis of phage or bacteriocin typing (Chapter 3). Many gram-negative bacteria produce bacteriocins (proteins coded for by plasmids), which can be bactericidal to other, closely related strains. Since bacteria are

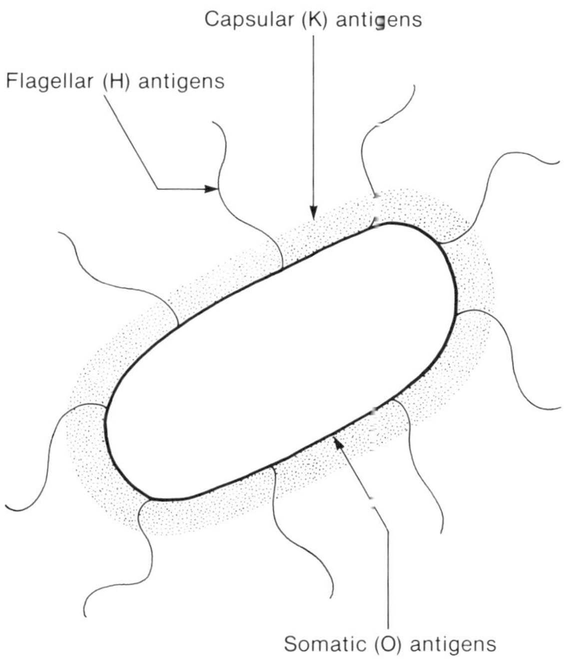

FIGURE 9-1. Antigens of gram-negative bacteria.

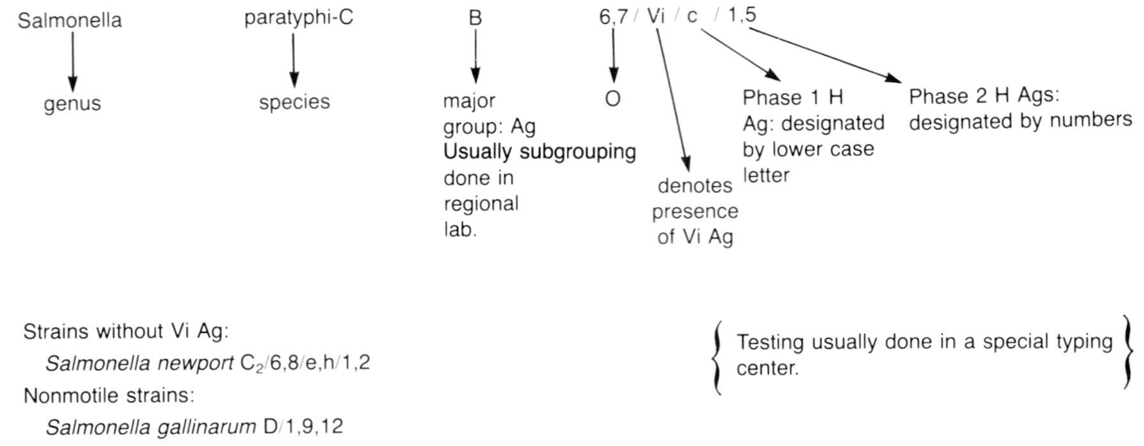

FIGURE 9-2. Typing *Salmonella.*

resistant to their own bacteriocins, these antimicrobial substances can be used to type strains. Bacteriocins have different names according to the type of organism producing them: *E. coli* produce colicins, pseudomonads produce pyocins, *Serratia marcescens* produces marcescins.

The fact that the enteric organisms share many common features and often have cross-reacting antigens is due in part to the active role of conjugation and transduction in genetic transfer between species. Thus, predominant serotypes vary within individuals and in certain geographic areas.

Because all gram-negative bacteria possess endotoxin, the clinical picture of endotoxic shock is very similar, regardless of the infecting organism (Table 9-4). The endotoxin is released when the bacteria lyse. It causes fever, poor perfusion of vital organs, shock, complement activation, and disseminated intravascular coagulation (DIC). Death often results despite antimicrobial therapy. Lymphocytes and macrophages play a primary role in host responses to endotoxin. The exact mechanisms by which endotoxin mediates host biologic responses in unknown. The fact that endotoxin is present in all gram-negative bacteria without causing severe consequences in the host indicates that other mechanisms of virulence besides endotoxin release are often present.

It is possible to detect the quantities of endotoxin in the body. The assay uses blood cells of the horseshoe crab species *Limulus polyphemus,* so the test is called the *Limulus* test. The test shows promise for use in diagnosing endotoxic shock and meningitis or sepsis caused by gram-negative bacteria. Repeated injections of small amounts of endotoxin in humans and animals have resulted in tolerance. Hence the development of an antitoxin is possible since the endotoxin produced by all strains of gram-negative bacteria is immunologically similar.

ENTEROBACTERIACEAE

Enterobacter

Enterobacter are motile and include five species: *cloacae, aerogenes, agglomerans,*

TABLE 9-4. Physiologic Effects of Endotoxin Release

Effect	*Mechanism*
Fever	Pyrogen released from neutrophils
Disseminated intravascular coagulation (DIC)	Activated factor XII (Hageman factor) in the clotting system
Hypoglycemia	Enhanced glycolysis
Impaired organ perfusion and shock	Dilatation of peripheral vascular system; increased vascular permeability
Hypotension	As above; release of vasoactive substances
Complement activation	Alternative pathway
Shwartzman reaction (in animals, I.V. injection of endotoxin 24 hours after intradermal injection causes necrosis at skin site; a second I.V. injection causes DIC and usually death)	RES unable to adequately remove subsequent doses of endotoxin; mechanism not completely understood
In small doses, increases resistance to radiation and infection	Enhanced activity of RES

sakazakii, and *gergoviae. Hafnia alvei* is also classified in this group. They are closely related to *Klebsiella* and are found in the intestinal tract, soil, water, and dairy products. *E. agglomerans,* which has 11 serogroups, has also been found with some frequency on hands, and is discussed in Chapter 10, "Commensals That Become Opportunistic Pathogens." The pathogenicity of *Enterobacter* is similar to that of other coliforms; they cause urinary tract and respiratory tract infections, sepsis as secondary or opportunistic pathogens, especially with hyperalimentation lines or instrumentation. About 69 percent of strains isolated in the United States are *E. cloacae,* 5 percent *E. aerogenes,* about 6 percent *E. agglomerans,* and the remainder are *Hafnia alvei* (which was formerly called *E. hafniae), E. sakazakii,* and *E. gergoviae. E. sakazakii* and *E. gergoviae* have been very recently described and very little is known about their clinical importance. Some strains of these species produce distinctive yellow pigments.

Escherichia coli

Biologic characteristics. *Escherichia coli* is an obligate gastrointestinal parasite that can survive only a short period of time outside the host. It is the major aerobic inhabitant of the colon. Most strains are motile. Over 160 O antigens, 50 H antigens, and 90 K antigens have been identified. There are three types of capsular antigens: about 30 L antigens, which are heat labile; about 26 A antigens, which are polysaccharide capsular material that is heat stable and can produce a Quellung reaction; and over 25 B antigens, which are heat labile like the L antigens but with the heated antigen still capable of absorbing antibody. Each *E. coli* strain may contain no K antigens or only one. Those possessing K antigens have increased virulence properties that include increased resistance to phagocytosis and antimicrobial agents. The synthesis of some K antigens is probably plasmid mediated.

Epidemiology. *E. coli* is the principle cause of about 20 percent of nosocomial infections, making the organism the leading cause of nosocomial infections in the United States. Two-thirds of these infections arise in the urinary tract (Fig. 9-3). *E. coli* is also the most frequently isolated organism from the urinary tract in infections that are not related to instrumentation or hospitalization. Enterotoxigenic or invasive strains are a common cause of childhood diarrhea in developing countries, as well as a common cause of "traveler's diarrhea" or "Montezuma's revenge." Though such strains seldom cause disease in the United States, outbreaks, primarily in newborn nurseries, have been documented since the 1940s. About 10 percent of diarrhea in infants is estimated to be caused by enterotoxigenic *E. coli* (ETEC).

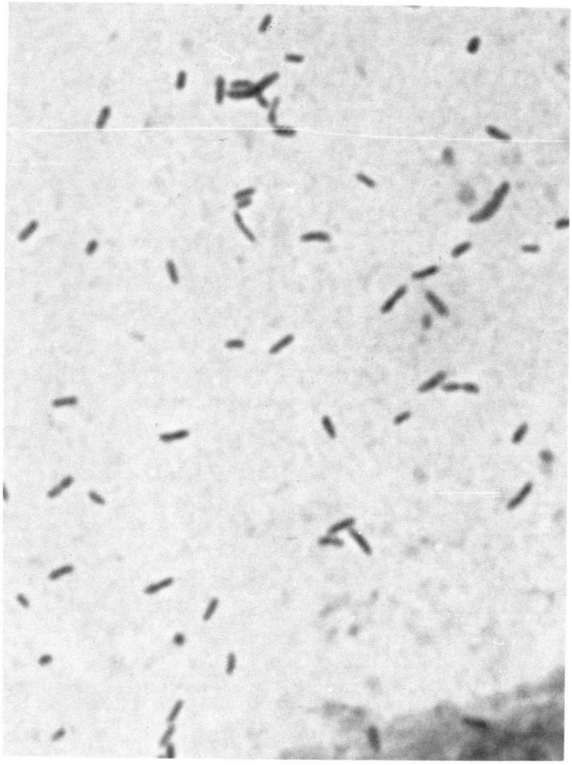

FIGURE 9-3. *E. coli* in urine. (Courtesy of Dr. James Plorde.)

Pathogenesis. *E. coli* is a part of the normal human gastrointestinal flora and does not usually cause disease. It can, however, cause pyogenic infections if introduced into other body sites. Extra-intestinal disease can occur, especially in the urinary tract, as a result of hematogenous, lymphatic, or direct spread of the *E. coli*. Most of the extra-intestinal disease is caused by a few—10 or less of the 160—serotypes, and none of these are enterotoxigenic strains.

E. coli may become pathogenic in the intestinal tract by two mechanisms: enterotoxin production or enteroinvasiveness. Enterotoxins are proteins that cause a secretory intestinal response. Each ETEC strain may produce one or both of two enterotoxins, a heat-labile (LT) and a heat-stable (ST) toxin that are compared in Table 9-5. To be fully virulent, ETEC must also possess surface pili, which aid in colonization of the bowel. These colonization factors related to the pili (CF I and II) as well as enterotoxigenicity and antibiotic resistance are plasmid mediated and can be transferred between strains. ST-producing *E. coli* can only be identified with an infant mouse assay, but LT can be detected in the feces with tissue culture and immunoabsorbent assays. The enterotoxin binds to receptor sites in intestinal cells and interferes with cellular metabolism, causing diarrhea and altered fluid and electrolyte balance. Hemolysins and colicins, also plasmid mediated, are produced by some strains. Colicins bind to re-

TABLE 9-5. Comparison of Heat-Labile (LT) and Heat-Stable (ST) Enterotoxins of *Escherichia coli*

Characteristic	*LT*	*ST*
Causes diarrhea	Yes	Yes
Destroyed by boiling for 30 minutes	Yes	No
Transferred by plasmid	Yes	Yes
Resembles *V. cholerae* toxin	Yes	No
Antigenic	Yes	No

ceptors on surfaces of other sensitive *E. coli* or related enteric strains and disrupt the cytoplasmic membrane. However, there is no strong correlation between pathogenicity and production of colicins or hemolysins.

Enteroinvasiveness is a separate mode of virulence possessed by a few strains of *E. coli,* but it is not common among strains found in the United States. It is probably genetically rather than plasmid mediated. Such strains can cause a severe dysentery like that caused by *Shigella* as a result of invasion of the intestinal mucosa. The mucopolysaccharide capsular antigen K1 is associated with increased invasiveness and strains that possess this antigen are frequently the cause of neonatal infections such as septicemia and meningitis. The K1 antigen cross-reacts with K antigens of group B streptococci and type B meningococci. Certain serotypes, not necessarily ETEC, are especially associated with neonatal outbreaks.

Host responses. Intestinal infections with *E. coli* are caused by ETEC or enteroinvasive *E. coli* (EIEC). In neonates the disease can be devastating, causing severe watery diarrhea, dehydration, and shock; outbreaks in nurseries occur intermittently. The organisms are transmitted usually by the fecal-oral route. Babies may be particularly susceptible to infection because they lack IgM protective antibodies (IgM antibodies do not pass the placental barrier). Winckel's disease is a neonatal hemorrhagic septicemia caused by *E coli*. The incubation period for diarrheal disease is 5 to 72 hours, depending on dose and strain virulence. These organisms can invade the blood stream and cause septicemia, meningitis, or pneumonia. Nonenterotoxigenic strains can also cause disease outside the gastrointestinal tract, especially in the urinary tract but at other sites such as the gall bladder, conjunctiva, and abdominal wounds as well (Fig. 9-4). *E. coli* is one of the most common causes of gram-negative meningitis and sepsis. *E. coli* is a major cause of urinary tract infection both in the general population and among hospitalized individuals.

Control and treatment. Immunity to individual serotypes develops, but since there

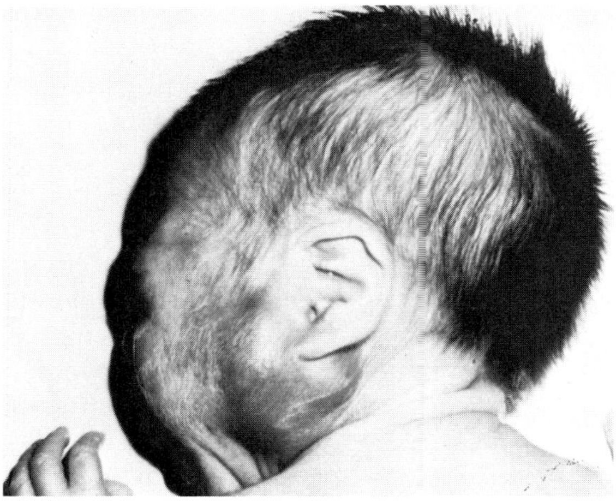

FIGURE 9-4. *E. coli* parotitis in a newborn. (Courtesy of University of Washington Clinical Microbiology Laboratory.)

are so many different serotypes, effective immunization is impractical. An antitoxin against enterotoxin is feasible, but is not currently in use. Adequate sanitary precautions can usually prevent spread of ETEC. The presence of *E. coli* in water supplies is used as an indicator of fecal contamination.

Outbreaks of diarrhea caused by *E. coli* in newborn nurseries have been controlled by implementation of strict handwashing and gloving procedures (see discussion in "Clinical Implications"). Infants with diarrheal disease are treated with oral colistin or neomycin though the efficacy of such therapy is unproven. Breast feading may provide infants with some passive protection due to IgA antibodies.

There is conflicting evidence regarding the efficacy of prophylaxis for traveler's diarrhea with streptomycin or doxycycline. Even if there is some benefit, the emergence of drug-resistant strains is of major concern. Agents such as diphenoxylate (Lomotil) and paregoric slow intestinal motility, interfering with one of the body's important defense mechanisms. They are therefore not recommended as prophylaxis for travelers. Bismuth subsalicylate (Pepto-Bismol) is also used by many travelers as prophylaxis against diarrhea. It may be of some use in blocking the action of enterotoxin.

Patients with urinary tract infection or systemic disease due to *E. coli* may receive any antimicrobial agent to which the organism is susceptible. Unless R plasmids are present, most *E. coli* are very sensitive to antimicrobials such as ampicillin, chloramphenicol, erythromycin, neomycin, tetracycline, streptomycin, and sulfonamides. Resistance, unfortunately, is becoming more prevalent, with more than half of strains now resistant to at least two major drugs. In one report from India, 80 strains of ETEC were examined for the presence of R factors. They were present in 63.7 percent. Another group studied antimicrobial resistance among *E. coli* isolated from normal individuals, hospital personnel, and hospitalized patients. Resistance was found in 38, 54, and 76 percent respectively. R plasmids were found in 77.3, 58.3, and 62.3 percent with more variety among hospitalized patients (Dutta et al. 1980; Echeverria and Murphy 1980; Zintel and Knox 1967).

Klebsiella

Biologic characteristics. *Klebsiella* species include *K. pneumoniae*, *K. oxytoca* (an indole-positive *K. pneumoniae*), *K. ozaenae*, and *K. rhinoscleromatis*, which is quite rare. *Klebsiella* are distinctive from *Enterobacter* in being nonmotile. They usually contain a large capsule and hence produce large mucoid lactose-fermenting colonies (Fig. 9-5). *Klebsiella* have O and K antigens. There are five O serogroups but the capsule hinders interaction between the O antigens and

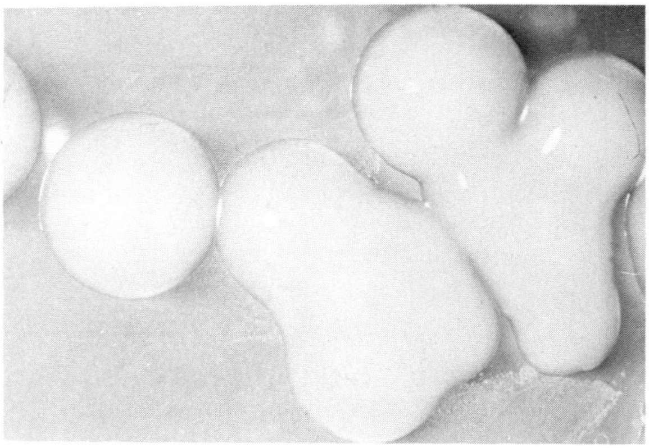

FIGURE 9-5. *Klebsiella* colonies on MacConkey agar. Large, mucoid colonies are due to capsular material. (Courtesy of Dr. James Plorde.)

antisera, so the organisms are not routinely typed by somatic antigens. There are about 70 polysaccharide capsular antigens that are identified by precipitation tests to make the capsule visible. Swelling of the capsule (Quellung reaction) does not occur. Virulence is not correlated with capsule type.

Epidemiology. Almost all (95 percent) clinical isolates are from the species *K. pneumoniae,* which is present in the intestinal and respiratory tracts of about 10 percent of healthy persons. *K. pneumoniae* causes about 0.5 to 5 percent of all pneumonia in developed countries. Members of the tribe Klebsielleae are found to frequently colonize the oropharynx and nasopharynx of the elderly or debilitated individual. Such colonization may lead to pneumonia or bacteremia. About two-thirds of *Klebsiella* infections are nosocomial, and type-specific capsular antisera as well as bacteriocin (klebocin) typing can be used in investigations of their source. *Klebsiella* and *Enterobacter* account for about 20 percent of nosocomial infections. In hospitals, food as well as personnel hands have been identified as modes of transmission for *Klebsiella.*

Pathogenesis. Some species produce a heat-stable toxin, but its role in pathogenesis is unknown. Its virulence is primarily due to the protective effect of its large capsule against phagocytosis.

Host responses. Like the other coliform, *Klebsiella* are normal flora and do not usually cause disease. However, *Klebsiella* are associated with upper respiratory disease and other septic complications of hospitalization (abscesses, meningitis, wound infections), especially as secondary invaders among individuals with chronic respiratory disease. They rarely cause septicemia. *Klebsiella* pneumonia causes multiple areas of lung necrosis and, eventually fibrosis and cavitation. There is considerable alveolar destruction, abscess formation, and bronchiole obstruction. Sputum production is profuse with thick, tenacious secretions. Even with antimicrobial therapy, lung damage results and may require surgical drainage or pulmonary resection. Case fatality rate without treatment is very high. Urinary tract infections are also common; *Klebsiella* is second to *E. coli* as the most common cause. *K. ozaenae* causes ozena and *K. rhinoscleromatis* causes rhinoscleroma, both rare destructive diseases of the nose. The strains that produce an enterotoxin have been isolated from persons with diarrhea and tropical sprue.

Control and treatment. Most antimicrobials except penicillin are effective against *Klebsiella.* Eighty percent of *Klebsiella* strains are sensitive to cephalosporins, whereas *Serratia* and *Enterobacter* usually are not.

Proteus-Providencia

There are six species in this group of highly motile gram-negative rods: *Proteus mirabilis, vulgaris,* and *morgani,* and *Providencia stuartii, rettgeri,* and *alcalifaciens.* They are part of the normal bowel flora and are also found in water, soil, and sewage. *Proteus* species have several unique characteristics. They produce urease, which allows them to use urea as an energy source. This results in production of ammonia. An alkaline pH of urine in an inidividual with a urinary tract infection usually indicates that the causative organism is *Proteus.* The urease also increases invasiveness of the organisms. Hence, persons with urinary tract infection due to *Proteus* are especially prone to the formation of urinary stones. *P. vulgaris* and *mirabilis* have numerous peritrichous flagella and grow on agar plates over the entire surface of the medium (Fig. 9-6). This characteristic swarming growth covers all other colonies that are growing, making it difficult for laboratory personnel to isolate other organisms. Both *Proteus* and *Providencia* cause primarily urinary tract infections, but wound infections, pneumonia, sepsis, possibly gastroenteritis, and other serious infections have been reported. *P. mirabilis* are the only enteric organisms that are commonly susceptible to penicillins. However, susceptibility testing is important as multiple resistance is common.

Serratia

Serratia are motile gram-negative rods in the same family as *Klebsiella* and *Enterobacter*. They generally do not ferment lactose. There are three species of *Serratia: S. marcescens, rubidea,* and *liquefaciens*. More than 90 percent of clinical isolates are *S. marcescens*. *Serratia* live primarily in water and soil, and do not normally inhabit humans. Some strains produce a characteristic bright red pigment, but this is unusual in those clinically isolated from patients. In the past, reports of miraculous "bleeding" from paintings and communion bread have probably been due to the production of red pigment by multiplying *Serratia*. *S. mar-*

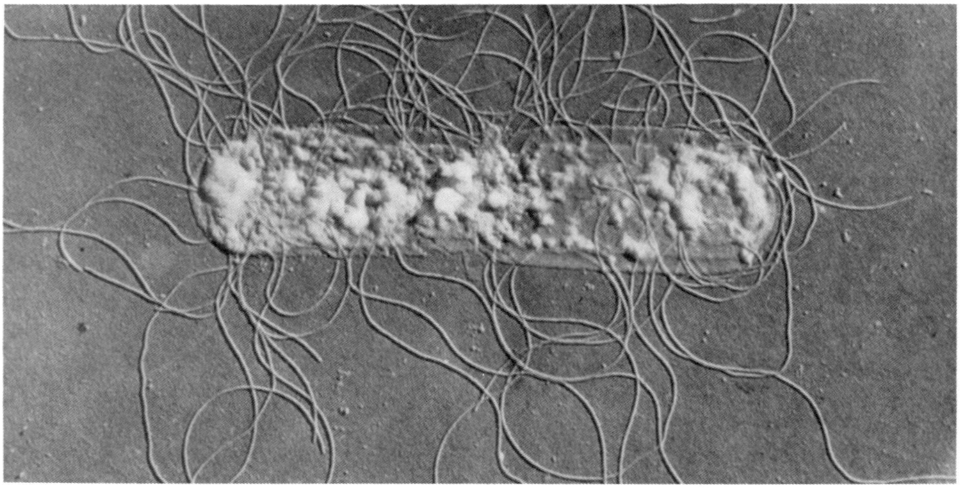

FIGURE 9-6. Electron micrograph of shadow-cast preparation showing the flagella of *Proteus vulgaris.* The bacteria were grown to the swarming stage on agar, refrigerated at 5°C for 16 to 20 hours, floated off in 5 percent formalin, and washed twice before mounting on collodion. The preparation is very transparent and flattens completely so that certain of the internal structures are demonstrable by shadowing. The flagella originate in small, spherical bodies about 100 nm in diameter, and their remarkably uniform size is clearly shown (van Iterson). (Reprinted with permission of the publisher from Freeman BF. Burrow's Textbook of Microbiology, 21st edition. Philadelphia: WB Saunders, 1979.)

cescens has 15 O and 16 H antigens known. It can be typed with bacteriophages or marcesins.

The organism is a significant opportunistic pathogen and common cause of urinary tract as well as other types of infections. Epidemics of infection caused by multiresistant strains of *Serratia marcescens* that spread between hospitals have been reported. *Serratia* are generally quite resistant to antimicrobials, again as a result of plasmid transfer. Gentamicin, carbenicillin, kanamycin, amikacin, or a combination of polymyxin B and rifampin may be used. Endocarditis, especially in drug abusers, septic arthritis, osteomyelitis, septicemia, and respiratory tract infections due to *Serratia* species occur.

Salmonella

Biologic characteristics. Unlike the enteric bacteria described previously, *Salmonella* as well as *Shigella* and *Vibrios* are *not* normal flora. Their isolation from human specimens indicates disease. Currently there are three recognized species of *Salmonella: S. choleraesuis, S. typhi,* which causes typhoid fever, and *S. enteritidis,* which has more than 1600 serotypes. Most of the organisms identified are not separate species but rather serotypes of *S. enteritidis* (e.g., *S. typhimurium, heidelberg,* and *newport*). About 95 percent of human isolates belong to serogroups A through E. Because of their highly complex antigenic structure, only certain typing centers can completely identify strains; most laboratories merely identify the group. *Salmonella* are motile with peritrichous flagella. In addition to H and O antigens, many strains have a specialized capsular antigen called "Vi" (virulent) antigen, which incurs increased virulence on those strains possessing it. Each O serogroup can be subclassified by one of two types of flagellar antigen: phase 1 H antigens, which are shared between a few species, and phase 2 H antigens, which are less specific. Typing schemes are illustrated in Figure 9-2.

Epidemiology. *Salmonella* are pathogens of humans and animals. *Salmonella* gastrointestinal infections occur especially in turtles, dogs, cats, parakeets, and many types of fowl. *Salmonella* are almost always acquired by ingestion, and almost any food product can be a source. As much as half of raw meats are contaminated with *Salmonella* and thus are a major source of human infection. In England from 1975 to 1979, turkeys were the vehicles of infection for 46 percent of food-borne outbreaks with one serotype of *Salmonella* (Centers for Disease Control 1980e). Eggs and egg products are also a common source, as are components of animal feed that perpetuate the infection among domestic animals. Because of the high prevalence of salmonellosis among small turtles, they are not sold commercially as pets. As many as 20 percent of *Salmonella* infections in some regions have been attributed to such pets.

Although most infections are acquired by eating or drinking contaminated items, person-to-person spread in dormitories, hospitals, and nurseries has occurred. Cases have also been traced to fomites such as contaminated suction tubing. In 1979, 31,123 isolates of *Salmonella* were received at the Centers for Disease Control (CDC), a slight increase over previous years. The rate was highest in infants aged 2 to 3 months (See Fig. 9-7). About 20,000 cases of *Salmonella* gastroenteritis are reported each year, though the incidence is probably closer to 100,000.

Only humans develop typhoid fever (*S.*

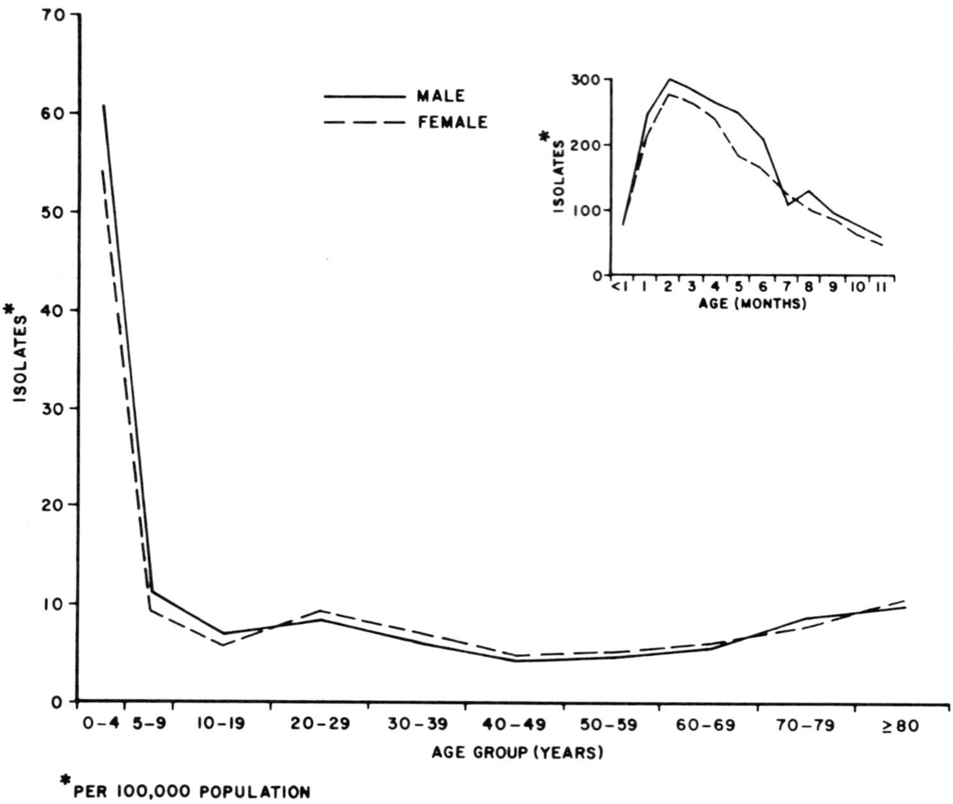

FIGURE 9-7. Rate of reported isolates of *Salmonella,* by age, United States, 1979. (Reprinted from Centers for Disease Control. Salmonella isolates—1979. Morb Mort Week Rep 1980;29:189-190.)

typhi infection), with sick persons or carriers serving as the only reservoir. Feces, which contain millions of organisms during the acute illness, serve as the major source of contamination of food or water, although the organism is also occasionally present in other body secretions. Raw seafood, contaminated ice, and even dust can contain sufficient organisms to cause disease. About 200 cases of typhoid fever per year are reported in the United States and in Great Britain. There are another approximately 200 known carriers of *S. typhi* each year in the United States. The incidence of typhoid has been steadily decreasing in developed countries where sanitation and public health measures control fecal contamination of potable (drinking) water, animal feeds, and human food. About 50 percent of cases in the United States are acquired abroad. The gallbladder in about 3 percent of cases becomes a site of bacterial multiplication so that organisms continue to be excreted from the gastrointestinal tract for long periods after recovery from the disease. Individuals who are chronic biliary carriers are usually without symptoms despite the fact that millions of organisms are shed in the stool. The majority of carriers are women over 50 years old. This carrier state can persist for months or years and may necessitate cholecystectomy to eradicate.

From 1968 to 1974 almost 270,000 isolates of *Salmonella* from blood and stool samples were received by the CDC. The most common serotypes were *S. typhimurium* (20.1 percent), *S. typhi* (16.3 percent), *S. heidelberg* (13.6 percent) and *S. enteritidis* (7.6 percent). In India, the most common type isolated from humans is *S. typhi*.

Pathogenesis. Salmonellosis results when organisms are ingested and invade the mucosa in the intestine. The organisms can survive in phagocytic cells, protected from antibody. *S. enteritidis* serotype *typhimurium,* the most frequent cause of *Salmonella* gastroenteritis, produces an enterotoxin that may be associated with virulence, but this has not been proven. Organisms may remain in the gut, or they can enter the lymphatic system and become disseminated throughout the body. In the lymphoid tissue, hyperplasia and necrosis occur. Hepatitis, liver necrosis, and localization and inflammation in the gall bladder often result from spread of organisms. A large infecting dose is necessary to produce typhoid fever: about 10^7 bacteria were necessary to cause disease in 50 percent of normal volunteers. The normal gut flora clearly plays a protective role against *Salmonella* infection, since individuals who have recieved antimicrobial therapy are at much greater risk of developing clinical salmonellosis.

Host responses. *Salmonella* are pathogenic for humans and many animal species and enter the host by the oral route. There are four clinical syndromes that result from *Salmonella* infection: acute gastroenteritis, enteric fevers, septicemia, and localized infection, as listed in Table 9-6. The most common form is gastroenteritis, which occurs from 8 to 48 hours after ingestion of contaminated food. *Salmonella* "food poisoning" may be mild or very severe with predominant symptoms being nausea, vomiting, diarrhea, and fever of 1 to 4 days duration. These symptoms are due to irritation of mucous membranes in the small and large bowel. The diarrhea is usually not bloody, and the organism can be cultured from the stool but only rarely from the blood.

TABLE 9-6. Clinical Manifestations of *Salmonella* Infection

Major syndrome	*Incubation period*	*Predominant symptoms*	*Duration*	*Common serotypes*
Acute gastroenteritis	8-48 hours	Nausea, vomiting, diarrhea, fever	1-4 days	*S enteritidis,* especially serotype *typhimurium*
Enteric fever	7-20 days	Fever, abdominal tenderness, rash, prostration, bacteremia	1-3 weeks	*S. typhi, paratyphi*
Septicemia	Variable	Spiking fever	Weeks to months	*S. cholerasuis*
Localized infections	Variable	Abscess formation	Weeks	Any serotype

Certain serotypes, especially *S. typhi,* if ingested in large enough doses, can cause enteric fever. The classic example of this is typhoid. Typhoid fever has a longer incubation period, 7 to 20 days, with an insidious onset of malaise, headache, and fever. The patient may suffer from extreme exhaustion and a rash on the trunk (called "rose spots") that lasts for a few days. Abdominal tenderness may be present due to accumulation of organisms, hyperplasia, and inflammation of Peyer's patches and other lymphoid tissues. This inflammation may also lead to gastrointestinal hemorrhage and bowel perforation in fatal cases. Enlargement of the spleen and bacteremia are common. Usually the disease lasts 2 to 3 weeks, though the organism may be excreted in the stool for several months (or longer with chronic carriers). Enteric fevers caused by *S. paratyphi* or other serotypes are milder and of shorter duration. The *Widal test* is a serologic test for agglutination in two serum samples, one early in disease and the second about 10 days later, to detect antibody rise.

Primary bacteremia caused by *Salmonella* frequently occurs without preexisting intestinal symptoms, especially among persons who are debilitated. The major symptoms are prolonged spiking fever and positive blood cultures, though organisms can be isolated only intermittently rather than constantly as in typhoid fever. Other symptoms of enteric fever are lacking and stool cultures are negative for *Salmonella.* In about 25 percent of cases localized foci of infection such as osteomyelitis, empyema, endocarditis, pyelonephritis, or meningitis occur. *Salmonella* are common secondary pathogens in patients with parasitic or other bacterial infections. *S. choleraesuis* is the most common species responsible for *Salmonella* septicemia.

Localized pyogenic infections can develop at almost any body site. They can occur without any other symptoms, or can be the result of bacteremia. Septic arthritis occurs and meningitis is most frequent among neonates. The clinical picture of *Salmonella* abscess resembles that of abscesses of any cause and diagnosis must be made by isolation of the organisms.

Control and treatment. Control of salmonellosis is by proper food handling and preparation, adequate sanitation of potable water, proper sewage disposal, milk pasteurization, and exclusion of carriers from food handling. It is theoretically possible to eradicate typhoid fever since humans are the only hosts and the disease can be controlled by adequate sanitary measures to prevent fecal contamination of food or water. Immunity results from the development of antibodies against O and Vi antigens. The anti-O antibody helps to opsonize *Salmonella* and enhance their phagocytosis; nonspecific cellular immunity also appears to play an important role. An attenuated parenteral typhoid vaccine is efficacious for use in travelers, contacts of cases, and military personnel. A new oral typhoid vaccine, tested in more than 32,000 children, has been found to provide 95 percent protection and many fewer side effects than the parenteral vaccine. It also has the advantange of stimulating gastrointestinal mucosal immunity (Wahden et al. 1982).

Salmonella gastroenteritis does not generally require antimicrobial therapy. On the contrary, use of antimicrobial agents is contraindicated since it increases the probability that resistant strains or the carrier state will emerge. The major treatment is the prevention of dehydration and electrolyte imbalances. The other clinical syndromes can be treated with chloramphenicol, ampicillin, or trimethoprim-sulfamethoxazole. Recently, however, a strain of *S. typhi* in the bowel of a patient with enteric fever developed resistance to both chloramphenicol and trimethoprim-sulfamethoxazole after the patient was treated with both. This plasmid-mediated antimicrobial resistance was shared by a *Klebsiella* also isolated from the same patient (Datta, Richards, and Datta 1981). All patients during acute phases of gastrointestinal illness should be isolated with enteric precautions since huge numbers of *Salmonella* are shed in the feces. Sometimes the chronic carrier state can be eradicated with ampicillin, but more often cholecystectomy is necessary. Carriers of the typhoid bacteria should not handle food.

Shigella

Biologic characteristics. There are four species of *Shigella* and 32 subtypes (Table 9-7). *S. dysenteriae* type 1 causes the most serious disease with high mortality and morbidity, but is very uncommon in the United States. *S. flexneri* and *sonnei* are the most common species; *S. boydii* infections are rare. Some *E. coli* strains (previously called *S. dispar*) cross react with *Shigella* antisera, but do not cause dysentery. *Shigella* are nonmotile and grow best aerobically. They have lipopolysaccharide O antigens, no flagellar antigens, and some smooth (S) strains have K antigens. Serologic agglutination or bacteriocin typing are used to identify subtypes. Characteristics of *Salmonella* and *Shigella* are compared in Table 9-8.

Epidemiology. Unlike *Salmonella, Shigella* have as their natural hosts almost exclusively humans and occasionally other higher primates. Transmission is by the fecal-oral route with vectors such as flies, as well as contaminated food and contaminated water used for swimming. Shigellosis is most common in children, 50 percent of cases occurring before age 4. The disease is worldwide in distribution, but is most common in areas where sanitary conditions are sub-optimal. Major epidemics in the United States occur very infrequently, except in institutions (Fig. 9-8). Most cases occur among travelers to foreign countries, especially to Mexico, Central America, and South America. There are about 15,000 cases per year

TABLE 9-7. Classification of *Shigella*

Species	*Serologic group*	*Subtypes*	*Frequency*	*Order of disease severity*
S. dysenteriae	A	10	Rare in U.S., common elsewhere	1
S. flexneri	B	6	Most common species in U.S.	2
S. boydii	C	15	Rare	3
S. sonnei	D	1	Very common in U.S.	4

TABLE 9-8. Comparison of *Shigella* and *Salmonella*

Factor	*Salmonella*	*Shigella*
Serotypes	More than 1600	32
Invasiveness	High; bacteremia common	Low; bacteremia rare
Clinical manifestations	Gastroenteritis, enteric fever, local infections, septicemia	Dysentery
Blood and mucus in stool	Very rare	Very common
Natural hosts	Humans, many cold- and warm-blooded animals, fowl and other birds	Almost exclusively humans
Mode of transmission	Fecal-oral; contaminated food or drink	Same
Reported annual U.S. incidence	About 20,000	About 15,000
Chronic carrier state	Yes	No
Size of inoculum to cause disease	Large, about 10^7	Small, about 500

reported in the United States. In 1979, the distribution of species causing shigellosis was:

S. sonnei	73.4 percent
S. felxneri	24.2 percent
S. boydii	1.8 percent
S. dysenteriae	0.7 percent

Pathogenesis. All species of *Shigella* are pathogenic. The infecting dose can be small; 200 organisms are sufficient. *Shigella* are much less invasive than *Salmonella. Shigella* penetrate the lining of epithelial cells and submucosa of the large bowel, but do not invade beyond the submucosa. Mucosal ulceration is produced. Epithelial cells are killed by enterotoxin and by the accumulation of metabolic products. Similar enterotoxins are produced by all species. *S. dysenteriae* also produces a distinct exotoxin that inhibits protein synthesis. It acts as a neurotoxin, causing death in laboratory animals. However, its role in human disease is not defined. Invasiveness is a more important determinant of virulence than is toxin production since noninvasive toxigenic strains do not cause disease while invasive nontoxigenic strains do.

Host responses. Shigellosis can be mild or it can be fatal. It is also called *bacillary dysentery* to distinguish it from amoebic dysentery. After an incubation period of 1 to 4 days, the usual course of disease is a sudden onset of severe dysentery with blood and mucus in the stool. Abdominal cramps are usually more predominant than with salmonellosis, and fever is also present. Symptoms usually subside after about 1 week. Though *Shigella* may be shed in the feces for weeks, a chronic carrier state such as that seen in salmonellosis is not present. Carriage is almost always for less than 1 month, with mean excretion time being

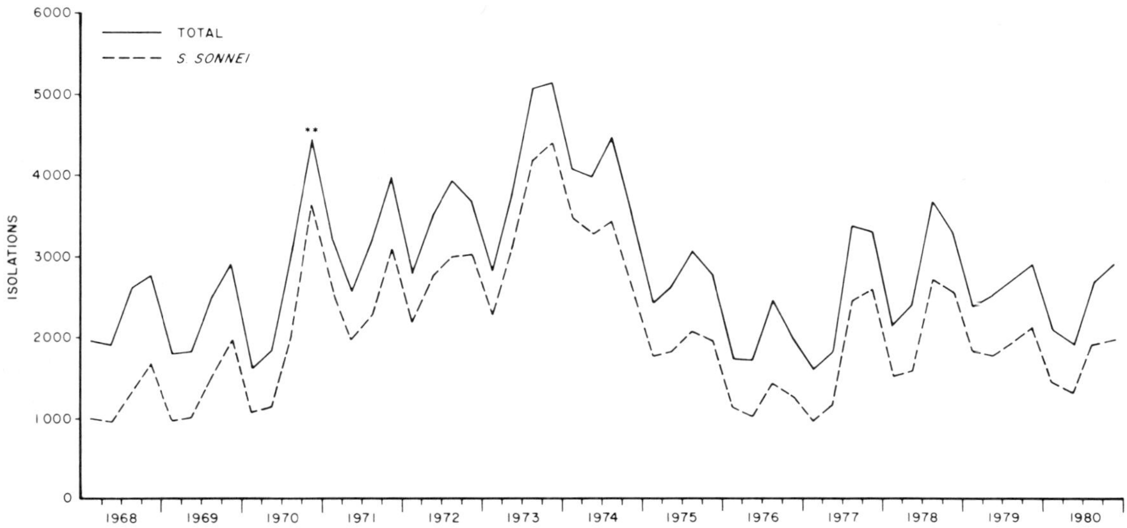

*NO REPORTS FROM CALIFORNIA OR THE VIRGIN ISLANDS AFTER 1969
**APPROXIMATELY 400 ISOLATIONS IN AUGUST 1970 COMMON-SOURCE OUTBREAK IN HAWAII

FIGURE 9-8. Reported isolations of *Shigella* from humans by quarter, United States, 1968 to 1980. (Reprinted from Centers for Disease Control. Annual Summary 1980. Morb Mort Week Rep 1981;29:76.)

about 27 days. The disease is generally self-limiting, but complications such as seizures (in about 10 percent of children), shock, and hyponatremia (low serum sodium) can occur from fluid and electrolyte loss. Rarely, arthritis, nephritis, disseminated intravascular coagulation (DIC), and even pneumonia have been reported. Mortality up to 20 percent has occurred with especially virulent strains of *S. dysenteriae*.

Control and treatment. Primary prevention involves improvement in sanitary conditions, prevention of fecal contamination of food and water, and elimination of vectors, especially flies. Travelers to endemic areas should avoid fresh unwashed food such as fruits and vegetables. Because *Shigella* rarely invade beyond the intestinal tract, there is minimal circulating antibody in individuals who have recently recovered from shigellosis. However, antibody is excreted in the feces. Reinfections occur, but a higher infecting dose is usually required. Though no vaccine is currently available, a live oral vaccine of noninvasive strains of *Shigella* is being developed. An antitoxin against *S. dysenteriae* toxin is available but is not in general use.

Early diagnosis is important. However, *Shigella* die quickly in the feces. Hence, the best culture methods are by rectal swab or by obtaining specimens from lesions by sigmoidoscopy. This is possible since many of the lesions are in the terminal ileus. Resistance transfer factors (plasmids) were first identified in *Shigella* in the 1950s. About half of *Shigella* isolates in the United States are now resistant to ampicillin.

For the same reasons as with *Salmonella*,

antimicrobial therapy of *Shigella* is not indicated except for very severe cases. Then the drug of choice is still ampicillin 100 mg/kg of body weight/day I.V. for hospitalized patients or 50 mg/kg of body weight/day orally for 5 days for outpatients. However, antimicrobial susceptibility testing should be done. Alternative drugs are chloramphenicol or trimethoprim-sulfamethoxazole. In Korea, *Shigella* are frequently resistant to these drugs as well. Gastrointestinal hypermotility and gastric acidity are important as host defenses against bacillary dysentery. Hence, drugs that decrease the diarrhea or stomach acidity are contraindicated and will make the disease worse. Likewise, chemoprophylaxis is not useful for contacts due to potential emergence of resistant strains. Hospitalized patients with shigellosis require isolation. Gloves should be worn for handling feces or for close contact with the patient during the acute phase of the disease.

Unusual Enterobacteriaceae

There is one species of *Edwardsiella, E. tarda,* which closely resembles *E. coli* but is much less common. It only rarely causes diseases such as meningitis or sepsis. *Citrobacter* are discussed in Chapter 10, "Commensals that Become Opportunistic Pathogens." *Arizona hinshawii* is closely related to *Salmonella,* is found rarely in the human gastrointestinal tract, and can be a rare cause of gastroenteritis and sepsis. At least one case of peritonitis has also been reported. Ingestion of contaminated dairy products and food handlers who are intestinal carriers of *Arizona* have been implicated in outbreaks. Ampicillin and chloramphenicol are drugs of choice for severe outbreaks.

PSEUDOMONAS

Pseudomonas are ubiquitous organisms that normally are harmless commensals. Though there are numerous species of *Pseudomonas* (Table 9-9), most human disease is caused by *P. aeruginosa. P. maltophilia, cepacia,* and *pseudomallei* also occur with some frequency in clinical specimens. Pseudomonads are motile by means of one or more polar flagella. They are obligate aerobes (not facultative, as are the Enterobacteriaceae), and grow readily on simple media. All species except *P. maltophilia* are oxidase positive. Most strains of *P. aeruginosa, P. fluorescins,* and *P. putida* produce water-soluble pigments that fluoresce under ul-

TABLE 9-9. Species of *Pseudomonas* Found in Humans

Species	*Pathogenic to Humans*
P. acidovorans	Probably not
P. aeruginosa	Commonly
P. alcaligenes	Rarely
P. cepacia	Rarely
P. diminuta	Questionable
P. fluorescins	Probably not
P. mallei	Occasionally
P. maltophila	Occasionally
P. maltophilia	No
P. multivorans	Occasionally
P. paucimobilis	No
P. pertucinogena	No
P. pickettii	No
P. pseudoalcaligenes	Probably not
P. pseudomallei	Yes
P. putida	Possibly
P. putrefaciens	Probably not
P. stutzeri	Rarely
P. testosteroni	Probably not
P. thomasii	Probably not
Pseudomonas sp. (numerous species)	Possibly

traviolet light. Pseudomonads are found in the water and soil, on plants, and to a lesser extent in the human gut and on skin. In hospitals these species are significant opportunistic pathogens and their propensity for growth in water sources (I.V. fluids, nebulizers, sinks, rinsing solutions, dialysate) has resulted in the nickname "water bugs." Since most species are highly resistant to conventional antimicrobial agents, the treatment of *Pseudomonas* infections is sometimes difficult.

Pseudomonas aeruginosa

Biologic characteristics. *Pseudomonas aeruginosa* produces several water-soluble pigments such as pyocyanin (blue) and fluorescein (green), which give its colonies a distinctive appearance and odor. These pigments also give rise to blue-green pus in wounds. It is the only gram-negative rod known to excrete pyocyanin. Because the species can use a large variety of carbon sources, it can multiply in distilled water and in the presence of moisture with just trace amounts of organic compounds. Serologic tests, phage and pyocin typing are used as epidemiologic tools to trace sources of infection. *P. aeruginosa* grows on simple laboratory media. A few strains have capsules.

Epidemiology. *P. aeruginosa* is a common inhabitant of soil and water throughout the world. It is also present in the bowel and on moist areas of skin in about 10 percent of persons. Moist environments are an important source of the organism, including eye drops, flower vase water, bathtubs, medications, incubators, antiseptic soaps, lotions, and solutions. Infections are usually among hospitalized patients who are debilitated, drug abusers, neonates, and the aged, and those who have received antimicrobial therapy, who are receiving invasive monitoring, or who have suffered extensive burns or injury. Children with cystic fibrosis often have encapsulated strains isolated from their sputum. *Pseudomonas* can be transmitted from person-to-person by human carriers as well as via fomites.

Pathogenesis. The organism may enter by the skin, directly into the blood stream with intravenous therapy, by the gastrointestinal tract, by the umbilical cord, or by the urinary tract. *P. aeruginosa* produces several toxic substances (listed in Table 9-10), which play a part in its pathogenesis. Exotoxin A, the action of which is similar to that of diphtheria toxin, is thought to be the most important component in toxigenicity. Injection of *P. aeruginosa* exotoxin into dogs causes death; many of the proteolytic enzymes elaborated produce hemorrhagic lesions when injected into rabbits.

Host responses. As mentioned, *P. aeruginosa* infections, like those caused by coliforms, are most likely to occur in patients who are debilitated, immunosuppressed, or otherwise have lowered defenses. The or-

TABLE 9-10. Toxic Substances Elaborated by *Pseudomonas aeruginosa*

Deoxyribonuclease
Elastase
Endotoxin common to all gram-negative bacteria
Esterase
Exotoxin A (inhibits protein synthesis)
Exotoxins, two others
Extracellular surface slime
Hemolysins
Lecithinase
Lipase
Phospholipase

ganism has caused epidemic diarrhea (in newborns), eye infections with corneal perforation, urinary tract infections, otitis media, mastoiditis, skin and wound infections, pneumonias, meningitis, endocarditis, and septic shock, especially in burn and trauma victims (Figs. 9-9, 9-10). Reports of skin infection due to *Pseudomonas* following use of contaminated hot tubs are increasing in frequency as hot tubs gain in popularity. Mortality following *Pseudomonas* bacteremia is about 75 percent.

Control and treatment. Preventive techniques in hospitals include careful attention to cleansing and monitoring of plumbing and humidifying equipment, bedpan cleaning devices, and other items that are moist. Active vaccination of guinea pigs with lipopolysaccharide *Pseudomonas* vaccine has proven to be protective against pneumonia and shows promise for use in humans. Hyperimmune globulin is also used in patients with systemic infections. Skin infections can be irrigated with 1 percent acetic acid, or topical colistin or polymyxin B can be applied. It is essential that wounds be kept debrided and clean. *P. aeruginosa* is notably resistant to antimicrobials and susceptibility testing is essential before treatment is

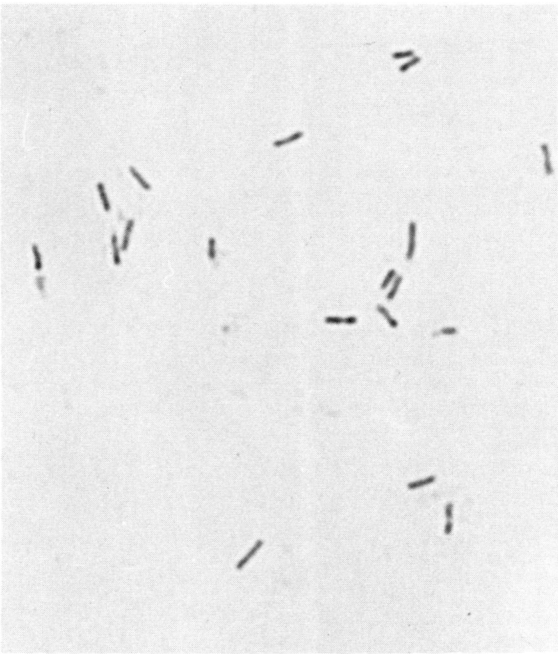

FIGURE 9-9. *Pseudomonas* species in urine. (Courtesy of Dr. James Plorde.)

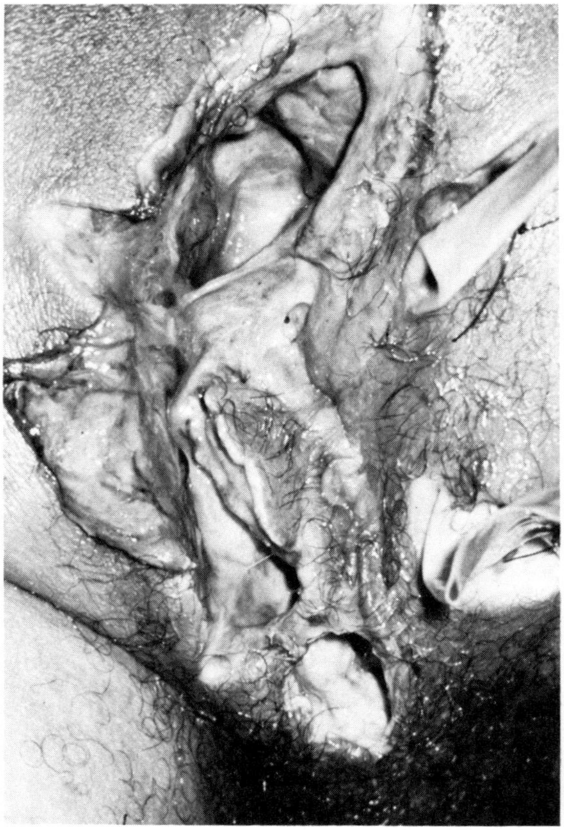

FIGURE 9-10. Perirectal abscess containing *Pseudomonas, Serratia, Peptostreptococcus* (aerobic and anaerobic organisms). (Courtesy of Dr. James Plorde.)

begun. Polymyxin B, colistin, carbenicillin, amikacin, tobramycin, or gentamicin may be used for susceptible strains.

Other *Pseudomonas* Species

Many other species of *Pseudomonas* are capable of causing diseases similar to those caused by *P. aeruginosa*. *P. maltophilia* is the second most common species isolated from clinical specimens (Fig. 9-11). *P. alcaligenes* and *pseudoalcaligenes* have been recovered from ponds, rivers, swimming pools, cosmetics, raw milk, and frozen fish, as well as from humans. They have been associated with empyema, eye infections, pneumonitis, and a few other clinical syndromes. Some of these species tend to be more sensitive to antimicrobial agents than does *P. aeruginosa*.

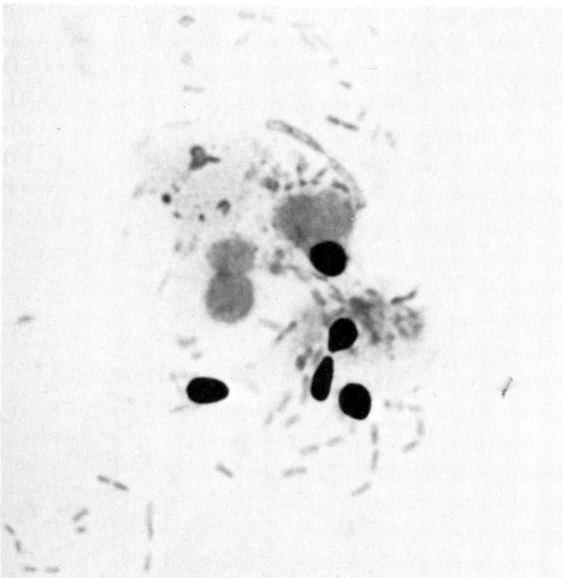

FIGURE 9-11. Sputum containing *Candida albicans* (a yeast) and *Pseudomonas maltophilia*. Such organisms usually cause clinical disease only in patients who are immunocompromised, debilitated, or receiving broad-spectrum antimicrobial agents. (Courtesy of Dr. James Plorde.)

P. mallei, a nonmotile species, causes a disease of horses called *glanders*. Glanders is occasionally transmitted to humans through skin abrasions or by inhalation, and can then be transmitted person-to-person. The disease has been eradicated in North America with no cases since 1938, but it occurs in Africa, Asia, South America, and the Middle East. It is often fatal, though it can be treated with sulfonamides. The incubation period is 1 to 5 days if infection is through the skin and 10 to 14 days if organisms are inhaled. If the organism enters by the respiratory tract a primary pneumonitis results. If the route of entry is the skin, necrotic nodules of the skin, lymphatics, and other organs ensue. Human cases must be isolated to prevent person-to-person spread.

P. pseudomallei is a small motile species isolated most often in southeast Asia. It causes *melioidosis* (meaning "a resemblance to distemper of asses"), which has a similar clinical picture to glanders. The only cases seen in the United States have been among those individuals (soldiers, for example) who have traveled in Vietnam, Cambodia, Laos, Thailand, and adjacent countries. The organism is found primarily in soil and causes disease in animals as well as humans. The primary route of entry is through skin abrasions (people working in rice paddies are at high risk). Ingestion and inhalation, as well as rare person-to-person and sexual transmission, are also possible. The incubation period may be as short as 2 days or the organism may remain latent for years before becoming clinically manifest. Subclinical infection occurs in about 10 percent of cases. Signs and symptoms are variable and include acute or chronic pustule formation, local abscesses, pneumonitis, septicemia, pharyngitis, diarrhea, or other manifestations mimicking various infectious syndromes. Diagnosis can be confirmed by isolation of the organism from lesions (the colonies have a characteristic wrinkled appearance after about 72 hours of incubation) and by serologic tests. Long-term antimicrobial therapy, sometimes for months, may be required. Untreated mortality is about 95 percent; even with treatment about 50 percent of individuals with septicemia die.

VIBRIONACEAE (VIBRIOS)

Vibrios are curved or comma-shaped gram-negative bacteria that are motile by means of a single polar flagellum. Vibrios exhibit characteristic rapid darting movement under darkfield examination. There are four major species or groups as listed in Table 9-11: *V. cholerae,* which causes cholera; halophilic (salt-loving) vibrios; *Campylobacter;* and noncholera vibrios. All are oxidase positive, like many of the other nonfermenting gram-negative rods. Non-O1 *V. cholerae* strains share identical H antigens with *V. cholerae* but have unique O antigens. Otherwise they are indistinguishable from *V. cholerae.* Differentiation of *Vibrio* types is based on differences in the polysaccharide O antigens.

Vibrio cholerae

Biologic characteristics. The cholera bacteria grow poorly on some enteric media such as eosin-methylene blue and *Salmonella-Shigella,* but grow well on MacConkey agar within 24 hours. *V. cholerae* are slow lactose fermenters, contain O and H, but no K antigens, and have three antigenic types: AB (Ogawa), AC (Inaba), and ABC (Hikajima). These types share one group-specific somatic antigen and also contain their own type-specific O antigens. In addition, there is a biotype called *El Tor,* named for the place where it was first discovered, that produces a hemolysin and is somewhat more

TABLE 9-11. Classification of Vibrios

Major groups	*Species*
Cholera vibrio	*V. cholerae* (comma)
	V. El Tor (biotype of *V. cholerae*)
Halophilic vibrios	*V. alginolyticus*
	V. group F (EF-6)
	V. metschnikovii
	V. parahaemolyticus
	V. vulnificus
Campylobacter	*C. fetus* subspecies *fetus*
	intestinalis
	jejuni
	C. sputorum subspecies *bubulus*
	sputorum
	C. fecalis
Noncholera vibrios (NCV)	*V. cholerae* non-O1, about 75 serotypes

resistant to treatment than other *V. cholerae* strains.

Epidemiology. Cholera is a disease that has been known throughout history and affects only humans. Epidemics have occurred over the centuries, the most famous of which was in England in 1855, at which time John Snow was first able to demonstrate epidemiologically that the responsible organism was spread by drinking water. By removing the handle of the Broad Street water pump, which contained cholera-infected water, he caused a reduction in the disease incidence. Pandemics have continued at intervals, the latest being in the 1970s. Cholera is still endemic in Asia and India. In 1979 there were 267 cases reported in Spain during a 4-month time period, though the disease was primarily mild or asymptomatic (Centers for Disease Control 1980c). Fish from contaminated waters was the major vehicle. The disease is most prevalent in areas that have poor sanitation.

Though cholera is quite rare in the Western hemisphere, 11 cases caused by El Tor biotype occurred in Louisiana in 1978 after these people ate steamed crab. In 1980 a case of cholera in Florida was traced to eating contaminated raw oysters. Cholera was also reported in 1980 to 1981 among Laotian refugees to the United States. They acquired the disease in a Thai holding camp (Centers for Disease Control 1980a, b). Person-to-person transmission is rare. The organisms are almost always transmitted by ingestion of food (especially seafood) or water that has been fecally contaminated. Most epidemics are water-borne. There is minimal danger of transmission from refugees. In endemic areas, the disease is most prevalent among children less than 5 years old. A chronic gallbladder carrier state as in typhoid is possible and may occasionally be a source of transmission.

Pathogenesis. After entering the gastrointestinal tract, *V. cholerae* colonize the duodenum and ileum. The organisms are not invasive and do not penetrate the epithelium. Rather, the pathogenic effects of cholera are due to the secretion of a potent protein enterotoxin called *choleragen.* There is no cellular response to infection, and the bowel remains histologically normal, but the enterotoxin causes a massive excretion of isotonic fluid from the small intestine. The cholera toxin causes the enzyme adenyl cyclase from the intestinal wall to be activated. Adenyl cyclase converts the cell's energy carrier, ATP, into cyclic AMP (CAMP), which causes cells in the intestinal wall to secrete chloride and water. This results in decreased absorption of sodium, resulting in massive fluid loss. The toxigenicity of *V. cholerae* is genetically rather than plasmid determined, which is different than the toxigenicity of *E. coli.* A large infecting does, about 10^9 organisms, is required to cause disease.

Host responses. The incubation period for cholera can be as short as 6 hours or as long as 5 days. The disease occurs abruptly and the predominant symptom is voluminous watery diarrhea called "rice water stools" because they are clear with flecks of mucus. Vomiting may also occur but there is usually no accompanying pain or nausea. The gastrointestinal fluid loss may be from 10 to 25 liters per day, resulting in profound dehydration, electrolyte imbalance, shock, and death, if untreated. The stool contains large numbers of vibrios that can be readily cultured. Without adequate fluid replacement, 50 to 75 percent of afflicted individuals die. This can be decreased to about 1 percent with therapy. The disease usually subsides within a week. There are, in addition, many subclinical or mild cases with a ratio of subclinical to clinical of about 10 to 1 in some epidemics.

Control and treatment. Adequate hygiene, sewage disposal, and public health measures to control sanitation of food and water are successful means of preventing the spread of cholera since the disease is usually caused by water contaminated by a patient or carrier. Immunity, either artificially or naturally acquired, is not complete. A vaccine made from killed *V. cholerae* is available for travelers to endemic areas and provides protection for 3 to 6 months but it does not prevent transmission of infection. Natural immunity following infection is protective for a maximum of about 1 year. Both circulating IgG and local secretory IgA against *V. cholerae* are produced by the infected individual, but the IgG may be destroyed by enzymatic digestion. This may explain the short duration of protection. A cholera toxoid, under investigation, may offer greater proteciton.

Malnourished and debilitated individuals are more susceptible to *V. cholerae*. Most persons with clinical disease are underweight. Gastric acidity also appears to provide some natural protection against cholera. Ingestion of raw or inadequately cooked seafood should be avoided, especially that obtained from the southeast coast of the United States (Florida, Louisiana, Texas, Alabama) where cases of cholera have been reported. Crabs and other shellfish require boiling for at least 15 minutes to ensure that any vibrios are killed.

The treatment of cholera is, of course, aggressive and rapid intravenous and oral fluid and electrolyte replacement. Antimicrobial therapy with tetracycline, chloramphenicol, or furazolidone may facilitate the elimination of organisms from the stool but is probably of little benefit after the first 24 to 48 hours of diarrhea. In 1979, strains of *V. cholerae* with plasmid-mediated resistance to five major antimicrobials were isolated in Bangladesh (Sarkar and Chowdhuri 1980). Patients with cholera should be isolated, and their stool disinfected before disposal.

Other Vibrios

In addition to *V. cholerae* there are several other *Vibrio* species that require a salty, alkaline environment. These halophilic species, especially *V. parahaemolyticus* and *alginolyticus* can proliferate in tidal estuaries where they contaminate seafood and swimming water. *V. parahaemolyticus* is one of the common causes of food poisoning in Japan. It is transmitted in a raw seafood dish called sashimi. The disease is becoming increasingly recognized in the United States as culture techniques amenable to isolation of vibrios are used. The resulting diarrhea is milder than that of cholera though still severe. It is accompanied by vomiting but little fever. Treatment is similar to that for cholera.

There were about 20 confirmed outbreaks of *V. parahaemolyticus* infection in the United States during the 1970s; most were due to poorly handled or inadequately cooked seafood. The mean duration of diarrheal disease is about 3 days. An enterotoxin is probably involved in pathogenesis, though none has been identified yet. Local infections and septicemia have also resulted from *V. parahaemolyticus*. Cases of *V. parahaemolyticus* infection should be reported to the local health department, as they usually result from improper food handling.

V. vulnificus is a rare cause of septicemia or wound infection after ingestion or handling of contaminated seafood. *Vibrio* group F (EF-6) has been shown to be linked to choleralike disease around the world, but a causative relationship has not been established. There is no evidence that *V. metschnikovii* causes disease in humans.

Noncholera vibrios (those not agglutinated by cholera O antiserum) are otherwise identical to *V. cholerae,* and can produce a toxin and a clinical syndrome like that of cholera. Such strains have been isolated in the United States as well as elsewhere in the world. Their taxonomy is not completely clearcut.

Campylobacter is a genus in the same family as the vibrios. It consists of three species: *C. (Vibrio) fetus* (with three subspecies), *C. sputorum* (with two subspecies), and *C. fecalis* (with no subspecies). *Campylobacter* are often curved or S-shaped and are motile with polar flagellum (Fig. 9-12a, b). *C. sputorum* is normal oral flora for about 5 percent of individuals and probably causes no human disease. Neither, in all likelihood, does *C. fecalis*. However, all three subspecies of *C. fetus* have been shown to cause disease in humans and animals. *C. fetus* subspecies *fetus* causes abortion and sterility in cows and has also been isolated from human clinical specimens. Both *C. fetus* subspecies *jejuni* and *intestinalis* have been increasingly recognized as human pathogens. Subspecies *jejuni* causes 3 to 11 percent of diarrhea worldwide, affecting normal as well as compromised hosts. It is also widespread in animals.

In one large study, *C. fetus* was isolated from 7.1 percent of patients with diarrhea and 66.7 percent of household contacts of index patients (Fig. 9-13). In only rare instances was it found in asymptomatic persons. Domestic animals had high carrier rates and could represent a significant reservoir for human infection. Carriage lasted 3 weeks maximum in patients with intestinal disease. Campylobacteriosis is now recognized as being as common as infection due to *Salmonella* or *Shigella*. The CDC is establishing a surveillance system for this disease.

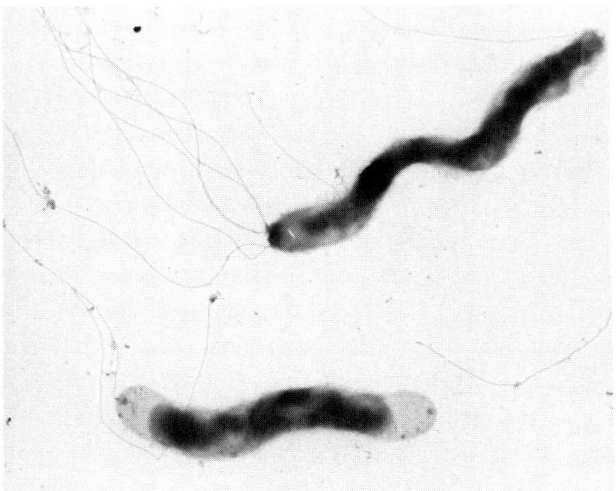

a

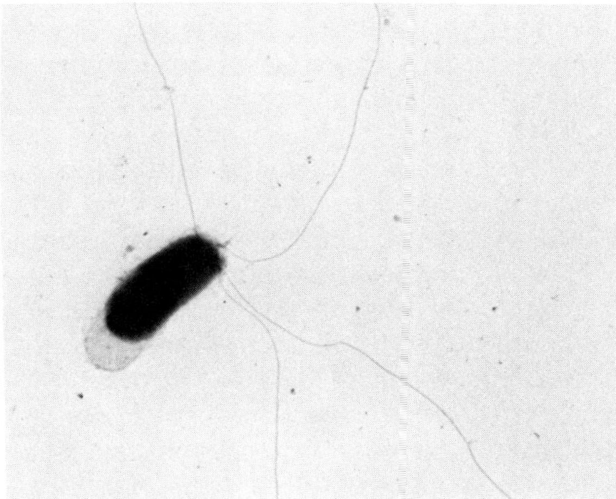

b

FIGURE 9-12. *(a)* and *(b)* Electron micrographs of *Campylobacter fetus.* Note S-shaped polar flagella. (Courtesy of Dr. James Plorde.)

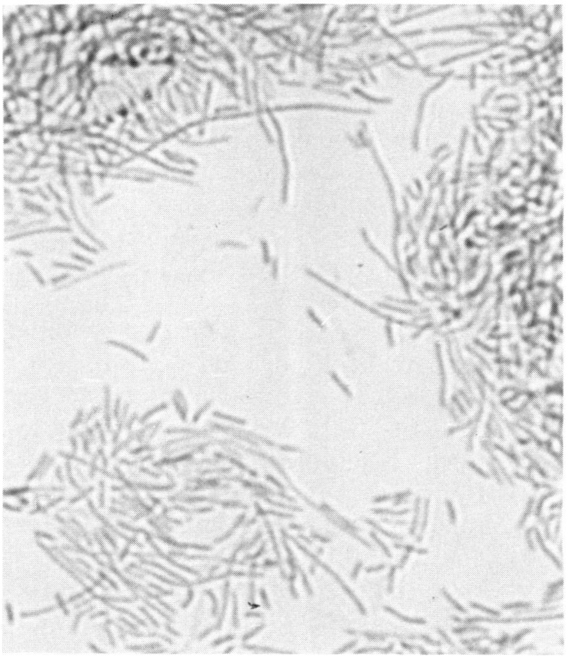

FIGURE 9-13. *C. fetus* in blood. (Courtesy of Dr. James Plorde.)

Campylobacter infections affect all ages; incidence peaks during summer months. *C. fetus* subspecies *jejuni* is isolated from fresh and salt water, raw milk, as well as many animal tissues. Contact with animals and poorly cooked meats may be routes of infection. The mode of transmission is fecal-oral; placental transmission from mother to fetus has been documented. From March 1977 until the end of 1980 there were 25,791 cases isolated in Great Britain (*Campylobacter* infections, editorial 1981). Diarrhea, abdominal cramps, and bloody stools can result from infection, but most young persons who were infected had no symptoms. Common source outbreaks were due to raw milk, water, chicken, and contact with sick animals. Three deaths were attributed during the time period to *Campylobacter* gastroenteritis. *C. fetus* subspecies *jejuni* was the causative agent in 13 milk-borne outbreaks. The organism can be readily eliminated from milk by pasteurization.

BACTEROIDES

Biologic characteristics. Organisms of the genus *Bacteroides* are a large group of usually nonmotile, often pleomorphic anaerobic gram-negative rods in the family Bacteroidaceae, which also includes two other genera: *Fusobacterium* (Fig. 9-14) and *Leptotrichia*. Motile species possess peritrichous flagella. *B. fragilis* is encapsulated

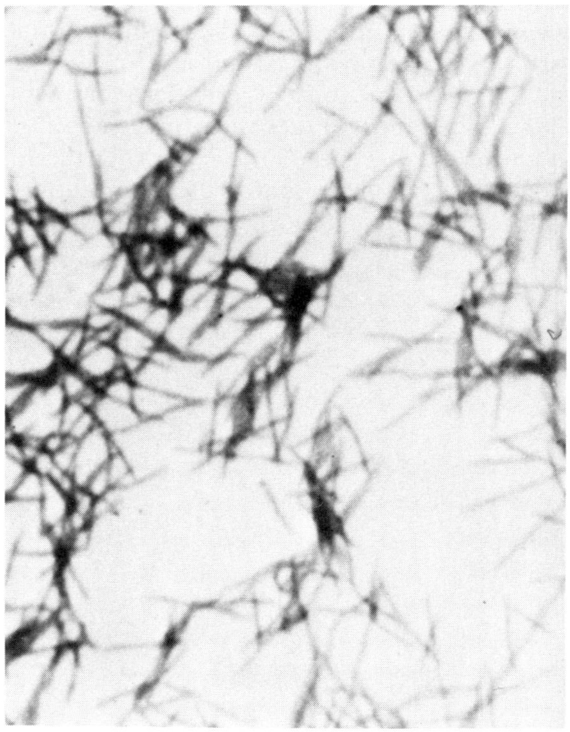

FIGURE 9-14. Gram stain of *Fusobacterium nucleatum.* Some strains of fusobacter do not stain and are called "ghost cells." (Courtesy of Dr. James Plorde.)

and has numerous subspecies. There are almost 40 species of *Bacteroides* and fusobacteria, the most common of which are listed in Table 9-12. Since they are strict anaerobes, they are somewhat difficult to grow and may be missed if specimens are inadequately collected or handled. Fluorescent antibody techniques are useful diagnostic tools. *Bacteroides* are sometimes called fusiform bacteria, as they may be slender rods, branching, or filamentous. Some strains of fusobacter do not stain and are called "ghost cells." *B. melaninogenicus* and *asaccharolyticus* produce tan or black colonies that fluoresce red under ultraviolet light. *B. fragilis* produce large, white, raised colonies.

Epidemiology. *Bacteroides* and the anaerobic bifidobacteria comprise 99 percent of the normal intestinal flora and are also normal in the mouth and female genital tract. Fusobacteria are normal flora of the mouth. It is probable that infections due to *Bacteroides* are more common than had been recognized in the past.

Pathogenesis. *Bacteroides* are not invasive and have low pathogenicity. They are primarily secondary invaders, especially in tissues that are necrotic and provide anaerobic conditions. The virulence of *B. fragilis* is related to its protective polysaccharide capsule and its resistance to bactericidal activity of serum.

Host responses. Under normal circumstances *Bacteroides* are harmless, but they occasionally cause abdominal, cervical, or brain abscesses, especially after perforation of the bowel or appendix. Genitourinary infections, primarily in females, also occur. These include salpingitis (usually combined with other organisms), septic abortion, and Bartholin's abscess. Ear infections, meningitis, arthritis, and empyema rarely occur. Abscesses with fusobacteria can occur fol-

TABLE 9-12. Species of Family Bacteroidaceae Common in Humans

Species	*Habitat*	*Associated diseases*
Bacteroides asaccharolyticus	Mouth, GI tract, GU tract	None
B. bivius	Mouth, vagina	None
B. disiens	GI tract	None
B. fragilis	Mouth, GI and GU tracts	Abdominal abscesses, kidney infections, postpartal infections
B. melaninogenicus	Mouth, GI and GU tracts	Rare abdominal abscesses, kidney infections, postpartal infections
B. orales	Mouth, vagina	Rare abscesses
B. pneumosintes	Nasopharynx	Probably none
B. thetaiotaomicron	GI tract	Rare intraabdominal abscesses
Fusobacterium nucleatum	Mouth	Ulcerative stomatitis (trench mouth), Vincent's angina
F. necrophorum	Mouth	Soft-tissue abscesses, ulcerative colitis, bacteremia
Leptotrichia buccalis	Mouth, vagina, GI tract	None

lowing human bites. Most infections caused by anaerobes are "mixed"; that is, they involve several species.

Control and treatment. Chloramphenicol, clindamycin, and metronidazole (which may be oncogenic) are used to treat *Bacteroides* infections. With *B. fragilis,* the most common pathogen, metronidazole and clindamycin are synergistic. Penicillins may be used if the causative organisms are susceptible. Fusobacteria are usually sensitive to penicillin. Proper cleaning and debridement of wounds and abscesses are an important aspect of treatment.

CLINICAL IMPLICATIONS

In underdeveloped countries where clean drinking water, proper sewage disposal and balanced food supplies are lacking, typhoid fever, bacillary dysentery, and cholera represent serious problems. In such countries diarrheal disease is still the major cause of death among small children. In developed countries the prevalence of severe intestinal infections associated with enteric organisms is lower. However, of major concern is the fact that gram-negative bacteria have become the predominant agents in hospital-acquired infections. During the 1950s and early 1960s gram-positive species such as *S. aureus* and *S. pneumoniae* caused most hospital infections. Today, however, *E. coli, Pseudomonas* species, *K. pneumoniae,* and *S. marcescens* are principle pathogens. Hence, the major clinical implications related to disease caused by gram-negative bacteria center around prevention and control of nosocomial infection.

Most nosocomial infections due to coliforms involve the urinary tract. Since 10 to 15 percent of patients admitted to hospitals receive indwelling catheters special guidelines should be implemented for urinary catheter care. Guidelines should also be adopted for care of intravenous catheters and for the handling and care of respiratory therapy equipment. Such guidelines are available from the Centers for Disease Control. Hospital surveillance and infection control committees are necessary to identify baseline levels of nosocomial infection in order to detect and manage epidemics. Ongoing surveillance of patients is necessary, noting not only those with infections or harboring resistant organisms but also those at high risk for acquiring infection, such as debilitated or immunosuppressed patients. Environmental control and adequate disinfection and sterilization techniques must be of concern to clinicians. Observation of patients is useful, though it is unnecessary to obtain microbiologic cultures from each patient. For example, in two years of surveillance, 1976 to 1978, for ampicillin- and gentamicin-resistant enteric bacteria in one newborn intensive care unit, colonization rates of infants did not correlate with either rates of clinical disease or with variations in antimicrobial usage (White et al. 1981). Far more important than culturing patients and the environment in search of colonizing strains is the prevention of spread of gram-negative bacteria by proper handwashing.

Transmission of Gram-Negative Bacteria By Hands

In more than 40 hospital outbreaks of infections caused by gram-negative bacteria over the past several years, carriage on hands of personnel has been implicated as the mode of transmission (Maki 1978). In two separate incidents of nosocomial infections among neonates, one caused by *E. coli* and the other by *Citrobacter freundii,* nurses' hands were found to be persistently colonized with the causative organisms. Intensive environmental surveillance uncovered no other potential fomites (Centers for Disease Control 1979; Salzman, Clark, and Klemm 1968). Another study compared the incidence of diarrhea for 35 weeks in two study day care centers and two control centers. In the study centers, handwashing was enforced for both staff and children before handling food, after toileting, and at other important times. The rate of diarrhea was half in those day care centers when compared with control centers (Black 1981).

During one year, 750 isolates of *Pseudomonas* from patients in a burn unit were studied. Extensive environmental cultures and cultures of nurses' hand flora suggested direct handling of patients by personnel to be the principle mode of bacterial transmission. About 24 percent of the nurses' hands yielded the same *Pseudomonas* species as were obtained from patients in the unit (Kominos, Copland, Grosiak 1972). *Pseudonomas* was also found to colonize 12 percent of the hands of hospital personnel working with patients with spinal cord injury (Montgomery and Morrow 1978).

Similarly, in a recent intensive care unit outbreak caused by *Acinetobacter,* 32 percent of 66 unit personnel had the epidemic strain isolated from their hands. Interestingly, the cultures were positive with the same frequency whether the individuals had just begun the work period or were just completing the shift. The organism was carried only on the hands of personnel. An inhalation therapist was found to be heavily colonized with the organism, and a wash cloth that he used each morning at his home yielded a pure growth of *Acinetobacter.* In this incident the organism apparently spread from one personnel member to patients, with subsequent contamination of other personnel hands (Burton et al. 1978).

An extensive investigation of the role of hand contamination of personnel in the epidemiology of gram-negative bacterial infections was carried out from 1974 to 1975. Samples from the hands were taken immediately after handwashing with solutions containing hexachlorophene or povidone-iodine. In 151 cultures from 13 nurses over a two-month time period, 86 percent yielded coliforms. Each nurse seemed to persistently carry a particular organism. For example, one nurse carried *Serratia,* another *Klebsiella,* another *Proteus.* These nurses had apparently healthy skin. From six of eight nurses the same organism was recovered after 2 to 5 days of vacation. In 27 of 47 sets of cultures (55 percent) colony counts of gram-negative rods were higher at the beginning of the work period than at the end. After use of a broad-spectrum antimicrobial agent within the work setting was discontinued, 101 samples from the hands (78 percent) of nine nurses over 3 weeks remained positive for gram-negative organisms. The author's conclusion was that hands were a true reservoir for these organisms, not just serving as a means of passive bacterial transfer between patients (Knittle et al. 1975).

During nonepidemic times, Maki found that 44 percent of 348 cultures from randomly sampled hospital personnel carried gram-negative bacteria on their hands; and Salzman and colleagues (1968) over a 9-month time period found antimicrobial-resistant gram-negative bacteria on the hands of over 20 percent of 118 hospital personnel. Neither, however, included a nonhospital comparison group to assess whether these numbers were unusual.

McBride and colleagues studied the hands of 50 nurses and 50 comparison subjects.

They found higher bacterial counts for all species or organisms among the control group, although the difference between the two groups for carriage of Enterobacteriaceae was not statistically significant (McBride et al. 1972, 1974). Polk and Lopez (1972) found that clinical pathogens such as *Pseudomonas* species, *Escherichia coli,* and *Proteus* species persisted on the hands of nurses, even after povidone-iodine scrub, for periods of one to three days. Reports of the prevalence of colonization of hands among hospital personnel with gram-negative bacteria range from 18 to 80 percent.

Effect of Handwashing Agents on Gram-Negative Bacteria

There have been many clinical tests using various sampling techniques to compare the efficacy of handwashing agents. Such comparisons are difficult, however, because of a lack of standardized testing procedures. The Food and Drug Administration accepts various types of efficacy tests. In addition, many agents, long in use, have not yet been adequately tested. Though there has been recently established a national plan for evaluating safety and effectiveness of handwashing agents, this plan has been found to exclude many suitable volunteers and to require more testing than necessary.

Lowbury and Lilly (1980), during an investigation to determine the efficacy of hexachlorophene (HCP) in decreasing *Staphylococcus aureus* transmission, found coliforms in large quantities on the HCP-washed hands of nurses who worked in a burn unit. Consistent with this were the findings of Bruun and Solberg (1973). Hexachlorophene handwashing as compared with bar soap was found to decrease *S. aureus* on the hands from 21 to 10.2 percent. A significantly higher rate of cultures yielding gram-negative bacteria was found after HCP handwashing than after washing with bar soap. In this study, as a result of serial sampling over a period of months, it was concluded that 13 percent of hospital personnel tested were persistent, not transient, carriers of gram-negative organisms on their hands.

Peterson and colleagues (1974), studied the effect of implementing strict handwashing with detergent containing HCP during an epidemic of *E. coli* infantile gastroenteritis. The handwashing regimen had no effect on either the epidemic or the prevalence of coliforms, including the etiologic agent, on the hands of nursery personnel. After personnel thoroughly washed their hands with either plain soap or HCP-containing agent, 33 to 89 percent of the hands sampled were positive for coliforms. To control this epidemic it was necessary to institute strict gloving techniques, at which time the prevalence of coliforms on the hands dropped to 19 percent. The investigators considered this to be the usual carrier rate, since gloving reduced exposure to transient contaminants.

Hexachlorophene is clearly more effective against gram-positive than against gram-negative organisms. Other antiseptics, particularly iodophors and chlorhexidine, have generally replaced HCP as handwashing agents used for patient care areas. Both povidone-iodine and chlorhexidine are more effective than HCP against gram-negative bacteria.

Comparative studies of povidone-iodine (polyvinylpyrrolidone iodine) have given contradictory results. Though some have found agents containing povidone-iodine to be more successful at removing gram-negative bacteria from hands than agents containing HCP, others, using more sensitive sampling techniques, have found little difference. Other problems with iodophors such as variable stability over time, reduced antibacterial potency due to binding of iodine in water-soluble complexes, loss of substantive effect when the product is washed off, have been identified (Crowder et al. 1967; Joress 1962; Polk and Lopez 1972; Spradlin 1980; Steere and Mallison 1975).

Agents containing chlorhexidine gluconate were approved for use in the United States by the Food and Drug Adminsitration in 1976.

Since it is a more recent product on the market, data are abundant and better standardized than that on iodophors or HCP. In a study of a 4 percent chlorhexidine detergent, 99 percent of *Serratia marcescens* artificially inoculated on the hands was removed after a 15-second wash. A similar finding has been reported for *Klebsiella* (Aly and Maibach 1976). Some investigators have demonstrated that 0.5 and 4 percent solution of chlorhexidine in alcohol were more effective than comparison preparations in removing gram-negative bacteria from hands. In some studies it appeared that the alcohol, with or without the chlorhexidine, was the effective antiseptic since aqueous solutions of chlorhexidine were comparable to plain soap in their effect against gram-negative organisms. Others have found, however, that a 4 percent chlorhexidine detergent produced a significantly greater reduction in skin flora than povidone-iodine or plain soap and had the added benefit of a stable substantive effect (Aly and Maibach 1976; Casewell and Phillips 1978; Rosenberg, Alatary, and Peterson 1976; Steere and Mallison 1975).

In summary, though studies of the relative efficacy of agents continue and new agents are being developed, we now have available several very acceptable options from which to choose a handwashing agent. In addition to being broad spectrum in their activity against bacteria, they often contain emollients to minimize skin drying and cracking, conditions which have been shown to increase the risk of bacterial colonization (Steere and Mallison 1975). Consistent use of antiseptic soaps in high-risk patient care areas will minimize the possibility of transmitting gram-negative bacteria among patients on personnel hands.

As shown in Table 9-13, hands can be a

TABLE 9-13. Types of Gram-Negative Bacteria Isolated From Hands of 153 Subjects

Bacteria	*Hospital personnel n = 103*	*Controls n = 50*	*Total n = 153*
Acinetobacter, Moraxella	89/209 (43 percent)	141/297 (47 percent)	230/506 (45 percent)
Klebsiella, Enterobacter	82/209 (39 percent)	116/297 (39 percent)	198/506 (39 percent)
Other *(Citrobacter, Pseudomonas, Serratia, E. coli)*	38/209 (18 percent)	40/297 (13 percent)	78/506 (15 percent)

Reprinted with the permission of the publisher from Larson EL. Persistent carriage of gram-negative bacteria on hands. Am J Infec Contr 1981;9:112-119.

primary (and not just a transient) source of gram-negative bacteria traditionally thought to inhabit the bowel. And, as shown in Table 9-14, the probability of colonization with such gram-negative bacteria is significantly correlated with frequency of handwashing. Thus, it behooves health care personnel to wash their hands not only to remove transient pathogens, but also to minimize the numbers of their own skin flora that can cause disease in patients.

SUGGESTED READINGS

Aber RC, Banks WV. An outbreak of nosocomial *Salmonella typhimurium* infection linked to environmental reservoir. Infec Cont 1980; 1:386-390.

Aly, R, Maibach HI. Effect of antimicrobial soap containing chlorhexidine on the microbial flora of skin. Appl Environ Microbiol 1976; 31:931-935.

Ayliffe GAJ, Babb JR, Quoraishi AH. A test for "hygienic" hand disinfection. J Clin Pathol 1978;31:923-928.

Berman R, Knight R. Evaluation of hand antisepsis. Arch Environ Health 1969;18:781-783.

Black RE. Handwashing to prevent diarrhea in day care centers. Am J Epidemiol 1981;113: 445-451.

Blaser MJ. *Campylobacter fetus* subspecies *jejuni:* the need for surveillance. J Infec Dis 1980;141:670-671.

Blaser MJ. *Salmonella* bacteremia; reports to the Centers for Disease Control, 1968-1979. J Infec Dis 1981;143:743-746.

Blaser MJ, et al. Reservoirs for human campylobacteriosis. J Infec Dis 1980a;141:665-669.

Blaser MJ, et al. *Salmonella typhi:* the laboratory as a reservoir of infection. J Infec Dis 1980b;142:934-938.

Blum D, et al. Failure of doxycycline prophylaxis of diarrhea in travelers to Mexico. Poster session, International Congress Antimicrobial Agents and Chemotherapeutics, New Orleans, 1980.

Bradley SG. Cellular and molecular mechanisms of action of bacterial endotoxins. Ann Rev Microbiol 1979;33:67-70.

Bruun J, Solberg C. Hand carriage of gram-negative bacilli and *Staphylococcus aureus*. Br Med J 1973;2:580-582.

Burton AE, et al. Nosocomial respiratory tract infection and colonization with *Acinetobacter calcoaceticus*. Am J Med 1978;65:507-513.

Campylobacter infections (editorial). 1977-1980. Br Med J 1981;282:1484.

Casewell M, Phillips I. Hands as route of transmission of *Klebsiella* species. Br Med J 1977;2:1315-1317.

Casewell M, Phillips I. Food as a source of *Klebsiella* species for colonization and infection of

TABLE 9-14. Colonization of Hands with Gram-Negative Bacteria by Frequency of Handwashing

Stated frequency per day	*Hospital personnel*	*Controls*	*Total*
< 8	10/30	33/40	43/70 (61 percent)
9-15	7/39	4/7	11/46 (24 percent)
> 15	5/34	3/3	8/37 (22 percent)
	$X^2 = 22.3$ $p < 0.001$		

Reprinted with permission of the publisher from Larson EL. Persistent carriage of gram-negative bacteria on hands. Am J Infec Contr 1981;9:112-119.

intensive care patients. J Clin Pathol 1978;31: 845-849.
Chun D, Seol SY, Suh MH. Transferable resistance to trimethoprim in *Shigella*. J Infec Dis 1981;143:742.
Cleary TJ, MacIntyre DS, Castro M. *Serratia marcescens* bacteremias in an intensive care unit. Am J Infec Cont 1981;9:107-111.
Clinical forum: *Salmonella*—unwelcome visitors. Nurs Mirror 1981;152:iv.
Centers for Disease Control. Nosocomial meningitis caused by *Citrobacter diversus*. Morb Mort Week Report 1979;28:249-250.
Centers for Disease Control. Cholera—Florida. Morb Mort Week Report 1980a;29:601.
Centers for Disease Control. Cholera in a Laotian refugee. Morb Mort Week Report 29:1980b;191-192.
Centers for Disease Control. Cholera—Spain. Morb Mort Week Report 1980c;29:324.
Centers for Disease Control. Human *Salmonella* isolates—1979. Morb Mort Week Report 1980d;29:189-190.
Centers for Disease Control. *Salmonella hadar*—England and Wales. Morb Mort Week Report 1980e;29:506-508.
Centers for Disease Control. Shigellosis—United States, 1979. Morb Mort Week Report 1980f; 29:201-202.
Crowder VH, et al. Bacteriologic comparison of hexachlorophene and PVP iodine surgical scrub soaps. Am Surg 1967;30:906-911.
Datta N, Richards H, Datta C. *Salmonella typhi* in vivo acquires resistance to both chloramphenicol and co-trimoxazole. Lancet 1981;1: 1181-1182.
Dutta R, et al. R-plasmids in intestinal *E. coli* of normal and hospitalized populations. Indian J Med Res 1980;72:794-801.
Echeverria P, Murphy JR. Enterotoxigenic *E. coli* carrying plasmids coding for antibiotic resistance and enterotoxin production. J Infec Dis 1980;142:273-278.
Fisher MC, Baluarte HJ, Long SS. Bacteremia due to *Bacteroides fragilis* after elective appendectomy in renal transplant recipients. J Infec Dis 1981;143:635-638.
Formal SB, Horrick RB. Invasive *Escherichia coli*. J Infec Dis 1978;137:641.
Gangarosa EJ. Epidemiology of *Escherichia coli* in the United States. J Infec Dis 1978;137:634-637.
Goldstein EJC, et al. Infections caused by *Klebsiella ozaenae:* a changing disease spectrum. J Clin Microbiol 1978;8:413.
Grehn D. Endotoxins of gram-negative anaerobic bacteria. Infec 1980;2:S 138.
Grieco MH, ed. Infections in the Abnormal Host. New York: Yorke Medical Books, 1980.
Johnson RH, et al. *Arizona hinshawii* infections. Ann Intern Med 1976;85:587-592.
Joress SM. A study of disinfection of the skin: a comparison of povidone-iodine with other agents used for surgical scrubs. Ann Surg 1962;155:296-300.
Knittle MA, et al. Role of hand contamination of personnel in the epidemiology of gram-negative nosocomial infections. Pediatr 1975;86:433-437.
Kominos SD, Copland CE, Grosiak B. Mode of transmission of *Pseudomonas aeruginosa* in a burn unit and an intensive care unit in a general hospital. Appl Microbiol 1972;23:309-312.
LeFrock JL, Ellis CA, Weinstein L The relation between aerobic fecal and oropharyngeal microflora in hospitalized patients. Am J Med Sci 1979;277:275-280.
Levine MM, et al. Duration of serum antitoxin response following *Vibrio cholerae* infection in North Americans: relevance for seroepidemiology. Am J Epidemiol 1981;114:348-354.
Linksay KL, Canawati HN. Spontaneous *Arizona hinshawii* peritonitis in cirrhosis with ascites. Gastroenterol 1981;81:349-351.
Lowbury EJL. Gram-negative bacilli on the skin. Br J Dermatol 1969;81:55-61.
Lowbury EJL, Lilly HA. Disinfection of the hands of surgeons and nurses. Br Med J 1960;2:1445-1450.
Maki DG. Control of colonization and transmission of pathogenic bacteria in the hospital. Ann Intern Med 1978; 89 (Suppl., Part 2):777-780.
McBride ME, et al. Microbial flora of nurses' hands. I. Quantitative differences in bacterial populations between nurses and other occupational groups. Intern J Dermatol 1972;11: 49-53.
McBride ME, et al. Microbial flora of nurses' hands. II. Qualitative differences in occupa-

tional groups. Intern J Dermatol 1974;13:197-204.

Michalek SM, et al. The primary role of lymphoreticular cells in the mediation of host responses to bacterial endotoxin. J Infec Dis 1980;141:55-63.

Michaud RN, et al. Improved experimental model for measuring skin degerming activity on the human hand. Antimicrob Agents Chemother 1972;2:8-15.

Montgomery J, Morrow J. *Pseudomonas* colonization in patients with spinal cord injury. Am J Epidemiol 1978;108:328-336.

Moseley SL, et al. Detection of enterotoxigenic *Escherichia coli* by DNA colony hybridization. J Infec Dis 1980;142:892-898.

Northey D, et al. Microbial surveillance in a surgical intensive care unit. Surg Gyn Obstet 1974;139:321-325.

Palmer SR, et al. Person-to-person spread of *Salmonella typhimurium* phage type 10 after a common-source outbreak. Lancet 1:1981;881-882.

Pennington JE, Kuchmay D. Mechanism for pulmonary protection by lipopolysaccharide *Pseudomonas* vaccine. J Infec Dis 1980; 142:191-198.

Peterson NJ, et al. Evaluation of skin cleansing procedures using the wipe-rinse technique. Health Lab Sci 1974;11:182-187.

Polk HC, Lopez JF. Bacterial ecology of hands of intensive care unit nurses cleansed with Betadine surgical scrubs. Medical and Surgical Asepsis with Betadine Microbicides. New York: The Purdue Frederick Company, 1972, pp 41-45.

Reid TMS, Porter IA. An outbreak of otitis externa in competitive swimmers due to *Pseudomonas aeruginosa*. J Hyg 1981;86:357-362.

Rennie RP, et al. Comparison of bacteriophage typing, serotying, and biotyping as aids in epidemiologic surveillance of *Klebsiella* infections. J Clin Microbiol 1978;8:638-642.

Robinson DA, Jones DM. Milk-borne *Campylobacter* infections. Br Med J 1981;282:1374-1376.

Rosenberg A, Alatary SD, Petersen AF. Safety and efficacy of the antiseptic chlorhexidine gluconate. Surg Gynecol Obstet 1976;143: 789-792.

Sack RB. Enterotoxigenic *Escherichia coli:* identification and characterization. J Infec Dis 1980;142:279-286.

Salzman TC, Clark JJ, Klemm L. Hand contamination of personnel as a mechanism of cross-infection in nosocomial infections with antibiotic-resistant *Escherichia coli* and *Klebsiella-Aerobacter*. Antimicrob Agen Chemo 1968;1967:97-100.

Sarkar R, Chowdhuri ANR. R-factors in enteropathic *E. coli* associated with diarrheal disease in children. Indian J Med Res 1980; 72:174-178.

Saxena SN, et al. *Salmonella* pattern in India. Indian J Med Res 1980;72:159-168.

Schaberg DR, et al. An outbreak of nosocomial infection due to multiply resistant *Serratia marcescens:* evidence of inter-hospital spread. J Infec Dis 1976;132:181-188.

Spradlin CT. Bacterial abundance on hands and its implications for clinical trials of surgical scrubs. J Clin Microbiol 1980;11:389-393.

Steere AC, Mallison GF. Handwashing practices for the prevention of nosocomial infections. Ann Intern Med 1975;83:603-610.

Thomas FE, et al. Sequential hospital outbreaks of resistant *Serratia* and *Klebsiella* infection. Arch Intern Med 1977;137:581.

Thoren A, et al. Antibiotics in the treatment of gastroenteritis caused by enteropathic *E. coli*. J Infec Dis 1980;141:27-31.

Wahden MH, et al. A controlled field trial of live *Salmonella typhi* strain Ty 21a oral vaccine against typhoid: three-year results. J Infec Dis 1982;145:292-295.

Wenzel RP, et al. *Providencia stuartii,* a hospital pathogen: potential factors for its emergence and transmission. Am J Epidemiol 1976; 104:170-180.

White RD, et al. Are surveillance of resistant enteric bacilli and antimicrobial usage among neonates in a newborn intensive care unit useful? Pediatr 1981;68:1-4.

Zintel HA, Knox WG. A comparative environmental investigation of surgeon's skin germicides. Ann Surg 1967;165:303-307.

10 OTHER GRAM-NEGATIVE BACTERIA

KEY POINTS

- The most important pathogens in this group of gram-negative bacteria include *Hemophilus influenzae, Bordetella pertussis, Neisseria gonorrhoeae* and *meningitidis,* and *Yersinia pestis.*
- *H. influenzae* is a major cause of meningitis in children; *N. meningitidis* causes meningitis in adults, as well as children. *B. pertussis* causes whooping cough, *N. gonorrhoeae* causes gonorrhea, and *Y. pestis* causes plague.
- Brucellosis, tularemia (caused by *Francisella tularensis*), and plague are zoonoses, primarily occurring as diseases in animals, with humans accidentally infected.
- *Brucella* species, *B. pertussis, N. meningitidis,* and *Hemophilus* form capsules that protect them from phagocytosis and add to their virulence.
- Many gram-negative bacteria *(Hemophilus, Neisseria),* as well as other types of organisms, have strains emerging that are resistant to the antimicrobial agents traditionally used for their treatment.
- Along with the staphylococci and the streptococci, the *Neisseria* are termed "pyogenic cocci."

Many of the gram-negative bacteria discussed in this chapter are closely related to each other and to the enteric organisms. *Yersinia* and *Citrobacter* for example are classified as Enterobacteriaceae, but are considered here because of their unique epidemiology and other characteristics. *Acinetobacter* and *Moraxella* are members of the family Neisseriaceae. The familes Brucellaceae and Neiserriaceae and the genera *Hemophilus* and *Bordetella,* are all very small gram-negative rods with closely related species. Commensals that become opportunistic pathogens are singled out because it is important for the health care practitioner to be aware of their role in clinical disease.

BARTONELLA BACILLIFORMIS

Bartonella bacilliformis is a small, slow-growing, motile, pleomorphic gram-negative rod that can be seen best with Giemsa stain. It has a single polar flagellum. Only one serotype is responsible for bartonellosis, a disease limited to certain areas of South America, particularly the mountains of Colombia, Ecuador, and Peru where the clinical syndrome is called Carrión's disease.

Bartonellosis may have two different manifestations: an acute fever and hemolytic anemia that develops in nonimmune individuals, called *Oroya fever;* or chronic skin nodules called *verruga peruana.* The organism is transmitted from the blood of infected hosts (who may be symptomatic or not) by a sand fly of the genus *Phlebotomus.* Many cases may be asymptomatic. In others, after an incubation period of about 20 days, the Oroya fever develops suddenly. *B. bacilliformis* adhere to the red blood cells and fill the cytoplasm of endothelial cells lining blood vessels, causing massive destruction with resultant severe anemia. Oroya fever has a high mortality rate—up to 40 percent—if untreated. Death is often due to superinfection with *Salmonella* because of impaired functioning of the reticuloendothelial system (RES). If the patient recovers it takes several months for the hematocrit to return to normal.

The skin disorder, verruga peruana, may occur 1 to 2 months after the hemolytic anemia or may occur without any preceding symptoms. Crops of nodular lesions develop, especially on exposed skin surfaces, and persist for up to several years. There is no bacteremia during the cutaneous stage. Successful treatment results with most antimicrobial agents including tetracycline, streptomycin, penicillin, and chloramphenicol. Control efforts are directed toward elimination of the sand fly vector.

BRUCELLA

Biologic characteristics. There are six species of *Brucella.* They are difficult to distinguish, but four are pathogenic to humans: *B. melitensis, B. abortus, B. suis,* and *B. canis.* These are pathogens of goats, cattle, pigs, and dogs, respectively. All *Brucella* are obligate intracellular parasites of the RES of humans and the genital tract of animals. In animals, infection causes abortion and infected milk production. In humans it follows a different course. The organisms are nonmotile, encapsulated aerobes. Smooth forms (with a capsule) are more virulent. There is cross reaction of antibodies of the species *melitensis, abortus,* and *suis;* they share two surface protein-lipopolysaccharide antigens called A and M.

Epidemiology. Humans become infected by contacting infected animal tissues (urine, feces, fetal, or other tissues) or ingesting nonpasteurized dairy products. *Brucella* can survive for weeks in fetal tissues and for days in unpasteurized butter. Brucellosis is rarely transmitted between humans. Various species occur worldwide as animal pathogens. Persons contacting animals are at high risk, especially those who slaughter swine. The current incidence in the United States is about 200 cases per year. Most cases are caused by *B. suis.* The disease was first described during the nineteenth century in British soldiers who were suffering from Malta fever (brucellosis) acquired from drinking goat's milk. Another synonym for brucellosis is Bang's disease, which was first described in Denmark as a cause of abortions in cows.

Pathogenesis. *Brucella* are intracellular parasites. After entering the body by direct contact, ingestion, or inhalation, smooth strains of *Brucella* are phagocytized and multiply in neutrophils and in monocytes in the regional lymph nodes. Rough strains, which are avirulent, do not multiply. If the host generates an adequate response, the infection is contained within the regional lymph nodes. Otherwise, the bacteria contained in phagocytes spread hematogenously. They aggregate in the liver sinusoids, spleen, kidney, and bone marrow (occasionally also bone, testes, or endocardium), where characteristic granulomatous areas similar to those of tuberculosis develop. In animals the organisms have a propensity to accumulate in the mammary glands and genital organs, hence the infectiousness of dairy products for humans. This attraction of *Brucella* to genital organs and especially to the pregnant uterus in animals is due to the presence there of erythritol, a polyhydric alcohol that enhances the growth of the bacteria. Erythritol is present in the fetal fluids of susceptible animals, but not of humans. *Brucella* do not produce exotoxins, though they seem to possess some factor that increases their virulence in vivo.

Host responses. After an incubation of weeks or months, the major clinical sign of brucellosis is prolonged, chronic, intermittent fever. The symptoms otherwise may be vague, resembling psychoneurotic disease, but diagnosis is suspected if there is a history of ingestion of raw dairy products or exposure to animals. Accompanying signs and symptoms include weakness, enlargement of the liver or spleen, lymphadenopathy, arthralgia, night sweats, and swollen testicles. Fever may last only a few days, or many months. Other manifestations include chronic abscesses of the liver or spleen, bone and joint infections, and skin infections. The organism can be cultured from the blood during the acute febrile disease if the specimen is retained for more than the usual few days to allow for its slow growth on laboratory media. However, most cases are diagnosed by agglutination tests. IgG antibody titers of greater than 1:250 indicate recent infection. There is also available an intradermal skin test for delayed hypersensitivity that indicates prior exposure to *Brucella*. Only about 15 percent of cases persist more than 3 months, even without antimicrobial therapy. The mortality rate is very low. Reinfection does occur.

Control and treatment. Control is similar to that for all such zoonoses: milk pasteurization, and careful handling of animal tissues. Vaccination of cattle also helps to limit infection. The safety of vaccination of humans is questionable; vaccination is practiced in some parts of the world but not in the United States. *Brucella* are sensitive to ampicillin, tetracycline, and streptomycin. Therapy must be long term because the organisms are intracellular.

CALYMMATOBACTERIUM (DONOVANIA) GRANULOMATIS

Calymmatobacterium granulomatis is a nonmotile, encapsulated, short, bipolar-staining gram-negative rod closely related to *Klebsiella*. It is an intracellular parasite, forming characteristic conglomerates, *Donovan bodies,* within macrophages that are found in cases of granuloma inguinale (Fig.

10-1). The disease is probably sexually transmitted, but the infectiousness of the organisms is very low. Its pathogenesis is not well understood. *C. granulomatis* has also been isolated from the gastrointestinal tract; the feces may be a source of infection. Though granuloma inguinale is endemic in the tropics, there are probably less than 100 cases per year in the United States.

Granuloma inguinale manifests itself about 1 month after sexual exposure with the development of granulomatous ulceration in the genital region (Fig. 10-2). The disease is chronic and progressive in nature, with local extension, spread via lymphatics, and even rare disseminated syndromes occurring. Extragenital lesions are sometimes present, especially around the face and mouth. Tetracycline is effective treatment, though, as is the case with all intracellular parasites, long-term therapy is needed and relapses may occur.

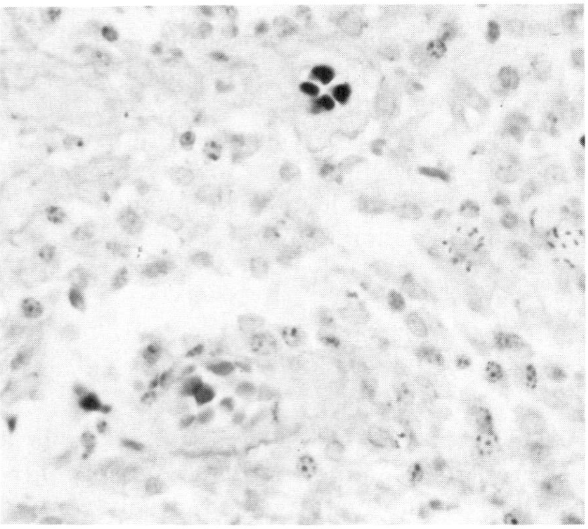

FIGURE 10-1. Granuloma inguinale Donovan bodies in tissue. Magnification 630×. (Courtesy of Dr. James Plorde.)

COMMENSALS THAT BECOME OPPORTUNISTIC PATHOGENS

Acinetobacter

Acinetobacter are aerobic, nonfermentative, nonmotile, gram-negative diplococci resembling *Neisseria* on a smear (Fig. 10-3). They are ubiquitous in nature; they are found in water, milk, and soil, and in the human gastrointestinal, genitourinary, and respiratory tracts as well as on the skin of about 25 percent of normal individuals. There is one species, *calcoaceticus,* and two variants: *Acinetobacter lwoffi* (former names: *Mima polymorpha,*) and *A. anitratus* (former name: *Herellea vaginicola*). *A. calcoaceticus* var. *anitratus* is much more common in humans

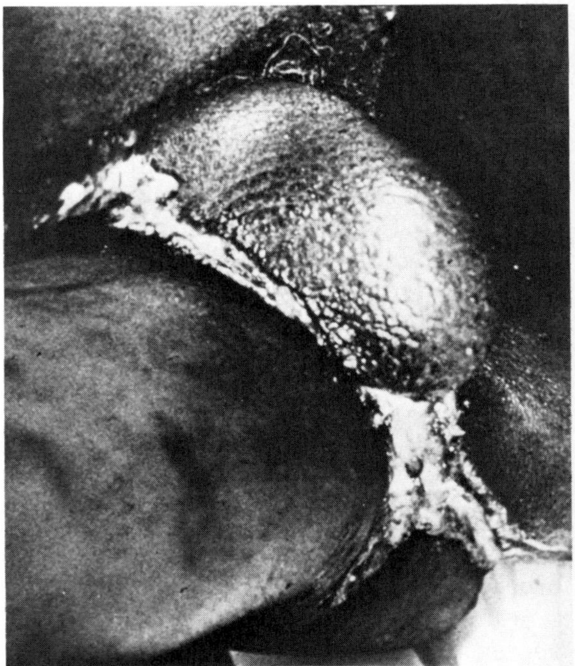

FIGURE 10-2. Granuloma inguinale. (Courtesy of University of Washington Clinical Microbiology Laboratory.)

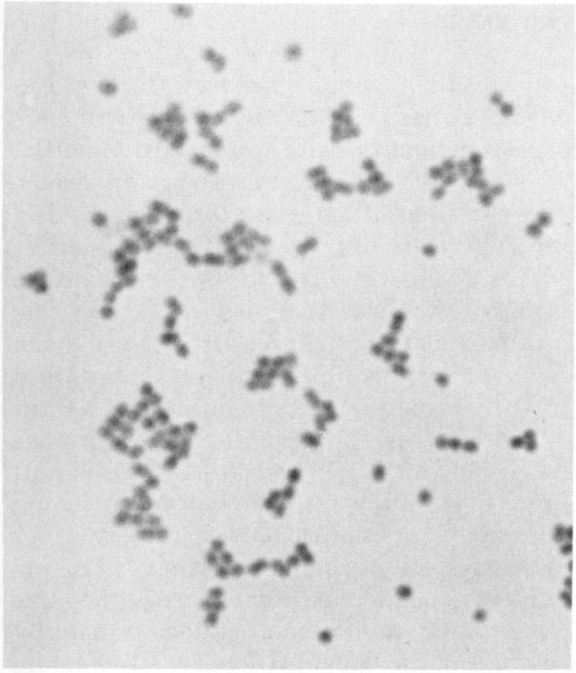

FIGURE 10-3. Pleural fluid containing *Acinetobacter.* (Courtesy of Dr. James Plorde.)

and is the species usually associated with opportunistic disease. In 1978, *A. calcoaceticus* var. *anitratus* caused nosocomial infection at a rate of 3.5 per 10^4 hospital patient discharges, a 14 percent increase over rates from 1974 to 1977. There was marked seasonal variation; infections peaked in August and were lowest in February. The most frequent site of infection is the lower respiratory tract. In one series, 15 percent of clinical infections were bacteremias. Pulmonary infections have been related to repiratory therapy equipment and to hands of personnel that were either contaminated or harbored large numbers of *Acinetobacter* as normal flora. *Acinetobacter* are relatively resistant to antimicrobial therapy, but usually sensitive to tetracycline, kanamycin, and gentamicin (Castle et al. 1978; Centers for Disease Control 1979a).

Aeromonas

Aeromonas hydrophilia is a motile gram-negative rod that primarily inhabits fresh water such as rivers, lakes, and sinks. It can survive on human skin up to 48 hours and can transiently colonize the gut or respiratory tract; it is occasionally isolated from stools in patients with diarrhea. *A. hydrophilia* produces enterotoxins, hemolysin, and cytotoxic extracellular enzymes. It is a rare cause of cellulitis, peritonitis, meningitis, wound infections, tonsillitis septicemia, pneumonia, and gastroenteritis. One recent report of a nosocomial epidemic of *A. hydrophilia* infections involved eight patients in one hospital during a 2-month time period. Patients involved were both immunosuppressed and normal. The spread of infection was significantly correlated with one team of physicians, with the apparent mode of transmission being wound dressing with inconsistent glove changing or handwashing between patients. It was thus demonstrated that person-to-person spread is possible, even of this saprophytic "water" organism. The organism is sensitive to tetracycline and aminoglycosides (Centers for Disease Control 1979c).

Citrobacter

Citrobacter, members of the Enterobacteriaceae, are antigenically similar to *Salmonella.* The two species *C. freundii* and *C. diversus* are aerobic, and are found in soil, water, and the intestinal tracts of humans and animals. In one series over a year's time period, two-thirds of 116 isolates from 77 patients were found to be hospital acquired and almost half of those were considered to be clinically significant. *Citrobacter* have been isolated from persons with diarrhea, urinary tract infections, and meningitis. *C. diversus* was the causative agent of five cases of neonatal meningitis in two hospitals during a period of several months. The only source of the organisms found was nurse's hands and stool, but it was not clear whether the nurses or the babies were colonized first (Centers for Disease Control 1979a; Hodges, Degener, and Barnes 1978).

Enterobacter agglomerans

The taxonomy of *Enterobacter agglomerans* is still undecided. It was until the late 1970s classified in the tribe Erwineae, which are primarily plant pathogens similar to members of the Enterobacteriaceae. It has also been called *Chromobacter typhiflavum*. It is currently classified in the genus *Enterobacter*. *E. agglomerans* is a motile gram-negative rod that is part of the normal skin flora. This was the causative organism for a nationwide epidemic of bacteremia due to contaminated commercial intravenous solution. There were 150 cases with nine deaths in that particular episode. In one general hospital over a 5-month time period, 13 patients had various clinical infections with *E. agglomerans*. In another hospital, on the other hand, there were only 63 isolates over a 5-year period and none were associated with severe disease. The organism has been responsible for abscesses, bacteremia, meningitis, wound and urinary tract infections. It is susceptible to chloramphenicol, colistin, kanamycin, and gentamicin (Geere 1977; Mackel et al. 1975; Ursing 1977).

Flavobacter

Flavobacter ("yellow bacteria") are gram-negative rods that may be motile by means of peritrichous flagella or may be nonmotile. They are aerobic or facultatively anaerobic and are ubiquitous in water, sewage, and soil. There are over 25 species. In one series over a 2-year time period, ten species were isolated as agents of clinical disease in hospitalized patients. *Flavobacter* have been found in hospital sinks, vials, catheters, and in the antiseptic chlorhexidine. *F. meningosepticum* is the primary opportunistic pathogen in the genus and has caused outbreaks of meningitis and bacteremia among normal and premature babies, and rare cases of septicemia after heart surgery. The organism appears to be much more pathogenic for infants; adult colonization is usually asymptomatic (Coyle-Gilchrist, Crewe, and Roberts, 1976).

Legionella

Biologic characteristics. In the family Legionellaceae there is one genus, *Legionella,* and five known species: *L. pneumophila,* the etiologic agents of Legionnaires' disease and Pontiac fever; *L. micdadei,* the etiologic agents of Pittsburgh pneumonia; *L. dumoffii;* and *L. gormanii. L. pneumophila* has about six serogroups. The name *L. pittsburgensis* has also been proposed for *L. micdadei.* The organisms are unique in their production of branched chained fatty acids (Fig. 10-4). Species are listed in Table 10-1. *Legionella* are small, flat, flagellated gram-negative rods. The Gram stain, however, is difficult to use with these fastidious organisms since they can only be seen after culture and passage in the laboratory. Instead,

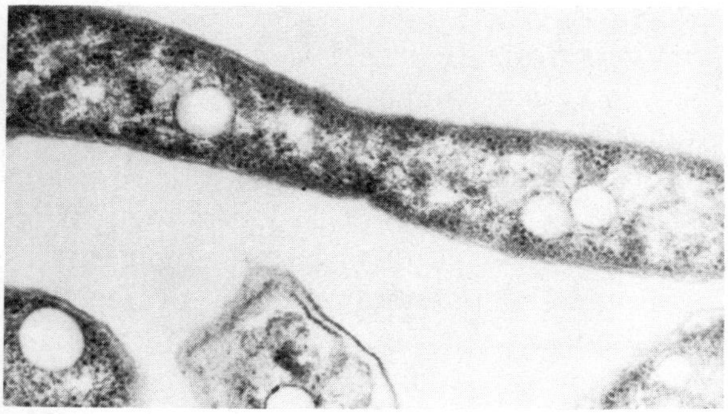

FIGURE 10-4. Electron micrograph of *Legionella dumoffii.* Notice characteristic round inclusion bodies of fat. (Courtesy of Dr. Jimmie Lara, Seattle.)

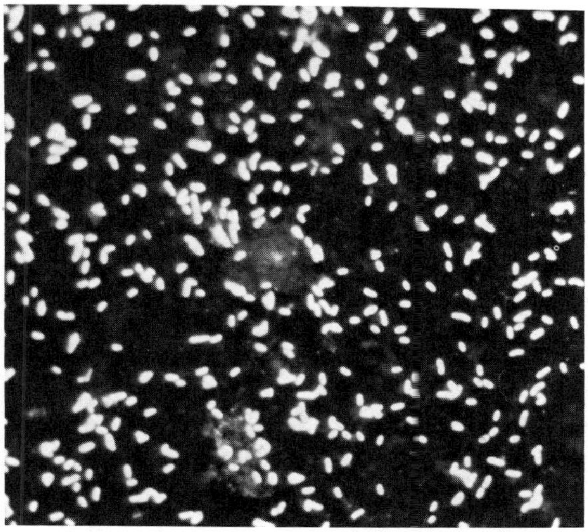

FIGURE 10-5. Fluorescent antibody positive for *Legionella dumoffii.* Magnification 40×. (Courtesy of Dr. James Plorde.)

TABLE 10-1. Species of *Legionella*

Species	*Associated with*
L. bozemanii	Probably no disease
L. dumoffii	Probably no disease
L. gormanii	Probably no disease
L. micdadei (*pittsburgensis,* or the Tatlock agent)	Pittsburgh pneumonia
L. pneumophila	Legionnaires' disease, Pontiac fever, other sporadic outbreaks since 1947

the Dieterle silver impregnation stain for spirochetes and fluorescent antibody (Fig. 10-5) are used. *Legionella* are very difficult to culture, grow slowly, and require rich media (they can be grown in yolk sacs of chick embryos). The antigenically distinct serogroups share a common O antigen and also have type-specific surface antigens.

Epidemiology. *Legionella* are normal inhabitants of soil and water, and are transmitted to humans primarily by aerosolization of contaminated water and in the soil during construction or excavation. Air conditioning systems, condenser water aerosols, and shower heads in large buildings have been sources of the organism in several outbreaks. *Legionella* survive for 12 months in tap water. No person-to-person spread has been documented. Outbreaks have thus far only been reported in Europe and in the United States, probably because of the difficulty in identifying *Legionella.* Because specimens had been saved from persons who were ill, it was possible to confirm that *Legionella* was the causative agent in a number of outbreaks of illness that occurred before *Legionella* was finally isolated and identified in 1976. Legionellosis is most common among those with decreased host defenses such as the aged and those who are hospitalized or immunosuppressed, or those who have lung disease. The disease has also occurred in children.

From October 1974 through September 1977, 263 nosocomial pneumonias among the hospitals participating in the National Nosocomial Infections Study (NNIS) were investigated by personnel at the Centers for Disease Control. Ten of these (3.8 percent) had Legionnaires' disease confirmed by examination of tissue specimens. The median age of cases was 59.5 years, with a range of 13 to 78 years of age. The disease was more common in men than women. Affected individuals had been hospitalized for a median period of 17 days; the range was 5 to 41 days. It occurred most frequently in summer. From these data, an estimated 950 cases of fatal nosocomial pneumonia yearly could be attributed to *L. pneumophila.* In one outbreak at a country club, a stream of air contaminated with *Legionella* from an evaporative condenser in the clubhouse was blowing directly toward an outdoor practice green where the organisms were apparently inhaled (Centers for Disease Control 1980c, 1980d, 1980e, 1981a; Cordes et al. 1980).

Pathogenesis. Though *Legionella* are phagocytized, they survive as intracellular parasites. After entering via the respiratory tract, the organisms remain localized. There are no known toxins produced. The disease is probably dose related and the normal staphylococcal and streptococcal pharyngeal flora have been shown to inhibit the growth of *L. pneumophila.* In guinea pigs the LD_{50} is 1.4×10^5 if given by aerosol and twice as much, 3.0×10^5, if injected intraperitoneally. The median infectious dose for guinea pigs is less than 129 organisms. Variations in virulence occur and may be due to differences in serotypes or to differences in host populations (Berendt et al. 1980).

Host responses. It is likely that many infections are subclinical or very mild. However, *Legionella* is probably responsible for about 10 percent of pneumonia that had been diagnosed previously as "viral" or in which an agent was not identified. The incubation period is 2 to 10 days, after which flulike symptoms or pneumonia may develop. The pneumonia is accompanied by high fever, lung consolidation, hypoxemia, and sometimes pulmonary cavitation. Many patients also have central nervous system symptoms, especially confusion or mental clouding. Acute renal failure can occur. The case fatality rate in pneumonia is about 10 percent. Diagnosis is made by immunofluorescence. If the patient's serum contains antibodies to *L. pneumophila,* they will be tagged on the organism.

Control and treatment. Prevention of legionellosis is difficult since the organism is widespread in nature. Treatment of the pneumonia is supportive and symptomatic. Erythromycin and rifampin, both active against intracellular parasites, are efficacious. Tetracycline is also used. The occurrence of nosocomial cases is an indication that an epidemiologic investigation to identify the source is in order.

Moraxella

The genus *Moraxella* is closely related to *Acinetobacter.* About eight species of *Moraxella* have been recovered from various clinical specimens including the nose, eye lesions, the upper respiratory tract, and the genitourinary tract. It has low pathogenicity in general. *M. nonliquefaciens, osloensis,* and *phenylpyrouvica* are part of the normal flora and may cause rare cases of urinary tract infection, septicemia, and septic arthritis. *M. kingii* and *lacunata* can cause conjunctivitis and occasionally, endocarditis. Other gram-negative rods that may be rare causes of opportunistic infections are listed on Table 10-2.

FRANCISELLA TULARENSIS

Biologic characteristics. *Francisella tularensis* was previously called *Pasteurella tularensis*. It is nonmotile, aerobic or facultatively anaerobic, nonspore-forming, nonencapsulated, and very small. Special media containing cysteine are necessary for growth, which is very slow. Even after up to 14 days the colonies are tiny, viscous, and gray to transparent. There is one serotype with three cell wall antigens: a polysaccharide, a protein, and an endotoxin. *F. tularensis* causes tularemia and is antigenically related to *Brucella* and *Yersinia*.

Epidemiology. Tularemia was first described in California in 1914. It is acquired by humans from small mammals, especially rabbits, but also rats, squirrels, muskrats, beavers, and ground hogs, either by an insect vector, by direct contact with infected tissues, by bites, by ingestion of infected meat, or by inhalation. The disease has been reported throughout Europe and in Japan as well as in the United States.

Pathogenesis. An endotoxin similar to that of other gram-negative bacteria is produced, but no exotoxins. *F. tularensis* is an intracellular parasite. After entering the body, the organisms are carried to the regional lymph nodes, which become painful, swol-

TABLE 10-2. Gram-Negative Organisms that are Rare Causes of Opportunistic Disease

Species	*Alternate names*	*Habitat*	*Human diseases*
Actinobacillus actinomycetemcomitans		Mouth	Abscesses of brain, soft tissue; endocarditis
Aeromonas shigelloides	*Pleisomonas shigelloides*	Water, soil	Diarrhea; abscesses; osteomyelitis; endocarditis; meningitis; wound infections
Cardiobacterium hominis	CDC Group IID	Upper respiratory tract	Endocarditis; liver abscesses; sepsis
Chromobacterium typhiflavum, violaceum		Water, soil	Abscesses; myocarditis
Eikenella corrodens	CDC HB-1; *Bacteroides corrodens*	Water, soil, mouth	Abdominal abscesses; endocarditis; meningitis; infections in human bites
Fusobacter fusiforme, gonidiaformans, necrophorus, sphaerophorus		Water, mouth, GI, GU tracts	Abscesses; endocarditis; conjunctivitis
Veillonella		Mouth, vagina	Intraabdominal abscesses; osteomyelitis

len, and may drain. Bacteremia results in spread to liver, spleen, and lungs. The organism can be recovered from lymph nodes for some time after clinical disease has subsided. Strains vary in virulence; the more virulent strains are from tick-borne disease of rabbits. Since a very small inoculum is needed to cause disease, the organism is dangerous in the laboratory.

Host responses. The incubation period of tularemia is 3 to 10 days. Clinical signs and symptoms vary by mode of entry as described on Table 10-3. The primary lesion is usually ulceroglandular with headache, fever, formation of abscesses, and areas of necrosis. The disease can become chronic or, because the organisms can survive well in the RES, can relapse for weeks. Eye disease, pneumonia, and a typhoidlike gastrointestinal disease can occur, as well as rare cases of endocarditis, peritonitis, meningitis, or osteomyelitis. Convalescence is prolonged. Without treatment the patient may be permanently, though mildly, debilitated. Case fatality rate with treatment is about 1 percent, in the untreated it attains 30 percent. Diagnosis is by fluorescent antibody, agglutination tests, and serologic tests. Agglutinins appear after about 10 days and remain for years. A rising titer (it continues to increase for 8 weeks after infection) confirms the diagnosis. Delayed hypersensitivity develops and a skin test is available as a sign of current or previous infection.

Control and treatment. Immunity is usually solid, though reinfection has occurred. Cross agglutination with *Brucella* sometimes occurs. Gloves should be worn for handling wild animal carcasses. An attentuated vaccine can protect individuals at high risk from virulent forms of the disease. Streptomycin is the best antimicrobial agent; recurrences occur with tetracycline and chloramphenicol since they are only bacteriostatic. Laboratory personnel who handle specimens are in danger of acquiring infection; therefore, specimens should be handled with care.

HEMOPHILUS-BORDETELLA ORGANISMS

Hemophilus and *Bordetella* are two genera of tiny gram-negative rods in the family Brucellaceae. They normally inhabit the human upper respiratory tract. There are eight species of *Hemophilus* and three species of *Bordetella,* all of which are closely related, and are known to cause disease in humans. *Hemophilus* species are listed on

TABLE 10-3. Clinical Manifestations of Tularemia According to Route of Entry

Route of entry	*Clinical manifestations*
Skin	Granulomatous nodules (buboes) of lymph nodes that form draining abscesses
Conjunctiva	Oculoglandular
Inhalation of infected droplets or hematogenous spread to lungs	Pneumonic, resembling plague
Ingestion	Disease resembling typhoid fever with fever, toxemia, gastrointestinal symptoms

Table 10-4. *Bordetella pertussis,* the etiologic agent of whooping cough, is the major pathogen in the *Bordetella* group. *B. parapertussis* and *bronchiseptica* can also cause upper respiratory tract infections resembling whooping cough. *B. bronchiseptica* is often isolated from the respiratory tracts of canines; disease results on rare occasion among persons with animal contacts. *B. bronchiseptica* is more closely related to *H. influenzae* and *Brucella* than to the other *Bordetella*.

Bordetella pertussis

Biologic characteristics. *Bordetella pertussis* is an aerobic, encapsulated, small, coccoid, piliated, nonmotile gram-negative rod that requires complex media for growth. Colonies are small and white, and take up to 5 days to grow. Often a special medium made of potato, blood, and glycerol, called Bordet-Gengou agar, is used. After several passages in artificial media in the laboratory, the smooth (S) encapsulated form loses its capsule and pili and becomes a rough (R) strain, which is nonvirulent. The states from fully encapsulated, virulent strains to nonencapsulated avirulence are phase 1 to phase 4. Vaccines produced from S (phase 1) strains are much more effective. *B. pertussis* contain both genus-, species-, and strain-specific antigens. Identification is by immunofluorescence or agglutination.

Epidemiology. Whooping cough occurs only in humans and is worldwide in distribution. Before the use of the DPT vaccine, whooping cough occurred in the United States in widespread epidemics, often devastating or fatal to children. There are now only about 2000 cases per year in this country (Fig. 10-6). The organism is spread by droplet and is highly contagious; over 80 percent of susceptible contacts develop the disease. The majority of unvaccinated individuals in endemic areas have had whooping cough by the time they are age 5. Morbidity and mortality from the disease, therefore, occurs primarily in infants. The organism may be spread by fomites, but this is uncommon since *B. pertussis* survives for only a short period of time outside a human host. The

TABLE 10-4. *Hemophilus* Species in Humans

Species	*Habitat*	*Diseases*
H. aegyptius	Conjunctiva	Conjunctivitis
H. aphrophilus		Rare endocarditis
H. ducreyi	Genital tract	Soft chancre
H. haemolyticus	Upper respiratory tract	Very rare subacute endocarditis
H. haemoglobinophiles (canis)	Dogs	Very rare endocarditis
H. influenzae	Upper respiratory tract	Upper respiratory infection; conjunctivitis; meningitis
H. parahaemolyticus	Upper respiratory tract	Pharyngitis; rare subacute endocarditis
H. parainfluenzae	Upper respiratory tract	Respiratory disease; rare endocarditis

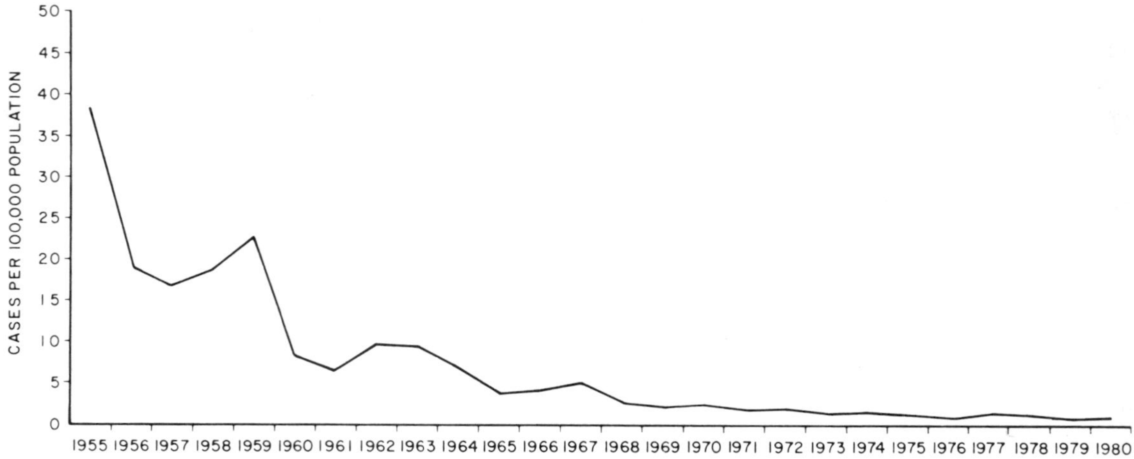

FIGURE 10-6. Pertussis (whooping cough). Reported case rates by year, United States, 1955 to 1980. (From Centers for Disease Control. Annual Summary 1980. Morb Mort Week Rep 1981;29:61.)

disease is most communicable during the early acute stages of illness, before the "whoop" begins. There is no carrier state, but mild, undiagnosed cases are major disseminators.

Pathogenesis. A number of factors determine the pathogenesis of *B. pertussis*. Since phase 1 virulent bacteria adhere better to ciliated epithelial cells than do avirulent phases, adherence may be important in the development of infection. A cell wall exotoxin is released on cell lysis and a lymphocytosis-promoting factor causes a severe lymphocytosis, often with a count of 20,000 to 100,000 with 80 percent lymphocytes. *B. pertussis* adheres specifically to epithelial surfaces of the trachea and bronchi, where it multiplies and causes localized inflammation and hyperplasia of lymphoid tissue. The organism does not spread in the blood stream, but remains localized.

Host responses. After an incubation period of 10 to 20 days, the disease starts with runny nose, mild cough, and inflammation of the eyes. These symptoms continue for 1 to 2 weeks. It is during this phase that infectiousness is greatest. Following this, the cough becomes more severe with paroxysms accompanied by an inspiratory noise or whoop, expectoration of thick sputum, and often vomiting. Fever is usually absent, but a profound elevation of leukocyte count is characteristic. There may be an increased bleeding tendency. The whooping phase may last up to 6 weeks. Secondary bacterial infections and bronchopneumonia occur in 5 to 10 percent of cases; encephalitis and seizures occasionally result. Death is uncommon, but occurs most frequently in infants. Adenovirus also causes a whooping-cough-like syndrome identical to that caused by *B. pertussis*.

Control and treatment. Though antibodies appear after about 20 days, immunity is not permanent. For that reason, newborns receive little passive protection from maternal serum and infants are highly susceptible to whooping cough, until they are immunized. However, subsequent attacks are milder than first ones. The primary control is with immunization beginning by age 8 weeks with killed virulent organisms, usually given in combination with diphtheria and tetanus toxoids (DPT). *B. pertussis* is an adjuvant for the two toxoids. Like protection after active infection, vaccine protection is also short lived. There is no evidence that hyperimmune pertussis globulin is effective in preventing or minimizing disease among susceptible contacts. However, low doses of booster vaccine may afford some protection.

Erythromycin is used to treat the disease and as a prophylactic measure among exposed susceptible individuals. It is of minimal use after the whooping stage has begun, however. Isolation of cases is of little value since the most contagious period often occurs before the diagnosis is made. Nevertheless, individuals with whooping cough should be segregated from high-risk individuals such as infants and the elderly.

Hemophilus influenzae

Biologic characteristics. *Hemophilus influenzae* is a gram-negative coccobacillus that is nonmotile, encapsulated, pleomorphic, and aerobic or facultatively anaerobic. There are six types, A through F, based on capsular polysaccharide and identified by capsular swelling tests. About 95 percent of human disease is caused by type B, which has a unique component in its capsule called polyribose phosphate (PRP).

H. influenzae has two distinct growth requirements that are useful in its identification, Factor X, hematin, and factor V, which can be replaced by NAD or NADP. *Hemophilus* species can be differentiated by whether or not they need factor X, factor V, or both for growth, and whether they are hemolytic on blood agar. Though red blood cells contain both factors, *H. influenzae* cannot grow until the blood cells in artificial media are lysed. Therefore, on sheep's blood agar plates, *H. influenzae* can only grow in proximity to hemolytic staphylococci that cause the release of factor V from the hemolyzed blood cells as they grow. This phenomenon of growth around staphylococcal colonies is called *satellitism* or the *satellite phenomenon,* and can be used for identification of *H. influenzae*. Chocolate (heated-blood) agar is used for isolation of *H. influenzae* because the red blood cells are then lysed and release factors X and V. Strains grown in the laboratory easily lose their capsules and, with them, their virulence.

Epidemiology. Humans are the only natural host of *H. influenzae*. Disease is spread by the respiratory route and occurs worldwide. About 20 to 30 percent of healthy newborns have nasopharyngeal (NP) colonization with avirulent nontypable strains, 3 to 5 percent with encapsulated type B. Colonization is about 80 percent after disease. The average annual incidence of meningitis due to *H. influenzae* in one study was 5.2 per 10^5 and of other types of infections due to *H. influenzae,* 3.5 per 10^5. Rates were highest in infants, black children were more susceptible to meningitis than white children, and the secondary attack rate in household contacts less than 2 years of age was about 1 in 24 in the first 2 months after infection. From 1971 through 1977, 68 percent of meningitis among Eskimos in Alaska was due to *H. influenzae*. The average annual incidence of all disease caused by *H. influenzae* in children under age 5 years was 491 per 10^5, for meningitis, 409 per 10^5. In this group, as well as in most populations, well over 90 percent of children affected were less than 18 months of age. It was concluded that the high incidence in the Eskimos was due to early exposure rather than to an unusual susceptibility since anticapsular antibody titers were 3 times those for children of similar age in the continental United States. Incidence is greatest in winter and early spring. Though there is some disagreement, there seems to be a rising incidence of meningitis caused by *H. influenzae* type B among children less than 5 years of age (Fraser 1980; Granoff and Basden 1980; Ward 1981).

Pathogenesis. The polyribose phosphate component of the capsule seems to be the major cause of virulence, as antibodies against PRP are protective against further clinical disease. Invasion of *H. influenzae* in the NP causes suppurative inflammation with an exudate of neutrophils, organisms, and fibrin. From the NP the organisms can infect regional lymph nodes, enter the circulatory system, or invade the meninges. There are no exotoxins produced.

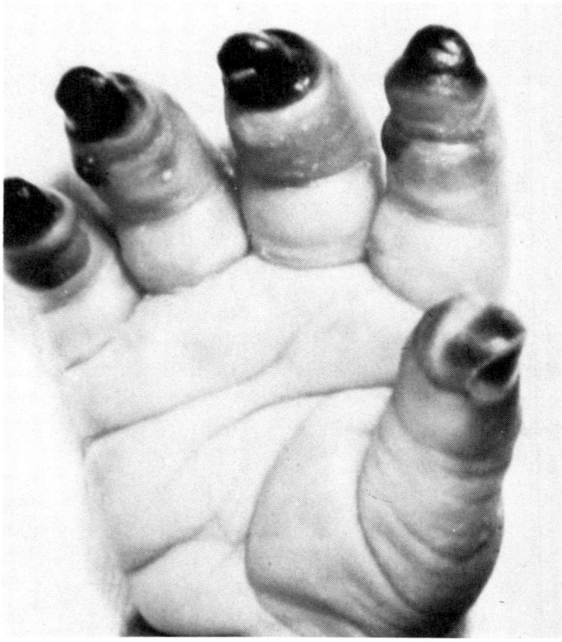

FIGURE 10-7. Peripheral gangrene accompanying *H. influenzae* sepsis in a child. Typical lesions of septic shock (not specific to *H. influenzae*). (Courtesy of University of Washington Clinical Microbiology Laboratory.)

Host responses. Asymptomatic carriers are common. However, in some individuals, serious diseases—primarily respiratory illness or meningitis—can occur. *H. influenzae* is a common cause of pharyngitis in children. Also occurring primarily in children are epiglottitis with "croupy" breathing, laryngotracheobronchitis with airway obstruction, pneumonia, bronchitis, otitis media, meningitis, cellulitis, and sinusitis. Laryngitis with epiglottitis can be a medical emergency in children, requiring tracheostomy. *H. influenzae* is second only to *S. aureus* as a cause of septic arthritis in children; other forms of sepsis also occur (Fig. 10-7). It is an occasional cause of genital infections such as epididymitis or salpingitis. Nontypable strains can occasionally cause otitis media, conjunctivitis, corneal ulcers (Fig. 10-8), and meningitis; however, 95 percent of *H. influenzae* meningitis is caused by type B. A few cases are caused by type A.

Almost all (about 97 percent) of *H. influenza* meningitis occurs in children less than 15 years of age. It is especially prevalent and devastating in children between the ages of 6 months and 2 years, during which time the organism is the most common cause of meningitis. About half of the cases are accompanied by bacteremia and are preceded by an upper respiratory tract infection. If untreated, the meningitis is almost

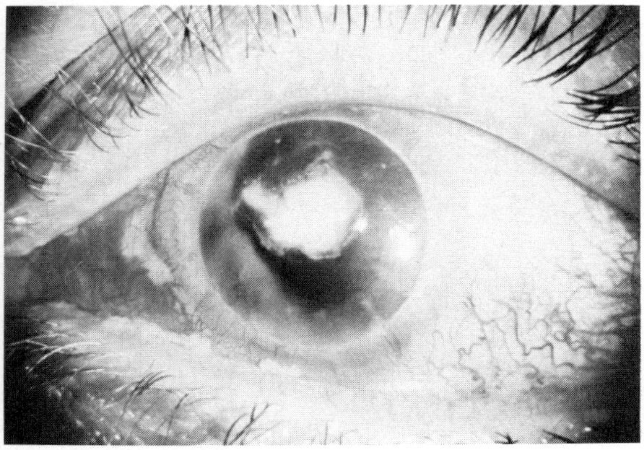

FIGURE 10-8. Corneal ulcer caused by *Hemophilus.* (Courtesy of Dr. John Chandler.)

always fatal. About 70 to 80 percent of adults have protective antibodies against type B. Nontypable strains may be more likely to cause disease in adults, especially those who are immunosuppressed. For example, in one series of 103 cases of bacteremia or meningitis due to *H. influenzae* in adults, 60 percent of cases were the result of pneumonia, 15 percent were the result of *H. influenzae* genital infections. Most infections in this series were due to nontypable serotypes (Wallace et al. 1981). *H. influenzae* is a frequent cause of exacerbation of bronchitis in persons with chronic obstructive pulmonary disease.

Control and treatment. Specific capsular polysaccharide antibody, especially secretory IgA, to *H. influenzae* type B occurs in high concentrations in breast milk and offers passive protection to infants for about 6 months after birth. No treatment for colonized infants is recommended, except perhaps for those carrying type B. Type-specific immunity develops after infection with the production of anticapsular antibody. *H. influenzae* meningitis is not infectious. Even with respiratory infections, isolation is not necessary. Close contacts who are less than 3 years of age (siblings, nursery school contacts) should be closely observed and may receive prophylactic ampicillin. However, there is an increasing frequency of ampicillin-resistant strains due to plasmid transfer of beta-lactamase production. Probably about 10 percent of strains are now resistant. Treatment consists of administration of ampicillin, chloramphenicol, or tetracycline early in the course of the disease. Treatment should be continued for 2 weeks. Currently no vaccine is available, though research in this area is underway.

Other *Hemophilus* Species

Hemophilus parainfluenzae may occasionally cause endocarditis, epiglottitis, or meningitis. *H. aphrophilus* rarely causes brain or soft tissue abscesses or endocarditis. *H. suis* played a significant role in the 1918 swine influenza epidemic. Concurrent infection with the swine flu virus and *H. suis* are synergistic in their disease-producing ability, causing a much worse syndrome together than either could separately. *H. aegypticus* is the most common *Hemophilus* strain that causes conjunctivitis or "pinkeye." It is highly contagious and is most common in winter when school children are in close quarters. *H. ducreyi* causes sexually transmitted genital ulcers called *chancroid.* The lesion of chancroid resembles the primary chancre of syphilis and the two conditions must be differentiated by laboratory techniques. A positive delayed hypersensitivity skin test called Ducrey's test remains for years after infection. Reinfections occur. Treatment is with oral erythromycin or trimethoprim-sulfamethoxazole.

NEISSERIA: THE GRAM-NEGATIVE COCCI

Neisseria are aerobic gram-negative cocci, some of which are normal oropharyngeal flora. Species found in humans are listed on Table 10-5. There are two important pathogens: *N. gonorrhoeae* and *N. meningitidis,* though occasionally other species can cause disease. *Neisseria* are piliated, but nonmotile. The cocci are kidney shaped; pairs tend to join on the concave sides, giving the appearance of doughnuts on Gram stain. The pathogenic species are often called pyogenic cocci and, like the gram-positive pathogenic staphylococci and streptococci, are resistant to phagocytosis. Though *Neisseria* are aerobes, they grow best in an atmosphere of 5 to 10 percent CO_2. They are all oxidase positive. *Acinetobacter* and *Moraxella,* discussed in the section "Commensals That Become Opportunistic Pathogens" are, along with *Neisseria,* in the family Neisseriaceae. Like *Hemophilus, Neisseria* grow best on chocolate agar because treating the blood removes the fatty acids and trace metals that inhibit their growth. Often, Thayer-Martin (also called VCN) selective medium is used for isolation of *Neisseria.* It contains vancomycin, colistin or polymyxin, and nystatin to inhibit the growth of many bacteria and yeasts that are normal flora.

Neisseria gonorrhoeae

Biologic characteristics. *N. gonorrhoeae* are gram-negative diplococci (Fig. 10-9). There are at least four distinct types, which can be identified by their colonial morphology. Types 1 and 2 are encapsulated, piliated, and are pathogenic; types 3 and 4 are nonpiliated and less virulent. All types seem to occur by mutation within a single strain. The organisms are antigenically heterogeneous and there is a large spectrum of ability among these organisms for growing on artificial media, with some types being much more fastidious than others. The virulent strains tend to lose their pili and pathogenicity after prolonged growth in the laboratory.

TABLE 10-5. Species of *Neisseria* Found in Humans

Species	*Habitat*	*Diseases*
N. catarrhalis (Branhamella catarrhalis)	Mouth, nasopharynx	Occasional endocarditis, meningitis, septicemia, bronchitis, otitis media
N. flava	Nasopharynx	Rare endocarditis, meningitis, septicemia
N. flavescens	Nasopharynx	Occasional meningitis
N. gonorrhoeae	Pathogen, GU tract	Gonorrhea
N. lactamica	Nasopharynx	Occasional endocarditis, meningitis, septicemia
N. meningitidis	Nasopharynx	Common meningitis
N. perflave	Nasopharynx	None documented
N. sicca	Nasopharynx	None documented
N. subflave	Nasopharynx	Rare meningitis

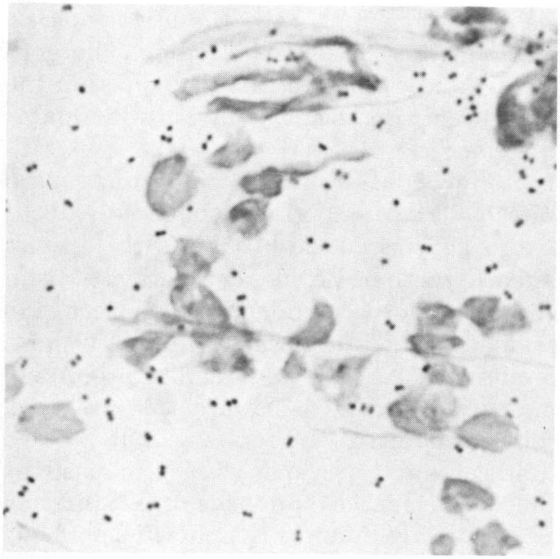

FIGURE 10-9. Extracellular gonococci in a urethral specimen. (Courtesy of Dr. James Plorde.)

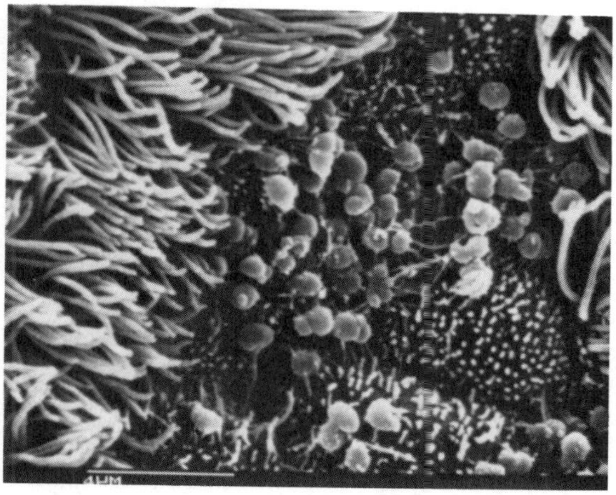

FIGURE 10-10. Scanning electron micrograph of normal human fallopian tube incubated with *Neisseria gonorrhoeae* (P+ +Tr phenotype) for 30 minutes. Attachment of the gonococci was to tips of microvilli of nonciliated columnar epithelial cells. Magnification 6750×; bar equals 4 mm. (Reprinted with permission of the publisher from Draper DL, et al. In vitro modeling of acute salpingitis caused by *Neisseria gonorrhoeae.* Am J Obstetr Gynecol 1980;138:996-999.)

Epidemiology. Humans are the only natural host of *N. gonorrhoeae*. The disease gonorrhea is highly prevalent worldwide. The organisms are almost always transmitted by sexual contact. Gonorrhea currently ranks first among reported communicable diseases in the United States. An estimated 5 million cases of genital infection occur each year, as well as about 3 babies per 100,000 born who develop gonococcal ophthalmia neonatorum, a congenital disease acquired by newborns as they pass through an infected birth canal. About 30 percent of babies born to women infected with gonorrhea congenitally acquire conjunctivitis or other gonococcal infection. Prolonged rupture of membranes may allow ascending infection to the fetus, as well as infection acquired during delivery. In some groups, the prevalence of genital infection is from 2 to 5 percent.

Pathogenesis. Strains of gonococci vary in virulence. There seem to be two major mechanisms for their pathogenicity. First, surface pili on types 1 and 2 enhance attachment to the genitourinary mucosal epithelial cells (Fig. 10-10). Second, their capsules prevent phagocytosis. Types 3 and 4 are readily phagocytized. The pathogenic *Neisseria* also produce a protease that can inactivate IgA. After attachment of pathogenic strains, neutrophils accumulate and cause a purulent exudate. Organisms are found both intracellularly and extracellularly. The bacteria penetrate epithelial cells and extend locally in the cervix, urethra, prostate, uterus, and sometimes to the ovaries and fallopian tubes. Dissemination in the blood stream also occurs. In vitro, the gonococcus can cause hyperplasia, focal necrosis, and sloughing of ciliated cells of the fallopian tubes.

Host responses. The primary disease in adults is a purulent urethritis, though asymptomatic disease in both sexes (but especially females) is common. About 30 percent of females are carriers without diagnosed disease; only about 5 percent of men are free of urethral discharge. In females, the symptoms of acute gonorrhea may mimic those of urinary tract infection. Dysuria and increased frequency of urination are prominent, along with increased vaginal discharge or abnormal menstrual bleeding, vaginitis, cervicitis, and urethritis. In about 20 percent of women the infection extends to the fallopian tubes and salpingitis (inflammation of the fallopian tubes) results. In males, there is almost always purulent urethritis after a 2 to 6 day incubation period. If untreated, it lasts several months and occasionally epididymitis results. However epididymitis is rare among those who are treated. Proctitis can also occur in males. When gram-negative diplococci are seen on smear from the urethra, it is often used as presumptive diagnosis of gonorrhea. The organisms can subsequently be isolated in culture.

Pharyngeal infections such as tonsillitis occur in both sexes as a result of oral exposure to the gonococcus. Both sexes can develop pelvic inflammatory disease, arthritis, endocarditis, meningitis, or perihepatitis (Fitz-Hugh-Curtis syndrome) from disseminated infection. Disseminated gonococcemia occurs in about 1 to 2 percent of individuals who are infected, whether or not they have previous symptoms. The disease is characterized by fever, pustular or hemorrhagic skin lesions (usually on the extremities), and synovitis. The bacteremia may last for about a week and subside, or septic arthritis can also develop. The gonococci can be isolated from the blood in the early stages and from synovial fluid in later stages of the illness.

Gonococcal ophthalmia neonatorum (GCON) is a conjunctivitis that occurs about 2 to 5 days after the birth of a child to an infected mother. Besides conjunctivitis, congenital gonococcal infection can result in prematurity, fetal loss, low-birth weight, premature rupture of membranes, pneumonitis, meningitis, anorectal infection, rhinitis, arthritis, and septicemia. It is likely that conjunctivitis can also be caused by contamination of the eyes from infected adults during infancy. Eye infections among adults in the military caused by penicillin-resistant gonococci have been recently reported. Adult eye infections, however, are rare (Fig. 10-11).

Control and treatment. Despite the devel-

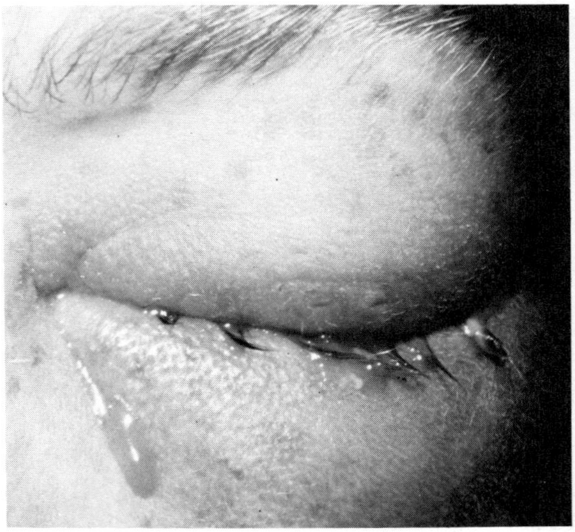

FIGURE 10-11. Gonococcal conjunctivitis in an adult. The syndrome is much more common in infants. (Courtesy of Dr. John Chandler.)

opment of antibodies, primarily secretory IgA, an adequate immune response does not ensue, so reinfection is common. Because of this, and because the rate of asymptomatic infection is so high, the tracing of sexual contacts of cases is very important. The use of condoms during sexual intercourse offers some degree of protection for both partners. Tests for syphilis and chlamydiae should be conducted for all individuals with gonorrhea, since these diseases often occur simultaneously.

Since its first use, penicillin has been the agent of choice in treating gonococcal infections. Unfortunately, in the past few decades, strains of *N. gonorrhoeae* that are capable of plasmid-mediated beta-lactamase production (see Chapter 26) have been recognized. These penicillin-resistant, penicillinase-producing *N. gonorrhoeae* (PPNG) strains were first discovered and are most prevalent in Southeat Asia and the western Pacific. The spread of R plasmids among *N. gonorrhoeae* is still rare in the United States, though PPNG account for 40 percent of gonococcal disease in the Philippines. Strains of PPNG have been reported from at least 27 countries. In four years (May, 1976 through April, 1980), 1022 PPNG were reported to the Centers for Disease Control, with a 66 percent increase for the first 4 months of 1980 (Fig. 10-12). Nevertheless, the treatment of choice in most regions of the United States is still aqueous procaine

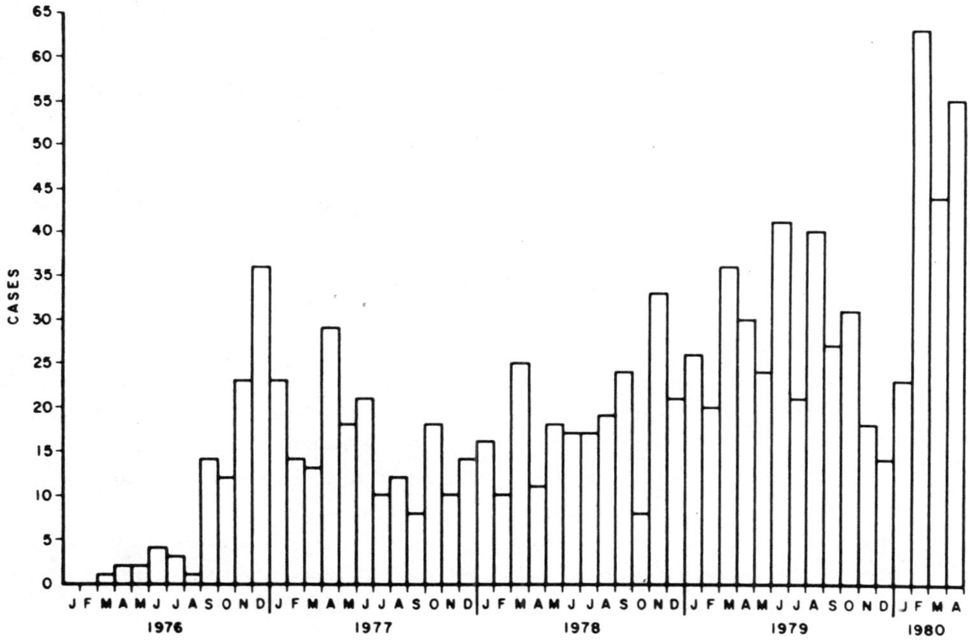

*Includes Commonwealth of Northern Mariana Islands, Federated States of Micronesia, Guam, Puerto Rico, and Virgin Islands.

FIGURE 10-12. Cases of PPNG, by month, United States, March, 1976 to April, 1980. (From Centers for Disease Control. An outbreak of penicillinase-producing *Neisseria gonorrhoeae.* Morb Mort Week Rep 1980;29:241-243.)

penicillin, 4.8 million units I.M. at two sites with 1.0 g of probenecid orally. The failure rate of this treatment is about 5 percent. Some of the failures are due to reinfection or to resistant strains. Alternatives to penicillin are tetracycline, cefoxitin, and especially spectinomycin (which should be reserved for use against PPNG only). Unfortunately, spectinomycin-resistant strains have recently been reported from England. Cultures should be taken again 3 to 7 days after therapy is complete. Cultures still growing *N. gonorrhoeae* should be tested for penicillinase production and treated with spectinomycin.

Preventive treatment of the eyes of newborns with 1 percent silver nitrate ($AgNO_3$) drops is well established. Though $AgNO_3$ is efficacious, 2 to 5 percent of infants with infected mothers still get GCON despite treatment. In addition, silver nitrate is not effective against *Chlamydia trachomatis,* another common sexually transmitted disease. This problem is discussed under "Clinical Implications" at the end of this chapter. Alternative eye prophylaxis is possible with 1 percent tetracycline or 0.5 percent erythromycin in single-dose containers, in the delivery room as soon after birth as possible. Tetracycline is probably more effective against the gonococcus than erythromycin, though efficacy is unproven for either drug. Penicillin, though effective, is not used because of the possibility of allergic sensitization in the newborns. Pregnant women should have cervical cultures taken prenatally. Those with intrapartum gonorrhea should receive 20,000 to 50,000 units I.V. or I.M. penicillin G. Gonococcal infections in babies are highly infectious. Affected neonates should be placed on secretion or wound and skin precautions for 24 hours following the initiation of effective treatment. Complicated or disseminated gonococcal infections in babies will require treatment for at least a week.

Neisseria meningitidis

Biologic characteristics. *Neisseria meningitidis,* commonly known as the meningococcus, is an aerobic, nonmotile, typical member of the genus *Neisseria.* Meningococci produce mucoid colonies due to large amounts of capsular polysaccharide. Only about 5 percent are piliated. The pili, contrary to those of the gonococci, seem to play no role in pathogenesis. The cocci are D-shaped, with one flattened edge. There are three major antigenic components of *N. meningitidis:* group-specific capsular polysaccharide, lipopolysaccharide cell wall components almost identical to the endotoxin of other gram-negative bacteria, and genus-specific proteins. There are seven serogroups defined by agglutination of susceptible bacteria with type-specific antisera against the various capsular polysaccharides. The groups are named A, B, C, D, X, Y, and Z. Like gonococci, the meningococci are best isolated on Thayer-Martin medium.

Epidemiology. Humans are the only natural host of meningococci. Transmission is by the respiratory route. In the first half of the century, epidemics occurred every 8 to 12 years, usually in winter, but there have been few major epidemics of meningococcal disease in the United States during the past 40 years (Fig. 10-13). Meningococcal infection is one of the most frequent infectious causes of mortality in healthy persons, with most cases occurring in children and youths. Infants less than 1 year of age have the highest frequency of disease occurrence of any age group. During the first 9 weeks of

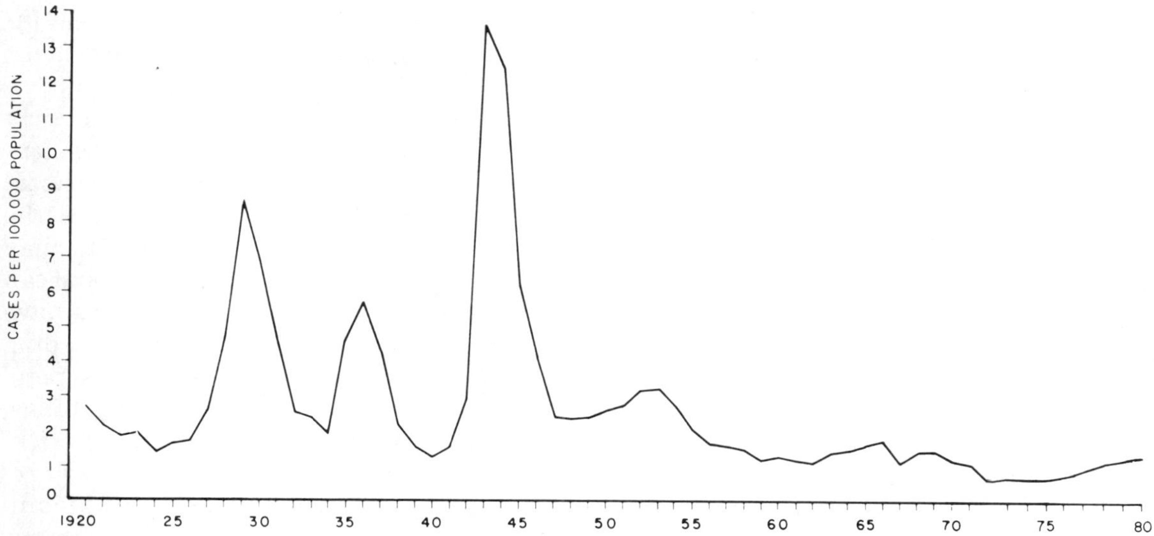

FIGURE 10-13. Meningococcal infections (total). Reported case rates by year, United States, 1920 to 1980. (From Centers for Disease Control. Annual Summary 1980. Morb Mort Week Rep 1981;29:54.)

1980, there were 893 cases of meningococcal disease reported in the United States, an estimated rate of 2 cases per 100,000 persons. Until the 1960s, group A meningococci had been the type primarily associated with epidemics. Groups B and C have become more prominent since that time. In New Zealand, however, most isolates before 1979 were commonly group B. In 1979 sulfonamide-resistant group A strains became predominant. Groups X, Y, and Z are frequently found the asymptomatic nasopharyngeal carriers but have caused only occasional scattered cases of disease. Y strains have caused epidemics of pneumonia (which is an unusual form of the disease) among military personnel. Group D infection is uncommon.

It is estimated that about 10 percent of healthy persons carry the meningococcus in their nasopharynxes. In one survey of more than 1200 asymptomatic Canadian children in a civilian population during nonepidemic times, 2.4 percent were carriers. No differences were found in carrier rates by sex, or by whether or not they were hospitalized. However, 35 percent of contacts of cases of meningococcal disease and carriers were colonized by the eighth week of surveillance. The mean duration of carriage was 15.2 weeks in disease-free families, 5.5 weeks in families in which there were persons with meningococcal disease. Serogroups B and C were most common. There were no secondary cases in 29 families of index cases (Marks, Frasch, and Shapera 1979). Most outbreaks occur in groups such as in military training camps and boarding schools, which have highly concentrated populations. Individuals in groups may be more prone to infection because there is a mass gathering of many people, with varying exposures, carrying many different strains. Nasopharyngeal carriage rates among individuals in these conditions can be 30 to 40 percent and during epidemic times, carriage rates exceed 90 percent. There is at least one reported case of nosocomial meningococcal group B meningitis in which the diseased patient required frequent nasopharyngeal suctioning. Transfer of the organism to another patient resulted from inadequately washed hands of hospital personnel (Rose, Lenz, and Sheth 1981). The meningococcus can occasionally also colonize the anus, cervix, or urethra.

Pathogenesis. *N. meningitidis* may remain localized in the nasopharynx or it may spread in the blood stream. The meningococcal capsule, by preventing phagocytosis, is the major factor in virulence of the organism. On the other hand, it has been demonstrated that the capsular polysaccharide impedes adherence, making host colonization more difficult. Thus, the nonencapsulated forms may be more likely to colonize the nasopharynx, yet less likely to cause disease. The lipopolysaccharide-endotoxin cell wall component is the toxic agent that causes blood vessel thrombosis, clotting, hemorrhage, and disseminated intravascular coagulation (DIC).

Host responses. Very few individuals who have nasopharyngeal colonization with *N. meningitidis* become clinically ill. Asymptomatic carriage can last for weeks or months, providing the primary source of spread of the disease to susceptible hosts. Thus *N. meningitidis* is a successful parasite: it lives well in the host while rarely causing damage. Protective antibodies are formed within 7 days of nasopharyngeal colonization. The absence of protective antibodies is the reason that meningococcal infections are most devastating and prevalent among children. During the nasopharyngeal colonization stage there may or may not be a mild respiratory illness. Occasionally *N. meningitidis* enters the cervical lymph nodes and from there invades the blood stream. This meningococcemia may cause acute, fulminating symptoms and can be fatal in less than 12 hours, or it can have a gradual onset. The organisms spreading through the body cause hemorrhagic petechial lesions in the skin, joints, heart, lungs, or adrenal glands (Waterhouse-Fredericksen syndrome). The accompanying rash results from blood vessel thrombosis. Meningococcemia is often a medical emergency with high fever, endotoxic shock, and DIC.

After the meningococcemia, the organisms may cross the blood-brain barrier to cause meningitis. In 20 to 40 percent of patients, meningitis occurs without overt symptoms of menigococcemia. The symptoms of meningococcal meningitis may be atypical in infants. Diagnosis is made by smear and culture of the organisms from lesions, blood, or spinal fluid. Serologic tests are used to identify the serotype. Meningococci can occasionally also cause sinusitis, conjunctivitis, and genitourinary diseases resembling those caused by gonococci such as cervicitis, epididymitis, vaginitis, and urethritis.

Groups A and C meningococci induce IgG or IgM antibody formation so that the organisms are opsonized, readily phagocytized, and killed. Group B capsular polysaccharide unfortunately is poorly immunogenic, so that hemagglutinating antibody titers are low even after infection. This is possibly because the capsular polysaccharide of group B strains contains largely neuraminic acid, which closely resembles that in human cell membranes. Thus, the host may not recognize the substance as foreign.

Control and treatment. Preventive measures are directed toward minimizing the carrier rate and vaccinating susceptible individuals. There are vaccines against the capsular polysaccharide of types A and C meningococci, but no effective vaccine against type B. The antibody response with group A vaccine has been shown to be markedly age-dependent. As age increases, there is a slower decrease in vaccine-induced antibody levels. Use of types A and C vaccines in military populations may be one reason why there is an increased incidence of group Y infections. Avoidance of overcrowding is useful to minimize the carrier state. The carrier state can be eradicated with rifampin that is adsorbed into nasopharyngeal secretions. There is no conclusive evidence that prophylaxis among household contacts or health care personnel with massive exposure to oral secretions is efficacious, but rifampin, ampicillin, and minocycline have been used for this purpose. Minocycline has side effects (such as vertigo, nausea, and vomiting) that limit its use.

Nasopharyngeal cultures for isolation of meningococci are best obtained with a swab attached to a bent wire to prevent contamination by normal mouth flora before reaching the posterior nasopharynx. Until the 1960s, treatment of choice for meningococcal disease was sulfonamides, but resistant strains have become increasingly common. Sulfadiazine-resistance among asymptomatic carriers of *N. meningitidis* in one series was present in 6.5 percent of group B and 39.4 percent of group C strains. Penicillin, erythromycin, and chloramphenicol are effective. Penicillin crosses the blood-brain barrier when there is inflammation present, but it is not effective in eradicating the carrier state.

PASTEURELLA

The genus *Pasteurella* has a number of species pathogenic to animals, but only one, *P. multocida,* is common in humans. *P. multocida* causes fowl cholera, septicemia, and pneumonia in animals with decreased host resistance. The disease is called "shipping fever" because of its prevalence when animals are herded and shipped. In humans, *P. multocida* and *pneumotropica* can cause infection and osteomyelitis after animal bites, especially from dogs or cats. Rarely, *P. multocida* causes respiratory or systemic disease in humans in the absence of a history of animal exposure. Lysozyme plays a major role in host defense against many bacteria, including *Pasteurella.* However, lysozyme may be detrimental as well as helpful because it causes increased release of endotoxin.

STREPTOBACILLUS MONILIFORMIS

Streptobacillus moniliformis, from the family Bacteroidaceae, is a nonmotile, short, nonencapsulated, pleomorphic, microaerophilic gram-negative rod that normally inhabits the upper respiratory tract of rats and other small mammals. It grows slowly on specialized media, either in the usual bacterial form or as L forms (see Chapter 2). In addition, stable L forms have been isolated from the blood of humans for weeks after infection. The organism causes one form of rat bite fever in humans (see Chapter 12 for discussion of *Spirillum minus,* the other cause of rat bite fever). The mode of transmission is usually by rodent bite, either in the wild or in the laboratory. Transmission by ingestion of contaminated dairy products has also been documented; this form of the disease is called Haverhill fever.

After a short incubation period, usually less than 3 days, there is sudden onset of fever, chills, headache, and malaise. There may occasionally be ulceration or swollen lymph nodes around the bite wound. In most individuals there is a macular rash especially on the extremities. Often there is arthralgia or arthritis. Recovery takes several weeks and relapses are uncommon. Untreated mortality is about 10 percent, and is due primarily to bacterial endocarditis. Effective treatment is penicillin or erythromycin administration. The clinical signs and symptoms of rat bite fever caused by *S. moniliformis* and *S. minus* are compared on Table 10-6.

YERSINIA

The genus *Yersinia* was reclassified in 1970 from the family Pasteurellae to the family Enterobacteriaceae because of its antigenic relationship to other enteric organisms. However, because prominent members of the genus are neither primary inhabitants nor major pathogens of the gastrointestinal tract, *Yersinia* is considered separately from the enteric bacteria. There are three species of importance in human disease: *Y. pestis,* which is the plague bacillus; *Y. pseudotuberculosis;* and *Y. enterocolitica*. The relatedness of the three species has been demonstrated by DNA hybridization techniques. There are probably also common virulence mechanisms.

Yersinia pestis

Biologic characteristics. *Yersinia pestis* is a nonmotile, aerobic encapsulated gram-negative rod that can be stained well with Gram, Giemsa, or Wayson (methylene blue and carbol-fuchsin) stains. The organisms have a characteristic bipolar staining pattern, which causes them to resemble safety pins. They can be grown rather easily on artificial media. On blood agar the colonies are brown. Because they grow well at room temperature (22°C), they can survive for weeks in excreta such as sputum or flea feces. *Y. pestis* produces a number of antigens and toxins, which are listed on Table 10-7. Two bacteriocins, pesticin I and II, are also produced that kill *Y. pseudotuberculosis* and some strains of *E. coli*.

TABLE 10-6. Signs and Symptoms of Rat Bite Fever Caused by *Streptobacillus moniliformis* and *Spirillum minus*

	S. moniliformis	*S. minus*
Latent period	Less than 10 days	Prolonged; 1-4 weeks
Bite site	Prompt healing	Recurrent inflammatory response
Fever	Sudden onset with chills; relapses rare	Sudden onset; relapses common
Other	Usually rash, arthralgias, arthritis	Usually without rash or joint symptoms

TABLE 10-7. Antigens and Toxins of *Y. Pestis*

Ag/toxin	*Characteristics*
Capsular Ag, Fraction 1 (F1)	Heat-labile protein produced at 37° C but not 22° C. Protects organism against phagocytosis
Protein V Ag	Cell bound
Lipoprotein W Ag	Excreted. V and W bind and protect organisms from phagocytosis in absence of capsule. V/W Ag produced at 37° C but not 28° C
Toxin A	Intracellular
Toxin B	Combines with A to form murine toxin
Lipopolysaccharide endotoxin	Similar to that of other gram-negative bacteria
Kunin Ag	Common Ag of Enterobacteriaceae

Epidemiology. Since the sixth century, massive pandemics causing millions of deaths have been caused by bubonic and pneumonic plague. The disease has been known as the "black death" because of the severe cyanosis of its victims. During the fourteenth century, about one-fourth of the population of Europe and the Middle East was killed by plague. There are two types: domestic plague, which is transmitted by rat fleas and causes epidemics; and wild (sylvatic) plague, which occurs among wild mammals such squirrels and rabbits and can serve as a reservoir for the plague bacillus. Since 1925, plague in the United States has been almost exclusively sylvatic. Implicated animal reservoirs include rabbits, wild rodents, domestic cats, prairie dogs, bobcats, coyotes, and wild mice.

When the infected flea (in domestic plague, the rat flea, though plague can be transmitted by many flea species) bites a human host, the bacteria are deposited in the wound in flea vomitus. Rats are also sickened and die from *Y. pestis*. If the rat population becomes low, rat fleas transmit the disease to humans. When these victims are subsequently bitten by human fleas, person-to-person spread by the human flea can also occur.

The incidence of plague in the United States has increased over the past two decades (Fig. 10-14). There were 105 cases reported in the 1970s, the largest number since the decade 1900 to 1909. Most disease occurred in persons less than 20 years of age (59 percent of cases), with ages ranging from 7 to 79 years. Most cases were in the southwest (California, New Mexico, Utah, and Colorado), and were isolated incidents. There were also six common source outbreaks among residents of the same household. The majority, 82 percent, occurred in May through September. Infected fleas have been found on squirrels and prairie dogs. Plague is a particular public health problem on Indian reservations in the southwest where the annual attack rate is 1.4 per 10^5 persons as compared to 0.1 per 10^5 persons for the rest of the United States population.

Pathogenesis. Virulence of *Y. pestis* is temperature dependent. When the organisms enter the human, they do not contain the F1 capsular or the V/W antigen because the flea-body temperature is 28°C. Only at

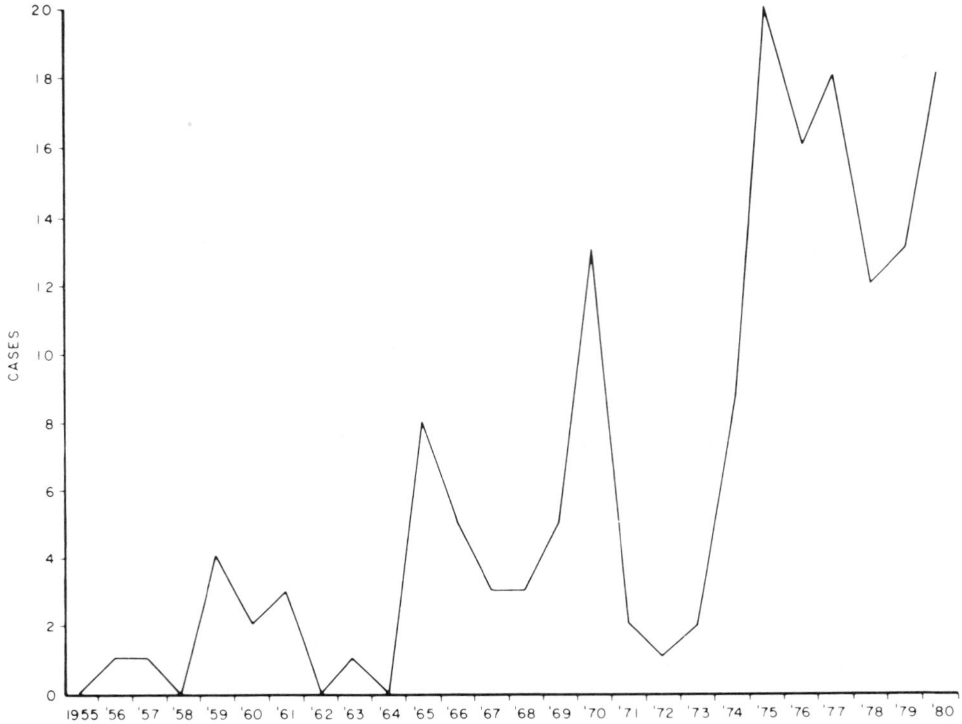

FIGURE 10-14. Plague. Reported cases in humans by year, United States, 1955 to 1980. (From Centers from Disease Control. Annual Summary 1980. Morb Mort Week Rep 1981;29:63.)

37°C are these substances formed. The F1 and V/W antigens are synthesized in macrophages and the organisms become virulent and resistant to phagocytosis. After the bacteria enter the skin by the flea bite, they travel to the regional lymph nodes, usually in the groin, where they cause swelling, pain, and inflammation. *Buboes* are the large, inflamed lymph nodes, hence the name *bubonic plague*. Fraction 1, a component of the capsule, is the major antigen, and the V and W antigens increase the organisms' resistance to phagocytosis, even in the absence of a capsule. If the bacteria are not killed (which is usually the case), septicemia occurs and the organisms spread to vital organs. *Pneumonic plague* occurs if the bacteria gain entrance into the lungs by way of the efferent lymphatics or blood stream. Two specific toxins, A and B, are produced, plus the endotoxin similar to that of other gram-negative bacteria.

Host responses. Plague may be asymptomatic or limited to the formation of buboes in the groin or axilla. The buboes there may become necrotic and eventually heal. In endemic areas this occurs with some frequency and the pharyngeal carrier rate of *Y. pestis* can be 10 percent. More often, however, after an incubation period of 1 to 6 days, the disease progresses and takes one of two clinical forms. Bubonic plague occurs when *Y. pestis* are disseminated throughout the body causing severe hemorrhagic lesions and disseminated intravascular coagulation. There may be meningitis; necrosis occurs in all parts of the body. The case fatality rate of bubonic plague without treatment is 50 to 75 percent. Pneumonic plague is even more devastating, with a case fatality rate reaching close to 100 percent. This form of the disease is characterized by pneumonia with massive hemorrhaging in the lungs. Death due to asphyxiation occurs rapidly. In the United States most cases of plague have been bubonic plague with a case fatality rate among treated individuals of 14 percent. Case fatality rates have not differed by race or sex, but the fatality rate is greater than 50 percent for those over 60 years of age.

Diagnosis is made by the finding of bacteria with a "safety pin" staining pattern from samples of sputum (for pneumonic plague) or buboes. Fluorescent antibody tests are also helpful, although serologic tests are not helpful for prospective diagnosis since antibodies develop late in the disease course. The presence of large numbers of *Y. pestis* in the blood is a late indicator of disease and may not provide diagnostic aid early enough to be useful.

Control and treatment. Preventive measures include raising the population's general standard of living, avoiding overcrowded conditions, and decreasing rat, rat flea, and human flea populations. There are attenuated and killed vaccines available that render short-term protection to high-risk individuals. Ships from endemic areas are quarantined for a period of time upon arrival in the United States. Drugs that are of use include streptomycin, chloramphenicol, and tetracycline. Penicillin is useless. In order to be efficacious, treatment must start as soon as possible, usually within 15 hours of onset. At that time mortality can be reduced to about 5 percent In addition, patients with pneumonic plague have highly infectious sputum and should be placed in strict isolation. Contacts of patients with pneumonic plague should receive prophylactic tetracycline, chloramphenicol, or streptomycin for 7 to 10 days.

Other *Yersinia*

Yersinia enterocolitica, previously called *Bacterium enterocoliticum,* is found in diseased or healthy animals, water, food, and raw milk. It was isolated from 16 of 495 small, wild animals trapped in Japan, whereas no strains of *Y. pestis* were isolated (Kaneko and Hashimoto 1981). In 397 feces samples collected from healthy Norwegian students, 20 strains were isolated from 10 (3 percent) individuals. Nine strains produced enterotoxin at 22°C, but not at 37°C (Kapperud 1980). There are 34 serotypes, but type 8 causes most of recognized disease in the United States, types 3 and 9 cause disease in other parts of the world. *Y. enterocolitica* are motile at 28°C, but not at 37°C. The mode of transmission may be ingestion of contaminated food, water, or milk, and possibly person-to-person spread. *Y. enterocolitica* causes an enterocolitis that can be easily mistaken for appendicitis or gastroenteritis due to other enteric pathogens such as *Salmonella.* The disease was first identified in Europe, but has become increasingly recognized in the United States, where several hundred cases are reported annually. There have been several outbreaks of *Y. enterocolitica*-caused enteritis reported in the United States since 1980. Often they can be traced to inadequately pasteurized milk. The largest such outbreak involved almost 200 individuals in three states. Common clinical syndromes include erythema nodosum, arthritis, Reiter's syndrome, and abscesses. Predominant symptoms of the enteric syndrome include abdominal pain and fever.

Y. pseudotuberculosis is a large rod that is motile by means of 1 to 6 peritrichous flagella at room temperature. It is not motile at body temperature. It is commonly isolated from healthy animals and birds, as well as from soil and milk. Human infections include ileitis and mesenteric lymphadenitis, but not gastroenteritis. It can also cause serious septicemia. *Y. pseudotuberculosis* resembles other enteric pathogens, and several antigens cross react with *Salmonella* species. There are 15 O antigens and six main serotypes, but almost all human disease is caused by type I. Treatment with kanamycin, tetracycline, and chloramphenicol is successful.

CLINICAL IMPLICATIONS

Once again, in cases of gonorrhea it is important for the clinician to help trace disease contacts and interpret to the patient the need for compliance with drug therapy. It is essential that treatment failures due to lack of compliance with therapy be separated from those due to reinfection or, more importantly, to the emergence of a strain of PPNG. Much gonococcal disease is transmitted by asymptomatic cases. Hence, all cases need to be questioned about sexual contacts so they can also be treated. Such tracing is the most important control measure for gonorrhea.

The best treatment for gonococcal conjunctivitis in newborns is still being investigated. The problem of treating the disease is complicated by the fact that currently the most common cause of neonatal conjunctivitis is *Chlamydia* rather than *N. gonorrhoeae,* and the ideal treatment agent for the two organisms is not the same. Problems with $AgNO_3$ and possible reasons for treatment failure are listed on Table 10-8. Though other antimicrobial agents may be more effective, all have disadvantages. Whatever treatment is used, the health practitioner is responsible for proper administration. The majority of treatment failures can probably be attributed to improper administration of the drug in the eyes.

The major sexually transmitted diseases

TABLE 10-8. Silver Nitrate ($AgNO_3$) as Prophylaxis for *N. gonorrhoeae* Ophthalmia Neonatorum

Problems with $AgNO_3$	*Possible reasons for treatment failure*
Does not provide perfect protection (2-5 percent treatment failure)	Improper instillation
Causes chemical conjunctivitis	Ascending infection, especially after premature rupture of membranes
Difficult to administer	Postpartal acquisition by autoinoculation
May cause eye damage if undiluted solution is inadvertently used	Failure to distinguish chlamydial conjunctivitis
May be unnecessary in some settings	Therapeutic agent is not totally effective
Other antibiotics are probably more effective	

(STDs) today, in addition to gonorrhea, include *Chlamydia,* genital herpes, cytomegalovirus infection, and trichomoniasis. Syphilis and hepatitis B are also transmitted sexually, but are less common. Each of these diseases are discussed in subsequent chapters. The clinician needs to be aware of individuals who are at increased risk for STD. These include persons with multiple sex partners, male homosexuals, and victims of rape or sexual abuse. Whenever an individual is diagnosed as having one STD, he or she should be screened for others as well. The recommendations of the Centers for Disease Control for screening of rape victims includes the following:

1. Cultures for *N. gonorrhoeae* and *Chlamydia trachomatis* from all potential sites of infection
2. Examination of vaginal specimens for *Trichomonas vaginalis*
3. Pelvic examination
4. Serologic test for syphilis
5. Obtaining a serum sample to freeze for possible future testing

There is recent evidence that vaginal spermicide contraceptives may be protective against sexually transmitted organisms. Further study regarding this issue is clearly indicated (New Developments in Infection Control 1983).

Obviously, any child who develops a STD should be considered a victim of sexual abuse and such instances should be reported by the nurse or physician to appropriate authorities.

Organisms such as *Hemophilus, Bordetella,* and *Neisseria* die quickly outside the human body. Specimens from body fluids, exudates, or the nasopharynx must be sent in transport medium immediately to the laboratory, or, ideally, inoculated directly onto laboratory medium at the patient's bedside. It is important that the clinician be aware of the fact that these organisms are very sensitive to temperatures cooler than body temperatures and to drying. Proper specimen collection is necessary for correct diagnosis. In addition, the clinican needs to educate patients about continuing the entire course of drug treatment even if symptoms subside. With diseases such as gonorrhea, the patient may feel better before the organisms are eradicated from the body.

The fact that so many opportunistic pathogens are emerging as causes for nosocomial infections is both encouraging and discouraging. It does mean that some of the major life-threatening infectious diseases of the past are controlled to a greater extent, and that our attempts at prevention and treatment are effective. It probably means that people are generally healthier today and their immune systems are more able to effectively combat pathogens. It is also clear that the virulence of certain organisms can become attenuated over time. This is certainly the case with some of the pyogenic cocci.

On the other hand, the emergence of opportunistic pathogens places an added burden on those who are providing direct patient care. It is clear that their own flora, and that of their patients, can cause significant infection. The care with which aseptic and clean techniques are used is closely reflected in patient infection rates.

When new disease-causing agents such as *Legionella* emerge, they are often first recognized in a major outbreak. Only when such organisms are characterized fully are their pathogenicity and virulence understood. Therefore, patients with legionellosis were initially treated with fear about their potential contagiousness and the virulence of the infecting organisms. It is now clear that person-to-person transmission of this disease is unlikely. Patients with legionellosis should be managed with secretion precautions or respiratory isolation, just as any other patient with a suspected or confirmed bacterial pneumonia.

SUGGESTED READINGS

Berendt RF, et al. Dose-response of guinea pigs experimentally infected with aerosols of *Legionella pneumophila*. J Infec Dis 1980; 141:186-192.

Bernstein G, Davis J, Katcher M. Prophylaxis of neonatal conjunctivitis. Clin Pediatr 1982;21:545-550.

Bramley HM, Dixon RA, Jones BM. *Haemophilus vaginalis (Corynebacterium vaginalis, Gardnerella vaginalis)* in a family planning clinic population. Br J Vener Dis 1981;57:62-66.

Brook I. *Yersinia enterocolitica* sepsis in an infant. NY State J Med 1981;81:778-779.

Castle M, et al. Outbreak of a multiply resistant *Acinetobacter* in a surgical intensive care unit: epidemiology and control. Heart Lung 1978;7:641-644.

Centers for Disease Control. *Acinetobacter calcoaceticus* infections—U.S. 1978. Morb Mort Week Rep 1979a;28:177-179.

Centers for Disease Control. Nosocomial meningitis caused by *Citrobacter diversus*. Morb Mort Week Rep 1979b;28:249-250.

Centers for Disease Control. Nosocomial acquisition of *Aeromonas hydrophilia*. National Nosocomial Infections Study Report. 1979c; Annual Summary 1979c: 23-25.

Centers for Disease Control. An outbreak of penicillinase-producing *Neisseria gonorrhoeae*. Morb Mort Week Rep 1980a;29:241-243.

Centers for Disease Control. Chromobacteriosis. Morb Mort Week Rep 1980b;29:613-615.

Centers for Disease Control. Legionellosis in a child. Morb Mort Week Rep 1980c;29:203-204.

Centers for Disease Control. Legionellosis in Italy. Morb Mort Week Rep 1980d;29:591-597.

Centers for Disease Control. Legionellosis-Västerås, Sweden. Morb Mort Week Rep 1980e;29:206-207.

Centers for Disease Control. Fatal nosocomial legionnaires' disease: a study in NNIS hospitals. National Nosocomial Infections Study Report. Annual Summary 1978. 1981a;23-25.

Centers for Disease Control. Gonococcal eye infections in adults—California, Texas, Germany. Morb Mort Week Rep 1981b;30:341-343.

Centers for Disease Control. Meningococcal disease. Morb Mort Week Rep 1981c;30:113-115.

Centers for Disease Control. Neonatal gonococcal infections: recommendation for prophylaxis and treatment. National Nosocomial Infections Study Report. Annual Summary 1978. 1981d;36-39.

Centers for Disease Control. Multi-state outbreak of yersiniosis. Morb Mort Week Rep 1982a;31:505-506.

Centers for Disease Control. Plague vaccine. Morb Mort Week Rep 1982b;31:301-304.

Centers for Disease Control. Sexually transmitted disease treatment guidelines, 1982. Morb Mort Week Rep 1982c;31 (Suppl).

Centers for Disease Control. Spectinomycin-resistant beta-lactamase-producing *Neisseria gonorrhoeae*—England. Morb Mort Week Rep 1982d;31:495.

Cordes LG, et al. Legionnaires' disease outbreak at an Atlanta, Ga country club: evidence for spread from an evaporative condenser. Am J Epidemiol 1980;111:425-431.

Coyle-Gilchrist MM, Crewe P, Roberts G. *Flavobacter meningoseptica* in the hospital environment. J Clin Pathol 1976;29:824-826.

Craven DE, et al. Adherence of isolates of *Neisseria meningitidis* from patients and carriers to human buccal epithelial cells. J Infec Dis 1980;142:556-568.

Denneberg T, et al. Glomerulonephritis in infections with *Yersinia enterocolitica* O-serotype 3. Acta Med Scand 1981;209:97-101.

Draper DL, et al. In vitro modeling of acute salpingitis caused by *Neisseria gonorrhoeae*. Am J Obstet Gynecol 1980;138:996-1002.

Ettensohn DB, Roberts NJ. *Yersinia enterocolitica* pneumonia. NY State J Med 1981;81:791-794.

Finkelhor RS, et al. *Gardnerella vaginalis* perinephric abscess in a transplanted kidney. New Engl J Med 1981;304:846.

Flesher AR, et al. *Legionella pneumophila:* growth inhibition by human pharyngeal flora. J Infec Dis 1980;142:313-317.

Fraser DW. Meningitis caused by *Haemophilus influenzae* type b (letter). J Infec Dis 1980;142:644.

Garner JG, et al. Bacteriological aspects of a case of *Brucella suis* infection. New Zea Med 1981;93:8-9.

Geere IW. *Enterobacter agglomerans:* the clinically important plant pathogen. Can Med Assoc J 1977;116:517-519.

Goodnow RA. Biology of *Bordetella bronchiseptica.* Microbiol Rev 1980;44:722-738.

Granoff DM, Basden M. *Haemophilus influenzae* infections in Fresno county, Calif: a prospective study of the effects of age, race, and contact with a case on incidence of disease. J Infec Dis 1980;141:40-46.

Harrison DN, et al. Commonality of a virulence factor among *Yersinia* species (letter). J Infec Dis 1980;141:413.

Henriksen SD. *Moraxella, Acinetobacter,* and the *Mimae.* Bacteriol Rev 1973;37:522-561.

Herbert DA, Ruskin J. Are the "nonpathogenic" *Neisseriae* pathogenic? Am J Clin Pathol 1981;75:739-743.

Hodges GR, Degener CE, Barnes WG. Clinical significance of *Citrobacter* isolates. Am J Clin Pathol 1978;70:37-40.

Holmes KK, Puziss M. Recommendations for the study group for research and training in sexually transmitted diseases. J Infec Dis 1980;142:639-642.

Jemsek JG, et al. Antimicrobial susceptibility testing of *Haemophilis parainfluenzae* by kinetic killing-curve method. J Infec Dis 1980;141:310-316.

Kaneko KI, Hashimoto N. Occurrence of *Yersinia enterocolitica* in wild animals. Appl Environ Microbiol 1981;41:635-638.

Kapperud G. *Yersinia enterocolitica* and *Yersinia enterocolitica*-like bacteria isolated from healthy humans in Norway. Acta Pathol Microbiol Scand B 1980;88:303-306.

Kaufmann AF, et al. Pontiac fever: isolation of the etiologic agent *(Legionella pneumophila)* and demonstration of its mode of transmission. Am J Epidemiol 1981;114:337-374.

Kaufmann AF, Boyce JM, Martone WJ. Trends in human plague in the United States: reports from the Centers for Disease Control. J Infec Dis 1980;141:522-524.

Käyhty H, et al. Serum antibodies to capsular vaccine of group A *Neisseria meningitidis* followed for three years in infants and children. J Infec Dis 1980;142:861-868.

Käyhty H, et al. Antibody response to capsular polysaccharide of groups A and C *Neisseria meningitidis* and *Haemophilus influenzae* type b during bacteremic disease. J Infec Dis 1981;143:32-41.

Kraus SJ. Incidence and treatment of gonococcal pharyngitis. Sex Trans Dis 1979;6:143-147.

Kuberski T, Papadimitriou JM, Phillips P. Ultrastructure of *Calymmatobacterium granulomatis* in lesions of granuloma inguinale. J Infec Dis 1980;142:744-749.

Mackel DC, et al. Nationwide epidemic of septicemia caused by contaminated intravenous products: mechanisms of intrinsic contamination. J Clin Microbiol 1975;2:486-489.

Marks MI, Frasch CE, Shapera RM. Meningococcal colonization and infection in children and their household contacts. Am J Epidemiol 1979;109:563-571.

New Developments in Infection Control. Vaginal spermicide contraceptives protect women from sexually transmitted disease. Infec Contr 1983;4:50.

Pasculle AW, et al. Pittsburgh pneumonia agent: direct isolation from human lung tissue. J Infec Dis 1980;141:727-732.

Perine PL, et al. Epidemiology and treatment of penicillinase-producing *Neisseria gonorrheae.* Sex Trans Dis 1979;6:152-158.

Pichichero ME, et al. Breast milk antibodies to the capsular polysaccharide of *Haemophilus influenzae* type B. J Infec Dis 1980;142:694-698.

Recalcitrant gonococci, plasmids, and antibiotics (editorial). Lancet 1981;1:816-817.

Rimler RB, Brown WE. *Pasteurella multocida:* plasma lysozyme in bacteremic and lipopolysaccharide-exposed turkeys. J Infec Dis 1980;142:614-617.

Rose HD, Lenz IE, Sheth NK. Meningococcal pneumonia: a source of nosocomial infection. Arch Intern Med 1981;141:575-577.

Rothenberg R: Ophthalmia neonatorum due to *Neisseria gonorrhoeae.* Sex Trans Dis 1979; 6:187-191.

Sparling PF. Treatment of gonorrhea: what effect will antibiotic resistance have in the future? Sex Trans Dis 1979;6:120-125.

Thompson SE. Treatment of disseminated gonococcal infections. Sex Trans Dis 1979;6:181-186.

Tramont EC, et al. Antigenic specificity of antibodies in vaginal secretions during infection with *Neisseria gonorrhoeae.* J Infec Dis 1980;142:23-31.

Ursing J. Characterization of *Enterobacter agglomerans* from clinical specimens. Acta Pathol Microbiol Scand B 1977;85:61-66.

Venkataramani TK, Sundararaj T, Madhavan HN, Sharma KB. Skin and mucous membranes as reservoirs of *Bacterium anitratum* and *Mima polymorpha* in patients. J Ind Med Assoc 1972;59:425-428.

Wallace RJ, et al. *Haemophilis influenzae* infections in adults: characterization of strains by serotypes, biotypes, and beta-lactamase production. J Infec Dis 1981;144:101-106.

Ward JI. *Haemophilus influenzae* disease in Alaskan Eskimos. Lancet 1981;1:1281-1284.

11 OBLIGATE INTRACELLULAR BACTERIA: CHLAMYDIAS AND RICKETTSIAS

KEY POINTS

- Chlamydias and rickettsias, though they are true bacteria, were originally classified as viruses because of their small size and the fact that they are obligate intracellular parasites.
- Chlamydias can cause a variety of clinical syndromes including trachoma, ornithosis, lymphogranuloma venereum, neonatal disease, and genital infections.
- Rickettsias are generally transmitted by arthropod vectors and cause fevers or rashes.
- The major rickettsial diseases are epidemic typhus, caused by *R. prowazekii;* endemic typhus, caused by *R. typhi;* Rocky Mountain spotted fever, caused by *R. rickettsii;* scrub typhus, caused by *R. tsutsugamushi;* and Q fever, caused by *Coxiella burnetii.*

The orders Chlamydiales (chlamydias) and Rickettsiales (rickettsias), as shown on Table 11-1, share a number of features in common. They are among the smallest eubacteria, and were mistaken in the past for viruses. Because of their minimal metabolic capabilities they require an external energy source for growth and are therefore obligate intracellular parasites. Though they are short rods related to gram-negative bacteria, they both have unique staining characteristics. The Gram stain is not helpful in identifying either chlamydias or rickettsias. The Giemsa, Giminez, or Castañeda stains are more useful in visualizing the organisms in host cell cytoplasm. Neither organism can be grown on artificial medium; they require yolk sacs of embryonated eggs

TABLE 11-1. Comparison of Chlamydias and Rickettsias

Similarities	*Differences*
Obligate intracellular parasites	Intracellular growth patterns differ
Can be grown in yolk sac of embryonated egg but not in artificial media	Chlamydias smaller (200-300 nm as compared to 300-500 nm for rickettsias)
Cell walls resemble gram-negative bacteria	Rickettsias primarily cause febrile exanthems; chlamydias primarily cause ocular and genital diseases
Staining characteristics	Rickettsias often transmitted by arthropods and human infection is incidental; chlamydias occur naturally in humans
Sensitive to many antimicrobials	Rickettsias yield lasting immunity; immunity does not usually develop after chlamydial infection
True bacteria that multiply by binary fission	Energy is required for adsorption into host cell by rickettsias; not by chlamydias
Very small	

or tissue culture to proliferate. Thus, their microbiologic identification is more difficult than that of other bacteria.

CHLAMYDIAS

In the order Chlamydiales there is one genus, *Chlamydia,* and two species, *C. psittaci* and *C. trachomatis*. Humans are the natural host for *C. trachomatis,* whereas *C. psittaci* infections are usually zoonoses. *Chlamydia* are closely related to the gram-negative bacteria, but they have a very unique life cycle. Since they are unable to produce ATP, they are dependent for reproduction and maturation on host cells. Extracellularly, *Chlamydia* exist as mature particles called *elementary bodies* (EB). An EB enters a host cell by specifically inducing its own phagocytosis, whereupon a vacuole is formed around it. Though the organism is intracellular, there is no eclipse phase as with viruses. Chlamydial antigens prevent lysosomal fusion with the phagosome, thereby protecting the *Chlamydia* from the host cell defenses.

In the vacuole, the EB changes over a period of 6 to 8 hours to a noninfectious larger particle called an *initial body* (IB). The IB is metabolically very active, rich in DNA, and cannot survive outside the host cell. Within about 15 hours binary fission of the IB begins and continues, forming an *inclusion body* that almost fills the entire host cell (Fig. 11-1). Maturation into new EB occurs within the inclusion body. Toxicity is due to multiplication of the IB in the host cell; a toxin is produced but has not been isolated. Host cell synthesis is inhibited during infection. The mature EB are released when the host cell lyses. The emerging EB are capable both of infecting new cells and surviving in the extracellular environment. The developmental cycle, summarized in Table 11-2, is rather long, lasting about 48 hours. Though *Chlamydia* cannot be cultured in cell-free media, the presence of inclusion bodies in tissue culture usually establishes the diagnosis. The

TABLE 11-2. Stages in Chlamydial Growth Cycle

Stage	*Reproductive events*
Mature particle, elementary body (EB)	Contacts receptors on host cell surface with resultant phagocytosis
Reticulate or initial body (IB)	Reorganization of elementary bodies in host cytoplasmic vacuole
Inclusion body	Contain numerous replicated initial bodies that mature and are released from the host cell as EB within 24-48 hours of initial infection

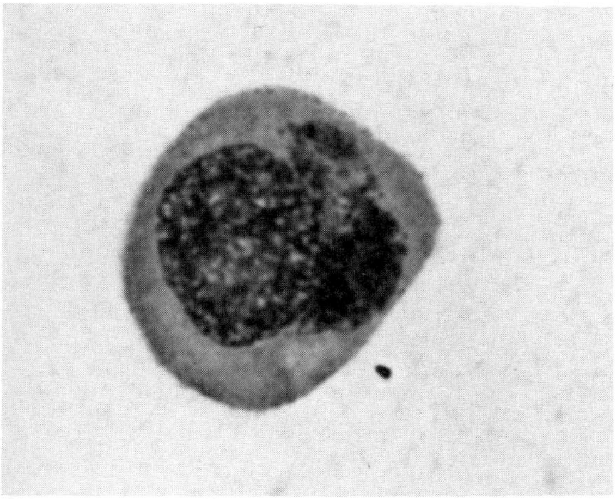

FIGURE 11-1. Chlamydial inclusion body (IB) filling almost the entire host cell. Conjunctival scraping from an individual with trachoma. (Courtesy of Dr. James Plorde.)

EB stain purple with Giemsa stain (cell cytoplasm is blue).

Chlamydia have two types of antigens, both cell wall components. A lipopolysaccharide group antigen that is heat stable and induces formation of complement-fixing antibodies is shared by all species. There are also species-specific protein antigens that can be used for typing strains. Complement fixation tests are available for the organisms of lymphogranuloma venereum and psittacosis, but not for other types of *C. trachomatis*. *C. psittacosis* has numerous serotypes. There are also at least 15 serotypes of *C. trachomatis:* three (L-1, L-2, L-3) cause lymphogranuloma venereum, four cause trachoma (A, B, Ba, C), and eight (D through K) cause genital and ocular diseases (Table 11-3). *C. trachomatis* has been known by the acronym "TRIC agent" (trachoma inclusion conjunctivitis) but this term is misleading because *C. trachomatis* causes several other syndromes besides conjunctivitis and because the term can be confused with *Trichomonas,* a parasite that also causes genital infections. *C. trachomatis* exist primarily in columnar epithelial cells of conjunctiva, cervix, urethra, or respiratory tract.

Antibodies are produced by infected hosts, but are ineffective in producing adequate immunity. The spread of EB to new host cells is inhibited by neutralizing antibody, but it is ineffective against intracellular initial bodies. Because of this inadequate immune response, latent, subclinical, and chronic infections are common with *Chlamydia.* The organism of lymphogranuloma venereum persists in the host for years. Persons who have been removed for years from areas where trachoma is endemic can develop acute disease in later life. Thus, treatment of chlamydial infections must be long

TABLE 11-3. Chlamydial Diseases

Species	*Serotypes*	*Diseases*	*Distribution*	*Mode of transmission*
C. psittaci	Multiple	Psittacosis, ornithosis	Worldwide, rare in U.S.	Usually contact with dried feces of avian carrier; rarely, person-to-person via sputum
C. trachomatis	A,B,Ba,C	Trachoma	Mostly Asia, Africa	Eye-to-eye
	D-K	Genital disease: urethritis, salpingitis, ectopic pregnancy, endometritis, PID, vaginitis, fetal death, proctitis, abortion Conjunctivitis Pneumonia	Worldwide, very common in U.S.	Sexual; for neonates, in birth canal of infected mothers; contact with infected genital secretions (in swimming pool, through handling by health care personnel)
	L-1,2,3	Lymphogranuloma venereum	Worldwide, especially in lower socioeconomic classes and in tropics	Sexual

term and is not always effective. The tetracyclines are usually the treatment of choice. Sulfonamides are inhibitory if the species synthesizes folic acid (usually *C. trachomatis* but not *psittaci*). Penicillin inhibits division and maturation of initial bodies.

Chlamydia psittaci

Biologic characteristics. *Chlamydia psittaci* is the causative agent of psittacosis (from parrots and parakeets) or ornithosis (from other birds). The two diseases are identical and are transmitted to humans in the same ways. *C. psittaci* is a prototype chlamydial organism, being intracellular, small, with the unique life cycle described previously.

Epidemiology. Psittacosis is primarily an avian disease with respiratory spread to humans by inhalation of infected bird feces or aerosol, by handling infected bird tissues, or by person-to-person spread via infected sputum. The disease is worldwide, but only 50 to 100 cases occur in the United States annually. Almost all bird species can be affected, but the turkey is the main reservoir in the United States. Other commonly affected birds include ducks, gulls, pigeons, and chickens. Asymptomatic avian carriers are common and their tissues and feathers as well as excreta are infectious.

Pathogenesis. After *C. psittaci* enter the human respiratory tract, the organisms settle in the alveoli and in cells of the reticuloendothelial system. Alveolar walls and lung tissue become thickened and occasionally necrotic or hemorrhagic. Patchy inflammation of the lungs, along with pneumonitis, may give flulike symptoms. In serious cases the liver, spleen, heart, or kidneys may become inflamed. Diagnosis is made by detecting the presence of cytoplasmic inclusion bodies in macrophages, and by serologic tests.

Host responses. After an incubation period of about 10 days, there is a slow or sudden onset of fever, malaise, sore throat, and headache. Psittacosis is often self-limited, but in severe cases central nervous system symptoms such as stupor or disorientation develop. Cough and epistaxis are common, but, otherwise, upper respiratory tract signs are not necessarily prominent. Infection in lower mammals may result in arthritis, abortion, enteritis, or pneumonitis. It is possible that human spontaneous abortions are sometimes related to infection with *C. psittaci*. A delayed hypersensitivity reaction develops and a skin test of this reaction is used for diagnosis in domestic birds. The case fatality is about 2 percent with treatment, but can be up to 20 percent in the untreated or elderly.

Control and treatment. Immunity after infection is only partial; individuals may carry the organism and excrete it in the sputum for up to a decade. Vaccines are also only partially effective, probably because of the large number of serotypes. Control therefore is directed at eliminating avian carriers. Imported birds are quarantined and flocks are tested for antibody to *C. psittaci*. Tetracyclines in bird feed are sometimes used to reduce the numbers of healthy carriers. Treatment for human disease is also with tetracyclines.

Trachoma

Biologic characteristics. Trachoma is a localized infection of the epithelial cells of the eye and occasionally of the nasopharynx. It is caused by four serotypes of *Chlamydia trachomatis*. The diagnosis is made by isolating the agent, by noting cytoplasmic inclusion bodies in scrapings of conjunctival cells, or by serologic tests of eye secretions.

Epidemiology. Trachoma is common in North Africa, the Middle East, Southeast Asia, and India. It is estimated that 400 million people have the disease, and about 10 percent of them will develop partial or total blindness. Trachoma is reputed to be the leading cause of blindness in the world. Humans are the only host and transmission is usually by direct contact with eye secretions infected with *C. trachomatis*. Is is also possible that fecal shedding of the organism may occur. In endemic areas, virtually every child is infected before the age of 2 years. A mild form of trachoma is seen in American Indians, with about 1000 cases per year reported.

Pathogenesis. Initially the organisms cause inflammation of the conjunctiva with corneal involvement. The disease is usually self-limiting, but sometimes corneal vascularization and infiltration of leukocytes (pannus formation) occurs. The cellular immune response may be partially responsible for pannus. With severe conjunctival scarring, distortions and inversion of the lids cause eye lashes to abrade and damage the eye, resulting in secondary bacterial infections and prolonged inflammation (Fig. 11-2).

Host responses. Trachoma is usually a mild disease. However, reinfection is com-

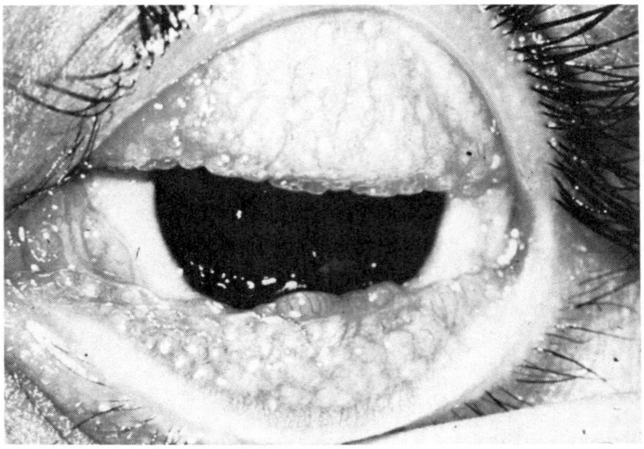

FIGURE 11-2. Active trachoma in the eye of a Southwest American Indian. (Courtesy of Dr. James Plorde.)

mon since partial immunity protects individuals for only a few months. By age 10 to 15 years the active disease usually disappears but eye damage may have occurred. Blindness may ensue 15 to 20 years after infection ceases, due to continued trauma, secondary bacterial infections, and inflammation.

Control and treatment. Control in endemic areas is extremely difficult because of the chronic nature of trachoma and the propensity for reinfection. General measures to improve hygiene and living conditions are important. Campaigns to prevent the blindness associated with long-term infection include measures such as topical antimicrobial administration and surgical repair of lid damage. Use of systemic antimicrobials, tetracycline and sulfonamides, is effective but has significant risks. Tetracyclines are contraindicated in children and pregnant women. Topical tetracycline is used, but must be administered for prolonged periods of time and compliance is often low. Vaccination is useless; hypersensitivity reactions have developed to some vaccines.

Inclusion Conjunctivitis ("Swimming Pool" Conjunctivitis)

Inclusion conjunctivitis or blennorrhea, also caused by *C. trachomatis,* is usually a self-limited purulent eye infection in which blindness is not a sequela. It is the equivalent of trachoma, but occurs in developed countries. After an incubation period of 5 to 12 days, the conjunctivitis lasts about 2 weeks and gradually subsides. Inclusion bodies in conjunctival scrapings are diagnostic. The reservoir is the human genitourinary tract and infection is acquired by contact with genital discharge. The name "swimming pool conjunctivitis" derives from the fact that many cases have been traced to exposure in swimming pools. The same disease can be acquired by neonates when passing through an infected birth canal (See "Neonatal Chlamydial Infections"), and occasionally by health care personnel caring for infected parturient women. Treatment is usually unnecessary but occasionally systemic tetracycline is indicated and must be given for 3 weeks.

Lymphogranuloma Venereum

Biologic characteristics. Lymphogranuloma venereum (LGV) is a sexually transmitted disease caused by *C. trachomatis,* types L-1, L-2, and L-3. These strains differ from other *Chlamydia* in at least three ways:

1. They are lethal on intracerebral injection in mice.
2. They are more invasive in vivo; spontaneous serial transmission in tissue culture occurs.
3. They are unaffected by diethylaminoethyl dextran (which increases infectivity of other trachoma agents) and by neuraminidase (which inhibits infectivity of other trachoma agents).

Epidemiology. LGV is uncommon in the United States, with about 600 cases reported annually, mostly among travelers who have had sexual exposure in Southeast Asia, Central or South America. Its occurrence is worldwide, especially common in tropical regions and among persons of lower socioeconomic classes. Humans are the only host. Since reporting of LGV is not mandatory, the exact incidence is unknown, but 20 to 40 percent of certain high-risk groups in the United States have positive serologic tests. There is a 20 to 1 sex ratio with a preponderance in males. The Frei test, a delayed hypersensitivity reaction using chlamydial organisms, was of only minimal usefulness in diagnosing LGV because of cross reactions with other *C. trachomatis* strains, and is no longer available in the United States. Diagnosis is usually differential; complement fixation or microimmunofluorescent studies are also useful.

Pathogenesis. LGV strains, the most invasive of the *Chlamydia,* spread in blood, tissues, and body fluids. However, symptoms are primarily localized to the genitalia with lymphatic invasion and inguinal and pelvic node enlargement.

Host reponses. After an incubation period of 3 to 20 days, a primary genital lesion develops. It is a painless vesicle or ulcer on the penis in males or on the labia or posterior vagina in females. It often goes unnoticed and heals rapidly. Secondarily, lymphatic spread of the organism can lead to lymphadenitis in the inguinal and femoral regions (bubo formation). The "inguinal syndrome" usually occurs within 2 months of infection. Small abscesses develop in the nodes, coalesce, and form large necrotic or draining areas. If untreated, the disease may even spread to the lower gastrointestinal tract where fistulae and strictures can develop. The disease is usually more serious in men, who more often have fever or other systemic symptoms. The disease causes leukocytosis, elevated levels of IgA, IgG, and IgM, the appearance of abnormal serum proteins, and a reversal of the albumin/globulin ratio. Occasionally, individuals develop a tertiary stage in which lymphatic obstruction causes chronic induration of the penis or vulva resulting in *genital elephantiasis.* The stages of LGV are listed on Table 11-4. Complement fixing antibodies are detectable after about 2 to 4 weeks and a delayed hypersensitivity reaction develops.

Control and treatment. As with all chlamydial infections, long-term therapy is often necessary. Often two or more courses for 3 weeks each of tetracycline doxycycline,

TABLE 11-4. Stages of Lymphogranuloma Venereum

Stage	*Pathology*
Primary	Painless genital vesicle forms on penis, vagina, cervix, urethra, or anus. After vesicle ruptures a shallow ulcer remains and heals.
Secondary	Inguinal and pelvic lymph nodes enlarge to form a painful bubo that may suppurate. Generalized symptoms include fever, muscle and joint aches.
Tertiary	Swelling and elephantiasis of penis, scrotum, clitoris, labia, rectum.

erythromycin, and sulfonamides are administered. Antimicrobial therapy does not expedite healing of lesions but systemic symptoms are terminated and the LGV organism can no longer be isolated from buboes. Buboes may require aspiration and elephantiasis may warrant surgical intervention.

Chlamydial Genital Infections

Biologic characteristics. *Chlamydia trachomatis,* serotypes D through K, are responsible for a number of genitourinary disease syndromes. These serotypes have been linked with about 20 different medical conditions. The organisms live in the columnar epithelial cells of the endocervix or the urethra.

Epidemiology. A tissue culture method developed in 1965 and a specific serologic test developed in 1970 have made it possible to more readily diagnose infection. Currently, an epidemic of genital chlamydial disease is occurring the United States, with an estimated 2.5 million cases per year. Chlamydial infections are several times more common than gonorrhea. Up to 40 percent of nongonococcal urethritis (NGU) in men is due to *Chlamydia.* About 70 percent of postgonococcal urethritis is also due to *Chlamydia.* In sexually active groups, 5 to 10 percent of individuals may carry *Chlamydia.*

Pathogenesis. Local growth and persistence of *C. trachomatis* occurs. Though a toxin is produced, it seems to be only minimally important in pathogenesis.

Host responses. Genital discharge often occurs within 2 to 3 weeks after infection. Asymptomatic infections of both sexes are common. However, the sequelae, especially in women of child-bearing age, can be serious. It is estimated that 10 to 20 percent of pelvic inflammatory disease (PID) may result from *Chlamydia.* In women, infection can cause salpingitis, ectopic pregnancy, cervicitis, vaginitis, and postpartal endometritis. In men, epididymitis and proctitis sometimes result. Both sexes can suffer from sterility, PID, conjunctivitis, perihepatitis manifested by right upper quadrant pain (Fitz-Hugh-Curtis syndrome), venereal Reiter's syndrome (for which a genetic predisposition may be present), and even pneumonia.

Control and treatment. A long (3 weeks) course of tetracycline is the treatment of choice, since no resistant isolates have been found as yet. Doxycycline, erythromycin, or sulfisoxazole, but *not* penicillins or ampicillin, may be substituted. To prevent reinfection, asymptomatic carriers and sex partners must also be treated. Increased screening of high-risk populations is recommended. The infection is difficult to treat successfully.

Neonatal Chlamydial Infections

Chlamydia trachomatis infections can be acquired by newborns during passage through an infected birth canal. About 50 percent of babies born under such circumstances develop inclusion conjunctivitis and about 10 percent develop chlamydial pneumonia. *Chlamydia* is now the most common cause of infant conjunctivitis, with about 100,000 babies per year or 1 to 5 per 1000 live births affected. Chlamydial pneumonia is more frequent than whooping cough, but the exact incidence is unknown. The diagnosis is difficult and, unless the lung infection is preceded by conjunctivitis, may not be made. *Chlamydia* have also been implicated as one possible cause of otitis media and nasopharyngeal problems in infants and young children.

The incubation period for conjunctivitis is about 5 to 14 days and it usually resolves spontaneously after several weeks. Rarely, conjunctival scarring can occur in a manner similar to trachoma. The incubation period for pneumonia is longer than for conjunctivitis. The disease is characterized by a cough, diffuse lung involvement, and absence of fever. Tetracycline is contraindicated for use with infants, therefore pneumonia is treated with erythromycin or sulfonamides.

Unfortunately, treatment failures for inclusion conjunctivitis are common. Silver nitrate, the prophylaxis employed in many hospitals primarily for gonococcol conjunctivitis, does not seem to be effective against *Chlamydia*. The efficacy of topical tetracycline and erythromycin is still under scrutiny. Treatment failures could be due to inadequate administration of eye drops or persistent reseeding of eyes with *Chlamydia,* as well as to antimicrobial ineffectiveness.

RICKETTSIAS

There are three genera in the tribe Rickettsieae, family Rickettsiaceae, and order Rickettsiales: *Rickettsia, Coxiella,* and *Rochalimaea*. Except for *Rochalimaea quintana,* the organisms have not been cultured in cell-free medium. All of the rickettsial diseases except Q fever are transmitted to humans, as an incidental host, by arthropod vectors. Rickettsias are small (about 300 nm), obligate intracellular parasites, and are nonmotile and nonencapsulated. Their cell wall structure resembles that of gram-negative bacteria. They are also gram-negative, but stain best with Giemsa or Castañeda stains. They actively penetrate the host cells, thereby expending energy. Thus they are more metabolically active than chlamydias, but are less active than most other bacteria. Their doubling time is about 8 hours. Unlike chlamydias, rickettsias do not have a good method for surviving outside host cells; they die very quickly in the extracellular milieu. Though they produce their own energy, they require host cofactors for survival.

All rickettsias produce a similar lesion, peripheral vasculitis, which is the result of infection of endothelial cells. Rickettsias multiply in the cytoplasm or nucleus of small blood-vessel endothelial cells and produce a species-specific endotoxin, a component of the rickettsial cell wall that helps destroy host cells. A hemolysin is also produced. However, virulence is probably related to other factors such as competition for host nutrients. Hemorrhage and thrombosis of small blood vessels causes a rash. When the rickettsias are in the endothelial cells, host defenses are of little use in fighting the infection. After rickettsias are released from lysed endothelial cells, they are readily phagocytized by macrophages. Their fate in the macrophages depends on the presence of circulating antibody. If no antibody is present, the rickettsial parasites multiply in the macrophages. In the presence of opsonizing antibody, however, rickettsias are readily killed.

Rickettsias have soluble antigens that are shared by major groups, and possess type-specific insoluble antigens as well. Group antigens are used in immunofluorescent and agglutination tests for diagnosis. The *Weil-Felix reaction* is also used to identify and type rickettsias. By chance, rickettsias share cross-reacting agglutinating antigens with some strains of *Proteus vulgaris*. Agglutination reactions with these *Proteus* O antigens help differentiate between species of rickettsias. The Weil-Felix test is almost always positive in epidemic typhus except in relapses (Brill-Zinsser disease) or when typhus occurs despite vaccination.

All rickettsial diseases can be successfully treated with chloramphenicol or tetracycline. Their growth is enhanced in the presence of sulfonamides. Immunity after infection is solid and long lasting. Major rickettsial disease of humans can be divided into three categories, as outlined in Table 11-5: typhus fevers, spotted fevers, and tsutsugamushi diseases.

Epidemic (Louse-borne) Typhus

Biologic characteristics. Epidemic typhus is the only rickettsial disease for which the natural reservoir is humans. The causative organism, *Rickettsia prowazekii,* is transmitted form person-to-person in the feces of the human body louse (Fig. 11-3), *Pediculus vestimenti,* which also becomes diseased and dies after 2 to 3 weeks. These rickettsias are not transmitted transovarially from generation to generation of lice. Infected louse excrement, which is deposited on human skin, is rubbed into the bite wound by scratching.

Epidemiology. Epidemic typhus is very rare in the United States, but occurs in Asia,

TABLE 11-5. Rickettsial Diseases

Species	*Disease*	*Distribution*	*Transovarial transmission?*	*Mode of transmission*
Typhus fevers				
R. prowazekii	Epidemic typhus	Asia, Africa, Central Europe	No	Human body louse
R. typhi (mooseri)	Endemic (murine) typhus	Southern U.S., Mexico, worldwide	No	Rat flea and louse, human louse
Spotted fevers				
R. rickettsii	Rocky Mountain spotted fever	The Americas	Yes	Wood, dog, or other ticks
R. australis, conorii, siberica	Tick fevers	Various parts of world	Yes	Ticks
R. akari	Rickettsialpox	Worldwide	Yes	Mite
Tsutsugamushi diseases				
R. tsutsugamushi	Scrub typhus, tsutsugamushi disease	Far East	Yes	Mite
Coxiella burnetti	Q fever	Worldwide	Rare	Inhalation
R. pediculi (quintana, wolhynica)	Trench fever	Mexico, Europe, rare in U.S.	No	Human body louse

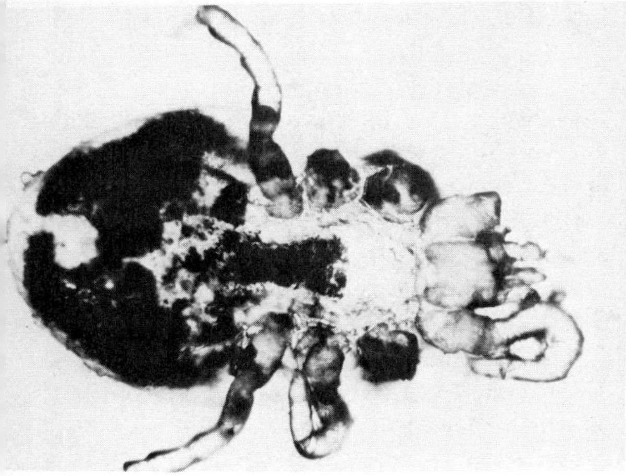

a

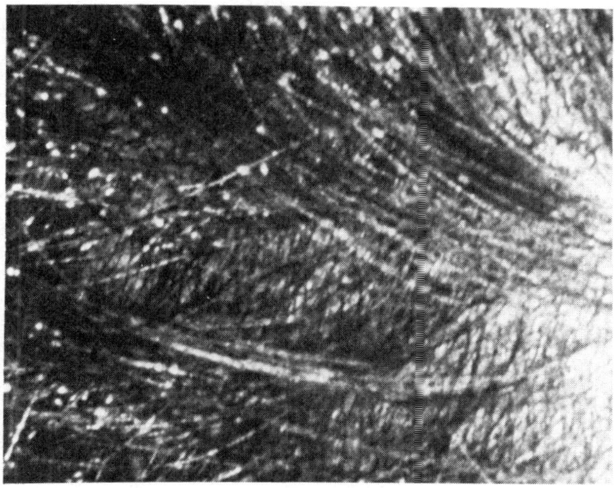

b

FIGURE 11-3. *(a)* A louse vector of epidemic typhus. *(b)* Head louse *(Pediculus humanus* var. *capitis)* infestation in scalp hair. Note tiny white dots. (*a* courtesy of Dr. James Plorde. *b* courtesy of Reed and Carnrick Pharmaceuticals.)

Africa, South America, and Central Europe. It is associated with overcrowding and unclean living conditions. There have in recent years been reports of louse-borne typhus attributed to flying squirrels in the southern part of the United States.

Pathogenesis. Vasculitis due to multiplication of the *Rickettsia prowazekii* in endothelial cells of small blood vessels is the cause of symptoms, and can lead to meningeal irritation, heart or kidney failure.

Host responses. Typhus is a Greek word meaning "hazy," which refers to the patient's mental condition. After an incubation period of 5 to 18 days, there is abrupt onset of severe headache, chills, and fever. The clinical course may be mild or fatal, with death rate being 10 to 60 percent in the untreated. The disease is more serious in adults than children. The accompanying rash begins about day 4 to 6 on the axilla, trunk, and extremities (Fig. 11-4). The palms of hands and soles of feet are usually unaffected.

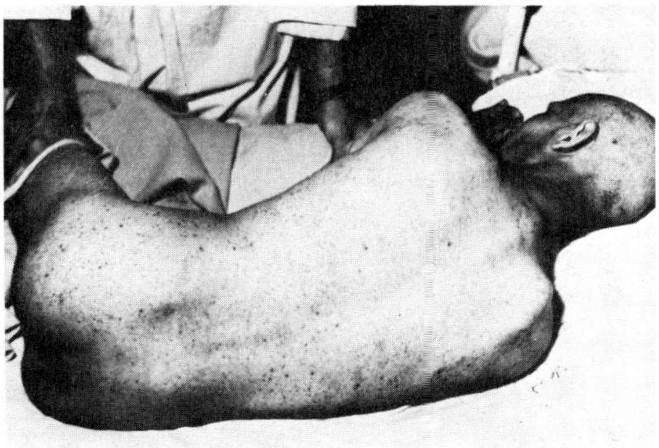

FIGURE 11-4. Epidemic louse-born typhus fever. Note diffuse petechial rash mostly on the back and buttocks about the eighth day of illness. (Courtesy of Dr. Theodore E. Woodward.)

Control and treatment. The main preventive measures relate to improving the general standard of living, and relieving overcrowding and the resulting lice infestation. During typus epidemics, DDT or other insecticides are used temporarily to kill lice (its toxic effects lasts 2 to 3 weeks). Immunity after infection is solid. A fairly effective killed vaccine is available that prevents or minimizes the disease. Vaccines are used during military campaigns. An attenuated live vaccine is also available but may produce mild symptoms. Treatment is with chloramphenicol or tetracycline, started as soon as possible after infection.

Endemic (Murine or Flea-borne) Typhus

The endemic, milder form of typhus is borne by the fleas and lice of rats. The organism responsible is *R. typhi* (also called *R. mooseri*) with which humans may become incidentally infected if they have contact with rats. There are about 30 cases per year with deaths occurring very rarely in the United States, most frequently in the southern states, especially Texas, and also in Mexico. In Ethiopia a significant proportion of fleas and lice obtained from the major rodent reservoir are positive for the typhus *Rickettsia*.

Endemic typhus has many names throughout the world including shop typhus, Moscow typhus, tarbadillo, red fever, and Manchurian typhus. It occurs worldwide, usually as an influenzalike illness, often without rash. It causes considerable morbidity among military troops. Control of rats and fleas is the major preventive mechanism. Rat runs can be dusted for fleas and rat populations can then be killed by poison. The fleas and lice (as well as the rats) must be killed so that they do not move from rats to another host.

Latent Typhus (Brill-Zinsser Disease)

Occasionally, *R. prowazekii* remain dormant in a host for many years and, under the influence of stress or lowered immune status, recur with acute disease. Usually the course is milder and shorter than primary epidemic typhus and no rash appears. This is due to the more rapid appearance of IgG antibodies, the classic secondary response. Latent typhus was common after World War II, but is extremely rare now. If the individual with latent typhus is also lice infested, the disease can be transmitted to others, especially family members, by lice bites.

Spotted Fevers

Biologic characteristics. The spotted fevers, also called "tick typhus," are all transmitted by endemic ticks. In the United States, *R. rickettsii* is the causative organism of Rocky Mountain spotted fever (RMSF). It resides in the dog tick *(Dermacentor variablilis)* and the wood tick *(D. andersoni)*. *R. rickettsii* is the only one of the nine or ten spotted fever *Rickettsia* in the United States that is of medical importance.

R. rickettsii live in the nucleus or cytoplasm of cells. They are transmitted from one generation of ticks to the next transovarially. Other tick-borne rickettsial diseases around the world are caused by *R. australis, R. conorii,* and *R. siberica,* all of which are antigenically similar to *R. rickettsii.* All diseases are similar, but vary in severity.

Epidemiology. Rocky Mountain spotted fever was named after the geographic location where it was first identified, but the identical disease occurs throughout North, Central, and South America. It is the only important rickettsial disease in the United States, with about 1000 cases occurring annually and the incidence increasing. The 1977 incidence was 5 times that of 1960 (Fig. 11-5), and the disease incidence has since continued at a higher rate. Mortality is about 10 percent despite therapy (Fig. 11-6). Highest attack rates occur in children ages 5 to 9 years, and in the southern United States. Other spotted fevers occur in the Mediterranean area and Africa (boutonneuse fever), Siberia (Siberian tick typhus), and Australia (Queensland tick typhus). Boutonneuse fever is milder than RMSF with only 1 percent mortality in the untreated.

Pathogenesis. Like other rickettsial diseases, pathologic changes result from multiplication of *Rickettsia* in endothelial cells of blood vessels. Endothelial cells are destroyed, inflammation and thrombosis occur throughout the body.

Host responses. Two to 10 days after the infected tick bites there is sudden onset of headache and fever. In another 3 to 5 days, a rash develops. The diagnosis is difficult without a history of tick bite. The disease is sometimes called "black measles." In areas where both are endemic, RMSF may be mistaken for typhus. However, the rash develops differently, appearing first on the ankles and wrists, then spreading to the trunk, palms of hands, and soles of feet (Fig. 11-7). Various rickettsial rashes are compared on Table 11-6. In some cases, mental confusion and other symptoms of central ner-

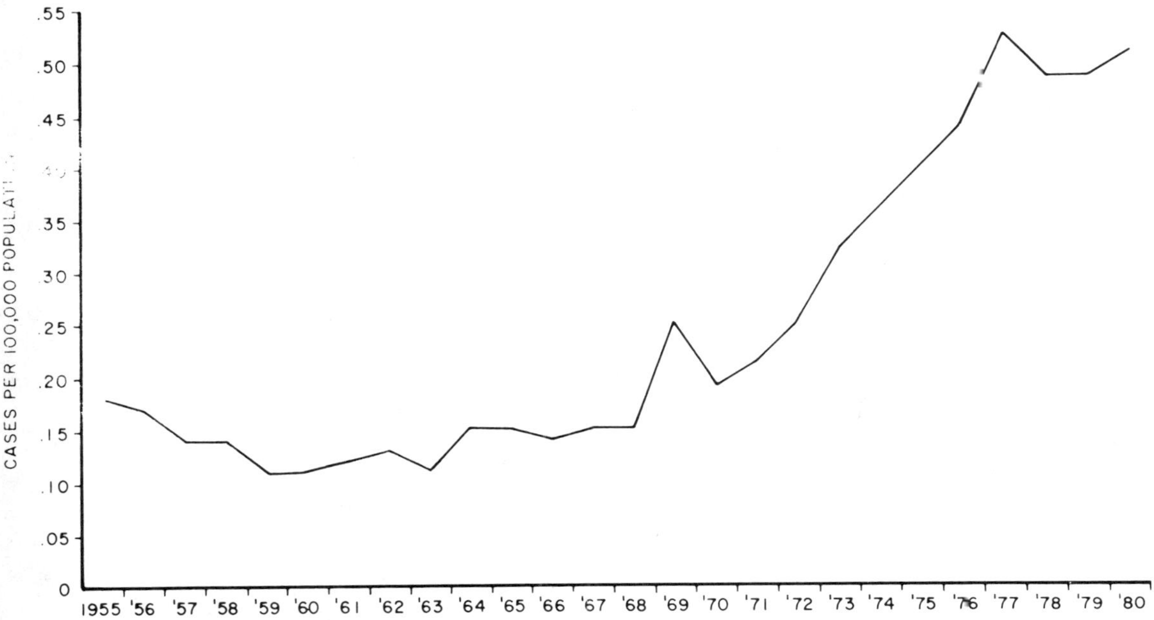

FIGURE 11-5. Typhus fever, tick-borne (Rocky Mountain spotted fever)—Reported case rates by year, United States, 1955 to 1980. (From Centers for Disease Control. Annual Summary 1980. Morb Mort Week Rep 1981;29:94.)

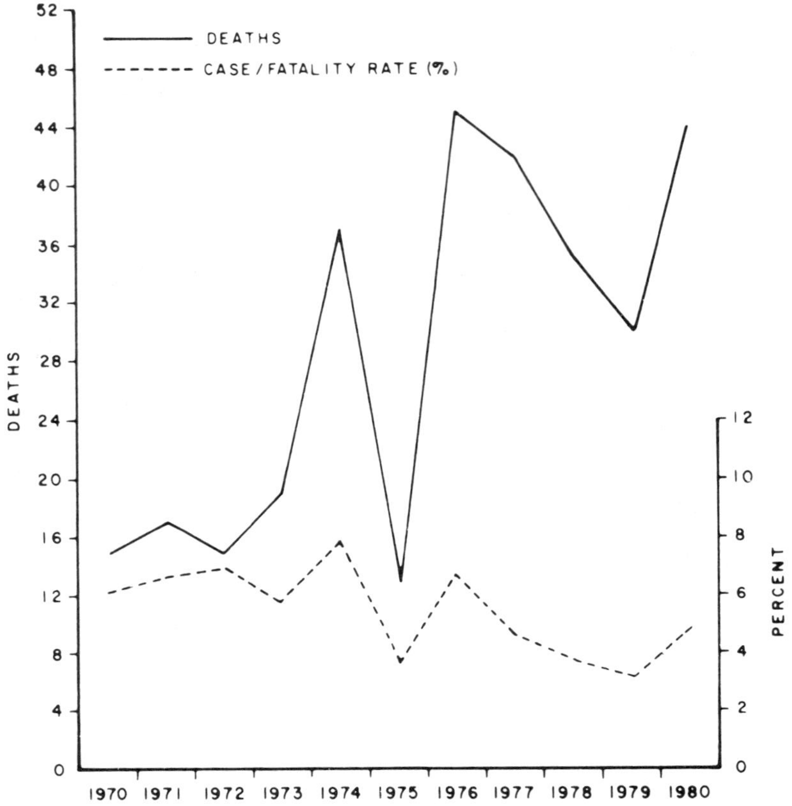

*Based on case-report form data; 1980 figure is preliminary.

FIGURE 11-6. Rocky Mountain spotted fever, death-to-case ratio, United States, 1970 to 1980. (From Centers for Disease Control. Rocky Mountain spotted fever. Mort Mort Week Rep 1981;30:319.)

vous system involvement occur. Myocarditis, shock, or renal failure can develop in fatal cases. Mortality occurs in about 3 to 8 percent of cases, despite antimicrobial therapy.

Control and treatment. Early and correct diagnosis with initiation of appropriate antimicrobial agents, tetracycline or chloramphenicol, is essential. In areas where ticks are prevalent, tick repellents and long clothing should be used. Solid immunity after infection develops; a vaccine is in the developmental stages.

Rickettsialpox

Rickettsialpox is a mild illness, often mistaken for chickenpox because of the varicellalike rash that develops. No fatalities are known to have occurred. The causative *Rickettsia, R. akari,* is transmitted by bloodsucking mites *(Allodermanyssus sanguine-*

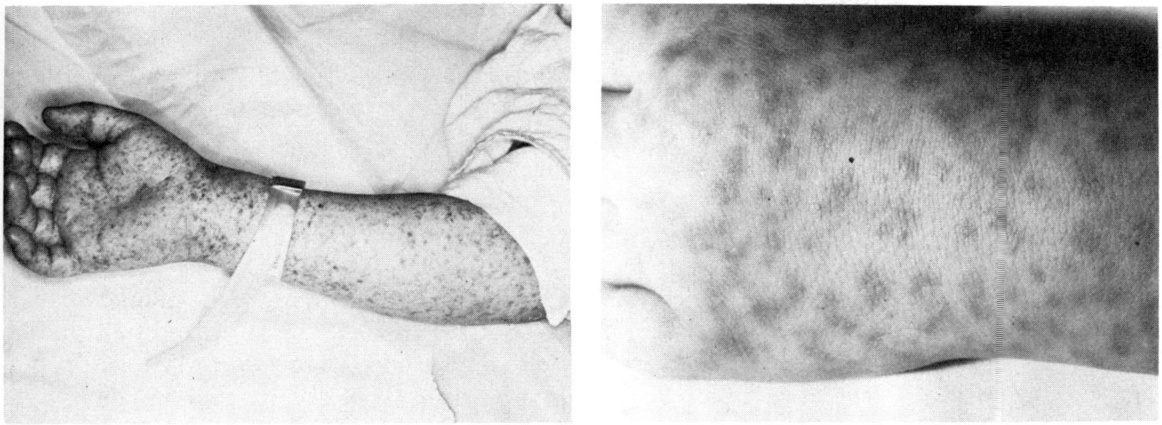

a b

FIGURE 11-7. *(a)* Rocky Mountain spotted fever in a child, age 4. Note petechial rash on palms and forearm with edema; about eighth febrile day. The patient recovered. *(b)* Close-up of wrist and hand of child with Rocky Mountain spotted fever. (*a* courtesy of Dr. Theodore E. Woodward. *b* courtesy of Centers for Disease Control, Atlanta, GA 30333.)

cis), which usually inhabit mice, so that the disease is common where people come in contact with mice. The disease is known to occur in the northeastern United States and in Russia. Possibly it occurs in other parts of the world as well. Since it is such a mild disease and usually not reported, the true incidence is unknown. A characteristic lesion at the site of the mite bite can help make the diagnosis. The lesion is a papule that develops into a vesicle with an eschar (called tache noir). It resembles the site of

TABLE 11-6. Rashes of Rickettsial Diseases

Disease	*Time of occurrence*	*Distribution*
Epidemic typhus	5-7 days after onset in all cases	Axilla, trunk, extremities; *not* on feet and hands
Endemic typhus	In about one-fourth of cases	As above
Brill-Zinsser disease	Only sometimes occurs	As with epidemic typhus
Rocky Mountain spotted fever	4 days after onset in all cases	Macular and petechial; ankles and wrists to trunk to palms and soles
Scrub typhus	At onset of disease	Generalized, maculopapular; distinctive eschar at site of mite bite, increasing to 1 cm
Rickettsialpox	At onset of disease	Varicellalike

a smallpox vaccination. The rash, fever, and headache usually last 1 to 2 weeks. *Rickettsia akari,* like *R. prowazekii,* are transmitted transovarially in the mouse mite.

Scrub Typhus (Tsutsugamushi Disease)

The causative agent of scrub typhus is *Rickettsia tsutsugamushi,* whose reservoir is also in mites (chiggers) that infest rodents. It can survive for many months in infected rat populations, and is transmitted transovarially in mites. People can be incidentally infected if infected mites attach to their skin and bite. Scrub typhus does not occur in the United States, but is very common in the Far East, South Pacific, Taiwan, Japan, Burma, India, and Ceylon. In Malaysia 20 percent of all acute febrile illness is due to scrub typhus. Scrub typhus was a major problem for military troops during World War II.

As in the spotted fevers, in scrub typhus there is frequently a lesion at the site of the chigger bite. As with other rickettsial diseases, scrub typhus is manifested after a 1 to 2 week incubation, by headache, high fever, and generalized rash. A unique feature of the disease is a lymphadenopathy. Mortality in the untreated can be up to 35 percent in more virulent strains. Because there are so many strains, development of vaccine is difficult. Immunity is not solid; it lasts only a few years. As in epidemic typhus, a latent form can persist for several months after recovery. Control with appropriate clothing and insecticides is helpful. A single dose of doxycycline has been used effectively for therapy.

Q Fever

Q fever is unique among rickettsial diseases in that it is usually transmitted to humans not by arthropods but by inhalation of products (urine, feces, birth tissue) from livestock infected with the causative agent, *Coxiella burnetii.* The disease is tick-borne to lower mammals. It is often latent until pregnancy and parturition. Then the organism is expelled with and multiplies in excreta, especially the placenta and other birth tissue. Those at high risk of the disease are persons who contact domestic livestock: farmers, veterinarians, and especially slaughterhouse workers. The disease occurs worldwide and can be epidemic in form, transmitted person-to-person. About 20 to 30 cases per year are identified in the United States. Q fever is unique in several respects when compared with other rickettsial diseases, as summarized in Table 11-7.

In humans, Q fever is an atypical pneumonitis resembling influenza. A rash does not develop. The incubation period is 14 to 26 days and the disease has a long course. Vaccines have caused severe local reactions, but are occasionally used in the military. In

TABLE 11-7. Differences Between Q Fever and Other Rickettsial Diseases

Other rickettsial diseases	*Q fever*
Weil-Felix reaction positive	Weil-Felix reaction negative
Arthropod transmission to humans	Inhalation transmission to humans
Vascular disease with rash	Pneumonitis without rash
Organisms unstable outside host cell	Organisms stable extracellularly

a recent study of over 3000 pregnant women in Maryland, cord blood samples were tested for antibody to *C. burnetii*. There was no evidence of fetal infection, though 3 percent of the samples had IgG titers consistent with contact with the organism. Immunity to Q fever seems to be cell mediated; immune serum has no effect on clearance of *Coxiella* in mice. However, spleen cells from immune animals result in accelerated clearance.

Trench Fever

Trench fever, caused by several different rickettsias, is transmitted by the human body louse. It is quite unusual in the United States, with 60 to 70 cases reported annually. It is endemic in Mexico and has caused serious morbidity among troops during wartime. Because the rickettsias may be carried in the blood, relapses may occur for years. Mortality, however, is very low. Symptoms resemble those of other rickettsial diseases: headache, fever, sweating, muscle aches, and rash.

CLINICAL IMPLICATIONS

The current epidemic of adult and infant chlamydial disease in this country is of major concern to the health professions. The fact that these diseases are chronic or subclinical in nature imposes a responsibility on clinicians to carefully educate their clients about the necessity to be persistent and consistent in complying with therapeutic regimens. Such compliance is known to be low, perhaps in part because many individuals are unaware of the potentially serious long-term sequelae of infection. The importance of adequate administration of ocular prophylaxis for newborns is also obvious.

Rickettsial diseases may be difficult to diagnose. The clinician can be helpful by soliciting a history of travel to areas where vectors such as ticks or lice are endemic. Because diagnosis must rely on clinical manifestations, and treatment must be initiated before laboratory confirmation is available, such observations are very important.

In endemic areas health care professionals can help in prevention of rickettsial diseases. Because ticks do not transmit organisms until they have been attached to the host for several hours, it is possible to prevent disease by frequently inspecting individuals for ticks and removing them when they are found. Ticks can be removed by grasping the tick with tweezers or fingers (which are protected with a cloth or tissue) as close as possible to their point of attachment to the skin and by pulling steadily and *slowly*. Because secretions from the tick can be infectious, the hands should be washed thoroughly after contact with the tick.

SUGGESTED READINGS

Arya OP, Mallinson H, Goddard AD. Epidemiological and clinical correlates of chlamydial infections of the cervix. Br J Vener Dis 1981;57:118-124.

Centers for Disease Control. Rocky Mountain spotted fever—U.S., 1981. Morb Mort Week Rep 1982;31:261-262.

Department of Social and Health Services. Report on chlamydial infections and other sexually transmitted diseases in Washington State. Olympia, Wa, 1981.

Hammerschlag MR, et al. Erythromycin ointment for ocular prophylaxis of neonatal chlamydial infections JAMA 1980; 244:2291-2293.

Holmes KK. The chlamydia epidemic. JAMA 1981;245:1718-1723.

Hunter JM, et al. *Chlamydia trachomatis* and *Ureaplasma urealyticum* in men attending a sexually transmitted diseases clinic. Br J Vener Dis 1981;57:130-133.

Lane RS, et al. Ecology of tick-borne agents in California. Am J Trop Med Hyg 1981;30:239-252.

Moe JB, Pedersen CE. The impact of rickettsial diseases on military operations. Military Med 1980;145:780-785.

Philip RN, et al. Rickettsiology conference. J Infec Dis 1980;141:112-118.

Schacter J. Chlamydial infections. Parts I, II, III. New Engl J Med 1978;78:428-435, 490-495, 540-549.

Siegel JD, et al. Single-dose penicillin prophylaxis against neonatal group B streptococcal infections: a controlled trial in 18,738 newborn infants. New Engl J Med 1980;303:769-775.

12 MYCOBACTERIA, *MYCOPLASMA*, AND SPIROCHETES

KEY POINTS

- Mycobacteria, *Mycoplasma,* and spirochetes are among the few bacterial agents *not* classified by Gram stain.
- These groups of bacteria contain both human normal flora and pathogens.
- The major pathogens in the group include *Mycobacterium tuberculosis* (causing tuberculosis), *Mycobacterium leprae* (causing leprosy), *Mycoplasma pneumoniae* (causing pneumonia), and *Treponema pallidum* (causing syphilis).
- Mycobacteria are "acid-fast" organisms that are very slow growing.
- *Mycoplasma* lack the usual bacterial rigid cell wall.
- Spirochetes are visualized by a special technique called darkfield examination.
- *T. pallidum* cannot be grown routinely on artificial laboratory media.

Organisms described in this chapter are ones for which the Gram stain is not a useful diagnostic tool. Paleontologic evidence indicates that tuberculosis and syphilis are ancient diseases known to humans for centuries. Leprosy too has been a scourge since recorded history began. Although the incidence of these diseases in this country has been generally on the decline for the past five or six decades, they continue to be leading causes of morbidity and mortality in developing countries. This chapter will discuss these diseases as well as those caused by other mycobacterial, spirochetal, and mycoplasmal species.

MYCOBACTERIA

Mycobacteria are aerobic, rod-shaped (sometimes spherical) bacteria. Some live in soil and water; some parasitize animals and humans. They are nonmotile, slow-growing, acid-fast, and occasionally form slender filaments. The Mycobacteriaceae are in the order Actinomycetales, along with the Actinomycetes (see Chapter 8, "Gram-Positive Rods"), but are treated separately here because of certain unique characteristics. Unlike the other families in their order, mycobacteria do not branch, and do not form hyphae or spores. In the family Mycobacteriaceae there are two genera, *Mycobacterium* and *Mycococcus*. None of the mycococci are pathogenic to humans; they live in soil and water. Medically important species of mycobacteria are summarized in Table 12-1.

Mycobacteria have the unique ability among bacteria of being able to resist decolorization by a solution of 95 percent ethyl alcohol with 3 percent hydrochloric acid. This acid-fast characteristic is due to the fact that mycobacteria are rich in many types of lipids. The high lipid content of mycobacteria is also responsible for some of the cellular response and necrosis in host tissues, and for the slow rate of growth of mycobacteria. Mycobacteria can be grown on artificial media, but growth is very slow, taking up to 8 weeks. Gram stain is not useful for identifying mycobacteria; the organisms are stained by Ziehl-Neelsen technique. Though the pathogenic strains are similar in appearance to nonpathogenic mycobacteria on stain, virulent strains do produce a characteristic pattern of parallel cord formation.

MYCOBACTERIUM TUBERCULOSIS

Biologic characteristics. *Mycobacterium tuberculosis* is a strict aerobe, about 0.3 μm wide, which thrives in the lungs (especially the apices where the oxygen content is high; Fig. 12-1). Its doubling time is about 12

TABLE 12-1. Species of Mycobacteria Causing Human Disease

Species	*Habitat*	*Diseases in humans*
M. avium	Birds, mice, rabbits, pigs	Lung infection, scrofula
M. bovis	Cattle, humans	Lung and opportunistic infections
M. chelonei	Dust, soil, water, animals	Lung infection
M. fortuitum	Soil, cattle, humans, reptiles	Lung infection
M. intracellulare	Dust, soil, humans	Lung infection
M. kansasii	Water, humans, soil	Lung infection, scrofula
M. marinum	Saltwater fish	Skin ulcers
M. leprae	Humans	Leprosy
M. scrofulaceum	Soil, water, humans	Scrofula
M. smegmatis	Soil, water, smegma	Probably none, but may cause false diagnosis of renal TB
M. tuberculosis	Humans	Tuberculosis
M. ulcerans	Humans	Skin ulcers
M. xenopei	Soil, water, humans	Lung disease

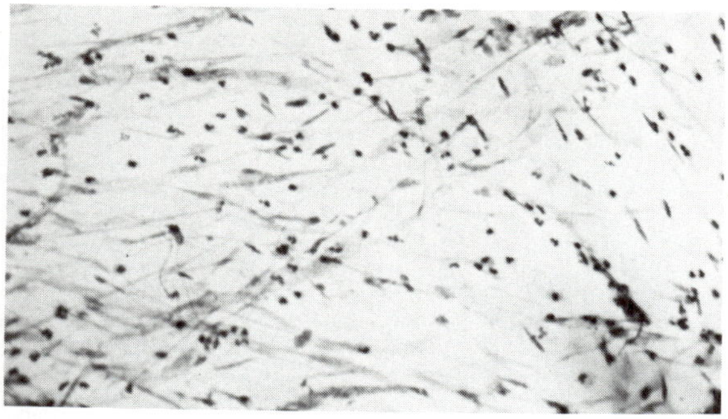

FIGURE 12-1. Sputum containing *Mycobacterium tuberculosis* stained with Ziehl-Neelsen technique. (Courtesy of Dr James Plorde.)

hours. The organism grows in moldlike clumps (hence the name "myco" or "fungus" bacteria). Its cell wall is complex and contains mostly lipids (60 percent) such as mycolic acid and true waxes. A cord factor is correlated with both virulence and serpentine growth. At least 20 distinct antibodies to *M. tuberculosis* have been described. However, the mycobacterial lipid coat may protect the cell wall glycoprotein antigens so that humoral immunity is poorly stimulated (though IgA levels are often increased in individuals with active tuberculosis).

Many patients with tuberculosis produce antibodies to antigens unique to *M. tuberculosis,* others that are restricted to slow-growing mycobacteria, and other antigens common for all species of mycobacteria. However, 20 percent of patients do not have elevated antibody levels to any of these antigen groups. Therefore, if purified soluble antigens specific for *M. tuberculosis* were available, serologic diagnosis of tuberculosis would still not be possible in all cases. Some work has been done in typing *M. tuberculosis* with bacteriophage.

In 1882 the German scientist Robert Koch discovered the cause of tuberculosis, then a prevalent and often devastating disease called consumption. It was based on his work with the tubercle bacillus that he first formulated "Koch's postulates," modified from earlier work by Henle (Chapter 4).

Epidemiology. The major host of *M. tuberculosis* is humans. The incidence and prevalence of tuberculosis (TB) have declined rapidly since the turn of the century. Whereas the majority of individuals in the United States prior to the twentieth century were infected by the time they finished their teen years, less than 1 percent in that age group are now infected. The current disease rates are about 12 per 100,000 persons annually. In 1980, 27,749 cases were reported in the United States, a slight increase over previous years due to an influx of Indochinese refugees. This reduction in TB is attributable primarily to the improved standard of living, but also to prompt recognition and treatment of cases. The prevalence of infection increases markedly with age. Between 1953 and 1979 there was an increase from 14 to 29 percent in the proportion of individuals with tuberculosis who were over age 65. This is because many elderly persons were infected at an earlier age (Fig. 12-2). Currently, the majority of tuberculosis occurs among persons who were originally infected in the distant past.

Despite its dramatic decline in developed countries, TB is still of major public health concern throughout the world. There is a clear association between disease and socioeconomic factors such as nutritional status. In one military study, overt clinical TB was three times more common among men who were at least 10 percent underweight than among those who were 10 percent overweight, even though infection rates were the same.

Nevertheless, TB is not highly commu-

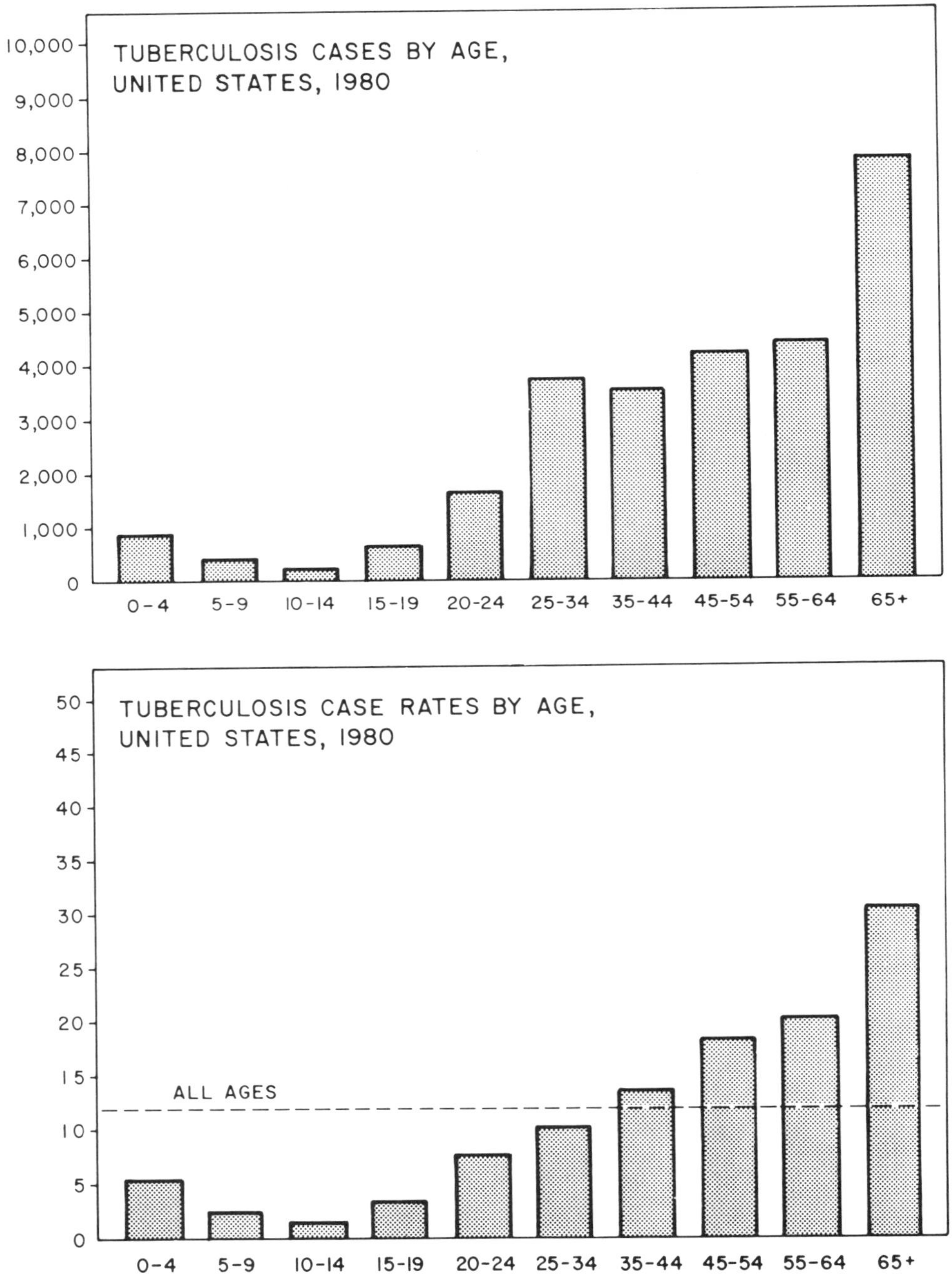

FIGURE 12-2. Tuberculosis. Reported cases and cases per 100,000 population by age group, United States, 1980. (From Centers for Disease Control. Annual Summary 1980. Morb Mort Week Rep 1981;29:88.)

nicable. Secondary attack rates among children aged 5 to 9 living with an individual with active TB have been shown to be about 50 percent, as compared with about 85 percent secondary attack rates for measles, mumps, or pertussis. TB rates in the United States are much higher among dark-skinned individuals, and higher for men of all races than for women. Despite the decline in pulmonary TB, there has been little change in the amount of extrapulmonary TB reported in recent years (Fig. 12-3). In Canada, one-sixth of cases from 1970 to 1974 were non-respiratory, including genitourinary, gastrointestinal, bone and joint disease, lymphadenitis, and meningitis (Enarson et al. 1980).

Pathogenesis. *Mycobacterium tuberculosis* most frequently enters the body by inhaled droplet nuclei, but it can also enter through the GI tract or by accidental inoculation. There is no toxin produced and no immediate tissue reaction, so the organisms multiply unabated. They are phagocytized and multiply in macrophages. Growth is slow, but eventually a mass of macrophages containing the tubercle organism forms in the lung. This growth is called a granuloma or tubercle, and is the reason the *M. tuberculosis* is called the tubercle bacillus. When multiplication of bacteria reaches a certain point after weeks or months, pulmonary symptoms occur. Dissemination from the pulmonary focus is via lymphatics.

The primary infection can result in death, or, more commonly, the infection will be overcome and the lesions will heal. The growth and spread of *M. tuberculosis* is pre-

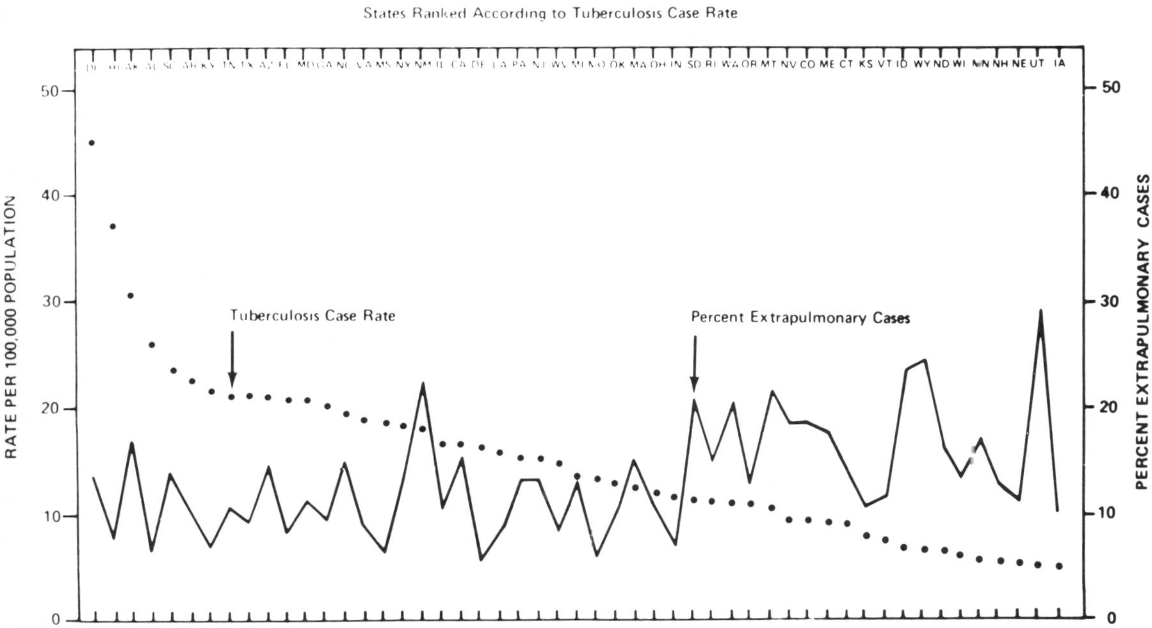

FIGURE 12-3. Tuberculosis case rates and percent extrapulmonary cases, United States, 1970 to 1974. (Reprinted with permission of the publisher from Farer LS, Lowell AM, Meador MP. Extrapulmonary tuberculosis in the United States. Am J Epidemiol 1979;109:205-217.)

vented by the development of acquired cellular immunity, especially an enhanced ability of activated macrophages to kill the organism. Along with cellular immunity, the infected individual also develops a delayed hypersensitivity reaction that causes a positive tuberculin skin test. Hypersensitivity develops as a response to the tubercle polypeptide, and can be passively transferred.

When immunity develops, granulomas and caseation necrosis develop in the pulmonary lesion, lymph nodes, and wherever the organism has spread. Necrosis results from the release of lymphokines from T cells that are lethal for the TB bacterium and for host cells as well. After primary infection, lesions usually heal and the necrotic area becomes fibrosed and calcified. Such healed primary lesions can be seen on chest x-ray. In most cases the infection becomes latent, the tubercle bacteria being present for life without causing clinical disease. The only symptom may be conversion of purified protein derivative (PPD) reaction from negative to positive.

Clinical TB can then develop in two ways: either directly by progression of the primary infectious process if host defenses are inadequate, or by recrudescence of the dormant lesion—by far the most common course in adults. Reactivation of tuberculosis can occur months or years after the initial infection as a result of conditions that lower host defenses (such as malnutrition, aging, or stress). Reactivation tuberculosis is also called secondary or adult tuberculosis. Spread of the organism occurs if the liquid caseous lung nodules (which contain numerous live bacteria) spill contents into the bronchi (bronchogenic spread) or if a blood vessel is eroded (hematogenous spread). After the liquid material from a caseous lesion is expelled, a cavity is formed.

Host responses. Among those infected with the TB bacteria, only about 5 to 10 percent develop clinical disease. The risk of developing clinical TB is greatest right after infection, about 1 in 30 in the first year. However, since the organism can remain viable in macrophages for years, the risk is life long. About 4 to 8 weeks after infection, the individual may experience malaise and fever, and the tuberculin test will become positive, denoting the development of delayed hypersensitivity.

Symptoms usually subside and the disease does not progress. In those with decreased host defenses, however, infection spreads in the lungs or circulatory system. This may occur immediately after initial infection or at any time throughout life. Pulmonary TB may be asymptomatic and discovered only on routine chest x-ray or PPD. Symptoms may include "night sweats," fever, excessive fatigue, weight loss, and cough. Sputum production is mostly in the morning; sputum is yellow or green and is odorless. As the disease progresses, cavitation and liquefaction necrosis resulting from inflammatory and immune responses occur. *M. tuberculosis* multiply and the sputum becomes infectious. Hemoptysis, pleurisy, bronchopleural fistula, empyema, laryngitis, and even spread to the peritoneum can occur.

Other body sites can be affected by hematogenous spread of the organism, including lymph nodes, kidneys, genital organs, bone, pericardium, meninges, and adrenals. Miliary TB, the most severe manifestation, occurs when a mass of liquid caseous material spills huge numbers of bacteria into the blood stream. A massive infection can occur that may be overwhelming and fatal unless promptly treated. Tuberculous spondylitis (Pott's disease), a frequent historical cause of "hunchback," is rare now. It may result from hematogenous spread and was commonly acquired in the past by nonpasteurized milk from tuberculous cows. In the past, if the individual was unable to generate adequate immune defense, TB was slowly but persistently progressive, and was eventually fatal.

The tuberculin test provides accurate information about whether an individual has ever been infected with *M. tuberculosis;* it is not an indication of whether the patient has active disease. However, it has been shown that individuals with the highest level of sensitivity to tuberculin, those who have strong local reactions, are at greatest risk of having or developing clinical disease.

Tuberculin is a protein fraction from broth in which *M. tuberculosis* have been cultivated. The preferred testing method is the Mantoux test. The method is to inject intradermally 0.1 μg of purified protein derivative in 0.1 ml (intermediate strength), which has been stabilized with a dispersing agent called Tween 80 (PPD-T) into the forearm.

Another test method, the *tine test,* uses a metal device containing multiple prongs impregnated with PPD. It is pressed into the skin. With the *Vollmer patch test,* a gauze saturated with PPD is placed next to the skin for several days. Neither of these tests are as sensitive or specific as the Mantoux test, so their use is limited to mass screening programs.

Control and treatment. Tuberculosis is generally well controlled in the United States (Fig. 12-4). Milk pasteurization has been a significant control measure, as TB was often acquired by children and young adults in infected milk. Improvement in the standard of living has also been essential, as has screening by chest x-ray or skin testing for high-risk groups (health care personnel, contacts of tuberculous patients, immigrants). Immunity to TB is primarily cellular and much of the pathogenesis is related to the immune response. Because the organism is latent, an individual once infected is never really immune to reinfection, although host defenses usually keep the organism in control. Less than half of patients with newly diagnosed TB have a detectable antibody response. Therefore, chemotherapy is necessary both to prevent disease in individuals who have recently become infected and to treat those with active TB. Chemotherapeutic agents increase antibody formation against the TB bacteria as well as having bactericidal effects.

Isoniazid (INH) is used as chemoprophylaxis for persons with newly acquired or dormant infections and is successful in preventing the progression of infection into active TB. Young children who have been in close contact with an active case may also be treated even if they are tuberculin negative. Once TB develops, chemotherapy is the same regardless of whether the disease is pulmonary or extrapulmonary. Because of the high frequency of organisms resistant to one or another chemotherapeutic agent, treatment with two or three bactericidal or bacteriostatic agents is standard. In 1976 there was an outbreak of TB caused by a *Mycobacterium* resistant to INH, streptomycin, and para-aminosalicylic (PAS). There were a total of 15 documented and 7 presumed multiply resistant, epidemiologically linked cases. Such resistance, fortunately, is quite rare, although most of the tubercle organisms isolated among Indochinese refugees are resistant to INH and streptomycin (Centers for Disease Control

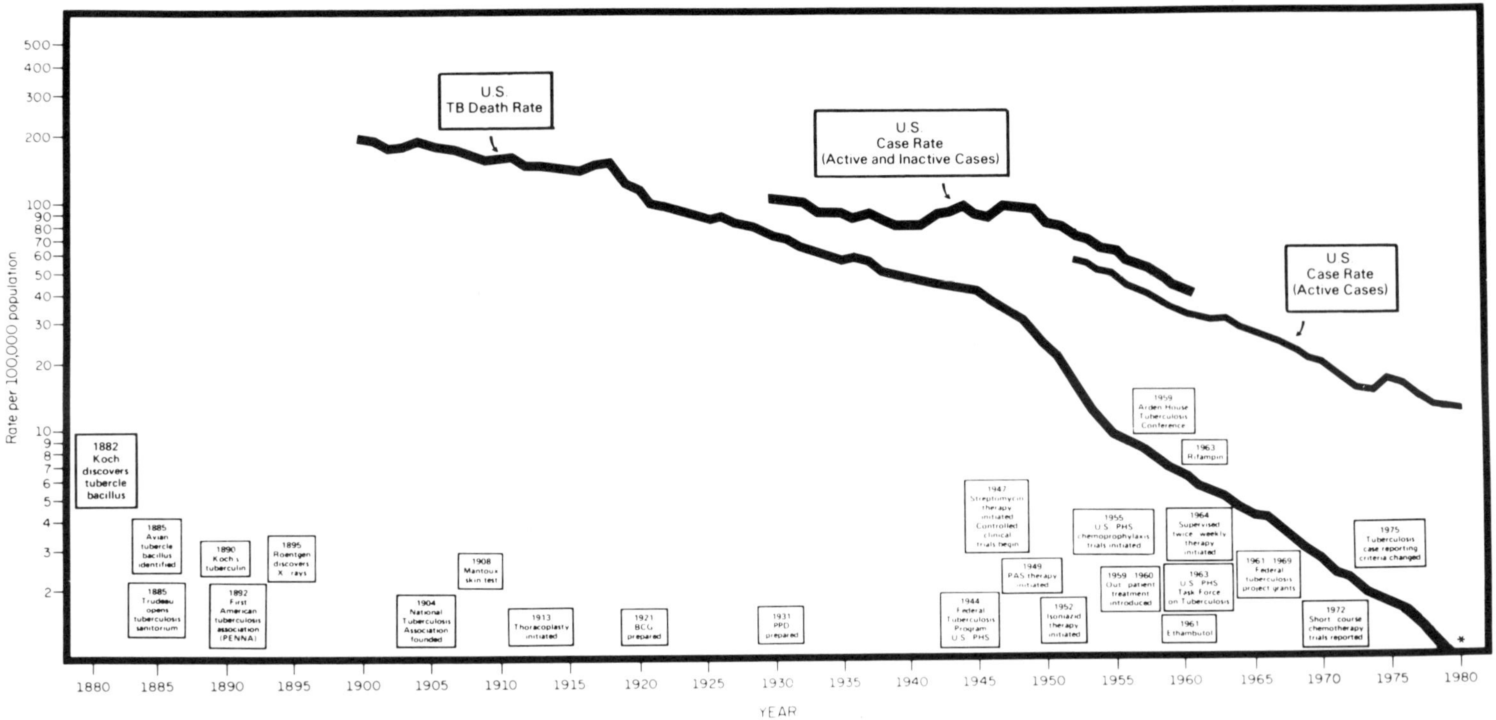

*Provisional data: 1979 = 0.9; 1980 = 0.8.

FIGURE 12-4. A century of advances in tuberculosis control, United States. (Reprinted from Centers for Disease Control. Centenniel: Koch's discovery of the tubercule bacillum. Morb Mort Week Rep 1982;31:122.)

1981a; Reves et al. 1981). Usually 18 to 24 months of therapy is necessary because the organisms replicate very slowly and the drugs are effective only against replicating organisms. The most effective agents are INH and rifampin (RIF). Others are also used and are listed on Table 12-2. If a second course of treatment is necessary, entirely different drugs should be used.

The days of the tuberculosis sanitarium

TABLE 12-2. Chemotherapy for Tuberculosis

Drug	*Dose*	*Comments*
Prophylaxis		
Isoniazid (INH)	Adults: 300 mg/day Child: 5-10 mg/kg/day for 12 months	Not recommended for individuals over age 35; risks of complications (hepatitis) outweigh risks of TB
Treatment of active TB* (in approximate order of usefulness)		
Isoniazid	300 mg daily for 2-4 wks; then 900 mg twice weekly for up to 24 months	Bactericidal; interferes with mycolic acid synthesis; may cause hepatitis
Rifampin (RIF)	600 mg daily for 2-4 wks, then 600 mg twice weekly for 6-12 months	Bactericidal; may produce liver dysfunction
Ethambutol	15 mg/kg/day	Not recommended for under age 13 due to possible visual toxicity; replacing PAS, fewer side effects
Streptomycin	1 g daily	Bactericidal; neurotoxicity can occur
Para-aminosalicyclic acid (PAS)	5 g 3 times per day	Not as effective, more side effects than other agents; useful in pregnancy and childhood; interferes with metabolism of *M. tuberculosis*, potentiates INH; side effects include anorexia, nausea and vomiting, diarrhea, allergic reactions
Cycloserine	500-750 mg daily	Used primarily in retreatment; inhibits cell wall synthesis; neurotoxicity can occur
Ethionamide	750-1000 mg daily in divided doses	Used primarily in retreatment; inhibits protein synthesis; causes gastric irritation
Aminoglycosides (kanamycin, viomycin)	1 gm I.M., 2 doses 12 hours apart, twice weekly	Limited effectiveness; used in retreatment
Thiocetazone, isoxyl	Not available in U.S. due to toxicity	Used in developing countries because of low cost, moderate effectiveness

*Two or three agents are given together to prevent emergence of a resistant strain. After several months of negative cultures, INH can be given alone.

are gone. Many tuberculous patients are not hospitalized but are treated as outpatients. The trend is increasingly toward drug therapy rather than long-term physical isolation of cases. In one study in India, TB cases receiving the same chemotherapy were randomly assigned to home or hospital care (because of lack of hospital beds). After 5 years, there were fewer TB cases among contacts of patients treated at home than among contacts of patients who were isolated in the hospital. Strict isolation is no longer recommended for hospitalized patients unless they are on a mechanical ventilator or are in other ways likely to disseminate infectious material in the hospital setting. Extrapulmonary TB without drainage is not infectious.

Even though the sputum is not cleared of *M. tuberculosis* for about 8 weeks after initiation of therapy, infectiousness rapidly decreases once chemotherapy is instituted. This may be due to decreased cough, or it may be due to the presence of the antimicrobial agents in expectorated material. Of compliant patients, 95 percent become sputum-negative within 3 months of treatment with INH and RIF.

A vaccine, Bacille Calmette-Guerin (BCG), contains an attenuated bovine strain of mycobacteria that is used in some countries where TB is highly endemic. In tuberculin-negative individuals it results in the development of tuberculin sensitivity in about 90 percent of cases. It provides about 80 percent protection for one or two decades against the development of clinical TB when natural infection occurs. It is of no value in the United States, except among those who cannot avoid exposure (babies with tuberculous mothers, health care workers entering an area of high endemicity, household contacts of active cases). It should not be used with the immunosuppressed.

Mycobacterium leprae

Biologic characteristics. *Mycobacterium leprae* is an acid-fast, weakly curved rod (Fig. 12-5). It is an obligate intracellular parasite living mainly in macrophages and Schwann cells (Figs. 12-6 and 12-7). It has not been cultivated on artificial media, but is propagated for study in the foot pads of mice (which are cooler than other body tissues) and in the nine-banded armadillo, which has a low body temperature and marked propensity for systemic infection with *M. leprae. M. leprae* causes one disease, leprosy (Hansen's disease), with several clinical manifestations. Leprosy is classified into four types according to its clinical presentation (see Table 12-3).

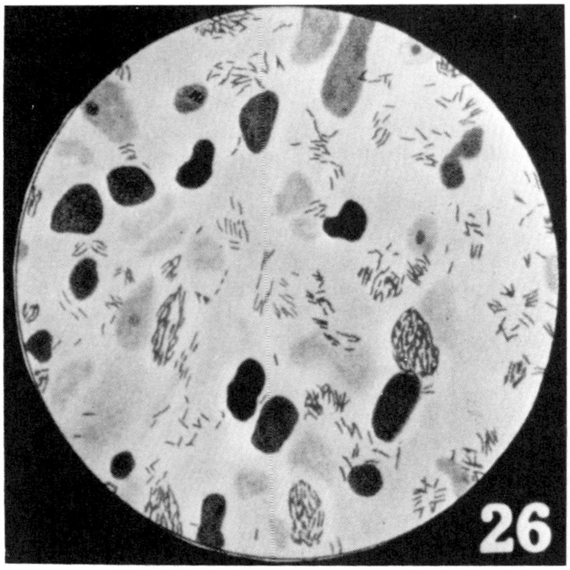

FIGURE 12-5. Photomic: *Mycobacterium leprae* (Hansen's bacillus of leprosy); from leprosy skin lesion; Ziehl-Neelsen technique. (Courtesy of Centers for Disease Control, Atlanta, GA 30333.)

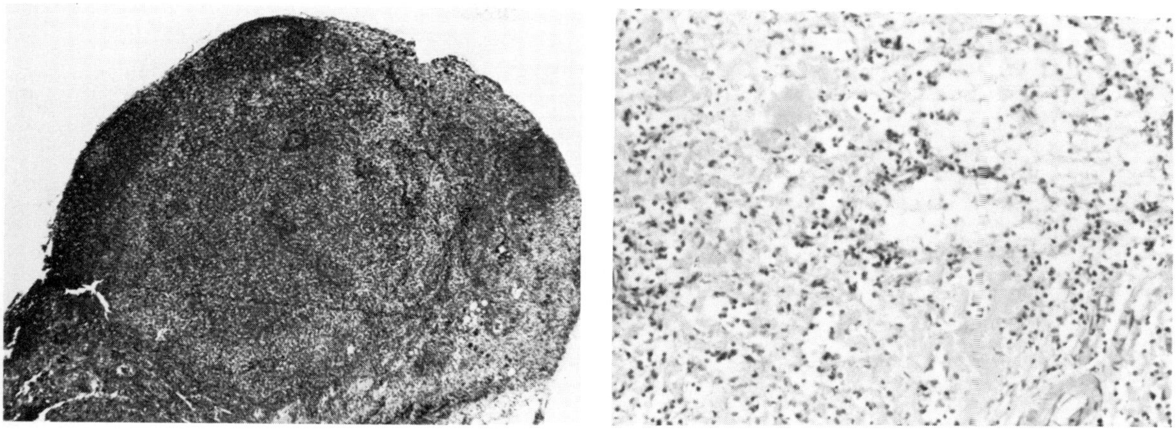

a b

FIGURE 12-6. *(a) Mycobacterium leprae;* lymph node. Magnification 50×. *(b)* Lepromatous leprosy. Spleen with nests of foam cells. (Courtesy of Centers for Disease Control, Atlanta, GA 30333.)

More than 40 antigenic components, many common to all mycobacteria and several proteins specific to *M. leprae,* have been demonstrated by immunoelectrophoresis. Antibodies cannot affect the organism inside the cell. Hence, cell-mediated immunity is considered to be of primary importance against leprosy. T cell-associated defects have been found in patients with lepromatous but not tuberculoid leprosy.

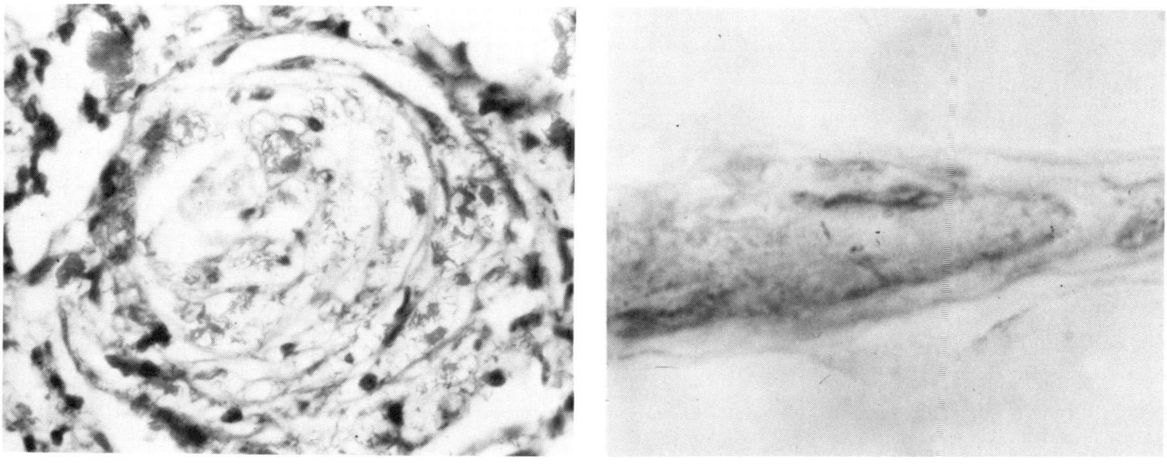

a b

FIGURE 12-7. Leprosy. *(a)* Cutaneous nerve invaded by innumerable bacilli. *(b)* Also cutaneous nerve. Occasionally, hours of searching are necessary to detect 2 or 3 bacilli. (Courtesy of Centers for Disease Control, Atlanta, GA 30333.)

TABLE 12-3. Clinical Classifications of Leprosy

Type	Characteristics	Usual lepromin reaction
Lepromatous	Extensive diffuse granulomatous reaction, bilaterally symmetrical; progressive, ultimately fatal	Negative
Borderline (dimorphous)	A combination of lepromatous and tuberculoid symptoms; disease may shift toward one form or the other	Positive
Tuberculoid	Skin lesions (papules or hypopigmented plaques) are few, often occur where other lesions had been previously; healing occurs; neurologic involvement can be severe	Positive
Indeterminate	Early stage, which may progress to any of above three	Positive

There is evidence that certain individuals have a genetic susceptibility for tuberculoid leprosy by means of recessive inheritance of HLA-linked markers.

As with TB, a delayed hypersensitivity reaction develops among individuals with leprosy. A skin-testing material called *lepromin,* composed of killed *M. leprae* from tissues of lepromatous patients, is available. It is not used for diagnosis, as most healthy individuals have a positive reaction. However, it can be used to differentiate lepromatous (negative reaction) and tuberculoid (positive reaction) forms of leprosy. A positive response, called the Misuda reaction, occurs 2 to 4 weeks after intradermal injection of lepromin. The Misuda reaction is the development of a granulomatous papule at the injection site.

Epidemiology. Humans are the only known reservoir for leprosy. The disease occurs primarily in the tropics, and less frequently in cooler climates such as in China and the United States. There are about 15 million cases currently in the world, about 3000 of these in the United States. About 100 to 200 new cases, mostly imported, are reported yearly in this country (Fig. 12-8). Epidemiologic studies are difficult in endemic areas because of the long incubation period (at least 3 years in most cases) and a lack of dependable tests to identify subclinical infections. The prevalence rate in India declined from 63 per 1000 persons in 1955 to 43 per 1000 persons in 1973; the lepromatous leprosy rates decreased from 12 to 5 per 1000 persons. However, the proportion of cases that were among children less than 15 years of age increased from 28 to 47 percent.

The portal of entry is either the skin or nasal mucosa and the disease is only spread by intimate, very long-term (years) exposure such as that occurring in families. The attack rate among untreated family mem-

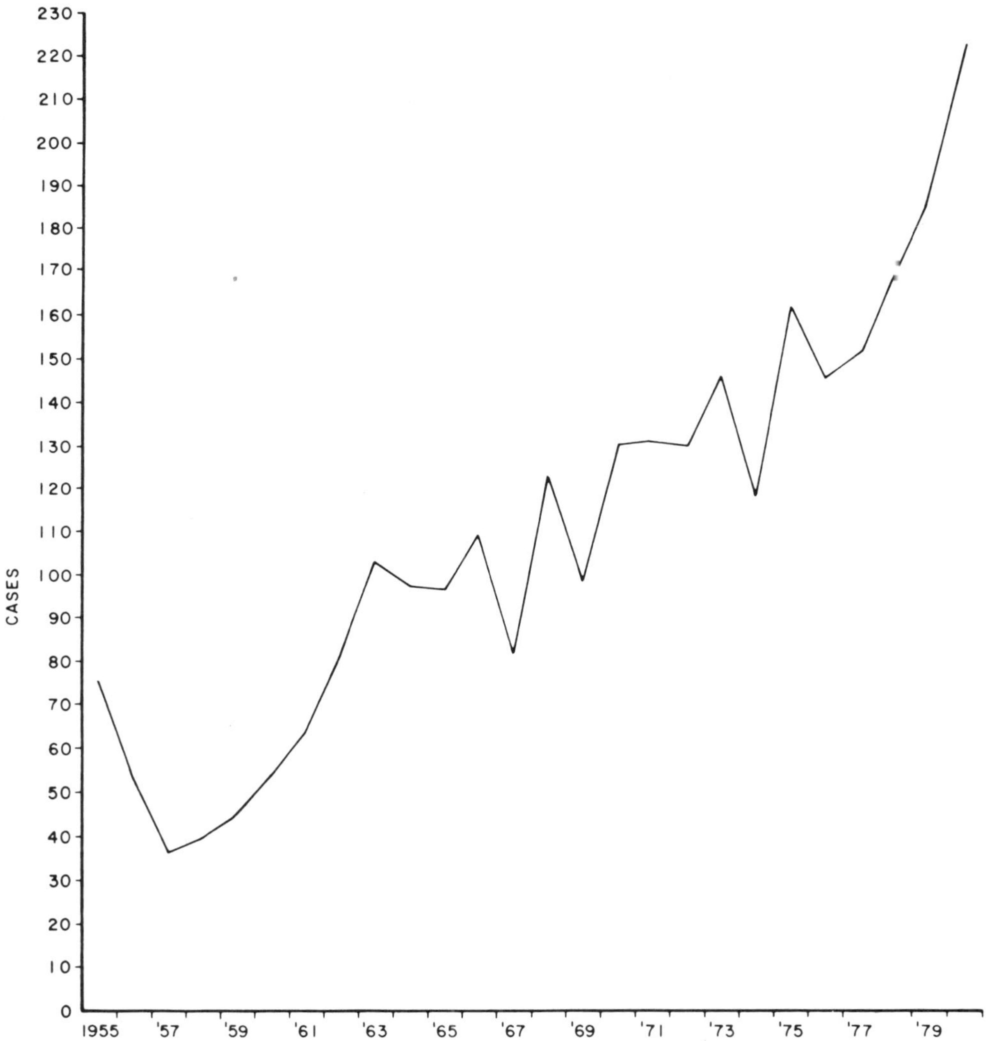

FIGURE 12-8. Leprosy. Reported cases by year, United States, 1955 to 1980. (From Centers for Disease Control. Annual Summary 1980. 1981;29:47.)

bers of individuals with leprosy is only 5 to 10 percent. Almost half of the children of parents with leprosy become infected, but the disease usually consists of a mild syndrome with single lesions that heal without treatment.

Pathogenesis. After entering the body, the organism deposits itself in the skin and peripheral nerves, probably because of its predilection for lower temperatures. The lesions produced are granulomatous and similar to those of TB (Fig. 12-9). T cell de-

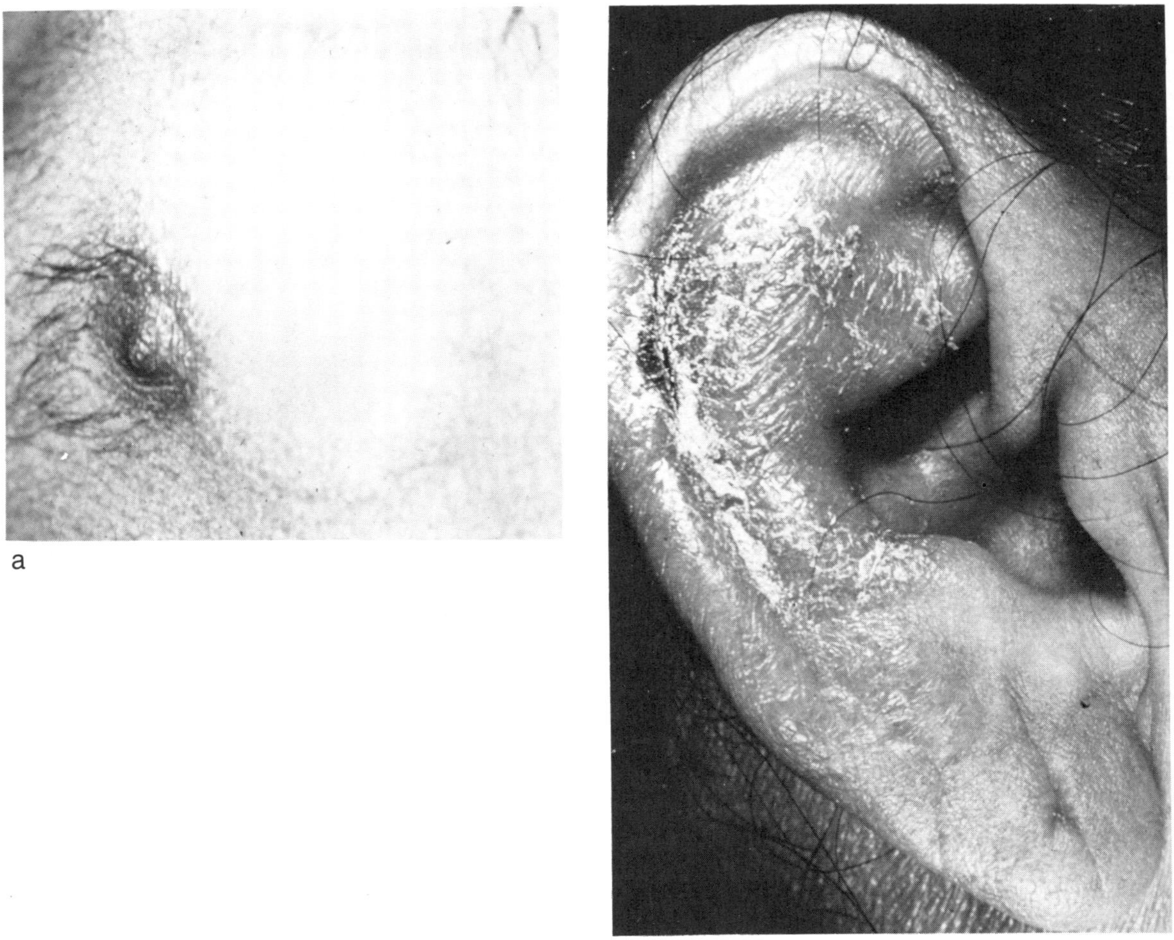

a

b

FIGURE 12-9. Leprosy. *(a)* Hypopigmented plaque with elevated scaly border (differential diagnosis: tinea). *(b)* Reddish-brown infiltration, external ear (differential diagnosis: *lupus vulgaris*, eczematous dermatitis). (Courtesy of Centers for Disease Control, Atlanta, GA 30333.)

ficiency in the host clearly plays a role in the progression of lepromatous leprosy, though whether the disease is the result or the cause of insufficient cellular immunity is unclear. In one study, the babies of 110 women with leprosy and 33 controls were compared for 2 years after birth. Babies of mothers with lepromatous leprosy weighed significantly less than those with tuberculoid leprosy or controls and had a trend of "failure to thrive," probably due to immunologic factors (Latapi et al. 1980).

Host responses. The incubation period for leprosy is variable from months to many years. As with TB, many infections may be dormant or subclinical. Most cases (84 percent in one series from India) in children are benign types (Latapi et al. 1980). Clinical manifestations include formation of nodules (lepromas) over the surfaces of the body and a change in facies so that the features are heavier (Fig.12-10). The organisms spread widely in the blood stream.

In the lepromatous type, cutaneous lesions contain many *M. leprae,* the disease is slowly progressive, and prognosis is poor. Affected individuals most frequently die from causes other than leprosy such as amyloidosis or infection, especially TB (Fig. 12-11). Peripheral nerve damage results in anesthesia of various body parts, especially fingers, toes, and elbows. This loss of sensation leads to injury and secondary infections, resulting in the disfigurement that is characteristic of the disease (Fig. 12-12).

Unlike lepromatous lesions, the lesions of tuberculoid leprosy are nonsymmetrical, fewer, macular, sharply demarcated, and heal slowly. Neurologic involvement results in pain, muscle atrophy, corneal ulcerations (Fig. 12-13), and loss of appendages. Diag-

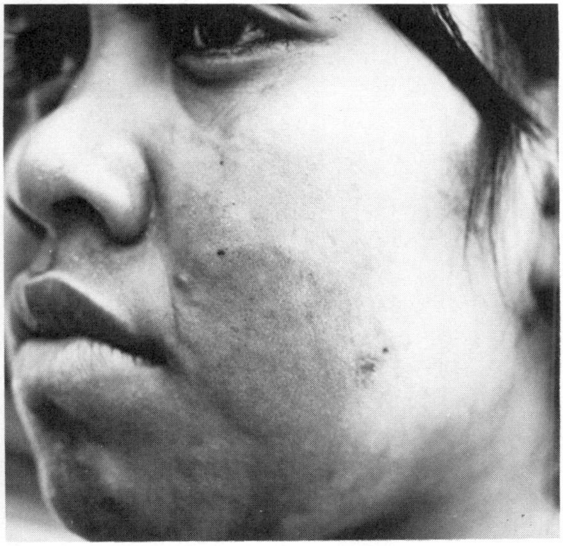

FIGURE 12-10. Face affected by leprosy *(Mycobacterium leprae).* (Courtesy of Centers for Disease Control, Atlanta, GA 30333.)

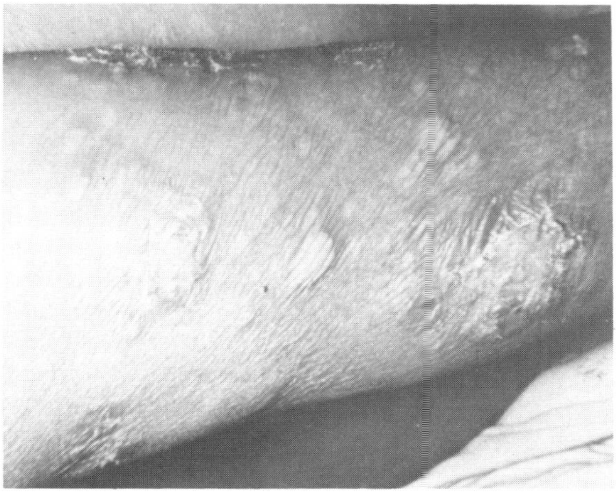

FIGURE 12-11. Lepromatous leprosy complication: erythema nodosum leprosum with pustulation. (Courtesy of Centers for Disease Control, Atlanta, GA 30333.)

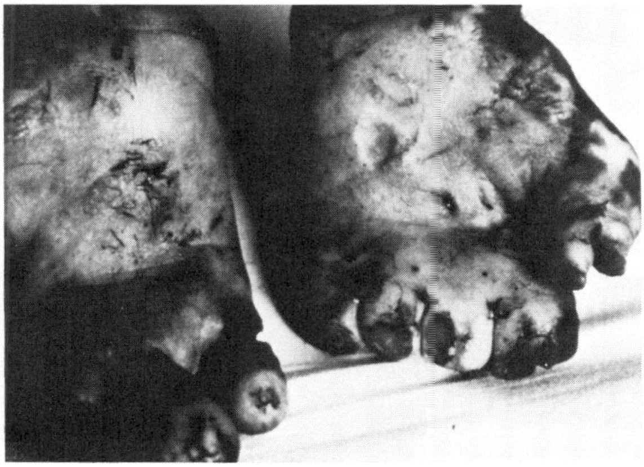

FIGURE 12-12. Lepromatous leprosy. Severe mutilations, both hands (often diagnosed as syringomyelia). (Courtesy of Centers of Disease Control, Atlanta, GA 30333.)

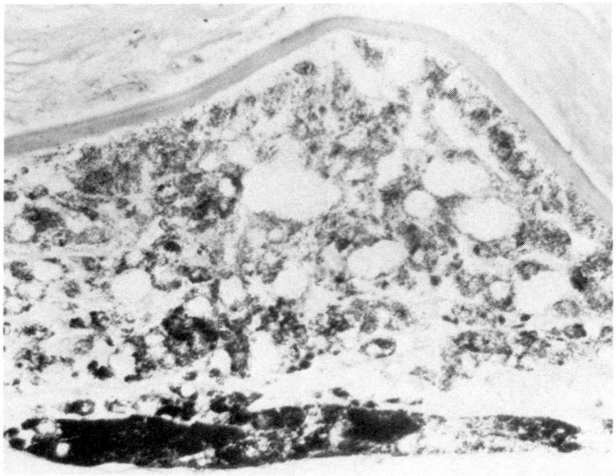

FIGURE 12-13. Lepromatous leprosy—eye. Severe destruction leading to blindness; innumberable lepra bacilli. (Courtesy of Centers for Disease Control, Atlanta, GA 30333.)

nosis is based on clinical findings accompanied by identification of acid-fast bacteria in tissue or lesion scrapings.

Control and treatment. In endemic areas, the establishment of clinics has been very helpful for enhancing early detection and treatment. It is unnecessary to remove individuals with leprosy from their homes if they are compliant with therapy. Routine skin examinations of family members are important, especially among children. Treatment of choice is diaminodiphenylsulfone (dapsone), though resistance may develop. Dapsone and alternative chemotherapeutic agents are listed on Table 12-4.

Other Mycobacterial Infections

At least ten other species of mycobacteria have been reported to cause clinical disease in the United States, the most common being *M. avium* and *M. fortuitum*. Organisms responsible for such infections were formerly called "atypical mycobacteria," but are now referred to as "mycobacteria other than tuberculosis" (MOTT). Pulmonary diseases resembling TB have been reported to be caused by *M. avium, bovis, chelonei, fortuitum, kansasii, intracellulare, xenopei*. Scrofula (cervical lymphadenitis) used to be quite common and still occurs occasionally. It has been caused by *M. avium, intracellulare, kansasii,* and *scrofulaceum*. Isolation of the organisms from aspirated pus is diagnostic. Skin ulcers can be caused by *M. marinum* and *ulcerans,* and several outbreaks of injection-related abscesses caused by *M. chelonei* and *fortuitum* have been reported. Surgical drainage may be necessary for localized infections. In England most strains of *M. kansasii* and *marinum* tested were sensitive to RIF and ethambutal, whereas all strains of *M. fortuitum* and *chelonei* tested were highly resistant (Yates and Collins 1981). Because of varying sensitivities, susceptibility testing of the isolate should be undertaken. Some mycobacterial strains can cross react with *M. tuberculosis,* interfering with the tuberculin test. Other nonpathogenic mycobacteria are common in nature and as normal human flora.

MYCOPLASMATACEAE

Mycoplasma are the smallest free-living cells identified and lack the rigid cell wall common to other bacteria. In the class Mollicutes, order Mycoplasmatales, family Mycoplasmataceae, there are two genera: *Mycoplasma* and *Ureaplasma*. Various strains cause pneumonia in cows, pigs, and rodents. *Mycoplasma* were previously called pleuropneumonialike organisms (PPLO) and were first identified as agents responsible for lung disease in cattle. *Ureaplasma* was previously called T (tiny) strains. There are

TABLE 12-4. Drugs Used in the Treatment of Leprosy

Drug	Resistant strains?	Daily dose	Comments
Dapsone	Yes	50 mg	Minimum inhibitory concentration (MIC) low, bactericidal; resistant strains common
Rifampin	Yes	600 mg	Bactericidal; works against dapsone-resistant strains, but also has resistant strains
Acedapsone	Yes	225 mg	Acts by releasing dapsone
Thiacetazone	Yes	150 mg	
Ethionamide/prothionamide	Yes	500 mg	
Clofazimine	Probably	100 mg	
Sulphamethoxypyridazine	Yes		Cross reaction with dapsone; much less effective

about 70 species of *Mycoplasma;* about 11 of which inhabit humans. Many species are mouth and genital flora (Table 12-5). *Ureaplasma urealyticum* is the only species of *Ureaplasma* and has about 10 serotypes. Other families in the class contain saprophytes or species that do not infect humans.

Mycoplasma are very pleomorphic due to their lack of a rigid cell wall structure. Replication may occur by budding; fission may lag behind replication, resulting in multinucleated filaments. The genome is several times smaller than that of other bacteria. Generation time is about 1 to 8 hours. Most *Mycoplasma* are facultative anaerobes. Complex media are necessary to supply nutritional needs, including sterol, a nutrient that only *Mycoplasma* need. This lipid is essential for development of the plasma membrane. On solid media *Mycoplasma* produce colonies that resemble fried eggs. *Ureaplasma* require 10 percent urea for growth. *Mycoplasma* have two antigens: a protein and a glycolipid. Species are classified by serologic tests. The lipid antigen is used most often for diagnostic serologic tests. The major human pathogen in this family is *M. pneumoniae,* but *U. urealyticum* and *M. hominis* are also associated with disease.

Mycoplasma pneumoniae

Biologic characteristics. *Mycoplasma pneumoniae,* the prototype *Mycoplasma,* was originally called the Eaton agent. It is not a common inhabitant of the human respiratory tract, as are some other *Mycoplasma,* but rather is considered to be a pathogen. *M. pneumoniae* lipid antigens cross react with human brain, heart, liver, and lung tissue. The organism is motile, possessing a terminal organelle that may play a role in attachment or motility. It produces nonspecific mitogenic activity in T and B lymphocytes.

TABLE 12-5. Species of *Mycoplasma*

Species	*Habitat*	*Frequency in humans*	*Diseases*
M. buccale	Respiratory tract	Rare	None
M. faucium	Respiratory tract	Rare	None
M. fermentans	Urogenital tract	Rare	None
M. hominis, types 1 and 2	Pharynx, urogenital tract	5-80 percent	Pharyngitis, spontaneous abortion, prematurity, PID, pyelonephritis, neonatal sepsis
M. lipophilum	Respiratory tract	Rare	None
M. orale, types 1 and 2	Mouth	Almost 100 percent	None
M. pneumoniae	Respiratory pathogen, not usually normal flora		Atypical pneumonia, myringitis, cardiac and neurologic involvement
M. primatum	Genitourinary tract	Rare	None
M. salivarium	Mouth	Almost 100 percent	None
Ureaplasma urealyticum (at least 8-10 serotypes)	Urogenital tract	5-80 percent	Nongonococcal urethritis (NGU), infertility, prematurity, endometritis, pyelonephritis

Definitive diagnosis of infection should include isolation of the organism from diseased sites (usually sputum, but also cerebrospinal fluid, blood, or tissues), and a 4-fold increase in antibody titer between acute (7 to 9 days after infection) and convalescent (3 to 4 weeks after infection) paired sera. *M. pneumoniae* infection is also characterized by the development of cold agglutinins (IgM antibodies that agglutinate certain red blood cells at low temperatures). Cold agglutinins are also present in viral, but not other bacterial, pneumonias. Hence, the test is not specific for mycoplasmal infection.

Epidemiology. Most infections by *M. pneumoniae* are subclinical or mild with only about 3 to 10 percent of infections progressing to overt pneumonia. However, 20 percent of all pneumonia in the general population is caused by *M. pneumoniae.* The disease is usually endemic without seasonal fluctuation, but epidemics also occur; for example, among military recruits. Highest attack rates are in children aged 5 to 9 years; the proportion of pneumonia due to *M. pneumoniae* is highest in teens (Fig. 12-14). The duration of illness increases with age. Occurrence is worldwide. Transmission is respiratory, but direct long-term contact as in families, schools, or military camps also seems necessary.

Pathogenesis. *M. pneumoniae* enters the respiratory tract, penetrates the mucociliary blanket, causing ciliostasis, and attaches by means of its terminal organelle to the surface of the respiratory epithelium.

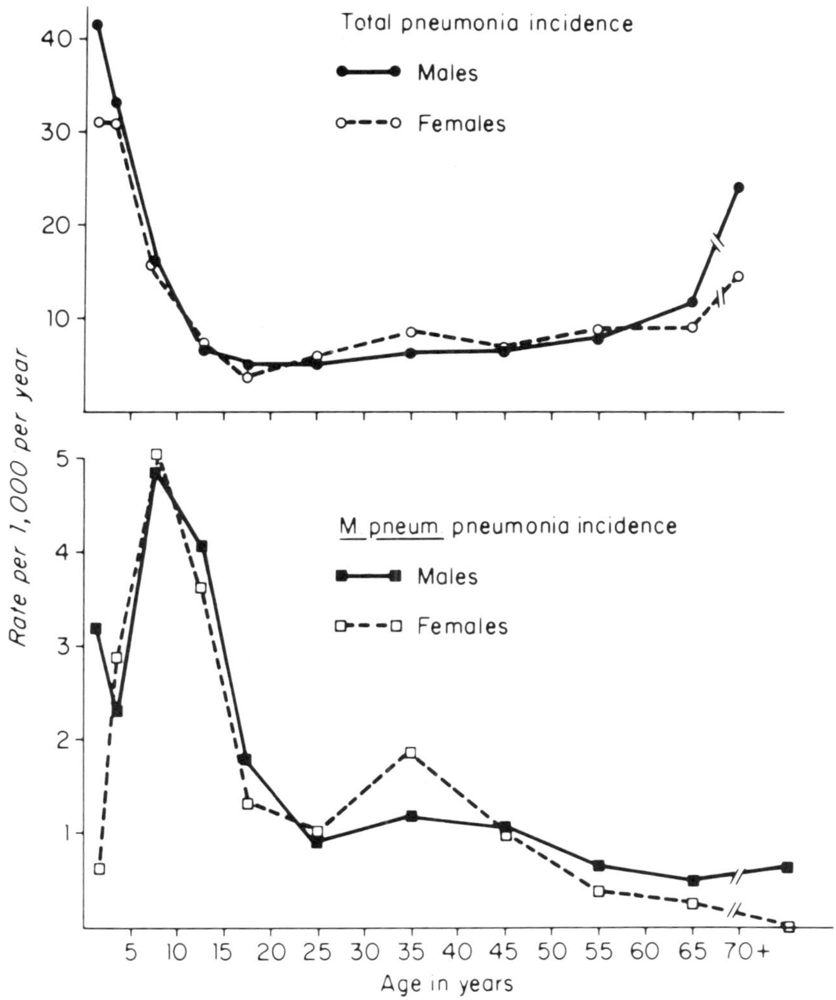

FIGURE 12-14. Incidence of *Mycoplasma pneumoniae* pneumonia by age and sex, Group Health Cooperative (Seattle, Wash.), December 1, 1963 through February 28, 1975. Infection was determined by isolation or 4-fold or greater rise of antibody titer. (Reprinted from Foy HM, et al. Long-term epidemiology of infections with *Mycoplasma pneumoniae*. J. Infect Dis 1979;139:681-687 by permission of the University of Chicago.)

In the respiratory tract, direct damage due to the *Mycoplasma* occurs. In addition, production of immune complexes and autoantibodies and the inflammatory reaction may well be responsible for much of the pathogenesis, especially for extrapulmonary complications. It is possible that toxins may also be produced.

Host responses. After an incubation period of 2 to 3 weeks, response to infection is often subclinical or results in mild coldlike symptoms. Occasionally, an atypical pneumonia develops with slow onset, no leukocytosis, many extrapulmonary signs, and radiologic findings more extensive than might be expected based on physical findings. Other manifestations of infection include swollen ear drums (myringitis), influenzalike illness, otitis media, laryngitis, and bronchitis. The disease course is long, 4 to 6 weeks, but recovery is almost always complete with only very rare fatalities.

Extrapulmonary symptoms also occur, usually from 1 to 21 days after the respiratory illness. In at least one case, a young adult with mycoplasmal pneumonia developed Guillain-Barré syndrome. There was evidence of direct neural infection rather than immunologically mediated disease because *M. pneumoniae* was isolated from blood and cerebrospinal fluid (Bayer et al. 1981). It has been estimated that 0.1 percent of patients infected with *M. pneumoniae* have central nervous system complications. *M. pneumoniae* has been occasionally isolated from brain abscesses, and pleural and joint effusions. Cardiac involvement may occur in about 4 percent of patients and mucocutaneous lesions in about 25 percent. Arthritis and glomerulonephritis have also been reported.

Control and treatment. Immunity lasts only 2 to 5 years and reinfection is well documented. Tetracycline and erythromycin are effective in decreasing length of illness though their efficacy in extrapulmonary complications is questionable. Penicillin is useless because the *Mycoplasma* have no cell walls on which the drug can work. Shedding of *Mycoplasma* can continue for several weeks after treatment. Patients acutely ill with mycoplasmal pneumonia should be isolated from patients who have decreased host defenses. No vaccines are available.

Genital Mycoplasmal Infections

About 30 percent of newborns have transient nasopharyngeal colonization with *M. hominis* and *U. urealyticum,* which disappears within a few weeks. The organisms reappear after sexual maturity and colonization patterns are consistent with sexual acquisition (low in nuns, high in sexually active persons). Colonization rates with *Ureaplasma* are so high among the normal population that it is difficult to establish a causal relationship with any disease. Nevertheless, *U. urealyticum* has been shown to cause about 10 percent of cases of nongonococcal urethritis (NGU). Both *M. hominis* and *U. urealyticum* have been implicated in postpartal endometritis, infertility, and arthritis. Cases of pelvic inflammatory disease, pyelonephritis, and neonatal meningitis with *M. hominis* have been reported, and *Ureaplasma* may be involved in the formation of urinary calculi. In one series, antibodies to *M. hominis* were present in 40 percent of women with acute salpingitis, compared with 8 percent in a control group (Mårdh et al. 1981). Mycoplasmal genital infections are similar in type and epidemiology to those caused by *Chlamydia.* Ureaplasmas have been shown to induce cell necrosis, ciliostasis, and other cytopathic changes in some cell lines. *M. hominis* causes cilial swelling. Strains of *M. hominis* and *M. urealyticum* have shown some resistance to tetracycline, but there is no clear indication that any treatment is necessary except in the case of NGU.

L FORMS

There are several other types of prokaryotes without a cell wall besides *Mycoplasma,* and they are often difficult to distinguish. These include protoplasts, spheroplasts, and L forms. *L forms* (named for the Lister Institute) are bacteria that have lost their cell wall as a result of the action of enzymes that cause cell wall lysis or of some antimicrobial agents such as the penicillins and cephalosporins (Fig. 12-15). Their colonial morphology ("fried egg" colonies) resembles that of *Mycoplasma. Spheroplasts* are irregularly shaped bacteria with defective cell walls. Some of the cell wall components may still be present. *Protoplasts* are bacteria with the cell wall components completely removed. L forms and protoplasts maintain the capacity to produce cell wall precursors and the forms can therefore be reversible; they may return to a cell-wall-producing state. L forms differ from protoplasts because they are more stable and variable in their growth capabilities and they may lose the ability to make cell walls.

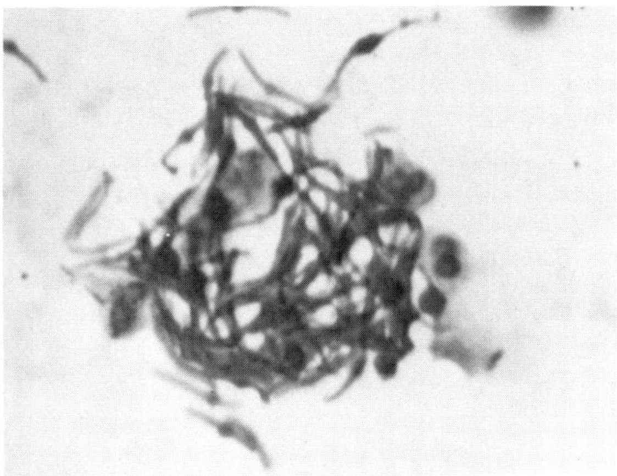

FIGURE 12-15. Gram stain of L forms. Magnification 1000×. (Courtesy of University of Washington Clinical Microbiology Laboratory.)

L forms stimulate an immune response and persist in vivo, especially in the urinary tract. They have a strong mitogenic effect on lymphocytes, demonstrating that the plasma membrane as well as cell wall lipopolysaccharide and peptidoglycan are antigenic. Induced *Pseudomonas* L forms have been shown to be more resistant to polymyxin, colistin, and gentamicin than their parent forms. They may be involved in the pathogenesis of pyelonephritis and endocarditis, and they may be important in allowing bacterial infections to persist and to cause relapses in human disease.

SPIROCHETES

The three genera of spirochetes that are pathogenic to humans are *Treponema, Leptospira,* and *Borrelia,* all from the family Spirochaetaceae. There are also two other genera of free-living nonpathogenic species. The spirochetes are flexible rods, that appear helical like a spring or appear to be undulating. Their cytoplasm is surrounded by three coverings: a cytoplasmic membrane, a layer of peptidoglycan that provides rigidity to the organism, and an outer layer of lipoprotein similar to that in gram-negative bacteria that serves as a semipermeable membrane (Fig. 12-16). All the spirochetes are difficult to visualize because of their thinness. A special microscopic technique called darkfield examination is used to increase the visibility of spirochetes.

Leptospira and *Borrelia* will be discussed in subsequent sections. The genera of treponemes is named "turning thread" because they rotate on their long axes. A number of

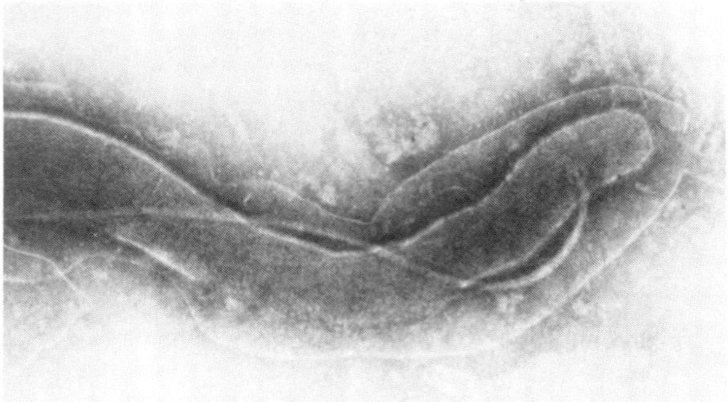

FIGURE 12-16. One end of a free-living spirochete, showing protoplasmic cylinder, axial fiber, and surrounding sheath. Magnification 84,000×. (Reprinted with permission of the publisher from Volk WA. Essentials of Medical Microbiology. Philadelphia: JB Lippincott, 1978.)

species are normal mouth flora (e.g., *T. microdentium*). The pathogenic treponemes have three axial filaments at each end overlapping in the middle. They have pointed, not blunt, ends, possibly to aid in attachment. Only nonpathogenic strains have been cultured in vitro. Pathogenic strains must be grown in laboratory animals as they will not grow in tissue culture, chick embryo, or artificial media. Treponemes are obligate anaerobes with a relatively slow division time of 4 to 32 hours. They reproduce by transverse fission.

There are four pathogenic syndromes caused by treponemes (see Table 12-6): syphilis, caused by *T. pallidum;* yaws, caused by *T. pertenue;* pinta, caused by *T. careteum;* and endemic nonvenereal syphilis, caused by *T. pallidum endemicum*. These treponemes are morphologically identical. It has not been possible to determine whether they are antigenically distinct. One hypothesis is that the various treponemal diseases are caused by interchangeable organisms whose characteristics are determined by environmental factors such as temperature. If such were the case, a single name such as treponematosis would be appropriate for all diseases cause by *Treponema* species.

Treponema pallidum

Biologic characteristics. Because of the difficulty with cultivating *Treponema pallidum* artificially, little is known about its growth requirements, life cycle, or antigenic structure. Recombinant DNA studies may be useful in this regard in the future. They are typical treponemes (Fig. 12-17), being obligate anaerobes and difficult to visualize except under darkfield, and having a slow generation time of about 30 hours.

TABLE 12-6. Treponemal Diseases

Organism	*Disease*	*Mode of transmission*
T. pallidum	Syphilis	Sexual
T. pertenue	Yaws	Direct contact with infected skin lesions
T. carateum	Pinta	Direct contact or fly vector
T. pallidum endemicum	Endemic syphilis (bejel, siti, belesh, dichuchwa)	Direct contact, especially sharing eating or drinking utensils

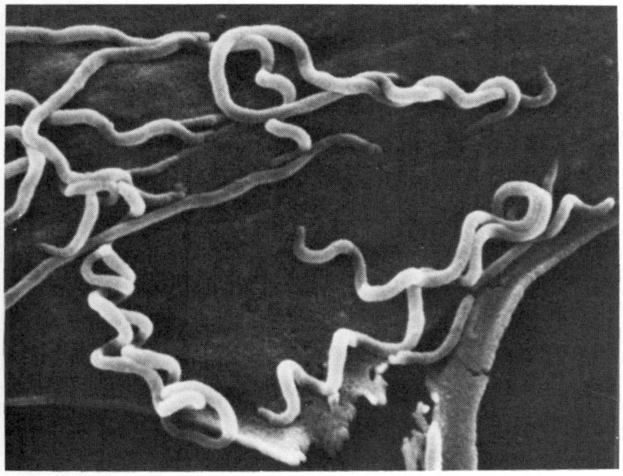

FIGURE 12-17. The three-dimensional qualities of scanning electron microscopy clearly reveal the corkscrew shape of cells of the syphilis-causing spirochete *Treponema pallidum,* attached here to rabbit testicular cells grown in culture. Magnification 8000×. (Reprinted with permission of the publisher from Volk WA. Essentials of Medical Microbiology, Philadelphia: JB Lippincott, 1978.)

Epidemiology. Almost all cases of syphilis are sexually acquired. Rarely, infection can result from contaminated fomites, other personal contact (at least one case acquired by a human bite has been described), blood transfusion, or transmission in utero. Both adult and congenital syphilis have decreased markedly (over 90 percent) in occurrence since the 1940s (Fig. 12-18). In the United States there are about 22,000 cases per year reported in adults and about 1000 cases of congenital syphilis. The frequency seemed to be remaining stable until 1980 when the rate increased by 26.3 percent from 1977 (9.5 to 12.0 per 100,000 persons). Fortunately, the serious late sequelae such as neurologic and heart disease are now quite rare. Peak incidence is in the sexually active years, ages 15 to 29 years. The risk of developing syphilis after sexual contact with an infected individual is about 50 percent.

Pathogenesis. Strains of *T. pallidum* seem to vary somewhat in virulence. After penetration through mucous membranes, the spirochetes enter the blood and lymph nodes within hours, causing a septicemia and making the blood infectious. The primary lesion occurs earlier when the inoculum of treponemes is large. There are no identified toxins produced; pathogenesis is due to invasion and multiplication of treponemes in multiple body sites and to the deposition (such as in the kidneys) of immune complexes. Delayed hypersensitivity reaction plays a role in the symptomatology of tertiary syphilis.

Host responses. If untreated, syphilis has three stages, listed on Table 12-7. The organism enters the genital tract during sexual intercourse through small breaks in the skin. After an incubation period of 2 to 8 weeks, during which time the organism is multiplying locally, the primary lesion called a *chancre* becomes manifest, accompanied by regional lymphadenopathy. The chancre becomes a small ulcer that heals spontaneously in about 2 weeks (Fig. 12-19). Treponemes can be found in phagocytic cells and in lymph nodes during this stage.

Within 6 weeks after the chancre heals, the secondary symptoms appear if the disease is untreated. These include a generalized maculopapular mucocutaneous skin eruption (Fig. 12-20), cerebrospinal fluid abnormalities in about 50 percent of patients, glomerulonephritis due to formation of immune complexes, and general lymphadenopathy. In this secondary stage, treponemes are present in body tissues in large numbers despite high antibody titers. In about one-fourth of untreated cases, re-

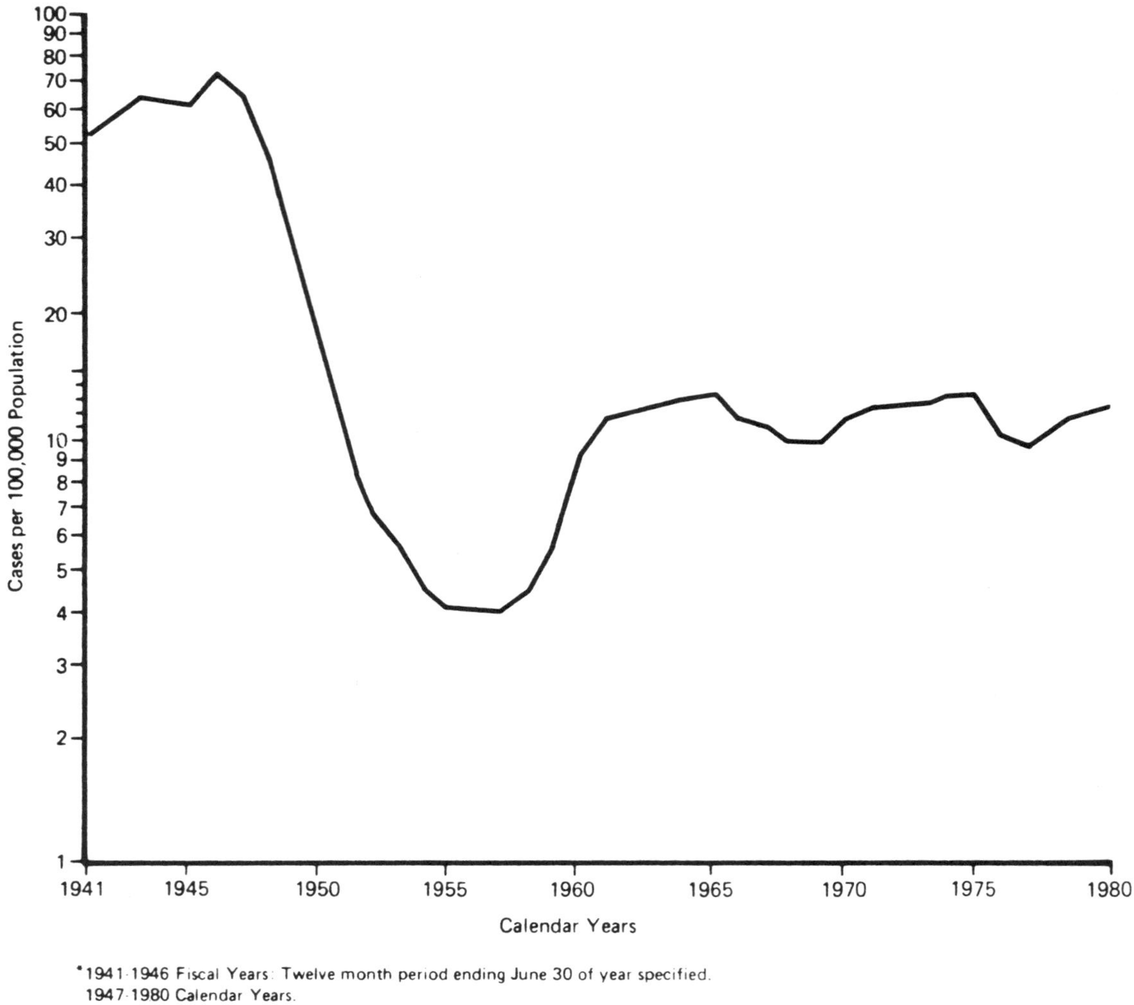

FIGURE 12-18. Syphilis (primary and secondary). Reported civilian case rates by year, United States, 1941 to 1980. (Notice the logarithmic scale, which makes differences at the lower end of the scale seem larger; hence, the decrease in syphilis is much greater than it may at first appear.) (From Centers for Disease Control. Annual Summary 1980. Morb Mort Week Rep 29:78, 1981.)

lapses of localized or generalized skin eruptions occur for years after infection.

Many years after the secondary stage, various clinical sequelae may occur. Tertiary syphilis may cause aortitis, gummas (granulomatous lesions), neurologic damage (neurosyphilis), and joint degeneration. Tabes dorsalis (manifested by ataxic gait, parasthesias, impotence, and areflexia) and general paresis are common central nervous system complications. Deaths from late syphilis are primarily a result of syphilitic cardiovascular disease.

Congenital syphilis also occurs. After the

TABLE 12-7. Stages of Syphilis

Stage	*Onset following infection*	*Symptoms*
Primary	2-8 weeks	Cutaneous ulcer (chancre) at site of inoculation of treponemes (usually penis or vagina, but may be mouth, rectum, hand)
Secondary	10-14 weeks	Generalized mucocutaneous eruptions, lymphadenopathy, glomerulonephritis, CNS abnormalities
Tertiary	10-30 years	Aortitis, gummas, neurologic damage, joint degeneration

fourth month of pregnancy, *T. pallidum* is transferred in about 90 percent of cases across the placenta from an infected mother to the fetal circulation. (If the mother has had untreated syphilis for 4 years or more, the risk of transmission to the fetus is much lower). In about 40 percent of cases, congenital infection results in stillbirth or prematurity. Early manifestations of congenital syphilis occur before the child is 2 years of age. These include skin lesions such as vesicles and pemphigus, osteitis, jaundice, and glomerulonephritis.

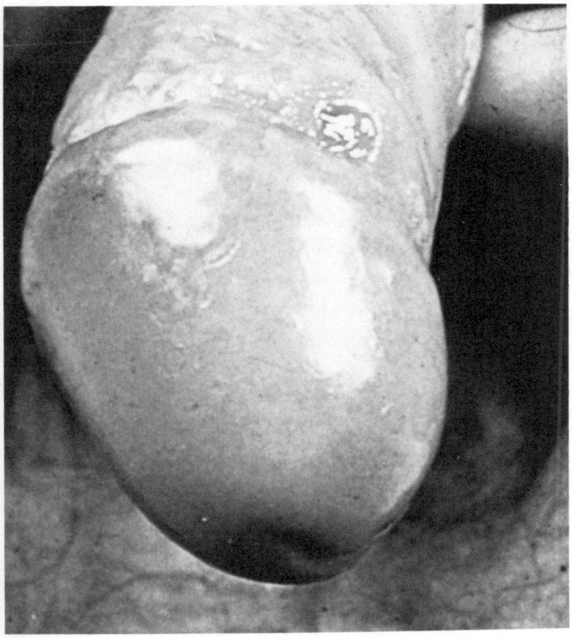

FIGURE 12-19. Primary syphilis chancre. (Courtesy of University of Washington Clinical Microbiology Laboratory.)

Affected infants are infectious and have a high mortality rate, often dying of hepatitis or thrombocytic disorders resulting in hemorrhage. However, only about one-fourth of congenital syphilis is recognized during the first year of life. Children who develop late manifestations are not infectious, and the majority of the cases are latent. Some children develop blindness, probably as a result of hypersensitivity. Others may develop deafness, central nervous system disorders similar to those seen in tertiary syphilis, notched teeth, abnormal facial configuration, and gummas. Syphilis is diagnosed by darkfield examination of motile treponemes in secretions from lesions and by nonspecific or specific antitreponemal antibody tests (Table 12-8).

Control and treatment. There is no natural resistance to *T. pallidum*. Treponemes are readily phagocytized and T cell immunity seems to play an important role in host defense. Resistance to reinfection occurs in un-

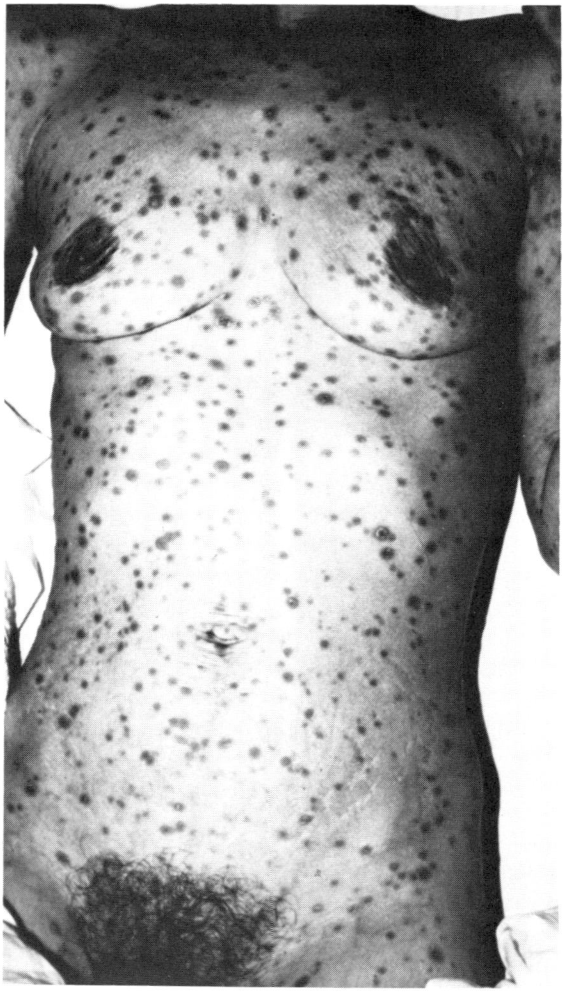

FIGURE 12-20. Skin manifestations of secondary syphilis. (Courtesy of University of Washington Clinical Mircobiology Laboratory.)

treated persons about 3 months after infection, but not in treated persons. Careful followup of all sexual contacts of cases is one of the most important control measures. Nonspecific tests (RPR or VDRL, see Table 12-8) are used for screening, followed by specific antitreponemal tests since the false positive rate is about 20 to 40 percent in nonspecific tests. If infected pregnant women are treated before the fourth month of gestation, the fetus will be protected. Routine serologic tests of all women during the first trimester are therefore recommended. Penicillin G is the treatment of choice, although other agents may be substituted if necessary. Treatment in the tertiary stage may need to be of longer duration, or may be too late to prevent serious complications. Asymptomatic neurosyphilis, however, can be treated with penicillin and symptoms may be prevented.

Other Treponemal Diseases

Yaws, caused by *T. pertenue,* is an ulcerative skin and bone disease endemic in the tropics. It is transmitted by direct contact with infected lesions and is particularly prevalent in children. Following an initial papular lesion (called a "mother yaw" or framboise (raspberry); Fig. 12-21) extensive

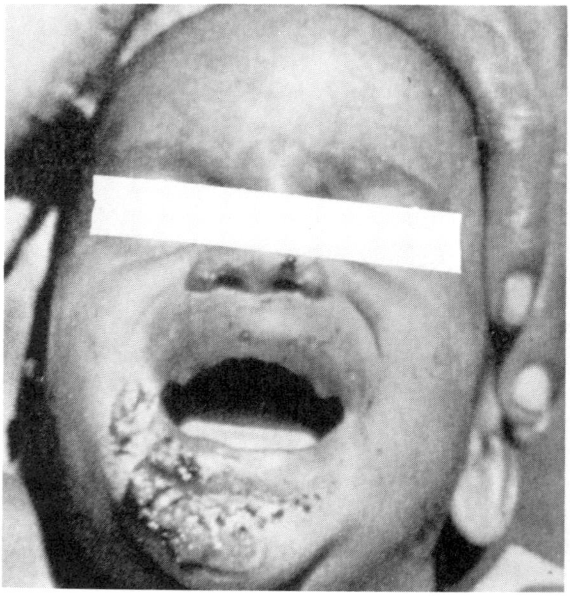

FIGURE 12-21. Early, papillomatous, stage of yaws. (Courtesy of Centers for Disease Control, Atlanta, GA 30333.)

TABLE 12-8. Diagnostic Tests for *Treponema Pallidum*

Test	Characteristics	Sensitivity (percent)		
		Primary	Secondary	Tertiary
Nonspecific Wasserman Ab (reagin) tests*				
Venereal Disease Research Lab (VDRL)	Slide microflocculation using extract of cardiolipin to test serum for treponemal antibodies; Ab titers can be determined with twofold dilutions	70	100	70
Rapid plasma reagin (RPR)	As above, but faster results and can be used where minimal lab facilities exist because it is not necessary to heat serum	50	100	70
Specific antitreponemal tests				
Fluorescent treponemal antibody absorption (FTA ABS)	Serum and *T. pallidum* are placed on a slide. If Ab is present in serum, it can be tagged with fluorescein dye. Serum is first treated with Reiter treponemes to prevent cross reaction with other treponemes. Results available within 24 hours	85	100	98
T. pallidum immobilization test (TPI)	Serum added to guinea pig complement; if individual has syphilis, treponemes grown in rabbit testes will become immobilized. Difficult to perform, not used commonly now	50	95	95
T. pallidum hemagglutination assay (TPHA)	Positive test causes clumping of treated RBCs by antitreponemal Ab to which treponemes are adsorbed	70	95	95
IgM-FTA ABS	Used to test for active congenital syphilis in newborns since IgG against syphilis is passively transferred from infected mother (maternal IgM does not pass the placental barrier)			

*Biologic false positives (BFP) occur with collagen diseases, some malignancies, aging, and heroin addiction.

secondary lesions resulting in pain and scarring develop (Fig. 12-22). These heal slowly and are infectious. Eventually, lesions become destructive and damage of nose or face similar to that occurring in leprosy results (Fig. 12-23). In some areas, the majority of young people will acquire the disease. For example, the incidence of yaws increased by 10-fold in Ghana from 1969 to 1979. For that reason, a major program funded by international health agencies such as World Health Organization was initiated in 1981 to prevent the spread of this endemic disease (Centers for Disease Control

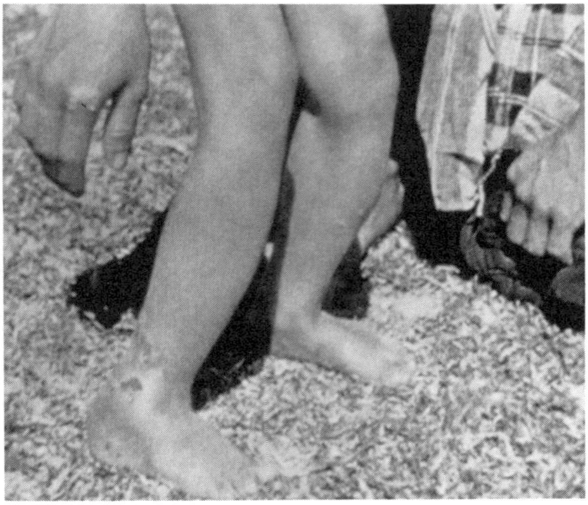

FIGURE 12-22. Early stage of yaws: bony involvement of legs. (Courtesy of Centers for Disease Control, Atlanta, GA 30333.)

1982). Penicillin, if available, is successful therapy.

T. carateum causes pinta, a disease similar to but milder than yaws. It too is a tropical disease transmitted by direct contact or by means of a fly vector. Within about 1 year after the appearance of the primary papular lesion, numerous red, deeply pigmented lesions appear all over the body. These eventually turn white and the skin develops a mottled appearance. Extensive bone involvement is rare.

Other diseases known by various names, but similar to yaws, occur throughout the world. Again, mucocutaneous lesions occur, often in the mouth, leading to periostitis, with late development of gummas and bone involvement. Though not sexually transmitted, these diseases are cumulatively known as endemic syphilis. Synonyms include bejel, siti, dichuchwa, and belesh. Transmission is by contact, especially sharing eating utensils.

Borrelia

Biologic characteristics. *Borrelia* are longer and thicker than treponemes and are the only spirochetes that can be readily stained and visualized with aniline dyes. They are microaerophilic, flexible spirals that move by twisting and rotating and can be cultured in media containing blood or tissue or in chick embryos. They are tapered at the ends. There are at least 28 species, some of which are pathogenic for various vertebrates. The species pathogenic to humans are either louse or tick-borne. *B. recurrentis* causes epidemic louse-borne relapsing fever and several species, including *B. hermsii, parkeri,* and *turicate,* cause endemic tick-borne relapsing fever. Division time for *Borrelia* is about 18 hours. Some species are found as part of the normal mouth flora (*B. duttonii, novyi*).

Epidemiology. Louse-borne relapsing fever has the same epidemiologic features as typhus, and epidemics of both may be simultaneous. *B. recurrentis* is transmitted from person-to-person by infected body lice that are crushed on the skin and enter the body through abrasions. Humans are the only host and the louse does not transmit the organism transovarially. The illness occurs in the Far East, South America, and parts of Africa.

Ticks of the genus *Ornithodoros* transmit an endemic form of relapsing fever. Many animal species, especially rodents and small mammals, may serve as reservoirs. Humans are infected by tick saliva when bitten, or by infected tissues when the tick is crushed. The infected ticks are not diseased and can transmit the *Borrelia* transovarially. Distribution of the disease is worldwide. Each year 10 to 15 cases have occurred among hikers and campers in the United

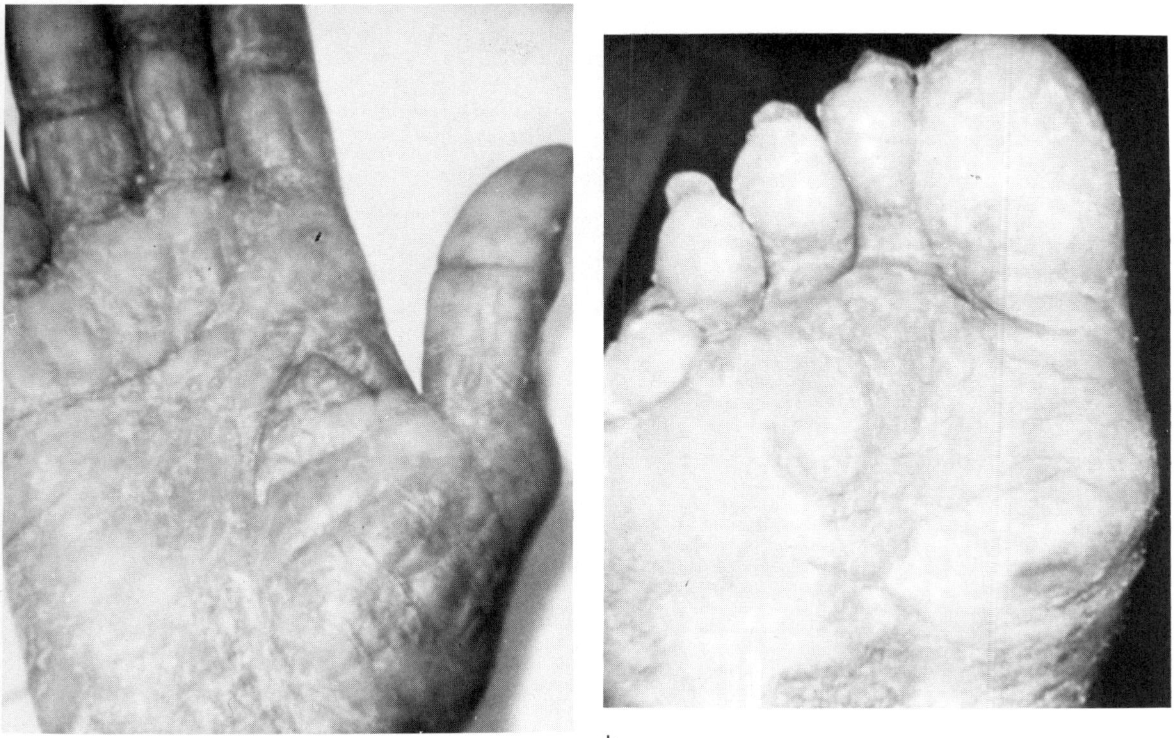

a b

FIGURE 12-23. *(a)* Late stage of yaws. *(b)* "Crab" yaws: foot. (Courtesy of Centers for Disease Control, Atlanta, GA 30333.)

States. Transplacental transfer in humans is possible.

Pathogenesis. *Borrelia* are extracellular parasites. Pathogenesis is due in part to a generalized bacteremia with invasion of multiple body organs and in part to an endotoxin that is produced. Relapses are the result of the emergence of several antigenic variants.

Host responses. Both louse- and tick-borne varieties of *Borrelia* infection are characterized by 3 to 10 relapses of fever and bacteremia. After an initial incubation period of 3 to 14 days there is sudden onset of fever that lasts 3 to 5 days, followed by 4 to 10 days during which the victim is free of symptoms. The number of subsequent relapses varies; usually there are more relapses of a less severe nature in the tick-borne as compared with the louse-borne fever. Despite the fact that antibody clears the organisms from the blood, new variants of the *Borrelia* are produced in the body of the host and emerge, requiring the production of new antibody. Recovery probably requires development of antibodies against several antigenic variants.

Besides fever, the affected individual experiences headache, anorexia, muscle aches, and sometimes a rash. In severe cases bleeding, splenomegaly, hepatomegaly, and cardiovascular collapse occur. The case fatality rate in tick-borne fever is about 5 percent, in epidemic louse-borne fever it is as high as 30 percent.

Control and treatment. There are no vaccines against *Borrelia* infection. Measures to control ticks and lice are useful. Treatment of relapsing fever is with penicillin, tetracycline, and chloramphenicol. However, a paradoxical worsening of the disease may occur in parallel with the killing of the *Borrelia*. This may be due to relapse of endotoxin from the dying bacteria.

Leptospira

Biologic characteristics. *Leptospira* are thin, helical spirochetes of two species, *L. interrogans* and *L. biflexa*. They are obligate aerobes with a single polar flagellum. The organisms can grow on a complex artificial medium. One end of the organism (sometimes both) is bent and resembles a hook. When it moves, it creates a corkscrew motion. *Leptospira* can barely be visualized on darkfield examination. *L. interrogans* has more than 170 serologic varieties (serovar) that are potentially pathogenic to humans and animals. The disease caused by *Leptospira* is called leptospirosis. About 30 serovar have been identified in animals in the United States to date. Common ones are listed in Table 12-9. *L. biflexa* also has more than 150 serovar, but these are not pathogenic to humans. These serovar overlap considerably in genetic structure. They all have a common lipopolysaccharide O antigen and varying agglutinating antigens. *Leptospira* are extracellular parasites that survive in water with alkaline pH for weeks.

TABLE 12-9. Some *Leptospira interrogans* Serovars Involved in Human Disease

Variety	*Source*
Autumnalis	?
Ballam	Mice
Bovis	Cattle, voles
Canicola	Dogs
Grippotyphosa	Rodents
Hebdomadis	Rats, mice
Icterohaemorrhagiae	Rodents
Mitis	Pigs
Pomona	Cattle, pigs

Epidemiology. Human leptospirosis is a zoonosis. The principle reservoir is the kidneys of carrier animals, including pigs, cattle, sheep, raccoons, dogs, and rodents. In animal hosts, the disease can be life long. The organism grows in the proximal tubules of the kidneys and is excreted in the urine. If the organisms are excreted in alkaline water, they can survive for weeks, and this water is the major source of infection to humans. The disease is rare in the United States; about 100 cases are reported annually. Leptospirosis, however, is prevalent in Australia and New Zealand. It is of economic concern to owners of domestic animals and livestock because the disease causes significant morbidity (including spontaneous abortion) and mortality. Infection also results after contact with contaminated animal tissues. At high risk are sewer workers, miners, swimmers in stagnant waters, and dog owners. In a Detroit-based survey of several hundred stray dogs, leptospiral agglutination was present in 38 percent of inner city and 19 percent of suburban animals. The most common serovar isolated from their urine was *L. icterohaemorrhagiae* (Thiermann 1980).

Pathogenesis. *Leptospira* are extracellular parasites that penetrate through abrasions of the skin or mucous membranes, or through the conjunctiva. They probably do not enter by ingestion. The organisms multiply in the blood and tissues during the first (bacteremic) phase. Lesions of small blood vessels, especially in kidneys, cause functional damage. A specific toxin may be produced, but it does not cause the lesions. A lipase and hemolysin are produced.

Host responses. The syndrome produced, which is biphasic, is similar for all serovar. After an incubation period of 5 to 12 days, there is sudden onset of flulike symptoms during the bacteremic phase. The individual has headache, photophobia, vomiting, oliguria, and a rash that lasts for 2 to 3 days. During the second phase various organs may become involved. Central nervous system involvement results in nonpurulent ("aseptic") meningitis. A syndrome of hepatitis, nephritis, and bleeding occurs in about 10 percent of cases and is called Weit's disease. It is usually caused by serovar *icterohaemorrhagiae.* Abdominal and muscle pain, conjunctivitis, pneumonia, arthritis, and other symptoms have also been reported. During the early stage, organisms can be cultured from the blood or visualized by darkfield examination. Subclinical cases in humans may be common, as they are in animals. Of diagnosed cases, mortality occurs in about 5 percent, often from renal failure.

Control and treatment. Antibodies are detectable within a few days after infection. Immunity due to agglutinating antibodies is serovar specific and long lasting. Vaccines comprised of the most common serovars are used for domestic animals and provide protection against spontaneous abortion and clinical disease for 6 to 12 months. Control in humans is primarily through avoidance of exposure by avoiding swimming in water contacted by animals, and by wearing rubber gloves for skinning or cleaning animals. Penicillin (tetracycline or erythromycin, as substitutes) may be efficacious in the early stages of disease, but is not totally satisfactory. Other treatment is supportive; i.e., renal dialysis for renal failure.

Other Spirochete Diseases

Rat bite fever (sodoka) is caused by *Spirillum minor,* a small circular organism that is motile by means of polar flagella. It occurs worldwide in rats and can cause human disease in areas where rat bites are common. Inoculation of the organisms via rat bite results in a local lesion, swelling of regional lymph nodes, rash, and a relapsing fever like that caused by *Borrelia.* Penicillin, tetracycline, and streptomycin are effective antimicrobials.

The fusiform anaerobic bacterial flora of the mouth can cause disease on rare occasions in compromised hosts. An ulcerative gingivostomatitis called trench mouth (or Vincent's stomatitis), lung abscesses, and infected leg ulcers can result. When trench mouth progresses it can lead to ulcerative tonsillitis called Vincent's angina. Predisposing factors are poor oral hygiene, simultaneous infection especially with herpes simplex virus type 1, and nutritional deficiency. Anaerobic spirochetes grow normally in the mouth and occasionally on normal genitalia *(B. refringens)* where they can be mistaken for the syphilis spirochete.

CLINICAL IMPLICATIONS

Drug resistance among mycobacteria can be minimized by regularity in taking medications. The clinician can help the patient appreciate the reasons for long-term treatment and the consequences of irregular compliance with drug therapy. Adequate collection of sputum samples is usually the responsibility of the health practitioner. To collect sputum for isolation of mycobacteria, a special acid-washed jar is necessary. The acid washing removes dead saprophytic acid-fast organisms that could give false positive results on Ziehl-Neelsen stain (the results of this stain form the basis for the first few weeks of chemotherapy). Special plastic containers are now available that are manufactured in such a way that dead mycobacteria are not present. Samples should be collected in the early morning when sputum production is greatest. Usually three to six samples on consecutive days are necessary since organisms are excreted only intermittently and sometimes in small numbers.

The clinician's role in the prevention of syphilis is similar to that for other sexually transmitted diseases (see "Clinical Implications," Chapter 9).

Mycoplasmal infections may be of particular concern to health practitioners working in community clinics, medical offices, or health maintenance organizations. Since the symptoms are often vague and insidious in onset, individuals with mycoplasmal disease are usually seen in such settings. Though mycoplasmal diseases are usually not fatal they can be debilitating for weeks, causing "walking pneumonia" or symptoms that are sometimes attributed to psychosomatic problems. Because *Mycoplasma* infection often goes undiagnosed, the symptoms can cause discouragement and depression to the infected individual. The clinician should be alert to young adults with upper respiratory symptoms that might be attributable to *Mycoplasma*.

SUGGESTED READINGS

Bayer AS, et al. Neurological disease associated with *Mycoplasma pneumoniae* pneumonitis. Ann Intern Med 1981;94:15-20.

Cassell GH, Cole BC. Mycoplasmas as agents of human disease. N Engl J Med 1981;304:80-88.

Centers for Disease Control. Bacteriologic conversion of sputum among tuberculosis patients. Morb Mort Week Rep 1982;31:454.

Centers for Disease Control. Drug resistance among Indochinese refugees with tuberculosis. Morb Mort Week Rep 1981a;30:273-275.

Centers for Disease Control. Syphilis trends in the U.S. Morb Mort Week Rep 1981b;30:441-444.

Centers for Disease Control. Tuberculosis—United States, 1980. Morb Mort Week Rep 1981c;30:325-326.

Centers for Disease Control. Yaws and yellow fever project—Ghana. Morb Mort Week Rep 1982;31:149-150.

Comstock GW. Frost revisited: the modern epidemiology of tuberculosis. Am J Epidemiol 1975;101:363-381.

Diesch SL. Leptospirosis: vaccination and titer evaluation. Mod Vet Prac 1980;61:905-908.

Drusin LM. Syphilis and other sexually transmitted diseases. Cutis 1981;27:286-287+.

Dustin P, et al. Generalized fatal chronic infection by *Mycobacterium scrofulaceum* with severe amyloidosis in a child. Path Res Pract 1980;168:237-248.

Dyckman JD, Wende RD. Comparison of serum and plasma specimens for syphilis serology using the reagin screen test. J Clin Microbiol 1980;11:16-18.

Eckman MR. Nontuberculous mycobacterial skin lesions (letter). Ann Intern Med 1981;94:414-415.

El-Najjar MY. Human treponematosis and tuberculosis: evidence from the new world. Am J Phys Anthropol 1979;51:599-618.

Enarson DA, et al. Non-respiratory tuberculosis in Canada. Am J Epidemiol 1980;112:341-351.

Faine S. Leptospirosis—here and now. Pathol 1981;13:1-5.

Farer LS, Lowell AM, Meador MP. Extrapulmonary tuberculosis in the United States. Am J Epidemiol 1979;109:205-217.

Foy HM, et al. *Mycoplasma pneumoniae* pneumonia in an urban area. JAMA 1970;214:1666-1672.

Fuimara NJ, Exner JH. Primary syphilis following a human bite. Sex Trans Dis 1981;8:21-22.

Gangadharam PR. Identification of mycobacteria by smear examination of the culture. Tubercle 1981;62:123-127.

Gilpin RW, Patterson SK, Knight RA. Quantitation of *Bacillus subtilis* L-form growth parameters in batch culture. J Bacteriol 1981;145:651-653.

Good RC. Isolation of nontuberculous mycobacteria in the United States, 1979. J Infec Dis 1980;142:779-783.

Grange JM, Gibson J, Batty A. The specificity of the humoral immune response to soluble mycobacterial antigens in tuberculosis. Tubercle 1980;61:153-156.

Gremillion OH, Mursch SB, Lerner CJ. Injection site abscesses caused by *Mycobacteria chelonei*. Infec Contr 1983;4:25-28.

Higgins R, Descôteaux JP, Degré R. Pathogenesis of experimental *Leptospira interrogans* serotype icterohaemorrhagiae infection in the guinea pig: possible role of endotoxin of intestinal bacteria in the development of lesions. Can J Comp Med 1980;44:304-308.

Hollander DH. Trepanematosis from pinta to venereal syphilis revisited: hypothesis for temperature determination of disease patterns. Sex Trans Dis 1981;8:34-37.

Holmes KK, Puziss M. Recommendations of the study group for research and training in sexually transmitted diseases. J Infec Dis 1980;142:639-642.

Houser WE, et al. Mycobacterial infection of a congenital bronchogenic cyst. Thorax 1980;35:312-313.

Jones WD. Typing *Mycobacterium tuberculosis* with mycobacteriophage Bo4. J Gen Virol 1980;49:319-322.

Judson FN. The importance of coexisting syphilitic, chlamydial, mycoplasmal, and trichomonal infections in the treatment of gonorrhea. Sex Trans Dis 1979;6 (Suppl 2):112-119.

Kaplan MH, Chase MW. Antibodies to mycobacteria in human tuberculosis. I. Development of antibodies before and after antimicrobial therapy. J Infec Dis 1980;142:825-834.

Kaplan MH, Chase MW. Antibodies to mycobacteria in human tuberculosis. II. Response to nine defined mycobacterial agents with evidence for an antibody common to tuberculosis and lepromatous leprosy. J Infec Dis 1980;142:835-843.

Latapi F, et al, eds. Leprosy. Proceedings of the XI International Leprosy Congress. New York: Excerpta Medica, 1980.

Mårdh PA, et al. Antibodies to *Chlamydia trachomatis, Mycoplasma hominis,* and *Neisseria gonorrhoeae* in sera from patients with acute salpingitis. Br J Vener Dis 1981;57:125-129.

Martin HH, Schilf W, Schiefer HG. Differentiation of Mycoplasmatales from bacterial protoplast L-forms by assay for penicillin binding proteins. Arch Microbiol 1980;127:297-299.

Metcalf JF, et al. *Mycobacterium fortuitum* pulmonary infection associated with antigen-selective defect in cellular immunity. Am J Med 1981;71:485-492.

Mornex JF, et al. Lung disease due to *Mycobacterium fortuitum* (letter). Lancet 1981;1:614.

Petrini B, Hellstrand P, Eriksson M. Infection with *Mycobacterium chelonei* following injections. Scand J Infec Dis 1980;12:237-238.

Powell KE, Faere LS. The rising age of the tuberculosis patient: a sign of success and failure. J Infec Dis 1980;142:946-948.

Reves R, et al. Transmission of multiply drug-resistant tuberculosis: report of a school and community outbreak. Am J Epidemiol 1981;113:423-435.

Schell RF, et al. Endemic syphilis: transfer of

resistance to *Treponema pallidum* strain bosnia A in hamsters with a cell suspension enriched in thymus-derived cells. J Infec Dis 1980;141:752-758.

Sopko JA, Fieselmann J, Kasik JE. Pulmonary disease due to *Mycobacterium chelonei* subspecies abscessus. A report of four cases. Tubercle 1980;61:165-169.

Thiermann AB. Canine leptospirosis in Detroit. Am J Vet Res 1980;41:1659-1661.

Toth A, et al. Subsequent pregnancies among 161 couples treated for T-mycoplasma genital tract infections. N Engl J Med 1983;308:505-507.

Tully JG, et al. A newly discovered mycoplasma in the human urogenital tract. Lancet 1981;1:1288-1291.

VonEden W, et al. HLA segregation of tuberculoid leprosy: confirmation of the DR2 marker. J Infec Dis 1980;141:693-701.

Yamamoto A, Homma JY. L forms of *Pseudomonas aeruginosa*. 2. Antibiotic sensitivities with L forms of *Pseudomonas aeruginosa*. Jpn J Exp Med 1978;48:355-362.

Yates MD, Collins CH. Sensitivity of opportunist mycobacteria to rifampicin and ethambutal. Tubercle 1981;62:117-121.

Zvetina JR, et al. *Mycobacterium kansasii* empyema. Tubercle 1981;62:135-138.

13 FUNGI

KEY POINTS

- Fungi are eukaryotes; bacteria are prokaryotes.
- Fungi have two morphologic forms, yeast and mold, and can reproduce sexually or asexually.
- Though there are more than 100,000 species of fungi, only a few (about 100) are potentially pathogenic for humans.
- The major pathogenic fungi that cause systemic human disease include *Blastomyces dermatitidis* (blastomycosis), *Coccidioides immitis* (coccidioidomycosis), *Histoplasma capsulatum* (histoplasmosis), and *Paracoccidioides brasiliensis* (paracoccidioidomycosis).
- The major fungi that can cause opportunistic fungal infections in debilitated hosts include *Aspergillus, Candida,* and *Cryptococcus.*

Despite the fact that there are more than 100,000 species of fungi, only a relatively few, about 100, have been known to cause human disease. There are even fewer true pathogens; most fungal disease are caused by opportunists in compromised hosts. Useful and harmful effects of fungi are listed on Table 13-1. The classification of fungi is based primarily on reproductive methods and on morphology. This differs from bacteria for which biochemical and immunologic characteristics play a major role in taxonomy.

Fungi vary in other ways from bacteria as well (Table 13-2). Fungi are eukaryotes rather than prokaryotes and as such have a much more complex microstructure than bacteria, including several (not just one) chromosomes, a true nuclear membrane, mitochondria, and endoplasmic reticulum. Their cell walls also vary structurally from bacteria. Many contain chitin, the same substance as contained in the exoskeleton of crustaceans. Others contain hexoses or polysaccharides. They are generally more complex and harder than bacterial cell walls. There are no strict anaerobes among the fungi.

Fungi may take one of two general forms: yeasts or molds. Yeasts exist as single cells

TABLE 13-1. Useful and Harmful Effects of Fungi

Useful	Harmful
Production of chemical products	Some cause disease (mycoses)
Antimicrobials	Poisoning (mycotoxicoses)
Corticosteroids	Possible carcinogenesis
Organic acids	
Use in food production	
Breads	
Cheese	
Alcoholic beverages	

that appear somewhat similar to bacteria on Gram stain, except that they are much larger and ovoid in shape (Fig. 13-1). They reproduce by budding from the parent cell (Fig. 13-2) and stain gram positive. Molds, the more common form of fungi, are characterized by twiglike, branching filaments called *hyphae,* which contain the cell's cytoplasm and nucleus and combine to form *mycelia* (Fig. 13-3). The hyphae, similar to branches on a tree, may be separated by walls *(septate)* or may be continuous and uninterrupted. In the nonseptate mycelia (called *coenocytic*), the cytoplasm and nuclei move freely among the filamentous branches. Such nonseptate mycelia are common among the phycomycetes (Fig. 13-4). Most fungi that are pathogenic to humans are dimorphic. That is, they can take either a yeast or a mold form depending upon environmental circumstances.

Because fungi are ubiquitous and often airborne, laboratory contamination of plates with fungi is common. However, any fungus that can grow at body temperature is a potential pathogen, and repeated isolation of

TABLE 13-2. A Comparison of Fungi and Bacteria

	Fungi	Bacteria
Type of cell	Eukaryote	Prokaryote
Chromosomes	Several	One
Rate of growth	Slow	Rapid
Antimicrobial therapy	Effective drugs often toxic to humans	Effective drugs usually comparatively nontoxic to humans
Reproduction	Sexual, asexual, parasexual	Binary fission
Taxonomy	Based primarily on reproductive methods and morphology	Based primarily on biochemical and immunologic characteristics
Oxygen requirements	All are aerobic or facultative anaerobes, no strict anaerobes	May be aerobic, facultative anaerobes, or strict anaerobes
Cell wall	Often contain chitin, more complex, firm, contain sterols	No chitin or sterols, less complex

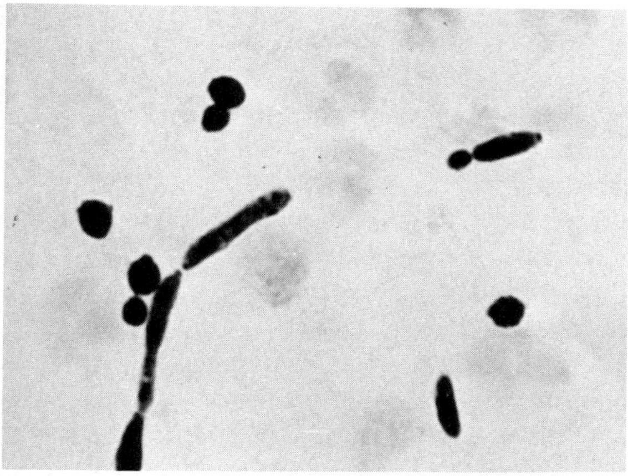

FIGURE 13-1. Gram stain of yeast, which is larger and more ovoid than a bacterium. (Courtesy of Dr. James Plorde.)

the same fungus, even a saprophyte, from the same type of clinical specimen may well indicate disease. Specimens plated for fungal isolation are retained for up to 6 weeks because of the slow growth of some potential pathogens. Sabouraud agar, which contains dextrose, peptone, pancreatic digest of casein, peptic digest of animal tissue, agar, and water, is used almost universally for isolation of most fungi. Sometimes anti-microbials are added to inhibit bacterial growth, though they may also be inhibitory to some fungi. Yeasts can be seen on Gram stain or by direct microscopic examination from clinical specimens emulsified on a slide in a drop of 10 percent sodium hydroxide (NaOH) or potassium hydroxide (KOH). Ideally, yeast specimens should be inoculated on Sabouraud agar, preserved with formalin, or dried on a slide at the patient's bedside so that the in vivo morphology can be observed.

REPRODUCTIVE METHODS OF FUNGI

Fungi reproduce by three methods: sexual, asexual, and parasexual (Table 13-3). There are several complex classification systems

TABLE 13-3. Reproductive Methods in Fungi

Method	*Mechanism*
Asexual	1. Sporulation with subsequent germination 2. Budding (yeasts) 3. Fission (certain yeasts only) 4. Fragmentation of hyphae (artificially induced; probably not important in nature)
Sexual	Donor cell haploid nucleus penetrates cytoplasm of recipient cell, fuses with recipient nucleus. This diploid nucleus produces four haploid nuclei by the process of meiosis. Spores are generated through successive divisions
Parasexual (mitotic recombination)	1. Haploid nuclei can inhabit the same cytoplasm when hyphae fuse; they can divide at the same rate 2. Rarely, heterozygous diploid nuclei can result from nuclear fusion 3. Rarely, somatic pairing of chromosomes results in crossing over, similar to meiosis

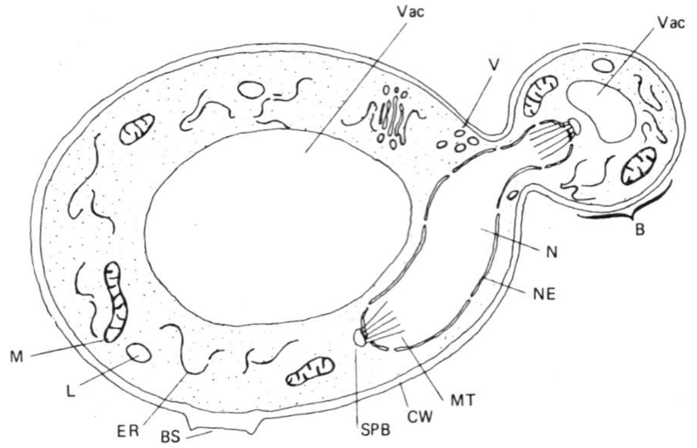

a

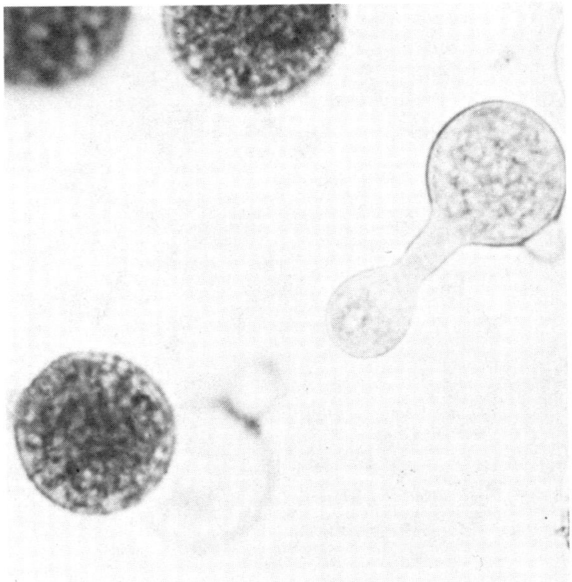

b

FIGURE 13-2. *(a)* Diagrammatic representation of a budding yeast. Magnification approximately 12,000×. B: developing bud; BS: bud scar remaining at a former site of budding; CW: cell wall; ER: endoplasmic reticulum; L: lipid globule; M: mitochondrion; MT: microtubules of the spindle apparatus; N: nucleus, NE: nucleur envelope; SPB: spindle pole body (centriole); V: vesicles; Vac: vacuole.
Note the development of a bud and the accumulation of vesicles at the point of budding, where most wall synthesis takes place. Also note the division of the nucleus and partitioning between the parent and daughter cell. The spindle pole bodies are equivalent to centrioles but lack the centriolar structure; they are typical of fungi in general, true centrioles being found only in the Mastigomycotina, where they are associated with the flagellar apparatus of zoospores. *(b)* A budding yeast cell. (*a* reprinted with permission of the publisher from Deacon JW. Introduction to Modern Mycology. New York: John Wiley and Sons, 1980. *b* courtesy of Dr. James Plorde.)

for fungi, but the most simple and generally accepted one, as shown in Table 13-4, defines four major classes of fungi on the basis of the type of sexual spores they produce: phycomycetes, ascomycetes, basidiomycetes, and deuteromycetes. Some taxonomists further delineate six subclasses of zygomycetes, but, medically, these are of minor importance since zygomycetes only rarely cause human disease. Basidiomycetes, the mushroom being a typical example, produce no infectious diseases in humans, though they can cause mycotoxicosis (poisoning on ingestion).

The mechanism of sexual reproduction is as follows: Nuclei of male (donor) and female gametes fuse after penetration of the donor cell haploid nucleus into the recipient cell cytoplasm. This diploid nucleus produces four haploid nuclei by the process of meiosis. Through successive divisions spores are formed. In ascomycetes, spores that re-

FIGURE 13-3. Typical branching filaments (hyphae) of the mold form of fungi that combine to form mycelia, producing a treelike appearance. *Oidiodendron.* Magnification 40×. (Courtesy of Dr. James Plorde.)

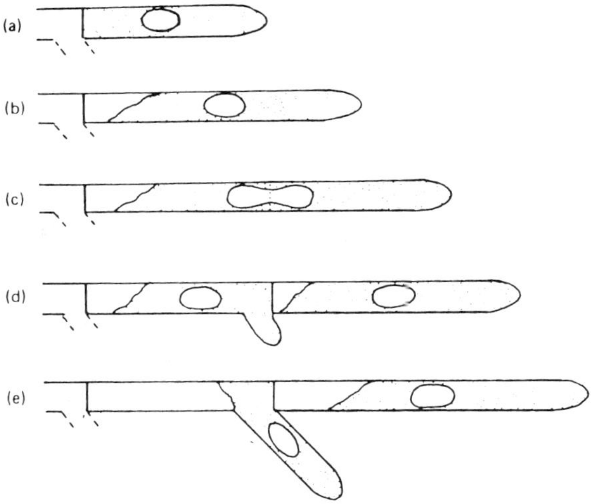

FIGURE 13-4. Branching pattern of *Basidiobolus ranarum,* a fungus with complete, unperforated septa. *(a)* Apical cell full of cytoplasm. Note the large nucleus. *(b)* Cytoplasm is drawn forward and more is synthesized during growth. *(c)* Nucleus divides, apparently when cytoplasm attains a critical volume. *(d)* A septum forms; the apical cell grows on, and a branch is initiated from the sub-apical cell. *(e)* Cytoplasm streams into the newly formed branch; the whole process will then be repeated in the two apical cells. Note the apparent relationship between (1) cytoplasmic volume; (2) nuclear division; and (3) number of apices. (Reprinted with permission of the publisher from Deacon JW. Introduction to Modern Mycology. New York: John Wiley and Sons, 1980.)

main in the egg are called *ascospores,* and the egg is then referred to as an *ascus.* In basidiomycetes, some spores leave the egg through tubes called sterigmata and then attach or fuse outside the egg. These are called *basidiospores,* and the egg is then called a *basidium.* Spores of zygomycetes result from the fusion of hyphae tips, meiosis, and the development of thick-walled spores called *zygospores.*

Fungi for which a sexual cycle has been observed are called "perfect" and those for which no sexual cycle is known are called "imperfect." Deuteromycetes, previously called fungi imperfecti, are species of fungi for which a sexual reproductive cycle has not been observed. For this reason, some genera may have species classified in the class Deuteromycota (if no sexual cycle has been observed) and other species in another class (Table 13-5). Some species of *Candida,* for example, which are known to produce sexual spores, are classified as ascomycetes in the subclass Hemiascomycetidae and other species of *Candida* are classified as deuteromycetes.

Asexual reproduction in fungi can occur

TABLE 13-4. Taxonomy of Fungi of Importance in Human Disease

Kingdom: Myceteae
Division: Amastigomycota

Class	*Subclass or order*	*Form*	*Typical genera*	*Usual environment*	*Human disease*
Phycomycetes	Mucorales	Nonseptate hyphae	*Absidia*	Soil, water	Phycomycosis
			Mucor		
			Rhizopus		
	Entomophthorales		*Basidiobolus*		
			Conidiobolus		
Ascomycetes	Onygenales	Septate hyphae, yeasts	*Ajellomyces*	Soil, water	Perfect (sexual) states of ringworm, histoplasmosis, blastomycosis
			Arthroderma		
			Emmonsiella		
			Nannizzia		
	Eurotiales		*Aspergillus*		Aspergillosis
			Penicillium		
	Hemiascomycetidal		*Saccharomyces*		Rare, perfect state of candidiasis
			Pichia		
Basidiomycetes	Agaricoles	Septate hyphae, yeasts	*Amanita*	Soil	Mushroom poisoning
	Ustilagenaceae		*Filobasidiella*		Perfect state of cryptococcosis
Deuteromycetes (Fungi imperfecti)	Blastomycetidae	Septate hyphae, yeasts	*Candida*	Soil, water, plants, animals	Candidiasis
			Cryptococcus		Cryptococcosis
	Hyphomycetidae		*Coccidioides*		Coccidioidomycosis
			Epidermophyton		Ringworm
			Paracoccidioides		Paracoccidioidomycosis
			Sporothrix		Sporotrichosis
	Dematiaceae		*Cladosporium*		Chromomycosis
			Phialophora		Tinea nigra

TABLE 13-5. Some Fungi that are Classified Both by Sexual (Perfect) State and by Asexual (Imperfect) States

Genus based on sexual (perfect) state	*Genus based on asexual (imperfect) state*
Ajellomyces	*Blastomyces*
Allescheria	*Monosporium*
Arthroderma	*Trichophyton*
Emmonsiella	*Histoplasma*
Filobasidiella	*Cryptococcus*
Nannizzia	*Microsporum*

in yeasts by budding or outgrowing of a daughter cell from the parent cell. Unlike cells produced by binary fission (simple splitting), the buds are smaller than the parent cell and will remain attached while slowly growing. If many such parent and progeny yeast cells adhere to each other, they form chains of "pseudohyphae." A few species of yeast can also reproduce by simple binary fission. A parent cell will produce many buds, which are indicated by budding scars on the surface of the cell (Fig. 13-5).

In molds, sporulation and germination is the primary method of asexual reproduction. There are seven major types of asexual spores (Table 13-6). Phycomycetes produce sporangiospores (Fig. 13-6). The asexual spores of other classes of fungi are together sometimes called *conidia* and a single fungus may produce up to four different types of asexual spores (conidia). Small, unicellular conidia are termed microconidia, and conidia that are large (comprised of many cells) are called macroconidia (Figs. 13-7 and 13-8). Because both sexual and asexual reproduction result from the formation of spores, it is important that the two types of spores not be confused (Fig. 13-9). Sexual and asexual reproduction are illustrated in Fig. 13-10.

A parasexual reproductive cycle can also take place in some fungi, though it is a rare

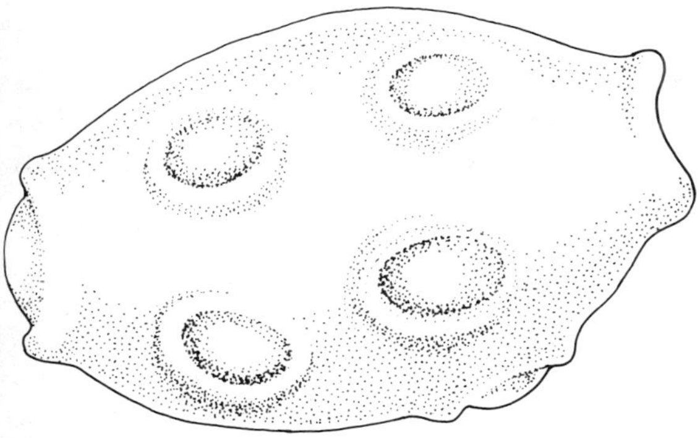

FIGURE 13-5. A schematic drawing of a yeast cell showing scars where cells have budded off during asexual reproduction.

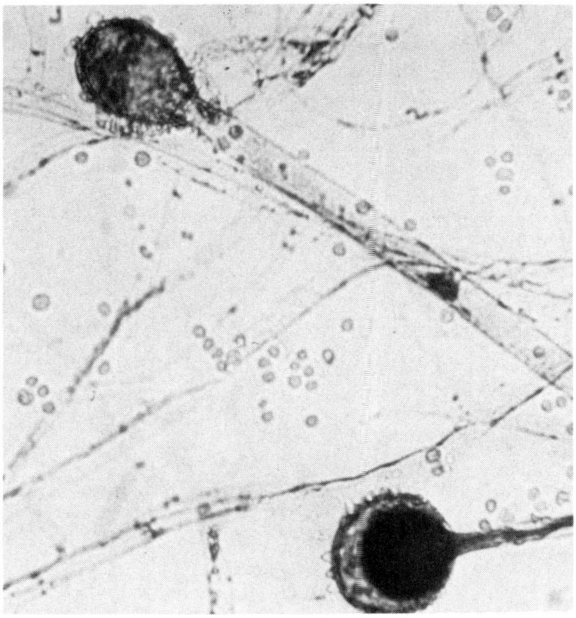

FIGURE 13-6. Sporangiospore and columella of *Mucor* species in culture (Courtesy of Dr. James Plorde.)

TABLE 13-6. Sexual and Asexual Spores of Fungi

Method of reproduction	*Description*	*Produced by*
Sexual		
Ascospores	Contained in the egg	Ascomycetes
Basidiospores	Attached to the outside of the egg	Basidiomycetes
Zygospores	Large, thick walled	Phycomycetes
Asexual		Example general
Conidia*		
Aleurioconidia	Single spores from blown-out ends of cells	*Blastomyces*
Annelloconidia	Cell elongates after production of each spore, leaving a ringlike scar (annulus)	*Scopulariopsis*
Arthroconidia (arthrospores)	Spores produced when hyphae fragment into individual cells	*Geotrichum*
Blastoconidia (blastospores)	From a bud of a portion of the parent cell	*Cladosporium*
Phialoconidia	Produced within the cell, but not causing an increased length as in annelloconidia	*Penicillum, Aspergillus*
Poroconidia	Produced through spores	*Drechslera*
Sporangiospores	Multiple spores produced within a sac called a sporangium	All zygomycetes

*Conidia is the general name given to all asexual spores *except* sporangiospores.

event. In such cases, mitotic recombination of DNA occurs without the sexual mating of gametes. Two nuclei fuse to form a diploid nucleus. Mitotic crossing over occurs and the diploid nucleus reverts to a haploid. This reproductive method, however, does not occur with any regularity.

MORPHOLOGY

As previously mentioned, most fungi grow from septate or nonseptate hyphae (branching filaments). Only a comparatively few genera, about 30, grow in the unicellular yeast form under normal environmental conditions. Yeasts are found primarily in three groups:

Ascomycetes	*Saccharomyces, Endomycopsis, Nematospora, Pichia*
Deuteromycetes	*Candida, Cryptococcus, Rhodotorula, Trichosporon*

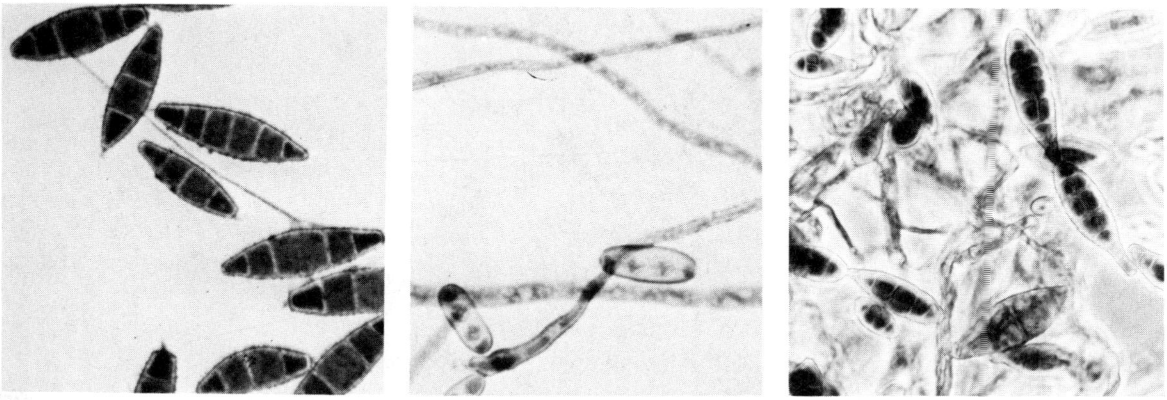

a b c

FIGURE 13-7. Various forms of macroconidia. *(a) Microsporum gypseum. (b) Helmintosporium. (c) Alternaria.* (Courtesy of Dr. James Plorde.)

Heterobasidiomycetes	*Filobasidiella, Leucosporidium, Syringospora*

Whereas the mold forms of fungi produce fluffy, huge, textured colonies on agar because of their aerial hyphae, yeast forms produce small, moist, creamy colonies resembling those of bacteria. Fungi are slower growing than most bacteria, taking from 2 to 7 days or up to 6 weeks for colonies to appear. Yeast forms can develop hyphae when "transitional cells" such as chlamydospores or rounded blastospores germinate. A chlamydospore is a thick-walled intercalary or terminal asexual spore formed by the rounding up of a cell. It is not shed. Many of the dimorphic pathogenic fungi, listed on Table 13-7, are in their yeast form when in the human at body temperature and in the mold form when in the soil or when grown at room temperature.

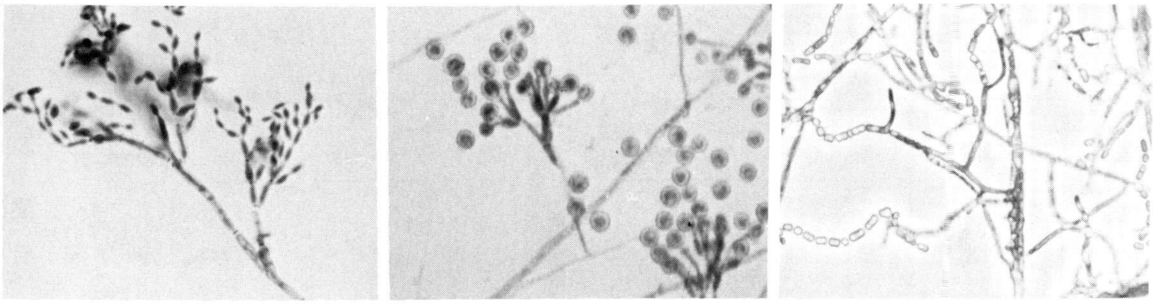

a b c

FIGURE 13-8. Various forms of microconidia. *(a) Cladosporium. (b) Scopulariopsis. (c) Oospora.* (Courtesy of Dr. James Plorde.)

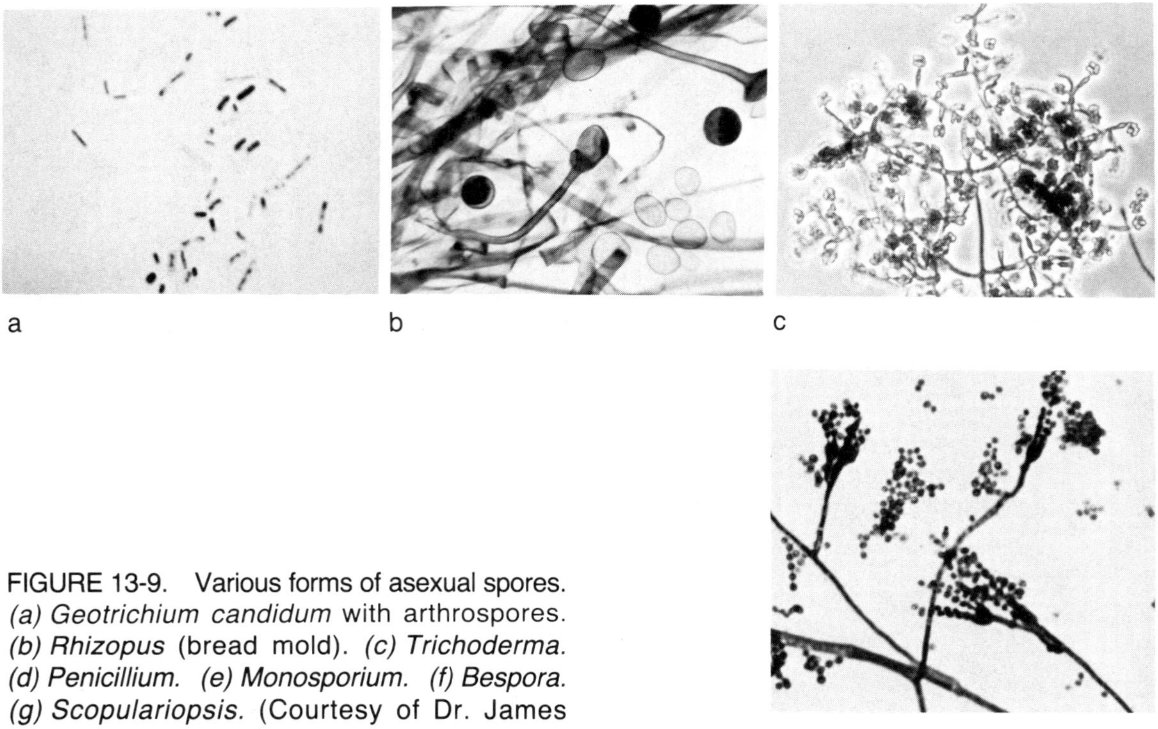

a b c d

FIGURE 13-9. Various forms of asexual spores. *(a) Geotrichium candidum* with arthrospores. *(b) Rhizopus* (bread mold). *(c) Trichoderma.* *(d) Penicillium.* *(e) Monosporium.* *(f) Bespora.* *(g) Scopulariopsis.* (Courtesy of Dr. James Plorde.)

CLINICAL SIGNIFICANCE OF FUNGI

Many fungal infections, though not life threatening, are very irritating and persistent. Most people at some time in their lives develop some form of fungal infection such as ringworm, athlete's foot, or thrush. Fungi that are true pathogens of humans, that is those causing serious systemic disease, are rare. They include organisms in only four genera: *Blastomyces, Coccidioides,*

TABLE 13-7. Forms of Fungi Causing Systemic Disease in Humans

Genus	*Dimorphic form: yeast at 37 °C, mold at 25 °C*	*Yeast form only*	*Predominantly mycelial form*
Aspergillus			X
Blastomyces	X		
Candida		X	
Coccidioides	X		
Cryptococcus		X	
Histoplasma	X		
Paracoccidioides	X		

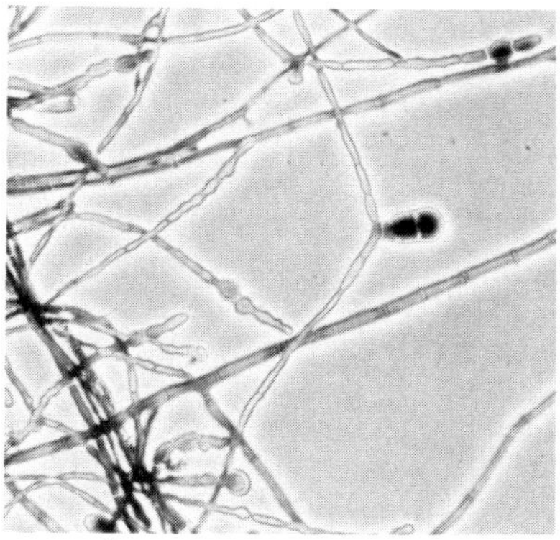

e

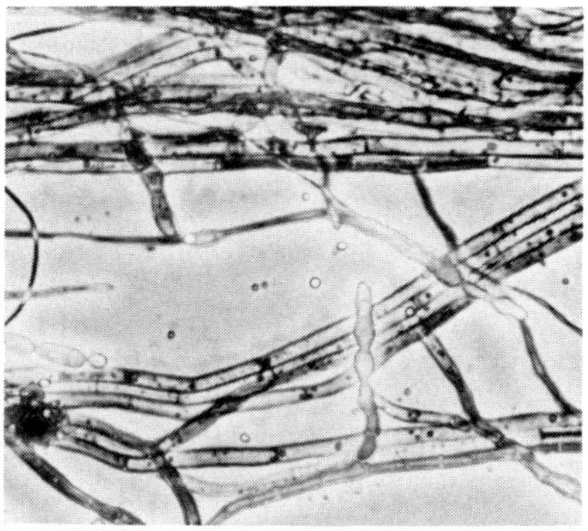

f

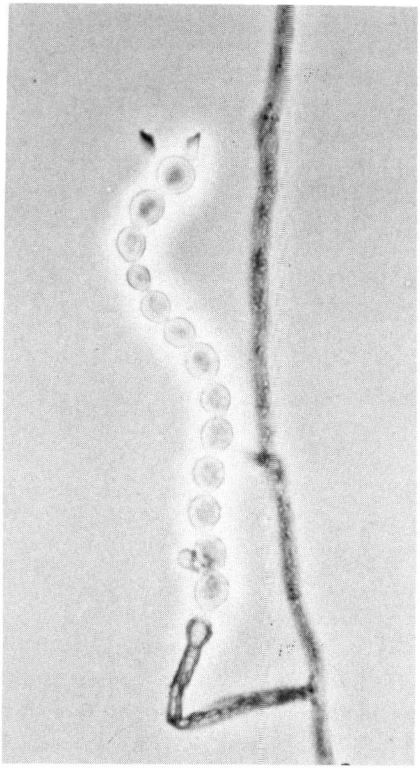

g

FIGURE 13-9. Continued

Histoplasma, and *Paracoccidioides.* All of these normally inhabit the soil and only incidentally cause human disease when their spores are inhaled. They are not infectious between humans. Serious disease in immunocompromised or debilitated hosts can also be caused by opportunistic fungi such as *Candida,* which is often acquired endogenously from the individual's own gut flora, or *Aspergillus* and some phycomycetes, which are acquired from the environment. Such opportunistic fungi are especially common in hospitalized patients or those with malignancies. They may gain entry into the body by inhalation, especially in the presence of mechanical ventilation, by intravascular lines, or by surgical wounds.

The pathogenesis of systemic fungi is similar to that of tuberculosis, due primarily to invasiveness and to the induction of hypersensitivity with resultant chronic granulomatous tissue reaction. With the exception of certain ingested mycotoxins (e.g., "poisonous" mushrooms), fungi do not produce toxins. About half of fungi clinically isolated are yeasts and about 10 percent of these cause clinical disease. Mycoses (infectious diseases caused by fungi) can be classified into four types: superficial, cutaneous, subcutaneous, or systemic (Table 13-8). After a discussion of treatment for fungal diseases, the remainder of this chapter will be devoted to a discussion of these four types of mycoses.

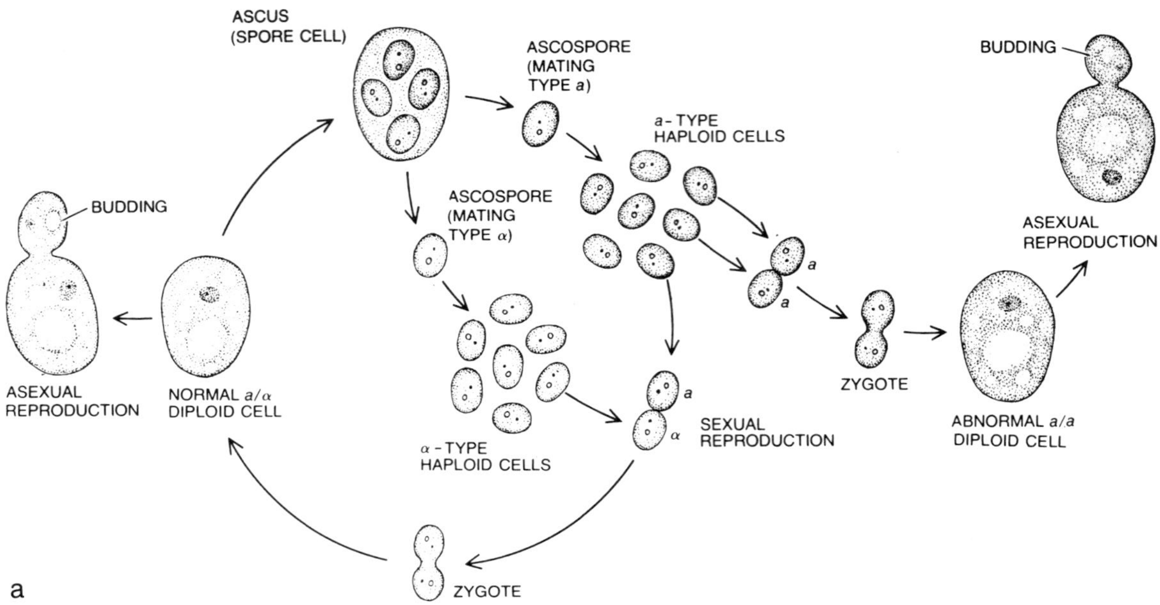

FIGURE 13-10. *(a)* Reproduction of yeast is normally asexual, proceeding by the formation of buds on the cell surface, but sexual reproduction can be induced under special conditions. In the sexual cycle a normal diploid cell (a cell with two sets of chromosomes and therefore two sets of genes) gives rise to asci, or spore cells, that contain four haploid ascospores (cells with one set of chromosomes and one set of genes). The ascospores are of two mating types: *a* and α. Each type can develop by budding into other haploid cells. The mating of an *a* haploid cell and an α haploid cell yields a normal *a*/α diploid cell. Haploid cells of the same sex can also unite occasionally

TREATMENT OF FUNGAL DISEASES

Mycoses are very resistant to conventional forms of therapy. Special chemotherapeutic agents and protracted, long-term treatment, often with highly toxic substances, is necessary. The agents used to treat fungal disease are listed on Table 13-9. Special precautions of therapy are discussed in "Clinical Implications."

SUPERFICIAL MYCOSES

Superficial mycoses affect only the outer nonviable layers of skin and hair. Hence they are of minor clinical importance. There are four such conditions that occur fairly commonly: black piedra, white piedra, tinea nigra, and tinea versicolor (Table 13-10). Only tinea versicolor is seen in the United States with any frequency and its importance is primarily cosmetic (it causes whitish or brownish scaly lesions of a chronic nature, especially on the trunk). Though the causative agent, *Malassezia furfur,* grows slowly on culture, it is usually not necessary to isolate the organism to make an accurate diagnosis. The organism fluoresces under a Wood's lamp to a reddish-orange. Skin scrapings can also be microscopically examined for mycelial and budding elements after being treated with 10 percent potassium hydroxide.

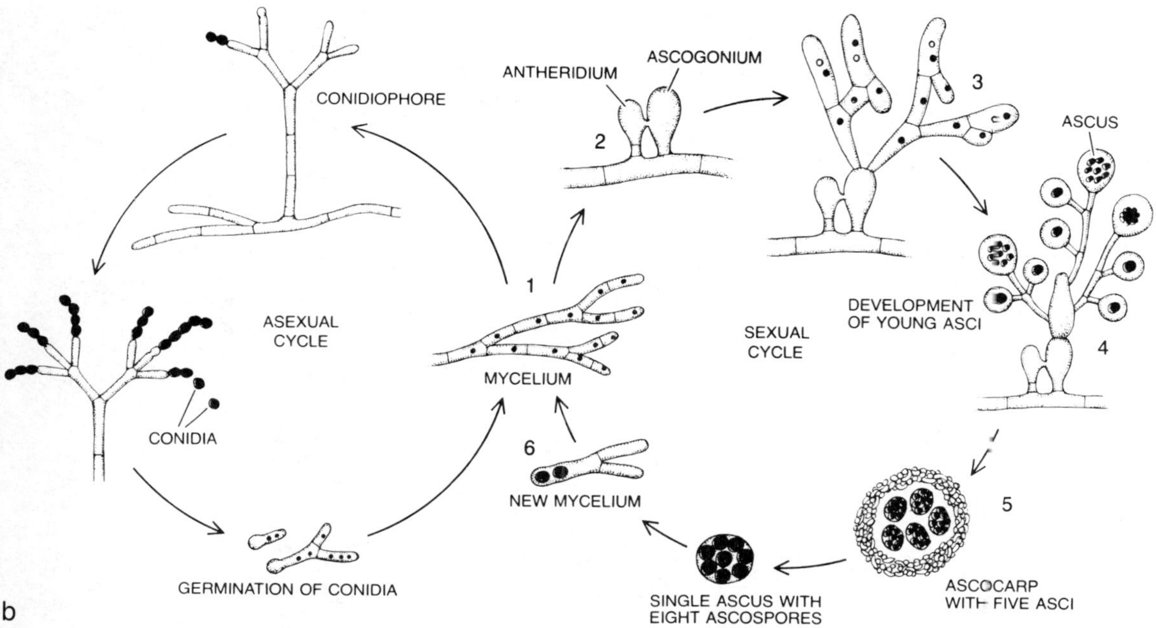

b

to form abnormal diploid cells (*a*/*a* or α/α) that can reproduce only asexually, by budding in the usual way. *(b)* Reproduction of a multicellular fungus, such as one of the higher Ascomycetes, can be asexual or sexual. The details vary with genus and species. The branched vegetative structure common to both reproductive cycles is the mycelium, composed of hyphae. (Reprinted with permission from Phaff HJ. Industrial microorganisms. Scien Am 1981; 245:76-89.)

TABLE 13-8. Types of Mycoses

Type	*Mode of transmission*	*Involved areas*	*Examples*
Superficial	Direct contact	Hair, skin; superficial layers only	Black piedra, white piedra, tinea nigra, tinea versicolor
Cutaneous (dermatomycoses)	Direct contact, fomites	Epidermis, hair, nails	Ringworm, athlete's foot
Subcutaneous	Implantation of spores in skin or mucous membranes	Subcutaneous tissue, fascia, bone	Chromomycosis, maduromycosis, sporotrichosis
Systemic	Inhalation of spores primarily; also via GI tract, skin	Internal organs and viscera	Aspergillosis, candidiasis, coccidioidomycosis, cryptococcosis, blastomycosis, histoplasmosis, paracoccidioidomycosis, phycomycosis

TABLE 13-9. Antifungal Chemotherapy

Drug*	Uses	Route of administration	Dosage	Precautions	Adverse reactions
Amphotericin B	Serious systemic mycoses	I.V.	0.25-1 mg/kg/day (never to exceed 1.5 mg/kg) in I.V. concentration of 0.1 mg/ml over 6 hours	Use must be balanced against many untoward effects; drug is highly nephrotoxic. Use strict aseptic techniques	Anorexia, phlebitis, fever, nausea, hypokalemia, vomiting (see "Clinical Implications")
Amphotericin B cream *(Fungizone)*	Cutaneous candidal infections	Topical	Apply 3% cream 2-4×/day for 1-3 weeks		Skin discoloration
Candicidin	Vaginal candidiasis	Vaginal	Insert tablet or ointment 2×/day × 14 days	May damage diaphragm; use other contraceptives	Rare hypersensitivity or irritation
Clotrimazole	Dermatophytes, vaginal candidiasis	Topical, I.V.	1% cream or lotion 2×/day, maximum of 4 wks; vaginal tablet daily × 7 days	Avoid during first trimester of pregnancy	Local hypersensitivity with urticaria, redness, blistering
Some fatty acids (undecyclenic acid)	Dermatophytoses	Topical	Apply powder or ointment 2-3×/day		
Flucytosine	Candidal, cryptococcal infections	Oral	50-150 mg/kg/day at 6 hour intervals	Close monitoring required if patient has renal, liver, or blood dysfunctions; may be teratogenic	Nausea, vomiting, rash, leukopenia

			daily Child: 10 mg/kg/day	of warfarin; contraindicated in hepatocellular failure, porphyria	sea, vomiting, candidiasis, granulocytopenia, photosensitivity
Iodochlorhydoxquin *(Vioform, Sterosan)*	Dermatophytes	Topical	As for fatty acids		
Ketoconazole	Candidiasis, systemic mycoses	Oral	200-400 mg/day	Possibly teratogenic; avoid during pregnancy	Usually mild; pruritus, gynecomastia, nausea, vomiting, increased liver enzymes
Miconazole	Severe systemic mycoses	I.V., intrathecal (I.T.), bladder irrigation, topical	200-1200 mg/infusion diluted in at least 200 ml IV solution over at least 30-60 min I.T.: 20 mg Bladder: 200 mg diluted	Rapid injection causes tachycardia; no renal or liver toxicity	Phlebitis, rash, nausea, vomiting, fever, flushing, thrombocytopenia
Nystatin	Oral, vaginal candidiasis	Topical, oral	Oral tablets or suspension, cream or vaginal suppositories for 48 hours after symptoms subside		
Potassium iodide	Sporotrichosis, zygomycoses	Oral	2-12 g/day for 2-3 months		
Tolnaftate *(Tinactin)*	Dermatophytes	Topical	1% cream, powder, liquid, or spray 2-3×/day × 2-3 weeks		Rare: local irritation and burning

*Brand names are indicated by *italics*.

TABLE 13-10. Superficial Mycoses

Disease	*Causative agent*	*Area affected*	*Signs and symptoms*	*Distributions*
Black piedra	*Piedraia hortae*	Scalp hair, rarely axillary and pubic	Hard nodules	Africa, Asia, Central America, South America
White piedra	*Trichosporon beigelii*	All body hair, especially on scalp or beard	Yellowish accretions around hair shafts, causes hair damage	Tropical and temperate zones
Tinea nigra	*Cladosporium werneckii*	Smooth body skin	Dark patches, primarily on palms, but can be on any smooth skin	Tropics; rare in U.S.
Tinea versicolor (pityriasis versicolor)	*Malassezia furfur (Pityrosporum furfur, P. orbiculare)*	Smooth body skin	Scaly patches on neck and trunk that spread. Chronic but only of cosmetic concern	Worldwide, mostly in warm climates

CUTANEOUS MYCOSES

Cutaneous mycoses are also called ringworm, tinea, or dermatophytoses. They are caused by a group of fungi called dermatophytes of which there are three genera (*Epidermophyton, Microsporum,* and *Trichophyton*) (Fig. 13-11) and 39 species. The dermatophytes affect the epidermis, and penetrate keratinized but not subcutaneous or deeper tissues. Each genus preferentially attacks either skin, hair, or nails (Table 13-11). Most dermatophytes have not had a sexual state observed and are classified as deuteromycetes. The ones that produce sexual spores are classified as ascomycetes. Their distribution is generally worldwide with marked host preferences. For example, *Microsporum canis* is preferentially found in dogs though human infection is also common. The host range may be related to inherent differences in physiologic factors (e.g., *M. audouinii,* which causes scalp ringworm, disappears after puberty because of fatty acid secretions from the scalp at that time) or to varying habits and exposures of hosts.

Cutaneous mycoses are classified by site of infection: tinea pedis (athlete's foot), tinea cruris (jock itch), tinea capitis (ringworm of the scalp), etc. (Fig. 13-12). Diagnosis is made by examination of skin, hair, or nail specimens under the microscope after preparation with 10 percent KOH. Hyphae and typical mycelial patterns can be observed. Rarely is culture necessary, though it can be done if there is need or interest in speciating causative agents. If isolation of the fungus is desired, growth takes about 4 weeks at room temperature on Sabouraud agar. Ringworm of the scalp fluoresces bright yellowish-green under Wood's light. Most species causing skin ringworm do not

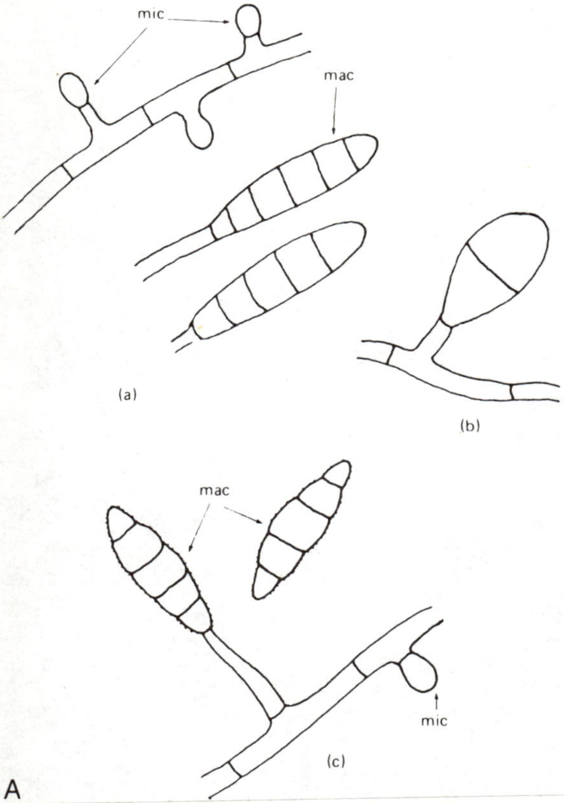

A

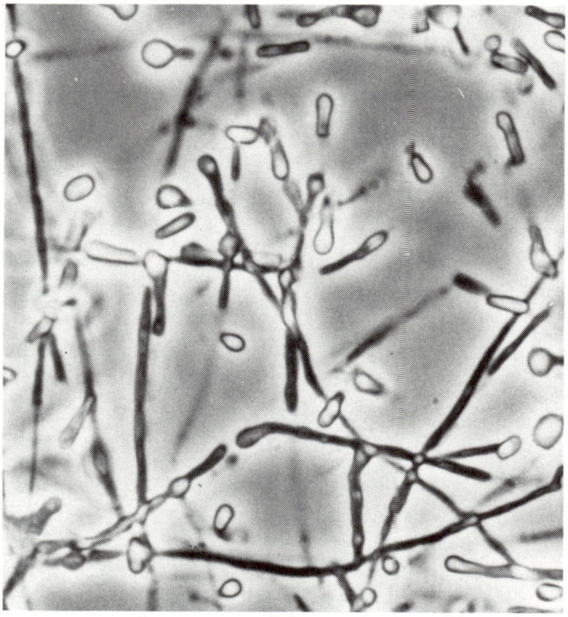

B

FIGURE 13-11. A. Dermatophytic fungi. (a) *Trichophyton,* showing macronidia (mac) and micronidia (mic). (b) Epidermophyton, macronidia (micronidia are not produced). (c) Microsporum, showing macro- and micronidia. All approximately 500X. (Reprinted with permission of the publisher from Deacon JW. Introduction to Modern Mycology. New York: John Wiley and Sons, 1980.) B. *Trichophyton tonsurans.* (Courtesy of Dr. James Plorde.)

TABLE 13-11. Classification of Dermatophytes

Genera	Species	Areas affected: Skin	Hair	Nails
Epidermophyton	*floccosum*	+	−	+
	stockdaleae	+	−	+
Microsporum	16 species: most common are *audouini, canis, gypseum*	−	+	−
Trichophyton	21 species: most common are *mentagrophytes, rubrum, schoenleinii, tonsurans, violaceum*	+	+	−

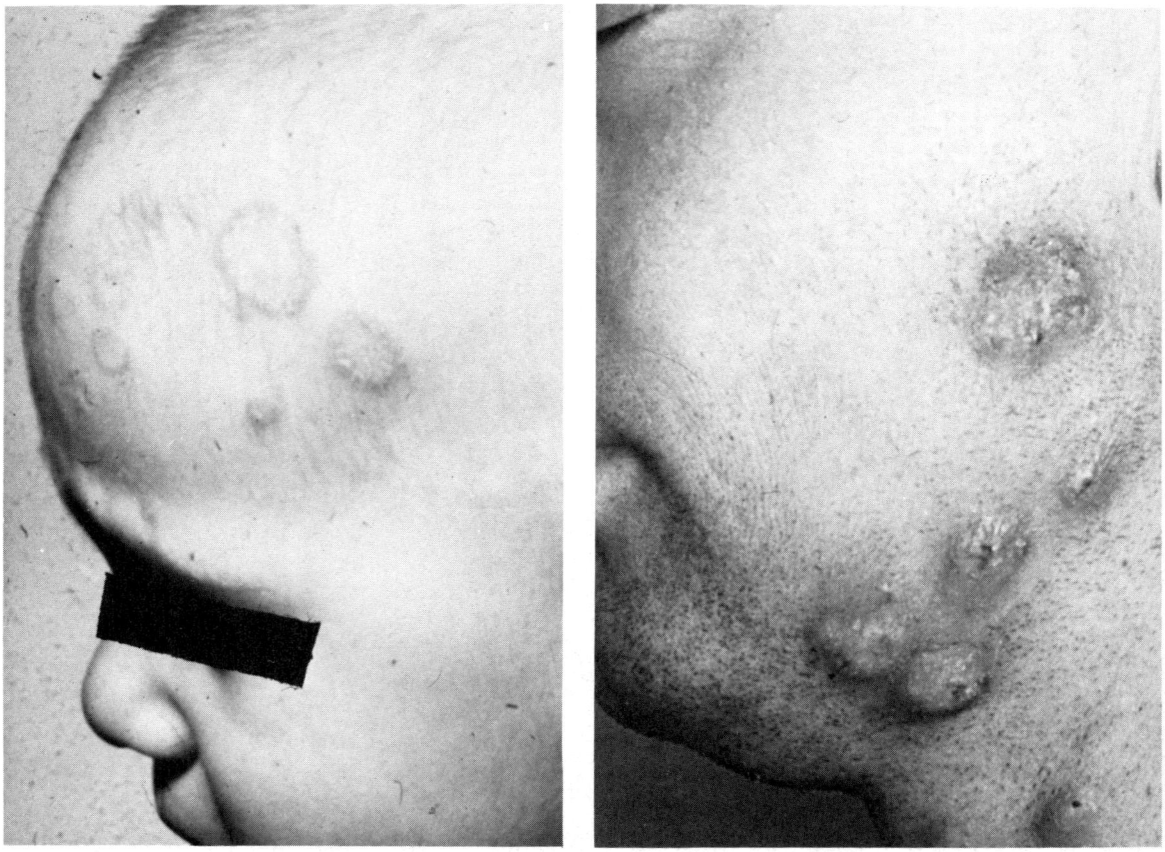

a b

FIGURE 13-12. *(a)* Ringworm in infant. *(b)* Ringworm in adult. (*a* courtesy of Dr. James Plorde. *b* courtesy of Centers for Disease Control, Atlanta, GA 30333.)

fluoresce. However, a bacterial skin infection called erythrasma (caused by *Corynebacterium minutissimum*) fluoresces orangish-red. The Wood's lamp is used to differentiate between these two skin conditions, which have a similar appearance and distribution.

Dermatophytes are some of the few fungi that are infectious between humans. Spores are transferred by direct contact with infected hosts and enter the body through small breaks in skin. Infection commonly occurs in shared showers or locker rooms where spores are present in sloughed-off epithelial cells. Invasions of intact skin may also be possible in very moist environments. Dermatophytes have the ability to degrade hard keratin for use as an energy source; hence their predilection for skin surfaces. The soil may also be a natural reservoir for dermatophytes. Infection is not serious but can be very annoying. It can persist for years, be difficult to treat, and be uncomfortable. In addition, allergic ractions to the fungi can result in inflammation and death of epithelium. Such allergic reactions called *dermatophytids* are usually in the form of skin vesicles on parts of the body other than those infected by the fungus (usually the hands).

Control of dermatophytosis includes the following:

1. Cleanliness of shared items such as combs and brushes, shower stalls, and headrests.
2. Careful removal of infected skin, hair, or nails.
3. Adequate aeration and drying of moist skin areas (athlete's foot is nonexistent among those who do not wear shoes). Persons with chronic athlete's foot should avoid wearing heavy stockings and closed shoes (such as tennis shoes).
4. Application of antifungal creams or powders such as miconazole, undecylenic acid, salicylic acid, usually for weeks.
5. In severe cases, use of oral griseofulvin. Weeks of therapy may be required. Overtreatment can increase the probability of developing hypersensitivity reactions (dermatophytids) to the infecting fungus.

SUBCUTANEOUS MYCOSES

Subcutaneous mycoses result from a different mode of fungal transmission. They are usually caused by soil saprophytes that enter the human host after an accidental puncture or break in the skin. There are three clinical subcutaneous syndromes caused by fungi: chromomycosis, maduromycosis, and sporotrichosis (Table 13-12). These will be discussed below. In addition, mycotic corneal ulcers (keratomycoses) can be caused by about 84 species of *Eumycota,* which are generally saprophytes. The most common causes are *Aspergillus fumigatus, Candida albicans, Culvaris* species, and *Fusarium solani.* Except for those caused by *C. albicans,* corneal mycoses are usually due to outdoor contamination in healthy people. They are characterized by elevated ulcers with an infiltrate that has hyphate margins (Fig. 13-13). Correct diagnosis is important to assure prompt antifungal therapy. Corticosteroids will worsen the condition.

Chromomycosis

Chromomycosis is a chronic skin disease characterized by warty, cauliflowerlike granulomatous subcutaneous growths. The causative agents are a group of dimorphic black molds, most commonly *Cladosporium carrionii, Fonsecaea pedrosoi,* and *Phialophora verrucosa,* which normally inhabit the soil and enter the skin through small abrasions or after trauma. The disease is most commonly seen on the feet, legs, or buttocks of agricultural workers in the tropics who

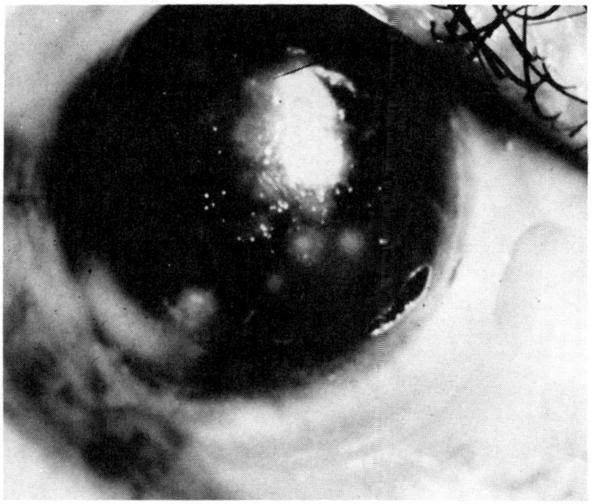

FIGURE 13-13. Corneal ulcer caused by *Candida tropicalis.* (Courtesy of Dr. John Chandler.)

TABLE 13-12. Subcutaneous Mycoses

Disease	*Hematogenous spread?*	*Causative agent(s)*	*Signs and symptoms*	*Mode of transmission*	*Distribution*
Chromomycosis	Yes	Dimorphic black molds (*Cladosporium carrionii, Fonsecaea pedrosoi, Phialophora verrucosa*)	Granulomatous skin lesions, occasionally brain abscess	Skin puncture with entry of soil fungi, primarily in feet and legs	Worldwide, especially in tropics
Maduromycosis (mycetoma)	No	About 15 species, especially *Monosporium apiospermum, Madurella mycetomi*	Granulomatous skin lesions with invasion of bone, and fascia	As with chromomycosis	Worldwide, especially in tropics (Sudan)
Sporotrichosis	Rare	*Sporothrix schenckii*	Local abscess or pustule with multiple subcutaneous nodules along lymphatic path	As with Chromomycosis	Worldwide, most common subcutaneous mycosis in U.S.

do not wear shoes. The lesions do not penetrate deeply or involve bone, though brain abscesses due to hematogenous spread can occur as opportunistic infections in debilitated individuals. The nodular skin granulomas develop over months and small dermal abscesses can crust and drain. Except when brain abscesses develop, chromomycosis is not fatal, though occasionally elephantiasis can result from lymphatic obstruction. The disease is very refractive to treatment, which includes oral flucytosine or thiabendazole and surgical excision of lesions.

Maduromycosis

Like chromomycosis, maduromycosis (or mycetoma) results from infection with certain soil fungi that enter the tissues, especially of legs, feet, hands, or back, after a traumatic injury. The most common causative fungi are *Monosporium apiospermum* and *Madurella mycetomi,* but there are about 15 different species that have been known to cause the disease. The filamentous bacteria, *Actinomycetes,* also cause a similar disease called mycetoma, though, not being caused by a fungus, the name is inappropriate.

Maduromycosis is a tumorlike granulomatous disease, but unlike chromomycosis, destruction of bone and fascia occur. Multiple fistulae are present with sinus tracts that erode and drain. The fungi do not spread in the blood stream, but secondary bacterial infections can cause death. Diagnosis is made by demonstrating the presence in lesions of fungal elements and granules, and by culture.

The epidemiology of maduromycosis is like that of chromomycosis. It is most common in the tropics (especially Sudan) among barefoot agricultural workers. Treatment, including surgical excision and possible amphotericin B (not well studied yet) is only partially effective and amputation of the affected limb may be necessary. *M. apiospermum* has also been known to cause arthritis, sinusitis, eye infections, and pneumonia.

Sporotrichosis

Sporotrichosis is the most common subcutaneous mycotic infection in the United States, where the causative dimorphic fungus, *Sporothrix schenckii* (Fig. 13-14), is common on certain plants, moss (sphagnum moss), and wood. The fungus enters the skin after a thorn prick or other trauma, and after 2 to 8 weeks causes a local infection, either a pustule, abscess, or small ulcer, that spreads via lymphatics along the affected limb, causing additional lesions (Fig. 13-15).

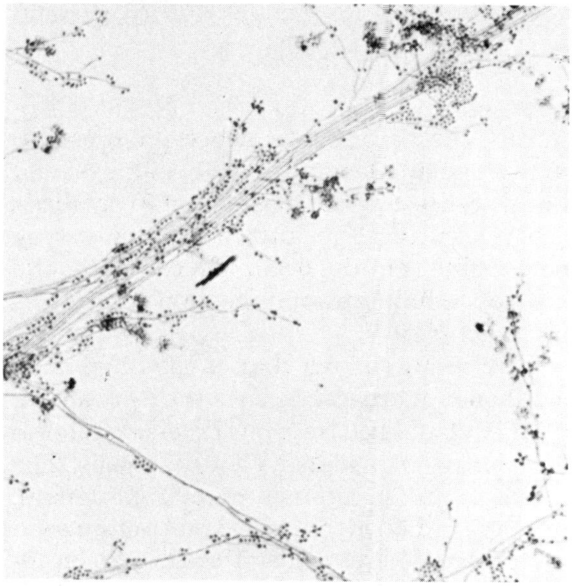

FIGURE 13-14. *Sporothrix schenckii.* Magnification 128×. (Courtesy of Dr. James Plorde.)

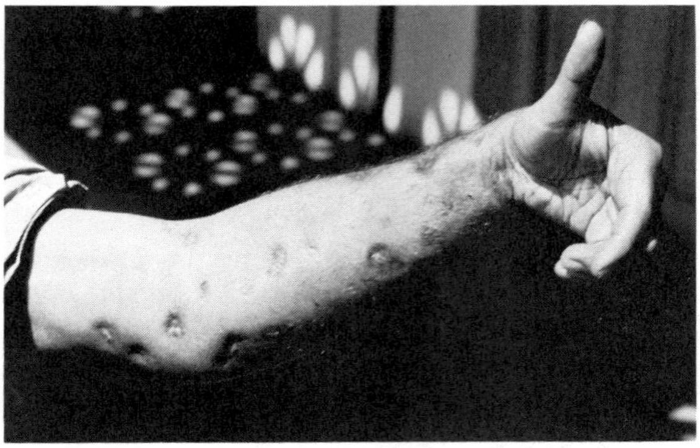

a

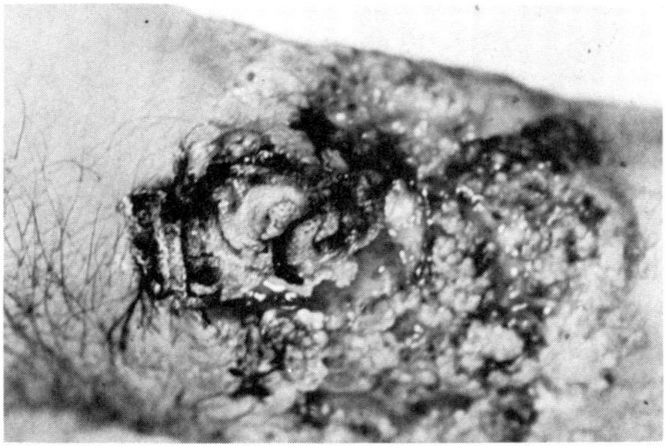

b

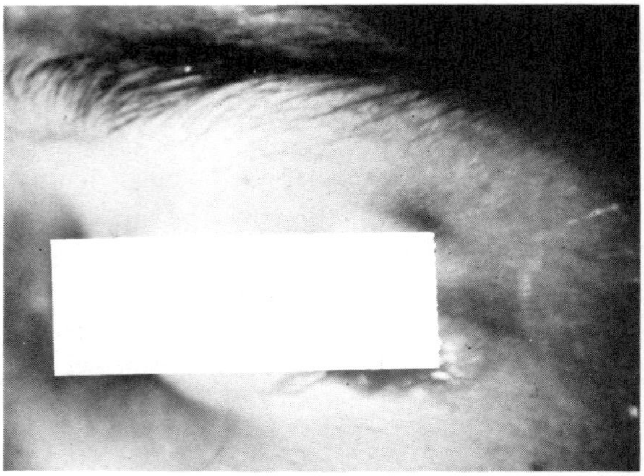

c

FIGURE 13-15. *(a)* Sporotrichosis of the arm. Note lesions along the lymphatic chain. *(b)* Sporotrichosis lesion on wrist. *(c)* Sporotrichosis of the eye. (Courtesy of Centers for Disease Control, Atlanta, GA 30333.)

Rarely the organism can become disseminated, causing lesions in bone, kidneys, and lungs. Since cases of disseminated sporotrichosis occur with no known trauma, it may be possible that the organism can enter the host by inhalation or ingestion as well as by puncture.

Unlike organisms that cause other subcutaneous mycoses, *S. schenckii* cannot be readily visualized from pus or lesions, therefore culture is necessary for diagnosis. The disease is most common among gardeners, miners, and farmers, and is most often seen on the hands and arms. Though person-to-person spread does not occur, outbreaks have been reported among industrial workers after exposure to plant materials containing a large growth of *S. schenckii*. The disease does not heal spontaneously, but is usually self-limited. Long-term treatment with potassium iodide and, for systemic spread, amphotericin B, is effective.

SYSTEMIC MYCOSES

Most fungi causing systemic disease do so only if the host is in some way immunologically compromised. T cell immunity plays an important part in defense against disseminated fungal disease, and the increased prevalence of some of these diseases among specific racial or ethnic groups probably reflects some deficiency in cellular immunity.

Most of these fungi grow on organic matter or soil, and human infection is incidental to the growth requirements of the organism. Therefore, person-to-person spread is quite rare. Spores are small enough (about 3 mm) to reach alveoli when inhaled.

There are two general types of systemic fungal diseases (Table 13-13): (1) those that can occur in otherwise apparently healthy individuals in specific geographic areas where the fungus is endemic ("pathogenic" fungi) and (2) those that are caused by ubiquitous, normally harmless fungi in compromised hosts ("opportunistic" fungi). The pathogenic fungi that cause blastomycosis, coccidioidomycosis, histoplasmosis, and paracoccidioidomycosis are normally soil inhabitants and are contracted by humans who inhale airborne spores. Each disease occurs primarily within a circumscribed geographic area. They are all caused by dimorphic fungi, occurring as spherical or yeast forms in the host at body temperature and in mycelial form in soil. Other factors besides temperature that affect dimorphism include high sugar levels *(Candida)*, anaerobic conditions plus high CO_2 or sugar levels *(Mucor)*, and high body temperature plus CO_2 (*Sporothrix* and *Histoplasma*). The opportunistic fungi, on the other hand, may enter the host in a variety of ways. These ways include inhalation, puncture of skin or mucous membranes, or ingestion. The opportunistic fungi are not limited to a particular geographic area. Many are part of normal flora *(Candida)* or are ubiquitous in soil or on plants (phycomycetes *Aspergillus)*.

Whether disease is caused by a pathogenic or an opportunistic fungus, the pathogenesis of all systemic fungal diseases have some similarities. All disease-causing fungi are facultative intracellular parasites that are phagocytized, grow in phagocytic cells, and are disseminated in the blood stream or lymphatics. Antibody formation plays little role in defenses against fungal disease. The most common systemic mycoses will be discussed in this section, though there are also other fungi that cause disease less commonly (Table 13-14; Figs. 13-16 and 13-17).

Blastomycosis

Biologic characteristics. *Blastomyces dermatitidis* is dimorphic. It grows as a thick-walled yeast at body temperature (Fig. 13-18) and in the mycelial form (producing aleuriospores) at room temperature. The yeasts produce only one bud at a time. The

TABLE 13-13. Types of Systemic Mycoses

	Caused by pathogenic fungi	*Caused by opportunistic fungi*
Diseases	Blastomycosis Coccidioidomycosis Histoplasmosis Paracoccidioidomycosis	Aspergillosis Candidiasis Cryptococcosis Geotrichosis Phycomycosis
Portal of entry	Inhalation of spores	Inhalation, skin puncture, ingestion, "endogenous"
Distribution	Limited to specific geographic areas	Worldwide

TABLE 13-14. Some Fungi Causing Rare Disease in Humans

Genus	Disease
Cephalosporium	Neonatal meningitis
Cladosporium	Chromoblastomycosis (Fig. 13-16)
Curvularia	Pulmonary or brain infection
Dermatophilus	Skin abscesses
Fusarium	Skin infection, fungemia
Protothera	Wound infections
Phialophora	Endocarditis
Rhodocrous	Fungemia
Torulopsis (classified by some in the genus *Candida*)	Fungemia (Fig. 13-17)

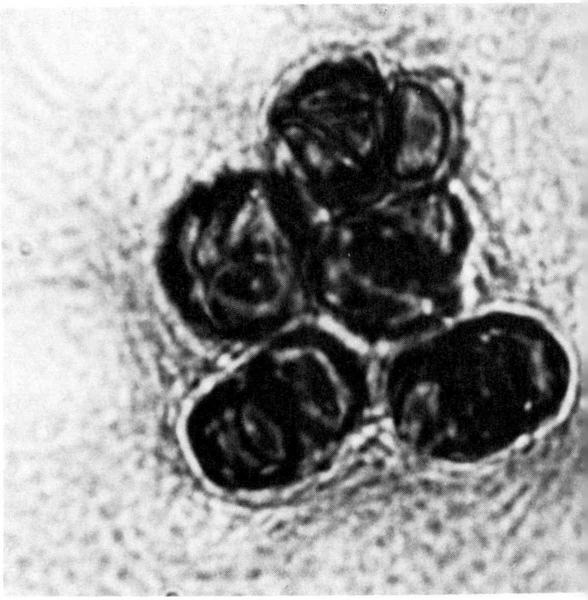

FIGURE 13-16. Chromoblastomycosis sclerotic body. Magnification 100×. (Courtesy of Dr. James Plorde.)

fungus grows well on Sabouraud agar and is fortunately not hazardous when grown in the laboratory (as is the organism causing coccidioidomycosis). There is a sexual stage, *Ajellomyces dermatitidis,* which is an ascomycete.

Epidemiology. Blastomycosis (sometimes called North American blastomycosis because of its prevalence in the United States and Canada) is endemic to the southeastern and central United States, the Mississippi and Ohio River Valleys, and occurs rarely in other parts of North America. Incidence of the disease is difficult to determine because it is usually mild or asymptomatic and because it is only optionally reported to state health departments. In 1980, 75 cases were reported in the United States.

B. dermatitidis has been isolated with less consistency from soil than have the fungi causing coccidioidomycosis and histoplasmosis. However, it is probable that the mode of transmission is the same for all three: inhalation of spores carried in dust and air currents. It is possible but not likely that infection from lower animals may be acquired by humans; person-to-person transmission has never been observed. The disease occurs more frequently in males, but there is no racial predisposition. Occasionally cutaneous blastomycotic nodules occur after accidental inoculation of spores into skin.

Pathogenesis. Spores are inhaled and spread in the blood stream or lymphatics to cause metastatic disease. The characteristic granulomatous or destructive lesions are a result of delayed hypersensitivity reaction, as in tuberculosis.

Host responses. Symptoms of blastomycosis may be varied, nonspecific, and the disease may be very difficult to diagnose.

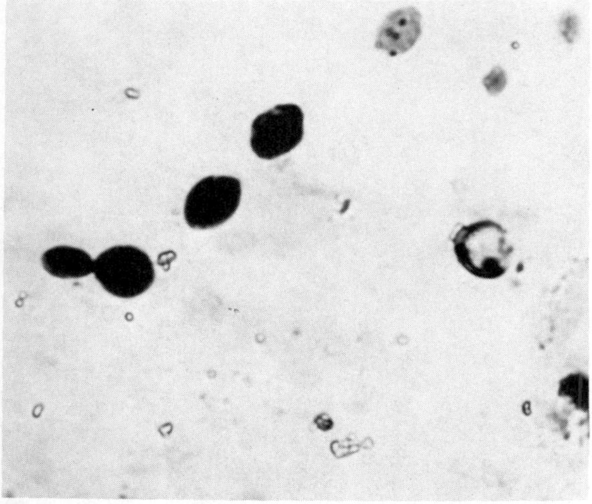

FIGURE 13-17. Stool specimen containing about 60 percent *Torulopsis* (yeast). (Courtesy of Dr. James Plorde.)

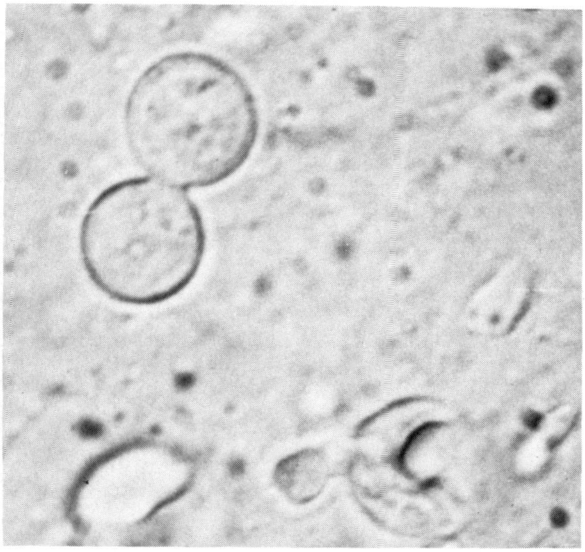

FIGURE 13-18. *Blastomycosis dermatitidis* in pus. Note budding cell. (Courtesy of Dr. James Plorde.)

As mentioned, progressive disease probably only occurs in individuals with deficient cellular immunity. The primary pulmonary infection is usually very mild. In the rare acute pulmonary form, an individual may suffer from persistent cough, low-grade fever, and weight loss with radiologic evidence of pneumonia. Lung cavitation occurs frequently. The most common manifestations are skin lesions, usually on exposed skin surfaces like the hands and face (Fig. 13-19). These are slowly developing painless papules that become raised, lumpy, and wartlike. The cutaneous lesions can remain localized for months or years, or disseminated disease can result in lesions in bone, in viscera (not the GI tract), in the central nervous system, in the male genital tract, and in the heart. Once dissemination has occurred, the mortality rate in untreated individuals is over 90 percent within 2 years. Since the 1960s, one or two persons annually have died of blastomycosis in the United States.

Unfortunately, serologic tests are not always useful in the diagnosis of blastomycosis because there is little circulating antibody. Complement fixation tests may be positive in some individuals. However, the organism can be isolated from infected tissues or sputum with relative ease, and the diagnosis is confirmed microbiologically. The blastomycin skin test is of minimal use because false positives can result from cross reactions with *Histoplasma* and *Coccidioides,* and because individuals with active blastomycosis may be nonreactive to the dermal test.

Control and treatment. No vaccine or preventive measures are known. Treatment of all systemic fungal infections is similar and consists of amphotericin B or other antifungal agents, all of which are quite toxic and require long-term administration (see "Clinical Implications").

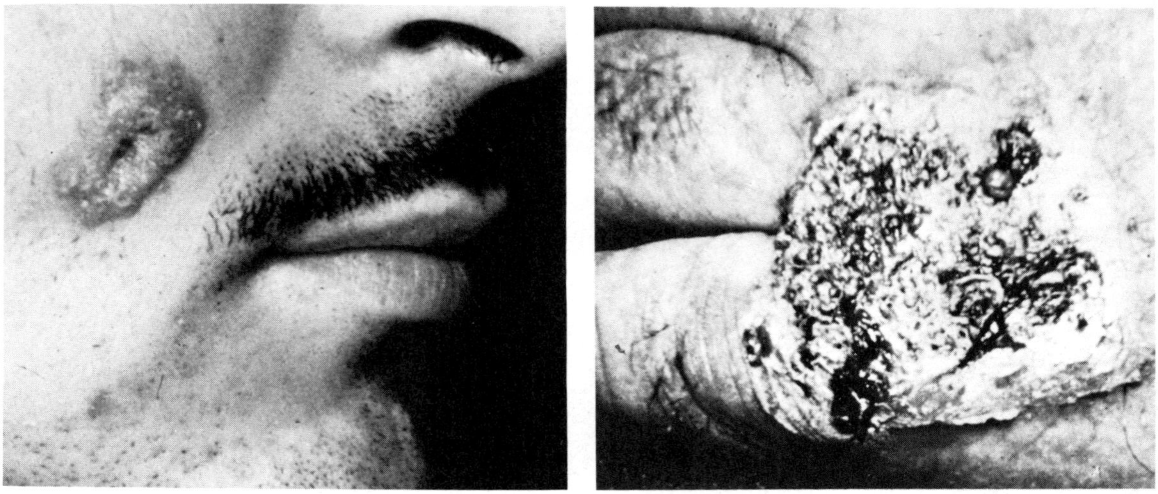

a b

FIGURE 13-19. *(a)* North American blastomycosis on face. *(b)* Closeup of blastomycosis on the hand. (Courtesy of Centers for Disease Control, Atlanta, GA 30333.)

Coccidioidomycosis

Biologic characteristics. Coccidioidomycosis is caused by a dimorphic fungus, *Coccidioides immitis,* that in soil or in the laboratory at room temperature produces fluffy septate mycelia with arthrospores and in the human host produces a spherule yeast form (Fig. 13-20) and endospores. Arthrospores (Fig. 13-21) growing on culture media are highly infectious and represent a significant hazard to laboratory workers, as they are readily aerosolized and inhaled. For this reason, diagnostic methods other than culture are used as often as possible for coccidioidomycosis.

Epidemiology. Coccidioidomycosis is also called San Joaquin Valley fever. The organisms, normally found in the soil in the southwestern United States, in northern Mexico, and in certain parts of Central and South America, enter the human host by inhalation. Like the other pathogenic fungi, humans are only infected accidentally and are not an essential part of the organism's life cycle. There are about 100,000 cases in the United States with less than 100 deaths reported annually. Most individuals in California, Nevada, and Texas have positive coccidioidin skin tests. The semi-arid climate in that region results in increased dust in which spores are transmitted. The disease is 50 times more common in individuals from the Philippines, 10 times more common in dark-skinned races than in Caucasians, probably as a result of differences in cellular immune function. In warm climates the spores survive winter, but the disease is more common in summer. Coccidioidomycosis also occurs in other animal species, but person-to-person or animal-to-animal transmission is rare. Spores can be transmitted long distances. Cases in other parts of the world are due to transmission of spores on inanimate objects.

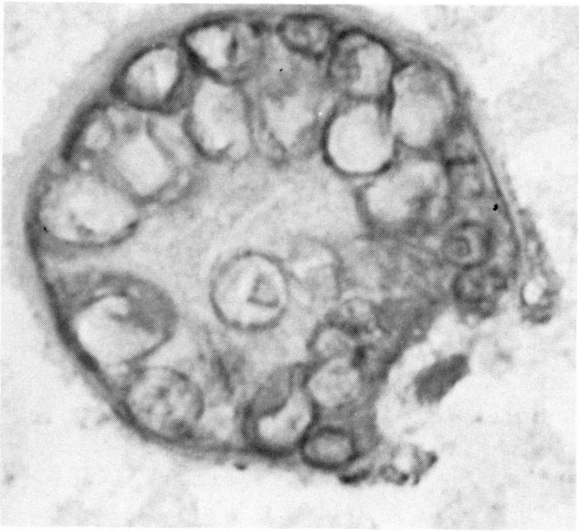

FIGURE 13-20. *Coccidioides* spherule. (Courtesy of Dr. James Plorde.)

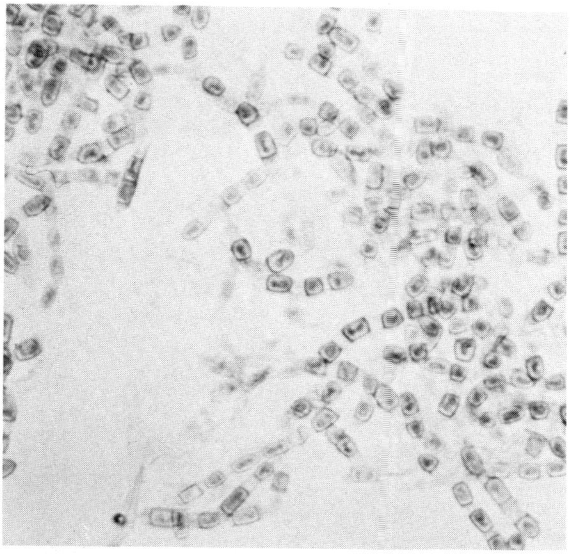

FIGURE 13-21. *Coccidioides immitis* arthrospores. (Courtesy of Dr. James Plorde.)

Pathogenesis. Arthrospores are phagocytized in the pulmonary alveoli. The organisms are usually killed, but since they are facultative intracellular parasites, they can proliferate in phagocytes. In 2 to 3 days they develop into spherules, causing pulmonary disease or becoming disseminated by blood or lymphatics. Cellular immunity usually stops the progress of the disease. Delayed hypersensitivity develops as in tuberculosis, and lung necrosis and cavitation can result in hosts who do not generate an adequate immune response. Strains of *C. immitis* differ somewhat in virulence.

Host responses. Following an incubation period of 1 to 3 weeks, at least 60 percent of primary infections exhibit no symptoms. In the other 40 percent, pulmonary symptoms may range from a mild flulike illness with a slight fever and dry cough to a severe pneumonia with abscess and cavity formation. Most cases resolve without treatment, but about 5 percent of individuals may develop residual pulmonary necrosis and fibrosis similar to that seen in tuberculosis. Development of allergic skin reactions such as erythema nodosum sometimes parallels the development of delayed hypersensitivity (Fig. 13-22). In less than 1 percent of cases the disease disseminates to meninges, skin, bones, liver, kidneys, and heart. Hydrocephalus can result. In addition to persons of dark skin, women in the third trimester of pregnancy and persons receiving corticosteroids are especially susceptible to disseminated disease. Without therapy, disseminated coccidioidomycosis is fatal. Diagnosis is usually made by direct examination of sputum or pus and by evaluation of epidemiologic evidence. Although the organism can be cultured easily in vitro, the procedure is to be avoided because of its high infectiousness. Detection of antibodies

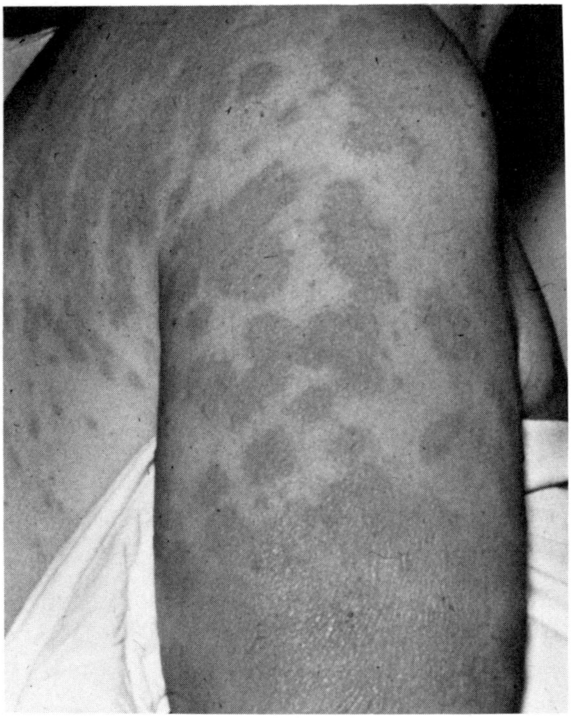

FIGURE 13-22. Coccidioidomycosis skin lesions (erythema nodosum) on arm of a patient. (Courtesy of Centers for Disease Control, Atlanta, GA 30333.)

and skin tests also have some usefulness. However, the skin test has the limitations described in Table 13-15.

Control and treatment. Infection induces solid immunity. No preventive measures are known. The only successful treatment is amphotericin B, though miconazole and ketoconazole are alternatives. Surgical excision of pulmonary cavities is sometimes necessary. Transfer factor to enhance cellular immunity has been trialed; vaccines are experimental.

Histoplasmosis

Biologic characteristics. *Histoplasma capsulatum,* which causes histoplasmosis worldwide, and *H. capsulatum* var. *duboisii,* which causes histoplasmosis in Africa, are dimorphic fungi that grow in reticuloendothelial cells (Fig. 13-23). The ascomycete, *Emmonsiella capsulata,* is the sexual stage. The fungus converts from yeast to mold and produces aleuriospores either as macroconidia or microconidia when grown at room temperature (Fig. 13-24). Handling the mycelial culture is less hazardous to laboratory personnel than handling of *Coccidioides,* but care is still warranted. The yeast form is not contagious.

Epidemiology. *H. capsulatum* flourishes in soil enriched by bird feces and is also carried in bat guano (feces). Birds are not infected with the disease because their high body temperature is not conducive to growth of the organism, but their droppings offer a rich culture medium in the soil.

Histoplasmosis occurs worldwide, but it

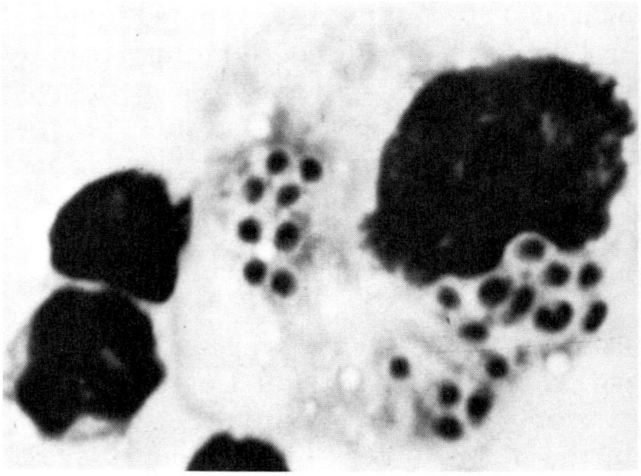

FIGURE 13-23. Histoplasmosis (Giemsa stain). (Courtesy of Dr. James Plorde.)

TABLE 13-15. Problems with Skin Tests for Systemic Fungal Diseases

Problem	*Blastomycosis*	*Coccidioido-mycosis*	*Histoplasmosis*	*Paracoccidioido-mycosis*
Nonspecific reactions (cross reactions with other fungi)	X	X	X	
Induces immune response (may cause misinterpretation of serologic tests)			X	
Skin tests often negative during acute or active disease	X	X	X	X
Only indicates past or present infection, does not indicate acute disease	X	X	X	X

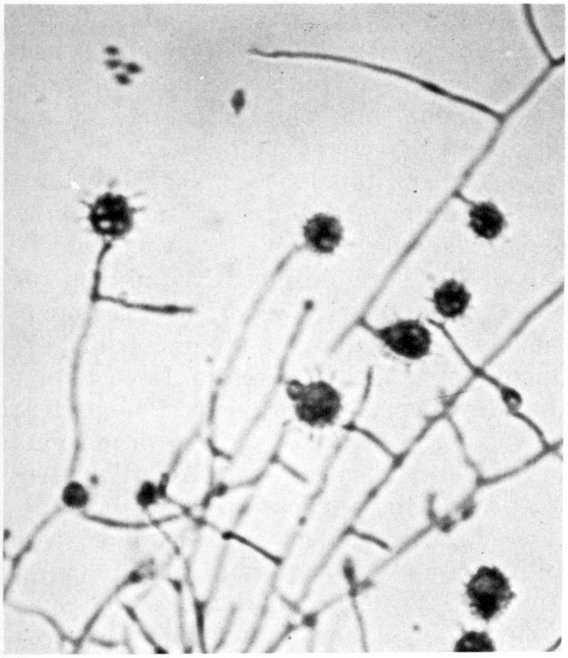

FIGURE 13-24. *Histoplasma capsulatum* with chlamydospores. (Courtesy of Dr. James Plorde.)

is especially common in certain areas such as the Ohio and Mississippi River Valleys (central and southeastern United States), where up to 80 to 95 percent of residents have positive skin tests. As many as half of the healthy children living in such areas have positive skin tests by 5 years of age, though the reactivity decreases over time. Histoplasmosis is endemic in at least 31 of the 48 contiguous states to some extent. There are an estimated 20,000 new cases each year with as many as 40 million people infected in the United States. Like tuberculosis, histoplasmosis is most common in young children (primary infection) and in persons 50 to 60 years of age (reactivation). Person-to-person spread is unknown. Most cases are acquired by inhalation of spores; rarely, accidental inoculation results in disease.

Epidemics of histoplasmosis occur and usually represent a focus of soil rich in bird droppings that has been disturbed such as in the demolition of an abandoned building or a canebrake. In 1968 a canebrake that was a blackbird roost was bulldozed. A subsequent outbreak of histoplasmosis occurred. After 7 days an individual developed histoplasmosis and died. Nine other people contracted the disease. Viable particles of *H. capsulatum* remained in the soil for 9 years. Of eighth grade students, 28 percent within a 20 kilometer radius had positive histoplasmin skin tests (Di Salvo and Johnson 1979). As a second example, 14 days after bat-cave exploring, a man and woman developed histoplasmosis. Twenty-three of 29 members of an exploring group had mild respiratory illness, cough, chest discomfort, fever, and dyspnea on exertion. The histoplasmin skin test was positive in 18 of 24 tested, as compared to a positive skin test prevalence in the area of about 7 percent (Lottenberg et al. 1979). Other animals are also susceptible to infection.

Pathogenesis. In experimental animals, microconidia are inhaled and germinate, releasing yeast forms about 3 to 4 days after infection. The organisms grow within phagocytes and an influx of macrophages occurs. By 2 weeks after infection, exudate has expanded and produced a patchy pneumonitis. Usually within 2 weeks vasculitis with local tissue necrosis occurs, resulting in hypersensitivity (positive skin test) and in most cases successful immunologic defense against further disease. The pathogenesis closely parallels that of tuberculosis. The yeast form of *H. capsulatum* also activates human platelets causing aggregation in the absence of complement, which explains the fact that thrombocytopenia often accompanies disseminated histoplasmosis.

Host responses. Three clinical syndromes occur after infection with *H. capsulatum:* an asymptomatic or mild respiratory disease, chronic pulmonary histoplasmosis, or disseminated disease. The incubation period after exposure varies, depending on the size of the inoculum and the condition of cellular host defenses, from 24 hours to many days. Though millions of persons are infected, most cases are subclinical or result in mild headache, cough, fever, and flulike symptoms that usually go unnoticed and resolve spontaneously. Cellular immunity and delayed hypersensitivity develop within a few weeks and the disease is arrested.

The respiratory syndrome, acute histoplasmosis, is usually only recognized when it occurs in epidemics. Rarely, chronic pulmonary histoplasmosis can occur, usually in individuals with underlying lung disease. It is characterized by numerous calcified areas and cavities in the lungs and lymph nodes. The signs and symptoms are indistinguishable from those of pulmonary tuberculosis or coccidioidomycosis.

In the debilitated (especially individuals with malignancies), disseminated histoplasmosis, similar to miliary tuberculosis, can occur. Disseminated histoplasmosis often presents with fever, hepatosplenomegaly, and hemolytic anemia. It may be mistaken for acute leukemia. In this stage the skin test is often negative. The patient may have pneumonia, skin rash, adrenal insufficiency, or meningitis. About 30 cases of *Histoplasma* endocarditis have been described, all in residents of the southeastern United States. The average length of time from initial infection to diagnosis for these patients was 9 months. The mortality rate of disseminated histoplasmosis, even with therapy, is about 50 percent. Reactivation histoplasmosis, like tuberculosis, can occur years after primary infection. The yeast cells can remain viable in the lungs for years.

Histoplasmosis is very difficult to diagnose because it mimics other diseases and because the causative agent may not be readily recovered from infected tissues. The skin test is to be avoided for reasons listed on Table 13-15, but especially because the injected antigen can generate an immune response, cause an increase in antibody titer, and interfere with the interpretation of serologic tests. Thus, repeated skin tests are not recommended. Serologic tests used for diagnosis include complement fixation, immunodiffusion, latex particle agglutination, and fluorescent antibody tests. Serologic tests are especially useful when serial samples are taken to follow the rise and fall of antibody titers. This is not the case with tuberculosis, in which there is no correlation between antibody level and disease state.

Control and treatment. There are no effective methods of ridding soil of *H. capsulatum,* and there is no vaccine for protection against histoplasmosis. Individuals, especially those with lowered resistance or pulmonary disease, should avoid endemic areas in which there are large volumes of bird or bat excreta. Usually no treatment is necessary, but in disseminated disease the only effective therapy is amphotericin B or ketoconazole. Even with adequate treatment, relapses occur.

Paracoccidioidomycosis

Biologic characteristics. *Paracoccidioides brasiliensis* causes South American blastomycosis or paracoccidioidomycosis. It is a dimorphic fungus that produces yeast cells characterized by multiple buds that resemble a crown of beads attached to the parent cell wall. It takes about 1 month to grow and, again, laboratory contamination with the mycelial form should be avoided.

Epidemiology. *P. brasiliensis* is a plant and soil saprophyte which, as its name suggests, is most common in South America, especially in Brazil, Colombia, and Venezuela. The portal of entry is most likely by inhalation, though pulmonary symptoms are usually subclinical. The disease is most common in adult male agricultural workers. Person-to-person spread does not occur.

Pathogenesis. After entrance into the alveoli, the organism does not establish itself in the lungs, but rather has a predilection for the face, nasopharynx, and oropharynx. From the nasal mucosa, lesions spread to the regional lymphatic system and into the blood stream. The basic lesions are granulomatous.

Host responses. Paracoccidioidomycosis is primarily focused in the skin and mucous membranes and only rarely causes pneumonia. The primary lesions are papillomas in the mouth, throat, and nose (Fig. 13-25). Purulent conjunctivitis, lymphadenitis, or subcutaneous abscesses may develop. The disseminated disease slowly progresses to involve any organ, including the spleen, gastrointestinal tract, lungs, bones, and central nervous system. If the disease is untreated, death usually occurs within 2 years.

Control and treatment. Spread of lesions can be prevented with amphotericin B. Miconazole, hydroxystilbamidine, and trimethoprim-sulfamethoxazole have also been used.

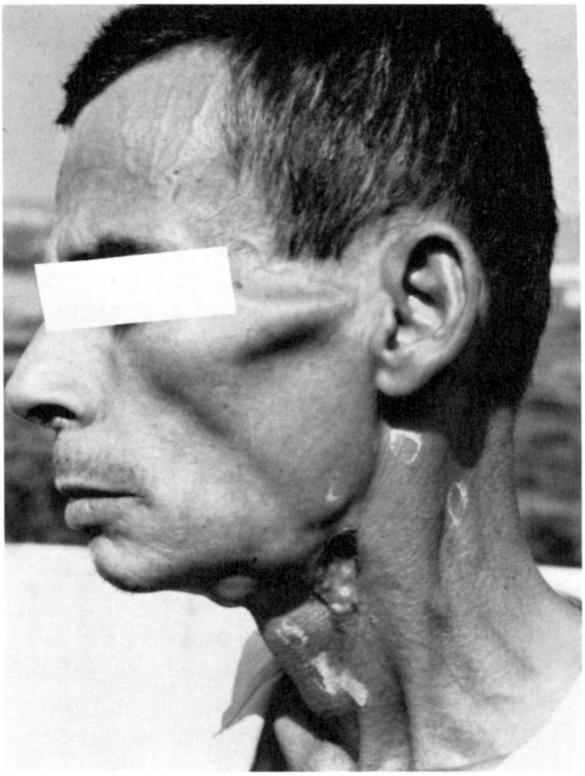

FIGURE 13-25. Paracoccidioidomycosis: close up of patient prior to treatment. (Courtesy of Centers for Disease Control, Atlanta, GA 30333.)

SYSTEMIC MYCOSES: OPPORTUNISTIC FUNGAL DISEASES

The fungal diseases discussed in this section almost always occur in hospitalized or markedly debilitated patients. They differ from the pathogenic fungi in that they may enter the body by many routes (not just by inhalation) and may even be endogenous infections, caused by the patient's own flora. Patients with these mycoses often have malignancies, are receiving immunosuppressive or antimicrobial drugs. For example, in 245 patients with chronic granulomatous disease recently studied, 20.4 percent developed clinically apparent fungal infections. The most common fungi involved were *Aspergillus* and *Candida*. Most of these patients had fungal pneumonia or disseminated disease; 18 percent of the infections were limited to soft tissue or bone. Without treatment, patients invariably succumbed; more than half of the treated patients were cured. Treatment consisted of antifungal drugs, surgical removal of infected tissue, and granulocyte transfusion (Cohen et al. 1981).

Aspergillosis

Biologic characteristics. There are at least 8 species of *Aspergillus*, a saprophytic mold, that have caused disease in humans. Though *Aspergillus* are classified as deuteromycetes, some species do produce sexual spores and are classified as ascomycetes. *A. fumigatus* is the most important species, causing the majority of clinical disease, with *A. carneus, clavatus, niger, nidulans, restrictus, terreus,* and *versicolor* also of some significance. *Aspergillus*, meaning "rough head," is characterized by the flowerlike appearance of its spore clusters (Fig. 13-26). The mass of spores grows at the end of a stalklike mycelium called a conidiophore. It is not dimorphic, but grows in segmented mycelial form at all times.

Epidemiology. Aspergilli are ubiquitous in the air and soil and because they thrive at higher temperatures (40-50°C), they are natural pathogens of birds, causing epidemics in penguins, turkeys, and chickens. The spores are particularly prevalent in fall and winter, especially in damp, warm areas such as piles of decaying hay, grain, or foodstuffs. The spores may also be spread in air filters or humidifiers, which can be a source of in-

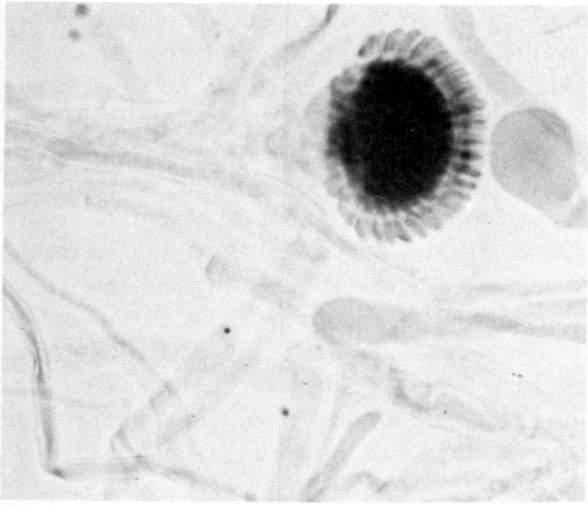

FIGURE 13-26. *Aspergillus,* which means "rough head"; note flowerlike appearance of spore cluster. Magnification 320 ×. (Courtesy of Dr. James Plorde.)

fection for hospitalized patients. The organisms usually enter the host by inhalation. *A. fumigatus* may be isolated from 10 to 15 percent of sputum cultures, but its presence is often nonsignificant.

Pathogenesis. After entrance into the lungs, the aspergilli form large masses of branching hyphae ("fungus balls" or aspergillomas) in preexisting cavities due to carcinoma, tuberculosis, or bronchiectasis. A second type of condition, hypersensitivity pneumonitis or rhinitis, is due to reaction against spore antigens. Invasion and thrombosis of blood vessels is common.

Host responses. The most common sites of infection are the lungs and the gastrointestinal tract. In the lungs, primarily among asthmatics who have developed antibodies to *Aspergillus* antigens, allergic aspergillosis (sometimes called the PIE syndrome—pulmonary infiltrates with eosinophilia) causes wheezing, cough, and pleurisy. Hypersensitivity pneumonitis (e.g., farmer's lung), often due to *A. clavatus,* also occurs in sensitized individuals. Aspergillomas cause severe hemoptysis, and acute pulmonary aspergillosis among individuals with malignancies is characterized by rapid, difficult breathing and fever. Invasive aspergillosis can result in brain or bone abscesses, endocarditis, gastrointestinal disease, or dissemination to any body part. Esophagitis, mediastinitis, and middle ear infections result from extension of local lesions. The mortality rate is high.

Diagnosis of invasive aspergillosis is difficult. Counterimmunoelectrophoresis and enzyme-linked immunosorbent assay (ELISA) are used and can minimize the risk of more invasive techniques such as surgical biopsy. Passive hemagglutination assay of antibody has also recently been used as a helpful diagnostic technique. *Aspergillus* must be isolated from a series of clinical cultures before one can be certain that it is actually causing disease since *Aspergillus* is so often a laboratory contaminant.

Control and treatment. Allergic aspergillosis is sometimes cured or alleviated by steroids which suppress the secretions in which the fungus grows. Amphotericin B, flucytosine, and pulmonary resection are used to treat pulmonary and disseminated aspergillosis.

Candidiasis

Biologic characteristics. *Candida* is a heterogeneous genus of deuteromycetes. *Candida* species are aerobic budding yeasts when found in human infection. A few species have been shown to reproduce sexually by ascospores, but most reproduce asexually with blastospores. *C. albicians* can also produce chlamydospores. *Candida* are faster growing than most fungi, with tiny white creamy colonies appearing within 1 to 2 days after inoculation on laboratory media. The most important human pathogen is *C. albicans;* however *C. guilliermondii, krusei, parapsilosis, pseudotropicalis,* and *tropicalis* also cause disease.

Epidemiology. *Candida* species are probably present in the intestinal tracts of 20 to 40 percent of normal individuals. They are normal flora of the mouth, intestinal tract, vagina, but usually in small numbers. They can become overgrown during antimicrobial therapy or in the presence of endocrine or immunologic defects. Candidiasis is often endogenous in origin when such factors decrease host resistance. In one hospital survey during a 3-year period, *Candida* species accounted for the majority (103 out of 136) of fungemias. In a second surveillance study of 37 bone-marrow transplant recipients and 52 patients receiving chemotherapy, 67 percent were colonized with *C. albicans,* 28 percent with *C. tropicalis,* and none with *Aspergillus.* Among those patients there were 21 cases of proven systemic fungal disease: 3 caused by *C. albicans,* 16 caused by *C. tropicalis,* and 2 by *Aspergillus* (Cohen et al. 1981; Meunier-Carpentier, Kiehn, and Armstrong 1981). There have been about 200 deaths annually attributed to candidiasis in the United States during the 1970s. Person-to-person transmission is rare, occurring primarily in newborns who acquire *C. albicans* in the birth canal. *Candida* also grow on organic material, especially fruits and juices.

Pathogenesis. *Candida* are normally susceptible to phagocytosis. Individuals receiving antimicrobial agents, corticosteroids, and those with high blood glucose (as in diabetes mellitus) are more susceptible to candidal infections because such conditions reduce the efficiency of phagocytosis. *C. albicans* can produce endotoxin, cause damage to mucous membranes, and invade the blood stream. However, its ability to cause disease is due to decreased host defenses and decreased competition from other organisms as a result of antimicrobial therapy. Adherence of *C. albicans* is enhanced in germinated cells by means of a surface protein that binds to an epithelial receptor.

Host responses. Candidiasis (a synonym is moniliasis) can take several forms: mucocutaneous infection, such as thrush in the mouth, vaginitis, esophagitis; or intertriginous lesions; or systemic infections, especially endocarditis. *Candida* is a common cause of diaper rash, which may spread extensively (Fig. 13-27). Vaginal infections cause a creamy vaginal discharge and considerable itching. Discomfort can be worse if bladder infection also occurs. Vaginitis and cystitis are common in young women taking oral contraceptives, during pregnancy, and in diabetics. Skin and nail infections similar to those caused by *Trichophyton* can be recurrent and are called chronic mucocutaneous candidiasis. Patients during intensive therapy with immunosuppressives or antimicrobial drugs can develop candidal esophagitis, with or without accompanying oral thursh. Oral thrush results in the formation of white, soft patches in the oropharynx that can be easily removed. Extension to the esophagus can cause chest pain and difficulty in swallowing.

Systemic candidiasis is much more serious. *Candida* is a frequent cause of sepsis associated with hyperalimentation solu-

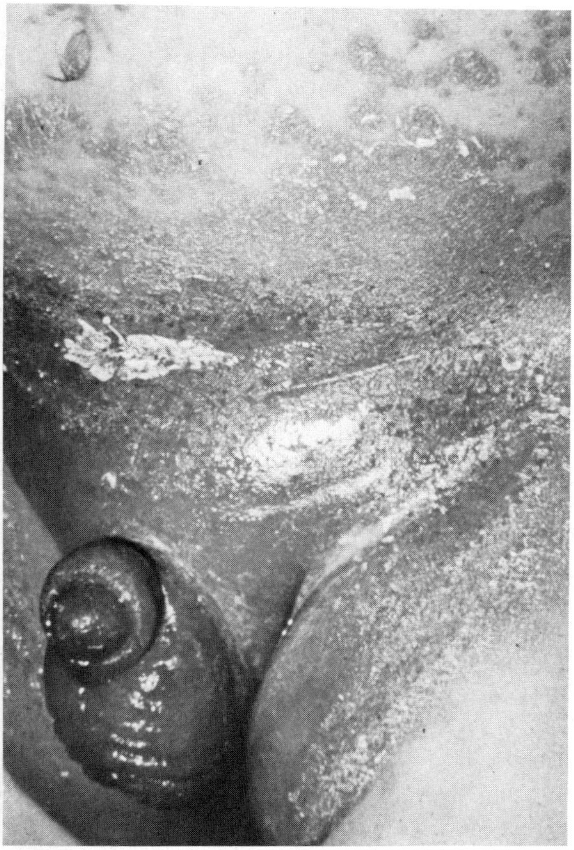

FIGURE 13-27. Candidiasis of skin on genitalia and lower abdomen of infant (severe diaper rash). (Courtesy of Centers for Disease Control, Atlanta, GA 30333.)

tions that are hypertonic. In a series of 131 patients studied prospectively during administration of hyperalimentation, 13 (9.9 percent) developed chorioretinal lesions due to *Candida,* probably as a result of hematogenous spread. Symptoms of disseminated disease may range from fever only to frank septic shock. *Candida* species can cause osteomyelitis, meningitis, pneumonia, and endocarditis as well as fungemia.

Diagnosis may be difficult and is usually based on repeated isolation of the organism from clinical samples. Serologic tests are of very little value because they are positive in many normal individuals. This is also the case with the skin test, which is positive for 90 percent of the population, but the test is sometimes used to evaluate the immunologic status of a patient (a negative test may indicate anergy). Latex agglutination tests may be used in diagnosis.

Control and treatment. Mucocutaneous infections should be treated topically, but are resistant to therapy. *C. albicans* is consistently susceptible to nystatin. Vaginitis is also treated with 1 percent gentian violet. Candidiasis will often persist as long as an individual is taking steroids or antimicrobials. It is sometimes necessary to discontinue such therapy in order to stop a systemic candidiasis. Flucytosine is used to treat systemic infections, though some *Candida* species are resistant, in which case amphotericin B is necessary. An infected heart valve usually must be removed.

Cryptococcosis

Biologic characteristics. *Cryptococcus neoformans* (formerly *Torula histolytica*) is always found in the nonfilamentous yeast form (it is not dimorphic) in tissues and culture. It does have, however, a sexual filamentous state, classified as a basidiomycete in the genus *Filobasidiella.* Though there are 12 species of cryptococci, all of which are encapsulated, *C. neoformans* is the only human pathogen. Its polysaccharide capsule distinguishes it from other fungi and on the basis of this polysaccharide, four strains (A through D) of *C. neoformans* have been identified. Type A infections are most common. The capsule causes the slimy mucoid appearance of cryptococcal colonies.

Epidemiology. Cryptococcosis is not limited to one geographic region, but is found throughout the world. *C. neoformans* has been occasionally found in milk, fruit juices, from skin and GI tracts of normal persons, but is most common in soil, especially if pigeon droppings are present. Though birds are resistant to cryptococcal infection because their higher body temperature is not conducive to growth of the organism, their droppings promote the growth of cryptococci by providing a rich alkaline medium. Infection is probably most often initiated by the respiratory route, though entry of the organism via the skin or intestinal tract may be possible.

The *Cryptococcus,* like *Candida,* is found on humans and causes disease only as an opportunist. Though the organism is quite common, only about 125 deaths per year due to cryptococcosis are reported in the United States. Individuals with malignancies of the RES and those taking steroids have an increased susceptibility to cryptococcosis. About 50 percent of patients with cryptococcal meningitis have some other underlying disease such as diabetes mellitus or lymphoma.

Pathogenesis. In contrast to other systemic mycotic diseases, cryptococcosis does not cause an intense inflammatory reaction or cellular response. Cryptococci have an affinity for the central nervous system. The cellular reaction develops slowly and the fungus, once it has established itself, proliferates rapidly so that in some cases visible masses of yeast cells grow in the nervous system, bone, or other body parts.

Host reponses. Despite the fact that the fungus enters by inhalation, respiratory symptoms are usually mild and the disease most commonly presents as chronic meningoencephalitis or acute menigitis. The individual's first symptoms are usually headache, vomiting, and diplopia without fever. The spinal fluid is nonpurulent, with increased protein and decreased glucose. About one-third of patients with cryptococcal meningitis also have cryptococci in their urine. Rarely, pneumonia, pyelonephritis, and dissemination to bone occur with or without other manifestations.

Diagnosis is made by direct observation or culture of cryptococci from the spinal fluid, sputum, or exudates, though relatively large volumes are often needed as the yeast cells may be infrequent. Most patients also have detectable cryptococcal antigens in their blood or spinal fluid that can be measured by latex fixation (latex particles coated with antibody to cryptococci are added to spinal fluid. Agglutination with visible clumping occurs if cryptococci are present in the CSF). Before treatment was available, most patients died within 3 months. Even with appropriate therapy, relapses are common. There is no skin test for cryptococcosis.

Control and treatment. There are no known preventive measures or vaccines for cryptococcosis. Amphotericin B, often in conjunction with flucytosine for about 6 weeks is required for treatment. Flucytosine alone, even though it penetrates the spinal fluid, is not recommended because resistant organisms emerge. For relapses it is sometimes necessary to administer amphotericin B intrathecally (directly into the spinal cord). This must be done with the strictest of precautions to prevent bacterial contamination and overdosage. Untoward effects of intrathecal injection include temporary paralysis of the bladder and legs and arachnoiditis.

Pneumocystis carinii

Pneumocystis carinii is a yeastlike agent that was previously categorized as a sporozoan parasite. It is ubiquitous in nature and causes lung infections in dogs, rats, and mice. Humans who are immunocompromised sometimes become infected by inhaling cysts from domestic animals. The organism can cause a life-threatening pneumonitis primarily in patients with acquired immune deficiency syndrome (AIDS; see Chapter 4) or other immulogic disorders, in those receiving immunosuppressive drug therapy, and in premature infants. Infection in most people, however, is asymptomatic. Though the organism can be cultured, diagnosis is usually made by biopsy of lung tissue or at autopsy. One tissue stage is quite characteristic: in pulmonary alveoli and bronchioles the organisms join to form a rosette shape. Treatment is with pyrimethamine or pentamidine.

Phycomycosis

Biologic characteristics. Phycomycosis (synonyms are zygomycosis and mucormycosis) is a disease caused by members of the order *Mucorales* in the class Phycomycetes, fungi that have nonseptate hyphae. Opportunistic pathogens are primarily from three genera: *Absidia, Rhizopus,* and *Mucor*. They are lower order terrestrial fungi producing sexual spores called zygospores and asexual sporangiospores. They do not have a yeast phase.

Epidemiology. Phycomycetes are abundant in soil and food such as bread (rhizopus is "bread mold"). Phycomycosis is rare, but can occur anywhere in the world. The rhinocerebral form occurs almost exclusively in persons with diabetes mellitus, especially in the presence of metabolic acidosis. Individuals with lymphomas and leukemia seem especially prone to pulmonary phycomycosis. The spores are usually inhaled, though phycomycetes can directly contaminate burns or open wounds.

Pathogenesis. After the organism enters the body via the respiratory tract it extends primarily into the nasal sinuses, the retroorbital tissues, and by means of thrombotic vessels, into the brain where infarcts result. The inflammatory response is minimal. Infection may spread to distant body sites when organisms are carried in the blood stream. Thrombosis and tissue infarction result.

Host responses. The most common form of phycomycosis is rhinocerebral, which presents with ptosis, loss of vision, periorbital swelling, facial palsy, and coma (Table 13-16). Lesions can, however, be present in the gastrointestinal tract, bone marrow, eyes, kidneys, or lungs. Diagnosis is confirmed by tissue biopsy, although the symptoms are so characteristic that diagnosis is not difficult. Untreated cases are almost always fatal; even with therapy mortality is about 50 percent. *Rhizopus* has also been known to contaminate elastic bandages, resulting in wound infections.

Control and treatment. There are no known preventive measures against phycomycosis except the careful management of diabetes mellitus to prevent ketoacidosis. Discomfort from rhinocerebral phycomycosis can be minimized with nasal irrigations of hydrogen peroxide or normal saline to decrease crusting. Amphotericin B is used for cerebral

TABLE 13-16. Physical Assessment in Rhinocerebral Mucormycosis

Area	*Technique*	*Positive Findings*
Nose	Inspection of nasal passages	Obstruction: black turbinates (easily mistaken for dried blood)
	Inspection of nasal discharge	Dark, blood-tinged discharge
	Palpation of perinasal area	Soft swelling progressing to pitting
	Inspection of skin	Discoloration
Eyes	Inspection of periorbital and orbital area	Discoloration, proptosis
	Palpation of periorbital area	Soft swelling progressing to pitting
	Test of cranial nerves II, III, IV, and VI	Ptosis, dilation and fixation of the pupil, loss of eye mobility, and no corneal reflex
Mouth	Inspection of oral cavity, especially the hard palate	Discolored palate (pale to black and necrotic)
Sinuses	Palpation of frontal and maxillary sinuses	Tenderness elicited or complaint of pain (especially headache)
	Illumination of sinuses with a penlight	Partial or no illumination of frontal or maxillary sinuses
Face	Test of cranial nerves V and VII	Facial weakness, numbness, or pain
Overall status	Assessment of level of consciousness	Progressive lethargy that continues when underlying condition, if any, is corrected

Reprinted with permission of the publisher from Triplett SJ. Rhinocerebral mucormycosis. Am J Nurs 1981;81:1163.

phycomycosis but debridement and surgical excision (enucleation of the eye) of infected tissue is often necessary. The surgery is quite disfiguring and the patient will need help adjusting to it.

MYCOTOXICOSIS

Some fungi, especially the basidiomycetes, contain species that contain toxins that, when ingested, cause damage to liver, kidneys, or bone marrow (Table 13-17). Others can also cause hallucinations and mental disturbances. These symptoms are commonly called "mushroom poisoning."

CLINICAL IMPLICATIONS

The collection of clinical specimens for fungal isolation is done somewhat differently than for bacterial isolation. Because fungi are frequent contaminants of laboratory specimens, careful collection will enhance the probability that a correct diagnosis can be made as soon as possible. It is often ad-

TABLE 13-17. Common Human Mycotoxins

Mycotoxin	Fungi producing	Common source	Toxic effects
Aflatoxin	*Aspergillus* sp.	Peanuts, other cereals and grains	Liver disease, including cirrhosis and cancer
Ochratoxins	*Aspergillus ochraceus*	Grains, fermented fish, other foods	Liver necrosis
Luteoskyrin and cyclochlorotine	*Penicillium islandicum*	Moldy rice	Liver disease, including cirrhosis and cancer
(Unnamed)	*Fusarium* and *Cladosporium*	Grains stored through the winter	Alimentary toxic aleukia (hemorrhage)

vantageous to collect specimens such as patches of "thrush" at the bedside since transport media should not be used when fungi are suspected. Collection techniques are summarized on Table 21-3c (Chapter 21). Often larger aliquots of clinical material are necessary to isolate fungi. This may be especially true with cerebrospinal fluid and sputum.

Patients who develop systemic fungal disease are often already very ill and may have been hospitalized and receiving medical therapy for long periods of time. The management of fungal disease is compounded by management of often complex physiologic problems. Needless to say, the mental attitude of such patients can be bleak. Psychological depression is intensified by the fact that antifungal therapy is prolonged, often lasting for months, and the side effects are dangerous and extremely uncomfortable. Thus, the clinical management problems of such patients can be immense.

Careful monitoring of toxic and side effects of amphotericin B is an important clinical function. Amphotericin B, unfortunately, can only be given intravenously, so the patient is usually required to remain hospitalized for the entire course of treatment, which may be up to 2 months. When initiating therapy, use the following guidelines:

1. Store the powder in the refrigerator protected from light.
2. Reconstitute with 10 ml of sterile water without preservatives or bacteriostatic agents. Use extreme care to maintain aseptic technique since the lack of preservative or bacteriostatic agents greatly increases the hazard of contamination.
3. Dilute to produce a concentration of no greater than 0.1 mg/ml.
4. Use the following dosage schedule:
   - Day 1: 1 mg test dose I.V. over 2 to 6 hours. Take patient's pulse and temperature every 30 minutes for 4 hours after the test dose has begun.
   - Day 2: 5 mg over 6 to 8 hours.
   - Day 3: Increase daily dose by 5 to 10 mg every other day to reach 0.5 to 1 mg/kg of body weight with a maximum of 1.5 mg/kg or about 50 mg.
5. Repeat the schedule in item 4 whenever therapy is interrupted for more than 7 days.
6. Do not add other medications to the

same I.V. line, as amphotericin is incompatible with most other drugs.
7. Do not use an in-line filter, as it removes the drug.
8. Mix the bottle occasionally while it is infusing to prevent precipitation.
9. Protect the infusion from sunlight or it will deteriorate, though it is not necessary to wrap the tubing.
10. Watch for adverse effects: fever, chills, headache, nausea, vomiting, thrombophlebitis at infusion site, nephrotoxicity, anemia, hypokalemia.
11. Corticosteroids (Solu-Cortef[R] or Solu-Medrol[R]) may be added to the infusion to minimize febrile reactions.
12. Intrathecal amphotericin B may be necessary to treat brain abscesses or meningitis since the drug given intravenously penetrates the spinal fluid poorly. *Aseptic technique is absolutely vital.* The usual intrathecal dose is 0.5 mg 2 to 3 times per week for 2 to 4 weeks.

The clinician needs to educate women who are being treated for candidal vaginitis. Though adverse reactions to such drugs as candicidin are rare, vaginal irritation can occur. The patient also needs to know that candicidin may cause damage to a diaphragm, so alternate contraceptive measures should be used during treatment for vaginal candidiasis. Candidal vaginitis, like the dermatophyte infections, can be very persistent and discouraging to treat because recurrences are common. To prevent recurrences, the woman can be instructed to clean the vaginal applicator used for drug administration between uses and to refrain from sexual intercourse or have her sex partner wear a condom. Because males can also become infected, the cycle of infection continues unless such precautions are taken.

SUGGESTED READINGS

Bayer AS. Fungal pneumonias: pulmonary coccidioidal syndromes (Part 2). Chest 1981;79:686-691.

Blair T, Raymond L. *Histoplasma capsulatum* endocarditis. Chest 1981;79:620.

Cohen MS, et al. Fungal infection in chronic granulomatous disease. Am J Med 1981;71:59-66.

Comer JB. Amphotericin B: ten common questions. Am J Nurs 1981;81:1166-1167.

Cope JR. Blastomycosis: diagnostic difficulties. South Med J 1981;74:284-287.

Deacon JW. Introduction to Modern Mycology. New York: John Wiley and Sons, 1980.

Delacrétaz J, Grigoriu D, Ducel G. Color Atlas of Medical Mycology. Chicago: Year Book Medical Publishers, 1976.

DesPrez RM, et al. Interaction of *Histoplasma capsulatum* with human platelets. J Infec Dis 1980;142:32-39.

DiSalvo AF, Johnson WM. Histoplasmosis in South Carolina: support for the microfocus concept. Am J Epidemiol 1979;109:480-492.

Gold JWM, et al. Diagnosis of invasive aspergillosis by passive hemagglutination assay of antibody. J Infec Dis 1980;142:87-94.

Goodwin RA, Loyd JE, DesPrez RM. Histoplasmosis in normal hosts. Medicine 1981;60:231-266.

Hawkins SS, Gregory DW, Alford RH. Progressive disseminated histoplasmosis: favorable response to ketoconazole. Ann Intern Med 1981;95:446-449.

Heggers JP. Medically important yeasts. J Am Med Technol 1980;42:73-78.

Henderson DK, Edwards JE, Montgomerie JZ. Hematogenous *Candida* endophthalmitis in patients receiving parenteral hyperalimentation fluids. J Infec Dis 1981;143:655-661.

Herman VS, Hurwitz S, Segal F. Multiple pulmonary abscesses resulting from pulmonary mucormycosis. Heart Lung 1980;9:514-517.

Holmberg K, Berdischewsky M, Young LS. Serologic immunodiagnosis of invasive aspergillosis. J Infec Dis 1980;141:656-664.

Hurwitz S, Solomon A. Aspergillosis in the lung. Heart Lung 1979;8:1127-1130.

Kauffman CA, Shea MJ, Frame PT. Invasive fungal infections in patients with chronic mucocutaneous candidiasis. Arch Intern Med 1981;141:1076-1079.

Lottenberg R, et al. Pulmonary histoplasmosis associated with exploration of a bat cave. Am J Epidemiol 1979;110:156-161.

Meunier-Carpentier F, Kiehn TE, Armstrong D. Fungemia in the immunocompromised host. Am J Med 1981;71:363-370.

Muchmore HC, et al. *Cryptococcus neoformans* serotype groups encountered in Oklahoma. Am J Epidemiol 1980;112:32-38.

Nobre G, Sobral T, Ferreira AF. In vitro susceptibility to 5-fluorocytosine and nystatin of common clinical yeast isolates. Mycopathol 1981;73:39-41.

Rees PL, Dixon DM. Opportunistic mycoses. Am J Nurs 1981;81:1160-1165.

Sanford GR, et al. The value of fungal surveillance cultures as predictors of systemic fungal infections. J Infec Dis 1980;142:503-509.

Sobel JD, et al. Adherence of *Candida albicans* to human vaginal and buccal epithelial cells. J Infect Dis 1981;143:76-83.

Triplett SJ. Rhinocerebral mucormycosis. Am J Nurs 1981;81:1162-1163.

Zeluff GW, et al. Chronic meningitis—benign and malignant forms. Heart Lung 1980;9:534-540.

14 ENTERIC VIRUSES

KEY POINTS

- Enteric viruses traverse the alimentary tract, are replicated therein, and are excreted in the feces.
- Enteroviruses cause a wide variety of clinical manifestations and cannot be diagnosed on the basis of symptomatology. For example, a single virus type may produce gastrointestinal symptoms in one person, rash in another, and aseptic meningitis in still another.
- The most important enteric viruses that cause disease are the polioviruses (causing polio), coxsackieviruses, echoviruses and enteroviruses (causing a variety of clinical syndromes), the reoviruses (causing infantile diarrhea), and the Norwalk-like agents (causing acute infectious gastroenteritis).

Enteric viruses enter and traverse the alimentary tract of humans. They then replicate there, and are excreted with the feces. Frequently the only indication of infection due to these viruses is a minor, self-limiting gastroenteritis or upper respiratory disease. This is particularly true of infection caused by members of the *Adenoviridae* and *Reoviridae* families. Although adenoviruses are commonly isolated from stools and sewage samples, they are most usually associated with upper respiratory disease and keratoconjunctivitus; therefore they will not be discussed further in this chapter. Only enteroviruses (in the family Picornaviridae), rotavirus (in the family Reoviridae) and the small Norwalk and Norwalk-like agents (probably members of the family Parvoviridae) will be addressed (Table 14-1).

ENTEROVIRUSES

The family Picornaviridae is divided into two major subdivisions, one composed of two genera of viruses that infect humans, the enteroviruses and the rhinoviruses, and a second subdivision composed of four groups of viruses that infect lower animals. For our purposes, only the enteroviruses of group I

TABLE 14-1. Viruses Causing Enteric Disease in Humans

Family	*Group*	*Serotypes*
Unclassified	Norwalk-like	Norwalk/Mc agents
		Ditchling/W agents
		Hawaii
		Cockles agent*
Picornaviridae	Polioviruses	1, 2, and 3
	Coxsackieviruses	
	Group A	1–22, and 24
	Group B	1–6
	Echoviruses	1–9, 11–27, 29–34
	Enteroviruses	68–71
Reoviridae	Rotavirus	1 and 2

*Antigenic studies inconclusive. At present there are either three or four serotypes.

(listed in Table 14-1) are of interest. Members of the picornaviruses, with the major exception of those that cause upper respiratory disease, are sometimes collectively referred to as enteroviruses. This term will be used throughout this section.

Biologic characteristics. The enteroviruses are composed of a single-strand RNA core encased in a protein coat (capsid). The capsid of the few enteroviruses studied appear to have a cubic symmetry (Fig. 14-1) of the icosahedral type with a 20 to 30 nm diameter (Mayor 1964). Proteins that make up the capsid provide the antigenic specificity of the virus, enabling it to adsorb to cells, and initiating the first step in the infectious process.

Although enteroviruses are sensitive to heat (50°C), thermal stability may be enhanced by the addition of bivalent cations such as magnesium. Poliovirus has been shown to be heat stable for 1 hour at 50°C in the presence of 1 M $MgC1_2$. Autoclaving does inactivate enteroviruses. However, at lower temperatures the virus is more stable. Enteroviruses remain viable for years at −70°C, for weeks at 4°C, and for days at 22°C.

Enteroviruses are unaffected by antibiotics and most disinfectants such as 70 percent alcohol, 5 percent Lysol, and 1 percent quanternary ammonia. They are sensitive to free chlorine at a level of 0.3 parts

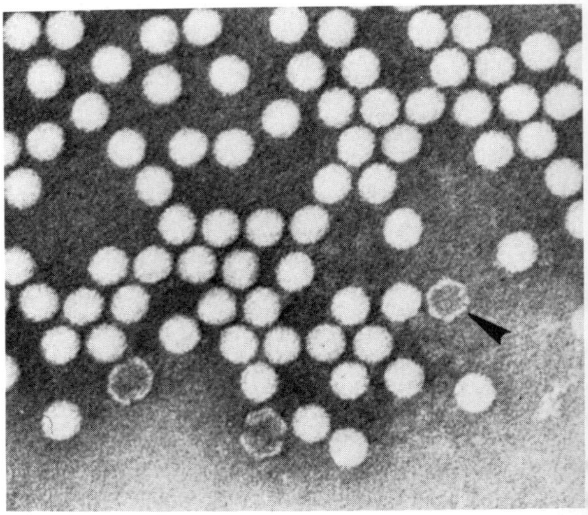

FIGURE 14-1. Poliovirus type 1. (Courtesy of Centers for Disease Control, Atlanta, GA 30333.)

per million (ppm) to 0.5 ppm, 0.3 percent formaldehyde, and 0.1 N hydrochloric acid, but may be protected from chemical inactivation by the presence of organic material. Detergents such as ether and deoxycholate, are ineffective because the enterovirus contains no essential lipids such as are found in the enveloped viruses (influenza, arboviruses, etc.). The enteroviruses are also stable over a wide range of pH (pH 2 to pH 11) for varying periods of time. These viruses must traverse the alimentary tract and in so doing are exposed to the acid stomach chyme, which reaches pH 2. Viability of the enteroviruses under these circumstances is necessary if replication within the intestinal mucosa is to proceed.

Differentiation of the various enterovirus types is based on the antigenic properties of the virus as determined by neutralization tests. If an isolate is completely neutralized by an antiserum to the prototype enterovirus (original isolate), the unknown is thus identified. However, if it is not neutralized by any of the prototype antisera, an antiserum is prepared against the new isolate. This is tested against the prototype strains. In some instances the prototype is completely neutralized by the antiserum. In such cases, the new isolate is referred to as a prime strain of the prototype. Many members of the echovirus group have recognized prime strains as does coxsackie A type 24.

Most of the prototype enteroviruses were isolated from human specimens. Poliovirus prototypes 2 and 3 were obtained from the spinal cords of individuals with fatal paralytic polio and prototype 1 was isolated from the stool of a patient with paralytic polio. All of the echovirus prototype strains were isolated from stools, 14 of which were obtained from healthy individuals. However, since those original isolations, all of the strains have been shown to be associated with disease. The other prototype strains were isolated from cases of aseptic meningitis (12 cases), diarrhea (7 cases), and fever of unknown origin (1 case).

All of the coxsackie group B prototypes were isolated from the stools of individuals with clinical disease with the exception of prototype 6, which was isolated from the stool of a healthy individual. Among the coxsackie A group, 7 of the 23 prototypes were obtained from specimens from healthy individuals. Two were isolated from environmental samples, coxsackie A type 4 from sewage, and coxsackie A type 12 from flies.

Since 1969, newly identified enteroviruses have been designated by a type number rather than being grouped as echoviruses or coxsackieviruses. To date, four prototypes have been recognized. Type 68 was isolated from a patient with lower respiratory illness, type 69 from a healthy individual, and type 70 from a patient with acute hemorrhagic conjunctivitis. Only type 71 was isolated from the stool of a patient with aseptic meningitis. Subsequently, strains of type 71 have been isolated from brain tissue from fatal encephalitis cases.

Epidemiology. The first recognized enterovirus disease was poliomyelitis. In 1840, the German physician Heine differentiated poliomyelitis from other diseases of the central nervous system (Holmes et al. 1975). Cases were relatively infrequent and for many years the infectious nature of the disease went unrecognized. It was not until a large outbreak occurred in Sweden (1031 cases) that the infectious nature of poliomyelitis was epidemiologically described by Ivar Wickman in 1913 (Howe and Wilson 1959). A characteristic of the disease in those early days was its propensity for producing paralysis in infants, thus it became known as infantile paralysis. It was not until 1909 that the causative agent of the disease was isolated by Landsteiner and Popper, who succeeded in providing evidence that infantile paralysis was indeed an infectious process. A bacteria-free homogenate of the spinal cord from a case of paralytic polio was inoculated intracerebrally into monkeys. Within 2 weeks, typical paralytic poliomyelitis developed. Further progress in research was very slow and exceedingly expensive because of the need to use baboons or monkeys for all experimental work.

In developed countries there was a gradual shift in the disease away from infections in infants to more and more infections in older children and adults. For the first time large epidemics became rather commonplace. This led to a change in name from infantile paralysis to poliomyelitis. The shift from infants to the older age group has been attributed to improvements in environmental sanitation (i.e., closed sewage systems replacing open sewage canals and septic tanks). Under these improved conditions, fewer infants were exposed to the poliovirus during their first 6 months of life—when they were protected by maternal antibodies. Exposure during this period would have led to the development of lifelong immunity. However, when exposure occurred later in life, many infections resulted in overt disease. As the pool of susceptible individuals increased, so did the epidemics. In the 1940s, money from the March of Dimes sponsored by the National Foundation for Infantile Paralysis provided sufficient monkeys for research. This permitted more stool specimens and environmental samples to be assayed for virus. Data from these studies confirmed the wide distribution of virus in the environment and the large number of subclinical infections that occurred.

During the same time period, Enders and his associates (1949) succeeded in growing poliovirus in cell cultures derived from human embryonic organs and monkey kidney cells. Data were accumulated at a rapid pace. Not only could more human specimens be tested for the presence of virus, but even more important, the cell cultures facilitated the eventual development of the vaccines. The use of cell cultures for virus isolation during the decade of the 1940s also led to the isolation of many previously unrecognized viruses. In 1948, Dalidorf and Sickles isolated the first coxsackievirus, named for Coxsackie, New York where the specimen originated.

During the 1950s, the use of cell culture became widespread and resulted in the isolation of large numbers of viruses from throat cultures and stools. Some of the isolates were identified as polio, some as coxsackieviruses, but many could not be assigned to any known groups. These unknowns were isolated frequently from apparently healthy individuals; the isolates failed to produce disease in newborn mice but did produce cytopathogenic effects in cell cultures. These agents were referred to as "orphan" viruses because they were not associated with a disease entity. This name was soon replaced with the coined name, echoviruses (enteric cytopathogenic human orphan viruses). Today they are no longer orphans because all of the isolates have been associated with disease.

In the 1960s the International Committee on Taxonomy of Viruses proposed that viruses be classified according to their biophysical and biochemical properties. Thus, today, the enteroviruses are classified in the family Picornaviridae, genus enterovirus. Again the name Picornaviridae was coined from the important characteristics of the viruses; that is, small (pico), ribunucleic acid (RNA) containing viruses (*picorna*viruses). The number of viruses comprising this group will probably increase in the future. In many laboratories today, viruses that cannot be typed are being isolated from stools and sewage samples. Only time will tell how many of these may represent new enterovirus prototype strains.

Enteroviruses are extremely infectious and intrafamilial transmission is common. Because the initial infection occurs in the nasopharynx, respiratory droplet spread occurs. Later in the disease, the fecal-oral route is dominant. Transmission may occur through flies and cockroaches that become contaminated by exposure to raw sewage and mechanically transfer virus to food or water. In recent years, sewage contaminated groundwater has been shown to be a source of enterovirus infection (Hejkal et al. 1982; Wellings, Mountain, and Lewis 1976).

Enteroviruses have been isolated from human specimens throughout the world. In temperate zones, infections usually occur during the summer months but in tropical and subtropical areas, they occur year-round. Children are more frequently infected than adults. Among lower socioeconomic groups over 90 percent of the children under 5 years of age may be immune to the endemic enteroviruses.

Pathogenesis. All enteroviruses have a similar pattern of pathogenesis. They enter by the oral route, and multiply in the oropharynx and alimentary tract for several weeks. Viremia causes the spread of the virus and there is multiplication in the reticuloendothelial system. The virus invades target organs such as the central nervous system, skin, or myocardium, and is excreted in the stool and from the pharynx for 1 to 2 weeks after infection. Poliovirus spreads along axons of peripheral nerves.

Host responses. Enterovirus infections may result in a spectrum of disease from subclinical to fatal. None of the enteroviruses produce a characteristic syndrome that would permit a specific diagnosis (with the possible exception of acute bulbar poliomyelitis). The fact that the poliovirus results in a wide spectrum of symptoms masked the infectious nature of the disease during early investigations. Even today, because of the protean expression of the enterovirus group, it is difficult to recognize an outbreak of a given virus type unless many of the individuals develop aseptic meningitis or encephalitis requiring hospitalization, and the specific virus strain is isolated and identified. For instance, a single virus type may produce a gastrointestinal upset in one person, an exanthem in another, and perhaps an aseptic meningitis in another.

Table 14-2 demonstrates the various symptoms produced by a single virus type by listing the symptoms produced by echovirus type 11 infections in 21 confirmed cases. The tentative diagnosis in these 21 cases included aseptic meningitis (15 cases), viral exanthem (3 cases), encephalitis (1 case), TORCH (toxoplasmosis, rubella, cytomegalovirus, herpes) syndrome (1 case), and enterovirus infection (1 case). Of particular interest was the newborn who had convulsions 2 days after birth and died at 9 days of age. Echovirus type 11 was isolated from tissues obtained at autopsy. Whether this was an intrauterine or nosocomial infection was not determined.

Enteroviruses in general are isolated most frequently from the stools of individuals with aseptic meningitis. Yet, the individual hospitalized for enterovirus infection may represent a very small percentage of the actual cases in the community. For example, in an outbreak caused by echovirus type 9 in Milwaukee in 1957, epidemiological studies indicated that for every person hospitalized in the epidemic, there were from 175 to 300 persons infected in the community. Thus, instead of only 149 cases of echovirus type 9 as represented by the hospitalized cases,

TABLE 14-2. Symptomatology Associated With Echovirus Type 11 Infections in Confirmed Cases

Month of onset	2	2	3	3	4	4	5	5	5	5	5	6	6	7	7	8	8	9	10	12	12
Age*	6 wk	5	5	26	10	NB†	6	3 mo	2 wk	1	5 mo	6	18 mo	9	9	14	1	3 mo	4	3 wk	1
Sex	M	M	M	M	F	F	M	M	M	F	F	M	M	F	M	F	M	M	M	F	M
Symptoms / Race	W	W	W	W	W	W	B	B	B	W	B	B	B	W	W	W	W	B	W	W	?
Fever	X	X	X	X	X	X	X	X	X	X	X	X	X	X	X	X	X	X	X	X	X
Headache				X	X							X		X		X			X		
Stiff neck				X					X		X			X		X			X		X
Vomiting	X				X							X	X	X			X		X		X
Diarrhea	X	X						X									X				
Lethargy	X										X	X		X			X				X
Irritability								X					X					X			
Abnormal reflexes																					
+ Babinski														X							
+ Kernig												X		X							
+ Brudzinski									X			X		X					X		
Mental confusion														X		X					
Convulsions					X																
Photophobia													X								
Cough										X	X										
Maculopapular rash	X	X								X					X					X	
Otitis																	X				
Rhinitis																	X				
Swelling of joints										X											
Bulging fontanelles			X																		

*Age given in years except where noted.
†Fatal infection in newborn.

approximately 45,000 residents were involved in the epidemic (Melnick and Sabin 1959).

Coxsackie B types 1 through 5, coxsackie A type 4, and echovirus types 1, 6, 9, and 19, have all been associated with pericarditis and myocarditis. In 1981, coxsackie A type 8 was isolated from autopsy specimens of the heart muscle and large intestine of a male 2 years of age. Myocarditis was evident on stained frozen section of the myocardium. Exanthems have been associated with coxsackie A types, 4, 5, 6, 9, 10, 16; coxsackie B type 5 and echovirus types 1 through 7, 9, 11, 12, 14, 16, 18, 19, 20. These cases are often tentatively diagnosed as atypical measles. In recent years it has become evident that the "failure to thrive" syndrome in infants up to 6 months of age is frequently related to an acquired enterovirus, rather than an intrauterine infection.

From the foregoing, it is evident that distinctive clinical syndromes are difficult to associate with any particular enterovirus. Initially, epidemic myalgia was considered to be caused by the coxsackie B group but subsequently, echoviruses have also been implicated. Therefore, distinctive clinical syndromes cannot be ascribed even for a particular enterovirus group.

The incubation period of enteroviruses varies from 2 to 35 days and the period of infectivity may be protracted. From the time the initial infection occurs in the nasopharynx until virus shedding in the feces ceases, live virus is being excreted. Infective oral secretions are predominant during the incubation period but decline rapidly following onset of disease. However, infectious virus is excreted in the stool for 2 to 4 weeks following onset of disease and may persist much longer.

Control and treatment. Infection with a particular enterovirus type leads to lifelong immunity to that type. Thus, susceptibility decreases with age in all populations. But, as with polioviruses, as overall environmental sanitation improves, the adult pool of susceptibles will increase.

There is at present no specific treatment for enterovirus. Palliative treatment should include forcing fluids, reducing the body temperature with cooling sponge baths, and maintaining good nutrition. The clinician's major role is to prevent virus transmission with scrupulous handwashing and careful disposal of excreta. This cannot be over emphasized because of the number of children admitted to nurseries and pediatric units who may be excreting poliovirus following live poliovirus immunization, or other enteroviruses, regardless of the admitting diagnosis. The incidence of poliomyelitis has dramatically decreased since the introduction of efficacious vaccine (Fig. 14-2).

ROTAVIRUS

Through the years, infantile diarrhea (ID) has posed a major public health problem in developed countries. Investigations of epidemic occurrences of ID failed to yield an etiological agent until 1973, when Bishop, et al, in Australia first visualized a 70-nm sized virus in the duodenal mucosa obtained from ID patients. Because of the spoked-wheel appearance of the agent, it was designated rotavirus from the Latin word *rota,* meaning wheel.

When the rotaviruses were first recognized they were referred to as reoviruslike, orbiviruslike, or as the infantile gastroenteritis virus. However, as more biochemical data were accumulated the International Committee on Nomenclature of Viruses established a new group, rotavirus, in the family Reoviridae. The rotaviruses have

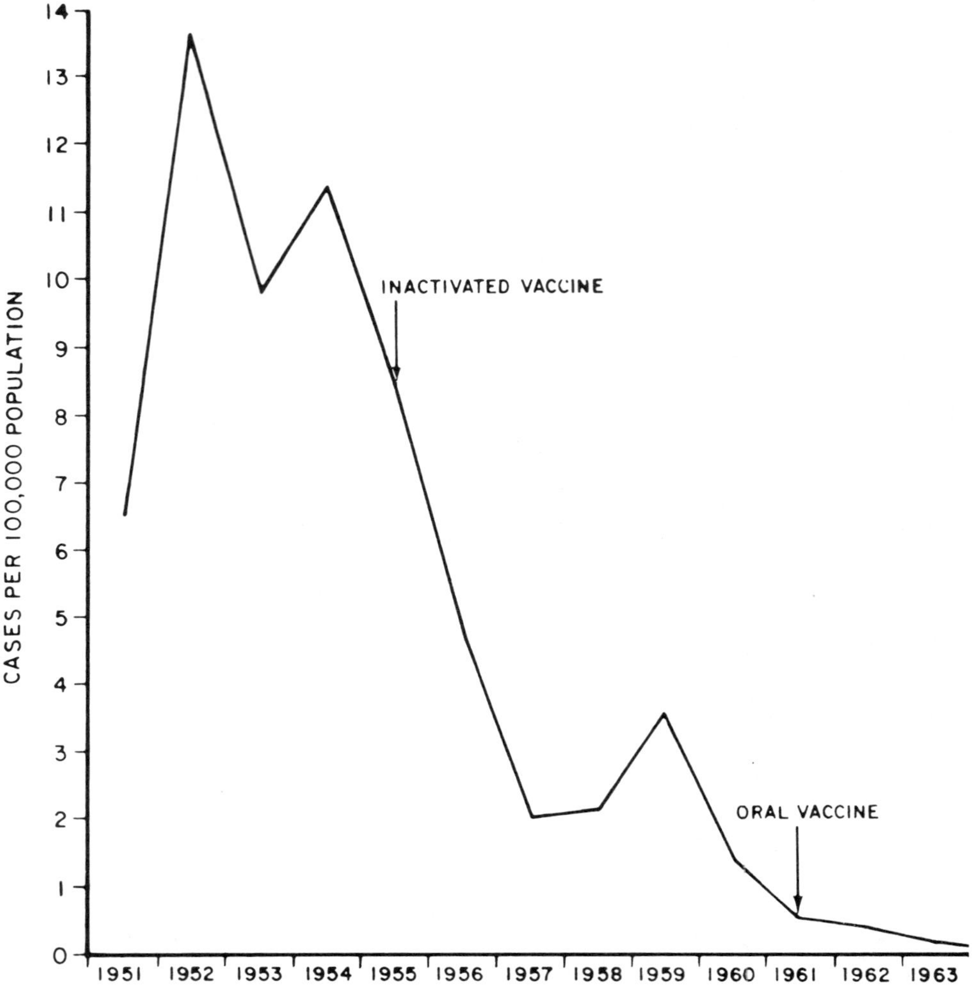

FIGURE 14-2. Poliomyelitis (paralytic). Reported case rates by year, United States, 1951 to 1963. (Reprinted from Centers for Disease Control. Annual Summary 1980. Morb Mort Week Rep 1981;29:64.)

been shown to be antigenically distinct from other viruses.

Biologic characteristics. The 70-nm sized rotavirus is composed of an inner and outer capsid connected by short "spokes" such as those seen in the old wagon wheels. The outer capsid is composed mainly of glycoproteins, at least one of which appears to be critical to the adsorption process (Fig. 14-3). The genome is composed of 11 segments of double-stranded RNA, which are quite distinctive in each of the rotavirus strains. At present, there are two antigenically distinct types of rotaviruses, but more may be identified in the future.

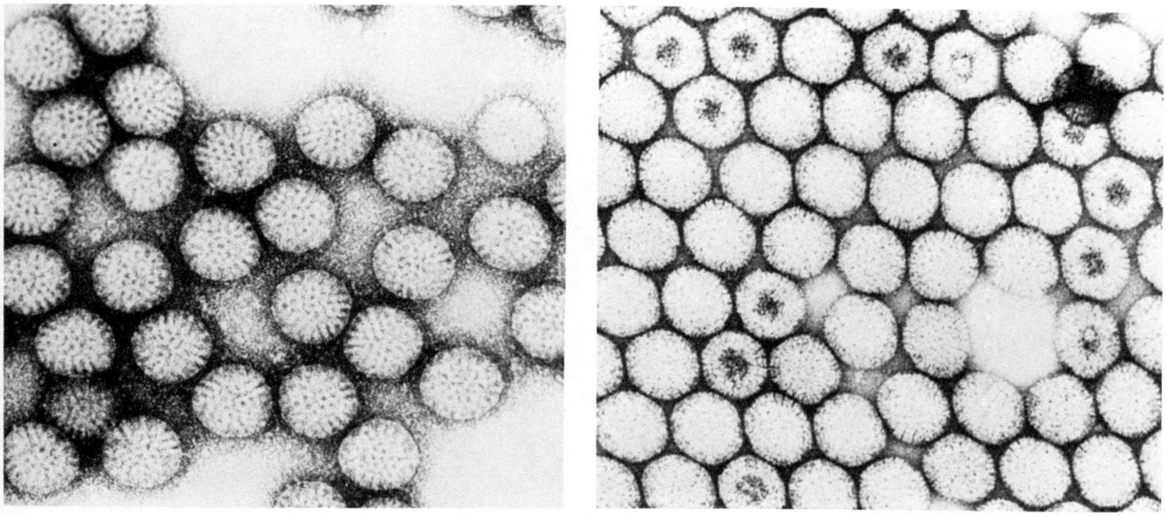

a b

FIGURE 14-3. *(a)* Rotavirus double-shelled particles. *(b)* Reovirus type 3. Prototype of family Reoviridae. (Courtesy of Centers for Disease Control, Atlanta, GA 30333.)

The rotaviruses are very stable structurally. After storage for 9 years at −20°C and over 30 years in the lyophilized state, samples of rotavirus were still recognizable by electron micrograph (EM). Whether or not these stored viruses are actually infectious has not been ascertained. Boiling and autoclaving will destroy the virus, probably through denaturation of the proteins present.

A confirmed diagnosis is possible by either visualization of the virus by EM or more recently, immune EM (IEM) in which immune serum is mixed with the preparation before EM examination. Other laboratory tests, such as fluorescent antibody assay, radioimmune assay, counterimmunoelectrophoresis and enzyme linked immunosorbent-assay (ELISA) have been used. Of these the latter is the most promising. (See Chapter 21 for a discussion of these techniques.)

Epidemiology. Symptomatic disease caused by rotavirus is usually confined to young children from 6 to 24 months of age with a peak in the 9- to 12-month age group. Mild gastrointestinal infections may occur rarely in older contacts and there is evidence that subclinical cases occur, particularly in adults exposed to infected infants (Kim et al. 1977).

In temperate climates, rotavirus infections usually occur during the cooler months of the year, unlike the enterovirus infections that occur during the summer. However, in semitropical and tropical countries cases occur year-round with possible predominance during the rainy seasons (Heiber et al. 1978).

Prevalence studies show the ubiquitous nature of the rotavirus. Serum antibodies to the agent are acquired early. The vast majority of adults and children past the age of 1 year have demonstrable antibodies to rotavirus. Over 50 percent of the hospital admissions for infantile diarrheas in temperate climates are due to rotavirus infections. Nosocomial spread of rotavirus among hospitalized pediatric patients and medical personnel has been well documented. An asymptomatic adult may be the source of infection for infants who lack contact with other children. Symptomatic outbreaks of rotavirus infection have occurred in adults with no known exposure to children. However, the very acute, life-threatening infection has been seen only in the infant.

Pathogenesis. Rotavirus causes damage to the absorbing mucoscal cells, but does not cause hypermotility of the GI tract. Malabsorption may lead to electrolyte imbalance and death.

Host responses. Infantile diarrhea is an acute disease characterized first by vomiting, followed by diarrhea. High temperature (39 to 40°C) and dehydration with associated acidosis are common. Both these factors determine the severity and the length of the illness, which may range from 2 to 10 days (an average of 5 to 7 days).

The incubation period is relatively short, ranging from 1 to 3 days followed by an abrupt onset of disease. The period of infectivity has not been firmly established, but, because of the high multiplicity of virus excreted in the feces, infectivity would be the greatest during the first few days of the disease. As the infection wanes, so does infectivity.

Control and treatment. Parenteral fluid replacement therapy is the only treatment in severe cases. Forcing the intake of oral fluids is contraindicated because of the malabsorption syndrome. However, in infants with only a 5 to 10 percent dehydration, oral rehydration with a glucose-electrolyte formula is an alternative treatment. Cooling sponge baths to reduce the temperature may be indicated. Prevention of virus transmission is critical because of the high level of virus present in the feces. Extreme caution and scrupulous handwashing is required.

NORWALK-LIKE AGENTS

Over the years, numerous enteroviruses were isolated from sporadic cases of diarrhea, but not during epidemics of acute infectious gastroenteritis (AIG). A viral agent was always suspected after bacterial and toxins were excluded. The first breakthrough occurred in 1968 when an epidemic of AIG was investigated in Norwalk, Ohio (Blacklow et al. 1972). The secondary attack rate was approximately 32 percent among family contacts, indicating the exceedingly infectious nature of the disease. To fulfill Koch's postulate that the etiologic agent was present in stools, human volunteers ingested bacteria-free and toxin-free filtrates of stools from symptomatic individuals and developed typical disease. However, it was not until 1973 that Kapikian et al actually visualized the 27-nm sized Norwalk agent by IEM (Fig. 14-4).

Since that time, several other Norwalk-like agents have been identified and shown to be antigenically related to the prototype Norwalk strain. The additional strains include the Hawaii agent and the MC agent, both isolated in two family outbreaks of AIG. A third serotype (Ditchling agent) was found in an AIG outbreak in a primary school in Ditchling, England. The fourth possible serotype (W agent) was identified in the stool from a boy who developed AIG during an outbreak in a boy's school and a similar strain (cockles agent) was isolated from stools of individuals who developed AIG after eating cockles.

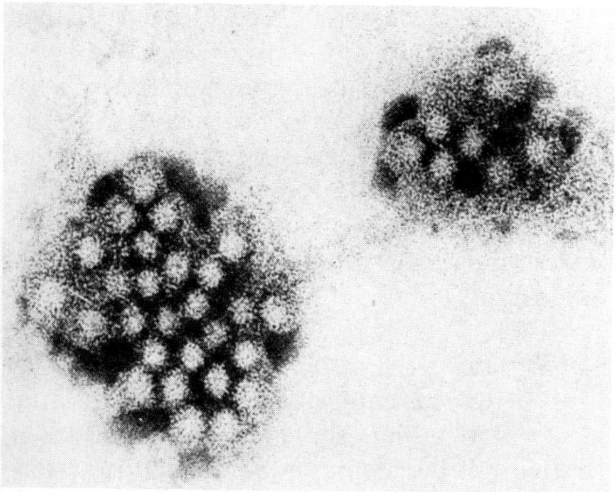

FIGURE 14-4. Immune electron micrograph, Norwalk-like agent. (Courtesy of Centers for Disease Control, Atlanta, GA 30333.)

Biologic characteristics. The virion is not perfectly round, measuring 25 nm to 27 nm in its shortest diameter and 32 nm in its longest. Neither the substructure nor the nucleic acid type has been determined. The Norwalk agent appears to be quite stable. It can withstand a pH of 2.7 for 3 hours at room temperature, 20 percent ether for 24 hours, 4°C for 18 hours and heating at 60°C for 30 minutes without loss of infectivity. Because of the inability of investigators to grow the virus in cell cultures, progress in determining the physiochemical characteristics of the Norwalk agent has been slow. At the present time, the Norwalk and Norwalk-like agents remain unclassified. They are morphologically similar to the picornaviruses or parvoviruses and hepatitis A virus. The nucleic acid content and number of proteins is unknown, therefore classification is not possible at present.

As discussed, antigenically related strains of the Norwalk agent have been studied. The Norwalk and Hawaii agents were shown by IEM and cross-challenge studies to be distinct, whereas, the MC and Norwalk agents show a relationship. IEM showed the Ditchling and W agents to be related but distinct from both Hawaii and Norwalk agents. The cockles agent studies were inconclusive. Thus, there appears to be at least three or four serotypes in the Norwalk-like group.

Visualization of the agent by IEM in stool specimens from infected individuals is diagnostic. Recently, radioimmune and the immune adherence hemagglutination assays have been adapted for virus identification and antibody titration of the Norwalk agent. Unfortunately, reagents are very limited because they must be derived from experimentally infected volunteers. Until these viruses can be cultured and reagents become more available, few laboratories will have the capability of identifying the virus or determining the antibody status.

Epidemiology. The AIG syndrome has been variously described as winter vomiting disease, epidemic collapse, viral diarrhea, epidemic diarrhea and vomiting disease, and acute infectious nonbacterial gastroenteritis. These represent descriptive terms applied to sporadic outbreaks of gastroenteritis in older children and adults. Infants and young children have not been involved.

As with other viruses, chlorination may not inactivate the Norwalk agent, particularly if solids are present. Shellfish that are eaten raw or partially cooked have given rise to AIG outbreaks. The fecal-oral route and fecal contamination of food and water are major routes of transmission. Droplet spread has not been documented. Maximal shedding occurs during the first 72 hours after onset of symptoms and occurs only infrequently thereafter.

Characteristically, outbreaks occur where large groups of people are together. Significantly, most of the outbreaks have occurred due to a breakdown in environmental sanitation (i.e., failure of equipment for disinfecting water and waste water or failure to follow accepted sanitary practices in food preparation). Outbreaks in resorts or summer camps are usually a result of sewage contamination of swimming waters or potable water. However, person-to-person spread is also significant as evidenced by the high secondary attack rate in families. Person-to-person spread may result from contact with vomitus rather than oral secretions, because virus has been noted in vomitus. There does not appear to be a sex or seasonal distribution of cases.

Studies of antibody prevalence indicate that Norwalk agents have a worldwide distribution. In developed countries serum antibodies usually develop during late adolescence and early adulthood. In tropical developing countries serum antibodies are present earlier in life. Epidemiologic studies are hampered by the lack of a simple, reliable laboratory test. A second deterrent is the lack of correlation between serum antibody titer and clinical disease. In epidemiologic studies, the pool of susceptible individuals is usually determined by establishing the number of people at risk based on the absence of serum antibodies. However, in human volunteers infected with the Norwalk agent, and challenged 20 to 42 months later, no protection was noted (i.e., those who became ill on first exposure also succumbed on challenge, even though serum antibodies were present). Volunteers who resisted the first challenge, resisted the second as well—yet failed to show serum antibodies. This unique situation precludes the defining of a pool of susceptible individuals based on serology. Two possible explanations have been proffered.

1. There may exist a genetic determinant.
2. Multiple exposures are required before an illness occurs or a serologic response is generated.

The latter explanation is compatible with the relative absence of antibody in young children, and with the gradual development of antibodies during early adulthood. This unique immune response is most unusual—additional research is necessary before it is explained.

Pathogenesis. Pathogenesis is apparently limited to the cells of the gastrointestinal mucosa in which the Norwalk agent(s) is replicated. Lacking a generalized infection, the immune response would be very limited, which may help to explain the need of numerous infections to generate a significant immune response.

Host responses. Symptoms resulting from the Norwalk-like agents are all similar but vary in individual cases. Some people develop only diarrhea, others experience only vomiting, while others have both. Of 604 cases studied, 85 percent had nausea, 84 percent vomiting, 62 percent abdominal cramps, 57 percent lethargy, 44 percent diarrhea, 32 percent fever, and 5 percent chills. Fever seldom exceeds 37.4°C. Anorexia, headache, and myalgias have also been noted. It is usually a self-limiting disease with symptoms subsiding in 24 to 48 hours. Extensive vomiting only occasionally leads to dehydration severe enough to require parenteral fluid replacement. The incubation period is roughly 48 hours. In volunteer studies, it was found to vary between 10 to 51 hours.

Control and treatment. Because of the self-limiting nature of the disease, little treatment is required. The use of bismuth subsalicylate has been shown to ameliorate the abdominal cramping but not affect the rate of virus excretion (Steinhoff et al. 1980). In the rare cases of extensive dehydration, parenteral fluid replacement therapy may be indicated.

CLINICAL IMPLICATIONS

Nurses or other clinicians are often responsible for obtaining patient specimens for viral culture. Serology may be used to demonstrate a 4-fold or greater rise in antibody titers between the acute and convalescent sera. Unfortunately, hospitalized patients are usually discharged before the convalescent sera must be obtained; this frequently precludes obtaining the necessary specimen. Some hospitals have established a recall system in which patients (or their parents) are notified when the specimen is due and patients are requested to have the blood drawn at the hospital or a health department laboratory. This has proven to be effective.

In the hospital setting, requests for virus isolation are most frequently submitted in cases of aseptic meningitis. This is probably due to the fact that once bacteria and toxins are ruled out as etiologic agents, a virus is suspected. Only then are specimens submitted for virus isolation. Unfortunately, this delay greatly decreases the probability of demonstrating a virus because of the decline in virus excretion over time.

Enteroviruses are most frequently sought as etiologic agents of enteric diseases in symptomatic patients, but because of the protean expression of enterovirus infections, laboratory data in conjunction with clinical and bacteriologic findings must be relied upon for establishing a specific diagnosis. To establish a diagnosis, enteroviruses must be actually isolated from feces, body fluids (CSF, blood, etc.), or tissues. If a virus is isolated, a 4-fold or greater rise in antibody titers against the isolate must be shown before the virus is considered the etiologic agent of disease. However, in fatal cases, the isolation of an enterovirus from autopsy tissues accompanied by confirming anatomic lesions is sufficient.

The acute serum for confirming serology should be collected during the acute phase of the disease. A convalescent serum should be collected 2 weeks later. These may be transported to the laboratory at room temperature, but to decrease possible bacterial growth, specimens should be stored at 4°C or −20°C until laboratory processing begins. Throat swabs should be placed in a transport medium containing a balanced salt solution (Hanks), or a bacteriologic broth. Autopsy specimens and spinal fluid may be collected in sterile containers or tubes, frozen at −70°C as soon as possible, and transported to the laboratory in the frozen state. The value of any laboratory test is dependent upon the timely submission and proper handling of the specimen.

At present, rapid isolation technology is not available. Specimens for virus assay are inoculated into newborn mice and cell culture. Both systems are required because none of the echoviruses will replicate in newborn mice, most of the coxsackie A viruses will not replicate in cell culture, but the coxsackie B group will usually replicate in both systems. Even the cell line used determines whether or not the unknown virus will be isolated. For instance, coxsackie A types 7 and 9 will grow well in monkey kidney cells but not in human cells. Therefore, the failure to demonstrate a virus in a specimen does not necessarily rule out the possibility of an enterovirus etiology.

The oral and injectable polio vaccines are discussed in detail in Chapter 27. Rare cases of clinical polio have occurred in nonimmune adults as a result of contact with a child who has received the attenuated oral polio vaccine. Therefore, parents or other adult contacts of children who are being immunized should be warned of this possible risk; they may also want to be immunized with the injectable killed virus vaccine.

SUGGESTED READINGS

Adler I, Zickl R. Winter vomiting disease. J Infect Dis 1969;119:668-673.

Albrey MB, Murphy AM. Rotavirus and acute gastroenteritis of infants and children. Med J Aust 1976;1:82-85.

Appleton H, Buckley M, Thom BT, Cotton JL, Henderson S. Virus-like particles in winter vomiting disease. Lancet 1977;1:409-411.

Appleton H, Pereira MS. A possible virus etiology in outbreaks of food-poisoning from cockles. Lancet 1977;1:780-781.

Bishop RF, Davidson GP, Holmes IH, Ruck BJ. Evidence for viral gastroenteritis. N Engl J Med 1973;289:1096-1097.

Bishop RF, Davidson GP, Holmes IH, Ruck BJ. Virus particles in epithelial cells of duodenal mucosa from chidren with acute gastroenteritis. Lancet 1973;2:1281-1283.

Bishop RF, Davidson GP, Holmes IH, Ruck BJ. Detection of a new virus by electron microscopy of fecal extracts from children with acute gastroenteritis. Lancet 1974;1:149-151.

Blacklow NR, Cukor G. Viral gastroenteritis agents. In: Lennette EH, Balows A, Hausler WJ, Truant JP. Manual of clinical microbiology, 3d edition. Washington: American Society for Microbiology, 1980, pp. 891-898.

Blacklow NR, Cukor G, Bedigian MK, et al. Immune response and prevalence of antibody to Norwalk enteritis virus as determined by radioimmunoassay. J Clin Microbiol 1979;10: 903-909.

Blacklow NR, Dolin R, Fedson DS, et al. Acute infectious nonbacterial gastroenteritis: etiology and pathogenesis. A combined clinical staff conference at the Clinical Center of the National Institutes of Health. Ann Intern Med 1972;76:993-1008.

Blacklow NR, Echeverria P, Smith DH. Serological studies with reovirus-like agents. Infect Immun 1976;13:1563-1566.

Bortolussi R, Szymanski M, Hamilton R, Middleton P. Studies on the etiology of acute infantile diarrhea. Pediatr Res, 1974;8:379.

Clarke SKR, Cook GT, Egglestone SI, et al. A virus from epidemic vomiting disease. Br Med J 1972;3:86-89.

Committee on Enteroviruses. Classification of human enteroviruses. Virol 1962;16:501-504.

Cubitt WD, Holzel H. An outbreak of rotavirus infection in a long-stay ward of a geriatric hospital. J Clin Pathol 1980;33:306-308.

Cukor G, Blacklow NR, Echeverria P, Bedigian MK, Puruggan H, Basaca-Sevilla V. Comparative study of the acquisition of antibody to Norwalk virus in pediatric populations. Infect Immun 1980;29:822-823.

Dalldorf G. The enteroviruses and paralytic disease. In Rose HM: Viral infections of infancy and childhood. Symposium of the Section on Microbiology. The New York Academy of Medicine, New York: Hoeber-Harper, 1960, pp 125-144.

Dalldorf G, Sickles GM. An unidentified, filtrable agent isolated from the feces of children with paralysis. Science 1948;108:61-62.

Dalldorf G, Sickles GM. The coxsackie viruses. In: Lennette EG, Schmidt NJ, eds. Diagnostic procedures for virus and rickettsial diseases, 2nd edition. New York: American Public Health Association, 1956, pp 153-168.

Davidson GP, Bishop RF, Townley RR, Holmes IH, Ruck BJ. Importance of a new virus in acute sporadic enteritis in children. Lancet 1975;1:242-245.

Dolin R, Blacklow NR, DuPont H. et al. Biological properties of Norwalk agent of acute infectious nonbacterial gastroenteritis. Proc Soc Exp Biol Med 1972;140:578-583.

Enders JF, Weller TH, Robbins FC. Cultivation of the Lansing strain of poliomyelitis virus in cultures of various human embryonic tissues. Science 1949;109:85-87.

Feinstone SM, Kapikian AZ, Purcell RH. Hepatitis A: detection by immune electron microscopy of a virus-like antigen associated with acute illness. Science 1973;182:1026-1028.

Flewett TH, Bryden AS, Davis H, Woode GN, Bridger JC, Derrick JM. Relation between viruses from acute gastroenteritis of children and newborn calves. Lancet 1974;2:61-63.

Flewett TH, Woode GN. The rotaviruses. Brief review. Arch Virol 1978;57:1-23.

Goldwater PN, Chrystie IL, Banatvala JE. Rotaviruses and the respiratory tract. Br Med J 1979;2:1551-1552.

Greenberg HB, Kapikian AZ. Detection of Norwalk agent antibody and antigen by solid-phase radioimmunoassay and immune adherence hemagglutination assay. J Am Vet Med Assoc 1978;173:620.

Greenberg HB, Valdesuso J, Kapikian AZ, et al. Prevalence of antibody to the Norwalk virus in various countries. Infect Immun 1979; 26:270-273.

Greenberg HB, Wyatt RG, Kapikian AZ. Norwalk virus in vomitus. Lancet 1979;1:55.

Greenberg HB, Wyatt RG, Valdesuso J, Kalica AR, London WT, Chanock RM, Kapikian AZ. Solid-phase microtiter radioimmunoassay for detection of the Norwalk strain of acute nonbacterial epidemic gastroenteritis virus and its antibodies. J Med Virol 1978;2:97-108.

Hejkal TW, Keswick B, LaBelle RL, et al. Viruses in a community water supply associated with an outbreak of gastroenteritis and infectious hepatitis. J Am Water Works Assoc June, 1982, 74:318-321.

Hieber JP, Shelton S, Nelson JD, Leon J, Mohs E. Comparison of human rotavirus disease in tropical and temperate settings. Am J Dis Child 1978;132:853-858.

Holmes IH, Ruck BJ, Bishop RF, Davidson GP. Infantile enteritis viruses: morphogenesis and morphology. J Virol 1975;16:937-943.

Howe HA, Wilson JL. Poliomyelitis. In: Horsfall FL, Rivers TM. Viral and rickettsial infections of man, 3rd edition. Philadelphia: JB Lippincott, 1959, pp. 432-478.

Kapikian AZ, Feinstone SM, Purcell RH, et al. Detection and indentification by immune electron microscopy of fastidious agents associated with respiratory illness, acute nonbacterial gastroenteritis, and hepatitis A. Perspect Virol 1975;9:9-47.

Kapikian AZ, Greenberg HB, Cline WL, et al. Prevalence of antibody to the Norwalk agent by a newly developed immune adherence hemagglutination assay. J Med Virol 1978;2:281-294.

Kapikian AZ, Kim HW, Wyatt RG, et al. Human reovirus-like agent as the major pathogen associated with "winter" gastroenteritits in hospitalized infants and young children. N Engl J Med 1976;294:965-972.

Kapikian AZ, Kim HW, Wyatt RG, et al. Reovirus-like agent in stools: association with infantile diarrhea and development of serologic tests. Science 1974;185:1049-1053.

Kapikian AZ, Wyatt RG, Dolin R, Thornhill TS, Kalica AR, Chanock RM. Visualization by immune electron microscopy of 27-nm particle associated with acute infectious non-bacterial gastroenteritis. J Virol 1972;10:1075-1081.

Kapikian AZ, Yolken RG, Greenberg HB, et al. Gastroenteritis viruses. In Lennette EH, Schmidt EH. Diagnostic procedures for viral, rickettsial and chlamydial infections. Washington, DC: American Public Health Association 1978, pp 927-995.

Kim HW, Brandt CD, Kapilian AZ, et al. Human reovirus-like agent infection: occurrence in adult contacts of pediatric patients with gastroenteritis. JAMA 1977;237:404-407.

Landsteiner K, Popper E. Ubertragung der Poliomyelitis acuta auf Affen. Z Immunitatsforsch 1909;2:377-390.

Martin ML, Palmer EL, Middleton PJ. Ultrastructure of infantile gastroenteritis virus. Virol 1975;68:146-153.

Matthews REF. The classification and nomenclature of viruses. Summary of results of meetings of The International Committee on Taxonomy of Viruses in the Hague, September 1978. Intervirology 1979;11:133-135.

Mayor HD. Picornavirus symmetry. Virology 1964;22:156-160.

Melnick JL, et al. Picornaviridae. Intervirology 1974;4:303-316.

Melnick JL, Shaw EW, Curnen EC. A virus from

patients diagnosed as non-paralytic poliomyelitis or aseptic meningitis. Proc Soc Exp Biol Med 1949;71:344-349.

Melnick JL, Sabin AB. The echo virus group. In: Rivers TM, Horsfall FL Jr. Viral and rickettsial infections of man, 3rd edition. 1959, pp. 547-569.

Melnick JL, Wenner HA, Phillips CA. Enteroviruses. In: Lennette EH, Schmidt NJ. Diagnostic procedures for viral, rickettsial and chlamydial infections, 5th edition. Washington, DC: American Public Health Association, 1979, pp 471-534.

Middleton PJ, Szymanski MT, Petric M. Viruses associated with acute gastroenteritis in young children. Am J Dis Child 1977;131:733-737.

Nalin DR, Levine MM, Mata L. Oral rehydration and maintenance of children with rotavirus and bacterial diarrheas. Bull WHO 1979; 57:453-459.

Palmer EL, Martin ML, Murphy FA. Morphology and stability of infantile gastroenteritis virus: comparison with reovirus and blue tongue virus. J Gen Virol 1977;35:403-414.

Parrino TA, Schreiber DS, Trier JS, Kapikian AZ, Blacklow NR. Clinical immunity in acute gastroenteritis caused by Norwalk agent. N Engl J Med 1977;297:86-89.

Petric M, Tam JS, Middleton PJ. Preliminary characterization of the nucleic acid of infantile gastroenteritis virus (orbivirus group). Intervirology 1976;7:176-180.

Rodger SM, Schnagl R, Holmes IH. Further biochemical characterization, including the detection of surface glycoproteins, of human, calf, and simian rotaviruses. J Virol 1977;24:91-98.

Rodriguez WJ, Kim HW, Arrobio JO, et al. Clinical features of acute gastroenteritis associated with human reovirus-like agent in infants and young children. J Pediatr 1977; 91:188-193.

Schnagl RD, Holmes IH. Characteristics of the genome of human infantile enteritis rotavirus. J Virol 1976;19:267-270.

Steigman AJ. Poliomyelitic properties of certain non-polio viruses: enteroviruses and Heine-Medin disease. J Mt Sinai Hosp 1958;25:391-404.

Steigman AJ, Lipton MM. Fatal bulbospinal paralytic poliomyelitis due to ECHO 11 virus. JAMA 1960;174:178-179.

Steinhoff MC, Douglas RG, Greenberg HB, Callahan DR. Bismuth subsalicylate therapy of viral gastroenteritis. Gastroenterology 1980;78:1495-1499.

Thornhill TS, Kalica AR, Wyatt RG, Kapikian AZ, Chanock RM. Pattern of shedding of the Norwalk particle in stools during experimentally induced gastroenteritis in volunteers as determined by immune electron microscopy. J Infect Dis 1975;132:28-34.

Thornhill TS, Wyatt RG, Kalica AR, Dolin R, Chanock RM, Kapikian AZ. Detection by immune electron microscopy of 26-27 nm virus-like particles associated with two family outbreaks of gastroenteritis. J Infect Dis 1977; 135:20-27.

Trask JD, Melnick JL, Wenner HA. Chlorination of human, monkey-adapted and mouse strains of poliomyelitis virus. Amer J Hyg 1945;41:30-40.

Tufvesson B, Johnsson T. Occurrence of reo-like viruses in young children with acute gastroenteritis. Acta Pathol Microbiol Scand 1976;84:22-28. (B)

VanKirk DH, Kapikian AZ, Wyatt RG, Kalica AR. Viral digestive tract infections. In: Seligson D. CRC Handbook Series in Clinical Laboratory Science. Section H: Virology and rickettsiology, Volume 1. Part 2 (section editors GD Hsiung and RH Green). West Palm Beach, FL: CRC Press, 1978, pp 211-233.

Von Bonsdorff CH, Hovi T, Makela P, Morttinen A. Rotavirus infections in adults in association with acute gastroenteritis. J Med Virol 1978;2:21-28.

Wallace C, Melnick JL. Stabilization of poliovirus by cations. Texas Rep Biol Med 1961;19:683-700.

Wallace C, Melnick JL. Cationic stabilization—a new property of enteroviruses. Virol 1962;16:504-505.

Wallace C, Melnick JL. Thermosensitivity of poliovirus. J Bacteriol 1963;86:499-504.

Wellings FM, Mountain CW, Lewis AL. Virus in groundwater. In: Proceedings of the Second National Conference on Individual Onsite

Wastewater Systems, Ann Arbor, Michigan: National Sanitation Foundation. 1976, pp 61-65.

Wenman WM, Hinde D, Feltham S, Gurwith M. Rotavirus infection in adults: results of a prospective family study. N Engl J Med 1979; 301:303-306.

Wyatt RG, Dolin R, Blacklow NR, et al. Comparison of three agents of acute infectious nonbacterial gastroenteritis by cross-challenge in volunteers. J Infect Dis 1974;129:709-724.

Wyatt RG, Kalica AR, Mebus CA, Kim HW, London WT, Chanock RM, Kapikian AZ. Reovirus-like agents (rotaviruses) associated with diarrheal illness in animals and man. In: Pollard M. Perspectives in Virology XX. New York: Raven Press, 1978, 10:121-145.

Wyatt RG, Kapikian AZ. Viral agents associated with acute gastroenteritis in humans. Am J Clin Nutr 1977;30:1857-1870.

Wyatt RG, Kapikian AZ. Viral gastrointestinal infections. In: Feigin RD and Cherry JD, eds. Feigin textbook of pediatric infectious diseases. Philadelphia: WB Saunders, 1980.

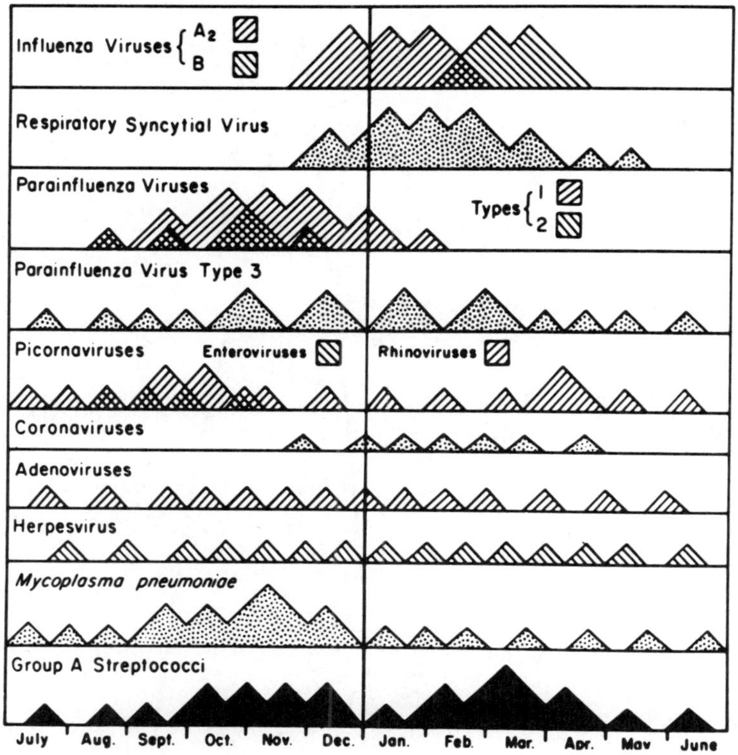

FIGURE 15-3. The seasonal occurrence of the major respiratory pathogens and a general indication of their associated morbidity. (Reproduced with permission of the publisher from Drew WL. Viral Infections. Philadelphia: FA Davis, 1976.)

Orthomyxoviruses

In the classification officially approved by the International Committee on Taxonomy of Viruses, influenza virus (type A and type B) was established as a genus within the family of Orthomyxoviridae. Type C was labeled a probable second genus.

Influenza Virus

Biologic characteristics. The three immunologic types of influenza virus are known as types A, B, and C, distinguished by the antigenic specificity of their ribonucleic protein (RNP). These viruses are smaller and more uniform in size than the other families of myxovirus and have a more narrow nucleoprotein of 6 to 9 nm, composed of six to eight segments contained in the nucleocapsid (Fig. 15-5). Small internal proteins that are associated with the RNP may function as virion polymerases. The antigens, hemagglutinin and neuraminidase, are separate, forming the glycoprotein spikes that stud the enveloped virion. Hemagglutinins (H) are located at the tip of the spike and are responsible for viral attachment to cell receptors. Neuraminidase (N) is the enzyme

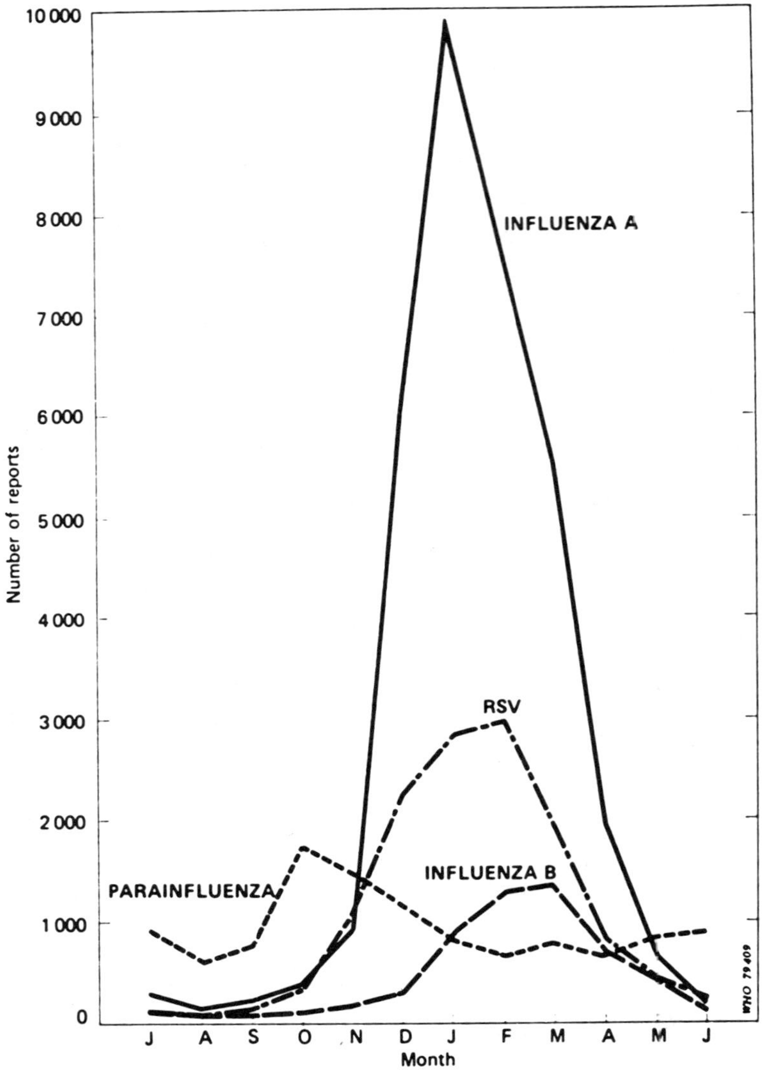

FIGURE 15-4. Respiratory diseases associated with influenza viruses A and B, parainfluenza virus, and RSV, by month, in the Northern Hemisphere. (Reproduced from World Health Organization. Technical report series #642. Viral respiratory diseases. Geneva: World Health Organization, 1980.)

that can free combined virus by destroying the host cell. The nucleocapsid is assembled intranuclearly and during replication the viral antigens are enveloped in the cytoplasm. Maturation or completion of the virion occurs at the cell periphery.

An example of the nomenclature of the subtypes is A/Hong Kong/1/68 (H_3N_2); strain

TABLE 15-2. Viruses that Cause Acute Respiratory Disease in Humans

Family	*Group*	*Nucleic acid*	*SS or DS**	*Symmetry*	*Envelope*	*Number of human serotypes*
Orthomyxoviridae	Influenza viruses	RNA	SS	helical	yes	3
Paramyxoviridae	Parainfluenza viruses	RNA	SS	helical	yes	4
	Mumps virus	RNA	SS	helical	yes	1
	Measles virus	RNA	SS	helical	yes	1
Pseudomyxoviridae	Respiratory syncytial virus	RNA	SS	helical	yes	1
Adenoviridae	Adenoviruses	DNA	DS	cubic	no	34
Picornaviridae	Rhinoviruses	RNA	SS	cubic	no	>110
	Enteroviruses	RNA	SS	cubic	no	>70
Coronaviridae	Coronaviruses	RNA	SS	helical	yes	6
Reoviridae	Reoviruses	RNA	DS	cubic	no	3

*Single-stranded or double-stranded nucleic acid.

TABLE 15-3. Characteristics of Myxoviruses

	Orthomyxoviruses	*Paramyxoviruses*	*Pseudomyxoviruses*
Characteristics	*Influenza*	*Parainfluenza Measles Mumps*	*RSV*
Particle size	Small (80-120 nm)	Large (125-250 nm)	120-200 nm
Diameter of internal helical core (nucleocapsid)	9 nm	18 nm	9-18 nm
Localization of nucleocapsid in host cell	Nucleus	Cytoplasm	Cytoplasm
Separate hemagglutinin and neuraminidase	+	0	0
Hemolysin	0	+	0
Prominent cytoplasmic inclusions	0	+	+
Syncytial formation	0	+	+

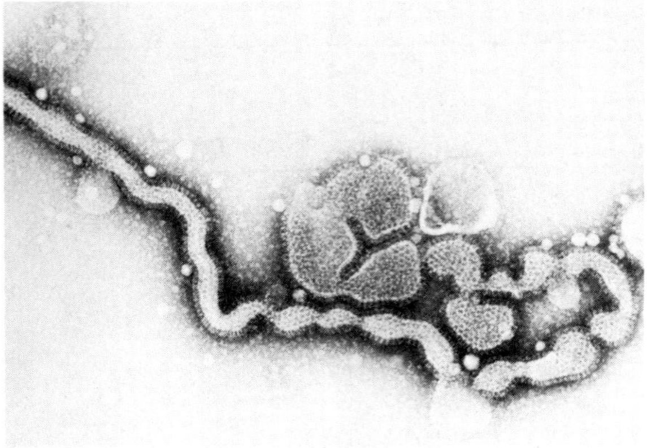

FIGURE 15-5. Influenza C virus particle. (Courtesy of Centers for Disease Control, Atlanta, GA 30333.)

designation of serotype/host of origin (humans unless stated otherwise)/geographic origin/strain number/year of isolation/(composition of hemagglutination and neuraminidase antigens).

Epidemiology. Influenza viruses are ubiquitous, affecting persons in all parts of the world. Influenza A virus repeatedly causes pandemics, characterized by a rapid spread from continent to continent (Fig. 15-6). Between the great pandemics, which occur approximately every 10 to 15 years, epidemics of type A occur at 2- to 3-year intervals (Fig. 15-7). Endemic occurrences and sporadic cases of type B irregularly surface during

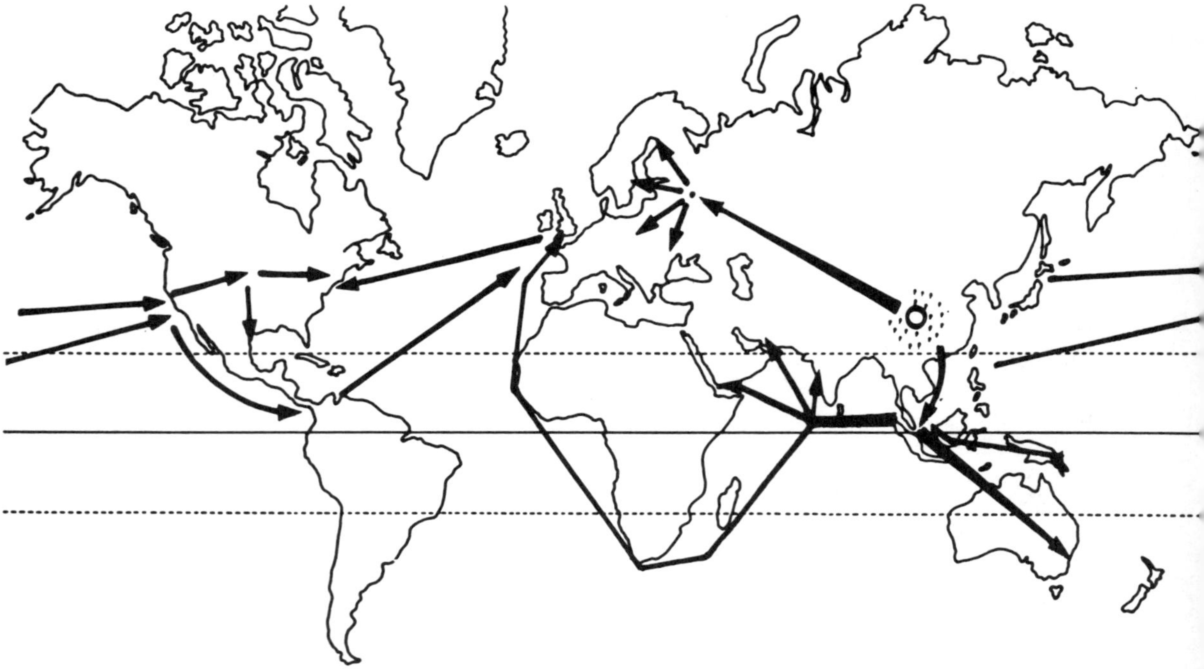

FIGURE 15-6. A diagram to indicate the spread of infection during the 1957 pandemic of Asian influenza. The origin of the epidemic in China is seen as a circle and dots. (Reproduced with permission of the publisher from Swain RHA, Dodds, TC. Clinical virology. Baltimore: Williams and Wilkins, 1967.)

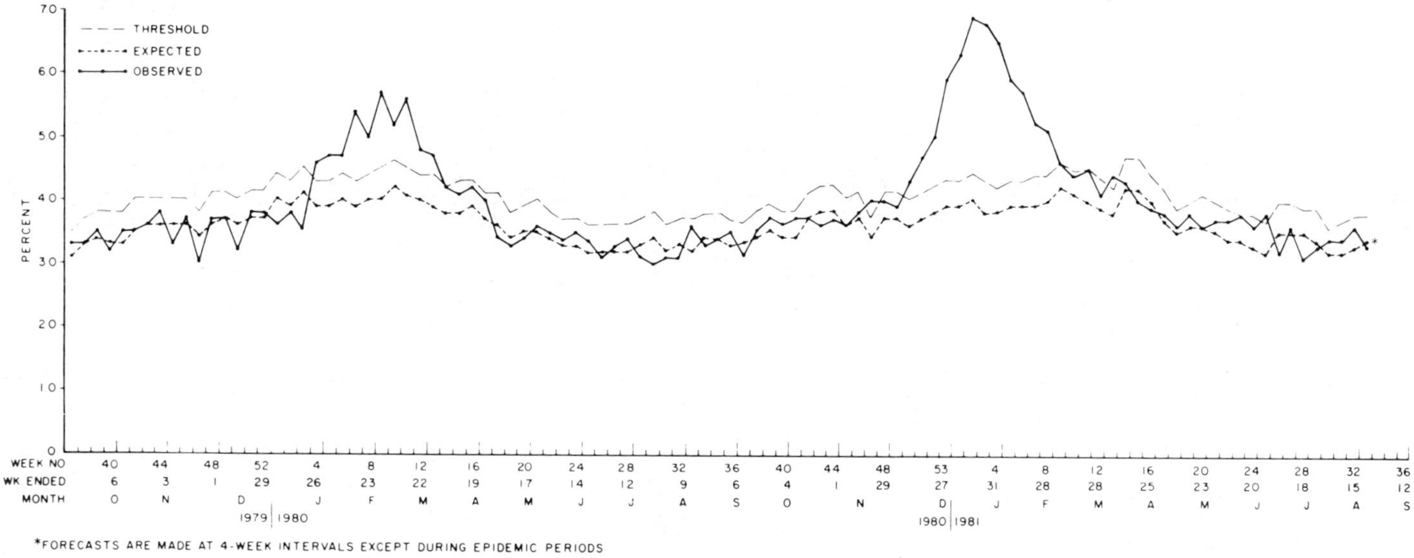

FIGURE 15-7. Pneumonia-influenza. Observed and expected ratio of deaths attributed to pneumonia and influenza in 121 cities in the United States, as determined by the time-series method, September, 1979 to August, 1981. (Reproduced from Centers for Disease Control. Annual Summary 1980. Morb Mort Week Rep 1981;29:12.)

the winter months. When epidemics of type B have been documented, they affect well-circumscribed populations, spreading particularly among schoolchildren, at approximately 4- to 5-year intervals. Influenza viruses types A and B often circulate concurrently in the same outbreak. It is rare that type C has been implicated in disease, but in some isolated cases it has been the causative agent for mild upper respiratory infection.

Influenza virus is easily transmitted by direct contact, by large droplet infection, or by articles contaminated by discharge from the respiratory tract of infected individuals. Because of its antigenic lability, which makes antibody induced by previous infection ineffective, this virus affects all age groups, causing a generalized febrile illness with respiratory tract involvement. The group most commonly affected by epidemics of influenza A is the population 5 to 14 years of age. The incidence declines with the population aged 20 to 25 years, and moderately rises until the age group 30 to 45 years. The distribution of type B according to age of victims is similar—a high attack rate in school-aged children with a decline in persons older than 20 years of age.

Pathogenesis. Influenza viruses cause mainly mild respiratory illness, but a high incidence of lethal pneumonia has been observed following some outbreaks. The duration of immunity following natural infection, as measured by failure to exhibit disease symptoms, lasts several years after a previous attack. Children have a more rapidly declining, less durable immune response. Subclinical reinfection with influenza probably occurs since antibodies are present in most adults in the United States.

The existence of independent segments composing the nucleocapsid allows the influenza RNA the capability of genetic recombination. Antigenic drift (see Fig. 15-9) or antigenic mutations are continually taking place within type A, and to a lesser degree type B, while type C appears to be antigenically stable. With the emergence of new antigenic characteristics on the surface of the virus, epidemics follow because the population lacks immunity to such mutated forms. The greater the deviation from previous antigens, the greater the pandemic proportions.

The influenza virus lodges in the respiratory tract. The hemagglutinin combines with the receptors of the mucoproteins in respiratory secretions. Viral neuraminidase thins the viscosity of the mucus, exposing bare surface cells. Since the mucus is thinned, the mucociliary escalator is compromised, and deposition of the virus is promoted to the dependent portions of the bronchial tree. Reproduction and synthesis follow penetration and adsorption of the virus.

The infective process is limited to those areas of the respiratory tract that are covered by ciliated epithelium: the larynx, trachea, and main bronchioles. Within 48 hours the epithelium of the respiratory tract is destroyed, leaving only the basement membrane. It has been shown that for several days following an influenza infection, the host's resistance to bacterial infection is reduced. Regeneration and repair follow in the succeeding days.

Host responses. Pulmonary virus titers peak in 24 hours, are maintained for 48 hours, then abruptly decline. The incubation period between contact and development of symptoms is one to two days. After the incubation period there is an abrupt onset of fever (39–41°C). The symptoms of influenza affecting the respiratory tract consist of dryness of throat and hoarseness.

Influenza virus A is an important cause of febrile convulsions in infants and may cause pneumonia and death in the very young. In older age groups, lower respiratory tract involvement is common, and most deaths from influenza occur in persons over 70 years of age. Influenza B causes a characteristic influenzalike illness, often accompanied by abdominal pain. The influenza C virus is probably responsible for mild illness of the upper respiratory tract, from which it is only infrequently isolated. In uncomplicated cases, there may be no clinical or radiologic evidence of parenchymal involvement and recovery occurs in a few days. Lung involvement may follow if a bacterial infection ensues. In recent years, staphylococcal pneumonia has been the most common bacterial superinfection. In older children, influenza virus has been implicated in laryngotracheobronchitis. Recently clear evidence has emerged for a strong correlation between influenza and a rare disease called Reye's syndrome (Fig. 15-8). Reasons for this are discussed in "Clinical Implications."

The diagnosis of influenza is often made clinically when typical symptoms are present during epidemic or pandemic periods. Laboratory diagnosis can be made by growth in tissue culture of virus from patient secretions. All strains are commonly identified by complement-fixation (CF) tests. Hemagglutinin inhibition (HI) tests reflect long-lasting antibodies and are highly strain specific.

Control and treatment. The treatment of influenza is aimed at palliation of symptoms and antibacterial therapy if secondary bacterial pneumonia develops. Amantadine hydrochloride, a systemic antiviral drug, has been found to be effective in preventing clinical disease with type A influenza, but not type B, measles, or any other RNA virus. It acts by blocking the penetration of the antigen to host cells. Its effectiveness is not well established but it is apparently necessary to administer the drug during the entire susceptible period, and it is ineffective if given after the disease is established. Optimal effectiveness is obtained when the drug therapy is instituted at least two days before exposure. Side effects are mostly those of the central nervous system: ataxia, nervousness, dizziness. It is recommended that prophylactic treatment to high-risk groups be given for one month prior to epidemic threats. Rimantadine is a drug with similar action that is effective for types A and B. This drug and its derivatives have not been recommended for use in children however. Ribavirin, a synthetic analogue of guanosine, has had inconclusive results against influenza viruses in clinical trials, and its therapeutic efficacy is not yet proven.

Vaccines developed for the control of influenza have been shown to be highly effective when applied to high-risk populations. In theory, mass immunization against influenza should prevent or halt the spread of epidemics. However, the prophylactic use of vaccines has been limited due to the genetic recombination capabilities of the virus and poor prediction of mutations (Fig. 15-9). The nearest approach to massive immunization was the national immunization program in fall of 1976 sponsored by the United States government that was to provide vaccination not only to those at high risk, but to the entire population. By following the patterns of influenza pandemics and by considering the continuous process of antigenic drift, scientists were able to predict the pandemic in late 1976 that was due to influenza A/New Jersey, a strain antigenically similar to swine flu, 1918. Though the strain of influenza did appear, disease did not occur in the epidemic proportions that had been an-

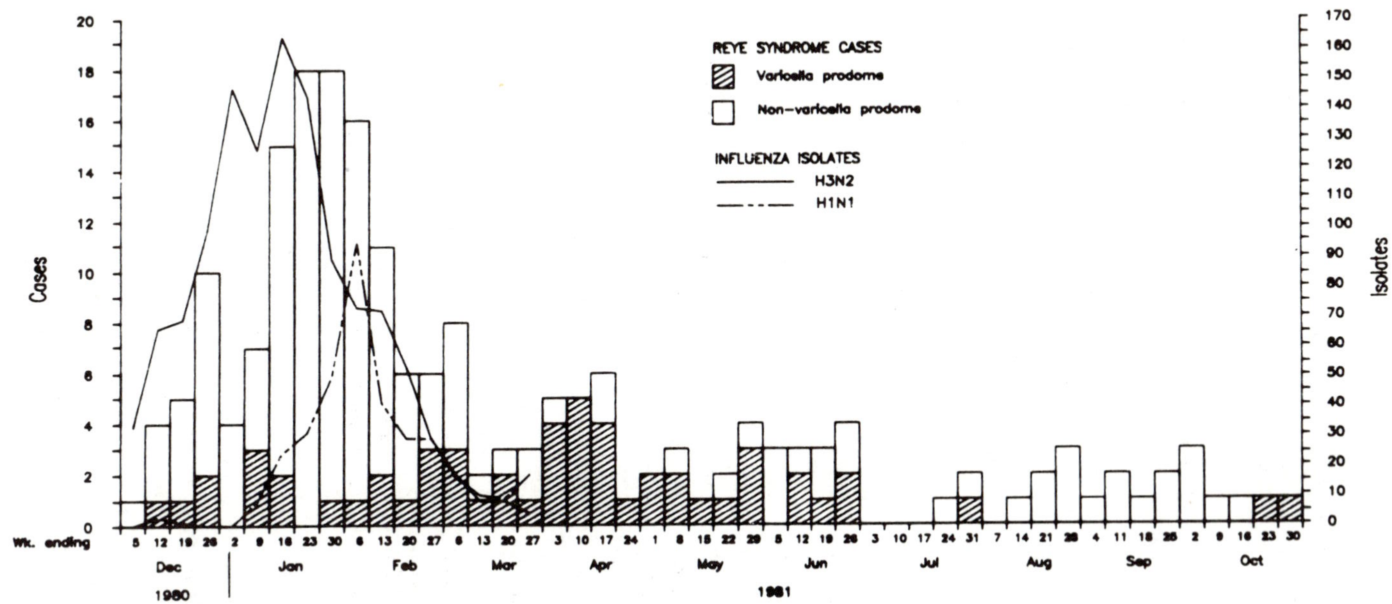

FIGURE 15-8. Reye's syndrome cases by week of hospitalization, and influenza isolates, by week of report, December 1, 1980 to October 30, 1981. (Reproduced from Centers for Disease Control. National surveillance for Reye Syndrome, 1981. Morb Mort Week Rep 1982;31:53.)

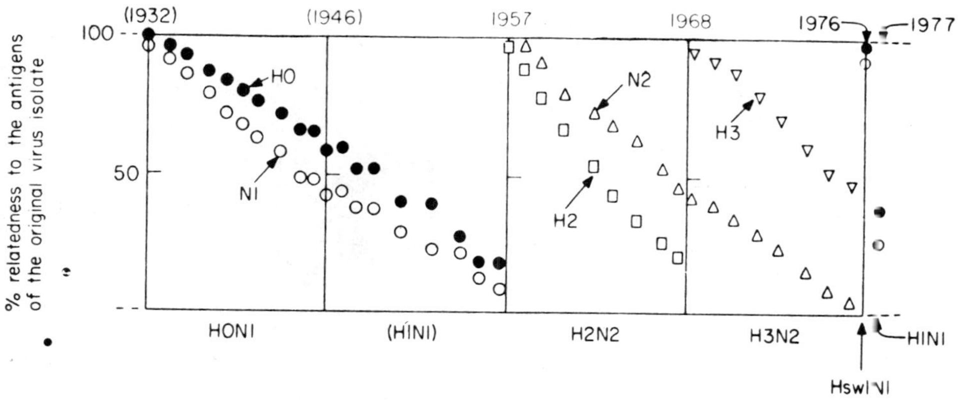

FIGURE 15-9. Diagram showing antigenic shift. In each of the years written in full, a hemagglutinin (•) or neuraminidase (○) appeared, which was antigenically unrelated to the strains of the preceding year. The hemagglutinin of the first human subtype isolated in 1932 is designated H0. Until recently it was thought that H0 and H1 (in parentheses above) were distinct serotypes. The current view is that these have arisen by antigenic drift from the Hswl of the strain causing the pandemic of 1918. The subtypes H2 and H3 are the result of antigenic shift, while the 1976 and 1977 strains may have resulted from recurrence of previous strains. (This is discussed in the text.) (Reproduced with permission of the publisher from Primrose SB, Dimmock NJ. Introduction to modern virology, 2nd edition. London: Blackwell Scientific Publications Ltd, 1980.)

ticipated. Unfortunately, epidemiologic data indicated that cases of Guillain-Barré syndrome were related to the vaccination for swine flu; about 1 in 100,000 vaccinated individuals developed this complication. Because of this unexpected complication, the mass immunization program was terminated.

So far, influenza vaccines have been used primarily in certain selected population groups—in industry to reduce absenteeism, and in public institutions to prevent interruption of needed public services. The elderly and individuals in any age group who have a known underlying chronic or debilitating disease should be selectively immunized because of the high risk of severe complications and fatal influenza infections in such individuals. Continuous monitoring is required to identify the antigenic nature of the viruses currently circulating in the population, any antigenic drift that might occur, and the appearance of new strains.

Paramyxoviruses

The three groups of viruses of clinical importance included in the family Paramyxoviridae are the parainfluenza viruses, *Myxovirus parotitidis* (mumps virus), and measles virus.

Parainfluenza Viruses

Biologic characteristics. As th
plies, parainfluenza viruses r
enza viruses. They too poss
tinin and neuraminidase, b

antigenic variation throughout their four types. The particle size as well as the nucleocapsid is considerably larger than that of influenza viruses. The capsid has a "herring-bone" appearance and is nonsegmented, therefore the nucleocapsid clearly differentiates the two families (Fig. 15-10). The helix of nucleoprotein is detected first in the cytoplasm and forms clusters of eosinophilic cytoplasmic inclusions. Replication occurs chiefly in the cytoplasm with some evidence of nuclear involvement of the host cell. Other properties of this family include hemadsorption or red cell adherence, cell fusion of adjacent cells, and hemolysis.

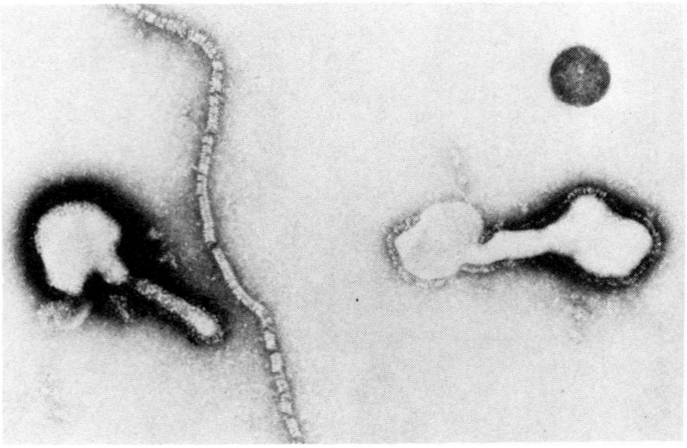

FIGURE 15-10. Parainfluenza virus. (Courtesy of Centers for Disease Control, Atlanta, GA 30333.)

Epidemiology. The four serotypes of the parainfluenza family are associated with different age groups, different epidemiologic manifestations, and different illnesses (Table 15-4). Type 1 is the principal cause of croup (laryngotracheobronchitis). Type 2 resembles type 1 but causes serious illness less frequently. Type 3 is a common cause of pneumonia and bronchiolitis in infants less than 6 months of age. In this respect type 3 approximates respiratory syncytial virus (RSV) in morbidity in this age group. Type 4 is infrequently associated with illness, therefore it is of little importance as a human pathogen.

All serotypes of parainfluenza viruses are highly communicable and have a short incubation period. Available evidence supports the fact that most individuals are infected during childhood. The virus spreads rapidly among children and families. Reinfections occur at a rate of 80 percent among children 0 to 4 years of age, 46 percent between 5 and 14 years, but only 22 percent among adults. Transmission is by direct person-to-person contact, or by large droplet spread.

Parainfluenza viruses are cosmopolitan in their distribution, but show differences in their epidemiologic manifestations. Parainfluenza type 2 characteristically appears

TABLE 15-4. Parainfluenza Viruses: Serotypes and Associated Age, Clinical Syndrome, and Epidemiologic Manifestations

Human serotype	*Major clinical syndrome*	*Peak age (months)*	*Epidemiologic manifestations*
1	Croup	6-24	Epidemic (fall)
2	Croup	6-24	Epidemic/endemic
3	Pneumonia, bronchitis	0-6	Endemic (all seasons)
4	URI	Unknown	Endemic

in epidemic form, with varying intervals between outbreaks. Type 1 virus produces epidemics at intervals of several years, but may entrench itself in an area and become endemic. Type 3 virus has an endemic pattern; cases occur at any time of year.

Pathogenesis. Little is known about the pathogenesis of parainfluenza viruses. Parainfluenza viruses are able to replicate in the respiratory epithelium by extrusion from the cell membrane without immediate epithelial cell destruction. This phenomenon allows continued release of viral particles for the life of the host cell. The development of cytoplasmic inclusions and vacuoles gradually robs the infected host cell of its structure. As with the influenza viruses, the ciliated epithelium is the chief site of involvement even though parainfluenza types 1 and 2 seem to have a predilection for the glottic tissue.

Although reinfection occurs frequently, children with high antibody titers before reinfection demonstrate milder illness and less virus shedding than those with few or no antibodies. There is a high positive correlation between immunity and the presence of antibodies in nasal mucosa. Those with high antibody titers suffer upper respiratory tract reinfections rather than more serious lower respiratory tract reinfections.

Host responses. Mild respiratory illness can result from parainfluenza viral infection in adults. Hoarseness is a frequent occurrence; pharyngitis, mucopurulent nasal discharge, nasal obstruction, and headache are also common. The adult patient may be slightly febrile. Despite the presence of a cough, there is little evidence of bronchial involvement.

The initial illness in children, which appears in the absence of antibody, may be a mild upper respiratory disease presenting as coryza or pharyngitis. Two other syndromes also occur. Croup, or spasmodic laryngotracheobronchitis, and more serious lower respiratory tract disease present clinically alarming pictures. Croup is commonly associated with parainfluenza virus type 2, and occasionally type 3. The laryngeal spasms cause stridor as air is inhaled with a high-pitched crowing sound. Some studies indicate that croup afflicts 40 percent of 0 to 4 year olds. The lower respiratory tract diseases due to parainfluenza may cause a multitude of symptoms, such as bronchospasm, retractions, copious secretions, dyspnea, rales, and rhonchi with temperatures greater than 30°C.

A determination of the etiology of sporadic episodes of acute respiratory diseases is usually made clinically. By process of elimination, infections with parainfluenza viruses may be suspected by grouping clinical manifestations with epidemiologic patterns. Serologic tests are difficult to interpret because the parainfluenza viruses share antigens with mumps virus and some other myxoviruses. Parainfluenza viruses grow well in laboratory animal tissues and their presence on the surface of the infected cells is detected by hemadsorption. Specific antisera inhibit hemadsorption and thus enable the virus to be identified. Complement-fixation tests are used for confirmation of the diagnosis, but are not yet fully reliable.

Control and treatment. Because of the severity of the lower respiratory tract illness produced by parainfluenza viruses in infants and very young children, efforts to develop suitable vaccines have been under way for some years. The poorly defined pathogenesis of the parainfluenza viruses retards the study of their prevention and control. Recipients of inactivated vaccines from virus derived from chick embryo, remain susceptible to infections. This may be due to the fact that local immunity linked to secretory nasal antibody is more important than circulating antibody. Reevaluation of parenteral vaccines and consideration of intranasal routes for live, attenuated virus, may prove to be rewarding. Studies are under way to develop temperature-sensitive mutants of parainfluenza virus types 1 and 3—the two types most frequently associated with severe lower respiratory disease.

Nosocomial infections with parainfluenza virus may occur readily if newborns and other susceptible infants are not kept isolated from children with viral disease. Infections with parainfluenza virus have been reported in 18 percent of children previously free of infection, after their admission to hospitals.

Mumps Virus

Biologic characteristics. The mumps virus varies in diameter between 80 and 240 nm, and its nucleic acid, RNA, is arranged in a helix. This virus has the typical paramyxovirus morphology, and also produces hemagglutinin, neuraminidase, and hemolysin. The mumps virus is antigenically homogeneous, although it shares general properties and some antigens with serotypes of the parainfluenza group.

Epidemiology. Humans are the only known reservoir for the mumps virus. It is transmitted by direct contact, by airborne droplets, or by fomites contaminated with saliva and possibly urine. More intimate contact is required for transmission of this virus than for measles or varicella. Infections occur endemically throughout the year, and are worldwide. The peak incidence is in winter and spring. The morbidity is high with this virus, but mortality is low even when there is central nervous system involvement.

Pathogenesis. The mumps virus has a marked affinity for glandular tissue. It affects the parotid glands most commonly, but may involve testes, breast tissue, and other glands, especially if infection occurs after puberty. The source of infection is the salivary secretions of infected subjects or carriers. The virus enters the mouth. It either passes directly along Stenson's duct to the parotid gland where a viremia follows, or it establishes a primary focus of replication in the superficial epithelium of the upper respiratory tract, followed by viremia and localization. The mumps virus has hemagglutination properties, but less than those of influenza virus. In high concentrations, the virus also causes lysis of red blood cells.

Host responses. The name "mumps" derived from the mumbling speech of patients due to painful jaw movements. There is a gradual onset of fever 18 to 21 days after exposure to the infection reflecting a long incubation period. With the fever there is a prodrome of malaise and anorexia. The first distinguishing sign is painful swelling of one or both parotid glands (Fig. 15-11); bilateral involvement occurs in 96 percent of cases. The swelling, with low-grade fever, reaches a maximum in 48 hours and persists 7 to 10 days. The sublingual and submaxillary glands may be involved with or without parotid involvement. If no complications ensue, the fever subsides in 3 to 5 days with full recovery 7 to 10 days later. Little tissue damage is associated with uncomplicated cases.

In 10 to 16 percent of cases there is involvement of the central nervous system

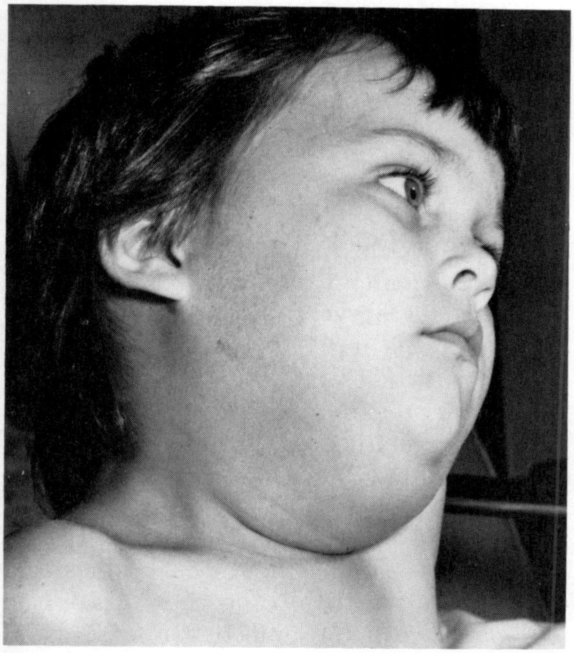

FIGURE 15-11. A child with mumps parotiditis. (Courtesy of Centers for Disease Control, Atlanta, GA 30333.)

with meningoencephalitis. Aseptic meningitis may occur without parotiditis, more commonly in males than females. Patients are infectious for 6 days after clinical onset of fever and swelling. The virus is shed in the saliva, urine, and cerebrospinal fluid for the first 5 days and may persist for 15 days. The virus may also be present in asymptomatic carriers. Immunity is permanent after infection. The diagnosis is not difficult clinically because of the distinguishing signs of parotid gland swelling in typical cases. Two complement-fixation antigens corresponding to the nucleocapsid and envelope can be identified. There is a skin test for mumps, but false negative tests are common.

Control and treatment. Killed virus vaccines have been used against mumps with varying success. The focus now is on administration of live attenuated vaccines. The product most commonly used combines live attenuated virus vaccines against rubella, measles, and mumps. Mumps vaccine is recommended for infants over age 1 year, as well as adolescents and adults who escaped infection as children. Mumps vaccine is contraindicated in all persons with sensitivity to eggs or neomycin, during pregnancy, and in all immunocompromised hosts.

In children uncomplicated mumps is usually a benign disease, and in the vast majority, no treatment is indicated. When treatment is indicated, therapy is directed to the relief of symptoms such as pain in the region of the parotid gland or other affected salivary glands, fever, and malaise. Management of complicated cases is directed at the system involved.

Measles Virus

Biologic characteristics. The measles virus is morphologically a member of the paramyxovirus family in the genus *Morbillivirus* but it lacks neuraminidase, and its hemagglutinin is nonreactive in humans. Like the other members of the family, the virion is large and spherical (Fig. 15-12). It contains single-stranded RNA, and is surrounded by a thick lipoprotein envelope with short spikelike projections. Only one serotype of measles virus has been isolated. The measles virus is highly temperature sensitive, and is rapidly inactivated at 37°C, or by refrigeration. It is inactivated by strong light, a pH of less than 5, and by proteolytic enzymes. Specimens for virus isolation must either be inoculated into tissue cultures shortly after collection or frozen at −70°C. Cell cultures are the only practical method for isolating the measles virus.

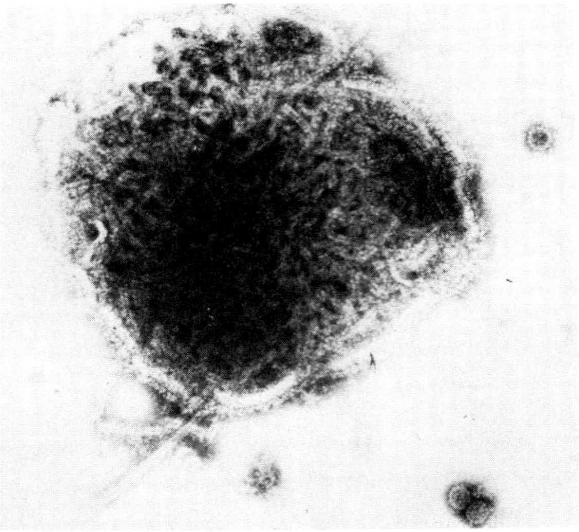

FIGURE 15-12. Measles virus (+SV 40). (Courtesy of Centers for Disease Control, Atlanta, GA 30333.)

Epidemiology. Humans are the only natural reservoir of measles virus, although when monkeys are removed from their natural habitat, they can acquire and shed the virus following contact with infected human beings. In the nonvaccinated population, measles is one of the most contagious of the viral diseases. The frequency is such that almost all human beings have had measles at some time. Those who appear to have escaped infection probably have had an inapparent form of the disease. The disease is endemic, with seasonal outbreaks occurring the world over. A few cases are usually reported each week in any large population, even when the epidemic trend is in the ebb phase.

The age at which measles is commonly contracted depends on the habits of the virus in a given geographic area. In economically advanced countries measles epidemics are closely related to the school year with the peak rising in early spring and declining abruptly in early summer. The age-specific incidence rate for reported measles was highest for 10 to 14 year olds in 1980, accounting for 60 percent of the measles cases that were correlated with age. Preschool children (age 0 to 4 years) had the second highest measles incidence rate (Fig. 15-13a). The age-specific rates are changing because of vaccination (Fig. 15-13b,c).

Due to its labile properties when exposed to light and low pH, the measles virus has a short survival time when drying on a flat surface or on exposure to an acidic environment such as the gastrointestinal enzymes. The virus does, however, survive drying in microdroplets, thus it is able to spread effectively as an aerosol. The virus is spread chiefly by aerosolation. Mucosal cells of the respiratory tract continue to release virus for several days after infection. Sneezing and coughing cause suspension and spread of the measles virus.

Pathogenesis. The direct action of the measles virus on tissues of the upper respiratory system is evident histologically by multinucleated cells that characterize virus multiplication. The great affinity of the virus for the upper respiratory tract explains in particular the associated coryza, cough, bronchitis, and conjunctivitis. The duration of the rhinopharyngeal shedding of the virus begins the day following inoculation and persists for about 9 days. Leukopenia (especially decreased eosinophils and lymphocytes) results from viral growth in leukocytes. The measles virus has been shown to grow in vitro in both T and B lymphocytes with consequent destruction, which may explain the transient anergy to a wide variety of antigens during the prodromal period of measles.

The hypothesis that explains the exanthem that appears so regularly 14 days after

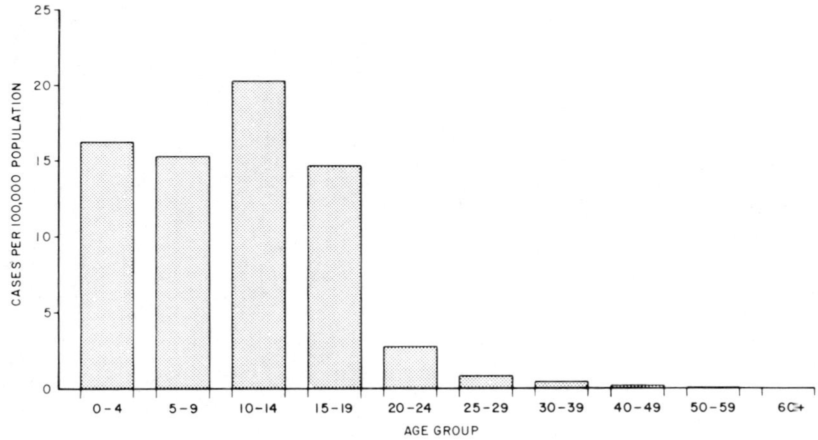

a *EXCLUDES UNKNOWN AGES.

FIGURE 15-13. Measles (rubeola). *(a)* Reported case rates by age group, United States, 1980. *(b)* Reported case rates by year, United States, 1955-1980. *(c)* Reported death rates by year, United States, 1955 to 1980. (All figures reproduced from Centers for Disease Control. Annual Summary 1980. Morb Mort Week Rep 1981;29:51,52.)

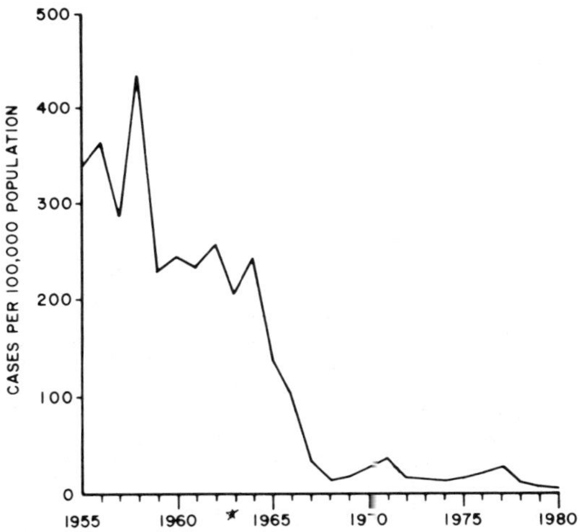

b

c *NOT AVAILABLE FOR 1979-1980.

exposure is that the rash results from a reaction between the virus and antibodies as the specific immune response begins; IgG and IgM antibodies begin circulating simultaneously with the appearance of the rash. Like the Koplik spots (small, irregular lesions with blue-white centers), the macular cells of the rash contain viral nucleocapsid. The mucous membrane and skin lesions show similar histologic changes, with greater inflammation associated with the exanthem.

Host responses. Measles is a cyclic disease, always passing through the same phases. The incubation period (or phase) lasts 10 to 11 days with little variation. A few discreet symptoms of upper respiratory infection may be noted concurrently with a low-grade fever, supposedly linked to the first phase viremia. The next phase, the invasion phase, persists for another 4 to 7 days after incubation and is marked by a series of characteristic symptoms; high fever (exceeding 39°C), conjunctivitis, and coryza first marked by sneezing, then abundant nasal discharge, giving the face a characteristic, blotched, edematous, and weepy appearance (Fig. 15-14), and a dry incessant, tiring cough. Koplik spots appear on the oral mucous membranes and occasionally on the conjunctiva or intestinal mucosa. The presence of the Koplik spots confirms the diagnosis as meales, but they appear after the fever, oculonasal catarrh, and cough. The time of onset of the exanthem, or maculopapular rash, is remarkably regular—14 days after infection. The fever usually subsides 2 to 3 days after onset of skin eruption.

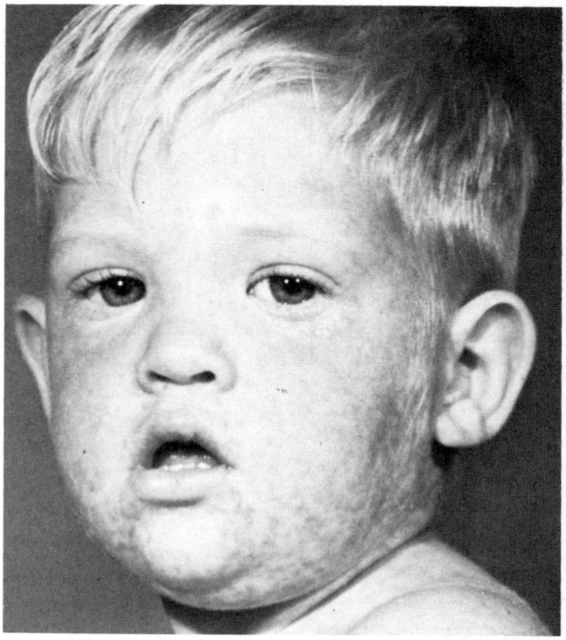

FIGURE 15-14. Big red measles (rubeola). Third day of rash. (Courtesy of Centers for Disease Control, Atlanta, GA 30333.)

The measles virus is present in the throat and serum during the days preceding the eruption and immediately after onset of eruption. The reliability for positive case identification of measles, on the basis of clinical assessment and history, is higher than for other infectious diseases due to its distinctive typical course. The rhinitis of measles occasionally takes a more complicated course. The purulent nasal discharge can erode the basal periphery of the nose and upper lip and ulcerate, resulting in breathing difficulties, especially in young infants. Often the pharynx is covered with mucus and the tonsils become red and edematous. Laryngitis may occur. Bronchopulmonary complications are rare and are less severe since chemotherapy has been used for treatment of bacterial superinfections. When measles occurs in an individual afflicted with a chronic respiratory disease, the condition is greatly aggravated; underlying asthma may be worsened, but the aggravation is transitory.

A measles virus infection in an individual with cystic fibrosis may be catastropic. The current mortality rate from all cases of measles worldwide has been estimated to be 2 per 100,000 cases, with fatalities being primarily among the very young, the very old, and those with underlying chronic illness.

A severe and rare complication of measles infection or of infection with a very closely related virus (the etiologic agent is still under dispute) is subacute sclerosing panencephalitis (SSPE). SSPE is a slow, progressive inflammation of the brain in children and young adults, and is almost always fatal. The natural measles virus or a variant is almost certainly causally involved in SSPE, but there may be other precipitating factors that result in chronic central nervous system infection. The disease is very rare; there are 30 to 40 cases per year reported in the United States (incidence less than 1 per million). It is worldwide in distribution, is more common in males than females, and is more common in rural than urban areas. It is most common in 5- to 9-year old children who contracted measles during infancy.

The onset of SSPE is insidious. Symptoms include loss of memory, poor school performance, and involuntary muscle movements progressing within a few months to coma and death. An atypical measles virus is often isolated from the brains of affected persons. Since those with SSPE have had measles at a much earlier age than usual, it is postulated that the onset of measles when maternal antibody may still be present and when immune systems may be immature may allow the measles virus to be established as a chronic infection. Measles vaccine may play a role in preventing SSPE by eliminating infection with unusual strains of natural virus. Evidence to date indicates that the risk of SSPE after vaccination is one-tenth of the risk after natural measles infection (see Chapter 16, section on "Slow Virus Diseases").

Control and treatment. Immunity to measles on subsequent exposure is directly correlated with the presence of serum antibody. Natural immunity is life long. A vaccine must give life-long protection to confer immunity comparable to that naturally acquired. The early immunizing agent, inactivated killed measles virus (KMV), offered temporary protection against the virus, but is no longer used because of serious hypersensitivity reactions with subsequent measles virus exposure after the temporary immunity ceased. Following KMV, early forms of live attenuated measles virus had to be further attenuated due to local and systemic reactions to the vaccine coupled with short-lived immunity.

The attenuated live measles virus (LMV) vaccine now in use produces minimal symptoms and reliable immunity. Since the introduction of LMV in 1963, more than 120 million doses have been distributed in the United States and at the same time there has been a rapid decrease in the incidence of measles reported to the Centers for Disease Control.

In 1978 the Measles Elimination Program was launched to eradicate measles from the United States. All states now have laws that require measles immunization prior to school admission. In 1981, 97 percent of children entering school for the first time were immunized. When compliance with immunization is maximal, most measles cases in the United States will be due to importation rather than to endemic foci. Unfortunately, the proportion of measles cases among college-aged individuals increased from 1.5% in 1980 to 27.2% in the first half of 1983. This indicates that compliance with immunization is still not optimal (Centers for Disease Control 1980, 1982, 1983).

Pseudomyxoviruses

Respiratory Syncytial Virus

Biologic characteristics. Repiratory syncytial virus (RSV) shares some biologic properties with the paramyxoviruses. It is a large, 120 to 200 nm, enveloped virus with a helix intermediate in size between the orthomyxoviruses and paramyxoviuses (Fig. 15-15). The nucleocapsid matures at the membrane of the infected cell. Like other members of this class the outer envelope is punctuated by spikelike glycoproteins, but includes no hemagglutinin. RSV is inactivated by ether and acidic pH. It is unstable at ≥ 37°C. RSV has no officially designated serotypes.

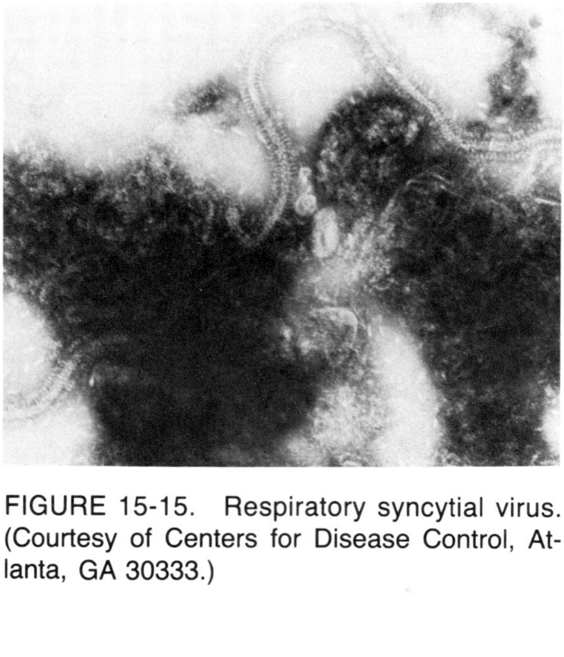

FIGURE 15-15. Respiratory syncytial virus. (Courtesy of Centers for Disease Control, Atlanta, GA 30333.)

Epidemiology. Since its isolation in 1956, RSV has emerged as the major lower respiratory tract pathogen of infancy and early childhood worldwide. RSV speads extensively in children during the winter months. In the United States epidemics occur in late fall, winter, and spring—but not summer. RSV is the only respiratory pathogen that produces a sizable epidemic every year in large communities. It is spread by person-to-person contact and airborne transmission of respiratory secretions.

Pathogenesis. Respiratory syncytial virus owes its name to its ability to form syncytial masses in growing cell monolayers. The virus-host cell interaction may undermine the protection of local immune defenses since syncytial formation allows recruitment of uninfected cells into the pathologic process. These pseudogiant cells enhance the pathogenic effects by spreading the virus. Cell-to-cell spread of virus accounts for spread of virus in a partially immune host. In order to persist for extended periods in eukaryotic cells, RSV uses the mechanism of transfection to form DNA intermediates infectious to susceptible cells. IgE mediated allergic reactions follow an initial sensitizing infection. Natural immunity appears to be minimal, since annual reinfections occur in the first years of life. Eighty percent of children possess serum neutralizing antibody for RSV by the time they are 4 years of age.

Host responses. Illness due to RSV begins in much the same way as illness due to parainfluenza infections; however, a cough soon becomes the predominant symptom and the fever—instead of subsiding quickly—continues for 4 to 7 days. There is a tendency in children for the inflammation of the mucous membranes of the respiratory tract to extend into the trachea, bronchioles and parenchyma. The younger the child, the greater the tendency for necrotizing bronchiolitis and pneumonia leading to atelectasis and emphysema. In older children milder infections of the upper respiratory tract with coryza and pharyngitis are common. Outbreaks in adults resemble the common cold. Aged adults and those with chronic lung disease are susceptible to pneumonia from RSV with serious consequences. This agent is the major viral cause of serious lower respiratory tract disease.

Serum neutralizing antibodies are known to circulate in the blood of children by age 4, but offer little protection. Older children and adults have less severe illness. The RSV is unstable and whenever possible, specimens should be inoculated directly into cell cultures without previous refrigeration. Neutralization and complement-fixation tests are used both to identify the virus and to estimate antibody titers.

Control and treatment. Although the severity of primary infection with RSV and the early age at which it occurs have not been fully explained, these characteristics demonstrate the need for vaccine. However, no safe, effective vaccine has been developed yet. If vaccine were to be developed, it would have to be safe for infants 1 month of age, since peak incidence occurs at age 2 months. Inactivated vaccines were recently proven ineffective in palliation and were found to cause more severe subsequent infections characterized by pneumonia, sufficiently severe to require hospitalization. Efforts to develop attenuated vaccines that induce nasal antibody formation are not widespread as yet. Research on methods of producing RSV vaccine must continue to be given high priority. Currently, treatment is limited to supportive care during the acute illness, but treatment may be lifesaving, particularly in individuals with airway obstruction.

PICORNAVIRUSES

The largest group of viruses that produce respiratory illness are the picornaviruses. The name is derived from their small size *(pico)* and from the fact that they have an RNA core (pico*rna*). They are classified as nonenveloped RNA viruses. This family includes two genera, *Rhinovirus* and *Enterovirus* (Table 15-5) that cause respiratory illness. The rhinoviruses are most frequently associated with the "common cold" syndrome, whereas the enteroviruses are most noted for infections of the gastrointestinal tract, but are also causes of respiratory diseases.

The rhinoviruses and enteroviruses share some basic properties—such as size, shape, nucleic acid composition, and ether resistance—but differ in their susceptibility to acidic environments. Rhinoviruses grow best in tissue culture at 33°C and lose their infectivity when held 1 to 3 hours at a pH of less than 3. Enteroviruses, on the other hand, grow well at 37°C and retain infectivity when kept at a pH of less than 3. These properties are believed to account for limitation of rhinoviruses to the nasal passages where low temperatures prevail, and

TABLE 15-5. Tissue Predilection of Picornaviruses

Group	Tissue predilection
Rhinoviruses	Nasal passages (common cold)
Enteroviruses	GI tract, upper respiratory tract
Echoviruses	Upper respiratory tract (common cold)
Coxsackieviruses	Upper respiratory tract (common cold), GI tract
Polioviruses	Spinal cord, brain

survival of enteroviruses in the acidic environment of the gastrointestinal tract. Similarities in the physical nature of the families may explain similarities in their epidemiologic behavior such as little seasonal variation, except possibly some increased frequency in winter and spring, and potential spread of the viruses by direct contact.

Rhinoviruses

Biologic characteristics. The number of serotypes accepted as antigenically distinct strains of rhinoviruses exceeds 110. The diameter of this diminutive group of viruses ranges between 15 and 30 nm. The capsomere has cubic capsids in 60 identical 4-chain subunits surrounding the RNA core. Rhinoviruses (RV) do not share a common antigen and there seems to be little antigenic cross reactivity between serotypes. Undoubtedly, this property accounts for the frequency of rhinovirus infections. Hemagglutinins and hemadsorption are not demonstrated within this family.

Epidemiology. Rhinoviruses have been associated with respiratory illnesses throughout the world. In terms of economic cost (days lost from work or school) and human discomfort, acute viral upper respiratory diseases represent a large health problem. Epidemiologic surveillance of RV infections has identified two relevant phenomena:

1. In each geographic area certain serotypes are more frequently encountered and tend to persist. Such common serotypes account for a disproportionate share of the infections.
2. There has been a clear increase over time in the proportion of RV strains isolated, suggesting that serotypes continue to emerge as the result of antigenic shift.

Infections may be detected in any season but are more prevalent in autumn or spring. Infections with RV affect a larger proportion of adults than children. There is no ready explanation for the age or seasonal incidence of RV infections.

The exact mechanism by which RV are passed from person to person is unknown. Virus shedding occurs primarily from the nose, 3 days after infection and continuing for up to 2 weeks. The concentration of virus is higher in nasal mucus than elsewhere in pharyngeal secretions; this is evidence that RV are not spread through the air by coughing or sneezing oral secretions. The potential for people with RV infection to contaminate the environment is high; the hands of infected individuals are contaminated with RV 50 percent of the time. The virus typically survives for about 3 hours on skin and other surfaces. The frequent finding of RV on the hands and its relative scarcity in secretions from simulated coughs and sneezes suggest that manual contact with nasal secretions may be the important mode of transmission.

Pathogenesis. The exact mechanism by which rhinoviruses produce disease is unknown. One mechanism is thought to be by direct cellular injury secondary to viral replication. Cell lysis and necrosis are followed by formation of multinuclear "giant cells" or syncytia. Hyperemia and edema of mucous membranes are accompanied by exudation of serous and mucinous fluid. Edema of subendothelial connective tissue with infiltration of neutrophils is evident histologically. Characteristic changes seen by electron microscopy include extrusion of goblet cells and areas of denuded epithelium, with the nasal passages being the major site of infection. These findings explain the rhinitis and depressed nasal mucociliary flow rates observed in infected individuals.

Serum neutralizing and nasal antibody titers rise with RV infections. Levels of these titers are inversely proportional to infection rates. The severity of illness may be modified with increased antibody titers.

Host responses. Illness associated with RV is usually afebrile and principal manifestations are rhinorrhea, nasal stuffiness, and sneezing and scratchy throat accompanied by nasal and pharyngeal mucosal inflammation. Increased production of mucus may lead to cough. The clinical syndrome associated with the RV is called the "common cold." The incubation period is short, 24 hours to 3 days. Lower respiratory tract involvement is unusual in adults, except in individuals with chronic bronchitis and asthma. The total impact of RV infections in children is minor compared to the impact of RV in adults. Complications of infections are rare. The relatively benign nature of rhinoviral illness usually precludes virologic studies in patients with this illness. The diagnosis can be suspected in an adult with an afebrile upper respiratory illness occurring in fall or spring. Diagnosis is confirmed by viral isolation from nasal washings, or by serologic tests.

Control and treatment. The high morbidity associated with RV infections, together with the associated economic loss, makes these infections of considerable importance. Preventive measures are thus desirable, but it is doubtful whether specific prophylaxis can be achieved in the near future. Studies on the etiology of "colds" have been motivated by the hope that, when the causative agent(s) are identified, effective vaccines can be developed. However, vaccines to RV have been ineffective because of the multiplicity of RV serotypes. Clinical trials with inactivated monovalent rhinovirus vaccines administered parenterally have shown such vaccines may ameliorate illness and reduce shedding of the virus, but not prevent infection. An effective vaccine will most likely be administered intranasally.

Enteroviruses

Enteroviruses of the coxsackie and echovirus groups are most commonly associated with intestinal symptoms (see Chapter 14), but can also cause respiratory disease. The echoviruses have been associated with mild febrile, upper respiratory tract disease in children, but rarely cause more severe respiratory illness. The viruses may spread via respiratory secretions or by fecal-oral contamination, including spread by vehicles such as contaminated water. The high communicability of these infections is most clearly evidenced in military camps and nurseries. Communal infections are due to the long period of virus shedding of the enteroviruses. The virus is regularly present in the throat and in the stools before the onset of illness. One week after onset, the presence of virus in the throat diminishes, but virus continues to be excreted in the stools for weeks.

ADENOVIRUSES

Adenoviruses were discovered when it was noted that adenoid tissue degenerates spontaneously in the laboratory due to the presence of a latent virus recovered from asymptomatic children following tonsillectomy. This virus was isolated from the excised adenoid tissue, hence the name adenovirus. A total of 33 serotypes have been isolated from human respiratory tissues and secretions.

Biologic characteristics. The viruses belonging to the adenovirus family are DNA viruses, medium sized, with a diameter of 60 to 90 nm. They lack an envelope. The hexagonal capsid consists of over 252 capsomeres arranged in icosahedral symmetry (Fig 15-16). The core, which contains DNA, is linear and double stranded, with the base composition of nucleic acid varying with each serotype. Chemically the adenovirus is relatively simple, consisting of only protein and DNA, 77 percent and 23 percent, respectively.

The number of structural polypeptides forming the protein portion varies with each serotype. The 34 human serotypes are divided into three groups according to their guanine and cytosine content of the DNA molecule. The viruses are ether resistant, and are stable at low temperatures and a fairly wide range of pH. Since the capsid units are soluble in denaturing solvents, it has been possible to develop methods to degrade the virion. All adenoviruses possess a common group-specific antigen, as well as subgroup-specific antigens.

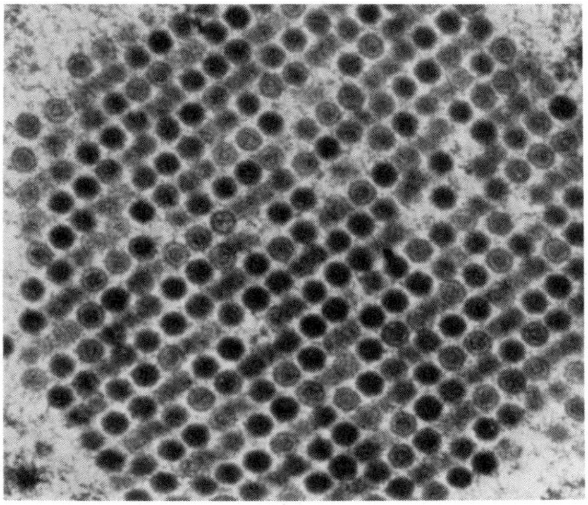

FIGURE 15-16. Adenovirus. Electron micrograph showing crystallike arrangement. (Reproduced with permission of the publisher from Moffet HL. Clinical Microbiology, 2nd edition. Philadelphia: JB Lippincott, 1980.)

Epidemiology. Adenoviruses are a ubiquitous group of viruses and possess diverse antigenic characteristics (Table 15-6). The "endemic" serotypes (1, 2, 5) circulate widely among young children, whereas the "epidemic" serotypes (3, 4, 7) are found more frequently in older children and adults. Adenoviruses cause less than 10 percent of viral upper respiratory tract disease in civilian groups. However, for unknown reasons, among military recruits the proportion of adenovirus respiratory infections exceeds 50 percent, while college students in the same age group rarely suffer from adenoviral infections. Epidemiologic surveys show that types 1, 2, and 5 are important causes of upper respiratory tract infections in children, and type 7 causes severe childhood pneumonia. Types 3, 4, and 7 have been found to be the major adenoviruses affecting military recruits.

Because of fecal shedding, the fecal-oral route may lead to infection, especially in children, though airborne transmission is the most common mode of spread. Although adenovirus infections are usually community acquired, a recent report suggests that these viruses may cause nosocomial disease as well. The "endemic" adenoviruses are isolated continuously, with no clear-cut seasonal distribution. The "epidemic" serotypes appear at irregular intervals. Outbreaks of adenovirus infection have been associated with swimming pools.

TABLE 15-6. Adenoviruses: Major Serotypes and Associated Clinical Syndrome, Age, and Epidemiologic Manifestations

Human serotype	*Major clinical syndrome*	*Peak age group*	*Epidemiologic manifestations*
1	Mild acute upper respiratory tract illness	Children	Sporadic, endemic
2	Mild acute upper respiratory tract illness	Children	Sporadic, endemic
3	Acute respiratory disease; minor respiratory disease in children	Older children and adults	Epidemic (military recruits)
4	Acute respiratory disease adenovirus pneumonia	Older children and adults	Epidemic (military recruits)
5	Mild acute upper respiratory tract illness	Children	Sporadic, endemic
6	Pharyngitis	All ages	Sporadic
7	Adult: acute respiratory disease; upper respiratory tract and severe pneumonia in children	Older children and adults	Epidemic (military recruits)

Pathogenesis. Human adenoviruses primarily affect the respiratory tract, and are sometimes accompanied by conjunctivitis and lymphoid tissue inflammation. These observations suggest a predilection for certain mucous membranes and lymph tissue, where they persist for long intervals. Infected host cells become rounded and enlarged, and clump to form grapelike clusters. Nuclei of the infected cells may contain characteristic inclusion bodies, which are the principal sites for viral maturation and assembly.

Host responses. The primary adenoviral infection usually occurs early in childhood with the production of a mild, acute respiratory tract illness. It is often accompanied by fever, malaise, headache, chills, pharyngitis, and cough. Acute febrile pharyngitis is perhaps the most common syndrome, but localizing signs may not develop and the illness may be thought to be a "fever of unknown origin." Sporadic cases cannot be distinguished clinically from other respiratory viruses, excluding the mumps and measles viruses. The incubation period for febrile pharyngitis is usually 5 to 7 days, with a gradual onset and short duration.

Acute respiratory disease (ARD), or febrile catarrh, is the disease that affects military recruits, but not civilian adults of similar age. The disease varies in severity, but usually presents as a mild, self-limiting illness characterized by fever, chills, headache, pharyngitis, and cough. Pneumonia sometimes occurs as a complication of ARD. Approximately 5 to 15 percent of patients with ARD develop radiologic evidence of pulmonary involvement. On rare occasions, these viruses have been known to cause serious illness, especially in infants and young children.

Isolation or cultivation of adenoviruses is

done by inoculation of tissue cultures. The virus may be recovered from collected nasopharyngeal secretions, eye exudate (Fig. 15-17), and rectal swabs. Specimens should be gathered a few days after onset of illness, but no later than 7 days. Group-specific and type-specific identification using conventional serologic technique is possible; however, it is a relatively slow process, requiring 1 to 2 weeks.

Control and treatment. No specific therapy exists for the treatment of adenoviral illnesses. Symptomatic, supportive treatment is usually sufficient. Because of fecal shedding, fecal-oral route spread should be minimized by adherence to strict aseptic technique when contacting secretions of an infected patient. Soon after the adenoviruses were isolated and shown to be etiologically associated with ARD in military recruits, formalinized vaccines for types 3, 4, and 7 were produced. The vaccines were administered as enteric-coated gelatin capsules to bypass the upper respiratory tract, the usual pathologic site, and produce a silent lower gastrointestinal tract infection. Recipients developed moderately high titers of protective circulating antibodies. The cost benefits of the vaccination program demonstrated its potential success, although the inability to consistently produce batches of vaccine, coupled with the occasional presence of a contaminating oncogenic component, have resulted in the discontinuation of vaccine production for type 3 adenovirus.

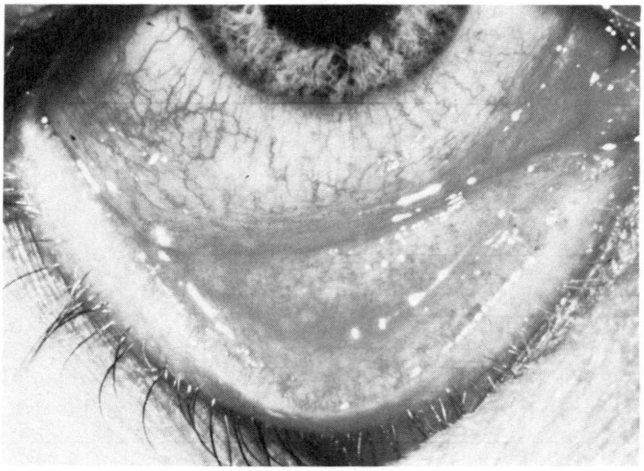

FIGURE 15-17. Adenoviral conjunctivitis. (Courtesy of Dr. John Chandler.)

CORONAVIRUSES

Biologic characteristics. Coronaviruses are in the family Coronaviridae. The external structure of the coronavirus particles is suggested by the designation: corona (Fig. 15-18). The particle is approximately 80 to 160 nm in diameter and has characteristic rounded projections, resembling a crown. The virus is inactivated by ether and an acidic pH of less than 3.0. The nucleic acid

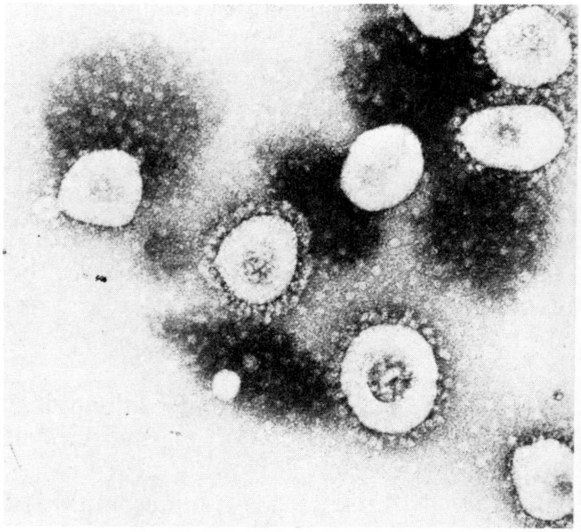

FIGURE 15-18. Coronavirus OC 43. (Courtesy of Centers for Disease Control, Atlanta, GA 30333.)

core contains RNA and is enveloped. Both hemagglutinin and complement-fixation antigens have been associated with the surface of the virion and are located in the projections. The antigenic characteristics of the isolated strains have not been defined, but antigenic relationships have been demonstrated to exist between some serotypes. Coronaviruses were previously called the human respiratory virus.

Epidemiology. Coronaviruses are worldwide in their distribution. All known human strains were isolated originally from the respiratory tract, most during upper respiratory illness. Evidence is mounting that the coronaviruses are of major importance in common respiratory infections of all age groups, but the total impact cannot be calculated because all serotypes have not been identified. The incubation period after coronavirus infection is short, 2 to 4 days from exposure to the development of symptoms. Illness persists for 4 to 7 days. The coronaviruses are transmitted by the respiratory route. It is likely that the viruses can be transmitted by aerosol and large droplet, which explains their ability to spread as rapidly as does the influenza virus.

Pathogenesis. The cytopathic effects of coronaviruses occur slowly, and usually do not involve the entire host cell layer. Virions develop by budding into cytoplasmic vesicles, utilizing the vesicle membrane to form an outer coat. Infected cells become elongated and "stringy" in appearance, and the cells eventually degenerate. Reinfection occurs in the presence of antibody titers, which may be explained by modification of the virus.

Host responses. Coronaviruses are a cause of the common cold, particularly in adolescents and young adults. Most illnesses associated with infection by coronaviruses are afebrile upper respiratory infections with profuse rhinorrhea. There are no documented cases of fatal human coronavirus infections.

Control and treatment. There are no prophylactic methods available for human strains of the coronavirus. Treatment is supportive.

OTHER RESPIRATORY VIRUSES

Adeno-associated viruses (AAV), which are ssDNA parvoviruses, are defective, and require a "helper" virus to grow. Though different than adenoviruses antigenically, they require simultaneous infection of the same cell with adenovirus to replicate. AAV antigens, hence, can only occur in cells that also contain adenovirus antigens. Likewise, no human possesses only AAV antibody unless they also have antibody to adenovirus. Herpesvirus can also act as a partial helper, since AAV DNA and some proteins are replicated in cells simultaneously infected with both AAV and herpesvirus, but no complete virions are formed. AAV are not known to cause disease.

Newcastle disease virus (NDV) is a typical paramyxovirus, which resembles parainfluenza and mumps viruses. Its primary host is poultry, in which it causes respiratory disease. Accidental infections among humans exposed to infected chickens or other birds result in mild conjunctivitis that resolves in less than 2 weeks. Because NDV has been used extensively in the laboratory to study paramyxoviruses, the disease also occurs in laboratory personnel. Antibodies to NDV cross react with those of mumps, hepatitis, and Epstein-Barr virus.

Reoviruses (*r*espiratory and *e*nteric *o*rphan viruses) are double-stranded RNA icosahedral viruses that have been isolated in young children with coryza, as well as in asymptomatic individuals. They may cause mild respiratory illness (see Chapter 14 for additional discussion).

CLINICAL IMPLICATIONS

Determining the proportion of respiratory diseases attributed to viruses is difficult because collecting and testing of suitable specimens is problematic. Clinically, viruses are only documented as the causative agent in 1 out of 4 actual cases. Since the advent of organic tissue cultures and electron microscopy, nearly 80 percent of suspicious cases yield viruses in the laboratory.

Respiratory tract infections are not precisely defined syndromes. Instead they are best described as a set of disorders, and present with a spectrum of clinical manifestations. Greater than 90 percent of upper respiratory tract infections are caused by viruses. These infections are usually associated with mild, localized illnesses that are accompanied by minor but discomforting symptoms.

In contrast, infections of the lower respiratory tract, 50 percent of which have viral origin, follow a more serious course. Although the exact mechanism is unknown, changes in the function of the large and small airways, evidenced by spirometry declination, airway resistance and decreased dynamic compliance, follow the third to fifth day of the lower tract viral infections. For the previously healthy individual, the pulmonary function changes are for the most part reversible, although some studies have demonstrated that an asthmalike tendency for airway hyperactivity develops after viral respiratory infections. This causes an increased responsiveness of the trachea and bronchi to various stimuli.

Viral respiratory infections place patients with chronic lung disease at high risk by aggravating their preexisting defects in pulmonary function, thereby causing acute exacerbation of their disease. Bacterial superinfections were traditionally implicated in acute exacerbative episodes of chronic lung disease; however, several studies show a direct relationship of such episodes to viral infections.

There are no antiviral drugs or vaccines in general use for the viruses that cause respiratory disease, though development is ongoing. Management of viral cases is aimed at reducing physiologic effects and providing palliative care. Often patients present a complex clinical picture that makes it difficult to localize sites and determine the origin of their illness as either viral or bacterial. In addition to a careful physical examination, chest x-rays and sputum specimens may be required. Diagnosis by virus isolation or serology may be indicated. In situations where there is considerable doubt as to the etiology, antimicrobial agents are often used. Most respiratory viral infections are self-limiting and only symptomatic treatment is desirable.

Patients with viral respiratory disease require careful nursing. Large amounts of fluids taken by mouth; sponging and antipyretics to control fever; moist, smoke-free air; protection from overheating and chilling; and adequate nutrition are essential. Hospital management is often necessary for children so that immediate assistance is available for acute respiratory distress, convulsions, delayed recovery, or lack of response to therapy. Transmission of viral agents may be reduced by proper personal hygiene, handwashing, and avoiding hand-to-mouth, hand-to-nose, and hand-to-eye contact. Nosocomial epidemics of RSV are common in hospitals during peak virus seasons. Hospitalized persons with influenza, parainfluenza, or RSV infections must have a private room, or be cohorted with other patients who have the same infection. Gowns are also recommended for personnel who care for patients with RSV or parainfluenza. Masks should be used by those caring for patients with influenza during the period in which the patient is symptomatic.

A discussion of the cause and prevention of respiratory viral infection is not complete without mentioning the health care team member as a possible source of transmission. Personnel should stay at home when symptoms of a head cold or respiratory illness begin, as this is the period of highest communicability. Health care personnel can provide maximum protection to their patients by maintaining their own proper health and thoroughly washing their hands before and after treating patients.

Since the 1960s a rare disease called Reye's syndrome has been associated with viral infections in children, especially influenza and chickenpox. Reye's syndrome causes death in about one-fourth of reported cases and can result in permanent brain damage among survivors. It was previously speculated that the infecting virus itself was the etiologic agent of Reye's syndrome, but careful case-control studies and additional evidence clearly implicate the salicylates (aspirin and related compounds) used to treat the viral illness rather than the virus alone as the agent correlated with Reye's syndrome. As a result of this evidence, the Surgeon General has advised against the use of salicylates for children with influenza or chickenpox.

SUGGESTED READINGS

Andrewes CH. The natural history of viruses. New York: WW Norton 1967.

Andrewes CH. Viruses of vertebrates, 4th edition. London: Baillière Tindall, 1978.

Attal C, Mozziconacci P. Measles: clinical features. In: Debré R, Celers J, eds. Clinical virology. Philadelphia: WB Saunders, 1970.

Belshe R. Viral respiratory infections. In: Conn HF. Current therapy 1979. Philadelphia: WB Saunders, 1979.

Centers for Disease Control. Measles prevention. Morb Mort Week Rep 1982;31:217-222.

Centers for Disease Control. Annual summary 1980. Morb Mort Week Rep 1981;29:54.

Centers for Disease Control. Measles—United States, first 26 weeks, 1983. Morb Mort Week Rep 1983;32:363-364.

Clyde W. Experimental models for study of common respiratory viruses. Environ Health Perspec 1980;35:107-112.

Debré R, Celers J, eds. Clinical virology. Philadelphia: WB Saunders, 1970.

Drew LW. Viral infections. Philadelphia: FA Davis Company, 1976.

Egerer RM, et al. Impact of swine flu vaccine on serum antibodies. Am J Epidemiol 1979;109:81-87.

Evans AS, ed. Viral infections of humans: Epidemiology and control. New York: Plenum Medical Book Company, 1976.

Fenner F, et al. The biology of animal viruses. New York: Academic Press, 1974.

Fox JP, Hall C. Viruses in families. Littleton, Mass: P.S.G. Publishing Company, 1980.

Knight V, ed. Viral and mycoplasmal infections of the respiratory tract. Philadelphia: Lea and Febiger, 1973.

McLean DM. Virology in health care. Baltimore: Williams and Wilkins, 1980

Moffit HL. Clinical microbiology. Philadelphia: JB Lippincott, 1980.

Pickin JJ. Effects of viral infections on small airways. Am J Med 1972;52:738-745.

Pingleton SK, et al. Type 3 adenoviral pneumonia occurring in a respiratory intensive care unit. Chest 1978;73:554-555.

Reichman RC, Dolin R. Viral pneumonia. Med Clin North Am 1980;64:491-506.

Ryden IC, Cobbs CG. Respiratory infections. Med Times 1981;109:80-93, 95.

Sarkkinen HK, Halonen PE. Detection of respiratory syncytial, parainfluenzae type II and adenovirus antigens. J Clin Microbiol 1981; 13:258-265.

Schonberger LB, et al. Guillain-Barré syndrome following vaccination in the national influenza immunization program, United States, 1976-1977. Am J Epidemiol 1979;110:105-123.

Shell G. Upper and lower respiratory tract infections. Nurs Clin North Am 1980;15:715-727.

Stuart-Harris CH. Influenza and other virus infections of the respiratory tract. Baltimore: Williams and Wilkins, 1965.

Taylor PE. Adenoviruses: diagnosis of infections. In: Kurstak E, Kurstak C, eds. Comparative diagnosis of viral disease. New York: Academic Press, 1977.

Tyrell DAJ. Approaches to the control of respiratory viral diseases, Volume 58. Geneva: World Health Organization, 1980.

Valenti VM, et al. Nosocomial viral infections. II. Guidelines for prevention and control of respiratory viruses, herpesviruses, and hepatitis viruses. Infec Cont 1980;1:165-178.

Wadell G, et al. Epidemic outbreaks of adenovirus type 7 with special reference to the pathogenicity of adenovirus genome type 7b. Am J Epidemiol 1980;112:619-628.

World Health Organization. Technical Report Series #642, Viral respiratory diseases. Geneva: World Health Organization, 1980.

Wistreich GA, Lechtman MD. Microbiology and human disease. Beverly Hills, Ca: Glencoe Press, 1976.

16 NEUROTROPIC VIRUSES

KEY POINTS

- The arthropod-borne viruses (arboviruses) usually cause febrile disease, often accompanied by bleeding or rash. They are usually geographically limited in distribution.
- Major encephalitis viruses are eastern equine encephalitis, western equine encephalitis, Venezualan equine encephalitis, Japanese B encephalitis, and St. Louis encephalitis. They are spread by mosquito vectors. Their natural hosts are wild birds or rodents.
- The tick-borne California encephalitis virus is the most common arthropod-borne encephalitis in the United States.
- Dengue and yellow fever are mosquito-borne febrile diseases that are widely distributed in warm climates. Both occur in epidemic proportions.
- Rubella virus is spread by the respiratory route from person-to-person; there is no arthropod vector. It is a very mild disease. The purpose of vaccination against rubella is not to protect the host from disease, but to prevent teratogenic effects in fetuses.
- Arenaviruses are not arthropod-borne. Rodents are the natural host, but humans can develop hemorrhagic fevers with agents such as the lymphocytic choriomeningitis, Lassa, Junin, or Machupo viruses.
- Rabies is transmitted by animal bite. It is fatal without treatment.
- Certain neurologic diseases such as Kuru and Creutzfeldt-Jakob disease are called slow-virus diseases. Their incubation periods last for months to years.

ARBOVIRUSES

Many of the viruses that cause primary central nervous system symptoms are from a group called arboviruses (*ar*thropod-*bo*rne). The diseases they cause are often difficult to diagnose because they cause clinical manifestations common to many other diseases. Arbovirus is not a taxonomic classification because the grouping is based upon the mode of transmission (by arthropod vectors) rather than any morphologic or biologic characteristics. The arboviruses do, however, share some characteristics (Table 16-1).

TABLE 16-1. Characteristics Common to Arboviruses

Transmitted by arthropod (insect) vector
RNA genome
Usually enveloped
Usually do not cause disease in arthropod vector
Mature by budding, multiply in host cytoplasm
May cause one or more of three clinical syndromes:
Benign fever, sometimes with rash
Encephalitis
Hemorrhagic fever
Limited geographically by location of vector and susceptible host(s)

There are more than 400 types of arboviruses, which have been divided on the basis of hemagglutination, complement fixation, or other serologic tests into over 50 antigenic groups. However, about 90 types still remain unclassified. Arthropod-borne viruses belong primarily to four viral families: Togaviridae, Bunyaviridae, Reoviridae, and Rhabdoviridae (Table 16-2).

There are several general principles of arbovirus infection. With the exception of yellow fever and dengue, which are named after the symptoms they cause, arboviruses are named for the geographic location where they were first isolated. For example, St. Louis encephalitis, Venezuelan equine encephalitis, Machupo virus, Junin virus. Because an acceptable vector, usually a tick or mosquito species, and a susceptible host must be available, most arboviruses are limited to relatively circumscribed geographic areas. The mosquito, *Aedes aegypti,* however, is present worldwide; therefore, the viruses transmitted by this species have the potential for being widespread.

A high proportion of arboviral infections are asymptomatic. If clinical symptoms occur, they are similar for all types: sudden onset of fever, headache, nausea and vomiting, drowsiness, stiff neck, stupor, loss of consciousness. Arboviruses cause primarily encephalitides and febrile illness, sometimes accompanied by genital rash or hemorrhage. Some may cause congenital defects. Vectors of arboviruses come almost

TABLE 16-2. Major Taxonomic Groups Containing Arthropod-Borne (Arbo) Viruses

Family	*Approximate number of arboviruses*	*Genera*	*Major diseases*
Togaviridae	85	*Alphavirus*	Western, eastern, and Venezuelan equine encephalitis
		Flavivirus	St. Louis encephalitis, yellow fever, dengue, Japanese B encephalitis
Bunyaviridae	115	*Bunyavirus*	California encephalitis
		Other	Phlebotomus fever
Reoviridae	39	*Orbivirus*	Colorado tick fever
Rhabdoviridae	29	*Vesiculovirus*	Vesicular stomatitis

TABLE 16-3. Vectors of Arboviruses

Type	Genus	Some diseases transmitted
Mosquito	*Aedes*	Dengue, urban yellow fever, California encephalitis, Chikungunya, Ross River fever
	Anopheles	O'nyong-nyong
	Culex	Eastern, western, Japanese B, St. Louis, Murray Valley encephalitis
	Culiseta	Eastern equine encephalitis
	Haemagogus	Yellow fever
Sandfly	*Lutzomyia*	Vesicular stomatitis
	Phlebotomus	Sandfly fever
Tick	*Haemaphysalis*	Powassan, Russian spring-summer encephalitis
	Ixodes	Colorado tick fever, Powassan fever
	Dermacentor	Colorado tick fever, Powassan fever

exclusively from a few genera of mosquitos, ticks, and sandflies (Table 16-3).

For most arbovirus infections, humans are an accidental host. (Yellow fever and dengue, which are diseases exclusively of humans, are the exceptions.) Common hosts include wild and domestic birds, rodents, reptiles, small mammals, and other vertebrates. Host responses to arbovirus range from completely refractory to the host acting as a disseminator (Table 16-4). This chapter is concerned about the few arboviruses (less than 100) for which humans can serve as hosts. In addition to arboviruses, this chapter will include discussions of other togaviruses and rhabdoviruses of the family Arenaviridae that are not arthropod-borne, and of the group of viruses known as "slow viruses."

TABLE 16-4. Levels of Host for Arboviruses

No involvement in transmission; host is refractory.
Incidental; susceptible but level of viremia is inadequate to infect vectors.
Maintenance or reservoir host; necessary for maintenance of the virus.
Amplifying host; same as above plus high population density.
Disseminating host; transfers virus between foci.

Adapted from Monath, TP. St. Louis encephalitis. Washington DC: American Public Health Association, 1980.

TOGAVIRUSES

Most of what were previously called arboviruses are now classified in the family Togaviridae, named for the fact that they all have an envelope (toga). They all contain single-stranded RNA and have a diameter of 45 to 75 nm with 32 capsomeres. The togaviruses can be grown with relative ease in cell culture, in embryonated chick eggs, and by intracerebral injection of newborn mice. They replicate in host cell cytoplasm and generally are released by budding through the plasma membrane.

On the basis of serologic tests, the togaviruses are divided in four genera: alphaviruses, flaviviruses, rubiviruses, and pestiviruses (Table 16-5). At present there are

TABLE 16-5. Togaviridae Causing Disease in Humans

Genus	Arthropod-borne?	Approximate number of types	Subgroup	Vector	Type of disease
Alphavirus	Yes	24	1	Mosquito	Western, eastern, Venezuelan equine encephalitis, Simbis
			2	Mosquito	Chikungunya, Mayaro, Ross River virus, O'nyong-nyong
Flavivirus	Yes	60	1	Mosquito	St. Louis, Japanese B encephalitis, West Nile virus, Murray Valley encephalitis, Ilheus
			2	Mosquito	Dengue
			3	Mosquito	Yellow fever
			4	Tick	Russian spring-summer encephalitis
Rubivirus	No	1		—	Rubella

identified about 24 alphaviruses, 60 flaviviruses, 1 *Rubivirus* and 3 pestiviruses. (The three pestiviruses are not arthropod-borne, affect only animals, and will not be discussed further.) Rubella, the only species in the genus *rubivirus,* is not arthropod-borne, but it causes an important disease affecting humans and will be discussed later in this chapter. Most of the alphaviruses and flaviviruses are arthropod-borne and thus are often termed arboviruses.

Togaviruses have surface projections that are hemagglutinins. Two immunologic subgroups of alphavirus and four of flavivirus cross react with hemagglutinating nucleocapsid proteins. The hemagglutination inhibition antibodies that develop after infection or immunization with one togavirus cross react and result in the formation of protective antibody against all togaviruses. The same virus, however, may be unique when isolated at different times. This is a problem in attempts to develop vaccines. Infection with togavirus results in permanent immunity. Neutralizing antibodies and hemagglutination inhibition antibodies last for years, complement-fixation antibody is of shorter duration, 2 to 5 years.

Encephalitides

Biologic characteristics. The majority of togaviruses that cause encephalitis in humans are spread by mosquito vectors. In addition, Russian spring-summer encephalitis and related viruses are group 4 flaviviruses and are tick-borne (Table 16-6). All are typical spherical togaviruses. The alphaviruses are produced and released from the cell in greater quantities (up to 10,000 per hour) than are the flaviviruses. The particles become infectious only when they have acquired their envelopes. The diseases are diagnosed by serologic tests: complement fixation, hemagglutination inhibition, neutralizing antibody, radioimmunoassay, and virus isolation.

TABLE 16-6. Some Encephalitides Caused by Togaviruses

Group	Virus	Usual host	Geographic distribution	Mortality rate (%)	Comments
Alphaviruses	Eastern equine encephalitis (EEE)	Birds	North America near large bodies of water, Caribbean, South America	50-75	Mostly in swamp birds
	Venezuelan equine encephalitis (VEE)	Rodents	Central, South America, southern U.S.	1	Economic problems resulting from loss of horses
	Western equine encephalitis (WEE)	Birds	Western U.S., Central America, northern South America	2-4	
Flaviviruses	Japanese B encephalitis (JBE)	Birds	Asia, Pacific	10-90	Sequelae are common
	Murray Valley encephalitis (MVE)	Birds	Australia	10	Sequelae are common
	Rocio encephalitis	Birds	Brazil		Often causes gait and equilibrium disturbances
	St. Louis encephalitis (SLE)	Birds	The Americas	0-37	25 percent have urinary symptoms
	West Nile encephalitis (WNE)	Birds	Africa		Encephalitis seen only in aged
Tick-borne viruses (type 4 flaviviruses)	Russian spring-summer encephalitis (RSSE)	Birds, rodents	Far East	30	Affects both sensory and motor functions
	Central Europe virus	Birds, rodents	Central Europe	1-8	
	Louping ill virus	Sheep	Great Britain	10	Rarely affects humans
	Powassan	Small mammals; mice, chipmunks, squirrels	North America, U.S.S.R.	10	

Epidemiology. As noted by their names, the togaviruses causing encephalitis vary in geographic location. It is interesting that the three alphaviruses of eastern equine encephalitis (EEE), western equine encephalitis (WEE), and Venezuelan equine encephalitis (VEE) cause epidemic encephalitis only in the Americas, despite the fact that other alphaviruses and vectors are present throughout the world. These three viruses cause disease in humans and horses (hence, the name equine). Natural hosts of these and other togaviruses that cause encephalitis are wild birds (EEE, WEE, Japanese B encephalitis or JBE, and St. Louis encephalitis or SLE) or rodents (VEE). That is, humans and horses are dead-end accidental hosts. They are not necessary for the viral transmission cycle (Fig. 16-1).

The mosquito-borne encephalitides and fevers are transmitted by the bite of mosquitoes that shed virus from their salivary glands. When a susceptible mosquito vector (primarily *Culex* and *Culiseta* species) takes a blood meal from an infected vertebrate, the virus replicates in the mosquito gut and after about 2 weeks travels to the salivary glands and nerve tissue where it is present in high titers and can be transmitted on next bite (Fig. 16-2). The mosquito does not

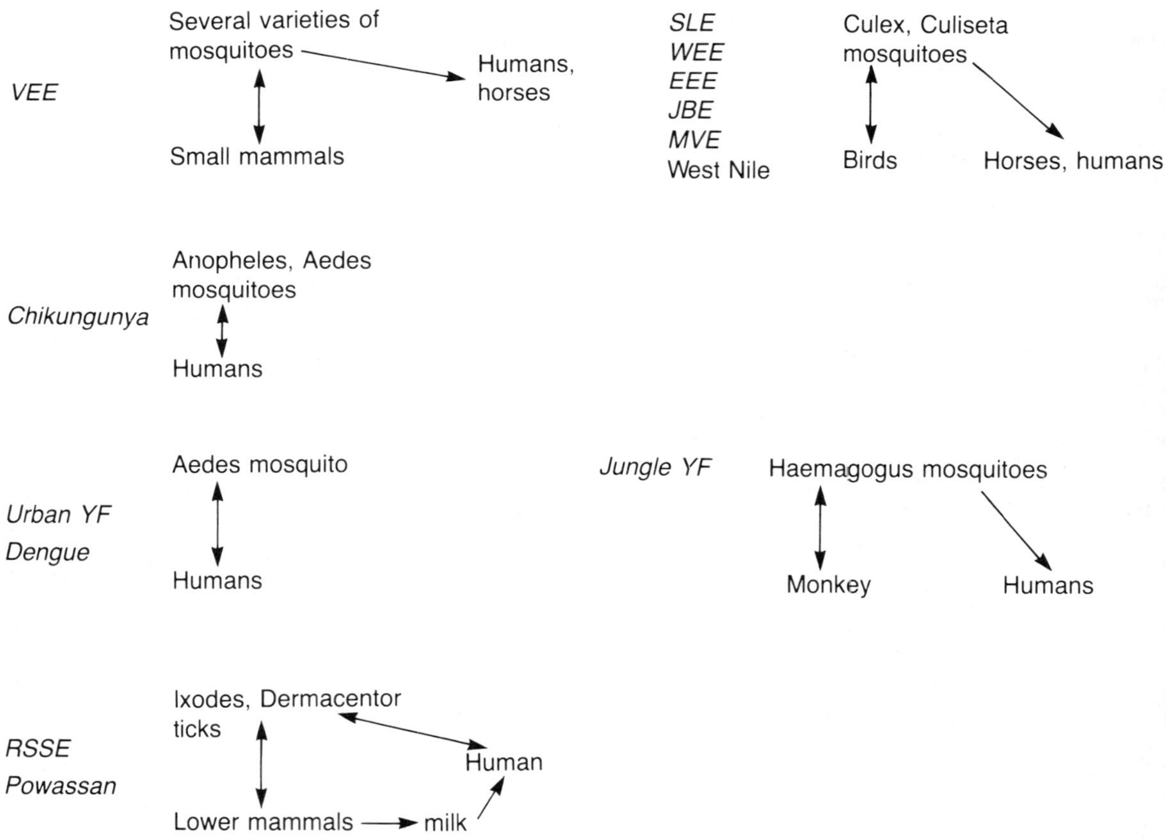

FIGURE 16-1. Transmission cycles of various arboviruses.

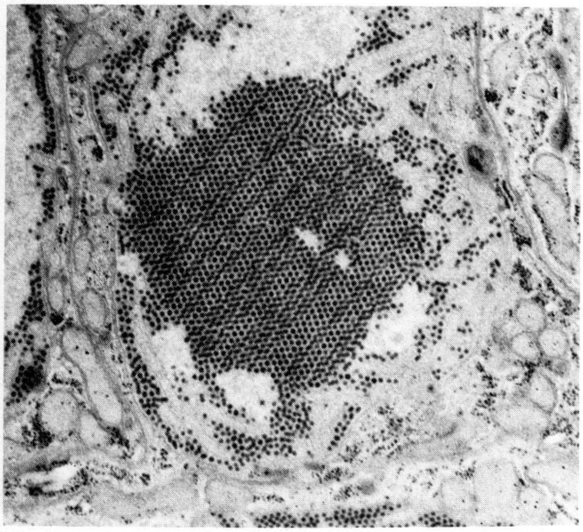

FIGURE 16-2. SLE virus infection in *Culex pipiens pipiens* salivary gland. By 25 days after infection of the mosquito, so many virus particles are present in the glandular lumen that crystalline arrays are formed which vary in appearance according to packing density and plane of section. Magnification 25,000×. (Reprinted with permission of the publisher from Whitfield SG, Murphy FA, Sudia WD. St. Louis encephalitis virus: an ultra-structural study of infection in a mosquito vector. Virol 1973;56:70-87.)

become ill, but stays infected and infectious for life (which lasts from weeks to months). Among mosquitoes no transovarial transmission occurs; hence, the method by which the viruses survive through the winter is speculative. They may survive in the mosquito vector, in birds, or possibly in reptiles and rodents.

The most common togavirus encephalitides in North America are St. Louis encephalitis, which causes 5 to 10 percent of all encephalitis in the United States during nonepidemic years (32 cases in 1979), eastern equine encephalitis, and western equine encephilitis, each of which cause less than ten cases annually in the United States. SLE is endemic in areas of California and Texas where the vectors, *Culex tarsalis* and *pipiens,* are abundant. It also occurs in tropical America (Fig. 16-3). The disease sometimes occurs in epidemics during summer (see Fig. 16-4).

The sporadic occurrence of outbreaks of SLE, as well as the other encephalitides, is a result of the fact that a constellation of conditions are necessary for the emergence of togavirus, including the efficient transference of the virus by a particular vector strain, the immunologic experience and competence of possible vertebrate hosts, and the competence of the viral agents. Attack rates can range from 5 to 20 per 100,000 persons in cities, to 800 per 100,000 persons in rural areas, where contact with infected birds (robins, sparrows, jays) and appropriate vectors is increased. There are many asymptomatic or mild infections (a ratio of 8 to 200 infections for every clinically manifest disease) with severe illness occurring primarily in the aged.

WEE, spread by the same mosquito vector as SLE, occurs primarily in western United States, Central America, and South America. EEE is more common on the Atlantic coast, in the Caribbean, and both coasts of South America. EEE virus is transmitted largely among wild swamp birds by mosquitoes that rarely bite humans or horses. Outbreaks of EEE in humans are often preceded by several weeks with outbreaks in horses. WEE and EEE, unlike SLE, affect primarily children, the incidence decreasing with age. In epidemic years, incidence can exceed 500 cases (for example, there were 815 cases of SLE in 1975), and it always peaks in July and August when mosquitoes are most abundant. Both VEE and EEE are fatal in about 98

FIGURE 16-3. Isolations of SLE virus in countries located partly or wholly in tropical America. (Reprinted with permission of the publisher from Monath TP, ed. St. Louis Encephalitis. American Public Health Association, 1980.)

percent of horses infected, making them economically devastating diseases. At least one epidemic of VEE among humans and horses has occurred in southern United States, but none since 1971. Other togaviruses that cause encephalitis have varying geographic distributions and vectors.

The tick-borne encephalitides caused by togaviruses are, as might be expected, more common in woodcutters, hunters, campers, or other persons with a high probability of exposure to ticks. There have been occasional outbreaks in Europe of tick-borne encephalitis. The viruses causing such disease are antigenically closely related, but are named for the region in which they are originally discovered. Only the Powassan virus occurs in the Americas, and fewer than 20 cases (one fatal) have been diagnosed. The incidence of tick-borne encephalitis peaks in May and June, with some cases through October. This is earlier than encephalitis caused by mosquito-borne viruses. The tick vectors remain infectious for life and can also transmit virus through eggs to their offspring. The milk of infected goats con-

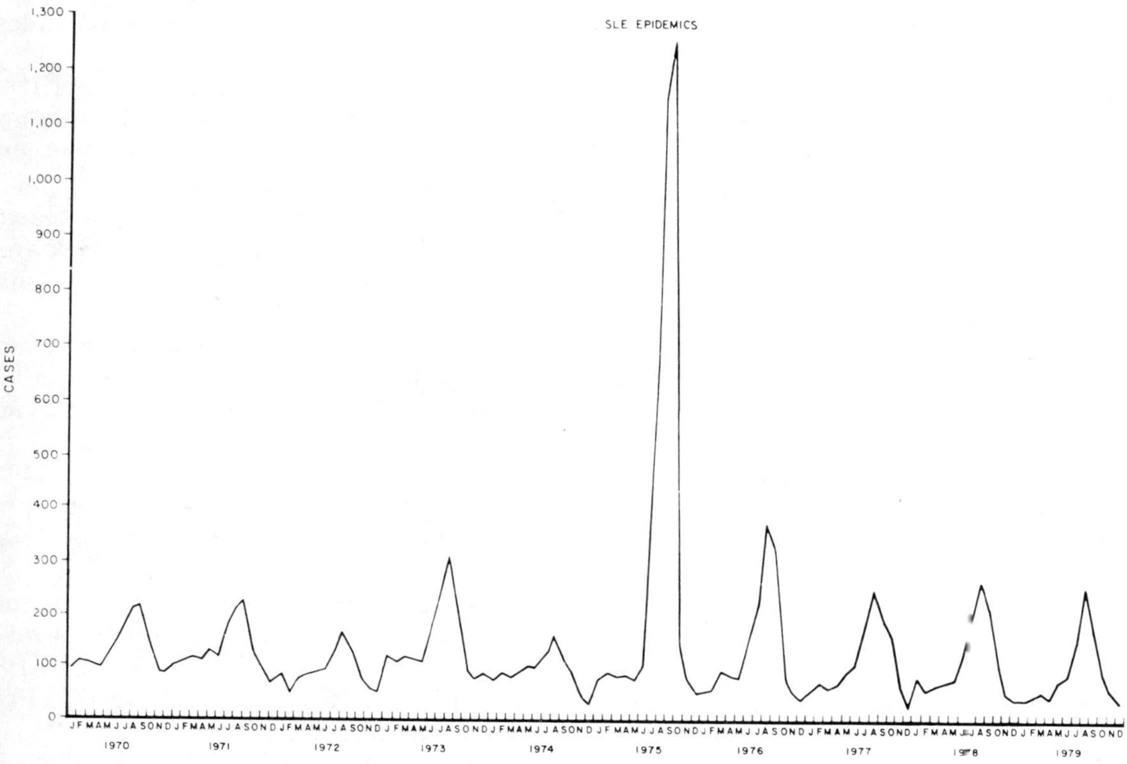

FIGURE 16-4. Encephalitis. Reported cases by month of onset, United States, 1970 to 1979. (Reprinted from Centers for Disease Control. Annual Summary 1980. Morb Mort Week Rep 1981;29:31.)

tains virus and milk-borne encephalitis has been documented with several viral species.

Pathogenesis. The encephalitis-causing viruses, after multiplying in the blood vessels, spread to the central nervous system in some way not completely explained (either via the peripheral nerves or more likely via the hematogenous route). The probability of invasion of the central nervous system is correlated with the level of viremia. In WEE and EEE, which affect infants less than 1 year old, there is massive neuroparenchymal destruction with large cystic lesions. Such lesions are uncommon in adults and never seen in SLE. WEE results in well-circumscribed foci of necrosis in the basal ganglia and thalamic nuclei. The brain stem, spinal cord, and cerebellum are usually unaffected. EEE results in a swollen congested brain with predominant polymorphonuclear infiltrates. Damage to basal ganglia, thalamus, and cerebral cortical gray matter ensue. SLE causes neuronal and glial necrosis and inflammation. In gray matter regions the most severely involved section is the substantia nigra of the midbrain.

Host responses. Most of the encephalitides, regardless of causative virus, are clinically indistinguishable. Diagnosis relies upon epidemiologic features and serologic tests rather than signs and symptoms. For those who develop clinical disease, an incubation period of 1 to 2 weeks is followed by abrupt onset of fever, headache, nausea and vomiting, changes in mentation, and convulsions (more common in children) that can advance to encephalitis with coma. The severity of SLE increases with the age of the host; encephalitis occurs in 56 percent of diagnosed infections in individuals under age 20, and in 87 percent of individuals older than 60. There is a predominance of lymphocytes in the cerebrospinal fluid. Because epidemiologic features, treatment, and preventive measures may differ, it is important to differentiate between encephalitis due to enteroviruses, arboviruses, or other viruses such as herpes simplex and varicella.

Among the togaviruses, EEE virus causes the most severe disease with a mortality rate of 50 to 75 percent as compared to 10 to 30 percent for WEE and SLE. Generally, the peak severity of disease is within 2 to 5 days of onset. VEE causes flulike symptoms in most people that are affected; only about 3 percent develop encephalitis. Sequelae such as seizures, mental retardation, or paralysis are uncommon after WEE and VEE, but may be present in about one-fourth of survivors of EEE. Meningitis may also be prominent, especially with the flaviviruses such as SLE and JE. Emotional lability and nervousness that may last for years are reported by 20 to 50 percent of individuals after SLE. Forty percent also continue to complain of headache, tremors, or memory impairment, though such manifestations are usually reversible. Individuals with SLE often have urinary tract symptoms such as incontinence, retention, or urgency, possibly due to basic kidney damage or changes in secretion of antidiuretic hormone. Recovery from the mosquito-borne encephalitides usually occurs within a month.

Russian spring-summer encephalitis causes sequelae in about one-half of cases. Because the virus has a predilection for neurons in the medulla and upper cervical spinal cord, the disease can result in damage to motor and sensory function of the neck and upper body, and even paralysis with respiratory failure. Recovery can take up to 6 months and mortality is about 30 percent. JBV has been transmitted transplacentally. Immunity after all infections is solid and long lasting.

Control and treatment. The prevention of arbovirus infections in humans has two components: vector control, such as aerial insecticide sprays; and early recognition and control among vertebrate hosts, such as use of effective vaccines for horses against EEE and VEE. There are apparently effective vaccines available for RSSE and JBV; attenuated vaccines against EEE and VEE are experimental. Milk-borne cases can be prevented by pasteurization. Once humans are infected, the only treatment is supportive.

Dengue Fever

Biologic characteristics. The dengue virus serves as a model for arboviral disease. There are four serotypes, 1 through 4. Complete immunity is type specific, though cross reaction occurs for 6 to 9 months following infection. The virus resembles that of yellow fever.

Epidemiology. Humans are the major host for dengue virus, though a monkey-mosquito cycle may occur in Malaysia. The virus is transmitted by *Aedes* mosquitos, usually *aegypti,* and distributed worldwide in warm climates. The mosquito breeds in sheltered clean (not dirty) water and the extrinsic incubation period in the mosquito is 8 to 11 days. About 50 cases per year are reported in the United States, almost always imported, though a type 1 strain was isolated in Texas in 1980. An outbreak of type 1 occurred in Puerto Rico in the fall of 1981 with about 7000 nonfatal cases. A type 4 strain caused five cases among individuals returning to the United States from Jamaica in the early 1980s. The risk of secondary spread from such imported cases is probably minimal since the mosquito vectors are likely to be different. Major endemic areas are Southeast Asia, West Africa, Micronesia, the Carribean, and South America. Outbreaks usually occur in children because adults in endemic areas are immune; 75 percent of individuals in some regions have antibody by age 2.

Though epidemics have been recognized for more than 200 years, the first pandemic that was confirmed by viral isolation occurred in 1963, a result of serotype 3. Hundreds of thousands of people have been affected during outbreaks, with attack rates of 20 to 50 percent of populations involved. A syndrome of hemorrhagic fever and shock accompanying dengue occurred in Manila in 1954 and has since been identified in South Pacific Islands and Southeast Asia. To date it has not been noted in Africa. It affects primarily children in ages 3 to 7, is caused by any one of the four serotypes, and seems to be more common among Oriental populations.

Pathogenesis. After reaching the blood stream the dengue virus multiplies in vessel endothelium. The viral multiplication and resultant infiltration of mononuclear cells causes vascular edema. Dengue hemorrhagic fever (DHF), a serious complication that occasionally accompanies dengue, occurs almost exclusively as secondary infection in those with preexisting antibody. It is hypothesized that the syndrome is an immunopathologic response among individuals (especially children who may have immature immune systems) previously sensitized to viral antibody. There is increased vascular permeability, disseminated intravascular coagulation, and thrombocytopenia. It is not likely that DHF is caused merely by a more virulent strain since several different subtypes may all cause the syndrome at one time.

Host responses. Classical dengue fever is also called "breakbone fever" because of the very severe joint and muscle pain and headache it causes. The incubation period is 3 to 14 days. Fever is always present for about 5 days and a maculopapular rash with petechiae and often itching is usual. The patient has leukopenia, an aversion to food, and swollen lymph nodes. Recovery without sequelae is the rule, though death rates of 5 to 10 percent occur. Adults have more severe symptoms than children.

DHF also a rapid onset of high fever. In addition, there is hepatic enlargement, bleeding manifestation, hypotension progressing to shock and circulatory collapse. The case fatality rate can attain 50 percent, but is usually about 10 percent. The virus can be recovered from the blood in only 5 to 10 percent of cases; more successful is recovery from peripheral blood leukocytes.

Control and treatment. Programs of mosquito eradication will control the spread of dengue fever, which remains a threat worldwide including the United States. Attenuated virus vaccines are available experimentally, but are not in widespread use, primarily because of the possibility that DHF could be induced by an immunologic response. As with the other arboviruses, no direct person-to-person spread occurs, but infected individuals should be screened for the first 5 days of illness to prevent biting by mosquitoes that could then spread the disease.

Yellow Fever (YF)

Biologic characteristics. The YF virus is the type species of *Flavivirus* subgroup 3, 22 to 38 nm diameter, with at least two different subtypes described. It grows well in chick embryos or tissue culture. There is no cross reaction between YF virus and dengue virus.

Epidemiology. YF virus, perhaps the best known and historically most devastating of the togaviruses, has caused endemic and epidemic disease in tropical climes since at least the seventeenth century. Though urban YF has been largely eliminated in North and Central America, outbreaks still occur in Africa (for example, there was a large epidemic in Ethiopia in the early 1960s). Despite the presence of the vector, *A. aegypti*, YF has not been reported in India or the Orient—and no outbreaks have occurred in the United States since the early 1900s. The last urban epidemic of YF in the Americas occurred in Brazil in 1928, with a few cases in Trinidad in 1954.

There were hopes of eradication of YF since humans were the only known reservoir of the disease. Such hopes have dimmed, however, with the discovery of a jungle focus of YF in which the transmission cycle is maintained in arboreal monkeys and species of mosquitoes other than *A. aegypti*. An infected individual can transmit the virus to vectors for about 5 days after onset of symptoms. The extrinsic incubation period is 12 to 14 days. Mosquitoes are infected for life; transovarial transmission in some mosquito varieties has been demonstrated. Humans are accidentally infected with jungle YF in forested regions of South America and Africa.

YF attacks all ages and races, and both sexes—though the jungle variety tends to occur more frequently in males of working age, who are more likely to be in contact with jungles and forests in the course of their employment as foresters or agricultural workers. Prior to the twentieth century, thousands of persons died annually of YF. Despite its disappearance in developed countries, however, the threat of YF remains as long as appropriate insect vectors abound. There are about 50 cases per year in the South American jungles due to monkey reservoirs and about 100 cases per year total in the Americas with a slight increase in the past decade.

Pathogenesis. After entering the host through the salivary glands of the infected mosquito vector, the virus enters the blood stream, multiplies in the epithelium of vessels (causing multiple small hemorrhages), and then travels to the liver (Fig. 16-5). A drop of blood may contain millions of viruses. YF virus multiplies in parenthymal hepatic cells, primarily in the mid-zone of liver lobules, causing characteristic damage. The cells become necrotic and contain eosinophilic bodies called "Councilman bodies." Viruses in the kidneys cause tubular necrosis. The virus also localizes in bone marrow, lymphatics, and spleen.

Host responses. The incubation period of YF virus is 3 to 6 days, followed by sudden onset of headache, fever, nausea and vomiting, and myalgia. (Many cases, however,

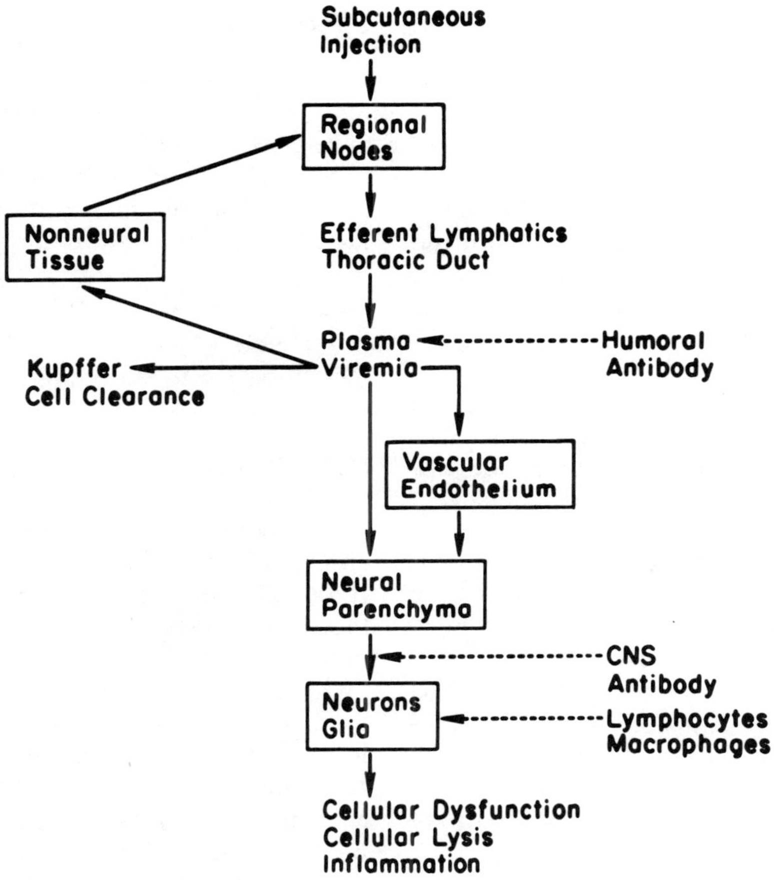

FIGURE 16-5. Sequential development of *Flavivirus* infections. *Boxes* indicate sites of virus replication, and *dotted arrows* indicate immunologic defense mechanisms. (Reprinted with permission of the publisher from Monath TP, ed. St. Louis Encephalitis, American Public Health Association, 1980.)

can be mild or asymptomatic, especially in children.) The pulse is relatively slow and there is leukopenia. Some individuals may have no further symptoms, but others (perhaps one-third) will develop jaundice, bleeding manifestations, and symptoms of renal damage such as albuminuria or oliguria about 2 to 3 days after disease onset. Bleeding is most common from the gastrointestinal tract: nosebleeds, hematemesis, or melena. The disease is usually of short duration, lasting only a few days. In most cases there is complete recovery without sequelae. Mortality due to heart failure, liver failure, or hemorrhage, is low (5 percent) in endemic areas, but has been up to 50 percent in epidemics.

Antibodies are present in the blood within

the first week after infection and diagnosis can be made by the measurement of acute convalescent titers or by isolation of virus from the blood of infected persons. Immunity is solid. Passively acquired antibodies in newborns last about 6 months. If YF occurs after infection with another *Flavivirus*, heterologous hemagglutination inhibition and complement fixation antibody titers are high, and a diagnosis based on serologic tests may not be possible.

Control. Control by eradication of *A. aegypti* mosquitos has been very successful for urban YF, but this is not possible for the jungle variety. Fortunately, an efficacious vaccine made from an avirulent viral strain (17D), which has been grown in chick embryo, is available. The protection is long lasting with a booster required after 10 years. However, because it takes 7 to 10 days for antibody to develop following vaccination, early recognition of outbreaks is essential for control. Insect repellents, protective netting, and clothing are recommended for nonimmune individuals in endemic areas.

Other Togavirus Fevers

Many other species of togavirus can cause fevers and encephalitis, especially in the tropics. The major ones are listed in Table 16-7. The hemorrhagic fevers cause clinical syndromes characterized by fever and bleeding and are similar on all continents, though viral agents are unique. Certainly more strains will be discovered in years to come, as many have only been identified since the 1950s. For example, a new *Flavivirus* named Rociovirus caused a large epidemic of encephalitis in 1975 to 1976. It was isolated from the mosquito *Psorophora ferox*. In an outbreak in the same country in 1978, 20 percent of 4000 persons were infected with Mayaro virus (group 2 Alphavirus) with a high proportion of overt disease. The greatest number of affected individuals lived close to the forest. Mayaro virus is probably mosquito-borne also. Epidemic polyarthritis is caused by Ross River virus in Fiji, where up to 90 percent of residents in some communities have antibody. It is endemic in Australia and New Guinea; transmitted by *Aedes vigilax* mosquito.

Rubiviris (Rubella)

Biologic characteristics. Like other togaviruses, rubella virus contains a single positive strand of RNA, is spherical and enveloped, and ether sensitive. It is readily inactivated by chemicals, ultraviolet light, and room temperatures. It is 50 to 70 nm in diameter with projections of 5 to 6 nm in length. It is the sole member of the genus *Rubivirus* in the family of togaviruses. It contains 3 to 8 viral specific proteins and a hemagglutinin. Rubella virus is antigenically distinct from all other togaviruses and does not multiply in arthropods. It can be grown in various cell cultures, though there is no convenient laboratory host. It replicates in host cell cytoplasm and matures upon budding through plasma membranes. Rubella virus was first isolated in 1960.

Epidemiology. Humans are the only natural reservoir for rubella and its occurrence is worldwide. Rubella is quite contagious, though less so than measles. Though the virus is present in blood, urine, feces, and genital tracts of infected individuals, transmission is probably by means of respiratory droplets or by direct or indirect contact. Fetuses become infected through the placenta

TABLE 16-7. Some Fevers Caused by Togaviruses

Group	Virus	Vector	Geographic distribution	Mortality rate	Bleeding complications?	Unique characteristics
Alpha-viruses	Chikungunya	Mosquito	Africa, Asia, Philippines	Less than 1%	Rare	Urban
	Kunjin	Mosquito	Australia	Less than 1%	Rare	Birds are usual hosts
	Mayaro	Mosquito	Americas	Less than 1%	Rare	Associated with jungle contacts
	Mucambo	Mosquito	Central and South America	Less than 1%	Rare	Rodents are usual hosts
	O'nyong-nyong	Mosquito	Africa	Less than 1%	No	Urban
	Ross River	Mosquito	Australia, New Guinea	Less than 1%	No	Polyarthritis-and-rash syndrome
Flavi-viruses	Dengue fever (4 serotypes)	Mosquito	Asia, Pacific Islands, Africa	Less than 1 %; 10% for hemorrhagic	Sometimes	
	Kyasanur Forest	Tick	India	5%	Yes	Serologically related to viruses of tick-borne encephalitides
	Omsk virus	Tick	U.S.S.R.	1-2%	Yes	Serologically related to viruses of tick-borne encephalitides
	Sepik virus	Mosquito	New Guinea	0%	No	
	Wesselsbron	Mosquito	Sub-Saharan Africa	0%	No	Usually in sheep farmers or lab workers
	Yellow fever	Mosquito	Worldwide in temperate climates	2-25%	Yes	Jaundice, renal damage common

if the mother has rubella during pregnancy. The occurrence of rubella has been altered significantly in developed countries where vaccine is in widespread use. In the absence of vaccine, the disease is endemic and occurs in epidemic peaks every 6 to 9 years, resulting in the development of immunity by the majority, about 85 to 95 percent, of the population. The prevalence of the disease is naturally highest among schoolage children, ages 5 to 9. Rubella occurs endemically even in sparsely populated areas such as Iceland.

Since vaccine was licensed in 1969, the incidence of rubella has decreased in the United States by over 90 percent (Fig. 16-6). For years after immunization was made available, age-specific incidence was high-

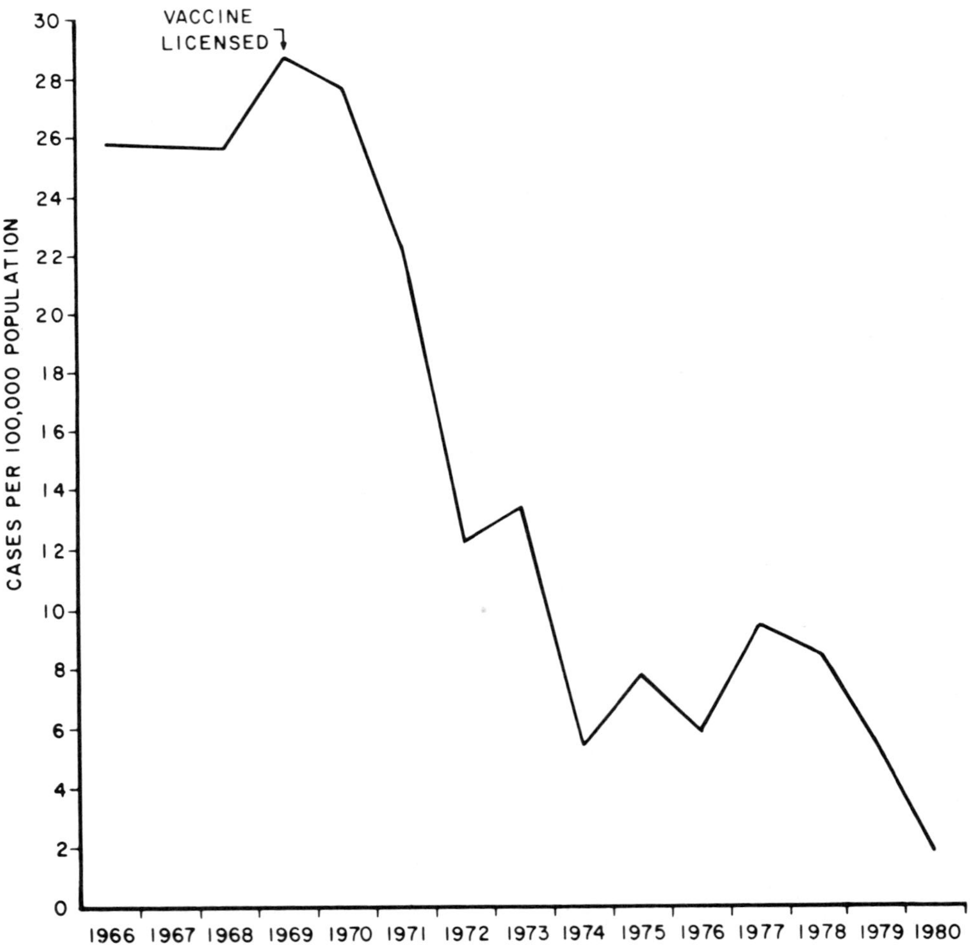

FIGURE 16-6. Rubella (German measles). Reported case rates by year, United States, 1966 to 1980. (Reprinted from Centers for Disease Control. Annual Summary 1980. Morb Mort Week Rep 1981;29:69.)

est in those over 15 years of age (70 percent). This group included adolescents who had missed immunization as children, but as of 1981 it was higher in the 0- to 4-year-old age groups. The current annual rate in the United States is about 2 per 100,000 persons (3904 cases were reported in 1980). There has not been an epidemic of rubella in the United States since 1964. Occurrence of the natural disease peaks in the spring in temperate climates.

Rubella is the only disease for which control is based on its effect on persons other than the diseased hosts. Of major concern is not the child or adult with rubella, since it is a very mild disease, but rather the unborn fetus of a pregnant woman who becomes infected in the first trimester. The correlation between a pregnant woman in the first trimester of pregnancy who has rubella and subsequent congenital anomalies in the newborn was first recognized and described by Gregg, an Australian ophthalmologist, in 1941. He noted that many cases of congenital cataracts occurred among newborns whose mothers had contracted rubella during an epidemic while they were in their first trimester of pregnancy. Subsequent prospective studies in several countries were able to demonstrate that up to 30 percent of babies with mothers who contracted rubella in their first trimester of pregnancy suffered some form of congenital malformation. Hence, vaccine development was aimed at preventing fetal complications rather than minimizing the clinical manifestations of the disease itself.

Though the incidence of congenital rubella syndrome (CRS) has declined since use of vaccine (attack rate was 3 per 100,000 in 1970 and 0.5 per 100,000 in 1978), there were about 20 to 40 cases per year of CRS reported between 1969 and 1980. Since it is theoretically possible to prevent all such cases, this incidence is clearly too high and is due to the fact that the age-specific incidence of rubella among individuals of child-bearing age has not declined sufficiently. In Oslo, a recent survey of almost 8000 pregnant woman identified 12.4 percent who were seronegative for rubella (Skaug, Ulstrup, and Stray-Pedersen 1981). In another recent report, a rubella outbreak among college students followed the introduction of the virus by some Japanese foreign exchange students, only 10 percent of whom had been immunized (Centers for Disease Control 1982f). Likewise, an outbreak of 23 cases among employees in an office building serves to demonstrate a lack of immunity among adults. It is estimated that about 15 percent of women of child-bearing age in the United States are still susceptible to rubella (Centers for Disease Control 1980i).

Pathogenesis. After entry by the respiratory route, the virus multiplies in the nasopharynx and local lymph nodes and then enters the blood stream (Fig. 16-7). The viremia results in dissemination of the virus to the viscera and skin; it can be recovered from skin biopsy. It is not clear whether the rash is due to viral damage or to antigen-antibody interaction. In utero infection occurs during the viremic stage in pregnant women. The viral infection can cause inflammatory infiltrates and lesions in any body organ and there is apparently arrested growth of the embryonic cells. Virus can be isolated from fetal viscera, eyes, liver, lung, brain, or heart.

Host responses. Rubella, also called German or 3-day measles, has a relatively long incubation period of 2 to 2 1/2 weeks. A rash is usually preceded by a mild prodrome of low fever and malaise. The rash is a typical

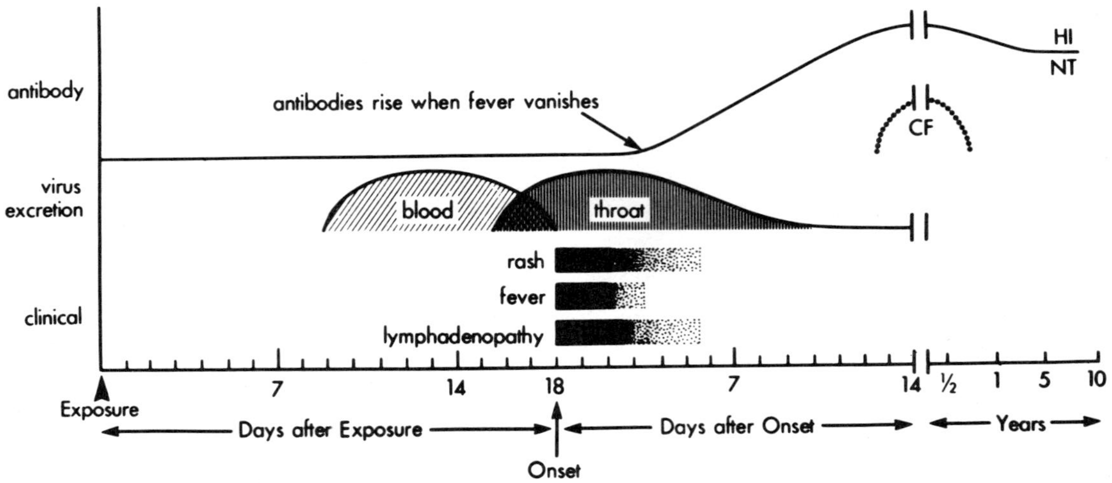

FIGURE 16-7. Pathogenesis of rubella: time course of clinical features, virus detection and antibodies. CF = complement fixing antigen; HI = hemagglutination inhibition; NT = neutralization. (Reprinted with permission of the publisher from McLean DM. Virology in health care. Baltimore: Williams and Wilkins, 1980.)

pink macular exanthem that appears first on the head and neck, and spreads to the trunk and limbs. It lasts for 2 to 3 days. Rash occurs in at least two-thirds of cases along with typcial lymph node swelling. It can readily be differentiated from measles (Table 16-8). About one-fourth of adults also suffer from joint pain or even frank arthritis. Recovery is usually complete; complications such as encephalitis and thrombocytopenia are extremely rare (about 1 per 5000 cases).

Hemagglutination inhibition antibodies appear at about the same time as the rash. The virus is shed in body secretions for as long as a week prior to development of the rash and up to 3 weeks after (from the throat only), though individuals are usually not considered infectious after about 6 days. Immunity is usually solid and enduring, though about 4 percent of individuals can develop subclinical reinfection (detected by a rise in IgG antibody). Even nuns who have been cloistered and without recent contact with

TABLE 16-8. Clinical Manifestations of Rubella and Measles (Rubeola)

	Occurring in	
Manifestation	*Rubella*	*Measles*
Coldlike symptoms, nasal congestion	Usually no	Yes
Generalized illness, constitutional symptoms	No	Yes
Blotchiness of rash	No	Yes
Enlargement of postauricular nodes	Yes	No

the rubella virus have been shown to possess persistent life long immunity to rubella (Rossier et al. 1981). Immunity correlates well with the presence of hemagglutination-inhibiting antibodies. Antibody is passively transmitted in utero and is lost by age 6 months.

The type and extent of fetal damage depends primarily on the gestational age of the fetus at the time of in utero infection (Table 16-9). If rubella occurs in the first 4 weeks of gestation, up to half of babies will develop CRS, 25 percent infected during weeks 5 through 8, 17 percent during weeks 9 through 12, about 3 percent in the second trimester, and essentially none in the third, although minor anomalies have been reported. The most common manifestations of fetal rubella infection are fetal wastage (miscarriage, spontaneous abortion, or prematurity), congenital heart disease (especially patent ductus arteriosus (PDA), and pulmonic stenosis), cataracts or glaucoma, mental retardation, deafness, microcephaly, and thrombocytopenia. Virus can be shed by congenitally infected babies for as long as several years (Table 16-10) after birth.

TABLE 16-9. Common Manifestations of Congenital Rubella Syndrome*

Week of gestation	*Anomalies*
First 8 weeks	Cardiac anomalies, especially patent ductus arteriosus and pulmonic stenosis; eye defects, especially cataracts; baby teeth defects
First 9 weeks	Multiple anomalies
First 16 weeks	Thrombocytopenia; low birth weight; hepatomegaly
5-15 weeks	Deafness

*See also Table 17-3.

TABLE 16-10. Usual Duration of Viral Shedding by Neonates with Congenital Rubella Syndrome (CRS)

Site	*Duration*
Throat	20 weeks, up to 2 years
Urine	8 weeks
Feces	2 weeks
Conjunctiva (in those affected with cataracts)	Up to 2 years
Spinal fluid	Up to 1 year
Leukocytes	6 months

Control and treatment. In countries where rubella vaccine is not available, every effort should be made to expose female children to rubella before puberty. In developed countries vaccination is the primary mechanism to prevent fetal infection with rubella. In the United States one live attenuated virus vaccine is available. One dose is adequate to induce an acceptable antibody response in about 95 percent of individuals. About 100,000,000 doses have been dispensed as of the early 1980s.

There are two different approaches to immunization. One is aimed at the target group, that is females who are just younger than childbearing age (12 to 14 years). This program, adopted in Great Britain and Canada, has been unsuccessful in stopping rubella epidemics, but it is decreasing rates of CRS. In the United States, the goal has been to provide mass immunization for all children and susceptible adults so that the pool of potential hosts is minimized and the spread of the virus is prevented. This program is based partly on the theory that "herd immunity" will result in potentiated protection. Unfortunately, outbreaks have occurred within groups with high levels of immunity, casting doubt on the effectiveness of herd immunity in protecting against rubella.

The current United States recommendation is that all schoolage children, nonpregnant women of child-bearing age, and individuals who contact children or pregnant women be immunized, including health care personnel (see "Clinical Implications"). Though the risk of CRS among babies of mothers receiving rubella immunization during pregnancy is probably minimal, it is recommended that any female of childbearing age use contraceptive precautions for several months before and after immunization. Among woman inadvertently immunized during pregnancy no congenital defects have occurred (Preblud et al. 1981).

The important features of the rubella vaccine are that the degree and duration of immunity are adequate, and that there is minimal shedding of virus, and hence no danger of spreading the disease by contacts. Though vaccine virus is shed from nasopharynx for several weeks after immunization, it has never been shown to be infectious. With prevention so easy, it is an unnecessary tragedy when a woman acquires rubella during the first trimester of pregnancy.

When a woman with unknown immune status becomes pregnant, it is possible to determine her immune status with the sensitive hemagglutination inhibition test. A titer of less than 1:8 indicates susceptibility to rubella, a titer of greater than 1:16 is indicative of immunity. However, reinfection has occurred in individuals with titers as high as 1:64. Fetal infection as a result of maternal reinfection (rather than primary infection), if it occurs, is rare. Therefore, it is vital if rubella occurs during early pregnancy to distinguish between primary and reinfection by testing for the early appearing IgM and the later appearing IgG antibodies. Both radioimmunoassay and hemagglutination inhibition with gel filtration tests are available for such evaluation. Susceptible pregnant women exposed to rubella have received immune serum globulin (ISG), but babies with CRS have been born to such women. Therefore, abortion is usually offered as an option to such women. Because babies with CRS are highly infectious, they should be segregated from potentially susceptible young women.

ARENAVIRUSES

Biologic characteristics. Members of the family Arenaviridae are pleomorphic, spherical, and contain many dense granules like sand, from which the name *arena* originated (Fig. 16-8). They are about 100 to 130 nm in diameter, are not associated with arthropod vectors, and contain three viruses that cause severe hemorrhagic fever: Lassa,

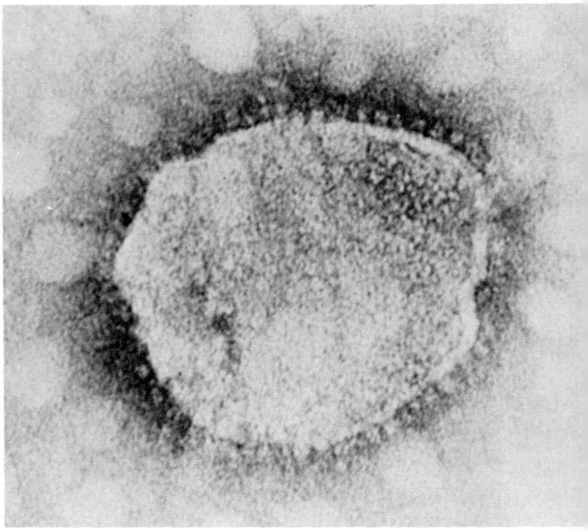

FIGURE 16-8. Arenavirus—Tacaribe. (Courtesy of Centers for Disease Control, Atlanta, GA 30333.)

Junin, and Machupo viruses (Table 16-11). Junin and Machupo viruses are the only members of a group of eight viruses called the Tacaribe groups that cause disease in humans. They are prevalent in the Western Hemisphere.

The prototype arenavirus is lymphocytic choriomeningitis virus (LCM), which, in 1934, was the first member identified. Like togaviruses, arenaviruses are enveloped. They contain a noninfectious (negative strand) single-strand RNA that codes for at least two glycoproteins. The viruses are released by budding and do not lyse their host cells. All viruses in this group are immunologically related. Junin and Machupo viruses can only be differentiated by specific neutralization tests. Other arenaviruses can be identified by fluorescent antibody or immunofluorescense.

Epidemiology. Rodents are the natural hosts of arenaviruses. The resultant diseases are characterized by a chronic carrier state in the urine or blood of their hosts. Life long infection in natural rodent hosts results from vertical transmission or from infection in the neonatal period. Infections that occur later in life are self-limited. Human cases of arenavirus infection usually occur as a result of accidental ingestion of food contaminated with infected rodent urine, feces, or oral secretions.

The virus of LCM, which probably naturally infects the house mouse, is distributed worldwide. The other agents of hemorrhagic fevers are more restricted geographically. The incidence of infection is difficult to determine since many infections may be mild or go undiagnosed due to lack of laboratory facilities. The prevalence of individuals with antibody to LCM has been found to be about 1 percent in selected populations. There do not seem to be any sex, age, or race predilections, but there is a slight seasonal peak in the winter, perhaps because people are indoors and more likely to contact fomites contaminated by mice. LCM infection is a special hazard to laboratory workers who contact animals. In ad-

TABLE 16-11. *Arenaviridae* that Cause Human Disease

Virus	*Disease*	*Geographic distribution*	*Discovered*	*Clinical manifestations*
Lymphocytic choriomeningitis virus (LCM)	LCM	Worldwide	1934	Benign aseptic meningitis
Junin virus	Argentine hemorrhagic fever	Argentina	1958	Hemorrhagic fever
Machupo virus	Bolivian hemorrhagic fever	Bolivia	1963	Hemorrhagic fever
Lassa fever virus	Lassa fever	Nigeria, Sierra Leone	1969	Hemorrhagic fever
Korean hemorrhagic fever agent (probable arenavirus)	Korean hemorrhagic fever	Korea, U.S.S.R.	1940	Hemorrhagic fever with pronounced kidney involvement

dition to mice, Syrian hamsters have been implicated as a source of human infection in at least two outbreaks.

Argentinean and Bolivian hemorrhagic fevers (caused by Junin and Machupo virus, respectively) predominate in rural agricultural workers during peak harvest times. Transmission from wild rodents may be through air or dust or through skin abrasions as well as by ingestion of contaminated food or drink. Person-to-person transmission is quite rare. In contrast, person-to-person transmission of Lassa virus is common. Lassa fever was first recognized after a small nosocomial outbreak occurred when an infected nurse was hospitalized in Nigeria. Two of the three secondary cases were fatal. Between 1970 and 1980 at least four outbreaks have occurred in Nigeria. The virus is also endemic in Liberia and Sierra Leone. As many as half of patients hospitalized with fever in Sierra Leone have documented Lassa fever and 50 percent of residents in some villages have antibodies. Reported outbreaks have usually been hospital-acquired cases following admission of a diseased individual; community outbreaks have been less common. Lassa fever does not have quite as marked seasonality as Argentinean and Bolivian hemorrhagic fevers. There have been two known importations of Lassa into the United States in 1969 and 1976 with no secondary cases. Transmission in nosocomial outbreaks is probably by droplets, direct contact, and accidental inoculation with infected material or fomites. The natural rodent reservoir, *Mastomys* species, is a mouse that inhabits dwellings in endemic areas. Intensive rodent trapping has reduced the transmission of Lassa virus to humans. Virus is shed from the nasopharyngeal secretions of infected individuals for several days after disease onset.

Another hemorrhagic fever which occurs in Korea and Russia and is called Korean hemorrhagic fever or hemorrhagic nephrosonephritis, has epidemiologic characteristics of an arenavirus. It appears to be associated with rodent vectors, but an etiologic agent has not been isolated. Its clinical picture is similar to the other viruses of the arenavirus family with the addition of pronounced renal involvement.

Pathogenesis. In their rodent hosts arenaviruses are able to establish carrier states by infecting immature animals before full immunologic status is attained. In humans, however, the disease is not due to immunologic response but probably to direct cytopathic effects of the virus. Inflammation and extensive damage of small blood vessels and capillaries occur and, in Lassa fever, hepatic necrosis.

Host responses. LCM in humans is usually a fairly mild disease with flulike symptoms or lymphocytic meningitis. The incubation period is about 1 to 2 weeks and asymptomatic infection can also occur. Thrombocytopenia and leukopenia are frequent, but more serious syndromes such as meningoencephalitis are rare. The patient is usually well within 1 to 2 weeks. Fatalities and sequelae such as paralysis or nervous disorders occur only as an exception.

Argentinean and Bolivian hemorrhagic fevers are clinically almost identical. Almost all cases are symptomatic after a 1 to 2 week incubation period. There is a slow onset of fever that remains steadily elevated for about 1 week. There may be rash and petechiae with bleeding manifestations in about one-third of cases. The liver and kidneys are usually involved, but jaundice is rare. Other common symptoms are fine tremor of tongue and hands, nausea and vomiting, headache, and muscle pain. Convalescence takes several weeks to months, and patients may lose their hair as a result of prolonged fever. Individuals either recover without sequelae or, in 5 to 30 percent of cases, die of uremia or hypovolemic shock.

Lassa fever presents a somewhat different clinical picture. Though it is probable that mild or asymptomatic infections occur, the disease was infamous after its discovery for its virulence and high infectivity. After a 1 to 2 week incubation period the disease has a slow onset with general symptoms of headache, nausea and vomiting, weakness, sore throat, and fever. Respiratory symptoms are prominent with pharyngitis, pneumonitis, pleural effusions, and white patches of oral deposits. Bleeding with petechiae occurs, as well as oliguria and somnolence. The acute disease lasts 1 to 2 weeks. Death due to hypovolemic shock and cardiovascular collapse occurs in about one-third of cases. Virus is present in the blood for about 2 weeks after onset of symptoms.

Control and treatment. Rodent control measures (which are not feasible in the jungles or fields) and avoiding contact with food, drink, or fomites contaminated with rodent excreta can successfully prevent exposure to arenaviruses. Laboratory infections of LCM can be avoided by monitoring hamster or mouse colonies for presence of virus or antibody. Vaccines developed from strains of the Tacaribe group have been successfully used to protect guinea pigs against infection with Junin virus, but efforts to develop vaccine for humans are still experimental. Hyperimmune serum has also been used for Bolivian hemorrhagic fever but its efficacy is as yet unestablished. An attenuated virus vaccine is being developed. Because of the difficulties with rodent control and the geographic restrictions of the viruses, vaccines appear to be the most promising control measures.

Since the virus of Lassa fever is more stable in body fluids, it presents a significant hazard for direct transmission between humans. Because of the probability of nosocomial outbreaks, rapid diagnosis is essential in order that strict isolation measures including gowning, gloving, and masking of staff can be implemented. The patient should be isolated to a single room. Such measures have been shown to prevent the spread of infection from cases imported to the United States. The virus also presents a hazard to laboratory workers and must be handled with extreme caution. Hyperimmune plasma from recovered patients has been used to minimize the severity of clinical disease. No vaccine is yet available. Ribavirin has been experimentally useful in minimizing clinical symptoms in monkeys.

BUNYAVIRUSES

Biologic characteristics. The family Bunyaviridae contains the largest number of arthropod-borne viruses; about 160 species contained in at least three genera. In the genus *Bunyavirus* (also called Bunyamwera super-group), there are more than 90 members in 12 antigenic groups. The three agents in this genus that cause human disease are the California encephalitis virus (CE), the group C arboviruses, and the Bunyamwera virus. There are about 70 other agents that will eventually be divided into at least two genera that are serologically distinct from members of the genus bunyavirus. These can be grouped into several major categories: the Turlock agents, Rift Valley fever, the tick-borne Ukuniemi group, Crimean-Congo hemorrhagic fever, the sandfly (phlebotomus) fevers, and the anopheles A and B groups (Table 16-12).

TABLE 16-12. Bunyaviridae Causing Human Diseases

Group	*Examples*	*Vector*	*Geographic distribution*	*Clinical manifestations*
Anopheles groups	Bwamba	Mosquito	Uganda	Headache, fever, muscle aches
Bunyavirus (Bunyamwera supergroup)	California encephalitis (CE, about 11 types)	Mosquito	North and South America, Trinidad, Czechoslovakia	Usually mild—fever, headache; CNS disturbances primarily in children
	Bunyamwera (at least 18 species)	Mosquito	North and South America, India, Africa, Finland	Headache, fever, muscle aches
	C group (about 11 species)	Mosquito	Florida, Trinidad, Brazil, Central America	Headache, fever, muscle aches
	More than 45 other species			
Crimean-Congo hemorrhagic fever		Tick	Africa, U.S.S.R.	Hemorrhagic fever
Phlebotomus	More than 20 species	Sandfly	Egypt, Italy, India, U.S.S.R., Mediterranean area	Febrile illness, papules
Rift Valley fever		Mosquito	Kenya	Fever, headache, muscle aches
Uukuniemi	At least 7 species	Tick	Finland	Fever, headache, muscle aches

The Bunyaviridae are larger than the togaviruses, measuring 90 to 100 nm in diameter. Their envelopes are covered with latticelike glycoprotein spikes with hemagglutinating properties that induce the formation of hemagglutination and neutralizing antibodies. The nucleocapsid is helical, containing a single negative strand of RNA. They contain, like the arenaviruses, an RNA-dependent RNA polymerase, probably a transcriptase, necessary for the transcription of genetic code. Their replication in the host cytoplasm resembles that of the orthomyxoviruses. Bunyaviridae can be grown in newborn mice and various cell cultures. They form by budding.

Epidemiology. Bunyaviridae have been isolated in every continent. They are spread to humans by inoculation from a blood-sucking arthropod; either a mosquito, tick, or sandfly. During most years, CE is the most common cause of arthropod-borne encephalitis in the United States. Its incidence is exceeded only during major outbreaks of SLE. In 1980, 139 of 177 cases (78.5 percent) of reported arboviral encephalitis was due to CE. It was first identified in mosquitoes in the San Joaquin Valley in 1943, though the disease is reported from all regions of the United States. Outbreaks occur from July to September. The natural reservoir is probably small mammals such as squirrels and rabbits.

Though the Bunyamwera group is named from the place in Uganda where it was first identified, it is present in many parts of the world. The main vector is *Aedes triseriatus* mosquito, in which transovarial transmission occurs. The group C viruses, about 11 species, are found in monkeys in Central America, Brazil, and the Everglades region. Human disease is accidental. Rift Valley fever is a disease primarily of sheep in Africa. Human secondary infection can occur, usually among exposed laboratory workers. The vector is a mosquito.

Phlebotomus (sandfly) fever is transmitted only by female sandflies (*Phlebotomus papatasi*) since males do not take blood meals. There is probably transovarial transmission. After ingestion of infected blood, the extrinsic incubation period in the sandfly is about 1 week. These agents are endemic in latitudes of 20 to 40°, especially in the Mediterranean area, Russia, Iran, Pakistan, India, Brazil, and Trinidad. Outbreaks occur among military troops or when groups from outside the endemic area are exposed. The agents of Congo (in the Congo and Uganda) and Crimean (in Russia) hemorrhagic fevers are antigenically identical. They are transmitted by tick bite and the natural hosts are small mammals such as rabbits.

Pathogenesis. Like other arthropod-borne viruses, bunyaviruses multiply in vascular epithelium and spread in the blood stream. Neural degeneration and cerebral inflammation occur.

Host responses. With one exception, the clinical diseases caused by bunyaviruses are mild and are very similar to each other. CE affects primarily children with symptoms of abrupt onset of fever, headache, and mild central nervous system involvement or encephalitis. Fatalities or long-term sequelae are rare. The ratio of inapparent to apparent infection is probably high and immunity is long lasting. The bunyamwera and group C agents cause rare mild febrile illness in humans with essentially no mortality.

Rift Valley fever is a denguelike disease with leukopenia and sometimes hepatitis. After an incubation period of 3 to 6 days, phlebotomus fever results in the development of itching papules, headache, fever, and stiff neck. There are no deaths; short-term specific immunity (about 2 years) develops for each antigenic type. Crimean-Congo hemorrhagic fever is the only serious disease in this group with mortality as high as 20 percent due to hemorrhage.

Control. Vector control, insect repellents, and screening or netting at night in endemic areas are the major mechanisms of control for bunyavirus disease. In general, the diseases are not serious or frequent enough to warrant major efforts aimed toward vaccine development.

REOVIRUSES

The family *Reoviridae* contains three genera that cause human disease: reoviruses (which may cause mild upper respiratory or gastrointestinal illness), rotaviruses (which cause infant gastroenteritis), and orbiviruses (which include the agent of Colorado tick fever; CTF). The agent of CTF is cubical, 75 to 80 nm in diameter, is nonenveloped, and contains double-stranded RNA. It is composed of 92 capsomeres and it replicates in the cytoplasm of the host cell. It is one of only two tick-borne viral diseases in North America (the other is Powassan viral infection, which occurs primarily in Canada). The natural hosts of CTF virus are rodents such as ground squirrels and chipmunks, and the vector is the tick *Dermacento andersoni*. It is prevalent especially in the Rocky Mountain region and accounts for as much disease as the other North America encephalitides combined (EEE, WEE, SLE, and CE). Infection is most common among those who have contact with wooded areas. Following an incubation period of 3 to 6 days the infected individual suddenly develops a fever, chills, headache, and severe myalgia without rash. Recovery is usually complete and rapid, and immunity is permanent. As with other such diseases, use of repellents and long clothing for hikers and others in contact with the outdoors in the western United States are adequate control measures.

RHABDOVIRUSES

There are two antigenically distinct genera in the family Rhabdoviridae: *Lyssavirus*, which includes the agent of rabies; and the *Vesiculovirus*, the vesicular stomatitis group. In addition, two viruses discovered since the late 1960s resemble the rhabodoviruses (Fig. 16-9) and can be placed, at least temporarily, into the rhabdovirus family. These are Ebola and Marburg agents (Table 16-13). There about 30 rhabdoviruses that have a unique bullet shape, are 150 to 180 nm in length, and 50 to 55 nm in diameter (Fig. 16-10). They are enveloped with surface projections, contain a single negative strand of RNA with a helical nucleocapsid, replicate in the cytoplasm, and bud from the endothelial reticulum in host cells. The

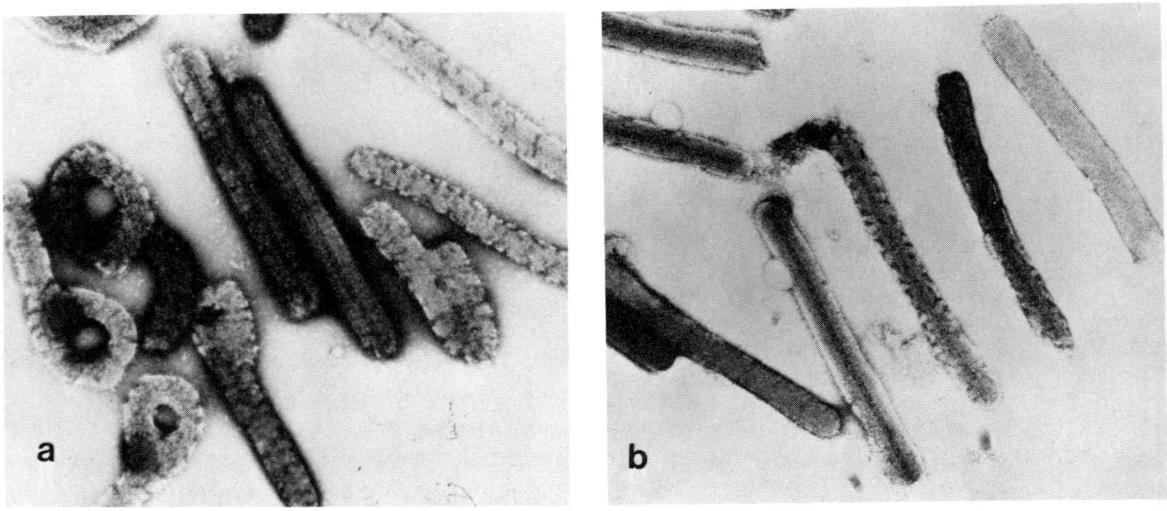

FIGURE 16-9. *(a)* Ebola virus. *(b)* Marburg virus. (Courtesy of Centers for Disease Control, Atlanta, GA 30333.)

TABLE 16-13. Rhabdoviridae Causing Human Disease

Genus	*Virus*	*Geographic distribution*	*Transmission*	*Clinical manifestations*
Lyssavirus	Rabies	Worldwide	Usually bite of infected carnivore; may be through mucous membranes, oral, or air	Neurologic damage, paralysis, death
Vesiculovirus	Vesicular stomatitis virus (VSV)	Americas	Inoculation of material from infected animal lesions through abrasions, especially on hands of farm workers	Influenzalike syndrome
	Piry virus	South America; in labs	Only lab infections by inadvertent inoculation have been reported in humans	Influenzalike syndrome
	Chandipura virus (CHP)	India	Bites of infected sandflies or mosquitoes	Denguelike fever
Unclassified*	Ebola	Zaire	Probably zoonosis; spread in nosocomial outbreaks is parenteral	Hemorrhagic fever
	Marburg	West Germany	Usually an infection of green monkeys; transmitted by lab accident, can be person-to-person	Hemorrhagic fever

*Possible members of Rhabdoviridae

RNA, being noninfectious, must be converted via a promoter, an RNA-dependent transcriptase, which allows the transcription of mRNA.

Rabies

Biologic characteristics. The rabies virus is a typical rhabdovirus. It is sometimes categorized as "street" virus, which is the wild, natural variety, and "fixed" virus, which is prepared by many hundreds of passages in the brains of rabbits to produce a strain with a much shorter incubation period. The fixed virus is historically used for vaccine purposes. There is one antigenic type. The agent was first demonstrated in the brains of infected animals. It can be grown in human diploid cells or by intracerebral inoculation of mouse brain with infected brain material.

Epidemiology. Rabies is transmitted by the bite of an infected animal or exposure of breaks in the skin or mucous membranes to infected salivary secretions. In several instances airborne transmission has occurred in laboratories and in caves inhabited by

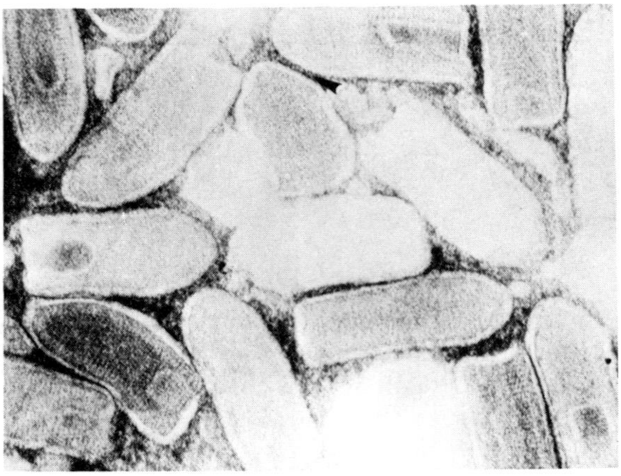

FIGURE 16-10. Vesicular stomatitis virus (Typical rhabdovirus). (Courtesy of Centers for Disease Control, Atlanta, GA 30333.)

rabid bats, but these are unusual occurrences. Contrary to popular belief, however, animals with rabies are not necessarily vicious. Most human cases in the United States are the result of exposure to an unaggressive animal. Abrasions or mucous membranes that become contaminated with infected saliva or brain tissue can result in disease. In fact, four of nine patients who died from 1978 to 1979 of rabies had no known history of animal bite. Though the saliva of an infected human is infectious, the only person-to-person transmission documented has been in two individuals who received corneal transplants from persons who died of rabies. Oral transmission of virus is also possible, but it is not important in human disease.

The disease is primarily one of carnivores; rodents and other small mammals (squirrels, hamsters, gerbils, rats, and rabbits) are rarely infected and have not been known to cause human disease in the United States. In countries where mass vaccination of domestic canines is not practiced, dogs are a primary source of rabies. In Ethiopia, for example, there are 30 to 50 human cases per year, and 98 percent are due to infected dogs. Jackals, hyenas, and donkeys have also transmitted the disease, and the mongoose is a common reservoir in Zimbabwe and South Africa.

In developed countries wild animals rather than domestic pets are the most important reservoirs of the rabies virus. In the United States the incidence of rabies has been decreasing, with about 2500 to 3000 laboratory-confirmed cases in wild animals, and about 400 cases in domestic pets yearly (Fig. 16-11*a*). However, there were 6421 confirmed animal cases in 1980, an 83.5 percent increase over the average for the preceding five years. Eighty-five percent of these cases were in wild animals; primarily skunks (63 percent), but also bats (11 percent), cattle and raccoons (6 percent each), dogs (4 percent), and cats and foxes (3 percent each). In Iowa, during 1981, cats were the most important source of human exposure (42 out of 57 cases). They in turn had probably been infected by skunks. In 18 European countries, during 1979, there were almost 17,000 confirmed animal cases, 70 percent in foxes. In Central and South America it has been estimated that as many as 1,000,000 cattle per year die of rabies as a result of bites by infected vampire bats.

There are about 1000 reported cases of human rabies in the world each year. A few areas of the world do not harbor rabies either because the virus has not been introduced or because of strict animal control and quarantine laws. These areas include Australia and New Zealand, Great Britain, Hawaii, Japan, and most of Scandinavia. There are 1 to 5 confirmed cases per year in the United States, although there were none in 1980 (Fig. 16-11*b*). There were four reported cases in Europe in 1979; however, much higher

a

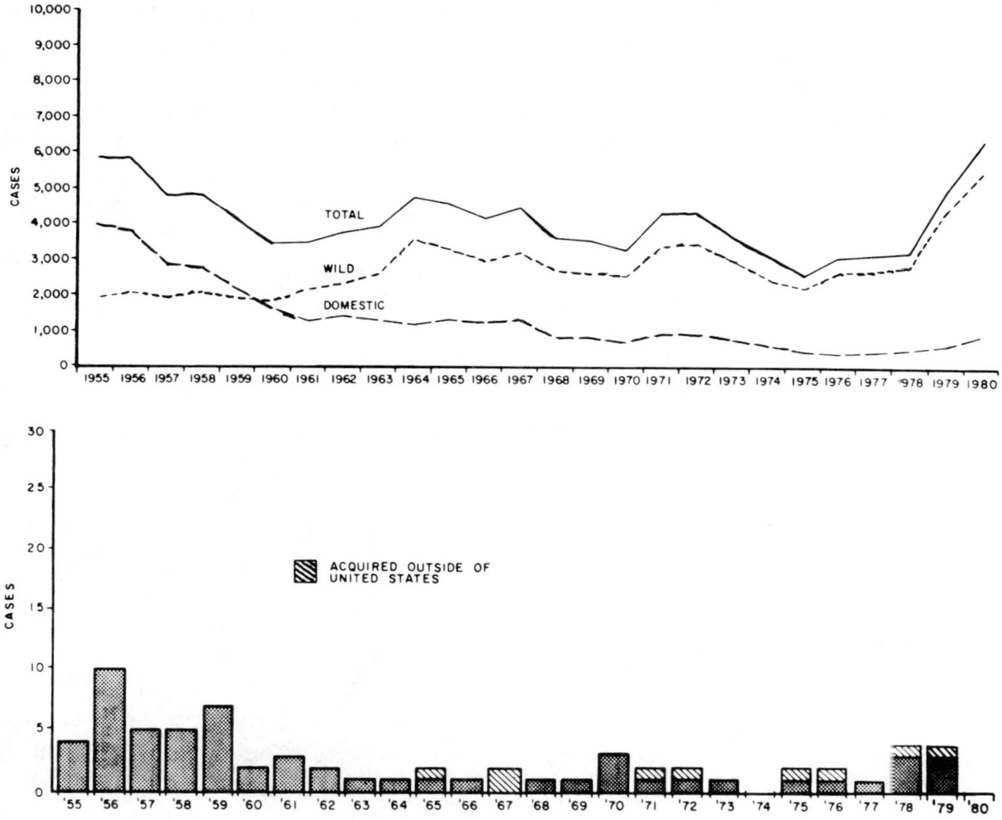

b

*Registered deaths, 1955-1961.

FIGURE 16-11. Rabies. *(a)* Reported cases in wild and domestic animals by year, United States, 1955 to 1980. *(b)* Reported cases in humans by year, United States, 1955 to 1980. (Reprinted from Centers for Disease Control. Annual Summary 1980. Morb Mort Week Rep 1981;29:67.)

rates exist in some parts of the world (for example, about 500 per year in Thailand). Rabies is more common in children during summer months, certainly a function of increased exposure to animals.

Pathogenesis. The rabies virus cannot penetrate unbroken skin. Even after bite by a laboratory-confirmed rabid animal, human rabies develops in only about half of untreated individuals. Thus disease is not an inevitable event. After entering the body the virus may multiply in muscle cells, but then migrates to the central nervous system along peripheral nerves. Most of the viral replication takes place in the central nervous system. In the neurons of the brain pathognomonic acidophilic intracytoplasmic inclusions, *Negri bodies,* are formed. These are large conglomerations of viral nucleocapsids. Negri bodies, detected by fluorescent antibody, are present throughout the brain and spinal cord; their presence is diagnostic of rabies. After replication in the brain the virus travels to the salivary glands, where it also proliferates and is excreted in massive quantities. Viral presence in saliva can be detected several days before onset of symptoms. During the disease, virus can also be detected in other organs of the body.

Host responses. The incubation period for human rabies is quite long, usually 3 to 8 weeks, though it can range from 6 days to 1 year. A short prodrome of 2 to 4 days includes irritability, discomfort around the bite site, and slight fever. Neurologic symptoms get progressively worse over the next few days with painful muscle spasms, difficult swallowing (the term "hydrophobia," meaning fear of water, comes from the fact that painful contractions of swallowing muscles can result from attempts to swallow—or even the thought of swallowing), excitable behavior, convulsions, and paralysis. Death ensues within a week, usually as a result of respiratory or heart failure. The mentation usually remains clear throughout the course of the disease. For the untreated, rabies is fatal.

Control and treatment. Rabies has been controlled to a great extent by the mandatory vaccination of pets and strict quarantine laws. Imported animals must be observed for 6 months to ensure that they are free of rabies. Wild animal reservoirs cannot be controlled, so there will always be a need for an effective vaccine. The first rabies vaccine was used by Pasteur in 1885, the newest vaccine was licensed in 1980. Because vaccination can be expensive, painful, and fraught with side effects, the decision to vaccinate an individual who has been bitten by an animal cannot be made lightly. Nevertheless, about 225,000 people per year are treated, partly because, to be maximally effective, vaccine must be given within 72, preferably within 24, hours of exposure and it is not always possible before that time to confirm whether the biting animal is rabid. However, prophylaxis has been given up to 7 months after exposure.

If a human incurs an animal bite, the following steps should be taken:

1. The wound should be thoroughly scrubbed and flushed with soap and water.
2. An antiseptic like alcohol or tincture of iodine should be applied.
3. Closure of the wound with sutures should be avoided.

This thorough cleaning is often effective in removing the rabies virus.

The decision about whether to provide specific prophylaxis depends upon the type and condition of the biting animal (Table 16-14). A healthy pet dog or cat that bites should be confined for 10 days and observed. If symptoms of rabies in the animal occur, the pet is killed, and the head is removed and shipped to the health department for viral isolation from the brain. Biting stray or wild animals should also be killed and the brain examined by fluorescent antibody.

Unless the patient has been previously immunized and has high antibody titers, treatment should include both rabies immune globulin (RIG) and one of 3 available vaccines. The globulin is necessary to provide passive protection and the vaccine to induce an active immune response. A fatal case of rabies in a 24-year-old American woman in Africa occurred in 1981 despite the fact that she received human diploid cell vaccine but no RIG.

The three available immunizing agents, duck embryo vaccine (DEV), human diploid

TABLE 16-14. Rabies Postexposure Prophylaxis Guide
The following recommendations are only a guide. In applying them, take into account the animal species involved, the circumstances of the bite or other exposure, the vaccination status of the animal, and presence of rabies in the region. **Local or state public health officials should be consulted if questions arise about the need for rabies prophylaxis.**

Animal species	*Condition of animal at time of attack*	*Treatment of exposed person**
Domestic		
Dog and cat	Healthy and available for 10 days of observation	None, unless animal develops rabies†
	Rabid or suspected rabid	RIG‡ and HDCV#
	Unknown (escaped)	Consult public health officials. If treatment is indicated, give RIG‡ and HDCV#
Wild		
Skunk, bat, fox, coyote, raccoon, bobcat, and other carnivores	Regard as rabid unless proven negative by laboratory tests**	RIG‡ and HDCV#
Other		
Livestock, rodents, and lagomorphs (rabbits and hares)	Consider individually. Local and state public health officials should be consulted on questions about the need for rabies prophylaxis. Bites of squirrels, hamsters, guinea pigs, gerbils, chipmunks, rats, mice, other rodents, rabbits, and hares almost never call for antirabies prophylaxis	

***All bites and wounds should immediately be thoroughly cleansed with soap and water.** If antirabies treatment is indicated, both rabies immune globulin (RIG) and human diploid cell rabies vaccine (HDCV) should be given as soon as possible, **regardless** of the interval from exposure.

†During the usual holding period of 10 days, begin treatment with RIG and vaccine (preferably with HDCV) at first sign of rabies in a dog or cat that has bitten someone. The symptomatic animal should be killed immediately and tested.

‡If RIG is not available, use antirabies serum, equine (ARS). Do not use more than the recommended dosage.

#If HDCV is not available, use duck embryo vaccine (DEV). Local reactions to vaccines are common and do not contraindicate continuing treatment. Discontinue vaccine if fluorescent-antibody (FA) tests of the animal are negative.

**The animal should be killed and tested as soon as possible. Holding for observation is not recommended.

Reprinted from Centers for Disease Control. Rabies prevention. Morb Mort Week Rep 1980;29:279.

cell vaccine (HDCV), and suckling mouse brain vaccine are discussed in more detail in Chapter 27: "Microorganisms in Public Health." Basically, the HDCV licensed in 1980 is the preferable agent whenever it is available because it is far superior in producing high antibody titers. Administration of one to three doses results in an antibody response in 90 to 100 percent of recipients and titers are 10 times those obtained following DEV. Also, 21 doses of subcutaneous DEV are necessary as compared with a 5 dose series of HDCV. Fewer systemic reactions, such as hives or anaphylaxis, have occurred after HDCV and, following the first 2500 doses, there have been no cases when it was necessary to discontinue treatment. Suckling mouse brain vaccine has been tested and found to yield satisfactory neutralizing antibody response with few side effects. It may be useful in Latin American and other endemic areas where HDCV is unavailable. Pregnancy is not a contraindication to treatment. Untoward reactions to vaccine, such as hypersensitivity and rash, can be treated with antihistamines, though anaphylactic reactions are a problem since corticosteroids to treat neuroparalytic ractions can inhibit the development of active immunity and should not be given. Pre-exposure vaccination is sometimes used for individuals at high risk of exposure: veterinarians, laboratory workers, or individuals with exposure to bats.

Other Rhabdoviruses

Several agents in the vesicular stomatitis group can cause flulike illness in humans after an incubation period of 1 to 2 days. Infection is transmitted in North and South America from domestic animals with oral lesions by mosquito vectors. Those at highest risk are cattle handlers or laboratory workers. The Chandapura virus has been associated with denguelike disease in India, and Piry virus has caused laboratory infections.

Ebola virus was discovered in 1976 in Zaire, first in a woman who died of severe hemorrhagic fever, and in subsequent hospital outbreaks in Zaire and Sudan, in which many hospital staff as well as patients were affected. From August to November of that year more than 500 cases occurred, resulting in 350 deaths. The virus contains RNA, has a lipid envelope, and is morphologically identical but antigenically different from Marburg virus (see below).

Ebola virus disease is characterized by sudden onset of sore throat, fever, rash, and severe abdominal pain and bleeding from multiple sites, especially the gastrointestinal tract. Though some complete recoveries occur, the fatality rate among identified cases is over 50 percent. There are, however, mild or asymptomatic cases since serum antibody titers to the virus were found in 7 percent of residents in endemic areas without history of disease. In the hospital outbreak previously described, women aged 15 to 29 years had a high prevalence of the disease related to attendance at a prenatal clinic where they received injections. The epidemic stopped when the use of sterile needles and equipment for parenteral injections was instituted.

Spread of Ebola virus is apparently by injection and by close contact. Hence, control measures include enteric isolation of cases and use of sterile equipment for parenteral injections. The natural reservoir of the virus is not known, but the endemic area is tropical rain forests or savannah, and the epidemiologic characteristics are consistent with a zoonosis.

Marburg virus was first isolated in 1967

TABLE 16-15. Slow Virus Diseases in Humans

	Disease	Implicated viral agent
Probable	Creutzfeldt-Jakob disease (CJD)	New agent ("viroid")
	Kuru	New agent (viroid)
	Progressive multifocal leukoencephalopathy (PML)	Papovavirus
	Subacute sclerosing panencephalitis	Measles, rubella
Possible	Alper's disease	?
	Alzheimer's disease	?
	Amyotrophic lateral sclerosis (AML)	?
	Multiple sclerosis (MS)	Associated with paramyxovirus in development of autoimmune response
	Pick's disease	?
	Schizophrenia	?

in laboratory workers in Germany who were handling tissues from African green monkeys. Seven of 31 died. The virus contains RNA. It is pleomorphic, usually rod shaped. It has one flat and one rounded end as in rhabdoviruses, though it is not antigenically related to any other rhabdovirus. Disease symptoms include sudden onset of fever, gastrointestinal disturbances, liver damage, rash, uremia, and hemorrhage. The mortality rate is 30 percent in hospitalized patients. The virus can also be spread person-to-person, therefore, affected patients must be isolated with special care taken for handling secretions (urine, blood, saliva). Many asymptomatic infections occur both in humans and monkeys. The agent can be grown in guinea pigs and cell culture.

SLOW VIRUS DISEASES

Biologic characteristics. Since the 1960s, four infectious chronic neurologic diseases have been recognized as probably of viral etiology, and speculations about the correlation between viruses and certain other chronic diseases have been made (Table 16-15). These diseases, known as slow virus diseases, share certain unique characteristics (Table 16-16): the absence of an inflammatory response or antibody formation, an extremely long incubation period of months to years, a propensity to attack neurologic tissue, and unusual resistance to conventional disinfecting methods. They are inactivated, however, by phenol and ether.

An etiologic agent for kuru and Creutzfeldt-Jakob disease has never been visual-

TABLE 16-16. Characteristics Unique to Slow Virus Diseases

Extremely long incubation period (up to 5 years)
No antibody produced
No inflammatory response (lack of cellular immunity)
Agents highly resistant to conventional disinfection techniques: heat, ultraviolet light, formalin, enzymes
Cause progressive neurologic diseases, usually fatal

ized by electron microscopy, despite the fact that particles have been retained by filtration and should be large enough to be visible. It is possible that the etiologic agents could be naked nucleic acid, biologic membranes or a new class of naked RNA agents called "viroids." Viroids are agents that are already known to cause a variety of plant diseases. The other slow-virus diseases—subacute sclerosing panencephalitis (SSPE) and progressive multifocal leukoencephalopathy (PML)—are caused by conventional measles, rubella, and papovaviruses or their variants.

Epidemiology. The epidemiology of kuru has been well described by Gajdusek, who studied the disease among the Fore tribe in New Guinea, the only place where it is known to exist. It was first described in 1957, primarily among women and children who had handled or eaten the brain tissue of relatives who had died of the disease. The incidence has markedly decreased since cannibalism was outlawed in this tribe.

Creuztfeldt-Jakob disease (CJD) is also rare, though it is not geographically limited (it occurs worldwide). The prevalence is 0.5 to 1 per 1,000,000 persons and it causes about 200 deaths per year in the United States. However, the rate in Libyan Jews is 30 times the average, possibly due to the custom among that group of eating beef and sheep brains and eyeballs (Neuget et al. 1979). Both CJD and kuru can be transmitted to chimpanzees by injection of infected brain material. There is evidence that CJD has also been inadvertently transmitted to a human by corneal transplant. There is sufficient clustering in the occurrence of CJD to implicate a genetic or infectious component to its transmission.

PML, first described in 1958, usually occurs in people with immunologic defects—those with Hodgkin's disease or leukemia. There have been fewer than 200 recognized cases from various parts of the world. Since the papovaviruses associated with PML are so common, it has not been established whether they represent a concurrent opportunistic infection or whether they have a causative role in the disease. PML could be the result of the first encounter of an immunodeficient host with a common agent.

There are about 40 new cases of SSPE reported annually in the United States. SSPE is worldwide in distribution. The disease is 3 times as common in males, and it usually occurs in children between ages 4 to 12, the majority of whom are from rural settings, especially in children who had measles as infants at an age much younger than normal. SSPE is also discussed in Chapter 15, "Respiratory Viruses."

Pathogenesis. Kuru and CJD have identical lesions as those seen in two animal diseases: sheep scrapie and mink encephalopathy. These four diseases may well be caused by the same agent. The primary lesion is intracytoplasmic vacuolation coalescing to form ballooning and destruction of neurons. PML is a demyelinating disease with white matter lesions and intranuclear inclusions in oligodendroglia. Two papovaviruses, SV40 and JC virus, have been isolated from infected tissue. SSPE is a sclerosing encephalitis. It is unknown why slow viruses almost invariably involve the central nervous system.

Host responses. Both kuru and CJD result in progressive neurologic degeneration, ataxia, tremors (Kuru means "shivering"), and dementia that lead to incapacitation and death. The duration of disease is 3 to 9 months for kuru, 8 to 12 months for CJD. CJD is also called transmissible virus dementia and subacute presenile dementia. There has been recorded survival from confirmed CJD, including recovery of mental functions (Behbehani 1979). PML and SSPE are also progressive, unrelentingly resulting in death within 4 to 12 months.

Control and treatment. The only successful treatment for slow virus diseases thus far has been the outlawing of cannibalism in the Fore tribe. Until more is known about the etiologic agents and modes of transmission, little can be done to prevent or control the diseases. However, the risk of SSPE after measles vaccination is about one-tenth the risk after natural measles infection, therefore, measles vaccination may be an effective control device for SSPE (Matthews 1981).

CLINICAL IMPLICATIONS

Because of the necessity for an arthropod vector, many of the diseases discussed in this chapter are not contagious between humans. The rare exceptions such as Marburg, Ebola, and Lassa fevers, however, remind us again of the need to be alert for potential infectious agents whenever a patient has a fever, leukocytosis, or other common indications of contagion. A number of cerebral malformations have been experimentally induced by viruses, and more knowledge is being added in this area.

One of the diseases in this chapter of special concern to health care personnel is rubella. A recent episode in which 56 pregnant women were exposed to an obstetrician and two other staff members with rubella (Fliegel and Weinstein 1982) emphasizes the importance of assuring that health care personnel working with pregnant women are immune to rubella. Likewise, health care personnel have contracted rubella from babies with congenital rubella syndrome. Such babies are highly infectious, excreting virus for up to 2 years. They require strict isolation and only immune personnel should provide their care while they are hospitalized. The Advisory Committee for Immunization Practices recommends requiring immunization of young adults as a prerequisite to employment in health care institutions. Many state health departments are setting policies to this effect (see example from Washington state, Fig. 16-12).

A second disease that has never presented a hazard in terms of person-to-person transmission but about which there is considerable concern is rabies. Because of high viral titers in the saliva of infected individuals, a person with rabies is potentially infectious. Thus, even though there has not been a case of known human-to-human spread, infected patients are placed in strict isolation with special care given to disposal of equipment such as syringes and suction catheters. It is important for the clinician to remember that even though the patient remains mentally alert, the prognosis is exceedingly grave and the family must be prepared for the event of the patient's death. Nursing care of patients with rabies at this time is extremely important since the only treatment possible is symptomatic and supportive.

Since the slow virus diseases are caused by agents as yet unidentified, it is prudent that caution is used when patients with diseases like Creutzfeldt-Jakob disease (CJD) are hospitalized. Recommendations from Centers for Disease Control published in mid-1982 include the use of needle precautions similar to those used for patients with hepatitis B infection. They suggest that patients with CJD do not require special isolation rooms, but that personnel take spe-

RUBELLA POLICY FOR HEALTH FACILITIES AND INSTITUTIONS

The facility/institution shall develop and implement written policies and procedures, reviewed by the appropriate governing authority to require, as part of the employee's medical clearance upon employment, and subsequently, that each employee is immune to rubella. Employees who cannot provide proof (including month and year) of a prior rubella vaccination, or of a positive antibody titer, shall be immunized with vaccine or shall have a rubella titer performed. If serologic screening is negative the employee shall be immunized. Proof of rubella immune status on each employee shall be kept on file at the facility/institution.

Each employee with negative or unknown immune status shall be informed in writing by the facility/institution that rubella immunization or rubella serologic screening is required.

Male employees with negative or unknown immune status shall receive rubella vaccine unless medical contraindications exist. Serologic screening to determine susceptibility prior to vaccination is not necessary.

Female employees with negative or unknown immune status may be serologically screened for susceptibility prior to vaccination, but serologic screening need not be a prerequisite for administering vaccine. Vaccine shall be administered to all seronegative females unless contraindications exist.

The facility/institution shall develop and implement policies and procedures to ensure that rubella vaccine is provided by the facility/institution to all employees requesting rubella vaccination. If the facility/institution does not have a medical director, or does not chose to provide vaccinations, the administrator of the facility/institution shall consult with the local health officer regarding the development and implementation of procedures for vaccinating employees.

The facility/institution shall ensure that a list is maintained of those employees who are seronegative and unvaccinated, because of contraindications, pregnancy or other condition, in the event that he/she is exposed to rubella and a determination is needed as to whether the employee can or cannot continue to work.

The facility/institution shall ensure that seronegative or unvaccinated employees will not be assigned to areas where he/she may have contact with patients/residents in their first or second trimester of pregnancy.

The facility/institution shall ensure that each employee's record contains documentation of a positive antibody titer or the date of vaccination with rubella vaccine, or, if applicable, of information provided to seronegative female employees of child-bearing age regarding rubella vaccination and pregnancy.

Immunization Unit/DSHS
July 1981

FIGURE 16-12. Sample statement: Rubella policy for health facilities and institutions.

cial precautions when handling blood, spinal fluid, or tissues, especially brain tissue. Contaminated objects should be carefully sterilized and surfaces should be cleaned with 5 percent sodium hypochlorite (household bleach) in addition to routine cleaning.

SUGGESTED READINGS

Aaskov JC, et al. An epidemic of Ross River virus infection in Fiji, 1979. Am J Trop Med Hyg 1981;30:1053-1059.

Anderson LJ, et al. Postexposure trial of a human diploid cell strain rabies vaccine. J Infec Dis 1980;142:133-138.

Behbehani AM. Slow virus disease. J Kans Med Soc 1979;80:600-604.

Brooks BR, et al. Slow viral infections. Ann Rev Neurosci 1979;2:309-340.

Centers for Disease Control. Adverse reactions to human diploid cell rabies vaccine. Morb Mort Week Rep 1980a;29:609-610.

Centers for Disease Control. Arboviral activity–U.S. Morb Mort Week Rep 1980b;29:353-355.

Centers for Disease Control. Dengue–Northeastern Mexico. Morb Mort Week Rep 1980c;29:407-408.

Centers for Disease Control. Dengue–Texas. Morb Mort Week Rep 1980d;29:451-452.

Centers for Disease Control. Human rabies—U.S., 1980. Morb Mort Week Rep 1980e;30:97.

Centers for Disease Control. Rabies. Morb Mort Week Rep 1980f;29:454-455.

Centers for Disease Control. Rabies prevention. Morb Mort Week Rep 1980g;29:265-271+.

Centers for Disease Control. Response of susceptible case of imported Lassa Fever. Morb Mort Week Rep 1980h;29:231-237.

Centers for Disease Control. Rubella outbreak in an office building. Morb Mort Week Rep 1980i;29:517-518.

Centers for Disease Control. Rubella–U.S., 1977-80. Morb Mort Week Rep 1980j;29:378-380.

Centers for Disease Control. St. Louis encephalitis. Morb Mort Week Rep 1980k;29:415-416.

Centers for Disease Control. Yellow fever—the Americas. Morb Mort Week Rep 1980l;29:281-282.

Centers for Disease Control. Nosocomial rubella infection. Morb Mort Week Rep 1981;29:629-630.

Centers for Disease Control. Human rabies—Rwanda. Morb Mort Week Rep 1982a;31:135-136.

Centers for Disease Control. Dengue fever in Puerto Rico—1981. Morb Mort Week Rep 1982b;31:103-104.

Centers for Disease Control. Cat rabies exposure in Iowa. Morb Mort Week Rep 1982c;31:67-68+.

Centers for Disease Control. Imported dengue type 4. Morb Mort Week Rep 1982d;30:622-623.

Centers for Disease Control. Rubella outbreak among foreign exchange students. Morb Mort Week Rep 1982e;31:104-105.

Chantler JK, Ford DK, Tingle AJ. Rubella-associated arthritis: rescue of rubella virus from peripheral blood lymphocytes two years post-vaccination. Infec Immun 1981;32:1274-1280.

Chaturvedi UC, et al. Transplacental infection with Japanese encephalitis virus. J Infec Dis 1980;141:712-715.

Cole AP, et al. Human and animal togavirus antibodies in congenitally deformed infants. J Med Virol 1981;7:79-90.

Coto CE, et al. Protection of guinea pigs inoculated with tacaribe virus against lethal doses of Junin virus. J Infec Dis 1980;141:389-393.

Coulombe L, Rosser WW. Can we prevent an increase in the incidence of congenital rubella syndrome in the next decade? Can Med Assoc J 1981;125:37-40.

Coyle PK, Wolinsky JS. Characterization of immune complexes in progressive rubella panencephalitis. Ann Neruol 1981;9:557-562.

Craddock-Watson JE, et al. Outcome of asymptomatic infection with rubella virus during pregnancy. J Hyg 1981;87 147-154.

Diaz AMO. Preexposure rabies immunization of man with suckling mouse brain vaccine. Am J Epidemiol 1982;115:274-277.

Evrard JR. Rubella, the trojan horse of obstetrics. Obstet Gynecol 1981;58:505-507.

Fekadu M. Rabies in Ethiopia. Am J Epidemiol 1982;115:266-273.

Ferguson CK, Roll LJ. Human rabies. Am J Nurs 1981;81:1175-1179.

Fliegel PE, Weinstein WM. Rubella outbreak in a prenatal clinic. Am J Infec Cont 1982;10:29-33.

Fox JP. Fetal perspectives of rubella vaccine efficacy and safety. JAMA 1981;246:1444-1445.

Francy DB, et al. Transovarial transmission of St. Louis encephalitis virus by *Culex pipiens* complex mosquitoes. Am J Trop Med Hyg 1981;30:699-705.

Gajdusek DC. Unconventional viruses and the origin and disappearance of kuru. Science 1977;197:943-959.

Hammon WM. Dengue hemorrhagic fever—do we know its cause? Am J Trop Med Hyg 1973;22:82-91.

Heymann DL, et al. Ebola hemorrhagic fever: Tandala, Zaire, 1977-1978. J Infec Dis 1980; 142:372-376.

Hoch AL, et al. An outbreak of Mayaro virus disease in Belterra, Brazil. III. Entomological and ecological studies. Am J Trop Med Hyg 1981;30:689–698.

Jahrling PB, et al. Lassa virus of rhesus monkeys: pathogenesis and treatment with ribavirin. J Infec Dis 1980;141:580-589.

Kangro HO, Jackson C, Heath RB. Comparison of radioimmunoassay and the gel filtration technique for routine diagnosis of rubella during pregnancy. J Hyg 1981;87:249-255.

LeDuc JW, Pinheiro FP, Travassos da Rosa APA. An outbreak of Mayaro virus disease in Belterra, Brazil. II. Epidemiology. Am J Trop Med Hyg 1981;30:682-688.

Lopes ODS, et al. Emergence of a new arbovirus disease in Brazil. Am J Epidemiol 1981; 113:122-125.

Matthews WB. Slow virus infections. J Roy Coll Phys Lond 1981;15:109-112.

McLean DM. Virology in health care. Baltimore: Williams and Wilkins, 1980.

Meyer HM. Rabies vaccine. J Infec Dis 1980; 142:287-289.

Moffet HL. Clinical microbiology, 2nd edition. Philadelphia: JB Lippincott, 1980.

Monath TP, ed. St. Louis encephalitis. Washington DC: American Public Health Association, 1980.

Morris CD, Zimmerman RH. Epizootiology of eastern equine encephalitis virus in upstate New York, USA. J Med Entomol 1981;18:313-316.

Neuget RH, et al. Creutzfeldt-Jakob disease: familial clustering among Libyan-born Israelis. Neurol 1979;29:225-231.

Parsonson IM, Della-Porta AJ, Snowdon WA. Developmental disorders of the fetus in some arthropod-borne virus infections. Am J Trop Med Hyg 1981;30:660-673.

Pattyn SR, ed. Ebola virus haemorrhagic fever. New York: Elsevier, 1978.

Pinheiro FP, et al. An outbreak of Mayaro virus disease in Belterra, Brazil. I. Clinical and virologic findings. Am J Trop Med Hyg 1981; 30:674-681.

Preblud SR, et al. Current status of rubella in the United States, 1969-1979, J Infec Dis 1980;142:776-779.

Preblud SR, et al. Fetal risk associated with rubella vaccine. JAMA 1981;246:1413-1417.

Rawls WE, Chan MA, Gee SR. Mechanisms of presistence in arenavirus infections: a brief review. Can J Microbiol 1981;27:568-574.

Roggendorf M, et al. Serological diagnosis of acute tick-borne encephalitis by demonstration of antibodies of the IgM class. J Med Virol 1981;7:41-50.

Rossier E, et al. Persistence of humoral and cell-mediated immunity to rubella virus in cloistered nuns and in schoolteachers. J Infec Dis 1981;144:137-141.

Schlesinger RW, ed. The togaviruses. New York: Academic Press, 1980.

Scott RM, et al. Isolation of dengue viruses from peripheral blood leukocytes of patients with hemorrhagic fever. J Infec Dis 1980;141:1-6.

Skaug K, Ulstrup JC, Stray-Pedersen B. A rubella epidemic in an unvaccinated pregnant population. Scand J Infec Dis 1981;13:11-15.

Smith H. Acute viral encephalitis. Br Med J 1981;283:935-936.

Taylor JL, Schoenherr C, Grossberg SE. Protection against Japanese encephalitis virus in mice and hamsters by treatment with carboxymethylacridanone, a potent interferon inducer. J Infec Dis 1980;142:394-399.

Tomori O, et al. Arbovirus infections among laboratory personnel in Ibadan, Nigeria. Am J Trop Med Hyg 1981;30:855-861.

17 DERMOTROPIC VIRUSES

KEY POINTS

- Important members of the herpes family include cytomegalovirus (a common cause of spontaneous abortion and congenital infection), Epstein-Barr virus (associated with Burkitt's lymphoma, nasopharyngeal carcinoma, and infectious mononucleosis), herpes simplex virus (causes "cold sores" and genital herpes), and varicella-zoster virus (causes chickenpox and shingles).
- The herpesviruses are characterized by their ability to remain latent in the host, causing recurrent disease.
- The human papilloma virus causes warts.
- Smallpox was declared by the World Health Organization to be eradicated from the Earth as of May 1980.
- Other diseases caused by poxviruses are relatively uncommon.

This chapter includes discussions of virus families that have major dermatologic clinical manifestations. These include the herpesviruses, papillomaviruses, and poxviruses.

HERPESVIRUSES

Herpesviruses are a family (Herpetoviridae) of double-stranded DNA, relatively large enveloped viruses of molecular weight 80×10^6 to 120×10^6 amu. All but two species (one that infects a fungus and one that infects an oyster) infect vertebrate hosts. There are more than 60 herpesviruses, but only a few cause human disease (Table 17-1). All herpesviruses have similar epidemiologic and clinical characteristics: some are transmitted sexually; and some cause encephalitis, keratitis, skin or genital lesions, and are disseminated neonatal diseases.

Herpesviruses share many other characteristics (Table 17-2). They multiply in the host cell nucleus and a glycoprotein-lipid envelope is added as the nucleocapsid passes through the nuclear membrane into the host

TABLE 17-1. Human Disease Caused by Herpesviruses

Virus	*Diseases*
Cytomegalovirus	Congenital defects Cytomegalic inclusion disease
Epstein-Barr virus	Burkitt's lymphoma Infectious mononucleosis Nasopharyngeal carcinoma (with other factors)
Herpesvirus simiae (B virus)	Ascending myelitis, encephaloymelitis
Herpes simplex virus	Oral and genital lesions Neonatal disseminated or localized disease Encephalitis or other disseminated diseases
Varicella-Zoster virus	Chickenpox Shingles

TABLE 17-2. Common Characteristics of Herpesviruses

Morphology	Double-stranded DNA viruses, glycoprotein-lipid envelope, icosahedral capsid with 162 capsomeres 180-200 nm
Replication	In host cell nucleus; most virions remain in host cell, some are released by budding
Antigens	Many; no common family antigen, but high degree of relatedness between strains, especially HSV-1 and HSV-2
Pathogenesis	Giant cell formation; gross disturbance in cell structure, ballooning, destruction of nucleolus; eosinophilic nuclear inclusions
Clinical manifestations	Characteristic vesicles; latency; dermatotropic, facultatively neurotropic; usually localized infection, however generalized can occur—especially in newborns and immunodeficient

cell cytoplasm. The envelope of the mature virion often has projections on its surface. Viruses contain about 30 polypeptides, six to eight in the envelope. The highly ordered icosahedral capsid has 162 capsomeres (Fig. 17-1). The virus infects its host by fusion with its cell membrane or by absorption and pinocytosis. Naked particles and nucleic acid as well as the mature enveloped virion can be infectious. The replicated virions often remain in the host cell, but may be released by budding.

Hybridization studies demonstrate a considerable degree of relatedness between herpes strains, the greatest being between herpes simplex types 1 and 2, which have 20 to 70 percent homology. Many antigens are present and are detected by immunofluorescence or immunodiffusion. Like all enveloped viruses, those of the herpes family are inactivated by lipid solvents such as ether and are readily killed by extremes in temperature and by low pH.

Host cells infected with herpesviruses are eventually killed by the infection. There is gross disturbance in cell structure with destruction of the nucleolus. Some herpes strains have a predilection for the nervous system, where they colonize axon tracts. Some cause proliferation of lymphoid cells where they can remain latent for years. Most strains are associated with giant cell formation. The cytoplasmic membrane of the host cell is affected by lipid substances in certain enveloped viruses so that cells merge

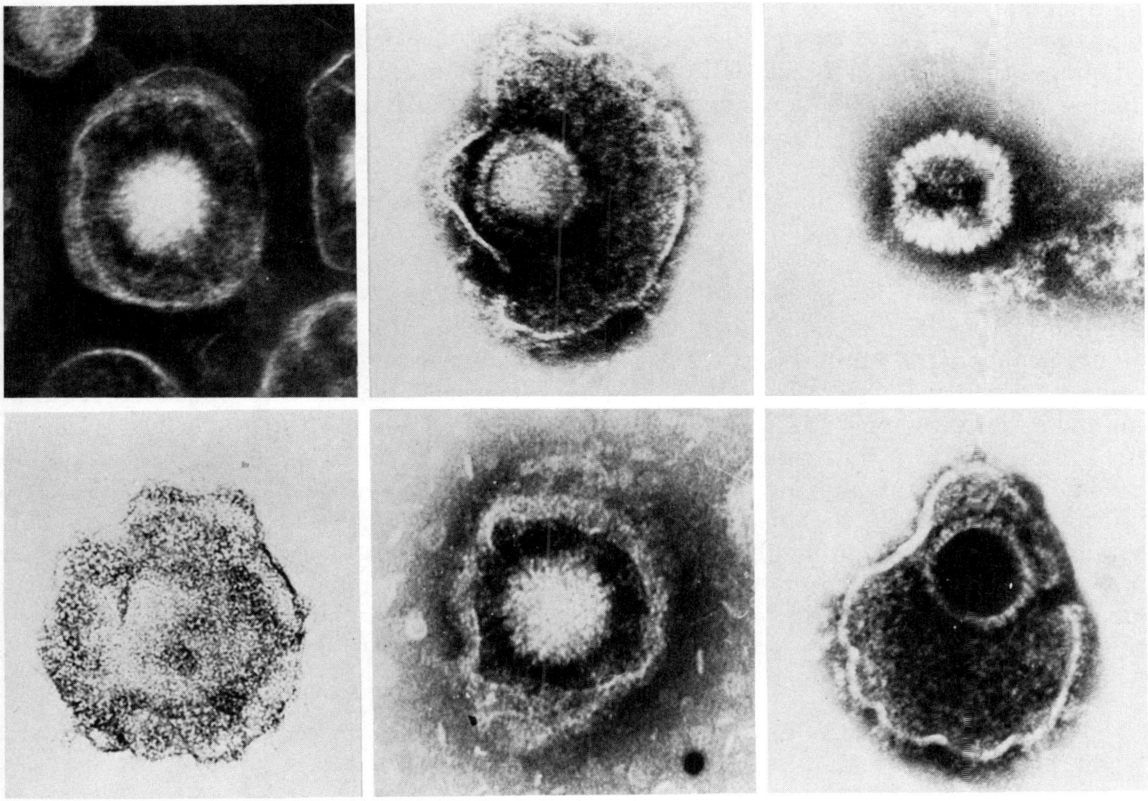

FIGURE 17-1. Various forms of herpes simplex virus. (Courtesy of Centers for Disease Control, Atlanta, GA 30333.)

to form a multinuclear syncytium. Giant cell formation is caused only by enveloped viruses, especially those of the herpes and paramyxovirus groups.

Probably the most unique and important clinical characteristic of the herpesviruses is their ability to maintain themselves in a latent state in the host for years, often for the lifetime of the host, even in the presence of circulating antibodies. The only other group of viruses that also result in latency are the adenoviruses, which are also DNA viruses that replicate in the host cell nucleus. It is not known how herpesviruses are able to be sustained for years in the host, but this characteristic causes difficult management and treatment problems. Recurrent infections are common, shedding of virus may occur intermittently for years, making control of spread very difficult, and latency is a concern in the development of vaccines, thus making prophylaxis difficult.

Cytomegalovirus (CMV)

Biologic characteristics. Cyto (cells) megalia (large) refers to large inclusion-bearing infected cells (Fig. 17-2). Disease caused by cytomegalovirus (CMV) has been called cytomegalic inclusion disease (CID) or inclusion body disease. CMV are typical herpesviruses. They grow only in human tissue and measure 180 nm in diameter. Replication of CMV is slower, and the degree to which the virus remains cell-associated is greater than with other members of the herpes group. CMV has only been isolated in culture from human fibroblasts despite the fact that in the human host CMV can grow in every type of cell. There is no animal model for growing CMV. Though there are strains of CMV that infect other animals besides humans, they are species-specific and human CMV can only be grown in human cells. There is probably more than one antigenic strain of human CMV; five subgroups have been identified by immunofluorescence, but these have not been classified into separate types.

Epidemiology. CMV is worldwide and ubiquitous. Humans are the only reservoir for human CMV and transmission can occur by direct or indirect contact with infectious secretions or tissues. Virus is excreted in saliva, urine, breast milk, vaginal secretions, semen, and blood, all of which can be sources of the virus. The most common methods of transmission are congenital or perinatal and by blood transfusions, though direct and indirect person-to-person spread has also been demonstrated.

The prevalence of CMV antibodies ranges in studies from 10 to 100 percent (Fig. 17-3), but it is apparent that most people do become infected during their lifetime and infection usually occurs in infancy or childhood. Eight to 60 percent of infants shed CMV in their urine during their first year of life. CMV can be acquired by neonates in three ways: either transplacentally, when passing through an infected birth canal, or

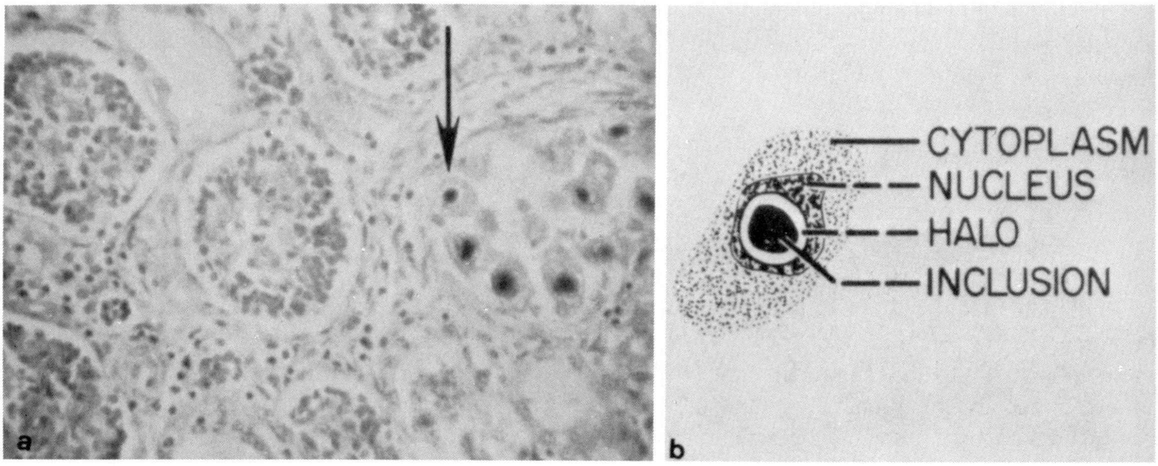

FIGURE 17-2. *(a)* Cytomegalovirus inclusions in renal tubular cells. Magnification 250×. *(b)* A diagram of the cell as seen in the urine. (Reprinted with permission of the publisher from Moffet HL. Clinical microbiology, 2nd edition. Philadelphia: JB Lippincott, 1980.)

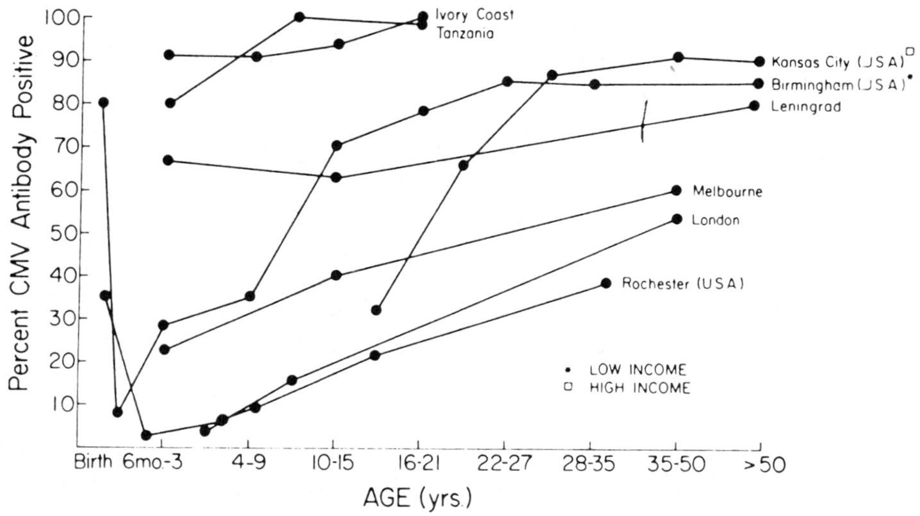

FIGURE 17-3. Seroepidemiology of cytomegalovirus infection. (Reprinted with permission of the publisher from Nahmias AJ, Dowdle WR, Schinazi RD, eds. The human herpesviruses. New York: Elsevier, 1980.)

after birth when being handled by the mother or other infected individuals. In a recent propsective study of 200 mothers and babies, the frequency of intrauterine infection was 2 percent, and 30 percent of the infants were infected around the time of birth. Perinatal infections were more common in babies whose mothers had IgM antibodies to CMV, indicating recent or active infection. Maternal antibody was not protective to the child. On the contrary, none of the babies whose mothers lacked antibody to CMV developed infection during the first year of life, evidence that the mother is often the source of infection for the child (Leiwikk et al. 1978).

After infancy the most common route of spread is respiratory. The virus is present in saliva and prevalence increases through the schoolage years into young adulthood. The higher incidence of infection among pregnant women may suggest increased susceptibility due to hormonal changes, or may suggest that CMV is also spread as a sexually transmitted disease. CMV excretion and high antibody titers coexist; though reinfection with CMV does occur, reactivation of latent virus is the most frequent source of recurrent symptoms.

There are no seasonal patterns to CMV infections, but there seems to be a higher prevalence of positive titers in tropical countries. This may in part be due to climate, but more likely to the general lower socioeconomic levels in these countries. Persons of lower socioeconomic status (in all countries) tend to acquire CMV infection earlier than do others. In a Canadian study, babies who acquired congenital CMV disease were not different from other babies in terms of mean gestational age, weight, or head circumference, but their mothers were younger, of lower socioeconomic status, and were three times more likely to be unmarried than the mothers of healthy children (Bryce et al. 1980). CMV infection frequently becomes clinically manifest following bone marrow or renal transplantation. In one series, one-fourth of patients with CMV antibodies had reactivation of symptoms following renal transplantation especially if they received a poorly matched graft and antithymocyte globulin (Pass et al. 1980). Blood transfusions also spread CMV. In a recent study, use of blood from donors seronegative for CMV decreased the prevalence of CMV excretion among transfused infants from 12.5 to 1.8 percent (Yeager et al. 1981).

Pathogenesis. The infectiousness of CMV is low, probably because the majority of viral particles remain cell-associated and because very few particles actually reach maturity. In culture, many particles (about 10^6) are necessary to result in infection. Infected cells such as those in the salivary glands are enlarged (up to 40 μm across), containing huge eosinophilic nuclear inclusions and basophilic granules in the cytoplasm. These cellular alterations are characteristic and can be seen with special staining techniques. CMV disease after organ transplantation or blood transfusion can be due to reactivation of host infection, or the transfused blood or transplanted organ itself can be the source of the virus. The former circumstance is probably most common. CMV is also potentially oncogenic, as transformation of human cells occurs.

Host responses. Though CMV is probably the most common cause of congenital infection of the human fetus, infections are usually subclinical and asymptomatic; less than 10 percent are clinically manifest. Nevertheless, CMV infection may account for 10 to 20 percent of fetal loss, a proportion of premature births, and a number of other congenital defects. CMV is the most common infectious cause of central nervous system damage in the fetus and newborn; about 15 percent of diseased infants are mentally retarded. Damage such as hyrocephalus can be present at birth and can continue to worsen after birth. Table 17-3 compares the clinical manifestations of CMV with those of the herpes simplex virus and rubella, which are also viral diseases occurring in neonates. Congenitally infected infants, whether symptomatic or not, shed virus for years and may be a source of infection to others.

Neonatal infections are commonly acquired at or near the time of delivery either from cervical secretions or, after birth, from breast milk. Resultant disease is of slow onset. Virus shedding, most common in the urine, begins at about one month of age and may continue for several years. It is possible to differentiate congenital from perinatal infection, since virus shedding in urine and saliva is present in the first week of life in infants infected before birth. Perinatal infections, like congenital infections, are usually asymptomatic; however, interstitial pneumonia, hepatosplenomegaly, and lymphadenopathy with rash can result. Long-term sequelae of infections acquired during birth are probably rare.

Clinical manifestations in older children and adults include hepatitis, mononucleosis, and mild respiratory, gastrointestinal, or central nervous symptoms (Table 17-4). The mononucleosis resembles infectious mononucleosis due to Epstein-Barr virus except that the heterophil antibody is negative and pharyngitis and lymph node enlargement are usually absent in CMV infection. Recovery within two months is the rule. CMV and toxoplasmosis have been reported to occur together with some fre-

TABLE 17-3. Frequency (Percent) of Symptoms in Neonates with Systemic Viral Diseases

	CMV	HSV	Rubella
CNS			
Encephalitis	50-75	50-75	20-50
Microcephaly	20-50	20	Rare
Other	Less than 20	Less than 20	Less than 20
Cardiovascular			
Structural defects	Not reported	Not reported	Up to 75
Peri-, myocarditis	Rare	Rare	0
Genitourinary	0	0	Less than 20
Eye			
Chorioretinitis	21-50	20	0
Conjunctivitis	Rare	20	0
Hematologic	Rare	20-50	20-50
Pulmonary	9-20	20	20-50
Liver	50-75	20-50	Less than 20
Long-term sequelae			
Mental retardation	15	Not reported	20-50
Hearing deficit	20-50	Not reported	20-50
Visual impairment	0-20	20	Not reported

Adapted from Galasso GJ, Merigan TC, Buchanan RA. Antiviral agents and viral diseases of man. New York: Raven Press, 1979.

quency in patients with neoplasms. Gastrointestinal disturbances, interstitial pneumonia, and chronic lung disease have also been reported. As previously mentioned, cytomegalic inclusion disease is much more common among those with malignancies, in post-transplantation, and following blood transfusions. Disseminated disease can be fatal.

Control and treatment. There are no known preventive measures nor is there any specific treatment for CMV disease. Interferons have not been shown to be efficacious. Vaccination of high-risk groups such as women of childbearing age and persons who will be receiving renal or bone marrow transplants

TABLE 17-4. Clinical Manifestations of Cytomegalovirus Infection

Type of victim	Manifestation
Congenital	Congenital defects (see also Table 17-3)
Perinatal	Pneumonia Lymphadenopathy Hepatosplenomegaly Rash
Children and adults	Mild respiratory tract, gastrointestinal, and CNS symptoms Hepatitis Mononucleosis Frequent simultaneous occurrence with toxoplasmosis

shows promise. Several live virus vaccines are being clinically trialled, as well as an inactivated virus vaccine in animal models. However, there are problems to be resolved before a successful program of vaccination is possible; especially the potential oncogenicity of CMV and the fact that many congenital defects are a result of reactivation rather than primary infection in mothers.

Epstein-Barr Virus (EBV)

Biologic characteristics. Epstein-Barr virus (EBV), discovered in 1964 by Epstein and Barr who were searching for an etiologic agent of Burkitt's lymphoma, is morphologically identical to other herpesviruses. It has four important biologic characteristics, including its ability to

1. Transform human lymphoid cells
2. Induce lymphoma in marmosets
3. Cause infectious mononucleosis in humans
4. Persist in an infected host (Epstein and Achong 1979).

The virus replicates in the host cell nucleus. Mature virions are housed in the host cytoplasm after they acquire their envelope from the nuclear membrane of the host.

EBV has four major antigens: viral capsid antigen (VCA), EBV-associated nuclear antigen (EBNA), EBV-associated early antigen (EA), and EBV-induced cell membrane antigen (MA) (Table 17-5). The presence of VCA and immunofluorescent tests are used to identify those who have been infected with EBV. At least three other distinct antigens have also been described. These antigens are unique from other herpesvirus antigens, but share some homology with three animal viruses that cause Lucke's adenocarcinoma of frogs, Marek's disease in chickens, and infectious bovine rhinotracheitis.

TABLE 17-5. Major Antigens of Epstein-Barr Virus

Antigen	*Characteristics*
VCA	Virus-specific capsid proteins; the reference test for infection with EBV is presence of VCA
EBNA	Fine granules in nuclei of infected cells
EA	Early antigen; may be formed with or without formation of virions
MA	Membrane antigen on cell surface; probably a component of the viral envelope; two serologically distinct types

Epidemiology. The distribution of EBV is worldwide. In developing countries most children by age 3 have antibodies to VCA, but in developed countries infection is often delayed until adulthood. Forty to 100 percent of children, even in developed countries, have antibody by age 5. The association of EBV with clinical disease is made primarily on epidemiologic evidence, and EBV infection has been implicated strongly as the causal agent (with or without other predisposing factors) in infectious mononucleosis (IM), Burkitt's lymphoma (BL), and nasopharyngeal carcinoma (Table 17-6). As can be seen in Figure 17-4, antibody titers to VCA are much higher in individuals with these diseases than in normal controls.

EBV is almost always detectable in African Burkitt's lymphoma, but not always in American Burkitt's, though a strain of EBV identical to that found in African BL has been isolated from American BL (Pizzo, Magrath, and Jay 1981). BL has a limited geographic distribution occurring primarily in the lower three-fourths of the African continent where it is the most common cause of cancer in 5- to 10-year-old children. In endemic areas the incidence is about 10 per

TABLE 17-6. Diseases Associated with Epstein-Barr Virus

Disease	Geographic distribution
Burkitt's lymphoma	Worldwide, but endemic in Africa and New Guinea
Infectious mononucleosis	Worldwide, but most frequently diagnosed in parts of the developed world, such as U.S. and Europe
Nasopharyngeal carcinoma	Certain parts of China, Algeria, Morocco, Tunisia

10^5 persons per year. The disease often occurs in those already infected with malaria. This seems to be a predisposing factor since BL is not endemic in malaria-free areas even when the weather and geography are similar to endemic areas. There is an excess of cases among males and there may be some genetic variation in susceptibility, though ethnic differences may be due to socioeconomic and geographic factors rather than race. The disease sometimes occurs in time and space clusters (epidemic pattern), and does occur in much lower rates throughout the rest of the world.

Though the evidence for a causative relationship between EBV and infectious mononucleosis (IM) is very strong, it is based on serologic tests, viral isolation from diseased individuals, and the fact that EBV experimentally transferred to humans and monkeys has been at least partially successful in producing IM. However, individuals do, on rare occasions, develop IM without developing high titers of VCA. Nevertheless we will proceed on the premise that EBV causes IM.

The incidence of IM has been estimated to be about 100 per 10^5 persons per year for the general population, about 1000 per 10^5 per academic year among college students (Fig. 17-5). The prevalence of EBV antibody among young adults varies from 25 to 100 percent; those in underdeveloped countries being more likely to be infected. Rates of clinically diagnosed IM are highest among 15 to 24 year olds. There is no sex predilection or seasonal variation, but whites are much more commonly diagnosed than are blacks. Perhaps persons of dark skin acquire EBV infection at an earlier age when clinical manifestations are much less severe.

Though the epidemiology of nasopharyngeal carcinoma (NPC) is very different than that of BL or IM, the serologic and epidemiologic evidence linking it to EBV is similar. NPC is very rare in most parts of the world (less than 1 in 10^5 persons per year) but in certain regions of China and in certain areas of Africa such as Algeria, Morocco, and Tunisia, rates are 20 to 30 times higher. The disease does not occur in epidemic patterns. There is an excess incidence in males, especially among certain occupational groups. The risk of NPC decreases among Chinese when they move from areas of high prevalence to areas of low prevalence of EBV and risk to Caucasians increases when they live in these same areas. This is evidence that environmental factors are determinants of risk. It is probable that EBV works in conjunction with other environmental variables to cause NPC.

For all diseases linked with EBV, the mode of transmission is presumed to be contact with infectious oropharyngeal secretions. However, the virus is not very contagious, which probably explains why other determinants in addition to EBV are usually necessary to initiate infection. The virus is also transmitted in blood transfusions.

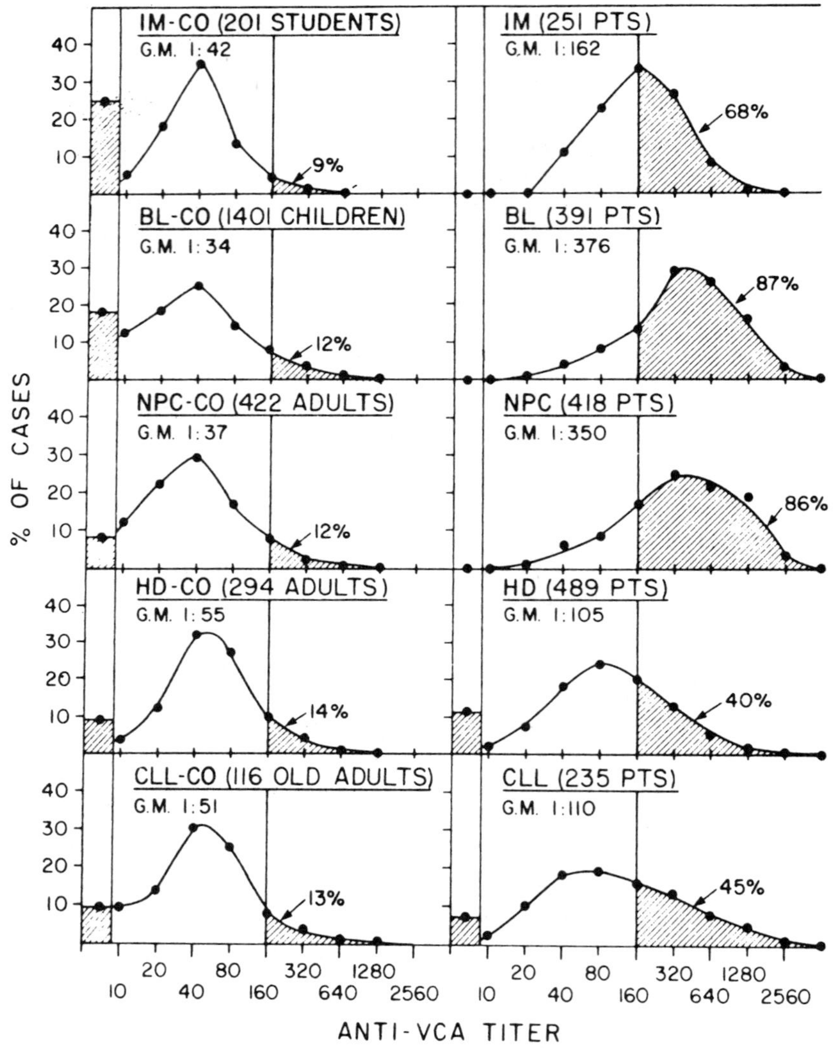

FIGURE 17-4. Distribution of antibody titers to VCA in patients with various diseases and appropriate control groups. IM = infectious mononucleosis; BL = Burkitt's lymphoma; NPC = nasopharyngeal carcinoma; HD = Hodgkin's disease; CLL = chronic lymphocytic leukemia; CO = control; GM = geometric mean. (Reprinted with permission of the publisher from Epstein, MA, Achong BG, eds. The Epstein-Barr virus. Berlin: Springer-Verlag, 1979.)

Pathogenesis. EBV can be cultured in lymphoid cells. Cells susceptible to EBV infection are primarily B lymphocytes, though EBV also is present in the nasal epithelial cells of persons with nasopharyngeal carcinoma. Some strains of EBV can change

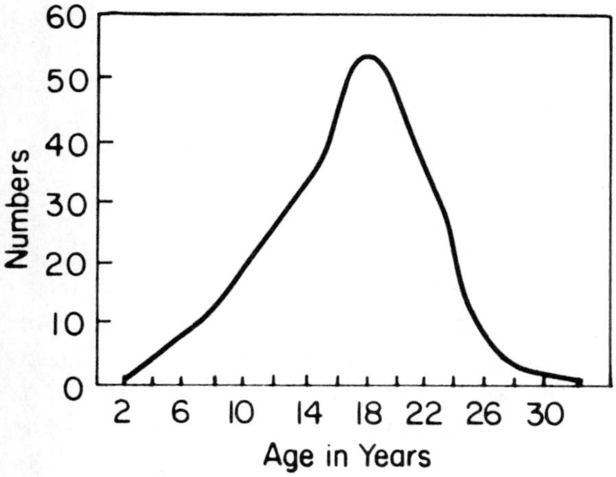

FIGURE 17-5. Typical age distribution of clinically diagnosed infectious mononucleosis.

human lymphocytes so that they are capable of rapid and persistent multiplication (called *immortalization*). EBV has not been shown to be cytotoxic to human cells. After establishment of EBV in B cells or the oropharynx, these cells are transformed and proliferate until up to 40 to 50 percent of local cells are infected. Infection may spread to other lymphoid tissue, especially the spleen, tonsils, and lymph nodes. There is lymphocytosis, appearance of atypical lymphocytes with emergence of suppressor T cells due to stimulation of cell-mediated immunity. Antibody is produced that neutralizes extracellular viruses, but persistent subclinical infection continues (same latency as with other herpesviruses) and the EBV persists in B lymphocytes. Clinical symptoms are due to destruction of infected cells, formation of immune complexes, or autoantibodies.

Host responses. Many EBV infections, probably 40 to 70 percent, are subclinical, especially when infection occurs in early childhood. The incubation period is unknown, but is probably long; from 1 to 3 months for person-to-person spread and 3 to 5 weeks if transfusion-transmitted. EBV, as opposed to CMV, causes only very rare infections in utero. It has also been linked with Hodgkin's disease, but no definite evidence for causation is available.

Burkitt's lymphoma is a tumor in the bones of the jaw that spreads primarily to the abdominal viscera, the meninges, the kidneys, and liver. It can be rapidly fatal in the advanced stages though spontaneous regressions sometimes occur. Without chemotherapy, 2-year survival is about 20 percent, but with treatment it is increased to 80 percent. Deaths occur only rarely after 2 years.

Infectious mononucleosis probably represents primary EBV disease among adolescents and young adults who escaped infection as children. The symptomatology is characteristic (Table 17-7). The most common symptom is sore throat with pharyngitis. Most people have low-grade fever and enlargement of cervical lymph nodes for up to six weeks. In children, abdominal pain is also common. Rarely, a rubellaform rash oc-

TABLE 17-7. Criteria for Diagnosis of Infectious Mononucleosis

Clinical	Slow onset of malaise, fever, sore throat; cervical lymph node enlargement, hyperplasia of pharyngeal lymphoid tissue
Hematologic	Absolute lymphocytosis lasting at least 2 weeks (4500/mm^3), at least 50 percent lymphocytes; atypical lymphocytes
Serologic	Heterophil antibody must appear in serum during illness*

*Serologic evidence of EBV infection without the other symptoms should be called primary EBV infection rather than infectious mononucleosis.

Adapted from Hoagland RJ. Infectious mononucleosis. New York: Grune and Stratton, 1967.

curs. However, infected individuals who are treated with ampicillin very frequently develop rash 5 to 14 days after its use; perhaps due to formation of abnormal antibodies. Central nervous system involvement is rare but does occur. Liver damage occurs in up to 90 percent of cases, jaundice in about one-third, as well as abnormal electrocardiogram, hyperuricemia, excessive autohemolysis, thrombocytopenia, and neutropenia.

Lymphocytosis (mononucleosis) is the hallmark of infectious mononucleosis, along with the presence of heterophil antibody. Heterophil antibody is antibody that is formed against antigens unrelated to the immunizing antigen. One that is exclusive to patients with infectious mononucleosis is called the Paul-Bunnell-Davidsohn (PBD) antibody. Though it does not react specifically with EBV, it is found almost exclusively in persons with IM. Occasional false-positive tests occur in those with other viral or malignant disease. The recovery period is long, but the mortality rate is less than 0.1 percent. Complications include splenic rupture, disseminated intravascular coagulation, hepatoencephalopathy, and pharyngeal hemorrhage.

Nasopharyngeal carcinoma causes death within a few months or slowly progresses for years. Presenting symptoms include cervical node swelling, nasal obstruction or bleeding, hearing impairment, or cranial nerve involvement. About one-third of cases are metastatic, and these have the longest survival rate. The invasive type, less than 10 percent of cases, spreads directly, and a combination of these two types occurs in over half of cases. Five-year survival with treatment is 70 percent.

Control and treatment. Many cases of BL may be prevented if effective malaria control exists. BL is very sensitive to chemotherapy, usually with cyclophosphamide (cytoxan). A vaccine is feasible, but presents the same problems as those discussed for CMV. Indeed, all herpesviruses present these same theoretic problems when vaccine development is considered.

There is no specific treatment for IM. Bed rest during fever, decreased physical activity, and corticosteroids shorten the duration of fever and hematologic abnormalities. Antibiotics are not useful, and ampicillin is especially contraindicated because of the high risk of rash development. Because infectiousness is very low and spread of the virus appears to require long-term, intimate contact with respiratory secretions, strict isolation procedures are unnecessary.

Though there are no means to prevent nasopharyngeal carcinoma, early diagnosis can decrease mortality. A painless lump in the upper neck of an adult from southern China is almost pathognomonic. Radiation therapy is the primary treatment.

Herpes Simplex Virus (HSV)

Biologic characteristics. The herpes simplex virus (HSV), also called herpesvirus hominis, is separated into two distinct serotypes: 1 and 2. They can be distinguished by type-specific antisera; type 2 has increased sensitivity to temperature and forms larger pocks on chorioallantoic membrane of embryonated egg. However, they cross react strongly. Both are isolated readily in tissue culture. They are about 150 to 200 nm in size, including their envelope and produce at least 49 proteins (Fig. 17-6). The term herpes (Greek: to creep) was first coined almost 2000 years ago by Herodotus to describe the spreading oral lesions most common with HSV-1.

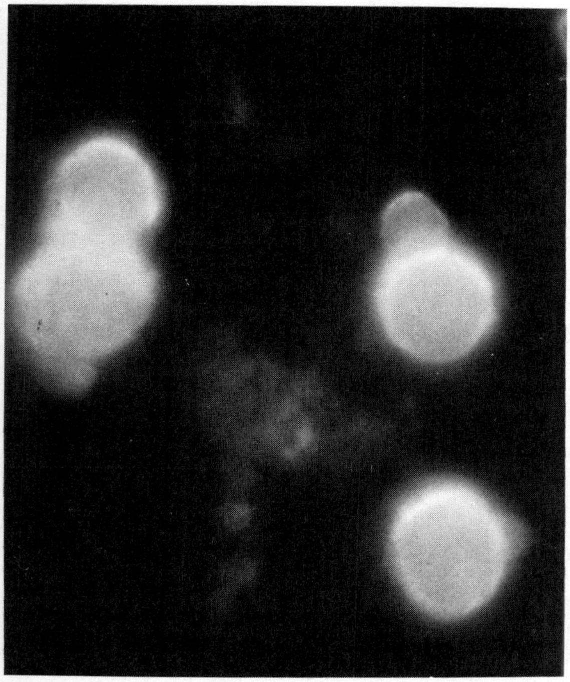

FIGURE 17-6. Fluorescent antibody stain of sputum containing herpes simplex, type 1. (Courtesy of Dr. Lawrence Corey.)

Epidemiology. Humans are the only reservoir of HSV. The epidemiology of HSV-1 and HSV-2 are very different (Table 17-8). Both types are distributed throughout the world, even in remote areas that have had minimal contact with developed civilizations. HSV-1 causes predominantly oral lesions: gingivostomatitis, cold sores, or fever blisters. HSV-2 causes predominantly genital lesions. However, genital herpes in some populations is also caused by HSV-1 (Table 17-9). Transplacental antibodies to HSV-1 are present in the blood of about half of newborns, but these disappear by 6 months of age.

Most people have encountered HSV-1 during their lifetime; about 50 to 100 percent of adults have circulating antibody (Fig. 17-7). However, primary infections are usually asymptomatic and usually occur in infancy, peaking at 6 to 18 months of age. Most primary infections occur between ages 6 months and 4 years. Transmission is by contact with infected oral secretions. Only about 10 to 15 percent of children infected develop symptoms, the usual clinical manifestation being gingivostomatitis (Fig. 17-8). There are no seasonal, sex, or geographic variations in incidence, though prevalence rates are somewhat higher among groups of lower socioeconomic status. The disease is highly endemic; it is not usually characterized by epidemic patterns (Fig. 17-9). However, hospital and institutional outbreaks of gingivostomatitis and skin lesions have occurred.

Despite the fact that there is a 4-fold or greater rise in antibody titer between acute and convalescent sera in individuals following a primary oral herpes infection, recurrences are common, occurring in 20 to 50 percent of cases. About 7 percent of the total population has more than two recurrences per year. Some events that precipitate recurrences are menses, exposure to sun, trauma, other infections, fatigue, and stress. Following bone marrow transplant, most patients who are seropositive for HSV, but only a few who are seronegative, develop HSV disease, indicating that reactivation rather than reinfection is the primary cause of such recurrences.

There are an estimated 200 to 2000 cases of HSV encephalitis per year in the United States; about 20 deaths are reported. This serious complication is most often associated with type 1, whereas a less serious meningitis occurs more frequently as a result of type 2 infections.

The incidence of genital herpes, 60 to 90 percent caused by HSV-2, appears to be in-

TABLE 17-8. Herpes Simplex Viruses, Types 1 and 2 (HSV-1, HSV-2)

	HSV-1	*HSV-2*
Host range	Humans only	Humans only
Mode of transmission	Contact with oral and respiratory secretions	Sexual
Usual age of primary infection	6-18 months	After puberty
Resultant diseases	Gingivostomatitis, herpes labialis (fever blisters, cold sores), keratitis (corneal infection), meningoencephalitis, chronic eczema (eczema herpeticum), whitlow	Genital herpes, neonatal herpes
Subclinical infection	Common	Common
Recurrences	Common	Common
Epidemiologic characteristics	Most people infected; only 10-15 percent develop primary clinical disease; recurrences induced by exposure to sun, temperature extremes, physiologic and emotional stress	Infection more common in persons of lower socioeconomic status, those with multiple sex partners; associated with cervical and vulvar cancer (though causative relationship not proven)

TABLE 17-9. Prevalence (Percent) of HSV-1 and HSV-2 as Causes of Clinical Herpes Syndromes

	HSV-1	*HSV-2*
Neonatal infection	20-40	60-80
Oral herpes	99	1
Dermatitis		
Above waist	95	5
Below waist	10	90
Genital herpes	10-40	60-90
Disseminated infection in children and adults	97	3

creasing over the past few decades (Fig. 17-10). There are an estimated 5 million individuals in the United States with genital herpes, but the prevalence varies, being much more common in individuals with multiple sex partners, in those who began having sexual intercourse at a young age, and in those of lower socioeconomic class. For example, the prevalence of genital herpes among patients being treated for other sexually transmitted diseases has been shown to be about 30 to 40 per 1000, in clinic patients, 3 to 10 per 1000, and in patients seen by private physicians, 0.3 to 0.9 per 1000. The incidence peaks in young adulthood and it has all of the epidemiologic features of a sexually transmitted disease. There appear to be no seasonal or sexual predilections to the disease.

About 80 percent of patients have recurrences after primary genital herpes infections, but the frequency of recurrences decreases over time. HSV can be isolated in

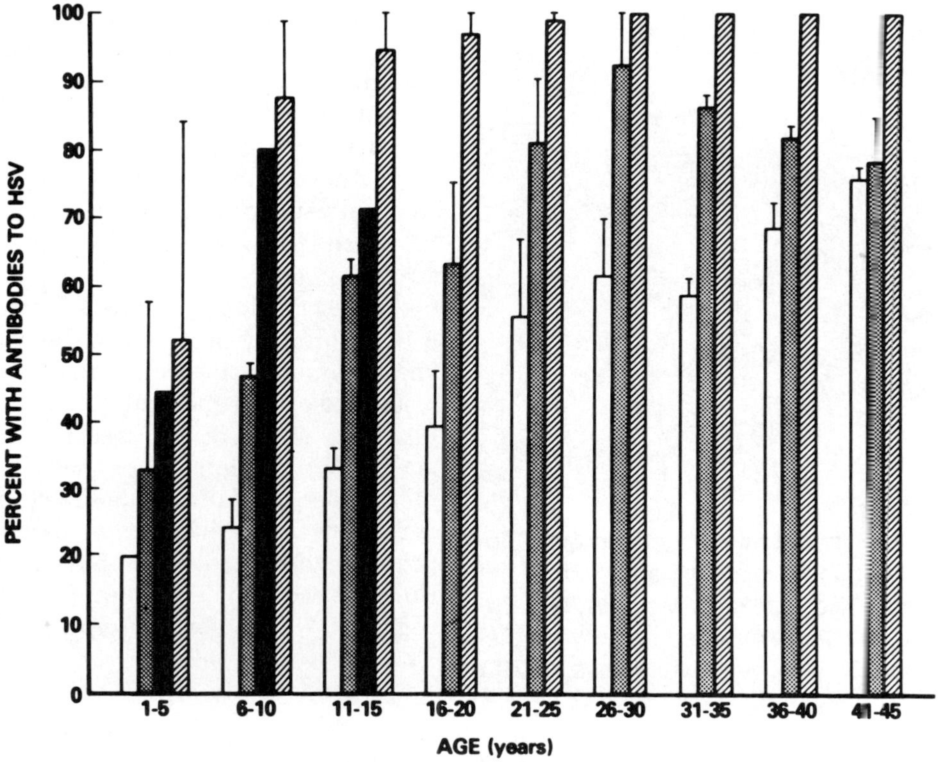

FIGURE 17-7. Age distribution of antibodies to herpes simplex virus among different populations. Columns with *slashed bars* represent mean value for three aboriginal Indian tribes; the *solid bars* represent values for children from low socioeconomic groups living in Uruguay. The *stippled bars* are means of values reported for blacks from lower socioeconomic class of Atlanta, Georgia, and Houston, Texas. The *open bars* are mean values for people from middle socioeconomic classes of Japan and Seattle, Washington. Antibodies to HSV-1 and HSV-2 were differentiated in some studies, but not in others. (Reprinted with permission of the publisher from Nahmias AJ, Dowdle WR, Schinazi RS, eds. The human herpesviruses. New York: Elsevier, 1980.)

about 85 percent of women with primary infection, about 10 percent with recurrences. Virus may also be shed when individuals are asymptomatic. Recurrences of genital herpes are not precipitated by fever as are those of oral herpes. However, hormonal changes may be important since the incidence of recurrences is 5 times higher in pregnant than in nonpregnant women. Onset of menstruation also triggers the development of lesions. Aside from the clinical disease itself, there are two major concerns associated with genital herpes in women: rare but serious neonatal herpes (about 120 cases per year reported in the United States) can be transmitted by infected mothers

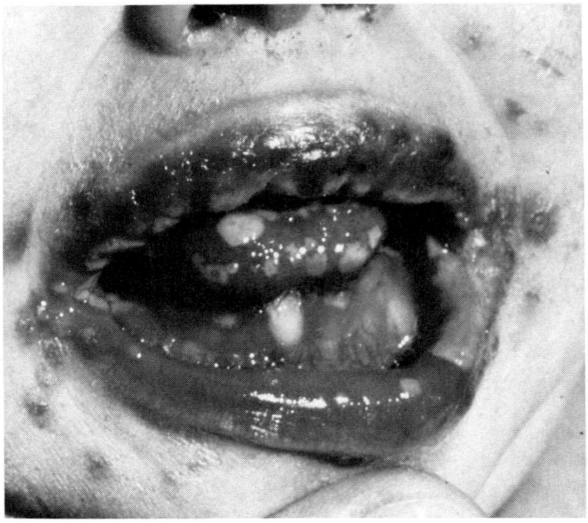

FIGURE 17-8. Infant with primary herpetic gingivostomatitis. (Reprinted with permission of the publisher from Galasso GJ, Merigan TC, Buchanan RA, eds. Antiviral Agents and Viral Diseases in Man. New York: Raven Press, 1979.)

transplacentally or at the time of birth, and there is a strong association between HSV-2 and cervical and vulvar carcinoma.

In 1966 an association between cervical cancer and HSV-2 was first noted. The epidemiology for both diseases is similar: common in those with multiple sex partners, and in those with early age of first intercourse. After at least 30 case-control studies comparing frequency of HSV-2 antibodies in women with and without cervical cancer, the cause and effect relationship is still unproven. Problems with serologic tests are

1. It may not be possible to identify past infection with HSV-2
2. HSV-1 and HSV-2 share many antigens, so HSV-2-specific antibodies are only a small portion of HSV antibodies produced after infection
3. It is very difficult to have subjects in study groups properly controlled for sexually related variables
4. Both HSV infection and cervical carcinoma could be related to particular epidemiologic patterns or sexual lifestyles, while being unrelated to each other

Nevertheless, 80 percent of women with cervical cancer are positive for latent infection with HSV-2, and the association is quite strong.

An association also exists between HSV-2 DNA-binding proteins in the cells of vulvar squamous cell carcinoma in situ. In nine of ten cases of cancer of the vulva, HSV-2 has been isolated. It is possible that there is a causal relationship or that HSV-2 could merely be a simultaneous infection that occurs more frequently in women who are also prone to develop carcinoma of the vulva. Further study of the role of HSV-2 in oncogenesis will be enlightening.

Pathogenesis. In oral herpes, the virus enters by the oral route; in genital herpes, the virus enters during sexual contact, and multiplies, causing local lesions. It may also spread to local lymph nodes and through the blood stream. Both HSV-1 and HSV-2 produce characteristic cell changes: intranuclear vacuoles, multinuclear giant cells, and host cell degeneration with ballooning. These cellular changes are demonstrable in the cervix with Papanicolaou (Pap) smear in infected women. The Pap smear is about two-thirds as sensitive for diagnosing genital herpes in woman as is viral isolation. HSV-2 (but not HSV-1) enhances permeability of infected cells.

Because the herpesvirus spreads directly from cell to cell, antibody, which can only reach organisms in the extracellular space, is probably of only minor importance in limiting infection. Persons with defects in humoral (B cell) immunity are not prone to more serious or frequent infections with HSV, also indicating that antibody plays lit-

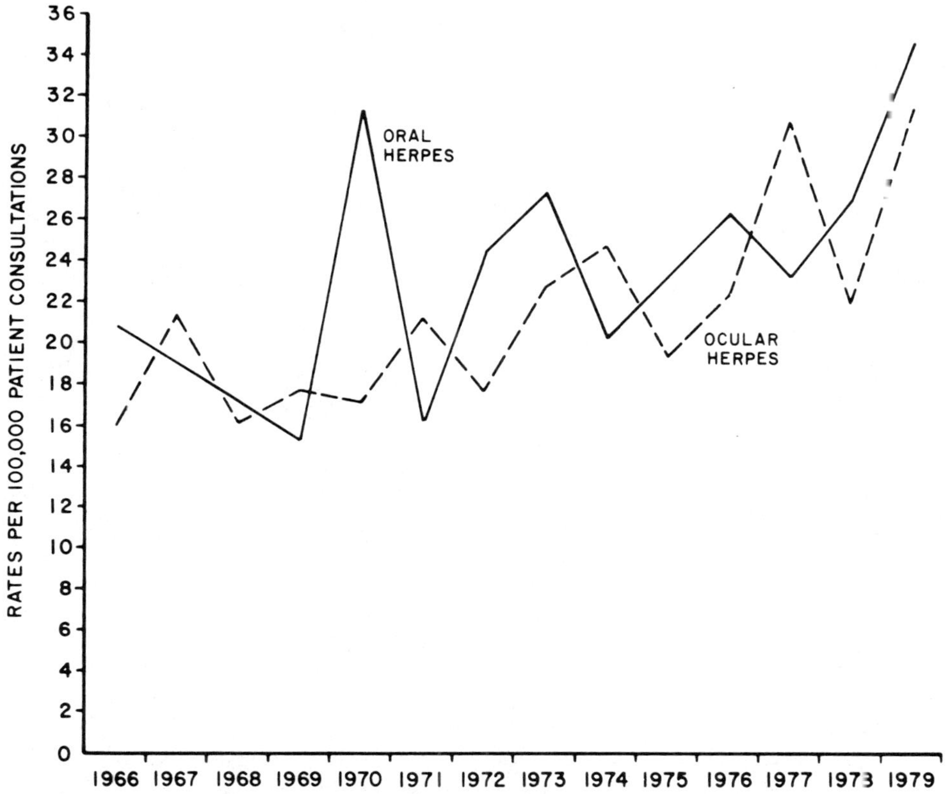

FIGURE 17-9. Estimated rates of patient consultations with private physicians for oral herpes infection and ocular herpes infection, United States, 1966 to 1979. (Reprinted from Centers for Disease Control. Genital herpes. Morb Mort Week Rep 1982;31:139.)

tle or no role in defense. However, T cell function seems to be critical, and recurrences more commonly result when there is impaired cellular immunity.

Despite circulating antibody, the virus takes up residence in a latent state in the neural senory ganglia; sacral (HSV-2) or trigeminal (HSV-1). The precise mechanism by which viruses are reactivated and again begin to replicate is unknown. There are two theories (Fig. 17-11). In the *ganglion trigger theory,* the viruses are in a nonreplicating, latent state in the ganglia until a stimulus causes them to again replicate and then travel down the intraaxonal pathway to the skin or mucous membranes. In the *skin trigger theory* there is a low, constant rate of viral replication in the ganglia, and it is stimulating factors at the skin itself that determine when clinical recurrences result. Virus shedding is greatest when there are clinical symptoms, but may also occur intermittently in the absence of symptoms for either herpes type.

Host responses. Asymptomatic infections with both HSV-1 and HSV-2 may be quite common. In either case, the primary infec-

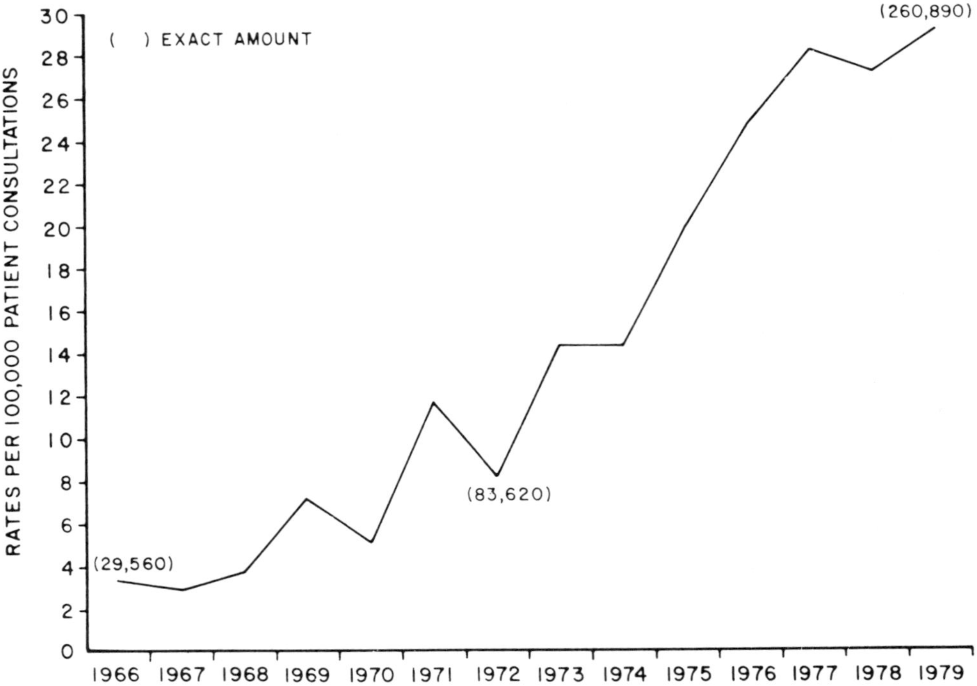

FIGURE 17-10. Estimated rates of patient consultations with private physicians for genital herpes infection, United States, 1966 to 1979. (Reprinted from Centers for Disease Control. Genital herpes. Morb Mort Week Rep 1982;31:138.)

tion is most likely to result in the most serious symptoms for both oral and genital herpes. Primary oral infection usually involves the entire oral cavity, whereas recurrences usually occur only on the lips or around the mouth. The incubation period for oral or genital herpes is 3 to 7 days. If gingivostomatitis occurs, it is characterized by sore mouth, anorexia, vesicular lesions that ulcerate within in a few hours and bleed readily, swollen lips, and enlarged cervical lymph nodes. Healing takes 2 to 3 weeks and virus is excreted for 2 to 5 weeks. Eighty percent of recurrences are single lesions, usually in the same or a proximal site as the previous lesions. Healing with recurrences is more rapid, taking about one week.

In genital herpes, 1- to 2-mm vesicles form, rupture, and cause painful erosions. Females may suffer from pruritus, dysuria, and pain with sexual intercourse. Lesions are most commonly on the labia, but sometimes on the cervix, buttocks, or thighs. The penis is the usual site of vesicles for the infected male. Fever and swollen lymph nodes may accompany the presence of vesicles. Lesions heal in about 3 to 6 weeks. Virus titers peak in the first few days after lesion formation and are shed for about 2 weeks after primary infection, and 4 days after recurrences. Recurrences occur as often as monthly. Though the risk of transmission is small with asymptomatic cervical infection, it is still possible that virus can be

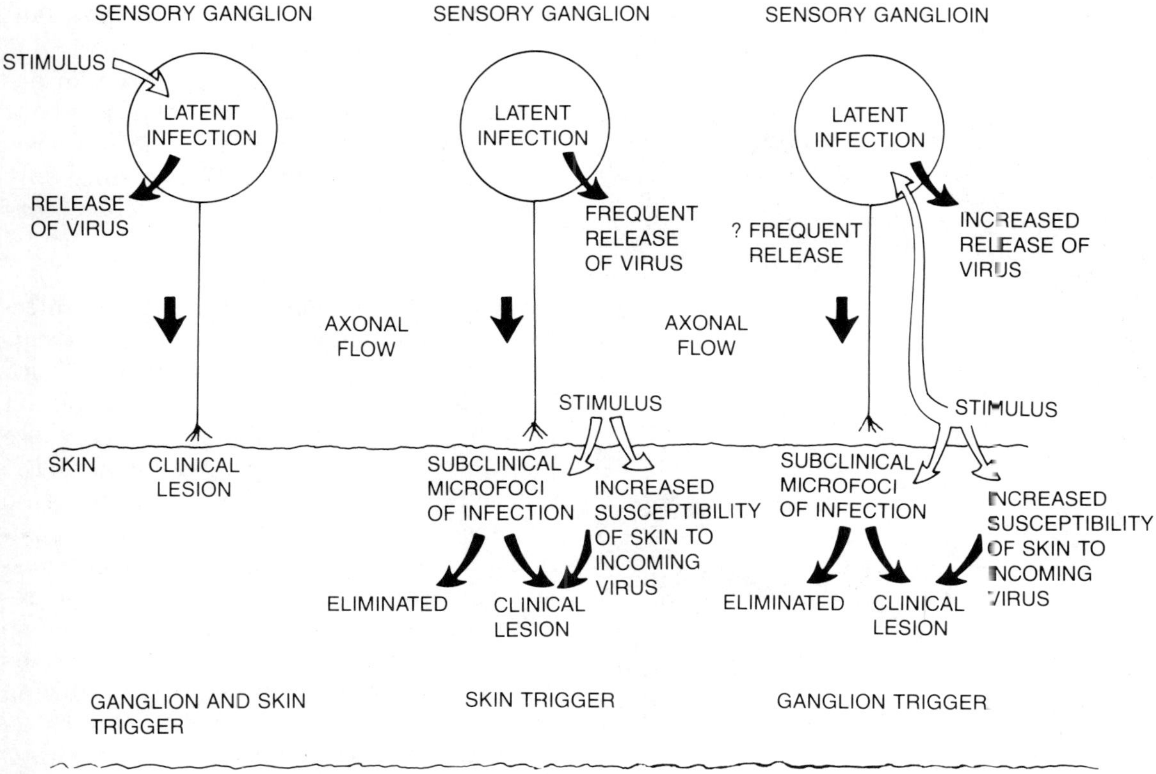

FIGURE 17-11. Possible mechanisms of induction of recurrent herpes simplex. (Reprinted with permission of the publisher from Nahmias AJ, Dowdle WR, Schinazi RS, eds. The human herpesviruses. New York: Elsevier, 1980.)

spread. Complications include cystitis, urinary retention (especially in women), bacterial superinfection, meningitis, sometimes rash, proctitis, endometritis, and whitlow.

Herpetic whitlow, or sores on the hands, occur when secretions infected with HSV contaminate a small abrasion on the hands. It is more commonly caused by HSV-1 and is an occupational hazard for health care personnel. In one recent report, there were five cases nosocomially acquired, three from a common source (Greaves et al. 1980). Herpes is also a major cause of keratitis and infectious blindness, and may cause dermatitis. Eczema herpeticum is an extreme form of viral invasion of the skin that usually occurs in individuals with atopic dermatitis or other diseased skin (Fig. 17-12). About three-fourths of victims are children. Large vesicles erupt on open skin areas, high fever results, and mortality is about 10 percent.

Neonatal herpes infection may include the spectrum of mild localized infection of eyes, skin, or mouth to severe disseminated fatal disease with liver and brain involvement. Unlike CMV, subclinical infections are not

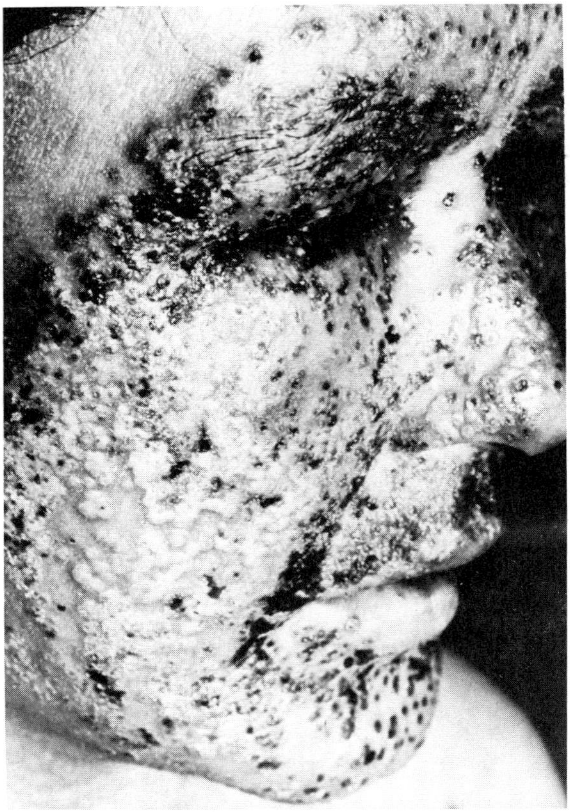

FIGURE 17-12. Eczema herpeticum; secondary infection. Two recurrences in a young man with atopic dermatitis. Some vesicles are hemorrhagic (type 1 infection). (Reprinted with permission of the publisher from Nasemann T. Viral diseases of the skin, mucous membranes and genitals. Philadelphia: WB Saunders, 1977.)

the rule with herpes simplex. In a small study, genital herpes in pregnant women has been associated with a 3-fold rise in spontaneous abortions and with increased prematurity. In another series, 70 percent of neonatal herpes occurred in babies whose mothers were asymptomatic. HSV-1 has caused infections in burn wounds and has caused hepatitits and disseminated intravascular coagulation after renal transplantation. HSV encephalitis, which can occur in neonates, the immunosuppressed, or apparently healthy people, is lethal in about 70 percent of untreated cases. The diagnosis is usually made by brain biopsy, since only rarely is the virus present in cerebrospinal fluid.

Control and treatment. Oral herpes infections are common and, in general, no treatment is necessary. Hospital personnel with active lesions should refrain from contact with high-risk populations such as those with dermatitis or burns, neonates, or the immunocompromised. Though there is no evidence that spread from personnel to patients is an important means of viral transmission, it is reasonable to make such restrictions. Corticosteroids are *contraindicated* for patients with herpes as they increase viral replication by inhibiting interferon. To prevent herpetic lesions on the hands, personnel giving oral or perineal care to patients with oral or genital herpes should wear gloves on both hands.

Treatment for genital herpes is primarily symptomatic: pain relief, lidocaine cream. In March 1982 acyclovir topical ointment was approved for use to shorten the duration of herpetic episodes, both genital and other mucocutaneous sites. Acyclovir has been shown to stop viral replication, shorten the duration of viral shedding, and in some cases, accelerate healing (Corey et al. 1982), but the emergence of resistant strains is a problem (Luby 1982).

Since the attack rate in susceptible individuals exposed to genital herpes is approximately 75 percent, individuals should refrain from sexual intercourse with partners who have lesions. Persons with genital herpes may also be at increased risk of other sexually transmitted diseases and should be screened for such. Probably about 40 percent of babies born to mothers with active lesions become infected. To prevent neonatal herpes, such babies are delivered by cesarean section. Weekly cervical cultures are recommended after 32 weeks gestation for those with a history of HSV-2 and for those whose sex partner has such a history.

Vaccination research is facilitated by the fact that there are experimental animal models for HSV infection. Hope for an acceptable vaccine remains poor, however, until the problems of latency and oncogenesis are resolved. Passive immunization, interferon, and substances like transfer factor and BCG that enhance immunity have all been trialled with no real effect to date.

Other drugs that are being tested include 5-iodo 2′ deoxycytodine (IDU or idoxuridine), trifluorothymidine (TFT), cytarabine (ara-C), vidarabine (ara-A). Three percent ara-A cream in genital and 10 percent in oral-labial (lip) herpes have not shortened the course of the disease, nor has 0.2 to 10 percent idoxuridine. However, in recent clinical trials a combination of IDU and dimethylsulfoxide (DMSO, an organic solvent) has shortened the duration of viral shedding, but not the rate of recurrence or duration of symptoms. All of this is experimental and the potential side or toxic effects have not been well delineated. Though no current therapy is efficacious topically, intravenous adenine arabinoside (ara-A or vidarabine) has been found to reduce mortality of disseminated neonatal infection and HSV encephalitis.

Herpesvirus Simiae (B Virus). The monkey is the natural host for B virus, but it can cause accidental infection in humans who contact monkeys or monkey cells. It produces acute ascending myelitis and encephalomyelitis, which is usually fatal. It is usually contracted by monkey bites though cases after merely handling monkeys have occurred. Initial healing appears to be without complication, but in 1 to 3 weeks generalized symptoms result. There is no definitive treatment and fortunately there is an average of only about one case per year.

Varicella-Zoster Virus (VZV)

Biologic characteristics. The virus that causes varicella (chickenpox) and zoster (shingles) is the same agent, varicella-zoster virus (Table 17-10). There is a single antigenic type containing at least five antigens. The virus is labile in the environment and has slight serologic cross reactivity with HSV. It can be grown in human and monkey cells, but has not been successfully grown in experimental animals, as has HSV.

Epidemiology. VZV causes disease only in humans and is worldwide in distribution. More than 85 percent of children have chickenpox by age 10; 95 percent of cases are in those less than age 15. There are about 400 per 10^5 per year cases of chickenpox in persons younger than age 15. Transmission is by droplet or direct contact with infected vesicles. Peak seasons in temperate climates such as the United States are winter and spring (Fig. 17-13). The attack rate among exposed susceptibles is about 60 to 90 percent, just slightly less than measles and influenza. Chickenpox can be considered an "occupational hazard" of school attendance. It occurs both endemically and in epidemic peaks.

Zoster (shingles), which represents a recrudescence of latent VZV infection among those who have previously had chickenpox, is most common in individuals past age 50,

TABLE 17-10. Comparison of Varicella and Zoster

	Varicella	*Zoster*
Causative virus	Indistinguishable from zoster	Indistinguishable from varicella
Age distribution	Usually young children; the primary form of disease	Usually adults; the secondary form of disease
Circulating antibody present?	No	Yes
Disease	Chickenpox	Shingles
Lesions	Painless vesicles occurring over entire body in crops	Identical vesicles, but very painful and occurring only along distribution of affected sensory nerve
Recurrences	No	Rare
Serious sequelae	Rare	Rare
Factors which increase disease severity	Steroid therapy, immune deficiencies	Same

though it has occurred in children. The prevalence of zoster increases as the numbers of elderly and immunosuppressed individuals increases. The annual incidence in the United States is about 3 to 4 per 1000 persons in the total population, but greater than 100 per 1000 in those over age 60. There are no seasonal variations in occurrence, and the only "outbreaks" noted have been among individuals whose immunologic defenses are compromised. Children can acquire chickenpox from contact with an individual with zoster. Likewise, exposure to children with chickenpox can cause an attack of shingles in an adult. There are no sex, race, or socioeconomic differences in prevalence. VZV infection cause about 100 deaths per year in the United States, most frequently from varicella pneumonia, hematologic defects, or encephalitis.

Pathogenesis. Despite the fact that the virus is quite labile and has not been recovered from the respiratory secretions of infectious individuals, the most likely route of VZV spread for chickenpox is by droplets. During the incubation period local replication of virus occurs, then a viremia with more replication, probably in the reticuloendothelial system. Histologic changes occur in the capillary endothelium and the virus spreads preferentially to the skin, in severe cases to vital organs as well. Successive crops of vesicles are probably due to intermittent viremia.

Zoster represents the emergence of the VZV in a host that has been previously sensitized and has partial immunity. It is not known if the virus after an episode of chickenpox is completely latent in the dorsal root ganglia, or if there is a small amount of virus continuously manufactured (see discussion of ganglion versus skin trigger theories in discussion on HSV). It is not known exactly why attacks of zoster are triggered in some individuals and not in others.

Recent evidence indicates that the proportion of persons positive for VZV antibody

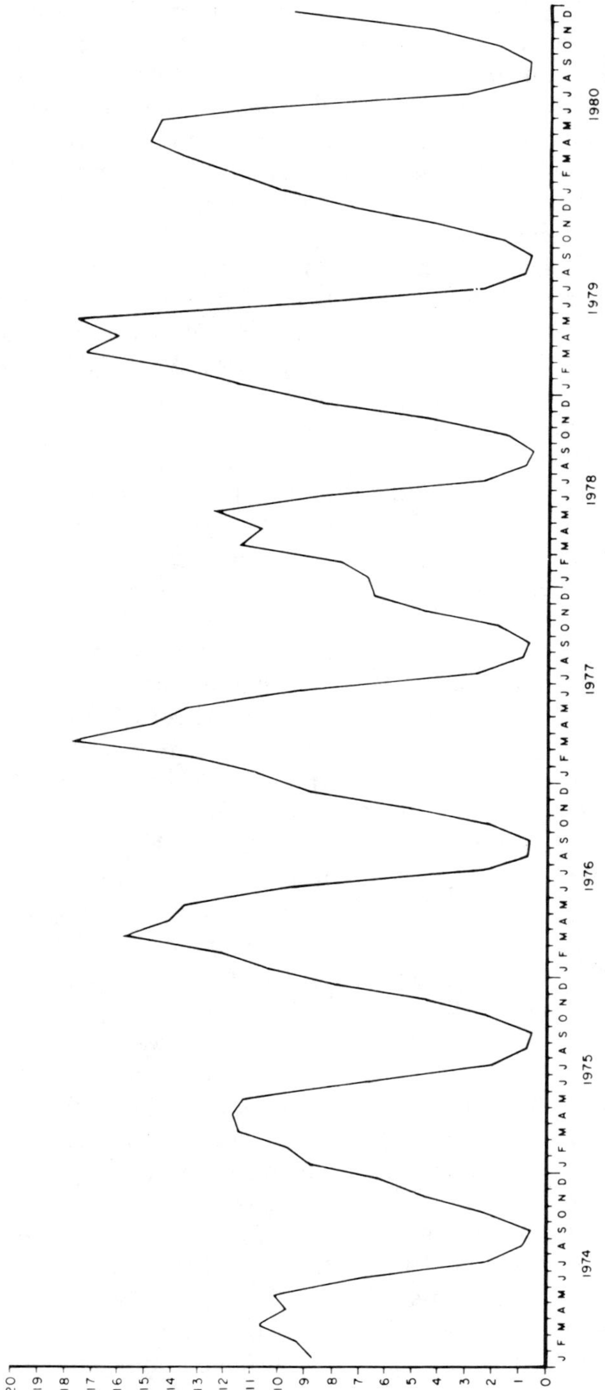

FIGURE 17-13. Chickenpox. Reported case rates by month, United States, 1974 to 1980. (Reprinted from Centers for Disease Control. Annual Summary, 1980. Morb Mort Week Rep 1981;29:28.)

rises until age 50. There is no fall in mean titers in the aged, so that decreased antibody levels do not account for increasing incidence of zoster in the elderly. Likewise, immunocompromised patients with disseminated zoster have been shown to possess the same or higher antibody titers as those with localized infection. Hence, humoral immunity seems to have little or no effect on the development or course of zoster infection. Cellular immunity is probably important, however, as reflected by the fact that clinical zoster and varicella are much more severe during pregnancy, in neonates, and in individuals with impaired host defences, those with malignancies, and those who underwent organ transplantation. It is possible that zoster can result from exogenous reinfection, but most cases are clearly reactivation of latent infection. VZV has been identified in nervous tissue. The typical pathogenic herpes changes in host cells occur; giant cell formation, intranuclear eosinophilic inclusions, and intracellular edema with cell ballooning.

Host responses. Unlike the other herpes infections, varicella is almost always symptomatic and recognized. Fewer than 5 percent of cases are missed. The incubation period is 10 to 20 days, and is usually about 2 weeks. There may be a 1- to 2-day prodrome with malaise, slight fever, and then the eruption of crops of red macular lesions that quickly evolve to papules, then to vesicles and pustules, which crust over. The crops appear over 2 to 4 days first on the scalp, then the trunk, with least involvement on the extremities. There may be lesions in various stages of evolution on any one body part (Fig. 17-14). Lesions are painless except on mucous membranes. Low fever, lymphadenopathy, and pruritus may accompany the rash. Recovery is within 2 weeks. Antibodies are present within about 5 days of onset of the rash.

Complications, less frequent than with other common diseases of childhood, include bacterial superinfection of lesions, especially with beta-hemolytic streptococci; otitis media; pneumonia; septicemia; and encephalitis. VZV antibody passes the placenta and may offer some protection to the newborn. Varicella can occur in neonates of mothers who acquire the disease late in pregnancy. Varicella in pregnancy has not been demonstrated to have adverse effects on the fetus.

Zoster lesions, which are like those of varicella, are usually unilateral and follow the pathway of the involved dorsal root or cranial nerve ganglion (Fig. 17-15). In addition, up to one-third of persons with zoster develop generalized cutaneous lesions. The emergence of zoster lesions is frequently preceded or accompanied by severe pain that can be quite debilitating. The individual with zoster may also have a low-grade fever and lymphadenopathy. Without complications, healing occurs in 2 to 4 weeks. However, hemorrhagic and gangrenous or necrotic forms can occur, resulting in prolonged healing and deep scarring. In addition, about 10 percent of individuals have some residual neuralgic symptoms for several months. Most cases are located in the region of the thoracic nerves, but lumbar, sacral, neck, and trigeminal regions are also affected. Eye damage is a potential danger if zoster occurs on the face. Fortunately, recurrences are an exception, occurring in less than 5 percent of cases.

Children with leukemia and adults with Hodgkin's disease are especially prone to develop serious clinical disease with VZV. Mortality in the general population is well below 0.1 percent, but may be increased considerably among those who are immuno-

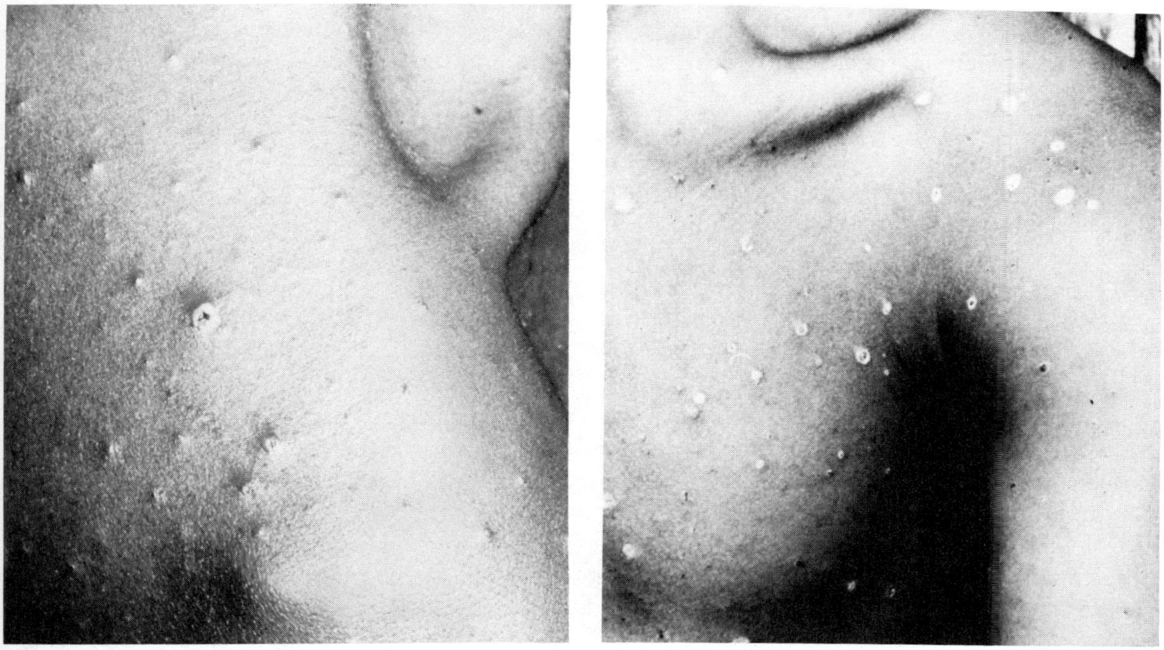

a b

FIGURE 17-14. Chickenpox. *(a)* Upper shoulder area (third day of rash). *(b)* Chest (eighth day of rash). (Courtesy of Centers for Disease Control, Atlanta, GA 30333.)

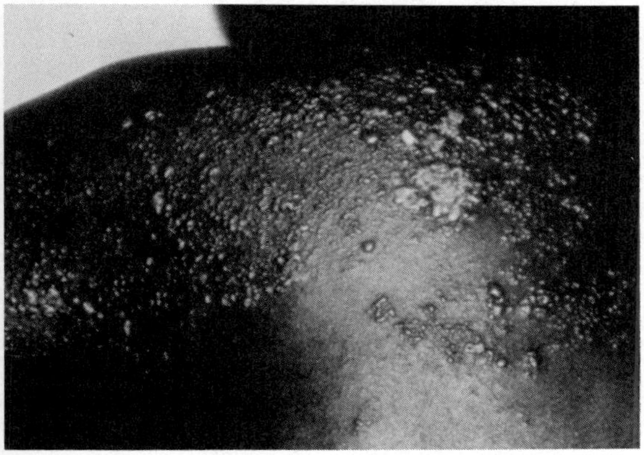

FIGURE 17-15. Typical dermatomal rash of herpes zoster in a normal host. (Reprinted with permission of the publisher from Galasso GJ, Merigan TC, Buchanan RA, eds. Antiviral agents and viral diseases in man. New York: Raven Press, 1979.)

compromised. Symptoms of varicella are worse in adults.

Control and treatment. A vaccine for chickenpox has been trialled with some success in Japan but is not currently recommended for use in the United States. Varicella-zoster immune globulin (VZIG) is available for use specifically in high-risk immunodeficient childen exposed to chickenpox (see Table 27-6 for indications for use). Human leukocyte interferon has been trialled with some success to treat children with malignancies who develop varicella. If given within 72 hours of the appearance of rash, disseminated disease is less likely to occur (Arvin et al. 1982). In general, however, no large preventive programs for chickenpox are possible or advisable.

Affected individuals are infectious from 1 to 2 days prior to emergence of skin eruptions until lesions are crusted and no new ones appear. Children should be restricted to home during that period. In hospital settings, isolation is necessary for patients with VZV because of the presence of many immunocompromised patients.

The zoster patient is less infectious than one with chickenpox, possibly because lesions are more circumscribed, adults are more careful with personal hygiene than children, or, most likely, presence of antibody at the time of the appearance of lesions causes partial neutralization of the virus. Nevertheless, both chickenpox and shingles are potentially infectious. See "Clinical Implications" for isolation guidelines. Idoxuridine with DMSO has been shown in clinical trials to decrease the duration of pain in zoster, cytosine arabinoside (ara-C) is not useful as it potentiates viral dissemination by decreasing host defenses, and adenine arabinoside (ara-A) is being tested with varying results. Acyclovir is the newest antiviral agent in which trials have demonstrated significant reduction in pain and improved healing; though recurrences are still common.

PAPOVAVIRUSES

Papovaviruses (*pa*pilloma, *po*lyoma, *va*cuolating viruses) are nonenveloped, icosahedral double-stranded DNA viruses, 45 to 55 nm in diameter, which are highly resistant to killing by formalin, heat, or acid. Multiplication occurs in the host cell nucleus. There are two major groups: papilloma virus and polyoma virus. The polyoma, SV-40, JC, and BK viruses are of primary concern because of their tumorigenic properties and will therefore be discussed in Chapter 19. The human papilloma virus, the etiologic agent of warts, will be discussed here. Other papilloma viruses infect other species such as rabbits, horses, dogs, hamsters, cattle, and deer.

Human Papilloma Virus (HPV)

Biologic characteristics. Human papilloma viruses (HPV) or wart viruses are small DNA papovaviruses that have never been grown in tissue culture. They produce latent and chronic infections in humans. They are round, about 46 nm, and replicate in the host cell nucleus (Fig. 17-16). There are four different serologic types of HPV, distinguished by complement-fixation tests and restriction enzyme analysis of DNA. The agent that causes condyloma acuminatum

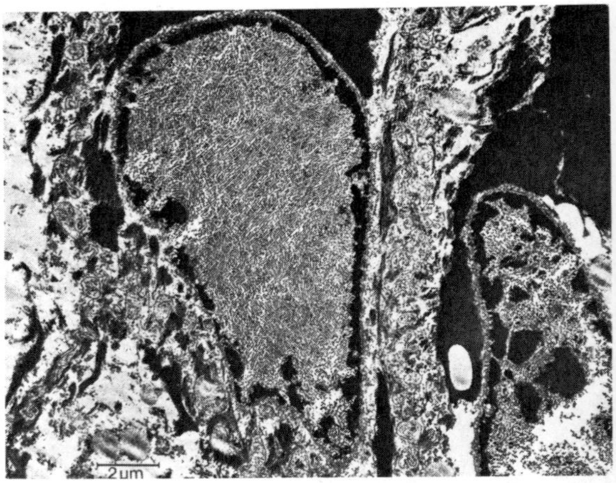

FIGURE 17-16. Thin section of plantar wart. Stratum granulosum of the epidermis. Virus crystals in the cell nuclei. Magnification 6500×. (Reprinted with permission of the publisher from Nasemann T. Viral diseases of the skin, mucous membranes and genitals. Philadelphia: WB Saunders, 1977.)

(genital warts) differs antigenically from the other HPV.

Epidemiology. HPV infections occur only in humans, affecting 7 to 10 percent of the population. The virus enters through minor skin abrasions following direct and indirect contact. Locker room and shower floors have been implicated in the spread of plantar warts. The occurrence of HPV infection (warts) is most common among children and manual laborers. Disease is more extensive and common among pregnant women and the immunocompromised.

Pathogenesis. HPV causes proliferation of certain layers of the epidermis with hyperkeratosis and parakeratosis, acanthosis, and vacuolated cells in the stratum granulosum, and the production of eosinophilic and basophilic intranuclear inclusions. These changes may be due to proliferation of the virus and infection of adjacent cells, or to transformation of the skin cells and passage of the virus to progeny. Malignant transformation of warts occurs only very rarely. Specific IgM and IgG antibodies develop in individuals who have warts.

Host responses. The incubation period for warts is long: 1 to 8 months. There are six morphologic types of warts (Table 17-11), which are common benign tumors of the skin. Verruca plana or juvenile warts are most common in children, are painless, and often occur in groups on the skin. Common warts results from hyperplasia of the papillary body and epidermis, and are most often located on uncovered skin surfaces, especially hands and knees. The lesions are solid and round with a rough surface, occurring singly or in groups (along a scratch line, for example). Mucous membrane warts can occur on the lips (often due to biting off other warts) or in the mouth, pharynx, or larynx. Laryngeal papillomas can become malig-

TABLE 17-11. Types of Papillomas (Warts)

Type	*Description*	*Location*
Verruca plana (juvenile or flat warts)	Flat, slightly rounded, usually skin colored papules; usually occur in clusters	Face, hands, arms
Verruca vulgaris (common warts)	Elevated, solid, round outcroppings of epidermal growth	Uncovered skin areas
Mucous membrane warts (larynx papilloma)	Hyperkeratosis, resembling common warts	Lips, inside mouth, pharynx, larynx
Plantar warts	Flattened lesions on the stratum corneum resembling common warts	Pressure points, especially bottoms of feet
Condyloma acuminatum (fig or genital warts)	Small round or pointed projections; may be sexually transmitted	Anogenital, other moist intertriginous areas
Epidermodysplasia verruciformis (generalized warts)	Hyperkeratosis, acanthosis, symmetric, numerous plane warts	Dorsum of hands and feet, spreading to any body area

nant. Plantar warts are papillomas occurring on the sole of the foot. They are very painful because they grow deeply, like a dagger into the foot.

Condyloma acuminatum, or genital warts, as the name implies, can be transmitted by sexual intercourse and occur as small papules in the genital regions. They thrive on the local moisture of intertriginous regions and are more common in the presence of vaginitis, urethritis, phimosis (when the foreskin of the penis cannot be withdrawn over the head of the penis), or in women who are pregnant or using contraceptives. Epidermodysplasia verruciformis (generalized warts) is more common among those with a family history of such condition. It may start in childhood or later, with the first lesions on the hands and feet, but it spreads over the entire body (Fig. 17-17). Fortunately, malignancy rarely ensues.

Control and treatment. Flat warts and common warts regress spontaneously at a rate of about 20 to 30 percent in 6 months. Individuals with plantar warts should wear shoes in public places to prevent spread of the virus and to avoid reinfection. The most accepted and successful treatment is cryotherapy (freezing with liquid nitrogen) or electrodessication with a fine needle. Caustics such as phenol or silver nitrate are also used, and have a cure rate of about 65 to 70 percent. Plantar warts are especially difficult to treat; recurrences are common. Surgical removal may be necessary, and then the recurrence rate is much lower (about 15 percent). Sensitization with dinitrochlorobenzene (DNCB), then weekly application of DNCB solution has resulted in an 80 percent cure rate with no recurrences in 6 to 12 months, but the treatment causes intense itching. 5-fluorouracil (5-FU) has also been used for warts. Its side effects include hyperpigmentation of the skin and erythema. Epidermodysplasia verruciformis does not spontaneously resolve and, even with treatment, a cure is rare.

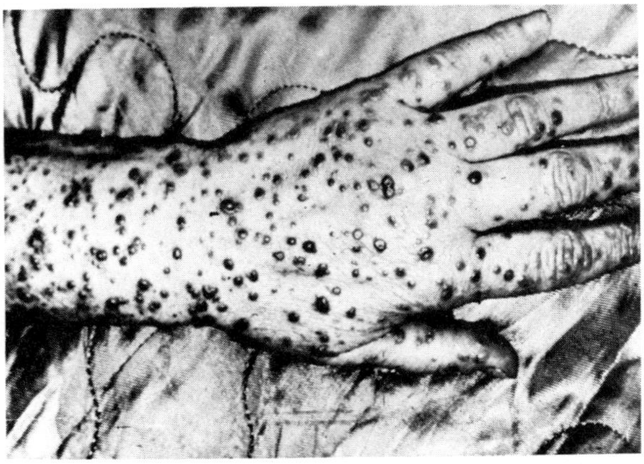

FIGURE 17-17. Epidermodysplasia verruciformis. (Reprinted with permission of the publisher from Nasemann T. Viral diseases of the skin, mucous membranes, and genitals, Philadelphia: WB Saunders, 1977.)

POXVIRUSES

There are poxviruses that infect monkeys, horses, cows, camels, pigs, many other mammals, and birds. In this section we discuss poxviruses which are etiologic agents of human diseases such as smallpox, alastrim, vaccinia, milker's nodule, cowpox, orf, and molluscum contagiosum. The family Poxviridae, named from the characteristic skin lesions or pocks usually produced by infection, contains the largest animal viruses. They are cuboid or brick-shaped cells of 250 to 450 nm (Fig. 17-18). There are six genera, two of which contain species that cause human disease and one unclassified poxvirus that causes molluscum contagiosum (Table 17-12). All poxviruses preferentially infect epidermal cells.

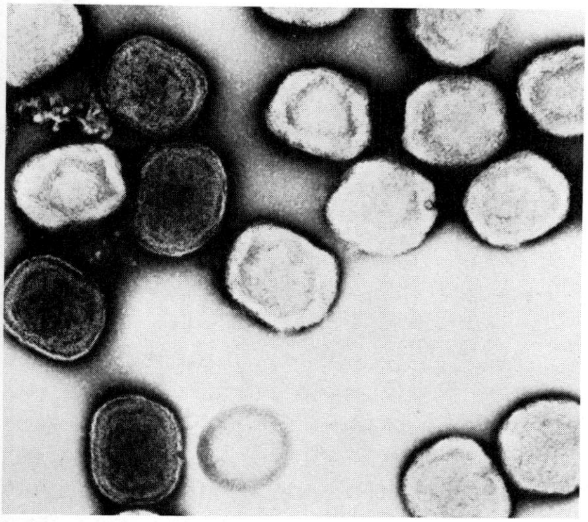

FIGURE 17-18. Vaccinia virus. (Courtesy of Centers for Disease Control, Atlanta, GA 30333.)

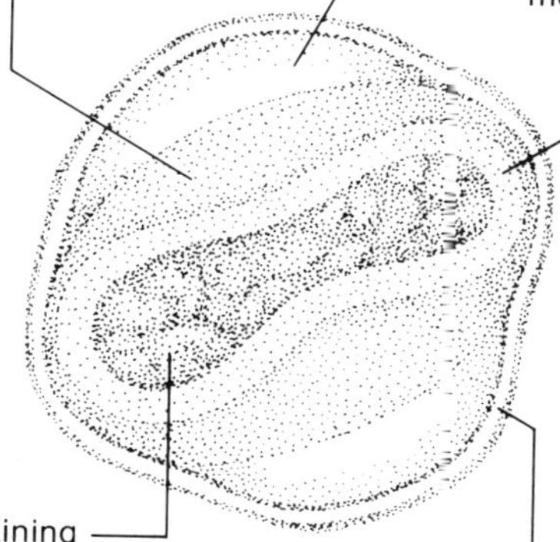

FIGURE 17-19. Poxvirus, schematic.

Poxviruses contain double-stranded DNA and have a complex structure—resembling a flattened, biconcave disc with two layers of protein membranes surrounding an internal core, all of which is covered by a ridged external covering. The internal membranes are shaped like a figure 8 (Fig. 17-19). The virus contains about 10 antigens, one of which is common to most species, and at least 30 structural proteins.

Poxviruses multiply in the host cytoplasm, producing eosinophilic intracytoplasmic inclusion bodies, which are "facto-

TABLE 17-12. Poxviruses

Genus	*Hosts*	*Human disease*
Entomopoxvirus	Invertebrates only	None
Orthopoxvirus	Humans, rabbits, monkeys, cattle, camels, raccoons	Vaccinia
		Smallpox (variola)
		Cowpox
		Variola minor (alastrim)
		Monkeypox
		Tanapox
		Yabapox
Parapoxvirus	Cattle, sheep, humans	Orf
		Milker's node
Capripoxvirus	Sheep, goats, cattle	None
Avipoxvirus	Birds	None
Leporipoxvirus	Rabbits, squirrels	None
Unclassified	Humans	Molluscum contagiosum

ries" of viral multiplication sometimes called *Guarnieri bodies*. Genetic recombination occurs within genera. Poxviruses also have the ability to reactivate each other. When there is simultaneous infection of a host cell with both an active infective virion and an inactivated one, the active virion produces an enzyme that frees the viral DNA from the inactivated viral coat, allowing replication of the inactive virion DNA.

Vaccinia and Variola

Biologic characteristics. Vaccinia (the virus used for smallpox vaccination) and variola (which causes smallpox and alastrim) are closely related and are often discussed together. They contain a double-stranded DNA molecule that can code for several hundred proteins. Because of its virulence, variola virus has not been studied extensively in the laboratory and only a few laboratories in the world are still housing the virus. However, vaccinia virus is well described, containing about 3 percent nucleic acid, 92 percent protein, and 5 percent lipid. DNA replication occurs entirely in host cell cytoplasm; no nuclear functions are necessary.

The virion enters the host cell by pinocytosis where uncoating occurs. Transcription takes place in three stages. In the first step of uncoating, the virus envelope is lost and the virion nucleoprotein core is freed in the cytoplasm. At this stage about 14 percent of the genome is transcribed. During second stage uncoating, RNA and protein synthesis are required to degrade the core so that the viral DNA is released. Then at least 30 polypeptides are produced by early transcription. Most of these components are part of the viral core. After this early transcription when DNA replication occurs, late transcription results in production of the remaining mRNA and protein.

Vaccinia and variola are very resistant to drying, freezing, ether, and to common disinfectants such as phenolics in dilute concentrations. Dried virus can survive 100°C for 10 minutes, but is killed by autoclaving, chloroform, and by one hour in 50 percent alcohol. Virus in inactivated by heat (60°C for 10 minutes) if it is not in a dry state. Virions have an internal nucleoprotein (NP) antigen common to all poxviruses and a surface antigen that has heat-labile and heat-stable components. There is also a lipoprotein hemagglutinin antigen. Complement fixation, precipitin, and hemagglutin-inhibition tests are used to diagnose infection.

Vaccinia virus was the first virus to be grown in tissue culture. Both vaccinia and variola grow well in chorioallantoic membrane of chick embryos and produce plaques or pocks. The lesions of variola are smaller, more dome-shaped, and less necrotic than those of vaccinia.

Epidemiology. Interest in both smallpox and vaccinia is primarily historical because the last naturally occurring case of smallpox was in Somalia in October 1977. The last laboratory-associated case occurred in England in August 1978. Laboratory-associated cases may occasionally continue to occur and it is possible that in future years smallpox or a similar disease may recur. However, the disease was declared by the World Health Organization in 1980 to be eradicated from the world.

Smallpox was apparently nonexistent during Biblical times. However, in the earliest records of the history of Europe, smallpox epidemics are noted, and the disease was endemic in many parts of the world by the eighteenth century. Smallpox occurs only in humans and is spread by the respiratory route and by contact with the lesions. Indirect contact is also a means of spread, especially since the virus can persist in linens and other fomites for months.

Contrary to past belief, the disease is not very contagious and spreads slowly. There were a number of characteristics of smallpox (Table 17-13) that made eradication possible. The eradication program was initiated by WHO in January, 1967 in western Africa. At that time smallpox was endemic in 30 nations; by 1972, 16 nations still reported cases. The total cost of the successful eradication campaign was $83 million, about one-half the cost for one year of vaccinations in the United States.

Vaccinia is a virus of unknown origin. It is closely related to both cowpox and variola, and is perhaps a hybrid of the two. In sheep and calves, it is the virus used for production of smallpox vaccination. It is transmitted through vaccination; persons may become infected by direct contact with the newly vaccinated individuals.

TABLE 17-13. Characteristics of Smallpox that Made Eradication Possible

Epidemiologic variables
Transmission almost always by direct contact (no vectors, rare vehicles)
Long incubation period (2 weeks) allows isolation of exposed
Seasonal variation enhances predictability and planning
Etiologic agent variables
Virus is antigenically stable and only one type
Agent is not highly contagious
Host variables
Humans are only natural host
No long-term carrier state
Immunity long term
Cases easily recognized
No subclinical infection
Treatment variables
Vaccine highly effective
Objective way to evaluate effectiveness ("take") of vaccine

Pathogenesis. After entrance via the upper respiratory tract, the smallpox virus multiplies locally and viremia causes spread of the virus through the reticuloendothelial system (bone marrow, liver, spleen, and lymphatics). After extensive replication, the virus preferentially attacks the epidermal cells where ballooning cells containing typical cytoplasmic inclusion bodies are found. Because all skin layers are involved in the disease process and necrosis occurs, scarring results.

Host responses. Smallpox was originally named to differentiate it from "large pox" (syphilis). There are two forms of smallpox: variola major, which is the classic serious, often fatal disease; and variola minor (alastrim), which is a benign form caused by a slightly different virus (alastrim virus).

After an incubation period of about 12 days, the individual infected with smallpox virus has 3 to 4 days of fever, muscle aches, and headache. Then skin eruptions appear, usually on the face and spreading to the entire body including the mucous membranes. Palms of hands and soles of feet often have many lesions. The severity of the disease is in direct proportion to the numbers of lesions. The eruptions progress from macules to papules to vesicles to pustules that crust over. After about 1 week when the pustular stage begins the patient often becomes worse. Complications include bacterial superinfection, pneumonia, otitis media, keratitis, nephritis, encephalitis, parotitis, and orchitis. After about 12 days of rash the pustules begin to dry, hair loss sometimes occurs, and deep scars remain.

Mortality rate varies from 10 to 60 percent; death is usually due to extreme prostration or complications. A hemorrhagic form is almost always fatal. Antibodies are present at about day 16 when the rash begins. Partially immune individuals can develop very mild disease with only a few lesions.

Alastrim is a much more benign disease with low fever, fewer vesicles, and much less damage to epithelial layers and only occasional scarring. Conjunctivitis and pneumonia can occur, as well as abortion in pregnant women. It is lethal is less than 1 percent of cases. The disease does not seem to become more virulent, even during epidemics.

Vaccinia is rare now because smallpox vaccination is not practiced routinely. It can occur by autoinoculation from a vaccination site, or from contact with another person's vaccination site. Generalized vaccinia with lesions over the entire body is rare, usually only in immunosuppressed individuals.

Control and treatment. As mentioned, routine vaccination is no longer necessary since smallpox is eradicatd. Vaccination may still occasionally be used among laboratory workers and travelers under special circumstances. If primary vaccination is administered to a susceptible individual, the response should be as outlined in Table 17-14. Vaccination is relatively safe but complications such as encephalitis and generalized rash do occur (Table 17-15).

Eczema vaccinatum, a disease very similar to smallpox, only occurs in the presence of pre-existing skin disease such as atopic dermatitis or if the individual is immunosuppressed. Patients with eczema or those immunocompromised should never be vaccinated. Vaccinia immune globulin is available for minimizing the severity of gener-

TABLE 17-14. Normal Response In a Susceptible Individual to Smallpox Vaccination

Day	*Skin condition*
1	Slight redness
3	Pink macule
4	Papule
6	Vesicle with central indentation
7-10	Pustule
14-21	Healing with scar formation

TABLE 17-15. Complications of Smallpox Vaccination

Complication	*Incidence per 10^6 vaccinations*	*Prevention/ treatment*	*Mortality*
Keloids		None; causes cosmetic problems only	0%
Postvaccinal exanthema (urticaria, macules, other rashes)	9	None; usually resolves in few days	0%
Postvaccinal encephalitis	2.9	Symptomatic	30-60%
Eczema vaccinatum	3	VZIG	30%
Fetal vaccinia	5	Do not vaccinate pregnant women	Fetal death in about one-fourth of victims if vaccine used in first trimester
Vaccinia gangrenosa	1.5	VZIG, methisazone	About 30%

alized complications. About 1 in 10^6 vaccinated individuals dies of complications of vaccination, though vaccination has not been practiced routinely in the United States since 1971. Methisazone, which inhibits growth of poxviruses, has a limited use in preventing smallpox among those exposed to known cases.

Poxviruses Acquired from Animals

The viruses that cause cowpox, milker's nodule, and orf are all morphologically similar and occur worldwide as an occupational hazard among individuals who contact domestic cows and sheep (Table 17-16). Both cowpox and milker's nodule are usually acquired by milking cows with infected teats. Cowpox results in the development of dark, hard nodules about the size of a large pea usually on the hands, arms, and face. The lesions are painful and can ulcerate, but usually heal within two months. Cowpox virus was the original agent used for vaccination against smallpox, but has been replaced by vaccinia virus.

Milker's nodule, caused by paravaccinia virus, is also acquired from infected cows. A wartlike, slightly painful papule is formed. It does not suppurate, as do the lesions of cowpox, and usually heals without scarring in about 6 weeks. The orf virus, closely related to paravaccinia, occurs naturally in sheep (especially young lambs) and goats and occasionally causes a pustular dermatitis in humans. Usually only one painless lesion appears on the fingers or face and heals in several weeks.

Three additional poxviruses that normally infect monkeys have been known to cause human disease: monkeypox, tanapox, and Yaba virus. All occur in Africa. Monkeypox resembles smallpox with generalized lesions, Yaba virus disease results in benign tumorlike growths on the head and arms, and, among children tanapox has caused small epidemics of febrile illness accompanied by one or two pocklike lesions.

TABLE 17-16. Poxviruses Acquired from Animals

Virus	*Normal host*	*Distribution*	*Human disease*	*Clinical picture*
Cowpox virus	Cows, possibly some wild animals	Worldwide	Cowpox	Dark nodules on hands, arm, face
Monkeypox virus	Monkeys	Africa, or in captive monkeys	Monkey pox	Generalized rash similar to smallpox
Orf virus	Sheep, goats	Worldwide	Orf	Pustular lesion on hands or face
Paravaccinia virus	Cows	Worldwide	Milker's nodule	Warty nodules on hands, arms
Tanapox virus	Monkeys	Kenya	Tanapox	1-2 pock lesions on upper body
Yabapox virus	Monkeys	Nigeria	Yaba virus infection	Benign tumorlike growths on head and arms

Molluscum Contagiosum

Molluscum contagiosum is caused by the largest poxvirus and it occurs only in humans. It results in wartlike nodular skin lesions with a diameter of about 2 to 5 mm. These lesions persist for months but are self-limited. The incubation period is 2 weeks to 2 months. Distribution of the virus is worldwide, and infections sometimes occur in small epidemics. Despite the presence of serum antibody in some people, specific immunity does not seem to develop. The disease is more common in children and adolescents. Lesions usually occur on the face, neck, eyelids, axilla, and genitalia. The central core of the lesions contain a soft mass that can be expressed for diagnostic and treatment purposes. It is apparently spread by direct contact, in swimming pools, and by fomites such as towels and clothing. Prognosis is good and recurrences are rare.

OTHER DERMOTROPIC VIRUSES

Many other types of viruses cause exanthems (Table 17-17, Figs. 17-20, 17-21). Two diseases that are almost certainly of viral origin, but for which an etiologic agent has not been identified, are erythema infectiosum (fifth disease) and exanthem subitum (roseola). Fifth disease is most common in children ages 6 to 15 and occurs in epidemics. Its incubation period is 7 to 14 days. It causes an itchy butterfly-shaped rash across the cheeks ("slapped cheek" appearance), which spreads to arms and legs within a few days. Healing takes about a week. Roseola occurs in infants from 6 months to 3 years of age. After an incubation period of 3 to 14 days, it results in 3 to 5 days of high fever, lymphadenopathy, and rubelliform rash. It is often confused with rubella. Recovery is the rule.

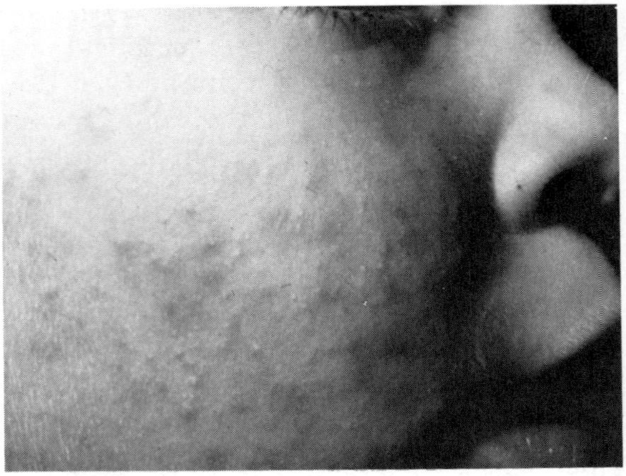

FIGURE 17-20. Echovirus type 9 infection on cheek of young girl. (Courtesy of Centers for Disease Control, Atlanta, GA 30333.)

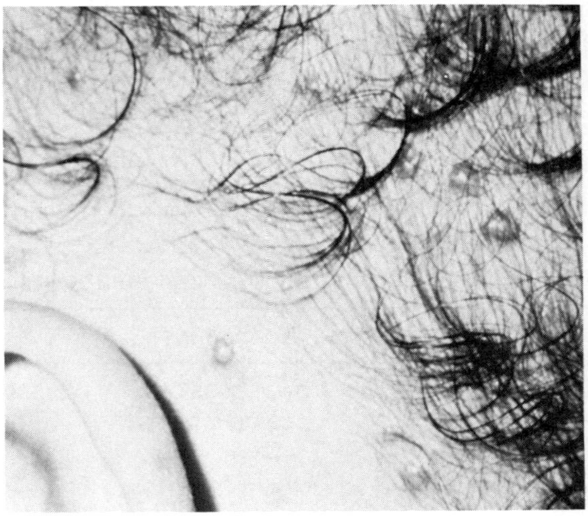

FIGURE 17-21. Coxsackie B virus. Four lesions in a young child. (Courtesy of Dr. Lawrence Corey.)

CLINICAL IMPLICATIONS

The increased recognition and probable increasing incidence of herpes infections demands from clinicians a commitment to educate patients about the agent and its

TABLE 17-17. Exanthems of Viral Diseases

Virus	Frequency of exanthem (percent)	Major site of infection	Description of rash
Adenovirus	9	Respiratory tract	Petechiae, macules
Arbovirus			
Dengue	70-90	Blood stream	Maculopapular on extremities, trunk; pruritus
Sandfly fever	Less than 50	Blood stream	Roseoliform, urticarial
Paramyxovirus			
Rubeola (measles)	100	Respiratory tract	Maculopapular, "Koplik's spots"
Mumps, influenza	Rare	Respiratory tract	Morbilliform, urticarial
Picornavirus			
Echovirus (especially types 6, 9, 16)	Depends on strain	Respiratory tract	Dark, maculopapular, may necrose (Fig. 17-20)
Rhinovirus (foot and mouth disease)	100	Respiratory tract	Vesicles in mouth, hands, feet (Fig. 17-21)
Coxsackie			
Herpangina	100	Respiratory, GI tracts	Grayish papulovesicles in mouth
Hand-foot-mouth disease	100	Respiratory, GI tracts	Oral vesicular eruptions
Rhabdovirus			
Vesicular stomatitis	100	Oral	Vesicles in mouth, hands, feet
Marburg virus	90	Oral	Perifollicular macules on face, trunk, limbs
Togavirus			
Rubella	Over 90	Respiratory tract	Pink macular exanthem
Erythema infectiosum* (fifth disease)	100	Probably respiratory tract	Raised butterfly erythema on face ("slapped cheek" appearance)
Exanthem subitum* (roseola infantum)	100	Probably respiratory tract	Small macular lesions, primarily on trunk

*Viral etiology probable, but agent not identified

clinical manifestations and a commitment to prevent infection in oneself and other health care colleagues. Herpes disease is discouraging to the host because of its recurrence at unpredictable times throughout life and because of its resistance to known therapy.

Genital herpes especially is often accompanied by mental depression and sexual dysfunction because of fear of infecting sexual partners and unborn babies. Persons with genital herpes should be cautioned to practice sexual abstinence when there are active lesions present, either in themselves or their partners. Expectant mothers with active lesions must have a cesarean section to prevent neonatal infection in the birth canal. Infants born to infected women require isolation, but mothers can be reassured that breast feeding is quite safe for the baby. To prevent postnatal spread of herpes to babies, infected mothers require instructions to wear gloves or to practice careful handwashing when handling the neonate. Even though infections with CMV, EBV, and HSV are not highly contagious, active herpes infections often require some degree of isolation while the patients are hospitalized (Table 17-18).

Nurses and other personnel giving oral or genital care to a patient with herpetic lesions can protect themselves by wearing gloves on both hands. Indeed, that is probably a safe rule in general, since patients may be shedding virus even in the absence of visible lesions. Hospitalized patients are more likely to have reactivation of disease and viral shedding because of the physiologic and psychologic stress of illness. And, of course, personnel with cold sores or herpetic paronychia (whitlow) should abstain from patient care in high-risk areas.

It is important for patients receiving acyclovir to recognize that the drug will not prevent transmission or reinfection with genital herpes. If the clinician is applying acyclovir topically, gloves or a finger cot should be worn. Reported side effects of topical application are mild; they include rash and transient burning. Intravenous administration, however, frequently causes phlebitis and may cause renal tubular damage. The patient should be well hydrated to minimize effects on the kidneys.

Much has been written in the popular press about the fact that herpes virus can survive for extended periods of time outside the body. Though it is true that virus has been recovered from gauze, toilet seats, and vaginal specula several hours after contamination, these items represent a very minimal hazard to the general public. It is a reminder, however, that disinfection and sterilization techniques are of major importance in health care facilities.

SUGGESTED READINGS

Andrewes C, Pereira HG, Wildy P. Viruses of vertebrates, 4th edition. London: Bailliere Tindall, 1978.

Arvin AM, et al. Human leukocyte inteferon for the treatment of varicella in children with cancer. New Engl J Med 1982;306:761-765.

Beltoli EJ. Herpes: facts and fallacies. Am J Nurs 1982;82:924-929.

Bray PF, et al. Progressive neurologic disease associated with chronic cytomegalovirus infection. Ann Neurol 1981;9:499-502.

Bryce Lurke RP, et al. Congenital cytomegalo-

TABLE 17-18. Types of Precautions for Hospitalized Patients with Herpes Infections

	Wound/skin	*Secretion*	*Strict*	*Gloves only*
HSV oral, primary or recurrent (cold sores, gingivostomatitis)		X		
HSV, genital				X
HSV disseminated, infants or adults			X	
Varicella (chickenpox)			X	
Zoster (shingles)	X			
EBV (infectious mononucleosis)		X		
Cytomegalovirus		X		

virus infection in an urban Canadian community. J Infec Dis 1980;142:647-653.
Centers for Disease Control. Chickenpox mimicking smallpox. Morb Mort Week Rep 1980;29:193.
Centers for Disease Control. Chickenpox—United States, 1979. Morb Mort Week Rep 1980;29:577-578.
Centers for Disease Control. Rubella outbreak in an office building. Morb Mort Week Rep 1980;29:517-518.
Centers for Disease Control. Rubella—United States, 1977-1980. Morb Mort Week Rep 1980;29:378-380.
Centers for Disease Control. Nosocomial rubella infection. Morb Mort Week Rep 1981;29:629-631.
Chang RS. Infectious mononucleosis. Boston; GK Hall Medical Publications, 1980.
Corey L, et al. A trial of topical acyclovir in genital herpes simplex virus infections. New Engl J Med 1982;306:1313-1319.
DeClercq E, et al. Comparative efficacy of antiherpes drugs against different strains of herpes simplex virus. J Infec Dis 1980;141:563-574.
Epstein MA, Achong BG, eds. The Epstein-Barr virus. Berlin: Springer-Verlag, 1979.
Evans AS. Viral infections of humans: epidemiology and control. New York: Plenum Medical Book Company, 1978.
Evans AS, Comstock GW. Epstein-Barr virus and Hodgkin's disease (letter). Lancet 1981;2:360-361.
Fiorilli M, et al. Cytomegalovirus excretion in children with antibodies (letter). Lancet 1981;2:422.
Fliegel PE, Weinstein WM. Rubella outbreak in a prenatal clinic: management and prevention. Am J Infec Cont 1982;10:29-33.
Galasso GJ, Merigan TC, Buchanan RA, eds. Antiviral agents and viral diseases in man. New York: Raven Press, 1979.
Garcia H, Edwards MS. Herpes simplex infections of burn wounds. South Med J 1981; 74:991-992.
Gershon AA, Steinberg SP. Antibody response to varicella-zoster virus and the role of antibody in host defense. Am J Med Sci 1981; 282:12-17.
Goldberg GN, et al. In utero Epstein-Barr virus (infectious mononucleosis) infection. JAMA 1981;246:1579-1581.
Greaves WL, et al. The problem of herpetic whitlow among hospital personnel. Infec Cont 1980;1:381-385.
Gurevich I, Cunha BA. Non-parenteral transmission of cytomegalovirus in a neonatal intensive care unit. Lancet 1981;2:222-230.
Hammer SM, et al. Temporal cluster of herpes simplex encephalitis: investigation by restriction endonuclease cleavage of viral DNA. J Infec Dis 1980;141:436-440.
Hamory BH, Luger A, Kobberman T. Herpesvirus hominis hepatitis of mother and newborn infant. South Med J 1981;74:992-995.
Hampl H, Schlehofer JR, Habermehl KO. Differences in the morphology of herpes simples virus infected cells. Med Microbiol Immunol 1981;169:209-223.
Herpes simplex encephalitis and vidarabine treatment (editorial) New Zeal Med J 1981;93;342-343.
Huang ES, et al. Molecular epidemiology of cytomegalovirus infection in women and their infants. New Engl J Med 1980;303:958-962.
Kaufman RH, et al. Herpesvirus-induced antigens in squamous-cell carcinoma in situ of the vulva. New Engl J Med 1981;305:483-488.
Leiwikk P, et al. Epidemiology of cytomegalovirus infections during pregnancy and infancy. Scand J Infec Dis 1978;10:165-171.
Luby J. Therapy in genital herpes. New Engl J Med 1982;306:1356-1367.
Marker SC, Simmons RL, Balfour HH. Cytomegalovirus vaccine in renal allograft recipients. Transpl Proc 1981;13:117-119.
Meyers JD, et al. Toxicity and efficacy of human leukocyte interferon for treatment of cytomegalovirus pneumonia after marrow transplantation. J Infec Dis 1980;141:555-562.
Meyers JD, Flournoy N, Thomas ED. Infection with herpes simplex virus and cell-mediated immunity after marrow transplantation. J Infec Dis 1980;142:338-346.
Nahmias AJ, Dowdle WR, Schinazi RS, eds. The human herpesviruses. New York: Elsevier, 1980.
Nasemann T. Viral diseases of the skin, mucous

membranes and genitals. Philadelphia: WB Saunders, 1977.

Nicholson KG. Coincident cytomegalovirus infection and toxoplasmosis in an uncompromised host. Br Med J 1981;282:2095.

Osborn JE. Cytomegalovirus: pathogenicity, immunology and vaccine initiatives. J Infec Dis 1981;142:618-630.

Pass RF, et al. Cytomegalovirus infection in patients with renal transplants: potentiation by antithymocyte globulin and in incompatible graft. J Infec Dis 1980;142:9-17.

Pizzo PA, Magrath IT, Jay G. Characterization of the Epstein-Barr virus isolated from a cell line derived from a patient with American Burkitt's lymphoma. Ca Res 1981;41:3161-3164.

Reeves WC, et al. Risk of recurrence after first episodes of genital herpes: relations to herpes simplex virus types and antibody response. New Engl J Med 1981;305:315-319.

Schwartz PE, Naftolin F. Type 2 herpes simplex virus and vulvar carcinoma in situ (editorial). New Engl J Med 1981;305:517-518.

Stagno S, et al. Breast milk and the risk of cytomegalovirus infection. New Engl J Med 1980;302:1073-1076.

Taylor RJ, et al. Primary disseminated herpes simplex infection with fulminant hepatitis following renal transplantation. Arch Intern Med 1981;141:1519-1521.

Thiers B. Therapeutic briefs. J Am Acad Dermatol 1981;4:249-252.

Thong YH. Chemotherapy for herpes simplex encephalitis. Med J Aust 1981;1:388-389.

Toplin MD, et al. Inactivated mouse cytomegalovirus vaccine: preparation, immunogenicity, and protective effect. J Infec Dis 1980; 142:569-574.

Tummon IS, Dudley DKL, Walters, JH. Genital herpes simplex. Can Med Assoc 1981;125:23-29.

Wade JC, et al. Intravenous acyclovir to treat mucocutaneous herpes simplex virus infection after marrow transplantation. Ann Intern Med 1982;96:265-269.

Yeager AS, et al. Prevention of transfusion-acquired cytomegalovirus infection in newborn infants. J Pediatr 1981;98:281-287.

18 HEPATITIS VIRUSES

KEY POINTS

- There are at least four hepatitis viruses: hepatitis A virus, which is primarily food- and water-borne; hepatitis B virus, and several non-A, non-B agents, which are primarily transmitted parenterally.
- Hepatitis A is a mild, self-limited illness. There is no carrier state.
- Hepatitis B virus causes more serious disease, and about 10 to 15 percent of individuals infected become chronic carriers of the virus.
- Hepatitis B (previously called serum hepatitis) is most prevalent in male homosexuals, I.V. drug abusers, health care personnel, and patients with exposure to blood (e.g., via hemodialysis or transfusion).
- The antigenic structure of hepatitis B virus is quite complex.
- Most transfusion-related hepatitis is caused by non-A, non-B agents.
- There is a safe, efficacious vaccine available against hepatitis B virus.

There are at least four distinct viruses that have a predilection for liver cells and cause similar clinical syndromes but which vary in their epidemiologic and morphologic characteristics. Hepatitis A virus, originally called infectious hepatitis virus, is the major cause of hepatitis in food- or water-borne outbreaks. Hepatitis B virus, previously called serum hepatitis virus, and the two or more agents causing non-A, non-B hepatitis are primarily responsible for transfusion-related and other parenterally spread cases of hepatitis. These viruses are not classified in a particular viral family, though hepatitis A virus could be considered an enterovirus similar to poliovirus. It is interesting that the agents of hepatitis A and hepatitis B, though totally unrelated, produce very similar clinical diseases.

The hepatitis viruses share a hardiness that surpasses that of other viruses; they are quite resistant to common disinfectants such as phenolics, to temperature extremes, to chlorine (they can survive in some treated water supplies), and to ultraviolet light. Special cleaning procedures may be necessary in an institutional setting to rid the environment of contamination. The general characteristics of the agents causing viral hepatitis are compared in Table 18-1, and nomenclature is listed on Table 18-2.

Most health departments currently require that hepatitis be reported, but cases are not always differentiated by etiologic agent (A, B, or non-A, non-B), especially in underdeveloped countries. It is also certain that considerable underreporting is the rule. Despite this, viral hepatitis is among the four most commonly reported infectious diseases in the United States.

Hepatitis A Virus (HAV)

Biologic characteristics. There is a single antigenic type of hepatitis A virus (HAV): it is 27 nm, nonenveloped, icosahedral, contains single-stranded RNA, and has epidemiologic characteristics similar to the enteroviruses. It can be grown in fetal rhesus monkey kidney cells, in marmoset liver cells, and, through a recent breakthrough, in human embryo fibroblasts. HAV infection can be diagnosed with complement-fixation, radioimmunoassay, or immune-adherence tests. The virus was discovered in 1973 by immune electron microscopy in the stools of individuals with clinical hepatitis. It is stable at 60°C, it is not killed by pasteurization, and is resistant to acid pH and ether. It contains four polypeptides and replicates in host cell cytoplasm. Blood tests for HAV and their interpretation are listed on Table 18-3.

Epidemiology. The HAV is ubiquitous worldwide. It is highly endemic in the developing world, where most children are infected early in life. For example, 97 percent of Thai adults older than age 16 years have been found to be seropositive for antibody

TABLE 18-1. A Comparison of the Viral Agents of Hepatitis

Characteristic	*Hepatitis A*	*Hepatitis B*	*Non-A, non-B*
Incubation period	15-45 days with average 25-30 days	50-180 days with average 70 days	15-180 days with average of 45 days
Age group	All ages, but especially ages 15-29	Especially ages 15-29; often associated with drug abuse, promiscuous sexual behavior, contact with blood of infected persons	All ages, transfusion-related
Seasonal incidence	Throughout the year with peak in summer and autumn	Throughout the year	Throughout the year
Route of infection	Predominantly fecal-oral. Occasionally transmitted parenterally or by contaminated water and food (i.e., shellfish)	Predominantly parenteral	Most frequently associated with blood transfusion—with incidence directly related to number of units received

TABLE 18-1. *(continued)*

Characteristic	*Hepatitis A*	*Hepatitis B*	*Non-A, non-B*
Occurrence of virus/antigen in blood	Present during late incubation and early acute phase: 2 wk before to less than or equal to 1 wk after jaundice	Present during late incubation and acute phase; occasionally persists for months and yrs	Probably similar to type B
Feces	Present during late incubation and acute phase: 2 wk before to 2 wk after jaundice	May be present but no direct proof	Unknown
Urine, semen	?	Infrequently present	Unknown
Saliva	?	Frequently present	Unknown
Fever greater than 39°C	Common; precedes jaundice	Less common	Less common
Duration of SGOT elevation	1-3 wk	1-6+ months	?
HBsAg	Not present	Present in incubation period and acute phase; occasionally may persist	Not present
Complications			
Chronic active hepatitis	−	+	+
Chronic persistent hepatitis	+, rare	+	+
Asymptomatic carrier state	−	+	Most likely
Mortality	Less than 0.1 percent	Less than 1 percent	?
Viral structure	27-nm single-stranded RNA	42-nm double-stranded DNA Dane particle with envelope	At least two antigenically different agents
Tissue culture	No	No	No
Passive immunization	NHIG	HBIG	NHIG of unknown efficacy
Jaundice	Rare in children; more common in adults	Same as type A	Same as type A
Type of onset	Usually acute	Usually insidious	Usually insidious
Immunity			
Homologous	Yes	Yes	?
Heterologous	No	No	?
Duration	Probably lifetime	Probably lifetime	?

Table developed by Christine Lima, RN, MN, Seattle, 1982.

TABLE 18-2. Hepatitis Nomenclature

Abbreviation	Term	Comments
Hepatitis A		
HAV	Hepatitis A virus	Causal agent of "infectious" hepatitis; probably an enterovirus; single sterotype
Anti-HAV	Antibody to HAV	Detectable at onset of symptoms; lifetime persistence
Hepatitis B		
HBV	Hepatitis B virus	Causal agent of "serum" or "long-incubation" hepatitis; also known as Dane particle
HBsAg	Hepatitis B surface antigen	Surface antigen(s) of HBV, detectable in large quantity in serum; several subtypes identified
HBeAg	Hepatitis B e antigen	Soluble antigen; correlates with HBV replication, high titer HBV in serum, and infectivity of serum
HBcAg	Hepatitis B core antigen	No commercial test available
Anti-HBs	Antibody to HBsAg	Indicates past infection with and immunity to HBV, passive antibody from hepatitis B immune globulin, or immune response from HBV vaccine
Anti-HBe	Antibody to HBeAg	Presence in serum of HBsAg carrier suggests lower titer of HBV
Anti-HBc	Antibody to HBcAg	Indicates past infection with HBV at some undefined time
Non-A, non-B hepatitis	Non-A, non-B	
NANBH	hepatitis	Diagnosis of exclusion; at least two candidate viruses; epidemiology parallels that of hepatitis B

Reprinted with permission of the publisher from Centers for Disease Control. Immune globulins for protection against hepatitis, Ann Intern Med 1982;96:193-197.

to HAV as compared to 13 percent in Scandinavian countries. The disease affects nonhuman primates and humans, with humans being the only important natural reservoir. Like enteroviruses, HAV is spread by the fecal-oral route and hence fluorishes where sanitation is poor.

Disease caused by HAV is more common in children since there is increasing immunity with age and since the possibility for fecal-oral spread is greater among children. Age-specific incidence rates in some countries indicate that there has been a rapid decline in infection in the past few decades. In Japan, antibody is present in less than 20 percent of people under age 30, and in more than 80 percent of those over age 40. In the United Sates the prevalence of hepatitis A antibodies is less than 10 percent in those under age 35; over 50 percent in those over age 40. In Greece, where the disease is hyperendemic with a prevalence rate of about 80 percent, sanitary conditions are poor. In contrast, among the isolated population of the Faroe Islands, Denmark, where high hygienic standards are maintained, the endemic rate of HAV infection between occasional outbreaks is very low. About 30 cases of hepatitis A per 100,000 population are reported every year in the United States (Fig. 18-1).

TABLE 18-3. Serologic Tests for Hepatitis A Virus

Test	*Characteristics*
Hepatitis A antigen (HA Ag)	Present in stools of patients with clinical hepatitis A
IgM antibody to hepatitis A virus (HAVAB-M)	Present from onset of symptoms for 2-3 months indicative of acute or recent infection
IgG antibody to hepatitis A virus (HAVAB-G or anti-HAV IgG)	Present 2-4 months after onset of symptoms; persists for life
Antibody to hepatitis A virus (HAVAB or anti-HAV)	Measures both IgM and IgG; indicates present or past infection with HAV

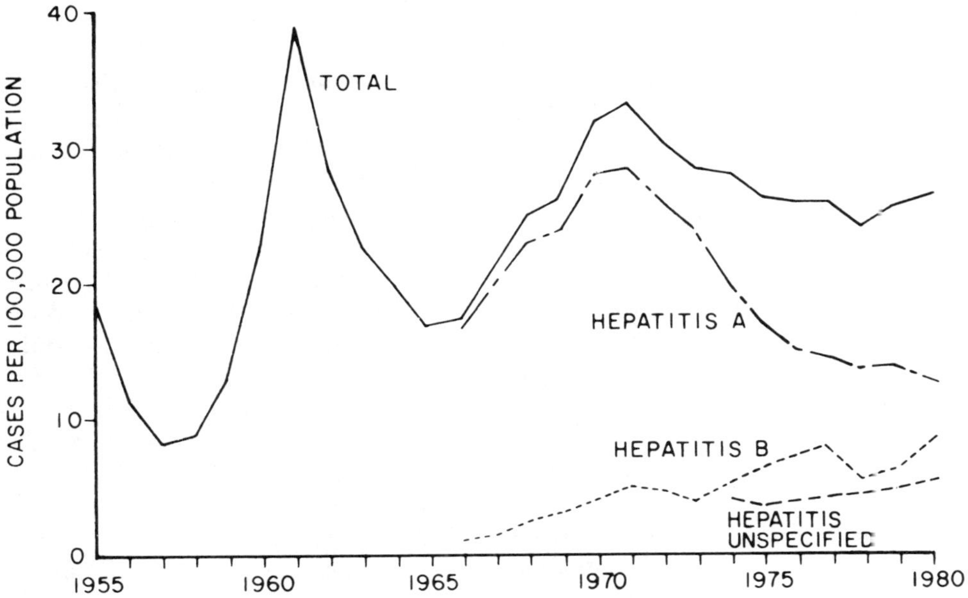

FIGURE 18-1. Hepatitis. Reported case rates by year, United States, 1955 to 1980. (Reprinted from Centers for Disease Control. Annual Summary, 1980. Morb Mort Week Rep 1981;29:39.)

Hepatitis A usually occurs in epidemics. It is transmitted most commonly in food or water contaminated by an infected individual. HAV is a predominant food-borne viral agent found in contaminated hot dogs, potato salad, ice, pastries, fish, and other foods. Raw shellfish such as clams, mussels, and oysters can become vehicles of viral spread if they are harvested from sewage-contaminated waters. Several outbreaks of HAV infection have had such an etiology.

Water-borne outbreaks are not common and when they occur they usually originate from wells rather than public water supplies. Among individuals who drank sewage-contaminated water during a large outbreak of gastroenteritis in 1975 at Crater Lake National Park, the rate of hepatitis A was 12 per 10^5 persons per year, which was not substantially different than the national rate.

In temperate climates there are usually summer and autumn peaks in incidence. HAV can be infective parenterally as well as orally but the parenteral mode of spread plays little role in the natural occurrence of the disease. The period of viremia is short and does not persist beyond the acute illness phase; transmission by blood transfusion has not been demonstrated.

Most cases of hepatitis A are asymptomatic (as many as 30:1), and only about 25 percent of persons infected have had known contact with another symptomatic person. Virus is shed in the feces mostly during the late incubation and early disease phases; that is, about 1 week before and after the onset of symptoms. During the first week after onset of symptoms, only about 45 percent of individuals still have virus in the feces, during week 2 that decreases to about 10 percent. There is virtually no viral shedding after the second week. Because the most contagious period of the disease is just before or at the time of onset of symptoms, the disease is often spread by food handlers or by direct contact before the infected person realizes he or she is ill. A second epidemic wave often occurs due to (direct) person-to-person contact.

Outbreaks in day care centers have been reported recently. Children less than 2 years of age are most likely to be the source cases and are frequently asymptomatic. Outbreaks are common also in institutions for the mentally retarded where it may be difficult to maintain sanitary conditions and prevent fecal-oral spread. The secondary attack rate among close contacts of those who develop clinical HAV infection is 10 to 15 percent. Fortunately there is no chronic carrier state for HAV (as there is for HBV), thus hepatitis A is a self-limited infection with no chronic reservoirs. Transmission requires serial passage of HAV, and the virus does not survive in small populations with highly sanitary conditions.

Pathogenesis. HAV enters the gastrointestinal tract, multiplies initially in the intestinal mucosa, and spreads hematogenously to the liver, spleen, and kidneys. Virus can be isolated in the stool, blood, and urine during the incubation and early disease phases. The pathogenesis of hepatitis A, B, and non-A and non-B is similar, and the clinical responses to infection involve a complex interaction of cellular and humoral factors. IgM antibody to HAV and antibody/antigen complexes are present at the time of onset of clinical disease, a fact that is consistent with an immune complex injury. That is, the cytopathic effects of HAV in the liver may in part be due to a secondary immune response and destruction of the infected host hepatocytes.

The hepatic morphologic changes are the same for all types of viral hepatitis. Hepatocytes become necrotic, and there is an infiltration of mononuclear cell inflammatory response, hyperplasia of Kupffer cells (liver macrophages), and bile stasis. Cell-mediated immunity appears to be very important to an individual's response to infection.

Diagnosis of HAV infection may be made by detecting hepatitis A antigen in feces, although, as mentioned, less than half of affected patients are still shedding virus at the time of disease onset. IgM antibody to HAV is present for just a few months after infection. IgG antibody is present for long periods, probably for life, and its presence is a sensitive indicator of past infection, though not necessarily of recent or acute infection.

Host responses. Most cases of HAV infection are asymptomatic or mild. Even if serious, they rarely result in chronic hepatitis or long term sequelae. After an incubation period of 15 to 50 days (averaging about 1 month), there is usually a fairly abrupt (more so than with hepatitis B) onset of fever, nausea, anorexia, abdominal discomfort, and leukopenia. Affected individuals express a loss of desire for food and, if they are smokers, for tobacco. In about half of adults and 20 percent of children, jaundice occurs. There is a marked increase in transaminase (SGOT and SGPT) levels and a slight increase in alkaline phosphatase. The disease is usually mild with recovery in 2 to 3 weeks. Recovery is complete in 3 months for 90 percent of individuals, though persons over 30 are prone to more severe or prolonged disease.

Chronic forms of hepatitis A are very rare; the fatality rate among symptomatic cases is less than 1 percent. Immunity following infection appears to be long lasting and complete. Symptoms of autoimmunity such as urticaria with immune complexes deposited in the blood vessels of the skin sometimes accompany the disease. Arthralgias and arthritis occur in about one-third of adults.

Control and treatment. Control of hepatitis A is directed at the interception of disease transmission. Persons who are handling raw fish should wear gloves, and individuals who ingest the fish should be sure that they are harvested from uncontaminated waters and are handled hygienically. Sanitation and food hygiene have resulted in decreasing rates of HAV infection in developed countries. The potential for development of a vaccine against HAV is great now that the virus can be cultivated in human diploid cells. Such a vaccine is likely to be forthcoming in the near future.

Normal human immune globulin (NHIG), if given within 1 to 2 weeks after exposure, can prevent or ameliorate clinical disease in 80 to 90 percent of individuals, and can be used for high-risk groups such as military personnel or other travellers to hyperendemic areas, individuals in homes for the retarded, and people who handle non-human primates, especially chimpanzees. The disease is modified by NHIG because of the high prevalence of antibody in normal serum. However, as the prevalence of the disease increases, efficacy of prophylaxis with NHIG may decrease and special donor pools (as those used to produce hyperimmune globulin for hepatitis B) may be necessary. Efficacy is greater than 90 percent for family contacts of patients with clinical hepatitis A. It is postulated that the long duration of effectiveness of NHIG, up to a year or more, is due to the mechanism of passive-active immunization. If one case of HAV infection occurs in a day care center or group living situation that houses diapered children, NHIG should be administered to other children and employees.

In hospitals, no patient-patient or personnel-patient transmission of HAV has been reported, though patient-personnel transmission has occurred, as have foodborne outbreaks. In general, a private room is unnecessary for a patient with diagnosed HAV infection unless the patient is fecally incontinent. Precautions such as gloving and careful handwashing are necessary when handling patient excreta (see "Clinical Implications").

HEPATITIS B VIRUS (HBV)

Biologic characteristics. Hepatitis B virus (HBV) is an antigenically complex, unusual, very small, double-stranded DNA virus that has not been classified with any other DNA virus. There is no cross reaction between hepatitis A and hepatitis B viruses. The circular DNA along with a viral-specific enzyme, DNA polymerase, are present in the viral core. The site and mode of DNA synthesis are not yet clear, but synthesis probably occurs in the nucleus of infected liver cells.

The virus has three morphologic forms, all of which contain the hepatitis surface antigen (HBsAg): spherical HBsAg particles, tubular or filamentous HBsAg particles (which are probably aggregates of the spherical particles), and the Dane particle (Figs. 18-2 and 18-3). The Dane particle is the hepatitis B virion, a 42-nm structure with a double shell that contains a surface antigen, an early (e) antigen, and a core antigen, along with the viral DNA. All three of these particle types are abundant in sera of individuals with hepatitis B. The spherical particles, being most numerous, occur in concentrations of 10^{11} to 10^{12}/ml of serum.

It may be possible that all three particles must be present in order for infection to result, or it may be that some of the particles merely represent incomplete virions. The

Appearance	*Particle*	*Size*	*Antigens*	*Description*
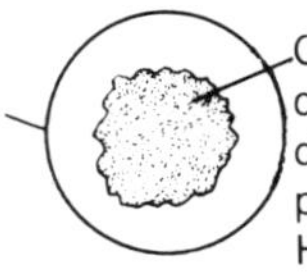 HBsAg; Core containing dsDNA, DNA polymerase, HBcAg and probably HBeAg	Dane particle	42 nm diam.	HBsAg, HBcAg	Enveloped, round, is probably the HB virion; core contains dsDNA
	Filamentous or tubular particle	22 nm dim; 50-230 nm long	HBsAg only	Probably aggregates of sphericle particles; contains no DNA
	Sphericle Particle	22 nm diam.	HBsAg only	Most common particles, found in abundance in serum of infected patients; no core or DNA

FIGURE 18-2. Structural components of hepatitis B virus.

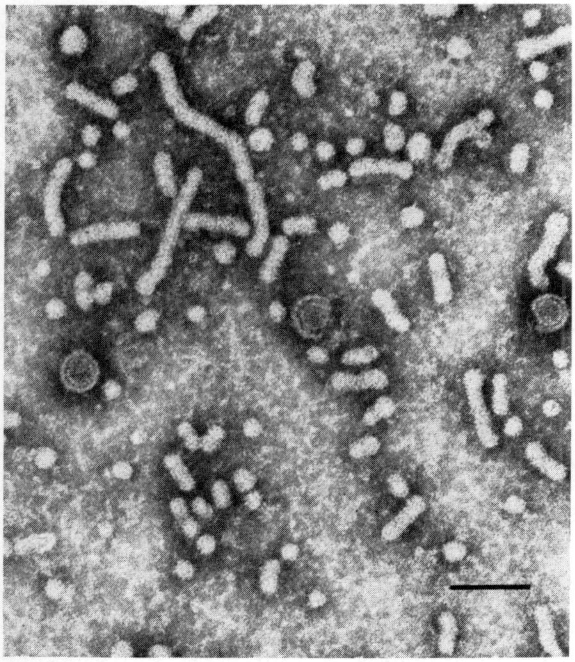

FIGURE 18-3. Electron micrograph of the three morphologic forms of particles associated with hepatitis B virus (HBV), purified from the serum of a chronic carrier of HBsAg. The bar represents 100 nm. Original magnification 132,000 ×. (Courtesy of Dr. Jules L. Dienstag. Reprinted with permission of the publisher from Koff RS. Viral hepatitis. New York: John Wiley and Sons, 1978.)

coding capacity of the short DNA is insufficient to explain the varied capacities of the HBV; thus, the Dane particle may be defective, it may be a multicomponent virus, or it may require a helper. The virus has not been grown in cell culture to date.

There are three major antigens to HBV: hepatitis B surface antigen (HBsAg), hepatitis B core antigen (HBcAg), and hepatitis B early antigen (HBeAg). HBsAg (previously called Australia antigen, AA, or hepatitis-associated antigen, HAA) was discovered by Blumberg in 1963 in the serum of Australian aborigines and was subsequently found to be a marker for infection with HBV. HBsAg is a lipoprotein containing a small amount of carbohydrate. It has a group-specific determinant a, and two mutually exclusive sets of antigens: d and y; and w and r. Hence there are four possible genotypes of the HBsAg: adw, adr, ayw, and ayr. These subgroups are only important as epidemiologic markers and seem to have no clinical significance.

Testing for the presence of HBsAg by radioimmunoassay, counterimmunoelectrophoresis, reverse hemagglutination, or immunodiffusion is the primary method for diagnosing HBV infection. HBsAg is abundant in membranes of endoplasmic reticulum, probably because the virus obtains its envelope by budding through the membranes. HBsAg is present in body fluids such as serum, saliva, urine (infrequently), breast milk, and semen of patients with acute and chronic forms of HBV infection.

The presence of antibody against surface antigen (anti-HBs) correlates strongly with immunity to HBV. Anti-HBs probably coats HBsAg and interferes with viral absorption. If anti-HBs does not develop, chronic HBsAg carriage ensues. There is strong cross immunity between subtypes of HBsAg. HBsAg is usually only present in the serum during the late incubation period and the acute illness phase, but in some individuals HBsAg remains present for months or years, indicating chronic infection.

HBcAg has a single antigenic type. It usually appears in the host cell nucleus. HBsAg appears in the cytoplasm. Antibody to hepatitis core antigen (anti-Hbc) appears at about the time that clinical symptoms occur and is a sensitive indicator of viral activity. As many as 10 percent of cases of acute hepatitis B may not be associated with the detection of HBsAg, and anti-HBc may be the only marker of infection. There is a separate antigen called the delta antigen (δ), different from but associated with the HBcAg, whose role is unknown.

IgM anti-HBc is a good indicator of acute HBV infection. The disappearance of anti-HBc IgM is an accurate indicator that HBV has been cleared from the host. Because IgG anti-HBc is measurable for years, perhaps life, it is an excellent marker for the prevalence of HBV, especially for the period of time when HBsAg disappears and before anti-HBs appears, which may be up to 3 months. Rarely, however, anti-HBc may disappear and anti-HBs may persist, so both antibodies should be measured to completely rule out HBV infection.

Hepatitis B early antigen (HBeAg), a structural, soluble antigen, is an indicator of acute infection and usually disappears early in the course of infection, before the disappearance of HBsAg. It is smaller than the other antigens and has two subtypes. The presence of HBeAg means that viral replication is persisting and people with circulating HBeAg are the most infectious, even in the presence of anti-HBe. If HBeAg titers are high (greater than 1:1000) or persist for more than 6 to 8 weeks after diagnosis, the possibility of developing chronic hepatitis is increased. Anti-HBe is found in healthy carriers of HBsAg, but rarely in patients with active disease. Significantly elevated levels of transaminase (indicating liver damage) have been found in individuals positive for HBeAg as compared with those positive for anti-HBe.

Individuals with chronic liver disease often continue to be positive for both HBsAg and HBeAg. In addition, vertical transmission from mother to fetus is most likely if the mother is positive for both HBsAg and HBeAg rather than just for HBsAg. The relative time periods in which the various HBV antigens are present in serum are illustrated in Figure 18-4.

Epidemiology. The HBV is epidemiologically quite unique from HAV. Though HBV can be isolated in the semen, saliva, peritoneal fluid, bile, feces, nasopharyngeal secretions, urine, and breast milk, the virus is spread primarily by the inoculation of contaminated blood or blood products into the skin or mucous membranes (eyes or mouth) or by very intimate contact such as sexual intercourse. Semen from HBsAg- and HBeAg-positive carriers causes infection in gibbons when given either subcutaneously or intravaginally. Saliva has been shown to be infectious when administered subcutaneously but not orally. HBV has also been experimentally transmitted by oral ingestion of infected serum. The virus is not transmitted in the feces as is HAV and hence is not food- or water-borne. Airborne transmission is not important either, though it is possible, especially in tropical climates, that insect vectors may play some role in transmission. HBsAg has been detected in numerous species of blood-sucking insects.

Humans are the natural host for HBV; there are no other known natural reservoirs. About 200,000 persons are infected each year in the United States, more than 10,000 of whom are hospitalized annually. About 250 people die each year. Like HAV, endemic rates of infection are high in lower socioeconomic groups, even in remote populations. Whereas hepatitis A is often a disease of children, HBV infection occurs primarily in adults. Carrier rates are higher for men in many populations, probably reflecting differential rates of exposure to the virus rather than a sexual predisposition to infection.

Persons at special risk of infection (Table

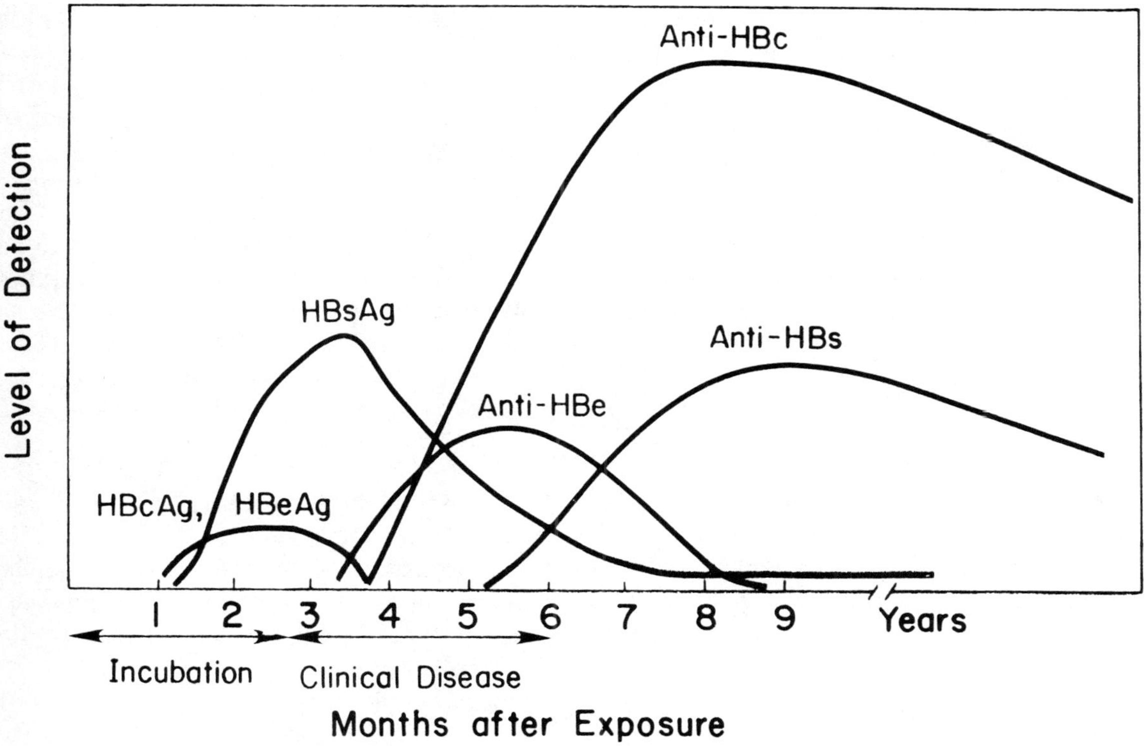

FIGURE 18-4. Usual time sequence of HBV serologic markers following exposure.

18-4) include male homosexuals, drug abusers, hospital personnel, persons receiving multiple blood transfusions, and patients on hemodialysis. As many as half the narcotic addicts have been infected with HBV. In general, HBV infection is about twice as frequent among health care personnel as compared to the general population and prevalence increases in direct proportion to their level of contact with blood products. In one study, phlebotomists had twice the incidence of serum positive for HBsAg or anti-HBs as secretaries at the beginning of the prospective observation period, but during the study developed infection at the same rate, probably due to increased care in blood handling as a result of attention focused on them by the study (Craig, Gribbie, and Suarez 1981). About 15 percent of hospital personnel are seropositive for HBV. The rate of hepatitis acquisition among staff on hemodialysis units is about 5 to 22 cases per 1000 employee months. On the other hand, one-third of 447 medical students in two

TABLE 18-4. Individuals At High Risk of Infection with HBV

Male homosexuals
I.V. drug abusers
Health care personnel, especially those who contact blood and blood products
Multiply transfused individuals
Patients receiving hemodialysis

years reported wounds contaminated by blood, but only 7 percent were exposed to HBV and none developed clinical hepatitis (Bock, Tong, and Bernstein 1981). HBV has also been spread by ear piercing, acupuncture, tattooing, and fiberoptic endoscopy.

Contrary to HAV, which is transmitted by individuals only for a short time period, HBV, in 10 to 15 percent of individuals infected, is chronically carried and transmission is often via such chronic carriers. Carriage is defined as having at least two positive tests for HBsAg at least 6 months apart. Such carriage, which may last for months or years, is usually asymptomatic but is associated with chronic active hepatitis in over 25 percent of carriers. The prevalence of HBsAg-positive persons in the United States is from 0.1 to 1 percent, but it may be up to 20 percent in some societies (Table 18-5). There are approximately 400,000 to 800,000 infectious carriers in the United States. HBsAg is identified in 4 to 62 percent of individuals with chronic hepatitis; the percentage varies by geographic area and by method of identification used.

TABLE 18-5. Prevalence of HBsAg Carriage

Location or part of population	*Percent carriers*
Asia, some parts	20
New Zealand	2.4
Panama	1-5
Polynesia	7
Senegal	10-12
Taiwan	15
Tokyo	3
United States	
General population	0.1-1
Hospital personnel	15
School for mentally retarded	6
Other high-risk groups: male homosexuals, multiply transfused, dialysis patients	40-70

Though infectivity of blood is correlated with the presence of HBeAg, the level of infectivity of blood positive for HBsAg requires further study. Currently, the blood of potential donors is routinely screened for HBsAg and, if positive, is considered to be infectious and not used. Such routine screening has decreased the numbers of transfusion-related hepatitis cases. Both anti-HBs and anti-HBc can also be passively transferred by blood transfusion (Carta 1982). HBV, however, accounts for less than 10 percent of transfusion-related hepatitis; the majority being caused by non-A, non-B hepatitis agents. HBV epidemics have occurred when contaminated blood products are pooled for vaccination purposes; otherwise hepatitis B is not an epidemic disease.

The development of the carrier state may be related to host immune status, and possibly to ethnicity. In a recent study of 25 Alaskan patients with clinical hepatitis B who were followed for up to 5 years, 15 were either slow in developing anti-HBs or never developed anti-HBs—despite the fact that only one patient became a chronic HBsAg carrier. The investigators postulated that the patients' ethnicity (Eskimo) may have contributed to their failure to develop antibody (McMahon et al. 1981).

HBV is also transmitted vertically, and umbilical cord blood is often positive for HBsAg. The infection rate among neonates born of mothers who are asymptomatic carriers of HBsAg in the United States is about 10 percent, for babies born to infected oriental mothers, the infection rate is 40 percent. This difference may be due to an increased proportion of oriental mothers who are positive for both HBsAg and HBeAg, rather than just for HBsAg. Neonatal infection may be acquired in utero or during the birth process. Perinatal transmission is of special concern because of a high probability of subsequent chronic hepatitis in affected babies. About half of the males and 13 percent of females born to mothers positive for HBeAg develop serious liver disease including hepatomas in adult life.

The fact that HBV infection is especially prevalent among individuals in underdeveloped countries, sexually promiscuous persons, male homosexuals, and about one-third of household contacts of persons with clinical hepatitis B is an indication that intimate contact as well as direct parenteral injection of the virus is an important mode of transmission.

Pathogenesis. Though the route of transmission for HBV is primarily parenteral, the viral replication site is unknown. It is clearly not the gastrointestinal tract (as with HAV). Within 5 to 8 weeks of exposure, viremia develops and the blood becomes infectious. Most knowledge of the pathogenesis of the HBV is speculative, though it is probable that the host's immune response to the HBV antigens rather than the cytopathic effects of the virus itself causes tissue damage. HBcAg levels are especially high in patients on immunosuppressive drugs.

Genetic factors may be important in determining whether chronic hepatitis ensues. "Ground glass" hepatocytes is a term applied to the appearance of infected liver cells. The ground glass appearance is due to the presence of HBsAg in the cell cytoplasm of chronic HBV infection, but is only rarely seen in cases of acute HBV infection. It is clear that persistent viral replication is related to hepatic cell necrosis, but not necessarily to liver cirrhosis. HBV DNA has been found integrated in hepatocytes in conjunction with chronic infection. Such integration precedes the development of neoplasm and there is a suggested correlation between hepatocellular carcinoma and chronic HBV infection.

Extrahepatic pathology can also occur. There is evidence that HBV can multiply in pancreatic as well as liver cells (Shimoda et al. 1981). HBV is associated with glomerulonephritis; HBsAg immune complexes are sometimes present in the kidney but not the liver. The persistence of antigenemia in HBsAg carriers could be due to failure of the host to produce anti-HBs, failure to lyse infected hepatocytes, or impaired cellular immunity with immunologic tolerance.

Host responses. The incubation period for HBV infection is longer than for other types of viral hepatitis, 1 1/2 to 6 months, averaging 2 to 3 months. The incubation period is shortened if the inoculum is greater. Most cases are asymptomatic. Clinical cases have a similar prodrome to that seen with hepatitis A, though the onset is often more insidious. Early symptoms include anorexia, nausea, abdominal discomfort, and liver and spleen tenderness and enlargement with absence of fever. Jaundice may or may not be present. The SGPT and SGOT both rise 5 to 10 days before or with the onset of symptoms. The SGPT is usually higher (mean levels of 300 to 800 units), reflecting significant hepatic damage. Generally, convalescence occurs over 2 to 6 weeks, jaundice (if present) gradually subsides, and recovery is complete within 6 months in 90 to 95 percent of affected individuals.

Extrahepatic symptoms due to autoimmunity and the presence of antigen/antibody complexes include a "serum sickness" syndrome, arthralgias and arthritis, and rash. The mortality rate in uncomplicated cases is less than 1 percent. Rarely, fulminant hepatitis can develop, causing mental derangement, increased intracranial pressure, and encephalopathy. About two-thirds of cases of fulminant hepatitis are caused by HBV (only a few cases are due to HAV, and the rest are attributed to non-A, non-B viruses). The mortality rate in fulminant hepatitis is 20 to 60 percent.

In 10 to 15 percent of cases, a chronic HBsAg carrier state persists following HBV infection. HBV persisting in the liver and blood will remain infectious for months to years and it is important to differentiate between past infection, the healthy carrier state, and chronic or active hepatitis. In 30 to 50 percent of individuals positive for HBsAg, the liver appears normal and there are no symptoms; others have some histologic indications of chronic hepatitis. Another 5 to 10 percent of carriers develop chronic persistent hepatitis (a protracted healing process with nonprogressive lesions associated with a good prognosis), and others develop chronic active hepatitis (continued hepatocyte damage and liver fibrosis).

Chronic carriage of HBsAg is estimated to be present in about 1 million Americans, about 120 million people worldwide, especially among the 20- to 40-year-old age group. HBsAg positivity for up to 20 years has been noted; the frequency of life long carriage is unknown. Both cirrhosis and hepatocellular carcinoma are correlated with chronic HBV infection.

Control and treatment. The use of volunteer blood donors, who have HBV infection rates 3 to 10 times less than paid donors, and avoidance of pooled blood products have decreased rates of HBV infection (but not of non-A, non-B hepatitis) associated with transfusions. Routine screening of potential donors for HBsAg and anti-HBc using radioimmunoassay techniques has also decreased the incidence of post-transfusion hepatitis.

Normal human immune globulin contains too low a titer of anti-HBs to be effective against HBV and is not recommended for postexposure passive immunization. The use of hepatitis B immune globulin (HBIG) is widespread in protecting individuals who have been exposed to blood potentially containing HBsAg, and it is at least partially effective in minimizing subsequent symptoms; however, in some cases it may merely prolong the incubation period. It does not prevent infection. In addition, it is quite expensive, so its use should be limited, whenever possible, to individuals exposed to blood proven to contain HBsAg, either as a result of accidental parenteral inoculation, or ingestion or splash into mucous membranes. In addition, it should be given to infants born to mothers with hepatitis B. The risk of developing clinical HBV infection after exposure to HBsAg-positive blood is 1 in 20; after exposure to blood of unknown HBsAg status, it is only 1 in 2000.

HBIG is most effective if given as soon as possible after exposure, but it is reasonable to test the blood of the contact for HBV antigens before giving specific prophylaxis, since test results can be available within 24 to 48 hours in most instances. HBIG is ineffective if administered longer than 2 weeks after exposure, its efficacy is questionable after one week (see "Clinical Implications") Pre-exposure prophylaxis using HBIG is neither necessary nor recommended.

A recently developed inactivated vaccine initially trialled in homosexual men has been found to be safe, with protective effectiveness of about 80 percent. It is made of inactivated surface antigen, purified from the plasma of chronic HBV carriers, and is recommended for high-risk groups: male homosexuals, drug abusers, clients and staff of institutions for the mentally retarded, and some health care personnel. To date, there have been no serious side effects associated with the use of the vaccine.

Enteric precautions for hospitalized patients are not justified in non-A hepatitis, although gloves should be worn when handling excreta. The patient with HBV infection can share a room with another patient if he or she is fecally continent. However, blood precautions are essential; special care should be given in cases where massive exposure to blood is likely (hemodialysis, labor and delivery, surgery). A diet high in protein, calories, and vitamins increases healing time. The use of drugs, especially those metabolized by the liver, should be avoided and bed rest is important to improve hepatic blood flow.

NON-A, NON-B HEPATITIS

Biologic characteristics. Non-A, non-B hepatitis (NANBH) is a term coined in 1975 when it became possible to rule out both HAV and HBV infection. It is a diagnosis of exclusion, the symptoms are similar to those of HBV, but there is an absence of serologic evidence of an etiologic agent. Though the agents have not been identified, there are at least two antigenically different ones. These agents do not cross react with HAV or HBV. Three good candidates for the agent of NANBH include

1. A 27-nm spherical virus isolated from the liver of chimpanzees.
2. A 37-nm spherical particle morphologically similar to HBV. It has been proposed that this agent be named hepatitis C.
3. The RJ agent, enveloped with radial projections similar to those on paramyxoviruses. The size of the RJ agent is 80 to 140 nm.

Epidemiology. Over 90 percent of transfusion-related hepatitis is NANBH. The incidence of acquiring NANBH following blood transfusion in the United States is about 9.8 percent. Comparative risks of developing hepatitis B or NANBH from various blood products is listed on Table 18-6. NANBH transmission is usually by the parenteral route, but may also be nonparenteral. The agents are responsible for about one-fourth of sporadic hepatitis in adults worldwide. Like HBV, NANBH is associated with drug abuse, hemodialysis, and can spread in institutions or among family contacts. It is more prevalent from blood of paid donors (40 percent versus 7 percent from volunteer donor blood) and tends to become chronic; 25 to 50 percent of patients have transaminase increases for at least 6 months after onset of disease.

Pathogenesis. Pathogenesis is the same as that of HAV and HBV infections.

Host responses. The incubation period is between that of HAV and HBV, averaging about 1 1/2 months. The clinical disease is slightly milder than that due to HBV. There is less jaundice, lower transaminase levels; however, almost one-fourth of diagnosed cases develop chronic active hepatitis. The disease must be differentiated from those caused by cytomegalovirus, Epstein-Barr virus, and *Toxoplasma,* as well as HAV and HBV.

Control and treatment. Screening blood donors for HBsAg has not decreased the incidence of NANBH. Until the etiologic

TABLE 18-6. Comparative Risk of Hepatitis B or Non-A, Non-B Hepatitis from Various Blood Products

Low or no risk	*Average risk*	*High risk*
Albumin	Factor VIII concentrates (from single donor)	Any products pooled from multiple donors
Globulins	Granulocyte concentrates	Cryoprecipitates
Plasma protein fraction	Packed red cells	Factor VIII concentrates
	Plasma, fresh frozen	Fibrinogen
	Whole blood	Platelet concentrates
		Prothrombin complex concentrates

Adapted with permission of the publisher from Koff RS. Viral hepatitis. New York: John Wiley and Sons, 1978.

agents are identified, very little can be done in the way of prevention. Increased alanine aminotransferase (ALT) levels in the blood of potential donors predict an increased probability of transmitting NANBH. Donors could be screened for this, but a cutoff level has not been set, and values differ with age and sex. Therefore, screening for this enzyme is not in widespread practice. The efficacy of NHIG in prophylaxis against NANBH is unknown. Patients with NANBH should be considered infectious and blood precautions implemented.

CLINICAL IMPLICATIONS

The infectivity of HBsAg is an important concern because so many hospital staff members are asymptomatic carriers. It may be a stigma in some settings to be labelled a "carrier," and health care personnel have been denied employment on that basis. It has been suggested that the low intensity exposure of hospital personnel to HBV results in the development of natural immunity and that many HBsAg-positive personnel may not be infectious (Dienstag and Ryan 1982).

Hepatitis B and NANBH are of considerable concern in hospitals and other institutions, while HAV is especially prevalent in day care centers for preschool children. Personnel in such settings must be involved in procedures to prevent the spread of hepatitis viruses. The spread of HAV is minimized by maintenance of scrupulous handwashing standards (especially after diapering young children and prior to handling food), by disinfection of diaper-changing surfaces with bleach (1:32 dilution), and by care that items such as vaseline or baby powder remain uncontaminated. In the case of an HAV outbreak, new admissions to day care centers should be temporarily suspended, parents should be discouraged from removing their children (to prevent spread to other centers), and HNIG should be administered to other children and staff.

The following precautions will minimize the transmission of HBV in hospitals:

1. The use of blood precautions for patients known or suspected of having HBV infection
2. Enteric precautions for infected patients with gastrointestinal bleeding or fecal incontinence
3. Since many infections are spread by

asymptomatic carriers, the use of golves for handling excreta (especially urine and feces), or careful handwashing with all patients
4. Routine screening for HBsAg carriage among high-risk groups (Table 18-7)
5. Vaccination among employees who handle blood or blood products frequently

If a mother has HBV infection when her baby is born, the infant should receive 0.5 ml HBIG within 30 minutes of delivery, or as soon as possible after that. HBsAg-positive patients undergoing hemodialysis should be dialyzed in a separate room and, if possible, attended by a nurse who is also seropositive for HBsAg. Staff caring for such patients should not provide care to seronegative patients during the same working period. All dialysis equipment used should be separated and carefully cleaned before use with any other patient. Ideally, equipment used for HBsAg carriers should be reserved specifically for that purpose. A recommended protocol for the care of health care employees with possible hepatitis exposure is outlined in Figure 18-5.

SUGGESTED READINGS

Aach RD, Wochspress JD. Post-transfusion hepatitis. Johns Hopk Med J 1981;149:45-49.

Aldershvile J, et al. Hepatitis Be antigen and antibody measured by radioimmunoassay in acute hepatitis B surface antigen-positive hepatitis. J Infect Dis 1980;141:293-298.

Aldershvile J, et al. The expression pattern of hepatitis Be antigen and antibody in different ethnic and clinical groups of hepatitis B surface antigen carriers. J Infect Dis 1980;142:18-22.

Anh-Tuan N, Novák E. Hepatitis B surface antigen circulating immune complexes (HBsAg-CICs) in patients with hepatitis B and asymptomatic HBsAg carriers. Clin Exper Immunol 1981;43:246-253.

Benenson MW, et al. A military community outbreak of hepatitis type A related to trans-

TABLE 18-7. Recommended Groups for Screening for HBV Infection

Group	*Frequency of testing*
Patients receiving hemodialysis in hospitals or using institutional equipment	Every 6 months
Patients receiving organ transplants, organ transplant donors	Prior to transplant
Multiply transfused patients, persons with history of parenteral drug abuse, patients from areas with high endemic rates (Asia, Africa)	On hospital admission
Patients with symptoms of hepatitis	At onset of symptoms
Hospital employees caring for high-risk patients (on dialysis units, in transplant surgery)	Every 6 months
Individuals who incur needle sticks or parenteral contamination with blood or blood products	Test both the individual contaminated and the person from whom the contamination occurred

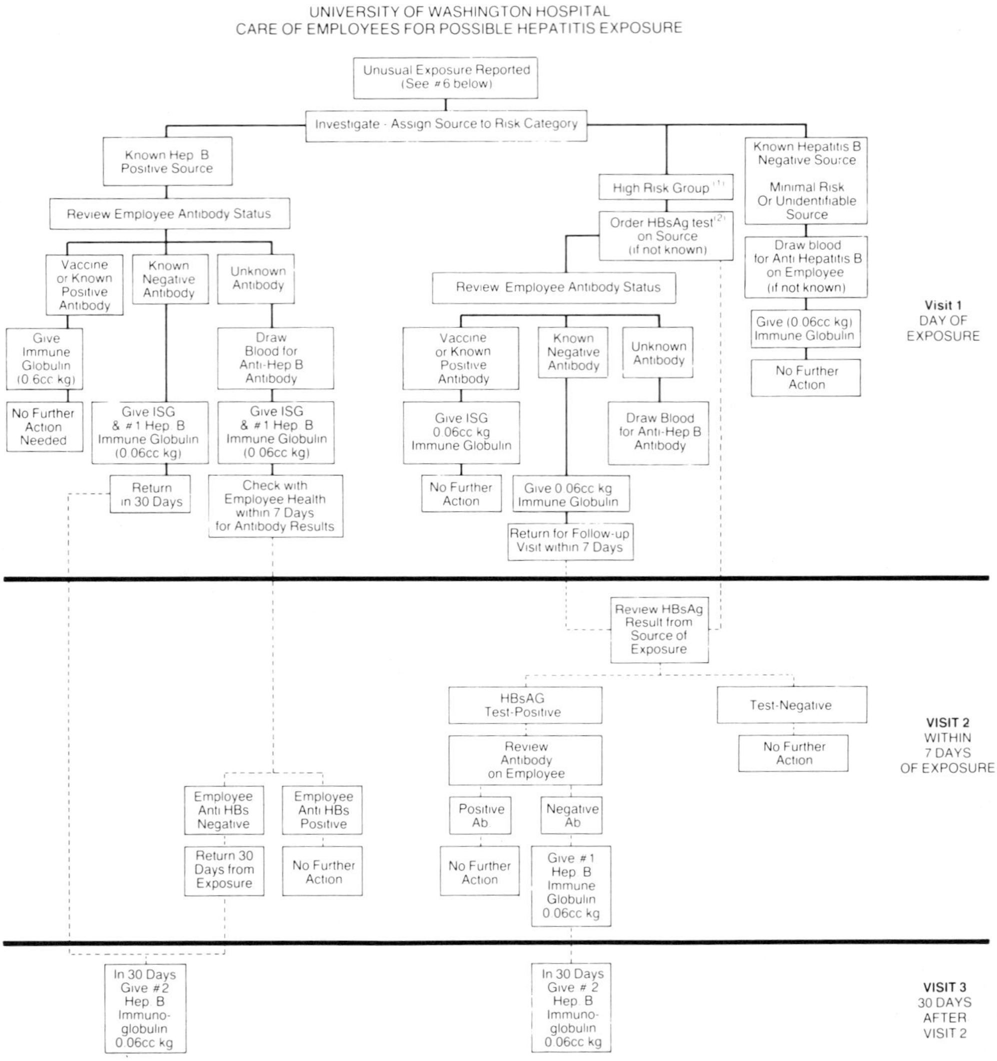

DEFINITIONS

(1) High Risk Patients for Hep. B

a. Has clinical hepatitis
b. On renal dialysis
c. Had renal or bone marrow transplant
d. Infants of Hep. B positive mothers
e. Had multiple past transfusions
f. Drug abusers
g. Active homosexual males
h. Persons from geographic areas with high endemic rates (10-20%)

(2) HBsAg - test for presence of Hep. infection
(3) Anti-HBs - test for immunity to Hep. B
(4) Hep. B immunoglobulin - passive antibodies to protect exposed persons from diseases
(5) Immune globulin - passive antibodies to protect exposed persons from Hep. A diseases and possible other agents. The degree of protection from Hep. B and non-A, non-B is not clearly established
(6) Unusual exposure includes contact with patient's body fluids blood, secretions or excretions through:
(i.) Broken skin or needle sticks
(ii.) Mucous membraines of eyes, nose or mouth

A more detailed list may be found in the University Hospital Infection Control Manual, Section III, G3.

FIGURE 18-5. Care of employees for possible exposure to hepatitis. (Reprinted with permission of the publisher from University of Washington Infection Control Manual. Seattle: University Hospital, 1982.)

mission in a child care facility. Am J Epidemiol 1980;112:471-481.
Bock KB, Tong MJ, Bernstein S. The risk of accidental exposure to hepatitis B virus via blood contamination in medical students. J Infect Dis 1981;144:604.
Burke DS, et al. Age-specific prevalence of hepatitis A virus antibody in Thailand. Am J Epidemiol 1981;113:245-249.
Carta JR. Presumed transfer of hepatitis antibody by blood transfusion to chronic hemodialysis patients. Am J Infect Control 1982;10:34-36.
Centers for Disease Control. Survey of viral hepatitis surveillance activity in state and local health departments. Morb Mort Week Rep 1981;30:164-169.
Centers for Disease Control. Inactivated hepatitis B virus vaccine. Ann Intern Med 1982;97:379-383.
Check WA. Looks like smooth sailing for experimental hepatitis B vaccine. JAMA 1981; 246:2111-2112.
Coulepis AG, et al. Detection of hepatitis A virus in the feces of patients with naturally acquired infections. J Infect Dis 1980;141:151-156.
Craig CP, Gribble C, Suarez K. Risk of hepatitis B among phlebotomists. Am J Infect Control 1981;9:11-14.
Deinhardt F. Predictive value of markers of hepatitis virus infection. J Infect Dis 1980;141:299-305.
Dienstag JL, Ryan DM. Occupational exposure to hepatitis B virus in hospital personnel: infection or immunization? Am J Epidemiol 1982;115:26-39.
Francis DP, et al. The prevention of hepatitis B with vaccine. Ann Intern Med 1982;97:362-366.
Frösner GG, et al. Antibody against hepatitis A in seven European countries. I. Comparison of prevalence data in different age groups. Am J Epidemiol 1979;110:63-69.
Gauss-Müller V, Frösner GG, Deinhart F. Propogation of hepatitis A virus in human embryo fibroblasts. J Med Virol 1981;7:233-239.
Gerlich WH, et al. Diagnosis of acute and inapparent hepatitis B virus infections by measurement of IgM antibody to hepatitis B core antigen. J Infec Dis 1980;142:95-101.
Gust ID, Lehmann NI. A seroepidemiologic study of infection with hepatitis A virus and hepatitis B virus in five Pacific islands. Am J Epidemiol 1979;110:237-242.
Hawkes RA, et al. The seroepidemiology of hepatitis in Papua New Guinea II A long-term study of hepatitis B. Am J Epidemiol 1981;114:563-573.
Hollinger FB, et al. Transfusion-transmitted virus study: experimental evidence for two non-A, non-B hepatitis agents. J Infect Dis 1980;142:400-407.
James JJ, et al. Serological markers for hepatitis types A and B among United States Army blood donors. Milit Med 1981;146:562-567.
Iwarson S, Lindholm A, Norkrans G. Hepatitis B and non-A, non-B in a Swedish blood center during ten years of HBsAg screening. Vox Sang 1980;39:79-82.
Kinoshita R, et al. Chronic hepatitis B: correlation between viral replication and clinical course. J Infect Dis 1981;144:303-311.
Koff RS. Viral hepatitis. New York: John Wiley and Sons, 1978.
Kuberski T, et al. Hepatitis B virus infections in Melanesians and Polynesians in New Caledonia. Am J Epidemiol 1981;114:355-361.
Lingao AL, Domingo EO, Nishioka K. Hepatitis B virus profile of hepatocellular carcinoma in the Philippines. Cancer 1981;48:1590-1595.
Mao JS, et al. Patterns of shedding of hepatitis A virus antigen in feces and of antibody responses in patients with naturally acquired type A hepatitis. J Infect Dis 1980;142:654-659.
Mazzur S, Bastiaans MJS, Nath N. Hepatitis B virus infection among children and adults in the Solomon Islands. Am J Epidemiol 1981;113:510-519.
McMahon BJ, et al. Delayed development of antibody to hepatitis B surface antigen after symptomatic infection with hepatitis B virus. J Clin Microbiol 1981;14:130-134.
Moss ALH. Hospital outbreak of hepatitis B (letter). New Zeal Med J 1981;94:65-66.

Mushahwar IK, et al. Radioimmunoassay for detection of HBeAg and its antibody. Results of clinical evaluation. Am J Clin Pathol 1981;76:692-697.

Osterholm MT, et al. IgM-specific serologic testing in an outbreak of foodborne viral hepatitis, type A. Am J Epidemiol 1980;112:8-16.

Papaevangelou GJ, Gourgouli-Goutou KP, Vissoulis HG. Epidemiologic characteristics of hepatitis A virus infection in Greece. Am J Epidemiol 1980;112:482-486.

Ray MB. Hepatitis B virus antigens in tissues. Baltimore: University Park Press, 1979.

Rosenberg ML, Koplan JP, Pollard RA. The risk of acquiring hepatitis from sewage-contaminated water. Am J Epidemiol 1980;112:14-22.

Scott RM. Experimental transmission of hepatitis B virus by semen and saliva. J Infect Dis 1980;142:67-71.

Shafritz DA, et al. Integration of hepatitis B viral DNA into the genome of liver cells in chronic liver disease and hepatocellular carcinoma. N Engl J Med 1981;305:1067-1073.

Shimoda T, et al. Light microscopic localization of hepatitis B viral antigens in the human pancreas. Gastroenterol 1981;81:998-1005.

Siebke JC, et al. Prevalence of hepatitis A antibodies in a normal population and some selected groups of patients in Norway. Am J Epidemiol 1982;115:185-191.

Skinhøj P, Mathiesen LR, Cohn J. Persistence of viral hepatitis A and B in an isolated caucasian population. Am J Epidemiol 1980; 112:144-148.

Tabor E, Ziegler JL, Gerety RJ. Hepatitis Be antigen in the absence of hepatitis B surface antigen. J Infect Dis 1980;141:289-292.

Tyeryai FJ. Summary of the U.S.-Japan hepatitis panel conference. J Infect Dis 1980; 142:471-475.

University of Washington Hospital. Infection control manual. Seattle: University of Washington, 1982.

Valenti WM, et al. Nosocomial viral infections. II. Guidelines for prevention and control of respiratory viruses, herpesviruses, and hepatitis viruses. J Infect Control 1980;1:165-178.

VanEtta LL, Peterson LR, Gerding DN. Asymptomatic hepatitis B carriers in a family (letter). N Engl J Med 1981;305:1093-1094.

Vladutiu AO, Leibowitz AI. Passive-active immunization after anti-HBs immunoglobulin? (letter). JAMA 1981;246:1546-1547.

19 ONCOGENIC VIRUSES

KEY POINTS

- Ocogenic viruses are those suspected, or proven, in the causation of cancer.
- The oncogenicity of viruses is very difficult to study because of the long lag period between exposure to virus and development of tumor; the fact that viruses may cause cancer only in some animal species; the long-term exposure to the oncogenic virus that might be required; and because of the complex nature of tumor formation.
- DNA viruses with oncogenic potential include some members of the papovavirus, adenovirus, herpesvirus, poxvirus, and hepatitis groups.
- RNA viruses with oncogenic potential include some members of the family Retroviridae, a group called oncoviruses. Tumor production by oncoviruses has been proven in some animal species, but not in humans.

Oncogenic (also called carcinogenic) viruses are viruses that are suspected or proven in the causation of cancer. Some viruses have been proven as cancer-causing agents in animals. Rous, for example, recognized in the early 1900s the correlation between the Rous sarcoma virus that infects chickens and the subsequent development of various types of tumor in these birds. Few viruses are actually suspected of causing cancer in humans (Table 19-1).

The study of oncogenic viruses is difficult for several reasons. First, there is a long lag period between exposure to a virus and the development of tumors. Second, the virus may only be oncogenic to certain animal species. Third, viruses may or may not cause tumors after superficial or one-time exposure; long-term, chronic, or frequent exposures may be necessary. Finally, though viruses may be an essential component in the production of some tumors, they are almost certainly not the sole component. Rather, cancer occurs as a result of a number of interacting factors; that is, immunologic, environmental, as well as microbiologic factors.

The mechanisms by which viral infection

TABLE 19-1. Human Viruses Exhibiting Oncogenic Properties

Virus	*Disease*	*Association with malignancy*	*Country*
Herpesvirus			
Herpes simplex virus 1	Predominantly oral lesions	Little	Worldwide, especially U.S.
Herpes simplex virus 2	Predominantly genital lesions	Cervical carcinoma Kaposi's sarcoma	Worldwide, especially U.S.
Cytomegalovirus	Inclusion disease	Prostatic cancer	Worldwide, especially the black population
Epstein-Barr virus	Infectious mononucleosis	Burkitt's lymphoma Nasopharyngeal carcinoma	Primarily Africa, China
Hepatitis B virus	Hepatitis	Hepatocellular carcinoma	Worldwide, especially U.S. and Africa
Oncovirus			
B type	—	Mammary tumors	Animals only
C type	—	Leukemias Lymphomas Carcinomas Sarcomas	Animals only

attributes to tumor formation are not well explained (Fig. 19-1). Oncogenic viruses cause host cells to become transformed and malignant. Transformed cells, or cancer cells, have several characteristics that differentiate them from normal cells. First, these cells lose their distinct shape and characteristics, having a different growth pattern and arrangement than normal cells. That is, they become *undifferentiated*. Second, they have an uncontrolled rate of growth. Normal cells will stop growing when their mass reaches a certain point, but transformed cells continue to grow. Third, cancer cells do not require an attachment to surfaces for growth as do normal cells. And last, cancer cells develop chromosomal changes, altered metabolism, and new antigens. Specific oncogenic viruses will be addressed separately, with emphasis on those suspected in human neoplasms.

DNA VIRUSES

The DNA viruses with oncogenic potential are contained in four groups: papovavirus, adenovirus, herpesvirus, and poxvirus.

Papovavirus

The papovaviruses with oncogenic potential incude the papilloma virus, polyoma virus (Fig. 19-2) and simian virus-40 (SV-40; also known as vacuolating virus), as well as several new human viruses, including the BK and JC viruses, which have similar structures as SV-40. (Notice that papovavirus consists of the first two letters of three of the viruses in the family.)

All of the papovaviruses have been found to cause tumors in animals. These tumors present as skin growths in rabbits, dogs,

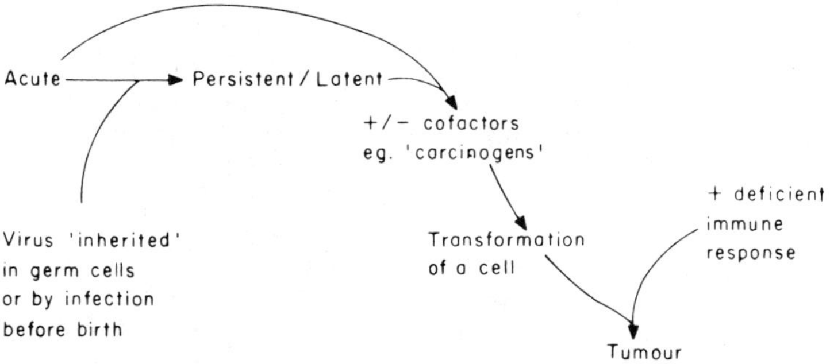

FIGURE 19-1. Possible relationship between virus infection and tumorigenesis. (Reprinted with permission of the publisher from Primrose SB, Dimmock NJ. Introduction to modern virology, 2nd edition. Oxford: Blackwell Scientific Publications Ltd, 1980.)

sheep, cattle, and humans. Unfortunately, papovaviruses are difficult to propagate, hence, they are difficult to study. These viruses are very large with a naked capsid. They most often contain double-stranded DNA present in circular molecules. The guanine and cytosine contents of papovavirus DNA and mammalian cell DNA are very similar, a fact that may play a part in their oncogenic potential.

Papovaviruses contain a large tumor antigen (T antigen), which is formed in the host cell nucleus during replication. The antigen is required to begin DNA replication of the virus. Tumor-specific transplant antigens (TSTA) have also been identified; they have been found to elicit immune responses to tumor introduction in some animals, aiding them in eliminating carcinogenic cells.

Papilloma Viruses

Papilloma viruses were first associated with wartlike tumors on both animals and humans in 1907. Since 1901 the papilloma virus family has been divided into four subgroups: HPV, HPV-2, HPV-3, and HPV-4. Studies show that HPV-4 has the ability to transform benign warts to malignancies in animals. Papilloma viruses have also been found in some women with mild cervical dysplasia. As yet, the viruses have not been detected in carcinoma in situ or in invasive cervical carcinoma, but research in this area continues.

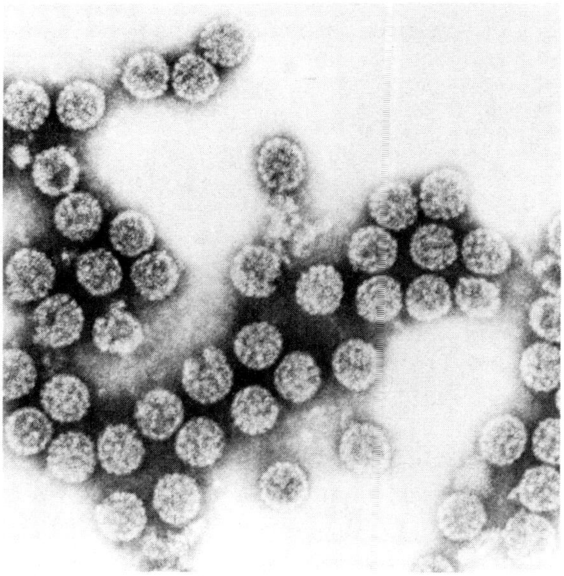

FIGURE 19-2. Polyoma virus (papovavirus.) (Courtesy of Centers for Disease Control, Atlanta, GA 30333.)

Polyoma Virus

Polyoma viruses have been isolated in newborn mice. They cause tumors in multiple sites in animals but have yet to be implicated in human neoplasms.

Simian Virus 40

Simian virus (SV-40) causes tumors in monkeys. Though it has been an inadvertent contaminant in certain human vaccines (such as the polio vacccine), no ill effects have been observed. SV-40 to date has not been found to cause tumors in humans.

BK Virus and JC Virus

The BK and JC viruses cause tumors in newborn hamsters; they can also tranform benign cells to malignant cells in vitro. The BK virus has been found in immunodeficient patients. The JC virus has been isolated from the brains of patients with multifocal leukoencephalopathy. Both BK and JC antibodies are common in humans but no definite link with cancer has been made.

Adenoviruses

Human adenoviruses are currently being studied for oncogenic effects, but none have yet been shown to cause cancer. They have been divided into two groups according to their potential ability to induce tumor formation: (1) highly oncogenic (types 12, 18, and 31), and (2) weakly oncogenic (types 3, 7, 11, 14, 16 and 21). The primary difference between the two groups is the guanine and cytosine content in the genetic make-up of each virus. Highly oncogenic viruses tend to have low guanine and cytosine content, while weakly oncogenic viruses have high guanine and cytosine content. Highly oncogenic viruses are infectious in tissue culture in hamsters and have caused sarcomas in newborn animals. In addition, lymphomas have been found in mice inoculated with adenovirus type 2.

Herpesvirus

The herpesviruses with potential oncogenic properties are simian herpesvirus, Marek's disease virus (MDV), herpesvirus of rabbit lymphoma, Lucke's herpesvirus of frogs (LHV), herpes simplex virus (HSV), and Epstein-Barr Virus (EBV). Simian herpesvirus consists of two strains: herpesvirus saimiri (HVS) and herpesvirus ateles (HVA). HVS has been known to induce lymphomas in primates. However, neither of these herpesviruses have been linked with human tumors.

Marek's Disease Virus

Marek's disease virus causes a highly contagious lymphoproliferative disease of chickens. Animal studies show that following vaccination against MDV, infection may still occur, but oncogenicity is lost.

Lucke's Herpesvirus

Lucke's herpesvirus causes renal adenocarcinomas in frogs. Studies show that its tumorigenic properties change as the frog's body temperature changes. Although both LHV and MDV have been shown to cause tumors in animals, neither has been connected with human tumorigenesis.

Herpes Simplex Virus

Herpes simplex virus type 2 was first detected through the Papanicolaou technique (Pap smear), in which epithelial scrapings are taken from the cervical wall and are inspected under the microscope. Pap smears detect the presence of active herpes infections as well as cancerous or precancerous cervical cells. A causative link between HSV-2 and cervical carcinoma is highly suspect for a number of reasons:

1. HSV-2 has been shown to transform benign hamster fibroblast tumors into malignancies.
2. HSV-2 antibodies are present more frequently in women with cervical carcinoma than in the normal population.
3. Rates of cervical cancer are higher in women with HSV-2 antibodies than without the antibodies.
4. HSV-2-induced tumor-specific antigens have been demonstrated in cervical cancer cells.
5. HSV-2 DNA particles have been found in human cervical tumors.
6. Epidemiologic studies suggest that both genital herpes and cervical carcinoma are caused by a sexually transmitted agent. Prostitutes, women with multiple sex partners, and women with first coitus at an early age are at high risk; virgins and nuns are at low risk of both diseases. In addition, the second wife of a man whose first wife died of cervical cancer is reported to have a higher risk of cervical cancer.

However, HSV-2 may not be the only causative agent of cervical cancer. Cytomegalovirus (CMV) as well as papilloma viruses are also being evaluated for possible carcinogenic abilities.

Epstein-Barr Virus

EBV is found in lymphoid cells, particularly in B lymphocytes that have EBV receptors. In vitro studies have shown EBV-infected lymphocytes transforming into malignant cells. EBV nucleic acid has been isolated from tumor cells. EBV is strongly correlated with two types of tumors. Burkitt's lymphoma and nasopharyngeal carcinoma (see Chapter 17). Some scientists believe that EBV may only be a passenger virus attached to the nasopharyngeal cancer cell instead of a causative agent. Others believe that genetic make-up or environmental factors may have some influence on incidence of nasopharyngeal cancer.

Hepatitis B Virus

Research studies in the early 1980s have uncovered strong evidence supporting the hypothesis that hepatitis B virus is associated with primary hepatocellular carcinoma. This carcinoma is found primarily in China, Asia, and Africa, but it occurs also in the United States. Several factors have been found to support a correlation:

1. Patients with primary hepatocellular carcinoma have a higher rate of HBsAg or anti-HBc positivity than do other individuals.
2. HBsAg and anti-HBc can be found in hepatocellular carcinoma cells or in cells directly adjacent to the tumor.
3. DNA sequences have been found in tumor cells of hepatocellular carcinoma patients.
4. Hepatocellular carcinoma is seen often in persons with cirrhotic livers. Within that group, there is a higher incidence of hepatocellular carcinoma among patients who are HBsAg-positive than patients who are HBsAg-negative.

While the evidence for the link with the hepatitis B virus is strong, it is still unknown if the virus is a causative agent, a promoter, or if it merely adds to the immunocompromised state that later leads to carcinoma. Research is ongoing concerning these issues.

RNA VIRUSES

Only the oncogenic group of RNA viruses, oncovirus, will be discussed in this chapter. Oncoviruses are all members of the family Retroviridae and possess the enzyme reverse transcriptase. Not all retroviruses, however, are oncogenic. Tumor production by oncoviruses has been proven in mice, chickens, vipers, and primates, and has been suggested in humans. Oncoviruses are divided as follows:

B type viruses
- Mammary tumor virus
  - Murine mammary tumor virus (MuMTV)
  - Mason-Pfizer mammary virus (MPMV)

C type viruses
- Leukemia viruses
  - Avian leukemia complex
  - Murine leukemia complex
  - Feline leukemia virus
  - Gibbon leukemia virus
- Sarcoma viruses
  - Murine sarcoma virus
  - Feline sarcoma virus
  - Whooley-monkey sarcoma virus
  - Rous sarcoma virus

The primary difference between B and C type viruses is changes caused in the host cell nucleoid. In B type infections, host cell nucleoids appear abnormal and contain protruding spikes; in C type infections, nucleoids appear normal and are located centrally in the cell. The RNA viruses will be discussed by animal host group, with emphasis on those important in human cancers.

Murine Viruses

Murine mammary tumor virus (MuMTV or Bitner's virus) causes mammary tumors in mice. To date, two types of MuMTV have been isolated: one is transmitted via the sex cells; the other is transmitted through breast milk. Newborn mice who ingest the milk-transmitted virus strain develop tumors early in life. Mice that acquire the virus sexually develop fewer tumors, and those occur late in life. Little is known about the MuMTV infection process. However, susceptible mice inoculated against MuMTV during the neonatal period exhibit a lower incidence of mammary carcinoma than those mice inoculated later in life. Tumorigenesis in these mice appears to increase when stress is introduced into their environment.

Despite active research, there is no conclusive evidence linking a virus with human breast cancer. However, nucleic acid sequences in the genetic coding of human tumors similar to MuMTV sequences do exist in human breast tumors. The fact that the incidence of breast cancer in populations where breast feeding is common is low and women who were breast-fed as infants are not increased risk for breast cancer argues against an oncogenic virus in breast milk.

Feline Viruses

Feline sarcoma and leukemia viruses cause fibromas and leukemias in cats, although only a small percentage of the cats that contract these viruses actually develop tumors. The viruses are transmitted both vertically and horizontally but little is known about any possible link with human neoplasm. Feline leukemia vaccines are currently being developed for possible use as models for human vaccines in the future.

Avian Viruses

Of all C type viruses being studied, only the avian leukemia viruses have been suspected in human cancer. Avian leukemia viruses and avian sarcoma viruses are antigenically similar. Avian leukemia virus and Rous sarcoma virus (RSV) are found in chickens. These viruses are transmitted vertically by females; males can be infected but will not transmit the infection. The incidence of tumors among infected chickens is low, but those found do contain infectious particles in high concentration.

RSV, the first virus to be connected with animal tumorigenesis, was first discovered by Rous in the early 1900s. It was shown that RSV could transform chicken fibroblasts to malignant cells at certain temperatures. RSV, however, has not been connected with any human cancer to date. It is primarily used in animal models for studies of cell transformation and vaccines.

Primate Viruses

Little is known about gibbon leukemia virus and wooley-monkey sarcoma virus, However, the Mason-Pfizer mammary virus has been isolated from rhesus monkey mammary tumors. It is horizontally transmitted and can transform benign cells to malignant cells in vitro. No conclusive evidence has been found that links primate oncogenic viruses with human neoplasms.

CLINICAL IMPLICATIONS

Although oncogenic viruses have yet to be proven in causation of cancer in humans, clinicians must be aware that this link may exist. Knowledge of clustering of certain viruses is important, especially if patients are from a geographic area in which specific viral agents or tumors are prevelant. If cancers could be detected earlier in these patients through more careful surveillance, disease could be better controlled. In cancer screening programs, health care personnel should ask questions such as:

Has the implicated virus ever been present in the patient?

Has the implicated virus ever been present in the patient's family?

Has the implicated virus ever been present in the patient's social group?

What is the patient's past medical history; does it include infectious mononucleosis? Malaria? Herpes simplex? Hepatitis B? CMV?

Where does the patient live? Is it an area where viral clustering has been observed?

What other factors are present that may predispose the patient to tumorigenicity? Age? Race? Sexual practices?

Individuals at risk for tumors of possible viral etiology should receive careful health teaching and should seek early and frequent checkups. For example, young women with frequent sexual partners should be instructed about ways to prevent infection with herpes simplex, and should be counseled to have yearly Papanicolaou smears.

Mass immunization against certain forms of cancer is a future possibility. As increasing evidence links viruses (e.g., hepatitis B virus) with cancer (e.g., liver cancer) immunization becomes more feasible as a preventive measure. Today prophylactic vaccination is successfully used against two neoplastic diseases in animals. However, in humans much research must still be done. A virus-cancer link must be firmly established, modes of transmission of the virus must be identified, and other environmental or synergistic factors eliminated or minimized. The safety and efficacy of vaccines must also be evaluated and their risks weighed.

In summary, oncogenic viruses are beginning to be linked with human cancers. Research is ongoing, therefore, practitioners must keep current with new developments. Evidence already available must be constantly evaluated, used in daily patient assessments, and implications for possible prevention must be recognized.

SUGGESTED READINGS

Beasley R, et al. Heptatocellular carcinoma and hepatitis B virus. Lancet 1981;11:1129-1132.

Davis S. The role of viruses in the etiology of cancer: an epidemiologic perspective. Wash Publ Health 1982;3:4-11.

deThe G. The epidemiology of Burkitt's lymphoma: evidence for a causal association with Epstein-Barr virus. Epidemiol Rev 1979;1:32-51.

Emmelot P, Bentvelzen P, eds. RNA viruses and host genome in oncogenicity. New York: American Elsevier, 1972.

Faras AJ. Virus genes and diseases. Minn Med 1981;64:215-220.

Gottlieb G, et al. A prelinary communication on extensively disseminated Kaposi's sarcoma in young homosexual men. Am J Dermatol 1981;3:111-119.

Harns J, et al. The properties of sarcomas induced in Wistar rats by RSV. Intern J Canc 1974;4:31-41.

Hillerman M. Human cancer virus vaccines. CA 1974;24:213-217.

Hynes R. Cellular location of viral transforming proteins. Cell 1980;3:601-602.

Kaufman R, Rawls W. Herpes genitalis and its relationship to cervical cancer. CA 1974; 24:258-265.

Levy J. Endogenous C type viruses in normal and abnormal cell development. Canc Res 1977;37:2957-2968.

Melnick J, et al. The causative role of herpesvirus type 2 in cervical cancer. Cancer 1974;34:1375-1385.

Menczer J, et al. Herpes virus type 2 in adenocarcinoma of the uterine cervix. Cancer 1981;38:1497-1499.

Meyskens F. Tumor virus and human cancer. Front Radiat Ther Oncol 1980;15:18-23.

Moore D, Chainey J. Breast cancer: etiology and possible prevention. Am Scient 1975;63:160-168.

Moore D, et al. Mammary tumor viruses. Adv Canc Res 1979;29:347-418.

Nomura A, et al. Presence of hepatitis B surface antigen before primary hepatocellular carcinoma. JAMA 1982;247:2247-2249.

McBride G, ed. Papilloma virus and cervical dysplasia. JAMA 1981;245:2483-2485.

Pitot H. Fundamental of oncology. New York: Marcel Dekker, 1981.

Potter C. Some aspects of the role of viruses in cancer. Postgrad Med 1979;55:150-158.

Rapp F. Virology and cancer. Preven Med 1980;9:244-251.

Safai B, Good R. Kaposi's sarcoma: a review and recent developments. CA 1981;31:2-12.

Sahnazarov N, et al. Cell transformation in vitro by herpes simplex virus. Virologie 1980; 31:231-239.

Siegel F, et al. Severe acquired immunodeficiency in male homosexuals, manifested by chronic perianal ulcerative complex lesions. N Engl J Med 1981;305:1439-1443.

Stone H. RNA tumor viruses. J Kans Med Soc 1979;80:417.

Sun L, et al. Isolation and identification of lymphocytic and myelogenous leukemia specific sequences in genomes of gibbon oncornaviruses. J Virol 1980;35:400-408.

Temm H. The cellular and molecular biology of RNA tumor viruses, especially avian leukosis-sarcoma viruses and their relatives. Adv Canc Res 1974;19:48-98.

Vrederoe D, et al. Concepts of Oncology Nursing. New Jersey: Prentice Hall, 1981.

Wentz W, et al. Induction of uterine cancer with inactivated herpes simplex type 1 and 2. Cancer 1981;48:1783-1790.

Wyke D. Oncogenic viruses. J Pathol 1981; 135:39-85.

ZurHaasen H. The role of viruses in human tumors. Adv Canc Res 1980;33:77-102.

20 PARASITES

KEY POINTS

- Parasites are of major health concern, particularly in developing countries.
- Protozoa are one-celled eukaryotes of which thousands of species are known.
- The major protozoa that cause infections include *Entamoeba histolytica* (which causes amoebiasis), *Giardia lamblia* (gastrointestinal disease called giardiasis), *Plasmodium* (four species of which cause malaria), *Toxoplasma gondii* (toxoplasmosis), and *Trichomonas vaginalis* (trichomoniasis).
- Helminths (which means worms) are classified into flukes (trematodes), tapeworms (cestodes), and roundworms (nematodes).
- Tapeworms are most commonly acquired from eating raw or inadequately cooked fish, beef, or pork.
- Important roundworm infections include ascariasis, pinworm (caused by *Enterobius vermicularis),* hookworms, trichinosis (caused by *Trichinella spiralis),* whipworm *(Trichuris trichiura),* and filariasis.
- Blood, liver, intestinal, and lung flukes all require water snails as intermediate hosts, and are therefore common where there are contaminated waters.
- Schistosomiasis is estimated to be second to malaria in worldwide public health importance.

Even countries with the highest standards of living are not free of endemic parasites, though there are differences between the developed and undeveloped countries both in types and in numbers of parasites. In the United States, two parasitic diseases—amoebiasis and trichinosis—are reportable to the Centers for Disease Control. Some states also request that cases of giardiasis and toxoplasmosis be reported. Parasitic infections, however, do not always produce symptoms, and the incidence is certainly higher than reported.

Parasites are generally divided into two major types (Table 20-1): protozoa and helminths. Protozoa are eukaryotes, that is, complex one-celled organisms with organelles that possess varied and distinct morphologies. They are ubiquitous in nature; there are and more than 45,000 species known (Table 20-2). The four most common protozoan infections in the U.S. are amoebiasis, toxoplasmosis, trichomoniasis, and giardiasis (see Neva 1974). Though malaria is not endemic in developed countries, it is discussed briefly here because cases may be seen among travellers to tropical and subtropical regions.

TABLE 20-1. Differentiation of Protozoa and Helminths

	Protozoa	*Helminths*
Structure	Unicellular eukaryotes; contain mitochondria and one or more nuclei	Complex, multicellular eukaryotes; with alimentary, excretory, and reproductive systems; several developmental stages: egg, larva, adult
Organelles of locomotion	Flagella, cilia, pseudopods, undulating ridges	Cilia, whiplike turning
Reproduction	Usually asexual by binary fission	Sexual
Classification	Four subphyla; over 45,000 species	Flukes (trematodes), tapeworms (cestodes), roundworms (nematodes)

TABLE 20-2. Major Groups of Protozoa

Subphylum	*Class*	*Human pathogens*	*Geographic distribution*
Plasmodroma	Mastigophora	*Giardia* (giardiasis)	Worldwide
		Leishmania (leishmaniasis)	Africa, Central and South America
		Trichomonas (trichomoniasis)	Worldwide
		Trypanosomes (sleeping sickness)	Africa, Central and South America
	Sarcodina	*Entamoeba* (amoebiasis)	Tropics, subtropics
	Sporozoa	*Plasmodium* (malaria)	Tropics, subtropics
		Toxoplasma (toxoplasmosis)	Worldwide
Ciliophora		*Balantidium coli*	Worldwide

Helminths (parasitic worms) can be simply classified for clinical purposes as flukes (trematodes), tapeworms (cestodes), and roundworms (nematodes) (Table 20-3). The helminths are the only organisms discussed in this text that are usually visible macroscopically. They range in size from about 5-mm long (pinworms) up to 20 to 30 feet long (some tapeworms). They are really classified as animals with eukaryotic, differentiated cells, and sexual reproduction. In the strict sense, then, the helminths do not belong in a book on microbiology; however, since they are traditionally included in such texts, they will be discussed briefly here. However, the reader with particular interest in parasitology or the clinician who plans to travel extensively or work in developing countries is referred to more detailed texts.

A surveillance report from the Parasitic Diseases Division of the Centers for Disease Control in 1976 stated that out of 414,820 stool specimens from 54 public health laboratories in the United States (including Guam, Puerto Rico, and the Virgin Islands), 15.6 percent contained at least one patho-

TABLE 20-3. Major Groups of Helminths (Parasitic Worms)

Phylum	*Class*	*Human pathogens*	*Geographic distribution*
Platyhelminthes	Cestodes (tapeworms)	*Diphyllobothrium latum* (fish tapeworm)	Worldwide, especially temperate climates like northern U.S. and Canada
		Taenia saginata (beef tapeworm)	Developing countries without meat inspection
		Taenia solium (pork tapeworm)	Developing countries without meat inspection
		Echinococcus granulosus (hydatid disease)	Worldwide
	Trematodes (flukes)	Intestinal, liver, lung flukes	Far East, Africa, Latin America
		Schistosomes (blood flukes)	Countries with untreated water supplies
Aschelminthes	Nematodes (roundworms)	*Ascaris* (ascariasis or hookworm)	Worldwide
		Enterobius (pinworm)	Worldwide
		Filarioidea (filariasis)	Tropics, subtropics
		Trichinella (trichinosis)	Warm to temperate zones
		Trichuris (whipworm)	Worldwide in warm, moist climates

genic or nonpathogenic intestinal parasite (Zaman 1979). Four of the most common helminthic diseases in developed countries are enterobiasis, trichinosis, ascariasis, and trichuriasis, with an estimated total of 54.2 million infections in the United States in 1972 (Warren 1974). In this chapter, the protozoan and worm infections most common in developed countries are briefly discussed.

PROTOZOA

Entamoeba histolytica

In the phylum Protozoa there is an organism responsible for a disease commonly called amoebiasis or amoebic dysentery. This disease is frequently confused with bacterial forms of dysentery. The trophozoite or mobile stage of *Entamoeba histolytica* is 8 to 30 μm long, has no rigid shape, and moves using pseudopodia (Fig. 20-1). A clear ectoplasm, dense endoplasm, and invisible nucleus are all characteristic of *E. histolytica*. The immobile stage, a cyst, measures 7 to 15 μm and contains stored glycogen. The life cycle begins when the mature cyst, the infective stage, which is ingested by a human host. The cyst is carried to the small intestine where it decysts, begins to divide, and travels into the large intestine. There the amoebae invade the mucosa by releasing proteolytic enzymes, remain in the lumen of the bowel without invasion, or again encyst.

Though other protozoa can cause intestinal disease (Fig. 20-2), *E. histolytica* is the

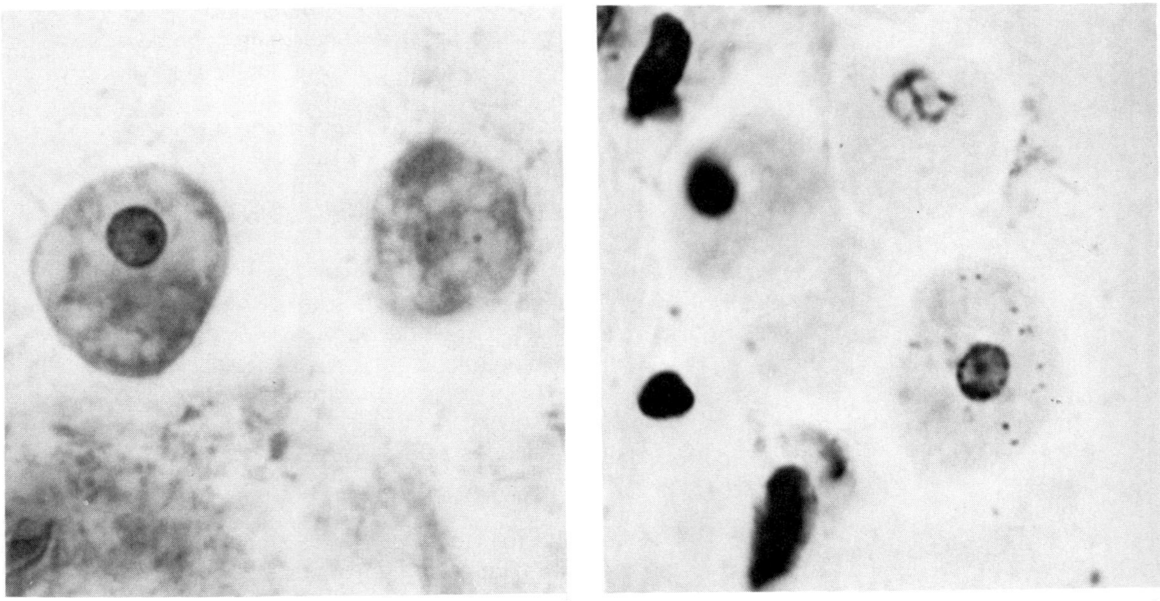

A B

FIGURE 20-1. Examples of trophozoites in tissue. A. *Entamoeba coli* B. *E. histolytica* (Courtesy of Dr. James Plorde.)

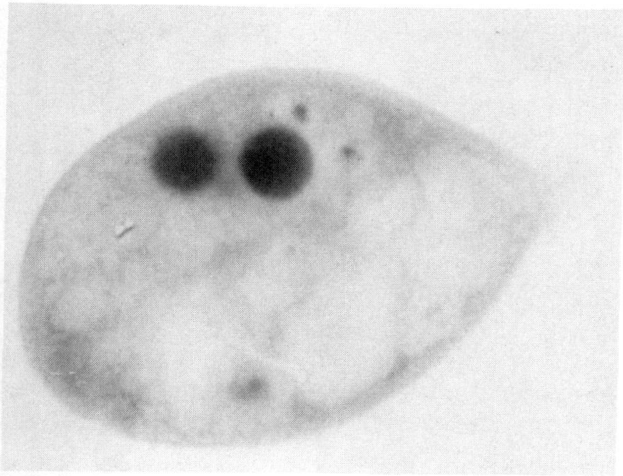

FIGURE 20-2. *Balantidium coli,* a ciliated protozoan, lives in the large intestine primarily of swine, and sometimes infects humans. It can cause diarrhea or dysentery. Infection results from ingestion of food or water contaminated by feces containing *B. coli* cysts. Tetracycline is the treatment. (Courtesy of Dr. James Plorde.)

most important. There are several thousand cases reported in the United States with about a dozen deaths. The incidence for both sexes is about equal, but young adults are more often affected. More cases were reported in the mid-Atlantic and Pacific states in 1979. In 1976, 0.6 percent of stool samples from public health laboratories in the United States were positive for *E. histolytica.* Transmission is by oral ingestion of cysts from direct person-to-person contact, or contaminated raw vegetables or water. The incubation period is 2 to 4 weeks. Some organisms can be hematogenously spread to various organs, most commonly the liver.

Most infections are asymptomatic. Loose stools, mild diarrhea, or severe dysentery, abdominal pain, cramping, flatulence, anorexia, and weakness are seen, and stools may be bloody. Symptoms of extraintestinal disease include right upper quadrant abdominal pain, liver tenderness and enlargement, and fever. Complications include bowel perforation, peritonitis, abscess, hemorrhage, anorectal fistula, and extraintestinal involvement. Diagnosis is made by identifying trophozoites or cysts in diarrheal stools, though endoscopy is sometimes necessary if stool samples obtained over several days are negative. An indirect hemagglutination test is useful in diagnosing 80 to 90 percent of those with active intestinal disease but only 10 percent of the asymptomatic cyst carriers.

There is no single drug that is completely effective in the treatment of amoebic dysentery. An amoebicide, metronidazole (Flagyl), is often used in combination with a broad-spectrum antibiotic. Since amoebae are symbiotic with normal intestinal flora, reducing numbers of enteric bacteria aids in the removal of amoebae. Metronidazole is much more effective in the treatment of amoebic liver abscess than of intestinal disease. The treatment course should be about 10 days. Several precautions should be mentioned about metronidazole: it has been shown to be carcinogenic in animals, it is not to be used during pregnancy, and if ingested simultaneously with alcohol, it can result in abdominal cramps, and nausea and vomiting. The Parasitic Disease Drug Service at CDC can advise on the most effective drugs available for treating amoebiasis.

As with all diarrheal diseases, it is important to monitor electrolytes, promote adequate nutrition and rest, and provide low-residue, bland foods. Follow-up stool examinations are needed to evaluate effectiveness of therapy. Recent contacts should also be examined. Control and prevention can be enhanced by sanitation, purified water supply, handwashing, and enteric isolation of cases.

Free-Living Soil Amoebae

Since 1965, infections due to free-living soil amoebae have been recognized in humans. To date, *Acanthamoeba* sp. and *Naegleria fowleri* have been reported as producing granulomatous amoebic encephalitis (GAE) and acute primary amoebic meningoencephalitis (PAM), respectively. Although no more than 150 cases of free-living soil amoeba infections have been reported worldwide, these infections may achieve greater prevalence and importance in the future due to increased use of immunosuppressive drugs and the ever increasing human population.

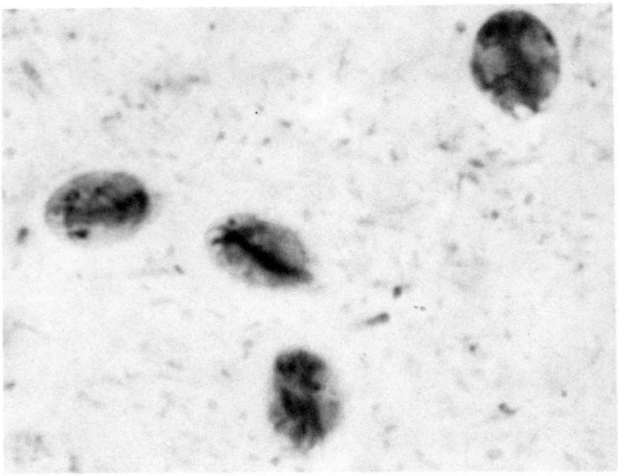

FIGURE 20-3. *Giardia lamblia* trophozoites. (Courtesy of Dr. James Plorde.)

Giardia lamblia

Giardia lamblia, a flagellated protozoan responsible for giardiasis, is from the same class as *Trichomonas vaginalis.* Functionally it has two stages, a trophozoite and a cyst. The trophozoite is uniquely shaped like a diagonally-cut pear. Two nuclei look like eyes and four pairs of flagella trail like long hairs (Fig. 20-3). The cysts, measuring 8 to 12 μm, are the infective form. Following ingestion, *G. lamblia* excysts in the duodenum. In the small intestine, a sucking disc enables the trophozoite to attach to epithelial cells. There the organisms live and multiply by binary fission. Later the trophozoite again encysts in the large intestine and the cysts are eventually passed in the feces.

There are about 5000 to 6000 cases annually reported in the United States; it is estimated that about 2 to 20 percent of individuals worldwide have been infected. Infants and children appear to be more susceptible. Lower socioeconomic status, poor hygiene, and immunodeficiency appear to correlate with increased frequency of infection. Transmission occurs primarily when water containing *G. lamblia* cysts is ingested, but may also be sexually transmitted. Cysts are viable up to 2 months in cold climates. The disease is common along hikers who drink stream water that has been infected by beavers and, as with other fecal-oral diseases, the disease is spread in day care centers and institutions.

General nonspecific intestinal symptoms can occur, but most infections are asymptomatic. In rare cases, biliary disease can result in adults, or steatorrheal diarrhea (loose stools containing large amounts of fat) in children. Epigastric pain and cramping, weight loss, and dehydration have been reported. Severe infection can result in malabsorption syndrome. Diagnosis is made by identification of trophozoites or cysts in the diarrheal stool. Serologic and skin tests are being developed.

Treatment is with one of three medications: quinacrine, metronidazole, or furazolidone. Precautions to prevent infection include cooking all food, boiling or iodinizing drinking water, and treating all pets with the disease. Enteric isolation is necessary for the duration of the illness.

Plasmodium

Four species of plasmodia cause human malaria: *Plasmodium vivax, falciparum* (these two causing about 85 percent of human malaria), *malariae,* and *ovale.* The plasmodia have a complex life cycle in the tissue and blood cells of vertebrates and in certain species (usually *Anopheles*) of female mosquitos. They are rarely transmitted by inoculation or transfusion of infected blood.

The saliva of an infected mosquito contains the infectious form of the parasite called *sporozoites,* which are carried via the blood stream to the liver of the person who has been bitten. The sporozoites then go through several asexual development stages called *schizonts* and *merozoites.* When the infected liver cells rupture, thousands of merozoites are released into the blood stream where they infect red blood cells. In the blood cells the merozoites become *trophozoites* and then, again, schizonts that in turn form new merozoites. This cycle may recur with continued infection of more red blood cells.

A second cycle in red blood cells can also occur in which male and female *gametocytes* are formed. Ingestion of these gametocytes by the appropriate species of female mosquito initiates the infectious cycle in the insect. In the mosquito stomach a *zygote* is formed when the male and female gametocytes mate. The zygote develops into an *ookinete* and then an *oocyst.* In the oocyst, new sporozoites are produced. Within 7 to 10 days sporozoites are present in the insect's salivary gland—from where infection of the next human host can occur (Fig. 20-4).

Malaria occurs primarily in Africa, Asia, and other tropical and semitropical areas of

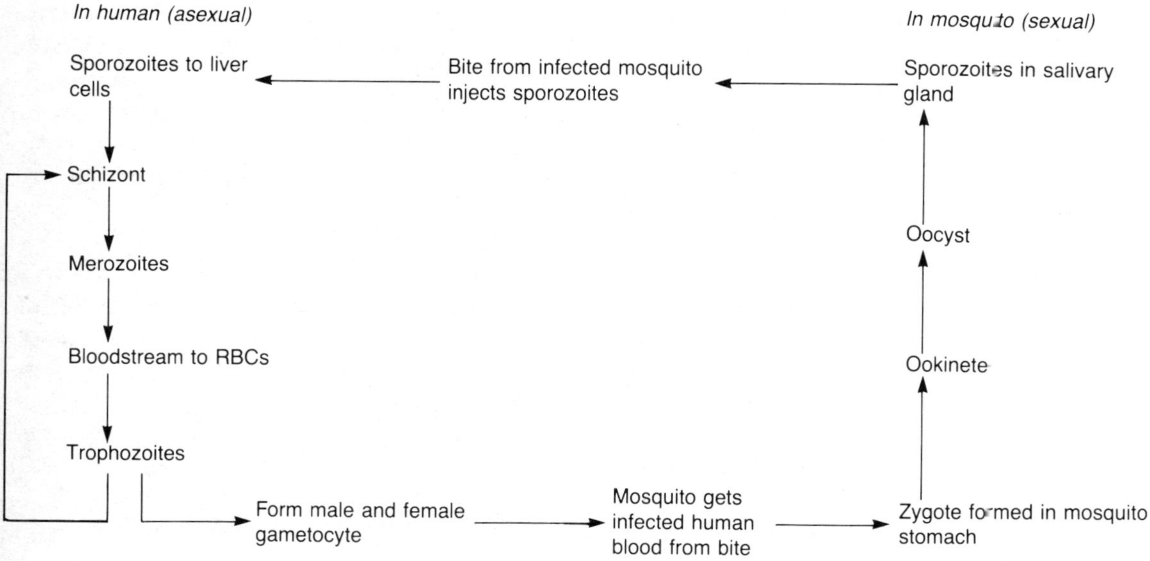

FIGURE 20-4. Life cycle of plasmodia.

the world. About 300 million people per year are infected, though the disease has been effectively controlled in the United States and other developed countries by elimination of the mosquito vectors. The most distinguishing clinical characteristic of malaria is fever that occurs paroxysmally in three distinct stages (Table 20-4) caused by the liberation of a new "batch" of merozoites into the bloodstream. These paroxysms occur every 48 to 72 hours for 3 to 12 weeks or even up to several years in untreated individuals.

Malaria caused by each of the four species of *Plasmodium* varies somewhat in clinical manifestations. Falciparum malaria is the most serious and can cause death. The other malarial forms are generally milder. In endemic regions, chloroquine and primaquine are used as prophylaxis. There is no vaccine, but infection results in the development of antibodies to the species causing the disease. Chloroquine is also used to treat cases, though some falciparum strains are resistant. Alternate treatment is with quinine, pyrimethamine, and sulfadiazine.

Toxoplasma gondii

Toxoplasma gondii is an intestinal protozoan parasite whose definitive hosts are members of the cat family. Humans, birds, or small mammals (such as mice) serve as intermediate hosts. Cats acquire infection congenitally or, more commonly, by ingestion of one of three infective stages of the organism, especially the oocysts. Millions of infectious oocysts are shed by infected domestic cats and in this way humans become infected. Oocysts can be present on food contaminated with cat feces or on uncooked meat. No oocysts are produced in humans. About 0.1 percent of human cases are acquired congenitally, and this only occurs if the mother becomes infected during pregnancy.

Though only about 300 to 400 human cases are reported in the United States annually, it is suspected that toxoplasmosis is one of the most prevalent pathologic protozoan diseases, with estimations that 25 to 40 percent of the population has been infected (Neva 1974). Toxoplasmosis has the highest mortality of any protozoan disease in developed countries. Among the immunosuppressed, those with Hodgkin's disease seem to be especially vulnerable. Congenital toxoplasmosis is estimated to occur in one or two infants per 1000 pregnancies.

The organism multiplies in the intestine where necrosis can occur. Lymphatic or hematogenous spread results in damage to other organs. *T. gondii* has a propensity for the retina, but cysts are also found in the central nervous system and in cardiac and skeletal muscle. Infection is not accompanied by an inflammatory response. Most human infections are subclinical but a wide range of clinical manifestations can occur,

TABLE 20-4. Stages of Malarial Paroxysms

Stage	*Duration*	*Symptoms*
Cold	30-60 minutes	Shaking chills, feeling of being cold
Hot	1-8 hours	High (up to 41°C) fever, headache, muscle and joint pain
Sweating	1-2 hours	Profuse sweating, exhaustion

most commonly retinochoroiditis and lymphadenitis (Table 20-5). Twelve percent of babies born alive with toxoplasmosis suffer from such problems as jaundice, chorioretinitis, hydrocephalus, microcephaly, mental retardation, or blindness (Johnston 1971).

Diagnosis is made by histologic confirmation of the tachyzoite (the actively growing stage) in tissue or by use of a serologic test called the Sabin-Feldman dye test. Pyrimethamine and sulfonamides act synergistically and are the treatment of choice. For the ocular effects, corticosteroids are prescribed. Pyrimethamine causes bone marrow depression and must therefore be given in conjunction with folic acid. Proper cooking of meat and good handwashing can help prevent the spread of toxoplasmosis. Infected cats need to be treated and their litter boxes sterilized daily. Pregnant women should avoid infected cats (Markell and Voge 1976).

Trichomonas vaginalis

Trichomoniasis is caused by a flagellated protozoan (Fig. 20-5). In the trophozoite stage, *Trichomonas vaginalis* has a characteristic gliding motion. Its body measures 15 μm and has a short undulating membrane. The posterior end of the organism attaches to tissues, causing inflammation and erosion.

TABLE 20-5. Most Common Clinical Manifestations of Toxoplasmosis

Congenital	*Postnatal*
Often asymptomatic	Often asymptomatic
Retinochoroiditis	Lymphadenitis
Stillbirth	Fever
Intracranial calcifications	Malaise
	Fatigue
Convulsions	Muscle pains
Hydrocephalus	Encephalitis (in immunosuppressed)

An estimated 10 to 25 percent of individuals in the United States are infected with *T. vaginalis*. Both sexes, but primarily females, are symptomatic, and the infection is more common in women with lowered vaginal acidity. Since cysts are not formed, person-to-person transmission must be direct, by contact with infected vaginal or urethral secretions during sex or birth, and possibly through contaminated personal articles. The incubation period is 4 to 20 days, and *T. vaginalis* is communicable for the duration of the infection.

Though the disease is nonlethal, it causes considerable discomfort in females. Symptoms include a creamy white, frothy vaginal discharge that is foul smelling, burning, and pruritus. Inflammation of the vulva and cervix may occur but the bladder is not affected. Males are usually asymptomatic but may present with urethritis or prostatitis. Diagnosis is by identification of the trophozoite in secretions from vagina, urethra, prostate, or urine. Drug therapy with metronidazole for 10 days is the recommended treatment. All sexual partners should also be treated and infected individuals instructed to abstain from sex or use condoms until adequately treated.

Other Protozoa

Other protozoa that cause disease primarily in developing countries include the trypanosomes, which cause sleeping sickness, and *Leishmania*, which has a visceral form called kala-azar—as well as a cutaneous form.

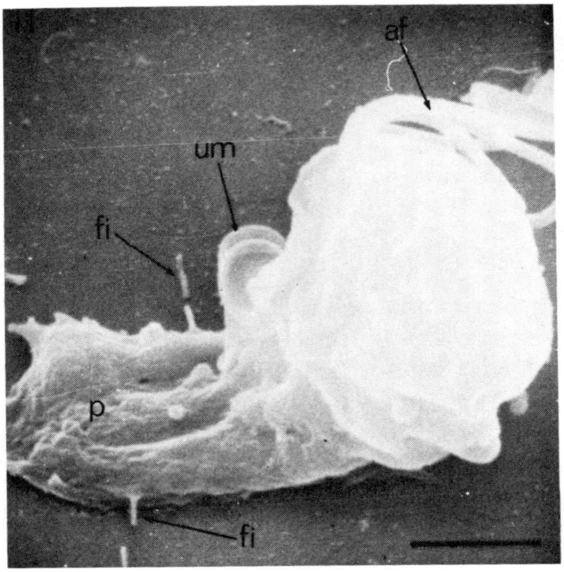

a

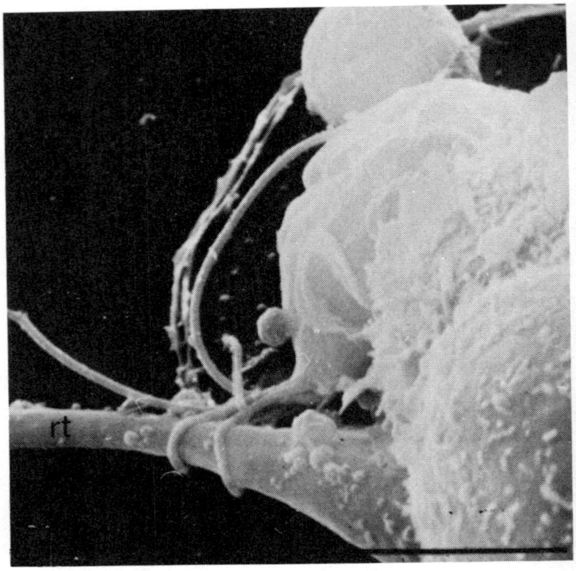

b

FIGURE 20-5.
(a) Scanning electron micrograph of an amoeboid *T. vaginalis* adhering to a glass coverslip in the center of a lesion. Note the broad pseudopodium (p) from which filopodia (fi) (which have fractured during preparation) extend down on to the glass surface. The anterior flagella (af) and part of the undulating membrane (um) emerge from the main body mass which projects up into the medium. Scale bar = 2 mm.

(b) T. vaginalis adhering to the side of a rounded epithelial cell. The trichomonad has looped its flagella around a fibre which remained behind as the peripheral parts of the epithelial cell retracted. Scale = 2.5 mm

HELMINTHS (PARASITIC WORMS)

Cestodes (Tapeworms)

There are four major cestodes that infect humans: *Diphyllobothrium latum* (fish tapeworm), *Taenia saginata* (beef tapeworm; Fig. 20-6), *Taenia solium* (pork tapeworm), and *Echinococcus granulosus* (Table 20-3). The adult tapeworms live in the small intestine of their hosts. They have a head, called a *scolex,* which is used to attach themselves to the wall of the intestine. The head is followed by 3 to 3000 body segments, *proglottids,* containing the male and female reproductive organs (tapeworms are hermaphrodites). The oldest body segments are at the proximal end of the worm. Because they have no digestive system, tapeworms obtain their nutrition through their segmental wall. They range from 60 mm to 10 meters in length.

Cestodes have complex life cycles requiring both an intermediary and a definitive host. Individuals are infected by eating inadequately cooked meat or fish containing

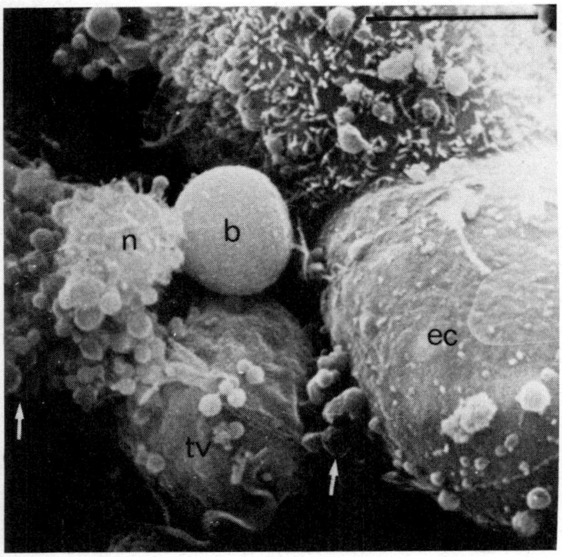

c

(c) Lysis of epithelial cells by *T. vaginalis.* A trichomonad (tv) lies beneath a collection of cytoplasmic constituents, including small vesicles (arrow), the nucleus (n), and bodies (b) of uncertain nature, released by the rupture of the plasma membrane of an epithelial cell. On the right two other epithelial cells are showing signs of injury; both are rounded and the lower cell has lost most of its microvilli. Scale bar = 2.5 Mm. (Reproduced with permission from Heath JP, et al. Behavior and pathogenicity of *Trichomonas vaginalis* in epithelial cell cultures. Br J Vener Dis 1981;57:109-117).

the tapeworm larvae. Fish, beef, and pork tapeworms all remain localized in the intestine where infection may be asymptomatic, or various gastrointestinal symptoms may be present. Anemia is most prevalent with fish tapeworm. Niclosamide is an effective and safe treatment.

E. granulosus (Fig. 20-7) is sometimes called the *hydatid worm* because it results in the formation of hydatid cysts in various organs, primarily the lungs and liver. It is transmitted to humans by contact with feces of domestic animals (usually dogs and sheep) that contain the worm eggs. Generally, the cysts cause no symptoms and take several decades to grow to a diameter of 15 to 20 cm. Symptoms occur if pressure is created on surrounding organs or if the cysts rupture and cause a toxic or anaphylactic reaction. Human infections are controlled by preventing dogs from eating infected carcasses and by personal hygiene to prevent transfer of eggs from dog feces to humans. The major treatment is surgical excision.

FIGURE 20-6. *Taenia saginata,* complete worm. (Courtesy of D.R. Seaton, Department of Tropical Medicine, Liverpool University.)

Nematodes (Roundworms)

Ascaris lumbricoides

Measuring in the female from 20-40 cm by 0.3-0.6 cm, *Ascaris lumbricoides,* these creamy white worms, are one of the larger

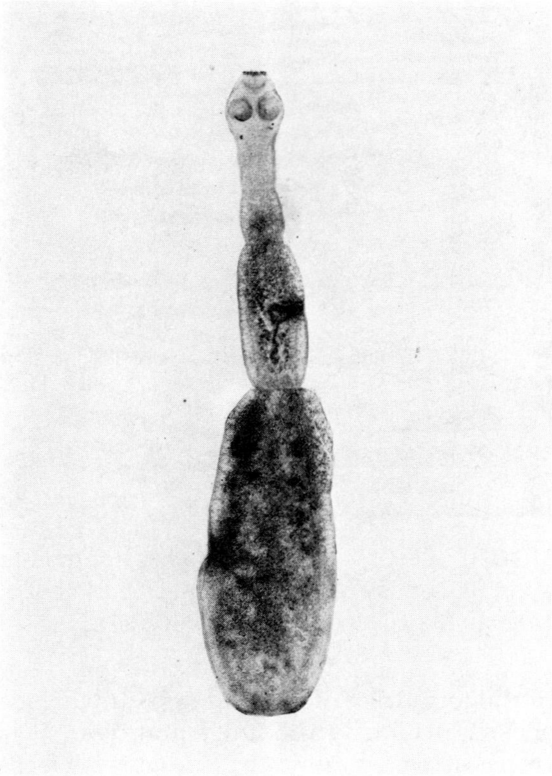

FIGURE 20-7. *Echinococcus granulosus*, complete worm (length 0.5 cm). (Courtesy of D.R. Seaton, Department of Tropical Medicine, Liverpool University. Reprinted with permission from Donaldson RJ: Parasites and Western Man. Baltimore, University Park Press, 1979.)

of the parasitic roundworms. The male measures 15 to 30 cm and has a curved tail and two copulatory spicules. In the infective stage the embryonated egg is ingested and later hatched in the intestine. Circulation carries the organism to the lungs where it migrates up the trachea to the pharynx. There it is swallowed and matures in the small intestine. Eggs are passed in feces but are not infective until they evolve into a second stage.

A. lumbricoides were found in 2.3 percent of stool samples in the United States public laboratories in 1976. They were especially prevalent in the southeastern states. Abundant in warm moist soil, *A. lumbricoides* contaminates vegetables and other food. During their migration in the respiratory tract, the worms can cause an allergic response as well as a block of the airway. Bile duct involvement also occurs. Death occasionally is caused by severe pulmonary invasions or obstruction of vital organs. Most cases are asymptomatic or mild, with the major danger of migration occurring when the worms become restless following ingestion of spicy food, medications, or after abdominal surgery. Pneumonitis can produce cough, wheezing, and dyspnea. Biliary involvement causes sudden colicky upper right quadrant pain with vomiting and jaundice. Abnormal liver tests are evident. Pharmacologic therapy includes piperazine or thiabendazole.

Enterobius vermicularis

Also called a threadworm or pinworm, *Enterobius vermicularis* is prevalent worldwide. There are about 45 million cases of infection in the United States, and 290 million throughout the world. The female measures 12 mm by 0.3 mm and is white with a long pointed tail. Its body appears distended by the gravid uterus. The male is smaller with a curved tail. The infective stage is the embryonated egg that—after entrance into the body—hatches in the duodenum, and penetrates and develops in the mucosa. The female exits the anus usually at night and deposits eggs in perianal and perineal areas. Eggs become infective after 6 hours.

Occurring in about 2 percent of the total population, *E. vermicularis* is considered to be the most prevalent intestinal parasite in the country, especially in young school age children. Transmission is by ingestion or inhalation of eggs, which are light enough to float in the air. Eggs are viable up to 2 days. Neither the adult worm nor ova are commonly found in the stool. Autoinfection through the rectum can also occur.

When the worms attach to the intestinal mucosa, mild inflammation without penetration occurs. Bacterial infections may result from scratching. The major symptom is anal itching, especially at night. Pruritus vulva with vaginitis also has been seen, and a low-grade eosinophilia may be present. Diagnosis is made by pressing a tongue blade with scotch tape against the anus to recover eggs.

Treatment of pinworm includes pyrantel pamoate, pyrvinium pamoate, mebendazole, or piperazine citrate in single doses that may have to be repeated in 2 weeks. The environment and clothing should be rid of eggs to prevent reinfection. All family members should be treated at the same time. Fingernails should be cleaned frequently and cut short to prevent accumulation and spread of eggs; gloves might be appropriate to discourage hand-to-mouth contact in small children. Each morning the rectal area should be washed and ointments applied. In addition snug underwear will help prevent spread of eggs. Linen should be handled with excretion precautions and air currents must be minimized.

Hookworms

Ancylostoma duodenale and *Necator americanus* are two species of hookworm that infect humans (Fig. 20-8). The larvae, which hatch from eggs deposited by infected feces in the soil, access the body by penetrating skin; hence its prevalence in the warm southeastern United States among individuals who frequently are barefoot. Local reactions are called "ground itch." If the larvae migrate, pneumonitis and/or considerable blood loss from intestinal hemorrhage can occur. The prevalence is less than 1 percent in the United States. Control depends upon good sanitation and wearing shoes or gloves when in contact with soil. The infection is susceptible to treatment with antihelminthic drugs such as mebendazole, thiabendazole, levamisole, and pyrantel pamoate.

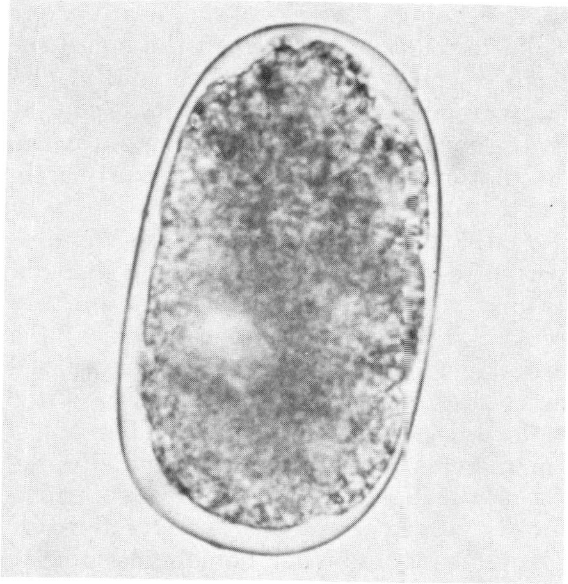

FIGURE 20-8. Hookworm egg. (Courtesy of Dr. James Plorde.)

Trichinella spiralis

Trichinella spiralis causes the disease known as trichinosis. The adult organism is cylindrical. The females are 2.8 mm by 0.06 mm and the males measure 1.1 mm by 0.03 mm. The infective stage is the encysted larvae, found in striated muscle of carnivores, especially the pig. After ingestion, there is an incubation period of 7 to 10 days. Then larvae are released in the small intestine where they mature to adults. The adults reproduce and the new larvae deposit in the mucosa. Later the larvae enter the circulation and become encysted primarily in striated muscle.

In 1979, 157 cases of trichinosis were reported to the Centers for Disease Control. Ingestion of uncooked or inadequately cooked pork results in ingestion of viable cysts. Incidence of trichinosis has decreased since the 1950s when it became prohibited to feed hogs uncooked garbage. The use of commercial feed probably also helps decrease the disease in the pig. Cases of trichinosis occur especially after home slaughtering and when homemade sausage is produced. Encysted larvae remain viable for several years.

Disease caused by *T. spiralis* imitates over 50 other diseases. Pathology in gastrointestinal disease is usually mild but in severe cases includes punctate hemorrhages, mucosal hyperemia, and edema. The more serious disease occurs when the cysts develop in extraocular, masseter, diaphragm, intercostal, and neck muscles. More *T. spiralis* are found at musculotendon junctions, however. Results are edema of fibers, shrinkage, replacement by fine granular tissue, and collapse of muscle wall. In addition, epicardial, lung, and CNS involvement occurs.

Gastrointestinal symptoms are usually mild and include malaise, mild fever, anorexia, and various GI complaints. Symptoms in the muscle phase depend on the area involved. The symptoms include severe muscular pain, dyspnea, pain with eye motion, edema of eyelids, muscular paralysis, and cardiac or pulmonary involvement. Diagnosis is by identification of larvae in muscle from biopsy. Treatment of choice is thiabendazole, though it is probably not effective against cysts. Side effects of the drug include nausea, vertigo, headache, and weakness. Bed rest and analgesics for muscle pain are indicated.

Prevention and control measures include education regarding the unhealthy effects of eating partially cooked or uncooked pork (including smoking, pickling, or spicing without cooking). Contamination of other meat products processed in the same area may occur. Garbage for hogs should be cooked. Official inspections of pork are important. The infected patient does not need to be isolated.

Trichuris trichiura

Commonly called whipworms, *Trichuris trichiura* resemble a buggy whip. The female is 40-mm long, slender in the longer esophageal region, and larger in its main body. Males are small with curved tails. The embryonated egg is the infective stage. After ingestion, larvae hatch in the intestine, penetrate, and grow in the mucosa. Adults appear in the cecum and deposit eggs, which later appear in the feces. Three to 6 weeks later in the soil or wherever deposited, these eggs mature to the embryonated infective stage. No multiplication occurs in humans, therefore more organisms are present in the intestine only with additional ingestion of eggs.

There are an estimated 2 million infections in the United States, most occur in the southeastern states. *T. trichiura* is primarily transmitted by hand-to-mouth from fomites or foods, especially raw vegetables that have been in contact with contaminated soil.

When the adult worm reaches the intestine it sucks blood (0.005 ml/day/worm) and causes small subepithelial hemorrhages, catarrhal inflammation, irritation, and edema. In severe cases, rectal prolapse may occur from straining against worm masses or from irritated nerve endings resulting in increased peristalsis. There are often no symptoms, but diarrhea, abdominal pain, vomiting, constipation, fever, flatulence, blood in the stool, and weight loss can occur. Diagnosis is confirmed by finding ova in feces. Medications used are mebendazole or thiabendazole. Washing of raw vegetables and good handwashing help prevent infections.

Other Nematodes

Other nematodes of medical importance throughout the world include *Toxocara canis* (in dogs) or *T. cati* (in cats), which can be acquired by humans by ingestion of food contaminated with feces that contain eggs. Larvae migrate to the lungs, liver, brain, kidney, and eyes, causing hemorrhage. A strong immune response results in the formation of granulomas. Most infections are asymptomatic. *Strongyloides stercoralis* occur worldwide in warm climates. Infection, once occurring, is life long. It can cause skin itching, and pulmonary and intestinal disorders if the lung or gut is penetrated. *Dracunculus medinensis,* Guinea worm, cause skin sores in Africa, Asia, and the Soviet Union. *Onchocerca volvulus,* in Africa, the Middle East, and tropical Americas, can cause skin nodules and blindness. Loa loa, or eye worm, can also result in blindness.

Filariasis is a disease caused by nematodes of the family *Filarioidea*. It occurs in tropical and subtropical climes. The parasite is spread by the bite of an infected mosquito. The filariform larvae migrate from the blood stream to the lymphatics where they mature and remain. After an incubation period of 3 to 12 months, the adult worms cause swelling and blockage of the lymph nodes and vessels with accompanying pain, fever, and redness. The inflammation in rare cases leads to elephantiasis. Diagnosis is usually based on symptoms and diethylcarbamazine is used as treatment. Mosquito eradication is the most effective control method.

Trematodes (Flukes)

Trematodes of medical importance include the blood flukes (schistosomes), liver and intestinal flukes *(Fasciola hepatica)* in temperate climates, *Clororchis sinensis* (also called the Oriental liver fluke), and others from the genera *Opisthorchis* and *Dicrocoelium*), and lung flukes *(Paragonimus westermanii)*. Other schistosomes present in natural waters in North and South America can cause dermatitis called *swimmer's itch.*

All trematodes have a similar life cycle with certain water snails as the intermediate hosts in which asexual reproduction takes place (Fig. 20-9). The fluke eggs are excreted in the urine, feces, or sputum of an infected human host. In the water the eggs hatch into larvae, called *miracidia,* which then penetrate the snail. Another developmental stage in the snail produces *sporocysts* and eventually *cercariae* (free-swimming infective larvae) are released from the snail into the water. These cercariae penetrate the skin of humans in the water, migrate to the liver and mature. In the case of nonschistosomal flukes, cercariae enter fish or shellfish; human infection occurs when these contaminated fish, or water plants contaminated with fluke larvae, are eaten. The eggs secrete an enzyme that can cause damage to host tissue and the formation of abscesses. Hence, the disease is more severe if more flukes infect the patient and additional eggs are produced.

Schistosomes (Blood Flukes)

About 200 million people in temperate and tropical zones are infected with one of the blood flukes, or schistosomes. Schistosomiasis is estimated to be second to malaria in

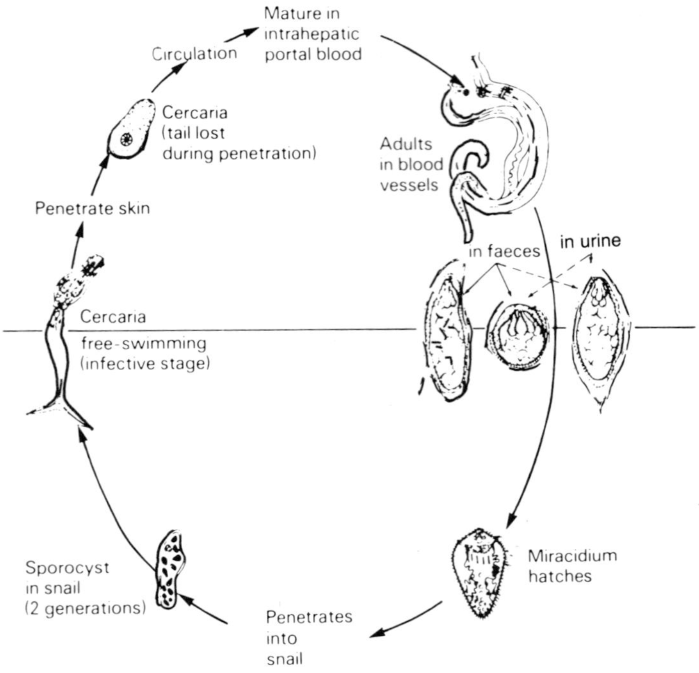

FIGURE 20-9. Life cycle of the schistosomes (blood flukes). (Reprinted with permission from Zaman V. Atlas of Medical Parasitology. Balgowlah, Australia: ADIS Press Australasia Pty Ltd, 1979).

worldwide public health importance. *S. japonicum* occurs in the Far East, *S. haematobium* in Africa and the Mediterranean, *S. mansonii* in Arabia and South America, and *S. mekongi* in Cambodia. The disease is common among rice growers or others who contact contaminated waters.

After maturing in the liver, the adult schistosomes migrate through the abdominal and pelvic circulation, depositing eggs (Fig. 20-10). The eggs can then be found in the bladder, the intestines, the liver. Foreign antigens in the flukes cause urticarial fever with headache, rash, fever, and chills, and sometimes death. Subsequent spread of the eggs causes necrosis, formation of polyps or granulomas, and liver cirrhosis. Diagnosis is made by positive complement-fixation or intradermal skin tests, and by finding eggs in urine or feces. Antimony compounds are an effective treatment.

Other Flukes

The liver fluke, *Fasciola hepatica,* is worldwide in distribution and causes disease of the liver and gall bladder. *Paragonimus westermanii,* human lung fluke, occurs mostly in the Far East, as does *Opisthorchis sinensis,* the Chinese liver fluke. These range in size from 2 to 30 mm. The types of

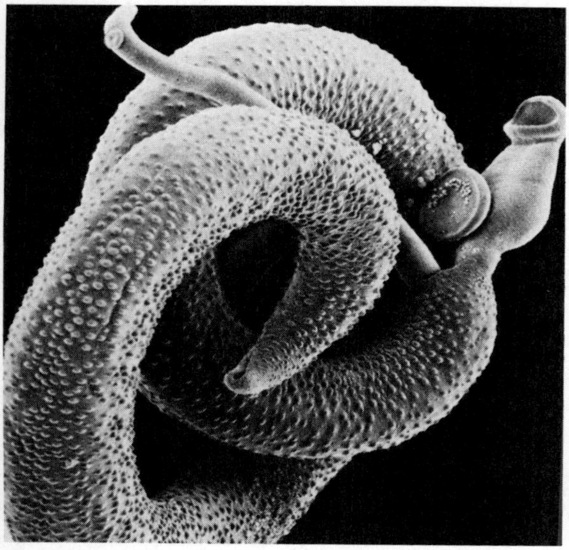

FIGURE 20-10. *Schistosoma mansonii* worms in copulation. The male has distinct tubercles on its cuticle, differentiating it from other species of schistosomes. The female appears as a thin tube-like structure emerging from the gynecophoric canal. (Scanning electron micrograph courtesy of Dr. M.M. Wong. Reprinted with permission from Zaman V: Atlas of Medical Parasitology. Balgowlah, Australia: ADIS Press Australasia Pty Ltd, 1979.)

symptoms occurs in humans depend upon the location of the organisms in the body: they can affect any system or organ, including the brain, heart, gastrointestinal tract, lungs, and liver.

Diagnosis is made by obtaining trematode eggs in stool or sputum. The eggs of lung flukes are present in the sputum about 3 months after ingestion of contaminated raw shellfish. The nonschistosomal eggs differ from those of blood flukes in that they are less mature when excreted (the schistosome eggs leave the human host as fully developed larvae or miracidia), they lack a spine (schistosome eggs possess one hook or spine), and each egg has a distinct lid, called an *operculum,* from which the larva hatches in the water. Treatment is with hexylresorcinol or tetrachloroethylene (for intestinal flukes), hetol or dehydroemetine (for liver flukes), or bithionol (for lung flukes).

CLINICAL IMPLICATIONS

Though parasitic disease in developed countries is, to a great extent, controlled by a high level of sanitation, the worldwide human toll taken by parasites is devastating. Clinicians who plan to practice in other countries should familiarize themselves with the endemic local parasites.

Successful control programs have two major components: (1) elimination of intermediate hosts such as snails or mosquitos, and (2) education of people in basic hygienic practices. Health care practitioners can play a very significant role in the education of individuals to prevent infection or reinfection with parasites. Persons living in areas where hookworm is prevalent should be taught to wear shoes, and gloves when digging in the dirt or handling excreta. The spread of pinworms and roundworms can be decreased with adequate handwashing and cleaning of fingernails. Parents must teach their children these practices, especially after touching the perineal area. Raw food such as vegetables and fruits in endemic areas should be washed before eating.

Trematode infection can be prevented by cooking crabs, crayfish, carp, and water plants if they are potentially contaminated. In addition, human excreta should be treated to destroy fluke eggs before it is used as fertilizer. The proper disposal of sewage is essential to the prevention of such parasitic diseases. To prevent malaria in endemic areas one must minimize the risk of mosquito bite by use of protective clothing, insection repellent, and netting.

There is need for the continued development of less toxic and more efficacious drugs to treat parasitic diseases (Tables 20-6 and

TABLE 20-6. Current Status of Parasitic Infections, Their Treatment, and Prevention*

Parasitic disease	Worldwide prevalence	Available drug(s)	Comments
Protozoan infections			
Amebiasis	10%	Metronidazole	Toxic, carcinogen (mutagen); inadequate efficacy (in dysentery, cyst passers, hepatic abscess).
Giardiasis	7%	Atabrine	Toxic, 10 days prescription; repeat prescription often needed.
Leishmaniasis	12 million	Toxic antimony compounds	Available only from Centers for Disease Control.
Malaria	300 million	Chloroquine, primaquine	Drug resistance; toxic.
Toxoplasmosis	30%	Pyrimethamine plus sulfadiazine	6 weeks prescription; unsatisfactory.
Trypanosomiasis, African	+ +	Melarsoprol	Very toxic.
American (Chagas')	+	Lampit® (a nitrofurantoin)	120 days prescription, toxic.
Helminthic infections			
Ascariasis	1 billion	Mebendazole	Effective, short course, relatively nontoxic, but rapid reinfection. Prescription 4 times annually for control.
Trichuriasis	800 million		
Hookworm	600 million		
Pinworm (enterobiasis)	>30 million in United States		
Toxocariasis (visceral larva migrans)	+	None	
Filariasis (Bancroft's, malayan), *Loa loa*	300 million	Diethylcarbamazine	Inadequate, allergic or Herxheimer-like reactions.
Onchocerciasis (river blindness)	+ + (West Africa)	Suramin	Very toxic; unsatisfactory.
Fasciola hepatica	Rare in humans	Emetine or chloroquine	Toxic, prolonged Rx.

TABLE 20-6. *(continued)*

Parasitic disease	*Worldwide prevalence*	*Available drug(s)*	*Comments*
Helminthic inflections			
Clonorchis (Chinese liver fluke)	Millions (Asia)	None	
Fasciolopsis (intestinal fluke)	Millions (Southeast Asia)	Hexylresorcinol	Toxic (not available in United States).
Paragonimus (lung fluke)	Millions (Asia)	Bithionol	Toxic, every 2 days × 10.
Schistosomiasis	200 million	Niridazole, Metriphonate	? Mutagenic and carcinogenic.
Schistosoma japonicum	30 million	Antimony compounds	Inadequate.
Strongyloidiasis	+	Thiabendazole	Difficult in the immunosuppressed.
Tapeworms (most)	+	Niclosamide	Effective.
Echinococcus (hydatid cyst)	+	Mebendazole	Needs confirmation.
Trichinosis	4% in United States	None	

*Courtesy of Kenneth S. Warren, M.D., Rockefeller Foundation: Personal communication, 17 March 1978.
Reprinted with permission of the publisher from Finland M. New developments in antibiotics. Ann Intern Med 1978;89(Part 2):849-53.

20-7). Many drugs cause serious side effects. For example, piperazine has caused central nervous system symptoms such as ataxia, numbness, muscle weakness or spasticity, and nystagmus, especially in children. Since this drug is used to treat pinworms and roundworms, parents should be warned to discontinue treatment if such symptoms occur. Quinacrine (atabrine), used to treat *Giardia,* tapeworms, and malaria, causes reversible yellowish discoloration of the skin and urine. Several drugs such as quinacrine and metronidazole (Flagyl) may be teratogenic and should not be given during pregnancy.

Proper techniques to obtain specimens for laboratory examination for parasites are outlined in Table 21-3d.

SUGGESTED READINGS

Baron S, ed. Medical microbiology. Menlo Park, Ca: Addison-Wesley, 1982.

Beck, WJ, Davies JE. Medical parasitology, 2nd edition. St. Louis: CV Mosby, 1976.

Brunner LS, Suddarth, DS. Textbook of medical-surgical nursing, 4th edition. Philadelphia: JE Lippincott, 1980.

Centers for Disease Control. Surveillance summary: intestinal parasite surveillance—United States, 1976. Morb Mort Week Rep May 19, 1978.

Centers for Disease Control. Surveillance summary, trichinosis—United States, 1978. Morb Mort Week Rep November 16, 1979, 541-543.

Centers for Disease Control. 1979 Annual MMWR Summary. Morb Mort Week Rep 1979.

Fozard G. Major helminthic diseases of North

TABLE 20-7. Major Antiparasitic Drugs

Drug	*Uses*	*Adult dosage*	*Precuations*
Metronidazole (Flagyl)	Giardia, amoebiasis	Oral: 250 mg, 3×/day for 7 days	Avoid in pregnancy; may be teratogenic
Piperazine citrate	Pinworm, roundworm	Oral for pinworms: 65 mg/kg/day × 7 days Roundworms: 3.5 g × 2 days	Side effects minimal; do not use with epileptics; CNS symptoms in some children include ataxia, muscle spasticity or weakness, nystagmus
Pyrantel pamoate		Oral suspension: 11 mg/kg × 1 dose	Avoid use in patient with liver disease, children under 2 years. Side effects minimal: nausea, vomiting, headache, diarrhea
Pyrvinium pamoate	Pinworms	Oral: 5 mg/kg × 1 dose	Red dye that stains clothing and feces; avoid use in patients with inflammatory bowel disease. Nausea, vomiting, diarrhea occur
Quinacrine HCl (Atabrine)	*Giardia*, tapeworm, malaria	Oral for *Giardia:* 100 mg 3×/day for 5-7 days	Use with caution in patients with psoriasis, alcoholism, liver or kidney disease; causes reversible yellowish decoloration of skin and urine; may be teratogenic
Thiabendazole	Hookworm, pinworm, roundworm, threadworm	Oral: 25 mg-1.5 g/kg (maximum of 3 gm)	Nausea, dizziness, headache, drowsiness; avoid use in patients with liver, kidney disease and children under 2 years

America: a review. Journal of Family Practice 1978;6:1195-1203.

Gilles HM. Protozoa. In: Donaldson RJ, ed. Parasites and western man. Baltimore: University Press, 1979.

Imperato PJ, et al. Parasitic infections in New York City: incidence and epidemiology. NY State Med J 1977;77:50-58.

Johnston TS. Diagnosis and treatment of five parasites. Drug Intelligence and Clinical Pharmacy 1981;15:103-110.

Keifer A. Waterborne giardiasis—California, Colorado, Oregon, Pennsylvania. Morb Mort Week Rep March 21, 1980;121-123.

Markell EK, Voge M. Medical parasitology, 4th edition. Philadelphia: WB Saunders, 1976.

Meyer EA, Jarroll EL. Giardiasis. Am J Epidemiol 1980;111:1-12.

Muller R. Nematodes. In: Donaldson RJ, ed. Parasites of western man. Baltimore: University Press, 1979.

Neva FA. Protozoal diseases endemic in the United States. Am J Trop Med Hyg 1974; 23:731-735.

Noble ER, Noble ER. Parasitology, 5th edition. Philadephia: Lea and Febiger, 1982.
Schmidt GD, Roberts LS. Foundations of parasitology, 2nd edition. St. Louis: CV Mosby, 1981.
Schultz MG. The surveillance of parasitic diseases in the United States. Am J Trop Med Hyg 1974;23:744-750.
Simons R. Amebiasis associated with colonic irrigation—Colorado. Morb Mort Week Rep March 13, 1981;30:101-102.
Tan JS. Common and uncommon parasitic infections in the United States. Med Clin North Am 1978:62:1059-1081.
Warren KS. Helminthic diseases endemic in the United States. Am J Trop Med Hyg 1974; 23:723-729.
Zaman V. Atlas of medical parasitiology. Philadelphia: Lea and Febiger, 1979.

III CLINICAL CONSIDERATIONS

21 DIAGNOSTIC LABORATORY METHODS: BACTERIOLOGY

KEY POINTS

- The correct diagnosis and identification of a microorganism is dependent upon appropriate specimen collection.
- The Gram stain is the most widely used staining technique for visualizing bacteria. Other important stains include acid-fast stains for mycobacteria, Giminez stain for *Rickettsia,* the Giemsa and Wright stains for parasites and intracellular microorganisms, and fluorescent antibody (FA) for identification of a variety of different organisms.
- Cultivation is the primary means for identifying microorganisms. Differential and selective media are used to isolate different organisms.
- Organisms can be directly visualized with wet mounts and by microscopy. Types of microscopy include routine light microscopy, phase and interference contrast microscopy, darkfield microscopy, and electron microscopy.
- Antimicrobial susceptibility testing is an important therapeutic and epidemiologic tool. The most commonly used technique is the disc diffusion method.
- Serologic tests are used to demonstrate antibodies in the blood, which is an indication of past or present infection. These tests include radioimmunoassay (RIA), enzyme-linked immunosorbent assay (ELISA), indirect fluorescent antibody (IFA), complement-fixation (CF), precipitin, agglutination, antistreptolysin O (ASO), and flocculation tests.

With his simple monocular microscope, Anthony van Leeuwenhook saw and marvelled at the abundance and variety of microorganisms ("animacules") in the environment and on and in humans. Since his time, more powerful microscopes have demonstrated that myriads of microbes inhabit the external—and some internal—body surfaces of humans and animals without causing disease. These organisms appear to be at equilibrium with the host and their microenvironment—they are referred to as normal flora. The composition of the normal flora is continuously changing due to the activities of the host in the environment. Host and microbial factors interact in determining whether an organism establishes itself as part of the normal resident or transient flora, or as a pathogen.

The discovery of microorganisms as causative agents of disease is one of the most interesting chapters in the history of medicine and microbiology. Semmelweis, Lister, Pasteur, and Koch laid the foundation on which rests the science of medical microbiology and the study of infectious diseases. The relationship between microorganism and host is termed *mutualism* if the association is beneficial to both, and *commensalism* if the microorganism benefits but does not harm the host. *Parasitism* is an association that is beneficial to the microorganism at the expense of the host. An organism that harms the host is termed a *pathogen.* All pathogens are parasitic and any parasite is capable of pathogenicity under certain circumstances.

Whether an infectious disease will occur once a pathogen enters the host tissue is determined by the host defense mechanisms. A simplistic division of microorganisms into those that live only in the environment, those that are medically important pathogens, and those that are nonpathogens has been made. This division is becoming increasingly unclear as medicine and chemotherapy advance. At present, clinical microbiologists have adopted a division of microorganisms into those that, when entering the host, will almost invariably cause pathology, and those that will cause disease only if there are special circumstances. Techniques are available for the health care professional to demonstrate and prove the infectious nature (microbial etiology) of disease, and these techniques are the subject of this chapter.

The most rapid, but often less definitive, means of looking for an infectious agent is by direct microscopy. Direct wet mount, darkfield, or phase microscopy may reveal characteristics such as morphology, motility, or location of a pathogen in the host tissue, which are so typical for certain organisms that their identity is assured. Many organisms, however, are similar to each other in morphology, locomotion, and habitat, making identity uncertain when direct microscopy techniques are used. Staining of a smeared preparation with differential dyes may add information that increases the probability of correct identification. Staining with antibodies tagged with a fluorescent or colored compound, cultivation, and biochemical characterization are the three most definitive means of establishing the identity of the infecting microorganism.

SPECIMEN COLLECTION

Infection affects many sites and may be due to a large variety of bacteria, either normal flora (Table 21-1) or pathogens (Table 21-2*a*), fungi (Table 21-2*b*), parasites (Table 21-2*c*), and viruses. The etiologic agent of an infectious process may often be hypothe-

(text continues on page 533)

TABLE 21-1. Normal Flora

	Mouth	*Upper respiratory tract*	*Lower respiratory tract*	*Skin*	*Conjunctiva*	*Ears*	*Vagina*	*Urethra*	*Intestine*
Bacteria									
Anaerobes									
Anaerobic streptococcus	+	+	−	+	−	−	+	−	+
Bacteroides sp.	+	+	−	−	−	−	+	+	+
Fusobacterium sp.	+	+	−	−	−	−	+	+	+
Eubacterium sp.	+	−	−	+	−	−	−	−	+
Clostridium sp.	−	−	−	−	−	−	−	−	+
Actinomyces	+	+	−	−	−	−	−	−	+
Propionibacterium sp.	−	+	−	+	−	−	+	−	+
Aerobes									
Staphylococcus sp.	+	+	−	+	+	+	+	+	+
Viridans streptococcus	+	+	−	+	+	+	+	+	+
Neisseria sp. (not *N. gonorrhoeae*)	+	+	−	−	−	+	−	−	−
Hemophilus sp.	+	+	−	−	−	−	−	−	+
Lactobacillus sp.	+	−	−	−	−	−	+	+	+
Enterobacteriaceae	+	+	−	−	−	−	+	+	+
Mycoplasma	+	+	−	−	−	−	+	+	−
Spirochetes	+	−	−	−	−	−	−	−	+
Yeasts	+	+	−	−	−	−	+	−	+

TABLE 21-2a. Selected Infectious Diseases, Causative Agent(s), and Transmission: Bacteria

Disease or infecting agent	*Site of infection/portal of entry*	*Treatment*	*Control of spread*
Actinomycosis *Actinomyces* sp.	Endogenous organism Enters damaged tissue	Penicillin	—
Anthrax *Bacillus anthracis*	Skin, respiratory tract Contact with infected animals, hair, hides, soil (spores) Systemic toxicity	Penicillin Erythromycin Chloramphenicol	Isolation of patient Vaccine Surveillance
Botulism *Clostridium botulinum*	Ingestion of preformed toxin Traumatic wound infections	Ventilatory assistance Antitoxin	Proper canning Cook canned foods Discard spoiled foods
Brucellosis *Brucella* sp.	Direct contact Dust inhalation Ingestion of unpasturized dairy products from infected animals	Tetracycline	Pasteurization of dairy products Vaccination of cattle
Diphtheria *Corynebacterium diphtheriae*	Pharynx, larynx, skin Direct contact with discharges of patient or carrier. Indirect contact with contaminated articles	Penicillin Erythromycin	Disinfection of contaminated articles Active and passive immunization Treatment of carriers
Gas gangrene *Clostridium perfringens*	Contamination of wound with dust or endogenous organism	Debridement Topical antibiotics Penicillin	—
Gastroenteritis *Escherichia coli* *Salmonella* sp. *Shigella* sp. *Campylobacter jejuni*	Fecal-oral route Animal products	Antibiotics	Careful handwashing, cohorting in nursery epidemics
Gonorrhea *Neisseria gonorrhoeae*	Vagina, urethra, pelvis Sexual contact	Procaine penicillin	Treatment of infection
Disseminated gonococcal infection	Birth	Spectinomycin Tetracycline	Silver nitrate instill in newborn's eyes
Impetigo *Streptococcus pyogenes* *Staphylococcus aureus*	Skin infection Direct contact with infected person	Penicillin Methicillin Nafcillin	Treatment of the infected person
Leprosy *Mycobacterium leprae*	Direct contact with infected person	Rifampin	Quarantine Handwashing Disinfection of families

TABLE 21-2a. *(continued)*

Disease or infecting agent	*Site of infection/portal of entry*	*Treatment*	*Control of spread*
Legionnaires' disease *Legionella pneumophila*	Respiratory tract Inhalation of contaminated air	Erythromycin	Control of air conditioning intake
Leptospirosis—Weil's disease: *Leptospira* sp.	Pneumonia Rats, mice, dogs, cattle Direct contact with contaminated urine or carcasses	Tetracycline Penicillin	Prevent contact
Listeriosis *Listeria*	Ingestion Inhalation, vagina, blood Transplacental passage Genital contact	Penicillin Ampicillin	Treatment
Meningitis *Neisseria meningitidis* Other organisms	Droplet inhalation Hematogenous spread from respiratory tract Dissemination of organisms from primary infected focus or normal flora	Penicillin Appropriate antibiotic	Prophylactic use of antibiotics
Pertussis—whooping cough: *Bordetella pertussis*	Inhalation of droplet from infected person	Adequate ventilation Erythromycin	Immunization Restriction of contacts
Pneumonia *Streptococcus pneumoniae*	Lung Droplet from infected person	Penicillin Tetracycline	Avoid contact
Mycoplasma pneumoniae	Same	Erythromycin	Same
Other organisms	Same or endogenous flora	Appropriate antibiotic	None
Plague—bubonic, pneumonic: *Yersinia pestis*	Systemic infection Direct or indirect (vector) Contact with infected animal	Streptomycin Tetracycline	Rodent control
Rat-bite fever, lymphadenitis *Streptobacillus moniliformis* *Spirillum minus*	Blood Bite of infected animal	Penicillin	Rodent control

TABLE 21-2a. *(continued)*

Disease or infecting agent	*Site of infection/portal of entry*	*Treatment*	*Control of spread*
Relapsing fever *Borrelia recurrentis*	Blood, bite of infected lice or ticks, reservoir: rodents	Tetracycline Chloramphenicol	Louse and tick control
Staphylococcal infections *Staphylococcus aureus* Food poisoning Abscesses	Gastrointestinal Ingestion of contaminated food Skin infection: Close contact hands	Methicillin Nafcillin	Adequate cooling and protection of food Handwashing Hygiene
Streptococcal infections *Streptococcus pyogenes*	Upper respiratory tract Direct contact with or droplet from infected person	Penicillin	Avoid contact
Syphilis *Treponema pallidum*	Sexual transmission Placental transfer	Penicillin	Treatment Sex education
Tetanus *Clostridium tetani*	Direct wound contamination	Debridement Muscle relaxants	Immunization
Trachoma, Lymphogranuloma venerum *Chlamydia trachomatis*	Direct contact Fomites, vectors Sexual contact	Doxycycline Sulfonamides Tetracycline	—
Tuberculosis *Mycobacterium tuberculosis*	Lung: Ingestion of infected droplets or dust, fomites Extra pulmonary Hematogenous spread	Triple therapy INH, PAS, Streptomycin, Rifampin Determine susceptibility	Treatment Isolation of shedder
Tularemia *Franciscella tularensis*	Lung, ulcers Direct inoculation or bites of infected fleas or ticks, rabbits, squirrels, cats, dogs. Inhalation, ingestion	Streptomycin Tetracycline Chloramphenicol	Control of infected animals
Typhoid fever *Salmonella typhi*	Intestinal tract Fecal-oral route Contaminated water Contaminated animals	Trimethoprim-Sulfamethoxazole Chloramphenicol	Control water supply Treatment and isolation of carriers
Typhus and spotted fever Rickettsiae	Skin, systemic Bite of infected tick, mite, body louse	Chloramphenicol Tetracycline Para-aminobenzoic acid	Control of vector

TABLE 21-2b. Selected Infectious Diseases, Causative Agent(s), and Transmission: Fungi

Disease or infecting agent	*Site of infection/portal of entry*	*Treatment*	*Control of spread*
Aspergillosis *Aspergillus* sp.	Lung, disseminates to other sites Inhalation of spores	Antifungal antibiotics	—
Blastomycosis *Ajellomyces dermatidis*	Lung, skin, bones Direct inoculation Inhalation of spores	Amphotericin B	—
Coccidioidomycosis *Coccidioidis immitis*	Lung, disseminates to other sites Inhalation of spores	Amphotericin B	—
Cryptococcosis *Cryptococcus neoformans*	Lung, disseminates to CNS Inhalation of contaminated dust, dust contaminated with pigeon droppings	Antifungal antibiotics	—
Histoplasmosis *Histoplasma capsulatum*	Lung, disseminates to other sites Inhalation of spores	Amphotericin B	—
Mucormycosis *Mucor* *Absidia* *Rhizopus*	Wound contamination, Dissemination Infected fruit, bread	Treatment of underlying disease Amphotericin B	—
Sporotrichosis *Sporothrix schenkii*	Lymphocutaneous Direct inoculation	Iodides	Protective clothing

sized from the clinical presentation and patient history. The proof, however, usually rests with the microbiologic examination. This examination is of value only if the specimen collected is appropriate for the infection suspected, and if there is sufficient quantity that is fresh and well identified in a sterile container adequate for the preservation of the specimen and organism.

Laboratory personnel must be advised as to the desired examination, the suspected pathogen, and any other relevant information such as antibiotic treatment, diagnosis and immune status of the patient. The laboratory establishes guidelines for specimen collection. It also reserves the right to refuse an improperly collected specimen or a specimen for which insufficient information is available to make an examination meaningful. Each specimen must be submitted with a written requisition. The ordering station must be noted on the requisition; otherwise, the report cannot be returned.

The following information should be present on each requisition and specimen container:

TABLE 21-2c. Selected Infectious Diseases, Causative Agent(s), and Transmissions: Parasites

Disease or infecting agent	Site of infection/route of transmission	Treatment	Control of spread
Amoebiasis *Entamoeba histolytica*	Proximal colon; disseminates Ingestion of cysts in food or water	Metronidazole Tetracycline	Chlorination Cooking Treat carriers
Giardiasis *Giardia lamblia*	Small intestine Ingestion of cysts in water	Metronidazole	Water treatment
Hookworm *Ancylostoma duodenale* *Necator americanus*	Small intestine Filariform larvae in soil burrow into skin, migrate	Adequate diet	Wearing shoes
Malaria *Plasmodium* sp.	Blood Bite of infected mosquito	Chloroquine Quinine	Prophylactic treatment in endemic areas
Tapeworms *Taenia saginata* *Taenia solium* *Diphyllobothrium latum*	Intestine Ingestion of infected beef, pork, or fish	Niclosamide	Well cooked meat and fish
Toxoplasmosis, congenital or acquired *Toxoplasma gondii*	Tissue, CNS Ingestion of cysts or congenital	Pyrimethamine Sulfonamides	Minimize contact with cat feces; Surveillance of pregnant women

Requisition	*Specimen container*
Patient's full name	Patient's name
Patient's number	Patient's number
Patient location and unit code	Date of specimen collection
Physician name and/or code	
Specific type/source of specimen	
Test requested (be specific)	
Tentative diagnosis	
Antibiotic regimen or other treatment (give dosage and time of each dose if antibiotic assay is requested)	
Date and time of specimen collection	
Suspected pathogen(s)	

Examples of methods of specimen collection are given in Tables 21-3*a-d*. These instruction should be considered as guidelines, not absolutes, since each laboratory must deal with problems peculiar to itself such as transport time, endemic pathogens, patient populations, and expectations of their physicians.

Swabs

Swabs are a convenient device for the collection of specimens from skin and mucous membranes for bacterial, viral, and fungal cultures. They are adequate for cultures of areas such as umbilicus in newborns, anterior nares, throat, or rectum. If several examinations are to be performed, separate swabs must be used since the quantity of

(text continues on page 551)

TABLE 21-3a. Specimen Collection: Routine Microscopy or Bacterial Culture

Specimen type	Patient preparation	Collection technique	Optimal specimen and volume	Collection container	Comments
Blood					
Peripheral	Skin decontamination with 70 percent alcohol followed by 2 percent tincture of iodine	Sterile venipuncture. Specimen drawn through catheter or cannulas are less satisfactory	10 ml (adults and older children), 1-2 (infants)	Sterile screw cap tube or vacutainer with SPS as anticoagulant or culture bottle for direct inoculation	Recommended number of specimens: for subacute bacterial endocarditis 2-3 within 24 hours; for fever of unknown origin, 4-6 within 48 hours. Send to laboratory immediately
Bone marrow	Same as above	Sterile percutaneous aspiration	1 ml or more	Same as above	
Body fluids or secretions (Other than blood, urine, or CSF) Joint fluid Hematomas Pericardial fluid Peritoneal fluid	Skin decontamination	Aspirate into syringe without air	1 ml or more	As above	Transport to lab without delay
Bile	Surgery	Aspiration with syringe during surgery, or from postop drainage site, or via nasogastric tube from duodenum	Several ml; first ml from post-op drain site often contains contaminants	Sterile tube or jar	
Breast milk	Skin decontamination	Gentle manual expression	Several ml	Sterile tube or jar	

TABLE 21-3a. *(continued)*

Specimen type	*Patient preparation*	*Collection technique*	*Optimal specimen and volume*	*Collection container*	*Comments*
Catheters (tips)					
Foley catheter			Not recommended		Foley catheters or tips should not be cultured except when used as sterility check
Vascular cannulae Central venous pressure lines, Umbilical or intravenous catheters	Skin decontamination	Cut off tip with sterile scissors or submit entire catheter	Tip is more relevant; may have infected clot	Sterile jar	Remove if sepsis or fever persists in spite of adequate antibiotic treatment
Central nervous system					
Brain biopsy	Surgical	Sample infected site rapidly. Transport to lab without delay	Tissue	Special anaerobic transport tube	Needs coordination with pathology and virus lab (herpes)
Cerebrospinal fluid	Skin decontamination	Sterile lumbar puncture. Ventricular or suboccipital tap	1-5 ml	Sterile, clean screw capped glass tubes	
Meningomyelocele	Same	Sterile aspiration through skin	Often only one specimen will be submitted	Same	Often contaminated or infected with skin flora
Shunt fluid	Skin and catheter decontamination	Sterile aspiration through shunt	Often only one tube	Same	Often contaminated or infected with skin flora
Ear					
Internal	Cleanse external canal with mild antiseptic	Use sterile funnel if collecting specimen from ear drum or beyond	Most often swabs; if volume allows, submit fluid	Sterile clean tube; transport media for swabs	If eardrum is perforated, specimen should be collected by physician

External	Cleanse external canal with mild detergent	Obtain specimen from active margin, preferably including fresh secretion from deeper areas	Swab, scrape, or aspirate fluid	Sterile tube. Transport media for swabs	
Eyes					
Internal	Surgical specimen	Surgical technique	Volume of specimen usually suboptimal	Sterile tube	Label carefully left and right eye. Speed in transport and care in handling are very important
External	Cleanse skin around eye with mild antiseptic. Remove ointment, etc. with sterile cotton and saline	Swabbing: pass moistened swab 2 times over lower conjunctiva. Avoid eyelid border lashes. Scrapings: use local anesthetic and platinum spatula. Rub spatula with scrapings gently over small area on slide. If too dry, use very small amount of sterile water	For most cases moistened swabs are used. For diagnosis of viral or chlamydial infections and for cytology, conjunctival or corneal scrapings are necessary	Sterile swabs in sterile tubes with a small amount of nutrient broth	Transport to laboratory immediately. Scrapings should be done by ophthamologist. Giemsa and Gram stains are most helpful
Gastrointestinal tract					
Duodonal aspirate	Through tube	Aspiration	Several ml	Sterile tube	Examine for bacterial overgrowth, for *Giardia lamblia*, and other parasites

TABLE 21-3a. *(continued)*

Specimen type	*Patient preparation*	*Collection technique*	*Optimal specimen and volume*	*Collection container*	*Comments*
Feces	None	Directly into any clean container	Several ml on 3 consecutive days	Waxed cardboard container with tight-fitting lid	Avoid contamination with urine, soap, etc. Transport to laboratory immediately
Gastric aspirate in infants	Nothing by mouth (NPO) for 10 hours before test	Through tube, to be done by physician	Any amount sufficient for smear and culture	Sterile container	During the first few days the infant's stomach fluid is not acidic
(In adults)					See under AFB and Fungal cultures
Rectal swab	If indicated: proctoscopy, sigmoidoscopy	Insert swab beyond anal sphincture (1 inch) and rotate once	On 3 consecutive days	Swab in transport media	Inferior to fresh stool
Genital tract female					
Amniotic fluid		Aspirate with syringe	Uncontaminated fluid	Sterile tube	May contain gonococcus
Cervix	Wipe cervix clean of vaginal secretion and mucus, use speculum, no disinfectant, no lubricants. Speculum may be rinsed with warm sterile water	Under direct vision, gently compress cervix with blades of speculum and use a wringing motion with swab, obtain exudate from endocervical glands	Endocervical secretion. Take 2 swabs if smears are to be made	As for wound cultures or special transport media	The use of ordinary transport medium or dry swabs as a holding medium for more than 2 hours results in loss of gonococcal viability

Cul de sac	Surgical procedure	Aspiration through posterior vaginal vault	Fluid, secretion	Sterile tube (special anaerobic transport device)	
Darkfield (for *T. pallidum*)	1-2 hour soak with sterile saline and gauze	Scrape base of lesion, express fluid, avoid bleeding. Collect secretion into capillary tubes	Aspiration into capillary tubes	Capillary tubes	Transport to lab immediately. Motility is seen only on warm material. Seal capillary tube with vaseline to protect against drying out and oxygen
Endometrium	As for cervix	If swabs are to be used, collection through a sterile tube sheath will avoid contamination	Curettings or aspiration	Sterile container. Anaerobic conditions	
Intrauterine device	Surgical	Surgical removal	Entire device plus secretion, pus	Same as above	
Lymph node	Skin decontamination	Aspiration of lymph node with syringe and needle	Biopsy or needle aspirate	Sterile container	
Products of conception (fetal, placenta, membranes)	Surgical	Select relevant or suspicious areas of tissues	Tissue or aspirates	Sterile container	
Urethra (for gonococcus)	Wipe clean with sterile gauze or swab	1 hour or more after urinating "milking" the urethra or use thin wire swabs to collect material from 2 cm inside urethra	Swab with urethral secretion	Chocolate agar transport medium	Discharge may be stimulated by gently stripping and massaging the urethra against the pubic symphysis through the vagina

TABLE 21-3a. *(continued)*

Specimen type	*Patient preparation*	*Collection technique*	*Optimal specimen and volume*	*Collection container*	*Comments*
Uterus, tubes, ovaries	Usually surgical specimen	If surgical specimen, representative sample should be cut and submitted	Tissue, aspirates or swabs	Sterile container	
Vagina	See cervix	Simple aspiration or swabbing. For Trichomonas, aspirates are required	Aspirate or swab	Sterile container, chocolate agar tube or plate	
Vulva (including labia, Bartholin's gland)	Skin prep for regular skin sites. Do not use alcohol on mucous membranes	Collect with swab or aspirate with syringe and needle	Swab or aspirate (Bartholin's gland abscess)	Sterile container	
Genital tract male					
Darkfield			As for female tract		
Lymph node			As for female tract		
Penis	Skin prep for regular skin sites. Do not use alcohol on mucous membranes	Vesicles must be opened and secretion submitted as aspirate, swab	Swab, aspirate, or secretion	Sterile container	
Prostate	——	Digital massage through rectum	Secretion for smear and culture	Sterile tube or swab	Helpful in some chronic urinary tract infections.
Urethra	Wipe clean with sterile gauze	Same as for the female; thin urethro-genital calcium alginate swabs are preferred	Swab with urethral secretion	Swabs in sterile tube or slide	In males the diagnosis of gonorrhea can often be made by microscopic examination of the gram-stained

Lymph nodes	Surgery	Aspiration biopsy or excision	Tissue or aspirate	Sterile container or syringe	Consider unusual organisms (e.g., LGV, viruses, *Actinomyces, Yersinia,* acid-fast bacteria)
Oral cavity					
Mucosal surface	Rinse mouth	Scrape, swab	Scrapings or swabs	Swab, tongue blade, in sterile container	Most useful for smears for diagnosis of Vincent's angina or yeast infections
Dental abscess	Rinse mouth, prep with dry sterile gauze	Needle and syringe aspiration	Fluid, pus	Anaerobic transport tube or syringe	Anaerobic bacteria including *Actinomyces* species; certain viridans, streptococci
Pus (abscesses)	Skin decontamination	Sample rapidly, insert in spec. gassed out transport tube	Coat swab well with pus	Anaerobic transport tube	To laboratory without delay
Respiratory tract					
Bronchotracheal secretions	Surgical	By intubation or transcutaneous aspiration	Secretions, bronchial brushings, tissue	Sterile tube	——
Epiglottis	Tongue blade, good light source		Swab, etc.	Swab in transport medium	Do not swab throat or use tongue blade in cases of acute epiglottitis unless prepared for tracheostomy
Lung	Surgical	Percutaneous or open surgery	Tissue or aspirated fluid	Sterile tube or container	
Nasopharynx		Swab is passed gently through nostril into nasopharynx. Stay near septum and floor of nose	Swab	Thin wire swab in transport medium	

TABLE 21-3a. *(continued)*

Specimen type	*Patient preparation*	*Collection technique*	*Optimal specimen and volume*	*Collection container*	*Comments*
Nose		Rotate swab	Swab, etc.	Swab in transport medium	Used routinely for *S. aureus* carrier only
Sputum	Careful instruction of patient	Expectoration. Assisted by saline nebulisation postural drainage, etc. if necessary	Morning specimen, no saliva	Sterile container	Respiratory therapist may be contacted to assist
Throat	Tongue blade to depress tongue	Swab inflamed areas, collect pus or piece of membrane	1 swab	Transport medium	Routine culture only practical for group A streptococcus
Tracheal aspirate		Through nose or mouth using sterile tube	Sputum without saliva	Sterile container	
Skin	To clean skin use 70 percent alcohol	Open vesicle, aspirate or swab or biopsy; punch biopsy for burn wounds	Needle aspirate or biopsy	Syringe or sterile container	
Tissue	Surgery	Surgery or needle biopsy	At least 1 cc piece unless needle biopsy	Sterile container for large piece; transport tube for small piece	Do not add any fluids

Urine					
Clean voided urine	Careful instruction or assistance. Follow standard protocol in nursing care manuals	Clean voided midstream technique: void–discard; void–collect; then empty bladder and discard	First morning urine; several cc	Sterile wide-mouthed jar with lid	May be refrigerated. If at room temperature for more than 30 minutes, specimen is useless for quantitative studies
Catheter urine (indwelling catheter)		Iodine prep of aspiration site on catheter	Same as above	Sterile container	
Bladder urine	Surgical	Suprapubic aspiration	Same as above	Sterile container	
Wounds	Skin decontamination	Swab active margin	Coat swab well	Sterile transport tube	

SPS = sodium polyanethol sulfonate

TABLE 21-3b. Specimen Collection for Culture of Acid-Fast Bacteria

Specimen type	*Patient preparation*	*Collection technique*	*Optimal specimen and volume*	*Collection container*	*Comments*
Acid-fast bacteria (AFB)					
CS fluid	As for other bacteria	Aseptically	5-10 ml or more	Plastic containers approved for AFB work	Skin test result helpful
Gastric washings	Patient should be fasting for 12 hrs before collection in the morning	Use chilled sterile rubber tube	10 ml or more; may have to be repeated	Plastic containers approved for AFB work	Send to laboratory immediately or neutralize specimen
Respiratory tract	Instruction and assistance to cough up sputum rather than to produce saliva	Expectoration, aerosol inhalation, nasotracheal suction, transtracheal aspirate	3-5 early morning samples; 5-10 ml each	Special plastic containers for AFB work	Raising the foot end of bed during the night may produce a better specimen the next morning
Urinary tract	As for clean-voided midstream technique	As for clean-voided midstream technique	Entire midstream specimen	Plastic jars approved for AFB work	
Wound tissue	As for other bacterial cultures	Aseptically	1 ml of aspirate or tissue	Same as above, or syringe	Swabs are *not* acceptable for AFB culture

TABLE 21-3c. Specimen Collection for Fungus Culture

Site	*Optimal specimen*	*Container*	*Comments*
Abscess	Aspirate of abscess contents. Swab base of abscess	Syringe or sterile container	Specimens should be cultured as soon as possible
Blood Peripheral	5-10 ml anticoagulated (SPS) blood or collect the specimen directly into culture media	Sterile tube	Unusual fungi may be encountered in immunosuppressed patients
Bone marrow	As much as is available, heparinized	Syringe, sterile tube	
Body fluids	Anticoagulated fluid	Sterile container or syringe	Process as soon as possible
Central nervous system CSF Brain biopsy (see Table 21-3a)	>2-3 ml	Sterile tubes	
Ear (otitis externa)	Scrapings in sterile tube or swab in transport medium	Sterile tube	For specimen from middle or inner ear use sterile funnel
Eye (corneal ulcer)	Use clean smooth spatula to obtain scrapings. Inoculate scrapings directly to culture media. Make smears; send to the laboratory		Inner eye infections (endogenous oculomycosis are most often caused by *Candida albicans*)
Oral cavity Mouth (thrush) Tongue	Tongue depressor scrapings or swab	Sterile tube	Mark the end of tongue depressor that has the specimen
Respiratory tract Sputum Transtracheal aspirate	Expectorated or aspirated specimen from the lungs. Tracheal aspirates, Bronchial washings and lung tissue	Sterile container	Early morning specimen is preferred. Have the patient brush teeth and rinse mouth well with water just prior to collecting the specimen
Nose Nasopharynx	Swab	Sterile tube or petri dish	
Skin system Nails	Nail clippings and scrapings	Sterile tube, petri dish, or pill envelope	Disinfect the infected area with 70 percent alcohol and use a sterile blade

TABLE 21-3c. *(continued)*

Site	*Optimal specimen*	*Container*	*Comments*
			to scrape the nail. Do not collect material from under the nail
Skin	Skin scrapings	Sterile tube or petri dish	Disinfect infected area with 70 percent alcohol and allow the alcohol to dry. Scrape from the periphery of one or more lesions. Include scrapings from the apparently healthy skin adjoining the lesion
Hair	Pluck several infected hairs	Sterile tube, pill envelope	Use a Wood's light to fluoresce the infected hair and pluck them for processing. Or pick the dull, brittle hair and follicles
Vesicles	Snip off the tips of several vesicles	Sterile tube or petri dish	Disinfect the vesicular area with 70 percent alcohol
Tissue			
Lymph nodes Liver biopsy	Biopsy material of walls and central area of tissue. Keep specimen moist (use sterile broth)	Sterile container	Culture as soon as possible
Ulcers	Saline aspirates from the ulcer base and wall. If biopsied, include aspirates of the ulcer base and wall	Sterile tube	A syringe with 0.25 gauge needle is convenient to use for collecting aspirates. Culture as soon as possible
Urogenital system			
Urine	10-50 ml urine	Sterile container	First morning specimen is preferred
Prostatic secretions Cervix, vagina	Swab well coated with secretions	Swab or secretions in either a sterile tube or small amount of sterile saline in a test tube (swabs only)	

TABLE 21-3d. Specimen Collection for Parasites

Site	*Parasite*	*Optimal specimen*	*Collection*	*Comments*
Blood	*Plasmodium* sp. (malaria)	Thin smears: capillary blood (i.e., finger or ear lobe, use disposable lancet). For thick and thin smears: 5-10-ml fresh anticoagulated blood. Thin smears are made the same way as hematology differential slides. For thick smears, place a drop of blood onto a clean glass slide, spread to size of a dime. Place the slide on a flat surface and allow the blood to dry gently and dust free before sending to the laboratory	Collect the specimen 12 hours after a chill. Then collect every 6 hours for 3 days. Smears may be made from the tube of anticoagulated blood up to 1 hour after it was collected	
	Trypanosome sp.	Place a drop of capillary blood from a finger stick onto a clean glass slide and cover slip. Alternatively send 5-6 ml of anticoagulated patient's blood to laboratory		To laboratory without delay
	Microfilaria	5-10 ml of anticoagulated blood "Skin snips" from thigh, buttocks, iliac crest for African onchocerciasis "skin snips" from scalp, buttocks, face for American onchocerciasis	*W. bancroffti* and *B. malayi:* collect between 10 PM and 2 AM. *Loa loa, A. perstans,* and *Manzonella:* collect between 10 AM and 6 PM.	
Bone marrow	*Leishmania*	Bone-marrow aspirates or thin air-dried smears		

TABLE 21-3d. *(continued)*

Site	*Parasite*	*Optimal specimen*	*Collection*	*Comments*
CNS	*Hartmanella*	Fresh spinal fluid	Aseptic collection. Do not refrigerate	Immediately examine by light or phase microscopy
	Acanthameba group	Fix in Schaudinn's if examination is delayed		
Intestinal system				
Biopsies Jejunal Duodenal	*Giardia lamblia* *Strongyloides*	Fresh biopsy material	Collect into a sterile jar or tube with a little saline	Examine immediately
Rectal	*S. mansoni* *S. japonicum*	For schistosomes: biopsy from the level of the dorsal fold (Houston valve): about 9 cm from the anus	In clean container	To lab without delay
Feces	*Entamoeba histolytica* Other amoebae *Trichomonas* sp.	3 freshly passed stools collected in the morning every other day	Clean container. If unformed fix in Schaudinn's or polyvinylalcohol reagent (PVA)	If unformed or diarrheal, the specimen should be examined within 15 minutes. Formed stools may be refrigerated until they are examined
	Giardia lamblia	3 freshly passed stools collected in the morning every other day	If the initial series of 3 specimens is negative, examine 3 more, 1 every week. Duodenal aspirates (Entero-test) may be necessary	See above
	Balantidium coli	Up to 3 stools collected daily	May be refrigerated	
	Trichuris trichiura	Up to 3 stools collected daily	May be refrigerated	
	Ascaris lumbricoides	Up to 3 stools collected daily	May be refrigerated	

	Hookworm	Up to 3 stools collected daily	May be refrigerated	
	Strongyloides	Up to 3 stools collected daily	May be refrigerated	
	Tape worms	Up to 3 stools collected daily	May be refrigerated	
	Flukes	Up to 3 stools collected daily	May be refrigerated	
	Enterobius vermicularis (pinworm)	Cellophane tape (Scotch tape) or anal swab	Allow for the patient to be resting and quiet for several hours before collecting the specimen. Usually in the morning before bowel movement or bath. Collect from area around the anus	
Sigmoidoscopy proctoscopy	*E. histolytica*	Fresh scrapings collected with a cuvette, Volkmann spoon, surgical instrument to snip off a piece of mucosa or a 1-ml serologic pipette with a rubber bulb to aspirate the lesion. Cotton-tipped swabs are not satisfactory	If the patient has not had a recent bowel movement before the sigmoidoscopy is begun, he should be purged. Wait for 2-3 hours before beginning procedure	The specimen must be either examined immediately or fixed for later examination
Respiratory tract				
Sputum	*Paragonimus westermani*	Fresh sputum	Instruct patient carefully as in Table 21-3a	
Aspirates (tracheal, bronchial)	*E. histolytica* *Strongyloides stercoraiic* *Echinococcus granulosus* Hookworm *Ascaris*	Any aspirated material or drainage material	The specimen should be examined as soon as possible or preserved for later examination	
Lung biopsy	All above plus *Pneumocystis carinii*	Lung tissue	Collect into sterile container. If needle biopsy add some sterile saline	For *Pneumocystis carinii* a lung biopsy is best

TABLE 21-3d. *(continued)*

Site	*Parasite*	*Optimal specimen*	*Collection*	*Comments*
Skin	*Onchocerca volvulus*	"Skin snips" from the thigh, buttocks, iliac crest (African cases) and face, scalp, and buttock (American cases)	For skin snips disinfect the skin with alcohol, insert a 25-gauge needle just under the epidermis, raise it, slice off a small piece of tissue, avoid bleeding	
	Leishmania *E. histolytica*	Aspirate of ulcer bed	Leishmaniasis: the parasites are typically found in the ulcer wall rather than in the pus	
	T. solium *E. granulosus*	Biopsy		
Urogenital system				
Vagina Urethra Prostatic secretions	*Trichomonas* sp.	Sterile swab in a tube with a small amount of sterile saline	Douche must not be used for 3-4 days before collecting the specimen. Send to the laboratory as soon as possible before the trichomonads lose motility	
Bladder	*S. haematobium*	Biopsy from area around the trigone		

material held on one swab is insufficient for multiple cultures. Swabs should not be used for pus or exudates, surgical specimens, culture for mycobacteria, and when anaerobes are sought.

Anaerobic Transport Tubes

Several types of anaerobic transport tubes are commercially available. Instructions for the proper use must be followed explicitly for dependable results.

Other Transport Devices

Many varieties of tubes, jars, and bottles are commercially available. Their sterility must be assured. They must be approved for bacteriologic work and their seals must be tight to prevent leakage of potentially contaminated materials. A sterile syringe may be used to aspirate pus or fluids. Plugged and properly labelled, a syringe can serve as a transport device. Antibiotic-containing transport media are provided by the laboratory for the collection and transport of specimens for viral, chlamydial, and *Mycoplasma* isolation. The laboratory may opt to supply outlying clinics and physician offices with special transport media that support the growth of *Neisseria gonorrhoeae*. Such media may include either of two commercially available products:

Trans-grow: A bottle containing a slant of modified Thayer-Martin medium and an atmosphere of carbon dioxide. The slant is inoculated with a swab from the infected site; the cap is replaced and incubated at 35°C. The bottle must be held upright during inoculation to minimize loss of carbon dioxide.

Jembec: A rectangular plastic plate containing medium and a carbon dioxide-generating tablet. After inoculation, the plate is sealed into a ziplock bag and incubated at 35°C.

DIRECT MICROSCOPIC EXAMINATION

Time is of the essence in care and treatment of a patient ill with infectious disease since many infectious processes progress quite rapidly. Direct examinations of stained, fixed smears and stained or unstained wet-mount preparations may provide useful information and sometimes definitive identification of the causative agent of an infectious process, thereby allowing initiation of specific antimicrobic treatment long before useful culture results and susceptibility studies are available or serodiagnosis is possible.

The most important instrument in clinical microbiology is the light microscope. It magnifies objects 1 to 3000 times, enabling one to observe microorganisms, cells, and structures otherwise invisible to the naked eye. The magnifying power of a microscope is determined by the power of the objective multiplied by the magnifying power of the eyepiece ocular. The objective is the lens system closest to the object viewed; the ocular is the lens system closest to the viewer's eye. Most microscopes used in clinical microbiology laboratories have three objectives of 10X, 40X, and 100X magnification and eyepiece of 10X power. Thus, the total magnification may be 100, 400, or 1000 times the original size of the object.

Gram Stain

The most widely used staining procedure is the Gram stain. Developed 100 years ago by Dr. H.C. Gram, the stain is the basis for classifying bacteria into gram-positive organisms that retain the primary dye (crystal violet; purple); and gram-negative organisms that lose the primary dye on treatment with alcohol, acetone, or a mixture of alcohol and acetone, and stain pink with the counterstain safranin. The factors responsible for the Gram reaction of an organism are not completely understood but include the composition of the cell wall: the gram-negative cell wall has a higher lipid content than the gram-positive cell wall. The following method is a modification of Gram's technique, which gives reliable results with most clinical specimens:

1. *Preparing the smear*
With a loop or swab, spread some material onto a clean and labelled slide. The film should be thin with an even distribution of material on a defined area of the slide. It is wise to mark the location of the material on the slide with a diamond pen. Air dry and then heat fix by passing the slide, smear side up, several times through a flame. Let cool.

2. *Primary stain (Crystal violet)*
Flood the smear with crystal violet solution and let stand for 10 to 30 seconds. Rinse with tap water.

3. *Iodine mordant*
Flood with Gram's iodine for 10 to 30 seconds. Rinse with tap water.

4. *Decolorization*
Decolorize with a 3:1 (v/v) mixture of ethanol and acetone. Decolorize the tilted slide until no more blue dye comes off of the thin section of the smear (this will only take a few seconds). Rinse with tap water.

5. *Counter stain*
Flood the smear with safranin for 10 to 30 seconds. Rinse with tap water; wipe the back of the slide clean and let the smear dry.

6. *Interpretation*
In a well-stained smear, most gram-positive microorganisms will stain dark purple and most gram-negative bacteria will stain pinkish-red. Cellular elements, particularly white blood cells, should stain various shades of red in the thin portion of the smear. Many fungi, especially *Candida,* will stain purple. The correct interpretation of many smears requires considerable experience and a laboratorian familiar with the microscopic examination of gram-stained material should be consulted with all critical specimens.

7. *Quantitation*
Gram smear and wet mounts:
Less than one organism per oil-immersion or high-powered field (HPF):
Very scanty = +
One to five organisms per oil-immersion field or HPF:
Scanty = + +
Five to 30 organisms per oil-immersion or HPF:
Moderate = + + +
Greater than 30 organisms per oil-immersion field or HPF:
Profuse = + + + +

Comments

1. Too much decolorization of thin portion of the smear is probably the most common error.
2. The thinner portions of the smear are usually the most fruitful areas for microscopic examinations.
3. Some gram-negative bacteria stain very faintly with this procedure, (e.g., *Legionella pneumophila*). Prolonged counterstaining with safranin for 2 to 3 minutes, or using carbolfuchsin (used for acid-fast bacteria) for 30 seconds, give better results.

Reagents

1. Crystal violet
Crystal violet: 10 g
Methanol absolute: 500 ml

2. Iodine solution
   Iodine crystals: 2 g
   Potassium iodide: 12 g
   Distilled water: 1800 ml
3. Decolorizer
   Acetone: 400 ml
   Ethanol 95 percent: 1200 ml
4. Counterstain
   Safranin: 10 g
   Distilled water: 1000 ml

Acid-Fast Staining

Mycobacterium tuberculosis, the causative agent of tuberculosis, and most other mycobacteria are extremely slow-growing organisms with generation times (the time an organism takes to make a copy of itself) of 24 hours and more. They stain poorly or not at all with Gram stain owing to the presence of mycolic acid in their cell walls. Several staining techniques were developed of which the Ziehl-Neelsen stain and the fluorochrome stain are the most popular.

Ziehl-Neelsen

Preparing stain

1. Prepare smears as for Gram stain.
2. Fix smears by heating to 65°C for 2 hours.
3. Flood slide with some carbolfuchsin; or heat to steaming (steam for 5 minutes).
4. Rinse with water.
5. Decolorize with acid-alcohol 2 to 5 minutes or until no more color runs off. If *Nocardia* is suspected, use aqueous hydrochloric acid (HCL).
6. Wash with tap water.
7. Counterstain with aqueous methylene blue.
8. Interpretation: Mycobacteria appear as bright red solid or beaded staining rods which may palisade or appear singly. Background and other organisms stain blue.

Comments

1. Occasionally mycobacteria are very scant in a specimen. It is therefore necessary to scan for 15 to 20 minutes before reporting a negative smear.
2. Personnel must be checked for color blindness before being allowed to read acid-fast smears.

Reagents

1. Basic carbolfuchsin
   Basic fuchsin: 3 g
   Ethanol 95 percent: 100 ml
   Aqueous phenol 5 percent: 900 ml

2a. Acid alcohol
   HCL conc.: 3 ml
   Ethanol 95 percent: 97 ml

2b. Aqueous HCL
   Water: 97 ml
   HCL conc.: 3 ml

3. Aqueous methylene blue
   Methylene blue: 0.3 g
   Distilled water: 100 ml

Auramine-Rhodamine (Fluorochrome) Stain

Mycobacteria that retain the fluorescent dye can be visualized using the 25× objective on the fluorescent microscope.

Note: Magnification of objective times magnification of eyepiece (ocular) = total magnification.

1. Prepare and fix smears as for Ziehl-Neelsen stain.
2. Flood with auramine-rhodamine, allow to stain 15 to 20 minutes.
3. Rinse with tap water.
4. Decolorize with 0.5 percent HCL for 2 minutes.
5. Rinse with tap water.
6. Flood with potassium permanganate, let stand 2 to 3 minutes.
7. Rinse with tap water. Air dry.

8. Use a BG-12 primary filter and an OG-1 ocular filter with 25 × objective. Use 40 × objective to scrutinize the morphology of the fluorescing structures.

Comments. Expertise is needed for the interpretation of fluorochrome stained smears.

Reagents

1. Auramine-rhodamine
   Auramine: 1.5 g
   Rhodamine: 0.75 g
   Glycerol: 75 ml
   Add melted phenol 10 ml in 50 ml distilled water; filter
2. Acid-alcohol
   Ethanol 95 percent: 70 ml
   Distilled water: 30 ml
   HCL conc.: 0.5 ml
3. Potassium permanganate
   Potassium permanganate: 0.5 g
   Distilled water: 100 ml

Other Stains

In addition to the Gram stain and acid-fast stain, most laboratories have a repertoire of staining procedures for those organisms for which Gram and Ziehl-Neelsen stains are not applicable and also for the purpose of more closely identifying pathogens. For *Rickettsia,* intracellular microorganisms, the Giminez stain (a modification of the Machiavelli stain) is commonly used. Giminez and the Dieterle stain make visualisation of the elusive legionnaires' bacillus possible.

A variety of staining methods are used to visualize fungi and parasites: hematoxylin for tissue parasites and fungi, trichrome stain for intestinal parasites, and Giemsa and Wright's stain for blood parasites and intracellular microorganisms. These polychromatic stains differentiate nuclear material from cytoplasmic or cell wall components and thus make it easier to recognize some of the identifying structures.

Antibody Stains

The aforementioned stains have some built-in technical difficulties and many laboratories prefer to use a fluorescent antibody (FA) technique or a peroxidase stain to look for specific pathogens. Both techniques employ specific antiserum or antiserum pools tagged with a fluorescent moiety or an enzyme. The organism (the antigen) will bind its homologous antibody with great specificity. The purer the antibody, the more specific the reaction. The organism now can be visualized under the microscope with ultraviolet or short-wave blue light if the antibody was tagged with fluorescin-isothiocyanate. If the enzyme method is chosen, the addition of a specific substrate allows direct visualization of the reaction or spectrophotometric evaluation.

The FA and immune-peroxidase stains have their greatest utility in demonstrating specific pathogens whose cultivation is difficult and time consuming. They are used routinely in the identification of rabies, plague bacillus, legionnaires' bacillus, *Rickettsia, Chlamydia,* and some viruses. Some large laboratories may use the FA technique to identify group A beta-hemolytic streptococci in direct throat specimen or as a confirmatory test of an isolate.

Wet Mounts

Stains have the advantage of permanence. They may be kept for years for teaching purposes or for medical or legal reasons. However, some are cumbersome to process and difficult to read and, of course, because the organism is fixed and its sometimes characteristic motility is lost. Wet mounts are a much used technique to rapidly determine the presence of fungi, parasites, or certain bacteria.

KOH (Potassium Hydroxide) Mount

For a KOH mount the specimen (sputum, skin scrapings, nail clippings, tissue fragments) is placed on a slide and a drop of 10 percent KOH is added. The preparation is sealed with a coverslip and gently heated to dissolve some of the tissue. Fungal hyphae will not dissolve, are refractile, and can be seen with the 10× or 40× objective under the light microscope.

India Ink

To prepare a mount using India ink, a small amount of specimen (sediment of spinal fluid) is mixed with a loop full of India ink; a coverslip is placed over it and examined with 10× and 40× objective of the light microscope. This mount is used to identify *Cryptococcus neoformans,* a yeast characteristically budding with a narrow isthmus. Its large polysaccharide capsule appears as a halo around the organism.

Saline Mount

To make a saline mount, a small amount of feces or vaginal secretion is mixed with saline to a slightly turbid suspension and a coverslip placed over it. The preparation must be examined immediately since some organisms rapidly lose their motility. To a second saline preparation, add a drop of Lugol's iodine. This will stain the internal structures of most parasites.

Optical Methods

Optical methods are available to enhance the image contrast, making it possible to observe structures and objects that would go undetected by routine light microscopy. Phase microscopy and darkfield microscopy require special optical equipment and observers trained in the use of these instruments and techniques. *Phase microscopy* is preferred for the study of internal and external structures of parasites and cells. *Interference contrast microscopy* serves a similar purpose and also offers the observer a three dimensional impression of the viewed object.

Darkfield is the most powerful means of enhancing image contrast. A special substage condenser prevents the light from reaching the objective directly. Light from the periphery of the condenser reaches the object on the slide, which in turn reflects the light into the objective. The field of observation is dark while the viewed object is bright yellow. This technique is used primarily when spirochetes are being sought. It is well suited for the observation of their typical morphology and motility.

Electron microscopy for the detection of microorganisms is mainly a tool for the virologist and experimental microbiologist (see Chapter 22).

CULTIVATION OF MICROBES

Koch's and Pasteur's early efforts to grow microorganisms in pure culture were hampered by the fact that only liquid media were used at that time (Fig. 21-1). By chance, Koch observed gray-, red-, and yellow-colored moist droplets on a left-over boiled potato. A stain made from these droplets revealed a tangled mass of microorganisms. Koch correctly concluded that these droplets arose from a single organism that had fallen onto the surface of the potato. For the first time, bacterial colonies, as these droplets were named, were observed.

FIGURE 21-1. Bottles of liquid nutrient broth. Note the "cloudiness," indicating bacterial growth. The presence of even one organism in a sample can result in such growth, making this technique very sensitive for such specimens as blood. However, contamination can readily occur, so the possibility of false positive results is approximately 3 percent. (Courtesy of Dr. James Plorde.)

Koch was quick to recognize the great potential of solid media for pure culture work. Meat extracts mixed with gelatin and coagulated serum were some of the early solid nutrients on which organisms were grown. Gelatin, of course, liquifies at 37°C, the temperature at which most human pathogens grow best, and the process of coagulating serum is a tricky one. Finally agar, a substance long known to homemakers, found its way into the microbiology laboratory. Agar has the unique property of liquifying at temperatures above 80°C and solidifying at temperatures below 52°C. Supplements such as blood, beef extract, vitamins, carbohydrates, minerals, peptones, and indicators can be added to agar to satisfy the growth requirements of even the most fastidious organisms.

Clinical microbiology laboratories have established protocols for the initial "plating"—seeding of the specimen onto agar plates—for each specimen type. Number and type of primary isolation media vary for each laboratory but should include the following: a growth medium (blood agar or chocolate agar or both), a selective or differential medium, and a fluid medium for the recovery of small members of organisms and dilution of inhibitory substances in the specimen. A prereduced growth medium and an anaerobic broth medium are included when anaerobes are suspected as a possible cause of infection.

A medium is considered a *growth medium* if it supports the growth of a fastidious organism; blood agar and chocolate agar are examples. *Selective media,* such as Thayer-Martin medium, are inhibitory to some organisms but favor the growth of others. *Differential media* are selective to some organisms and, in addition, give some information about the physiology of an organism. MacConkey agar, for example, allows many gram-negative rods to grow and at the same time gives some information about the organism's ability to utilize lactose.

Growth characteristics, nutritional requirements, atmospheric requirements, microscopic morphology, and Gram reaction were the criteria for classification of bacteria in the early days of microbiology. At present, relatedness of organisms is determined by DNA and antigenic composition, physiologic activity, and the metabolic pathways used to satisfy their energy requirements. The clinical microbiology laboratory, however, does not routinely perform DNA homology to delegate an organism to its proper position in the bacterial hierarchy: order, family, tribe, genus, species, and biotype. In the clinical microbiology laboratory the following parameters are considered in establishing the identity of an organism:

1. *Atmospheric requirements. Aerobes* require oxygen for growth while *anaerobes* will grow only in the absence of oxygen. A *facultative* organism can grow under either condition. A *microaerophilic* organism is one that grows well anaerobically, but prefers just a little oxygen for growth, while an *aerotolerant* organism is an anaerobe that is not killed by oxygen, but does not grow in it. *Capneic* organisms are those requiring increased carbon dioxide for growth.
2. *Temperature requirements.* Most human pathogens grow best at 35°C (±2°C). Some grow well at room temperature, 23°C (±2°C), while others thrive best at 42°C. *Yersinia* and *Listeria* are enriched at refrigerator temperatures of 6°C (±2°C).
3. *Growth and action on agar.* The growth pattern of the organism on nutrient, selective, and differential media is a helpful clue to its identity. Some organisms will grow only on very rich nutrient media such as chocolate agar, while others are less fastidious. Colonies may be large and luxurious, small, or pinpoint in size. Organisms may produce hemolysis, complete, or partial lysis of red cells on blood agar. Complete red blood cell lysis will produce a clearing around the colony—beta hemolysis. Incomplete lysis produces a green halo around the colony—alpha hemolysis. Nonhemolytic (gamma hemolysis) organisms do not disrupt the red blood cells in the agar.
4. *Microscopic morphology and Gram stain reaction.* The shape and size and Gram stain reaction are the parameters that allow the microbiologist to make an intelligent choice for additional testing procedures to arrive at the identity of the organism. Examples include
   - Aerobes
     - Large (1 × 3 μm) gram-negative rods: possible Enterobacteriaceae
     - Small (0.5 × 2.5 μm) gram-negative rods: possible Pseudomonadaceae
     - Pleomorphic gram-negative rods: probable *Hemophilus*
     - Large gram-positive rod: *Bacillus*
     - Small gram-positive rod: probably *Lactobacillus*
     - Pleomorphic gram-positive rod: *Corynebacterium*
     - Gram-positive cocci in clusters: *Staphylococcus*
     - Gram-positive cocci in chains: *Streptococcus*
     - Gram-negative diplococcus: *Neisseria*
   - Anaerobes
     - Pleomorphic gram-negative rods: *Bacteroides*
     - Tapered gram-negative rods: *Fusobacterium*
     - Tiny gram-negative cocci: *Veillonella*
     - Large gram-positive rods: *Clostridium*
     - Small gram-positive rods: *Actinomyces*
     - Gram-positive cocci in clusters: *Peptococcus*
     - Gram-positive cocci in chains: *Peptostreptococcus*
   - Microaerophiles:
     - Curved gram-negative rod: *Campylobacter*
5. *Utilization of and sensitivity to some biochemicals.* Some frank human pathogens are members of the families and genera mentioned above. They must be differentiated from commensals in the same family or genus, which are harmless most of the time.

The modern clinical microbiology laboratory may use conventional tubed biochemicals and observe changes due to the growth of the organism for identification. It may use commercially available kits with a binomial code system that translates into the name of the organism tested, or it may use automated equipment to identify the pathogen. A simplified dichotomous key accompanies each section of organisms.

Gram-Positive Cocci (Fig. 21-2)

Anaerobes. *Peptostreptococcus* sp. forms chains. *Peptococcus* sp. occurs in clusters and are frequently found in brain abscesses, dental and oral abscess, aspiration pneumonia, and deep wound infections.

Aerobes. *Staphylococcus aureus* produces the enzyme coagulase, which causes plasma to clot. It is the causative agent of furuncles, postoperative wound infection, pneumonia, sepsis, osteomyelitis, scalded skin syndrome, and toxic shock syndrome. *S. epidermidis* and *S. saprophyticus* are two species of coagulase-negative staphylococci that cause urinary tract infections and catheter-associated sepsis.

Streptococci are classified as nonhemolytic, alpha hemolytic, or beta hemolytic. The Lancefield classification of streptococci recognizes 19 distinct species and serotypes of beta-hemolytic streptococcus. *Streptococcus pyogenes* or group A beta-hemolytic streptococcus, is the causative agent of streptococcal sore throat, scarlet fever, rheumatic fever, and puerperal sepsis.

Streptococcus agalactiae (group B streptococcus) is a frequent cause of neonatal sepsis. Group C contains three species and has been implicated in pharyngitis. Group D is

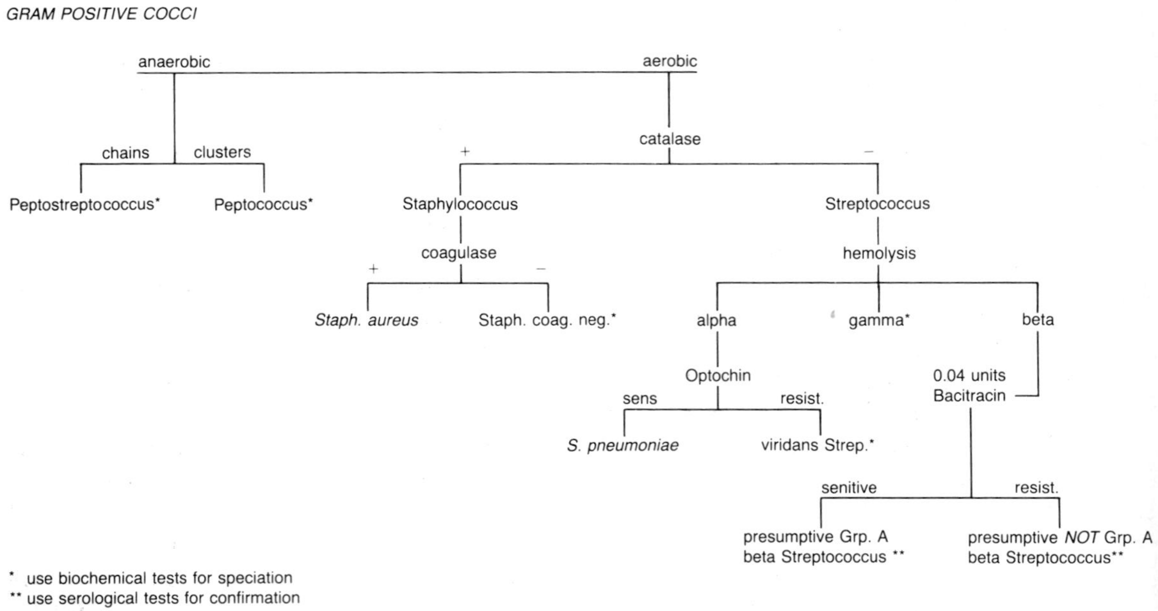

FIGURE 21-2. Gram-positive cocci.

divided into enterococcal streptococci that are frequent causes of urinary tract infection, and nonenterococcal streptococci that occasionally cause bacteremia. The presence of *Streptococcus bovis* (nonenterococcal) in the gut has been associated with a higher incidence of colon cancer. Members of other serogroups are recognized pathogens of animals and infrequently associated with human pathology.

Streptococcus pneumoniae is alpha hemolytic, optochin sensitive, and bile soluble, which serves to differentiate it from the viridans streptococci. *S. pneumoniae* is a frequent cause of meningitis, sepsis, and otitis media during the cold seasons of the year. It causes a particularly devastating infection in asplenic patients. *S. pneumoniae* is found in the throats of 10 to 30 percent of the normal healthy population. Several species of viridans streptococci (alpha-hemolytic streptococci) have been isolated from cases of subacute bacterial endocarditis. Their speciation is mainly based on carbohydrate utilization and glucan production.

Gram-positive rods (Fig. 21-3)

Anaerobes

Clostridium species are inhabitants of the normal human gut. *C. perfringens* produces a striking double zone of hemolysis on blood agar, which is due to its alpha and theta toxins. When introduced into deep tissue, *C. perfringens* causes necrotizing fasciitis and gas gangrene. Food poisoning results from ingestion of large numbers of *C. perfringens* organisms. In leukopenic patients or patients whose intestinal integrity is impaired, sepsis due to *C. perfringens* may occur.

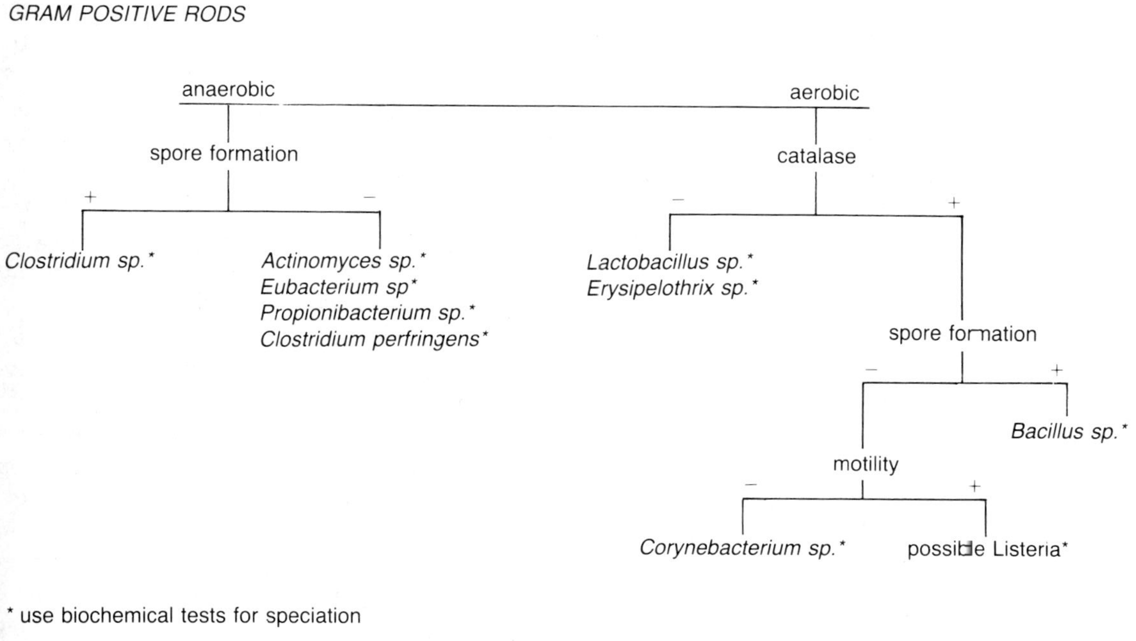

FIGURE 21-3. Gram-positive rods.

C. tetani produces a powerful neurotoxin when introduced into deep tissue. *Clostridium botulinum* toxin is the most powerful neurotoxin known. The preformed toxin is ingested with spoiled food. Wound botulism occurs when the spores are introduced into a deep wound. Seven antigenic types of *C. botulinum* are recognized. *C. difficile* is a relative newcomer to the list of human pathogens. *C. difficile* is now recognized as the cause of pseudomembranous colitis. Laboratory methods for diagnosis include isolation of the organism on specially formulated media and demonstration of *C. difficile* toxin. Other clostridial species have been found in intra-abdominal sepsis and wound infections. These organisms usually originate (endogenously) from the patient's own gut flora or from contamination with dust or soil. The species of *Clostridium* are identified using biochemical tests and specific antitoxins.

Actinomyces and *Propionibacterium* are normal inhabitants of the skin, intestinal tract, and the mucous membranes of the oral cavity and the vaginal tract. They may find their way into wounds and circulation via a break in the natural barriers. Identification is by use of biochemical tests, their morphology, and appearance in tissue. *Propionibacterium acnes,* a pleomorphic rod, is a member of the skin flora. Catheters become contaminated with this organism and it can cause meningitis (shunt infection) and sepsis.

Actinomyces are true filamentous bacteria. Several species are normal inhabitants of the oral cavity of humans and animals. *A. israelii* has not been found outside the human body. It is the etiologic agent of actinomycosis (the name "mycosis" was established when *A. israelii* was thought to be a fungus). Lumpy jaw, mandibular and maxillary abscesses, and osteomyelitis are often attributable to *A. israelii.* Actinomycosis of the uterus associated with intrauterine devices has been reported.

Aerobes

Lactobacilli prefer acidic pH for growth, thrive poorly on blood agar, and cause alpha or no hemolysis. They are normal inhabitants of the oral cavity, intestine, vagina, and skin. These organisms are occasionally isolated from blood cultures of patients following cesarean section, but this bacteremia usually clears spontaneously. In humans *Erysipelothrix insiduosa* causes an acute skin infection, though it is primarily a pathogen of animals.

Only two of the many *Bacillus* species are considered human pathogens. *B. anthracis* is the causative agent of anthrax. In culture *B. anthrasis* produces a "medusa head" colony. The organism is nonmotile, occurs in chains, and is encapsulated. Identification can be made by FA stain and biochemical tests. *B. cereus* when ingested in high numbers causes a food poisoning syndrome. Drug addicts using nonsterile utensils may become septic with any of the *Bacillus* species found in the environment. *Listeria monocytogenes,* a small gram-positive rod with a typical tumbling motility, is found in rodents and a variety of other animals. It infects primarily compromised hosts, leading to sepsis and meningitis. Neonatal meningitis may be due to intrauterine infection or contact with the organism during delivery.

An efficient immunization program has made infections due to *Corynebacterium diphtheriae* a rare occurrence. However, diphtheria is still endemic in some "skid row" populations in this country. Only toxigenic strains produce disease. Toxin production is phage mediated. The respiratory tract is the most common site of infection but skin diphtheria also occurs. Laboratories in endemic areas use Telluride or Tinsdale agar media (selective and differential media) to screen for this organism. Identification is by biochemical tests and antitoxin test. Other species of this genus are found in human disease. The J-K diphtheroid is a nosocomial pathogen causing sepsis in leukemic patients. Its exact taxonomic status is not yet resolved.

Gram-Negative Cocci (Fig. 21-4)

Anaerobes

The anaerobic gram-negative cocci in the genus *Veillonella* are normal inhabitants of the oral cavity. They are occasionally isolated from specimens from the lower respiratory tract in cases of aspiration pneumonia.

Aerobes: Neisseria species except *N. gonorrhoeae* are members of the normal flora of the respiratory tract of humans and some animals. Epidemics of *N. meningitidis* have been associated with closed populations such as boot camps, but sporadic cases do occur. The infection may spread via the blood stream to other systems in the body. The result of hematogenous spread is vasculitis, adrenal hemorrhage, hypotension, and vascular collapse leading to death in a high percentage of patients.

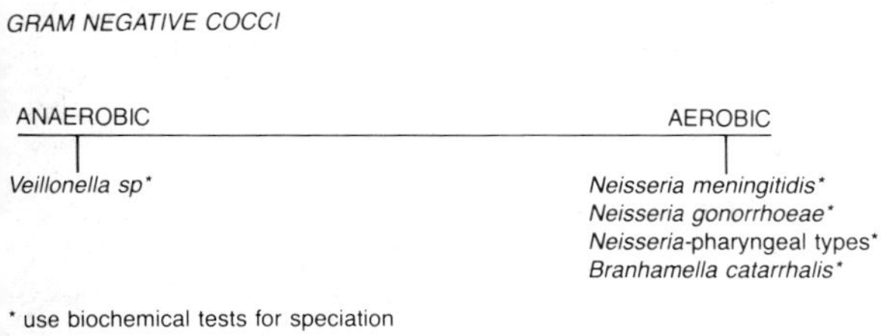

FIGURE 21-4. Gram-negative cocci.

Neisseria gonorrhoeae is the cause of the sexually transmitted disease, gonorrhea. Symptoms include vaginitis, cervicitis, urethritis, pelvic inflammatory disease, gonococcal arthritis, and sepsis. Gonococcal ophthalmitis in the newborn if untreated leads to blindness. In certain populations oral and rectal carriage and infection occur. *N. gonorrhoeae* isolates resistant to penicillin are reported with increasing frequency.

Branhamella catarrhalis is part of the normal flora of the upper respiratory tract. It has been reported as a possible causative agent of pneumonia in Belgian coal miners. While most *Neisseria* and *Branhamella* species grow quite well on blood agar, *N. gonorrhoeae* requires chocolate agar for growth. Several selective media are available for the isolation of *N. gonorrhoeae* from contaminated sites. The most widely used is the Thayer-Martin medium, a chocolate agar that contains vancomycin to inhibit gram-positive organisms, colistin to inhibit the enteric gram-negative rods, and nystatin to inhibit saprophytic yeasts. This medium inhibits all *Neisseria* except *N. gonorrhoeae* and *N. meningitidis*. Speciation of *Neisseria* is by biochemical tests, FA, or coagglutination tests.

Gram-Negative Rods (Fig. 21-5)

The gram-negative rods constitute by far the most numerous and diverse group of microorganisms associated with the human host.

Anaerobes

The anaerobic gram-negative rods are part of the normal flora of the intestine, the oral cavity, and the vaginal tract. Infection in deep tissue or sepsis occurs when the physical barriers are violated and the organisms

GRAM NEGATIVE RODS

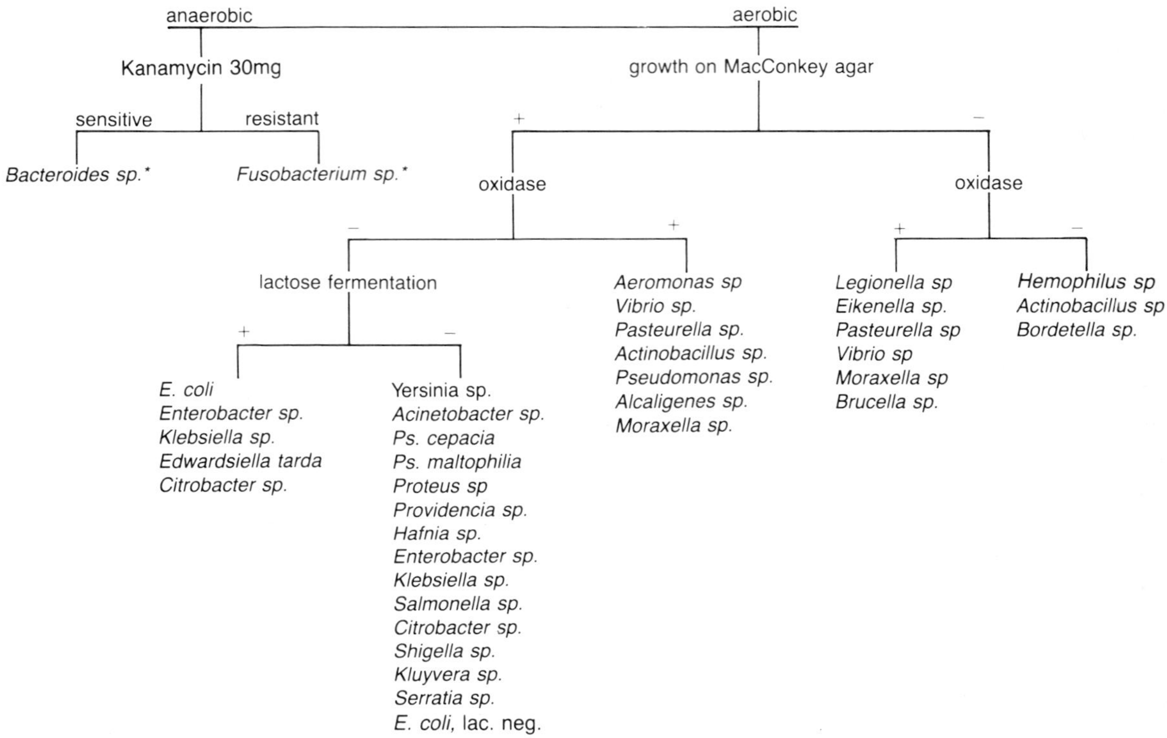

* use biochemical tests for speciation

FIGURE 21-5. Gram-negative rods.

gain access to normally sterile tissue. When anaerobes are considered as the cause of an infection, specimen collection and transport are very important. Specimen contact with air must be minimized and the use of special anaerobic transport devices is advisable. Most anaerobes require hemin-enriched media for growth. The medium should be prereduced in an atmosphere of carbon dioxide for at least 2 hours before use in order to drive off oxygen, which is detrimental to most anaerobes. Cultures for growth and isolation of anaerobes are incubated in special anaerobic jars in an atmosphere of 10 percent carbon dioxide, 85 percent nitrogen, and 5 percent hydrogen. Identification of anaerobic gram-negative rods is by biochemical tests and gas chromatography. *Bacteroides fragilis* is highly resistant to many antimicrobial agents and therefore warrants sensitivity testing when isolated from a significant site.

Aerobes

Members of the Enterobacteriaceae belong to the group of organisms collectively referred to as "enterics." Exclusive of *Salmonella, Shigella,* and *Yersinia,* enterics are part of the normal intestinal flora. *E. coli* is the most frequent cause of urinary tract infection but all other species are equally capable of infecting the upper and lower urinary tracts. Wound contamination, particularly of decubitus ulcers, with enteric organisms is common. Pneumonia and sepsis caused by *Klebsiella pneumoniae, E. coli,* and *Serratia marcescens* occur.

Salmonella, Shigella, and *Yersinia enterocolitica* are sought for in all stool cultures and cultures from rectal swabs since they cause diarrhea and enterocolitis. Enterotoxin-producing *E. coli* and *Campylobacter* must also be considered as causes of diarrhea. All species of Enterobacteriaceae grow well on simple media. Selective and differential media such as MacConkey, Hektoen enteric agar (HE), xylose lysine deoxycholate agar (XLD), and deoxycholate (DC) agar aid in the isolation of the enteric pathogens. Identification of enteric gram-negative rods and enteric pathogens is by biochemical and serologic tests.

Pseudomonas species are opportunists that colonize and infect the compromised host. Burn patients and patients on respirators are especially at risk to acquire an infection with *P. aeruginosa,* a particularly ubiquitous and resistant species. *P. aeruginosa* is also reported as a causative agent in hot-tub-associated folliculitis. *Aeromonas* and *Plesiomonas* are rarely isolated from humans. They are found in wounds contaminated with grass and soil. *Pseudomonas, Aeromonas,* and *Plesiomonas* can be readily isolated on simple agar media. They are identified by biochemical tests.

Pasteurella and *Yersinia* are animal saprophytes and pathogens; humans are accidental hosts. Respiratory tract infections are most common, but other sites may become infected. *Y. pestis* is the causative agent of plague.

Vibrios are comma-shaped rods. Several species are pathogenic to humans. *V. cholerae* is endemic and epidemic in countries with poor sanitation. *V. parahemolyticus* is found in the oyster beds in the coastal waters of the United States and Japan. *V. parahemolyticus* causes a gastroenteritis and occasionally infects wounds contaminated by sea water. *Pasteurella* and *Yersinia* grow well on routine laboratory media. Vibrios prefer alkaline pH or increased salt concentrations. Special media are available for their isolation.

Brucella, the causative agent of brucellosis (undulant fever), requires blood agar and increased carbon dioxide for isolation. If *Brucella* is being considered as the possible etiology in a febrile illness or if there is patient history of ingestion of unpasteurized milk, the laboratory must be notified so that culture conditions can be met. Infection with *Bordetella pertussis* precipitates a respiratory illness commonly known as whooping cough. The presence of this organism is usually identified by FA stain. Culture is only successful in the early stages of the disease on a specially formulated media that contains 20 percent blood. Other members of the *Bordetella* genus are animal saprophytes accidentally infecting human respiratory tracts.

All species of *Hemophilus* are quite fastidious. They require special factors—NAD, heme, vitamins, yeast extract—in their culture media for growth. Growth factor requirements are used to speciate this group of organisms. Capneic variants of *H. influenzae* occur. *H. influenzae* is one of the most frequent causes of meningitis in young children. It often leaves the patient with life long neurologic deficits. Epiglottitis, otitis media, and sinusitis are other infectious diseases in which this organism is implicated. The high frequency of superinfection with *H. influenzae* after influenza A virus infection is a curious unexplained paradox. *Hemophilus* species have been isolated from brain abscesses, lower respiratory tract infections, and cases of septicemia. *Eikenella corrodens* gained notoriety as an etiologic agent in abscesses associated with administration of methylphenidate (Ritalin). The organism is a member of the normal flora of the mouth.

Legionella pneumophila and related organisms long eluded detection since they do not stain with conventional dyes nor grow on the usual laboratory media. Most laboratories now will routinely culture pleural effusions, bronchial washings, and lung biopsies for *Legionella*. Since the organism grows very slowly even on specially formulated media, a direct fluorescent antibody (DFA) test is desirable for speed of diagnosis.

Acid-fast bacteria (Fig. 21-6)

Nocardia preferentially infects the respiratory tract and urinary tract of the compromised host. *Mycobacterium cheloni* has been a contaminant in porcine heart valves used in repair of human valves. It also has been recovered from catheters of patients on dialysis. *M. gordonae* and *M. kanasii* cause respiratory illness similar to *M. tuberculosis*. *M. tuberculosis* infections appear to be on the increase in the United States due in part to the influx of refugees. The acid-fast bacteria other than *Nocardia* and the rapid growing bacteria require albumin and oleic

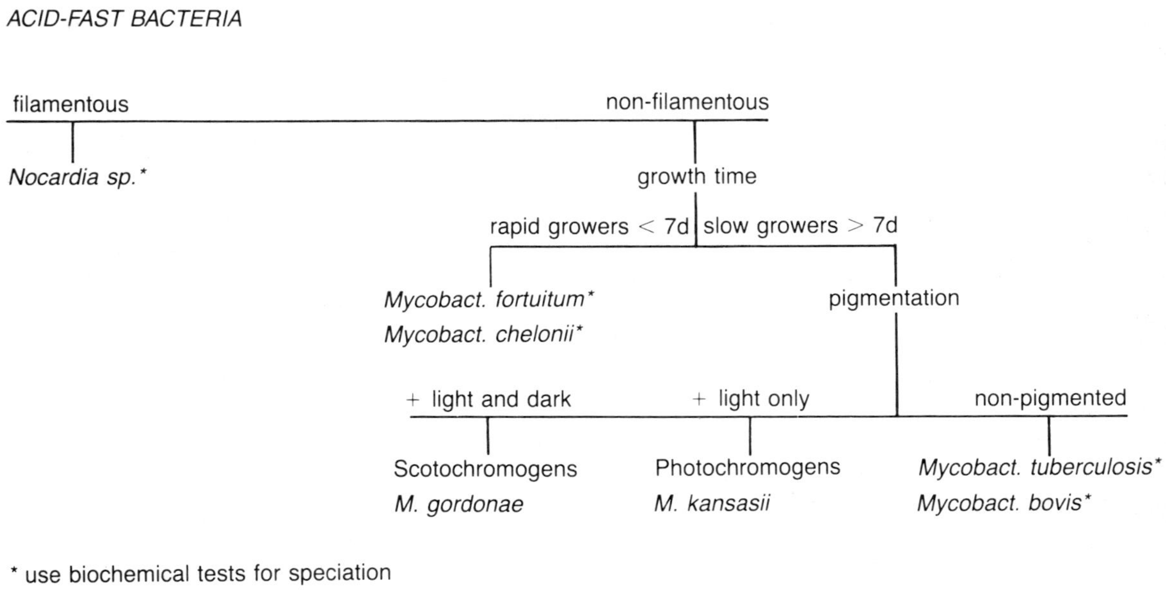

FIGURE 21-6. Acid-fast bacteria.

acid-enriched medium for growth. They are speciated by a battery of biochemical tests or gas liquid chromatography.

Fungi (Fig. 21-7)

Yeasts grow readily on most media. They are identified by use of biochemical tests and by their macroscopic and microscopic morphology. *Candida albicans* is an endogenous organism often infecting the oral cavity and vaginal tract. It is the frequent cause of systemic mycosis in patients affected with diabetes and in the immunocompromised host. Other species of *Candida* produce infection as well but with lesser frequency. *Cryptococcus neoformans* is a yeast surrounded by a huge polysaccharide capsule. This organism has a predilection for the central nervous system, but respiratory tract infections do occur. The organism is found in soil enriched with pigeon droppings.

The phycomycetes—*Mucor, Rhizopus, Absidia*—cause devastating infections in the burn patient. Their mycelia have no or very few cross walls (septa). The reproductive structures are typical for each species.

The ascomycetes and fungi imperfecti constitute a large number of organisms, of which *Aspergillus fumigatus, Coccidioides immitis, Histoplasma capsulatum, Ajellomyces dermatididis,* and *Sporothrix schenkii* cause clinical disease, especially infecting the compromised host. All grow well on Sabouraud's agar. Their identification requires considerable skill. Growth characteristics in tissue and in vitro and their microscopic morphology must be considered.

Parasites

The macroscopic or microscopic morphology of the adult parasite, or any stage of maturation or reproduction, is studied for the proper identification of parasites. Culture methods are available for some amoeba, trichomonas, and trypanosomes.

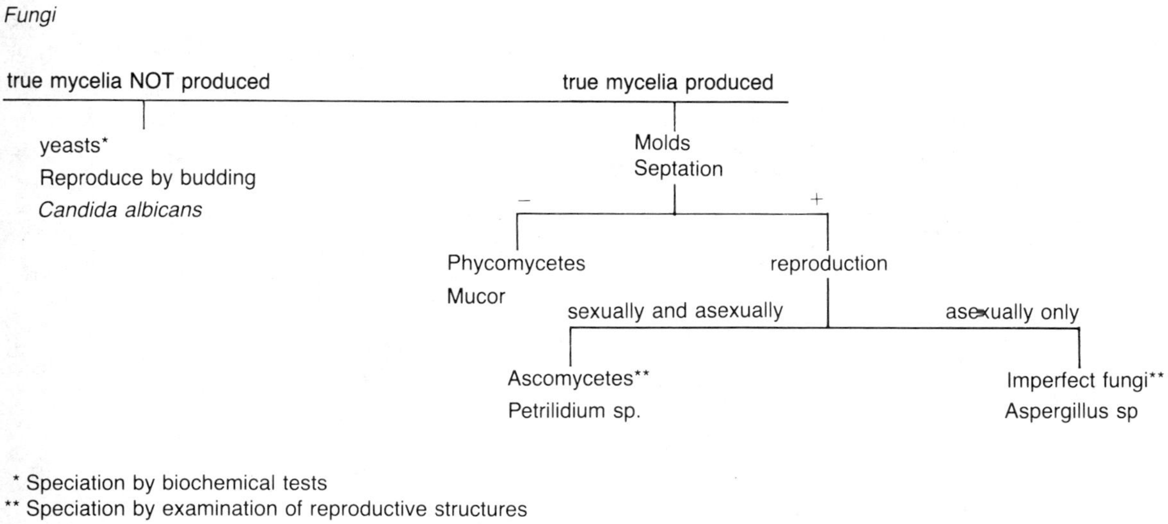

FIGURE 21-7. Fungi.

SPECIMEN PROCESSING

The prerequisite of most identification procedures is the growth and isolation in pure culture of the suspected pathogen. Each microbiology laboratory has its rules and written protocols that determine what happens to a specimen from the time it arrives in the laboratory to the time the final report is issued.

Bacteria

Table 21-4 shows the initial plating of specimens for bacterial culture and the minimum time required for completion. The cultures are inspected at approximately 24-hour intervals. Organisms that require biochemical testing or susceptibility testing are subcultured so that these tests can be performed on a guaranteed pure culture.

Fungi

Specimens for fungal isolation are plated on specially formulated media usually of low pH to promote their growth. Cultures are generally held from 3 to 4 weeks before being discarded as negative.

Mycobacteria

Most mycobacteria have defined growth requirements. The medium must contain egg white or albumin or lipidlike substances. Cultures for mycobacteria are incubated for at least 8 weeks in 5 percent carbon dioxide.

Quantitation

The laboratory will attempt to inform the clinician, not only of the quality of the microbial flora present, but also of the quantity. This, of course, is at best an approximation since in most cases the laboratory has no control over specimen collection, which influences the number of organisms initially sampled and the number and type of survivors. Quantitative elution of the specimen and dilution is not a feasible procedure for a routine laboratory. Most laboratories have adopted a quantitation of 1+ to 4+ depending on the number of colonies present in each quadrant of a culture plate or on a slant.

1+ = 10 or fewer colonies in the primary streak area
2+ = greater than 10 colonies in the primary streak area
3+ = greater than 5 colonies in the second streak area
4+ = greater than 5 colonies in the third streak area (see Fig. 21-8).

Some laboratories have adopted a scale of quantitation 1+ to 10+.

Urine and, in some laboratories, blood are plated to allow reasonably exact quantitation. A measured amount of urine (usually 1/1000 ml and 1/100 ml) or blood (usually 1 ml) is plated and the number of colonies arising from that amount are counted and multiplied to result in a report of colonies per ml. Colony counts of $> 10^5$ are considered clinically significant in voided urine specimens, while the presence of any bacteria in a catheterized specimen or bladder tap specimen generally correlates well with the diagnosis of urinary tract infection. Recently Stamm et al (1982) reported a sign-

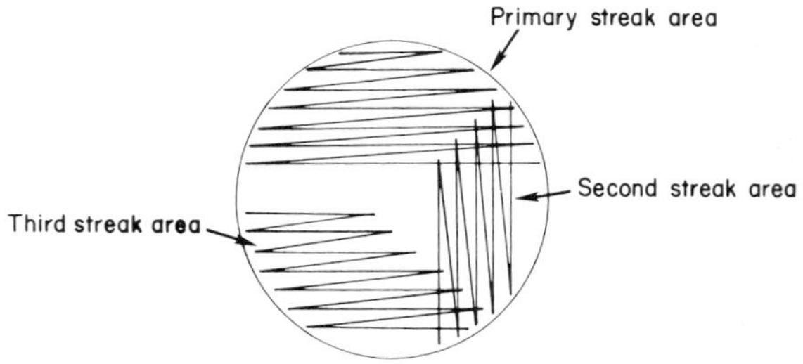

FIGURE 21-8. Common streaking method for the isolation of bacteria.

TABLE 21-4. Initial Plating of Specimens for Bacterial Culture

	Blood agar	Choc-o-late agar	Mac Con-key	He-XLD	TSY broth*	BHI broth†	C. diffi-cile agar	Brucella blood agar	Thayer-Martin	Aerobic CO_2	An-aero-bic	Temp (Cel-cius)	Turn-around time mini-mum	Comments
Upper respiratory														
Throat	X									X	X	37°	2d	
Nose	X									X		37°	2d	
Lower respiratory														
Expect, spu-tum	X	X	X							X		37°	2d	
Transtracheal	X	X	X					X		X	X	37°	2d	
Bronchial washing	X	X	X		X	X		X		X	X	37°	2d	
Lung biopsy	X	X	X		X	X		X		X	X	37°	3d	
Genitalia r/o *gonorrhea*		X							X	X		37°	2	
Gen. culture	X	X	X					X	X	X	X	37°	2	
Urines	X		X							X		37°	1 or 2	
Wounds	X		X			X		X		X	X	37°	2d	
Body fluids	X	X	X		X	X		X		X	X	37°	3d	
CSF	X	X			X	X				X		37°	3d	
Stool			X	X			X	X			X	37°	3d	Special media and 42°C for *Campylobacter*
Blood					X							37°	7d	Special broths also used

*TSY = Trypticase soy yeast broth
†BHI = Brain-heart infusion broth

ficant correlation between lower bacterial counts in urine and urethritis. Quantitative cultures are of prognostic value in burn biopsies.

ANTIMICROBIAL SUSCEPTIBILITY TESTING

Susceptibility tests measure in a qualitative or quantitative way the effect of an antimicrobial agent on a particular microorganism. An organism that is killed by concentrations of an antimicrobial well below the attainable serum or tissue level is considered *susceptible* or *sensitive* to the antimicrobial agent. If the antimicrobial concentration necessary to inhibit the growth of the organism approaches the concentration of the attainable serum or tissue level, it is termed *intermediate sensitive*. If the inhibitory effect is achieved by concentrations exceeding the attainable serum level or is not inhibited by an concentration, it is considered *resistant*.

Because the physiology of the host tissue cannot be duplicated in the test tube or petri dish, in vitro susceptibility tests are an approximation of the action of the drug in the host tissue. When antimicrobial therapy is being selected, the pharmacology of the drug is one important aspect to consider. This includes distribution of the drug and penetration to the site of infection, half-life of the drug in the body (a function of biodegradation by enzymes and the rate of excretion), route of elimination, and toxicity. Other considerations are the underlying disease of the patient and concomitant treatment with other drugs.

The nature of the infecting microorganism also influences the decision for susceptibility testing. Certain "bug-drug" combinations have predictable outcomes: the organism may be either explicitly susceptible (as is the case with *Streptococcus pyogenes* and penicillin) or quite resistant (as is *Pseudomonas aeruginosa* to most antimicrobials except the aminoglycosides, polymyxins, carbenicillin, and third generation cephalosporins). Much greater is the number of pathogenic microorganisms with unpredictable behavior towards many antimicrobial agents. Usually it is the microbiologist handling the culture who decides for which isolate antimicrobial studies are appropriate.

In spite of some acknowledged pitfalls in in vitro susceptibility testing, the tests remain invaluable in the management of infectious processes. Most clinical microbiology laboratories perform antimicrobial susceptibility tests. Several well standardized procedures are available. The choice of one versus the other is often personal preference, consideration of costs, predominance of isolates tested, and turn-around time for results.

Disc Diffusion Test

The disc diffusion test was first described by Ericson et al in 1954 and was standardized by Bauer et al at the University of Washington, in Seattle in 1967. A modification was made by Barry et al. The pathogen to be tested is grown in pure culture in broth, from which a standardized dilution is made. Mueller-Hinton agar in a petri plate is seeded with this dilution so that the agar surface is completely covered. Discs impregnated with a measured amount of antimicrobic agent are then placed onto the inoculated agar surface and the plate is incubated 18 to 24 hours at 35°C. After incubation the plate is inspected for areas of clearing (inhibition of growth) around the disc (Fig. 21-9). Minimum zone sizes have been defined for organism-drug combinations that allow qualitative interpretations of sensitive, if the zone size is met or exceeded; and intermediate (indeterminant), if the zone size is between the mimimum for susceptibility and maximum zone size for resistance. The zone-size diameter can be extrapolated to a quantitative estimate of inhibitory concentrations.

The disc diffusion method is easy to perform, gives reliable and reproducible results if media requirements are met, and requires no sophisticated equipment. It is relatively slow in producing results and is limited to rapid-growing organisms.

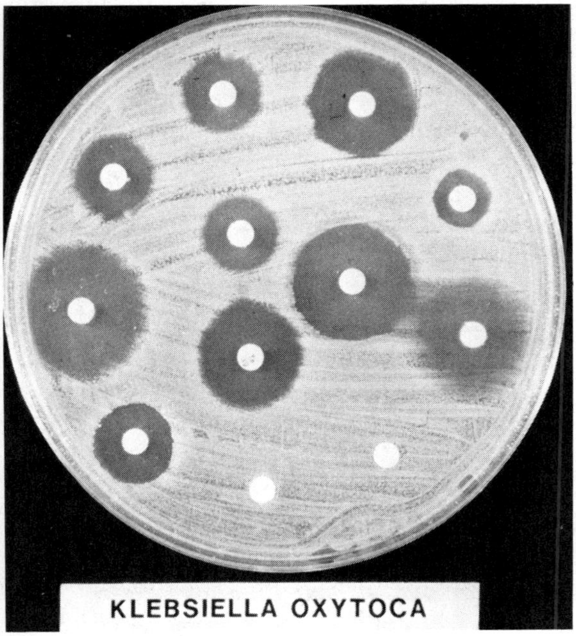

FIGURE 21-9. Antimicrobial susceptibility testing using the disc diffusion method. Notice areas of clearing (inhibition of growth) surrounding some discs, which have been impregnated with specific antimicrobial agents. This "zone of inhibition" indicates that the bacteria growing on the surface of the agar medium is susceptible to the specific agent. Bacterial growth surrounds other discs, indicating that the organism is resistant to the antimicrobic in that disc. (Courtesy of Dr. James Plorde.)

Minimum Inhibitory Concentration

The correctness of the extrapolations to numerical quantitative results from zone sizes is dependent on the care with which the inoculum and the agar are prepared and the zones are measured. More exact quantitative procedures make use of measured antimicrobial concentrations in broth or agar. A standardized suspension of the pathogen is added to dilutions of the antimicrobial. The concentration of antimicrobial, which prevents the organism from multiplying after 4 to 24 hours (depending on the test method), is considered the *minimum inhibitory concentration* (MIC) of the drug for this organism. The concentration of drug capable of killing the organism is determined by placing a measured amount of fluid from the "no growth" tubes of the MIC test to agar and observing for growth of the subculture in 24 hours. The concentration of antimicrobial that kills 99.99 percent of the inoculum is termed the *minimim bactericidal concentration* (MBC), or *minimum lethal concentration* (MLC). Automated equipment has made it possible to apply principles of MIC testing and of a variant to the disc diffusion test: the disc elution test. This test is recommended for sensitivity testing of anaerobes.

Synergism Testing

In some cases neither the disc diffusion tests nor the broth dilution MIC test yield sufficient information. It may be desirable to treat a patient with two drugs. Testing of combinations of drugs for possible synergistic effect is done in a microtiter plate using a checkerboard arrangement of the dilutions of the two drugs. Synergy is proven if the organism is killed by lower dilutions of the combined drugs than of the single drugs. New procedures and equipment are being developed to increase the reproducibility and speed of antimicrobic susceptibility testing.

Which Antimicrobials to Test

A great number of antimicrobials are already on the market, however, the search for others continues. In general, the laboratory will test one antimicrobial typical of its class for all classes of antibiotics: penicillin G, for the penicillins; tetracycline, for its class, and gentamicin, for the aminoglycosides. The choice, however, is harder in the cephalosporin class because the antimicrobial activities of the different cephalosporins vary considerably. In most hospitals the infection control committee and the infectious disease service will make recommendations to the laboratory as to which drug should be tested. Table 21-5 shows the recommendations of the National Committee for Clinical Laboratory Standards (NCCLS) of antimocrobial batteries for testing of certain groups of organisms.

Antimicrobial Assays

The effectiveness of an antimicrobial in treatment of infections is a consequence of the pharmacology of the antimicrobial, the site of infection, the immune status, underlying disease, and age of the patient, and the susceptibility of the organism. The aim of any drug treatment is to achieve the desired therapeutic effect without side effects or toxicity. Whether the outcome of drug treatment has therapeutic or toxic effects is related to drug dosage. Some antimicrobials have a low therapeutic index (TI = lethal dose [LD]/effective dose [ED]). If the secretion or metabolism of the drug is hindered due to hepatic or renal insufficiency, continuation of therapy with normal dosage will lead to toxic levels with often disastrous outcome. Gray baby syndrome due to chloramphenicol, renal damage due to amphotericin B, ototoxicity due to aminoglycosides, and hepatic toxicity due to erythromycin are examples of overdose-related events in the patient. Therefore, under certain circumstances, it is desirable to establish how a patient metabolizes a certain drug

TABLE 21-5. Suggested Groupings of Antimicrobics that Should Be Considered for Routine Testing and Reporting by Clinical Microbiology Laboratories[d*]

Staphylococci	*Enterobacteriaceae*[b]	*Pseudomonas*	*Enterococci*[c]
Amikacin[d]	Amikacin	Amikacin	Ampicillin[g]
Cephalothin[j]	Ampicillin	Carbenicillin or Mezlocillin[d] or Ticarcillin	Vancomycin[g]
Chloramphenicol[d,h]	Carbenicillin[d]	Cefoperazone[d]	Erythromycin[j]
Clindamycin	Cefamandole[d]	Cefotaxime[d] or Moxalactam[d]	Nitrofurantoin
Erythromycin	Cefotaxime[d] or Moxalactam[d]	Chloramphenicol[d,h,k]	Tetracycline[j]
Gentamicin[d]	Cefoperazone[d]		
Kanamycin[d]			

TABLE 21-5. *(continued)*

Staphylococci	*Enterobacteriaceae*[b]	*Pseudomonas*	*Enterococci*[c]
Methicillin or Nafcillin or Oxacillin	Cefoxitin[d]	Gentamicin	
Penicillin G	Cephalothin	Netilmicin[d]	
Tetracycline[d,l]	Chloramphenicol[h]	Piperacillin[d]	
Vancomycin[c]	Gentamicin	Tobramycin	
	Kanamycin	Sulfisoxazole[j,k]	
	Mezlocillin[d] or Piperacillin[d]	Tetracycline[j,k,l]	
	Netilmicin[d]		
	Tetracycline[j]		
	Tobramycin		
	Trimethoprim-Sulfamethoxazole		
	Nalidixic Acid[j]		
	Nitrofurantoin		
	Sulfisoxazole[j]		

a. Selection of the most appropriate antimicrobics to test is a decision best made by each clinical laboratory in consultation with their medical staff, infectious disease practitioners, and the pharmacy. The lists presented are antimicrobials of proven clinical efficacy for that organism group and additionally having acceptable in vitro tests performance.

b. Also applies to Acinetobacter sp and other similar organisms *not* recognized as a Pseudomonas on primary isolation.

c. Antimicrobics *not* listed in this column such as the cephalosporins, clindamycin, and the aminoglycosides, should not be tested against the enterococci because the reporting of their results can be dangerously misleading.

d. Represents secondary antimicrobics which may require testing as a primary agent in those institutions harboring endemic or epidemic resistance to one or more of the primary drugs (especially in the same family; e.g., beta-lactams or aminoglycosides) or as an epidemiologic aid. This designation was applied to some drugs in the pseudomonas list due to lack of long-term clinical experience with that antimicrobic or in the case of the cephalosporins, their first clinical application.

e. Ampicillin susceptibility may be used to predict the susceptibility to penicillin G, ampicillin analogs, and amoxicillin, to which enterococci are also "moderately susceptible." However, combination therapy of penicillin G or ampicillin, plus an aminoglycoside is usually indicated for serious enterococcal infections.

f. The cephalothin disc test *cannot* be relied upon to detect cephalosporin resistance in methicillin-resistant staphylococci.

g. Vancomycin is often used for serious enterococcal infections in patients with significant penicillin allergy.

h. *Not* routinely applied to organisms isolated from the urinary tract.

j. Only tested with organisms recovered from urinary tract infections.

k. May be indicated for testing some *Pseudomonas spp.* other than *P. aeruginosa,* or for other nonfermentive gram-negative bacilli.

l. Doxycycline or minocycline may be tested in place of or in addition to tetracycline for some isolates of S. aureus and nonfermentative gram-negative bacilli; e.g., *Acinetobacter* spp.

*Note: These tables are updated every 6 months. Therefore, this table serves as a *sample* and may not be current.

Permission has been granted to use portions of the text of the Second Informational Supplement (M2-A2S2) to the "Performance Standards for Antimicrobic Disc Susceptibility Tests, Second Edition," by the National Committee for Clinical Laboratory Standards. NCCLS is not responsible for errors or inaccuracies. Copies of the complete standard may be obtained from NCCLS, 771 E. Lancaster Avenue, Villanova, PA 19085.

and what peak and trough levels are achieved with a given dosage regimen.

Antimicrobial assays may be performed by either chemical methods, gas liquid chromatography, enzyme assay, or bioassay. Whatever the choice of procedure may be, the collection and handling of the specimen requires attention. Trough levels should be drawn 30 minutes before the next dose is given, and peak levels should be obtained 1 hour after intramuscular injections or oral dose, and 30 minutes after completion of an intravenous dose. Blood should be drawn into glass tubes without anticoagulant or preservatives and allowed to clot. The specimen must be promptly transported to the laboratory. If this is not possible, the serum must be separated from the clot and then frozen. The laboratory must be informed of the dosage regimen, when the specimen was collected, and whether the patient is on any other antimicrobial agents or drugs. The presence of other antimicrobials may interfere with the assay method the laboratory is using and may invalidate the test.

Serum Inhibition Assay (Schlichter Test)

The serum inhibition assay is frequently used when antimicrobial assays cannot be performed due to interfering substances. It is also used to follow patients on long-term therapy for endocarditis. Dilutions are made of the patient's serum at times of peak and trough antimicrobial levels, and a measured dilution of the organism against which the therapy is directed is added. After 24 hours of incubation the inhibitory dilution is noted. The "no growth" tubes are subcultured and reincubated. The next day the serum cidal dilution is determined. Dosage is considered adequate if the patient's serum is bactericidal for the organism at a dilution of 1:8 or 1:16.

Plasmids

The development of resistance of microorganisms to antimicrobial agents sometimes occurs in a stepwise fashion and is probably due to an adaptive mechanism in the organism's genetic makeup, the chromosomal DNA. Other organisms acquire resistance suddenly. This sudden change in behavior toward antimicrobials is plasmid mediated. Plasmids were long thought to be transmitted only among organisms that are closely related. In recent years it has become evident that intergenus exchange does occur. The plasmid encoding for ampicillin resistance in *Hemophilus influenzae* was passed to this organism by a member of the Enterobacteriaceae. Plasmid analysis has great utility in the epidemiology of infections, particularly infections with multiresistant organisms.

SEROLOGIC TESTS

The etiology of an infectious disease is best established by cultivation, isolation, and identification of the infecting agent. As desirable as this approach may be, it is often difficult to accomplish. The inability of an organism to grow on conventional laboratory media, antimicrobial treatment of the patient prior to sampling (which partially or completely inhibits growth of the offending microorganisms), small numbers of organisms present in the sample that fail to grow, or sampling late in the course of the disease are the major obstacles to successful isolation of the pathogen. The detection in the serum of a patient of specific antibodies produced during the course of the infection may be the only proof for the diagnosis.

Serologic tests allow us to demonstrate the types and numbers of antibodies and other components in the blood that contribute to immunity. More important than mere detection of antibody is demonstration of the type of immunoglobulin and change in the titer of the patient's serum. Titer is defined as the highest dilution of the test serum that will give an observable reaction with a specific antigen. Blood samples should be obtained during the acute phase of the illness and again 2 to 3 weeks later in adults (3 months later in infants). The two sera should be tested simultaneously to ensure similar test conditions. Single-serum titers usually have no significance unless they are extremely high. Exceptions may be taken in those cases where the acute onset of infection is insidious, as in some fungal and parasitic infections, or where the prevalence of a certain disease and, therefore, the prevelance of natural occurring antibodies is zero.

Interpretations of serologic test results must take into account the following factors:

1. Prevalence of disease and naturally occurring antibodies in the population.
2. Age and immune status of the patient. Presence of IgG antibodies in the newborn reflects transplacental passage of mother's antibodies. Presence of IgM antibodies indicates the baby's own immune response to infection, as IgM does not pass through the placenta.
3. Previous infection or vaccination with the same organism.
4. Known cross reactions. Antibodies produced in response to infection with one species of organisms sometimes react with unrelated antigens.
5. Anamnestic response. For poorly understood reasons, antibodies to a certain antigen can rise in response to infection with a different antigen.
6. Testing method. Antibodies may be detectable with one testing system and will not react or be demonstrated with another test system. The sensitivity and specificity of different methods for detecting antibodies vary greatly. A test of high specificity expresses antibody titers to one type of organism or a product of one type of organism and does not cross react with any other antigen.

Methods

A great number of testing methods are now available to the serologist. The choice may often be limited by availability of equipment, high-grade reagents, and sophistication of laboratory personnel.

Radioimmunoassay (RIA). RIA is the most sensitive testing method; it is approximately 1000-fold more sensitive than counterimmunoelectrophoresis. The test has its greatest utility in detecting small amounts of antigen but is also useful in detection of antibody. The principle is the same for antigen or antibody detection. A labelled component (antigen or antibody) is reacted with antibody (serum) or antigen (specimen in soluble form). Immune complexes are formed and then precipitated from the unreacted material. Radioactive counts in the precipitate are converted to titers by comparing them to a standard curve. The technique is used in the detection of antiviral antibodies or antigen. Shortcomings of the test are the need for expensive equipment and potential danger of the radioactive materials that are used.

Enzyme-linked immunosorbent assay (ELISA). ELISA is a very sensitive testing method with many variations to assay for either antigen or antibody. Antigen is adsorbed into wells of a microtiter plate, and test serum is added and allowed to react. Excess serum is rinsed off, enzyme conjugated antihuman globulin is added, excess conjugate is removed, and enzyme substrate is added to the complex. Degradation of substrate is dependent on the amount of labelled antihuman globulin present, which in turn is related to the amount of antibody that reacted with the antigen. ELISA is used in the detection and measurement of antiviral, antibacterial, antifungal, and antiparasitic antibodies. A major shortcoming of the assay is that some antigens or antibodies will not adsorb onto plastic surfaces and will not retain immunologic activity. In addition, special equipment and sophisticated personnel are needed to perform the test properly.

Indirect fluorescent antibody (IFA). IFA is a moderately sensitive test with good specificity. In this test, antigen is allowed to react with serial dilutions of patient serum. Fluorescin isothiocyanate-tagged antihuman globulin is then added to the antigen-antibody complex. The highest dilution of patient serum that still shows fluorescence is the IFA titer of that serum. It is used for antiviral, antibacterial, and antiparasitic antibodies. The principle of the IFA test is applied in the fluorescent treponemal antibody (FTA)-absorbed test. This test measures antibodies that are produced in response to infection with *Treponema pallidum,* the causative agent of syphilis. Special optical equipment and expertise in reading these tests are needed.

Complement-fixation test (CF). Complement fixation is a useful but complex test for detection of antibodies. Antigen, antibody in various dilutions, and complement form a complex to which a hemolytic system (sensitized red blood cells from sheep) is added. If all complement is bound, no lysis of the red cells will occur. If complement remains unfixed due to lack of antibody or antigen, lysis will occur. Hence, a positive test is one in which no effect is detectable; that is, no lysis occurs. The CF titer is that dilution of serum that completely fixes complement—the dilution in which no lysis occurs. The test has been adopted to microtiter technique. It is used for detection and measurement of antibacterial, antifungal, antiparasitic, and antiviral antibodies. Unfortunately, the test is complex and the reagents used are unstable.

Precipitin tests. Precipitin tests work on the principle that soluble antigen and antibody react and form a visible precipitate. The tests can be performed in capillary tubes, test tubes, or agar gel. In agar gel the test is called immunodiffusion (ID). Variations of the agar gel ID tests are a number of electrophoretic techniques. Counterimmunoelectroosmophoresis (CIE), rocket electrophoresis (RE), and crossed immunoelectrophoresis make use of the fact that most antibodies do not move in an electrical field at pH 8.6, while antigens will move. The electrical current carries the antigen molecules toward the antibody. Where the optimum proportion of antigen and antibody is met, a precipitin line will form. While ID test requires 48 to 96 hours for completion, electrophoretic tests may be completed in 90 minutes to 4 hours. These tests are used for the demonstraton of antigen or antibodies—either viral, bacterial, fungal, or parasitic. A prerequisite is the electrophoresis equipment.

Agglutination tests. A particulate antigen reacts with antibody and agglutination occurs. This simple test can be performed on a slide or in a tube. Dilutions of patient's serum are tested against a constant amount of antigen. The highest dilution of serum that still shows agglutination is the agglutination titer of the serum. Variations of agglutination tests are the latex agglutination test, in which the antigen or antibody is coded onto an inert particle, and the coagglutination test, in which the Fc end of antibody reacts with protein A (a special protein produced by a specific strain of *Staphylococcus*), and the Fab end of antibody binds to antigen. Agglutination tests are useful in identifying antigens and also in measuring antibody titers in some bacterial, fungal (latex), and parasitic infections. "Febrile agglutinins" is a collective name for a diverse group of antibodies that are produced following febrile illnesses caused by *Salmonella* (typhoid), *Brucella* (undulent fever), *Rickettsia* (typhus), and *Franciscella* (tularemia).

Antistreptolysin O test (ASO). ASO test is a unique test developed to demonstrate antibodies produced in the host in response to a metabolic product of group A beta-hemolytic steptococcus. Streptolysin O has hemolytic activity on human red blood cells. This activity is inhibited by antibodies to streptolysin O. The ASO test gives an estimate of these antibodies by means of dilution of test serum that can inhibit red blood cells hemolysis in the presence of constant amounts of streptolysin O. Other streptococcal tests measure the antibodies to such streptococcal metabolites as deoxyribonuclease B, hyaluronidase, and fibrinolysin. Antideoxryibonuclease B test (ADNB) appears the best single test for serologic detection of streptococcal infection.

Flocculation tests. Alkaline solutions of dry antigen prepared from beef heart are employed to coat cholesterol crystals that act as specific antigens in the reagin test. The nontreponemal antibody reagin is produced in response to infection with *T. pallidum*. The VDRL and RPR tests are modifications of the original rapid plasma reagin test. Nonspecific chemical flocculation tests (bentonite flocculation, Chopra's antimony test) are used in diagnosis of parasitic diseases. These tests are based on changes in plasma protein.

No single value can be assigned to represent normal, suspicious, or significant titer in antibody testing. Each test system must have its interpretative values based on the factors outlined on page 573. It is the responsibility of personnel in the laboratory performing the test to provide information on the testing method used and the significance of the results.

CLINICAL IMPLICATIONS

The correct identity of a pathogen must be established so that the effective antibiotic, chemotherapeutic, or surgical regimen may be chosen for the patient. The nature of the pathogen may also necessitate isolation procedures to prevent the spread of the organism to other susceptible hosts. Testing for specific antibodies in the patient's blood is an indirect means of establishing the identity of an infectious agent. It is the least desirable way, since it takes 5 to 14 days following infection for antibodies to rise to a significant titer. Serologic tests are frequently difficult to interpret because of errors in timing of specimen collection, anamnestic rises of related antibodies, or poor sensitivity or specificity of the testing method.

The clinician's primary function in the facilitation of microbiologic diagnosis is to collect specimens appropriately. Guidelines for specimen collection are produced in this chapter, but if questions arise, a local microbiology laboratory should be consulted prior to specimen collection. Delay in specimen transport may cause death of fastidious organisms or overgrowth with normal flora or rapid growing organisms. This results in errors in quantitative and qualitative interpretation of the microbiologic examination. The health care practitioner is also responsible for providing the laboratory with information regarding the exact type of specimen and any suspected infections.

To properly assess any microbiologic finding, the health care team (i.e., physician, nurses, and microbiologists) must consider the following factors:

1. The site of the infection and the normal flora of the site (see Chapter 23 and Table 21-1).
2. The integrity of the physical barriers. Skin, mucous membranes and their secretions, and ciliated epithelium prevent the entry of potential pathogens into the underlying tissue. Any break in these barriers predisposes the host to infection with microorganisms.
3. Underlying disease and immunologic competence of the host. The newborn is at high risk for infection due to its inability to muster an adequate immune response in the early weeks of life. Impairment of the natural immune response due to genetic error or intentional intervention (tumor or transplant immunochemotherapy) may prevent the host from producing antibodies (humoral defense), or activating cytotoxic cells (cell-mediated immunity), or both. Infections and microbiology of the immunocompromised host has been the subject of several books.
4. Prevalence of certain microorganisms in the environment. Living habits and the socioeconomic status of the host may play a role in acquisition of infections with certain unusual microorganisms.

Infection due to bacteria or viruses may result from direct contact with contaminated hands or instruments or indirect contact with contaminated droplets, air, or water. While droplets and air most often contaminate and infect the respiratory tract, contaminated water and food are the vehicles for infections of the gastrointestinal tract. Sepsis and systemic disease often follow injection or infusion of drugs and blood products contaminated with bacteria or viruses. If an illness is preceded by an animal bite, infections with organisms endemic in an animal population must be considered.

Certain organisms have a particular propensity for spreading in a hospital setting and cause nosocomial infections. It is of great importance to the infection control practioner to ascertain that the organism being transmitted is the same without reasonable doubt and therefore, may originate from some index person or vector. *Staphylococcus aureus* is long surpassed by gram-negative rods as the scourge of hospitals. Yet, nosocomial outbreaks of staphylococcal infections still occur. Phage typing is an effective tool to ascertain the relatedness of staphylococcal isolates. The test makes use of the fact that some strains of *Staphylococcus* can be parasitized by bacterial viruses called phages. The isolates are exposed to the phages known to infect staphylococci. If the same phage infects and lyses the isolated strains, these strains are considered identical.

The spread of *Pseudomonas aeruginosa* in burn centers is particularly troublesome. Pyocin typing is a testing method which allows one to trace a strain of *Ps. aeruginosa*. Pyocin is a metabolic product of this organism with several unique antigenic variations for which one can test.

Other methods to establish relatedness of organisms causing nosocomial infections are serotyping, bacteriocin typing, and plasmid analysis. Plasmids may confer multiple resistance to organisms. It is now established that plasmids are passed from organism to organism, not only within a genus, but also between unrelated genera. Prevention of spread of plasmids would limit the spread of multiresistance in microorganisms.

SUGGESTED READINGS

Allen JC. Infections and the compromised host. Baltimore: Williams and Wilkins, 1976.

Hoekelman RA, et al. Principles of pediatrics: health care for the young. New York: McGraw-Hill, 1978.

Hoeprich PD, ed. Infectious diseases, 2nd edition. Hagerstown, Md: Harper and Row, 1977.

Lechevalier HA, Soloterovsky M. Three centuries of microbiology. New York: McGraw-Hill, 1965.

Lennette EH, et al, eds. Manual of clinical microbiology, 3rd edition. Washington DC: American Society for Microbiology, 1980.

Lorian V, ed. Antibiotics in laboratory medicine. Baltimore: Williams and Wilkins, 1980.

Rose NR, Friedman H, eds. Manual of clinical immunology, 2nd edition. Washington DC: American Society for Microbiology, 1980.

Sausker F, et al. Pseudomonas folliculitis acquired from a health spa. JAMA 1978; 239:2362-4365.

Schmidt GA. The microbiology of respiratory tract infections. Techniques. Seattle, Olympic Medical Corporation 1977;7:1.

Stamm WE, et al. Diagnosis of coliform infection in acutely dysuric women. N Engl J Med 1982;307:463-468.

22 DIAGNOSTIC LABORATORY METHODS: VIROLOGY

KEY POINTS

- Successful isolation and identification of a virus is dependent upon proper specimen collection.
- Viruses are more difficult to cultivate than bacteria. They require living cells (not artificial media) to propagate—in the form of embryonated hen's egg; cell, tissue, or organ culture; or living laboratory animals, primarily mice.
- Diagnostic tests to identify viruses can be grouped into four categories: microscopic examination of specimens to observe for specific changes indicative of viral infection, viral culture, detection of virus or viral antigens in patient specimens using methods other than culture, and serologic studies.
- Methods for rapid viral detection and identification include use of electron microscopy, fluorescent antibody tests, and solid phase immunosorbent assays.

From the time of their discovery, viruses were surrounded with an aura of mystery. The classical techniques employed in the study of bacteria were found to be unsuccessful until viruses could be visualized by the electron microscope. Thereafter, studies soon revealed that the physical structure, biochemical composition, and biologic properties of viruses differed vastly from organisms composed of cells. In fact, the mode of viral replication, together with the discovery that viruses in highly purified state could be crystallized, led to extensive scientific and philosophical debates about whether viruses were living or inanimate. These debates continued for some time until it was finally recognized that the preconceived concepts of life itself had to be redefined. Finally, viruses have become the objects as well as tools for sophisticated molecular studies. In the process, a new vocabulary has developed, which undoubtedly has helped to perpetuate the aura of mystique for the uninitiated—and even for some virologists.

From the clinical point of view, viral diseases are the most common infections experienced by humans. The virulence of some viruses as well as their ability to spread rapidly between populations, even those separated by seemingly insurmountable geographic barriers can be frightening. It is not surprising then that the efforts of viral laboratories from their earlier beginnings were directed toward the surveillance, control, and prevention of viral diseases that were of significant concern for public health. In some cases, efforts in epidemiologic control were extraordinarily effective. For example, smallpox is the first infectious disease to become eradicated by vaccination and control methods, and many experts believe that measles might also become extinct, at least in industrialized nations.

Many viral diseases, although not eradicated, are under control; for instance, yellow fever and poliomyelitis. On the other hand, some important viral diseases, influenza type A, in particular, seemingly under control at some times, may resurface at any given time in explosive fashion. Although a recent strain, Russian influenza, was relatively benign, once it really started to spread, it circled the entire globe practically in a single season with the attack rates as high as 85 percent in secluded, susceptible populations. "Old" viral diseases can suddenly become a major concern of public health, as is currently happening with the explosive spread of sexually transmitted herpes simplex virus type 2 in the United States. Finally, new viral diseases might emerge whose etiologic agents are not yet known. Consequently, surveillance and detection methods must be kept at competent levels.

During the last few years, a new focus on viral diseases is the concern and management of viral infections in the individual patient. Within a short period of time, it was recognized that in humans different viruses, and even the same virus, can behave in an amazing variety of ways, including asymptomatic infections, acute disease, chronic disease of recurrent and persistent nature, or diseases of extraordinary complexities in the interaction between host and virus. Asymptomatic infections in humans do occur with most viruses, the notable exception being rabies virus. The explosive onset of influenza is a dramatic example of acute viral disease. Examples of recurrence and persistence in chronic viral infections are seen in herpes simplex virus and hepatitis B virus infections, respectively. Subacute sclerosing panencephalitis (SSPE), an autoimmunelike disease, occurs as a sequelae in some measles infections many years after exposure, and is an example of the complex relationship between a virus and its host. Lastly, several viruses have been strongly associated with human cancer, especially HSV type 2 with cervical carcinoma and EB virus with Burkitt's lymphoma. The ever-expanding list of clearly defined diseases attributable to a specific virus has demanded and hastened the developed of diagnostic tests. With the advent of rapid diagnostic procedures that also offer greater sensitivity, diagnostic virology has drastically changed and to a large extent has shed the stereotypic image of being slow, low-yield, clinically irrelevant, and not cost-effective.

From the point of view of the clinician, an accurate diagnosis of many viral diseases cannot be made on clinical grounds alone. A specific diagnosis, on the other hand, is indispensible in the accurate prognosis of the disease that, ultimately, is more important to the patient than treatment. Obviously, only the virology laboratory can provide the clinician with the necessary information. In many instances, a competent diagnostic viral laboratory can provide services for patient management comparable to those in diagnostic bacteriology. This evolution should be hastened, even mandated, by the development of more numerous and more specific antiviral agents.

The rationale in the choice of methods for diagnosis of viral infections is based on facts gathered from clinical observation: characteristic changes that occur in the cells of infected hosts; infectious virions and viral antigens that are produced and can be detected; and a specific antibody response that ensues and can be measured. The various diagnostic tests used can be grouped into the following four categories:

1. Microscopic examination of specimens from the patient for the presence of viral inclusions in infected cells, or characteristic pathologic changes in host tissues
2. Isolation (culture) of virus from the infected host
3. Detection of virus or viral antigens in specimens from patients—independent of cultivation
4. Serologic studies to detect the presence or development of virus-specific antibodies in the patient's serum

The importance and popularity of the methods in each category have changed over time. Presently, methods to diagnose viral infections rapidly and independently of viral culture are growing in popularity and importance. They are the direct outgrowth of significant advances in molecular biology, in particular monoclonal antibody and gene cloning techniques. They undoubtedly will affect profoundly the approach to viral diagnosis by laboratory methods.

From a practical and functional point of view, the sophistication of a particular diagnostic viral laboratory depends largely on its resources, which in turn are in most cases directly related to the needs of the community. It is important to remember that even rather sophisticated virus laboratories cannot and should not attempt to diagnose all possible viral infections. The agent of foot and mouth disease, due to its tremendous impact on livestock in the United States, or Lassa, Marburg, and Ebola virus, rarely encountered in the United States but of extremely high potential hazard, can only be handled in maximum containment facilities that are very expensive to build and to operate. On the other end of the spectrum, even relatively simple laboratories can expand their services by cooperating with larger institutions. Finally, regardless of what means are employed, the most significant steps in laboratory diagnosis of viral diseases is the proper selection and handling of viral specimens.

SPECIMEN COLLECTION

Viral diagnosis is a complex process that requires many decision-making steps. At each step, pertinent and adequate information must be available to reach the correct decision. Therefore, close communication and cooperation between the clinician and the laboratorian are essential factors in the proper utilization of the virology laboratory for patient care. The physician, who sets in motion the process of viral diagnosis, must be informed about what tests are available and appropriate and what specimens should be collected. The nurse must know when and how specimens should be collected and transported and what information is needed by the laboratory to provide appropriate testing.

As a rule, studies should be directed toward a specific diagnosis, but should never be undertaken as an afterthought when all other tests have failed. Virus shedding in most viral diseases occurs early in illness, and is usually of only short duration. Therefore, virus isolation or most rapid diagnostic procedures attempted late in the course of the illness usually result in false negative reports. Early initiation of studies might also avoid confusion regarding whether a viral isolation should be attributed to the illness that required admission into the hospital or to a nosocomially acquired infection.

As for bacteriologic and fungal diagnosis, the selection of sampling sites, collection methods, and handling of specimens are, without doubt, the most important parameters for successful viral diagnosis by laboratory methods. What could be confusing in virology is the fact that many different viruses may cause the same disease syndrome, and the same virus may infect many organs. A guide for selection of specimens when only the clinical picture is known, but a specific virus is not suspected, is shown in Table 22-1. If a specific virus is suspected, the most suitable sources for culture and diagnosis are presented in Table 22-2.

Clearly, specimens are similar to those used in bacteriologic diagnosis (many hospitals use the same specimens for both) and are usually obtained from the site of infection. The great majority of specimens are obtained by using swabs, and, therefore, it is very important to remember, that viruses replicate within cells. Consequently, sampling techniques must strive to collect as many cells as possible. Also, in cases of suspected viral meningitis, throat and stool swabs or fecal specimens are submitted in addition to cerebral spinal fluid to increase the likelihood of viral isolation. Several urine specimens should also be submitted when cytomegalovirus or mumps virus are suspected. On the positive side, there are no viral anaerobes to contend with, and surgical wound specimens or catheter tips are never cultured.

Types of Specimens

Throat, Nasopharyngeal Swab, Washings

Nasopharyngeal washings are superior to swabs for the recovery of viruses (optimal for coronaviruses) because the viral sample obtained is larger (a swab holds 0.1 to 0.15 ml), and a larger surface area is being sampled. However, swabs are more convenient and generally less unpleasant to the patient. Nasopharyngeal swabs are the preferred specimens for recovery of respiratory syncytial virus and rhinoviruses.

Throat swab specimens are best obtained by swabbing with a dry cotton-tipped swab over both tonsillar fossae and finally the posterior pharyngeal wall. In suspected cases of mumps, the inflamed orifices of Stenson's ducts should be swabbed. Swabs should be rubbed vigorously over any lesion in the mouth or throat. Swabs must be immediately placed into viral transport medium, usually supplied in vials by the laboratory. The applicator stick should be cut or broken off, leaving the swab in the medium, and vials should be capped. Specimens should be transported to the laboratory on ice (4°C).

(text continues on page 590)

TABLE 22-1. Appropriate Specimens for Viral Diagnosis in Common Syndromes

Clinical diagnosis	*Nasal throat swab*	*Rectal swab*	*Urine*	*CSF*	*Lesion*	*Buffy coat*	*Comments*
Respiratory disease	++++	+ (Adenovirus)	+ (Mumps) (CMV)				Rectal swab useful if enteroviruses or adenoviruses are suspected. For respiratory syncytial virus nasal washes are better than throat swabs
Enteric illness	+++	++++					Stool is best specimen for rotavirus
Vesicular exanthema	++	+++			++		
Maculopapular exanthema	++	+++					Maculopapular lesions not cultured
Aseptic meningitis	+++	+++	+ (Mumps)	++			
Myocarditis, pericarditis, pleurodynia	+	+					Acute and convalescent sera for coxsackie B and influenza serologies
Newborn baby with suspected intrauterine infection	+++	+++	+++	++	++++	++	Conjunctival swabs also appropriate
Orchitis, parotitis	+++		+++				Include diagnosis with urine specimens for appropriate work-up
Immunocompromised host	+++	+++	+++		+++	++	
Conjunctivitis							Swab of conjunctival exudate or conjunctival scrapings are recommondod
Infectious mononucleosis	+++		+++			+	Culture for CMV if heterophil and EBV negative; serum specimen for EBV

TABLE 22-2. Procedures for Diagnosis of Viral Infections

Virus	*Seasonal prevalence*	*Disease*	*Appropriate specimens*	*Clinical diagnosis*	*Detection w/o culture*	*Serologic diagnosis*	*Comments*
Adenovirus, 31 serotypes	Year round, especially fall	Conjunctivitis, keratoconjunctivitis pharyngitis, pneumoitis, hemolytic uremia	Eye, throat, urine, stool,	—	IFA, EM	CF, SN, HI	Diagnosis questionable if isolated from stool only in low titer
Arenavirus		Hemorrhagic fever	Blood, tissue	Fair	IFA	IFA, SN	Only handled in special laboratories; extremely hazardous
Lassa	?						
LCM	?						
Coronavirus, 2 types	Fall, winter, spring	Upper and lower respiratory infections	Nasal wash	—	IFA, EM	SN, CF; HI (one type)	Not routinely done
Coxsackie A virus, 23 serotypes	?	Herpangina, hand-foot-and mouth disease, exanthema, aseptic meningitis, paralytic syndromes, URI, FUO	Throat, rectal, CSF	—	—	SN	Most serotypes can only be isolated in suckling mice
Coxsackie B virus, 6 serotypes	Summer/fall	Aseptic meningitis, exanthemas, hemolytic uremia	Throat, rectal, CSF, urine	—	—	SN	

Coxsackie B virus, 6 serotypes	Year round	Myocarditis, pericarditis, pleurodynia Bornholm's disease	Serum	——	——	SN	Serodiagnosis (paired sera taken 2 wks apart) recommended for these syndromes; consider also influenza and perform CF tests
Cytomegalovirus	Year round	Congenital infection, interstitial pneumoitis, hepatitis, heterophil negative mononucleosis, pneumonia in compromised host	Urine, throat, tissue, buffy coat (occasionally)	Poor	EM	CF	Isolation may take up to 3 wks
Echovirus, 32 serotypes	Summer/fall	Aseptic meningitis, exanthema, FUO, URI	Throat, rectal, CSF	——	——	SN	Serodiagnosis usually done at state health labs
Epstein-Barr virus	None	Infectious mononucleosis, usually heterophil positive	Serum	Good	——	IFA	Capsid antigen should be done on acute sample first: convalescent speciment not needed, if positive

TABLE 22-2. *(continued)*

Virus	*Seasonal prevalence*	*Disease*	*Appropriate specimens*	*Clinical diagnosis*	*Detection w/o culture*	*Serologic diagnosis*	*Comments*
Hepatitis A virus	All year, peaks in summer and autumn	Hepatitis	Serum (not plasma)	Fair	EM	CF, RIA	RIA for antibody only; can differentiate IgM vs IgG; plasma may give false positive results
Hepatitis B virus	None	Hepatitis	Serum (not plasma)	Fair	RIA		RIA for surface antigen (HBsAg) and for antibody (HBsAb); RIA for antibody to other components (HBcAb, HBdAb, HBeAb) plasma may give false positive results

Herpes simplex virus, 2 types	Year round	Gingivostomatitis, keratitis, congenital infection, genital infections, aseptic meningitis, encephalitis	Lesions, throat, conjunctival scrapings, CSF in aseptic meningitis, brain biopsy in encephalitis	Good	IFA	SN, IAHA	Subtyping of isolates now possible; type 1 antibody can be differentiated from type 2
Influenza virus, many serotypes	Winter	Influenza (headache, fever, myalgias, myositis)	Throat, sputum, tracheal aspirate, nasal washes	—	IFA	HI, CF	—
Mumps virus	Winter/spring	Mumps, aseptic meningitis	Throat, urine, CSF	Good	—	CF, HI, SN	Virus is very unstable, must be processed immediately
Parainfluenza virus, 4 serotypes	Year round, especially wintor to early spring	Bronchitis, croup, broncho-pneumonia	Throat tracheal aspirate, sputum	—	IFA	CF, HI	—
Poliovirus, 3 serotypes	Summer	Poliomyelitis	Throat, rectal, CSF	Good	—	SN	Often isolated following oral vaccination

TABLE 22-2. *(continued)*

Virus	*Seasonal prevalence*	*Disease*	*Appropriate specimens*	*Clinical diagnosis*	*Detection w/o culture*	*Serologic diagnosis*	*Comments*
Respiratory syncytial virus	Winter/spring	Bronchiolitis, croup, bronchopneumonia	Nasopharyngeal swab or wash, tracheal aspirate throat swab inferior	—	IFA	CF, SN	Very severe disease in compromised children; nosocomial spread common
Rhinovirus, over 100 serotypes	Fall, winter/spring	Common cold	Nasal swab or washing	—	—	SN	Serodiagnosis not commonly done in clinical laboratories
Rubella virus	Winter/spring	Rubella, congenital infection	Throat, nasal, urine, placental or fetal tissue	Good	IFA	HI, CF	Difficult to culture, requires special cell lines; notify laboratory
Rubeola virus	Winter/spring	Measles, SSPE (subacute sclerosing panencephalitis)	Throat, nasal	Good	—	HI	Notify lab if rubeola is suspected; special ID procedures required

Rotavirus	Winter/spring	Nonbacterial gastroenteritis	Stool; do *not* put into transport medium	——	EM, ELISA	——	Nosocomial spread common; stool specimen better than rectal swab
Togavirus	Depends on arthropod vector	Encephalitides: St. Louis, eastern equine, western equine, Venezuelan equine	Blood, tissue	——	——	CF, HI, SN	Only handled in special laboratories; extremely hazardous
Varicella-zoster virus	Winter/spring	Chickenpox, shingles	Lesion swab, throat swab	Good	IFA	SN, IHA	——

CF = complement fixation
ELISA = enzyme-linked immunosorbent assay
EM = electron microscopy
HI = hemagglutination inhibition
IAHA = immune adherence hemagglutination
IEM = immune electron microscopy
IFA = immunofluorescence assay
RIA = radioimmune assay
SN = serum neutralization
FUO = fever of unknown origin
URI = upper respiratory infection

Nasopharyngeal Swab
Specimens are collected by gently inserting a dry, special nasopharyngeal swab through the nares until resistance is met at the level of the turbinates. The entire cavity at that level is gently swabbed while rotating the swab. This process is repeated on the other side. Swabs are placed into transport medium and if respiratory syncytial virus is suspected, they are sent immediately to the laboratory on ice.

Nasal Washings
Nasal washings are obtained by instilling 5 ml of sterile saline into each nostril with the head hyperextended. The subject is instructed not to swallow. The head is then tilted forward, allowing the fluid to run out of the nares into a sterile container.

Sputum and Tracheal Aspirates
Specimens from the upper respiratory tract, which are of little value in bacteriology and mycology, can be quite useful in the diagnosis of viral infections of the lower respiratory tract. Herpes simplex virus is the only virus that can be recovered from the upper respiratory tract of a signficant number of "normal" adults. The specimens, collected in sterile containers and transported to the laboratory on ice, may be used for rapid diagnosis.

Rectal Swabs and Stool Specimens
The viruses most commonly associated with gastroenteritis (rotavirus, Norwalk-like agents, parvovirus, some adenoviruses) cannot be cultivated in cell culture. These agents may be demonstrated by electron microscopy or antigen detection methods. Stool samples are preferred for these analyses. Rectal swabs are adequate for other agents, are easily obtainable, and are preferred by laboratories.

Rectal swabs. Gently insert a dry cotton or Dacron swab several centimeters into the anal canal (just enough to dirty the swab) and gently twist. Remove, place into transport medium, and transport to the laboratory (maintain on ice).

Stool specimen. Collect 5 to 10 grams of stool in a clean container and maintain and transport to the laboratory at 4°C.

Urine
Urine should be collected in a sterile container. First morning, cleanly voided urine usually contains highest virus concentration. Urine should be kept at 4°C, but, unlike some other viral specimens, it should never be frozen, even if delivery is much delayed. Mumps, adenoviruses, and cytomegalovirus are the viruses most commonly recovered. The recovery of CMV is increased significantly when three sequential urine specimens are tested. Examination of the urine sediment for cytomegalovirus inclusions in cells is not a sensitive method for diagnosing CMV infection.

Cerebrospinal Fluid and Other Body Fluids
Cerebrospinal fluids and other body fluids should be collected in sterile containers without transport medium, maintained at 4°C, and transported to the laboratory as soon as possible. At least 0.8 ml of CSF are needed for complete viral testing.

Eye
The most common causes for viral conjunctivitis are adenovirus and herpes simplex virus infections. Varicella-zoster, cytomegalovirus, and enterovirus 70 have also been isolated. Conjunctival swab specimens should be collected by placing a sterile swab in the lower conjunctival sac so as to occlude

the nasolacrimal duct and allow the swab to become saturated with fluid. Corneal and conjunctival scrapings might be superior for isolation of herpes simplex virus and can be used for rapid diagnosis. They should be obtained by an ophthalmologist or other trained person. Both types of specimens should be placed into transport medium, maintained at 4°C, and should transported to the laboratory as quickly as possible.

Lesions

Viruses can often be isolated from vesicular-pustular lesions (fluids and cells) but rarely from macular lesions. Vesicular lesions should be opened with a 25-gauge needle and vigorously swabbed with a dry swab. The swab should be placed into transport medium, maintained at 4°C, and transported to the laboratory as soon as possible. Lesion scrapings collected in this manner provide material for isolation and fluorescent antibody examination for herpes simplex virus and varicella-zoster virus. Most laboratories prefer to make slides from the lesion scrapings in the viral transport medium.

Tissue Specimens

Small pieces of tissue to be used for tissue specimens should be placed in viral transport medium. Large pieces may be transported in dry sterile containers, as long as one is certain the specimen will be moist on arrival at the laboratory. Alternatively, they may be maintained in sterile saline at 4°C.

Buffy Coat

For buffy coat, whole blood should be collected in vacutainer tubes containing anticoagulants and immediately transported to the laboratory. Hemolyzed blood is unsuitable.

Serum for Serologic Studies

For serologic studies, blood should be drawn in tubes without anticoagulants, and sent to the laboratory. For shipment, only serum should be sent. If the patient is receiving anticoagulants, the laboratory must be notified since this might be a source of false positive results in hepatitis testing.

Other Specimens

Occasionaly, pleural fluid, semen, urethral and cervical secretions are submitted. If sufficient fluid can be aspirated, it is the most satisfactory specimen. Alternatively, a swab can be used to absorb secretions. The swab is placed into transport medium and submitted for culture.

All specimens must be properly labelled with the patient's name, age (child or adult), sex, site of culture, and date obtained, and should be submitted with a request slip. A brief statement describing the clinical signs and symptoms (such as croup, bronchiolitis, parotitis, meningitis, encephalitis, rash, postoperative fever) is needed to assist the laboratory in selecting the most rapid and appropriate procedures. For serologic studies of serum specimens, this clinical information, plus a history of past vaccination (if known) is extremely important. A summary of procedures for collection and transport of the most common viral specimens is presented in Table 22-3.

Transport Media

From a molecular point of view, viruses consist basically of unique units of nucleic acid, DNA or RNA, which are wrapped into coats of varying complexities. In their simplest form, coats consist of proteins; in their most complex form they consist of intricate structures composed of proteins and lipids. They

TABLE 22-3. Summary of Procedures for Collection and Transport of Viral Specimens

General
Collect specimens early in illness; immediately place swabs into viral transport medium; do not freeze specimens

Throat
Swab vigorously with dry swab to obtain cells; swab all lesions

Nasopharynx
Use nasopharyngeal swab or get washing; swab vigorously; if RSV is suspected specimens must be transported to laboratory immediately

Stool
Rectal swabs are adequate and often preferred; except in suspected rotavirus infections

CSF
Complete work-up requires 0.8 ml of fluid

Eye
Scrapings best for HSV; should be done by trained person

Lesions
Vesicles yield virus more often than ulcers; scrape base of vesicle vigorously to obtain cells

Urine
Mostly for suspected CMV and mumps cases; early morning specimen contains highest virus titers; three successive specimens optimal for CMV; diagnosis from cellular inclusion bodies very insensitive; *never freeze urine*

Biopsy material
Submit large specimens in a clean container; place small specimens into transport medium

Blood culture
Submit at least 3 ml of blood containing anticoagulant (heparin, versene)

Serologic studies
Require 5 ml of clotted whole blood; submit both acute and convalescent specimen; specify specific tests desired; send serum if longer transport time expected

Transport
Maintain and transport specimens in viral transport medium on wet ice or refrigerate (4°C); *never freeze specimens;* transport to laboratory as fast as possible

serve as a protective shield for the viral genome and as a vehicle for cell entry. Conditions that affect the structure and integrity of the coat, such as adverse physical and chemical environment or enzymatic attacks, are also detrimental to viral viability. Several types of media have been used to stabilize viruses in specimens during transport to the laboratory. These media consist mainly of buffered, balanced salt solutions with a pH indicator; they are supplemented with a protein source (albumin or gelatin) to stabilize viruses, and antimicrobial substances to inhibit bacterial growth. In order to maintain antimicrobial potency, transport media are best stored at −20°C and thawed just before use.

Drying and temperature, two other factors that fortunately can be controlled, also greatly affect viral stability. With rare exceptions (hepatitis B antigen and poxviruses), viruses are very sensitive to dehydration. Swabs, therefore, should be immediately placed into viral transport medium. Attacks from proteolytic enzymes and instabilities of protein-lipid structures are markedly decreased at 4°C. Thus putting viral specimens on ice, even if the specimen is in transport for only short periods, is a most desirable practice. It is also the preferred method for shipment and storage up to several days duration.

Only if very long transit or storage periods are to be expected should viral specimens be frozen. Freezing should be carried out immediately and quickly, but only at −70°C (dry ice), and never at −20°C (as in freezer compartments of refrigerators); the frozen specimens must be maintained at −70°C during shipment and storage. It is important to know that freezing, even at optimal temperatures and conditions, is often associated with decreased viral infectivity. In contrast, freezing and storage at −20°C is adverse to most viruses, drastically reducing the viability of many clinically important viruses (HSV, for example) and almost surely abolishes the chance for isolation of CMV, RSV, and VZ viruses. In virology, in fact, $-20°C$ freezers are useful only for the storage of virologic reagents and serum specimens.

CULTIVATION METHODS

Mammalian cells must be propagated and maintained under physiologic conditions controlled for osmotic pressure, pH, and organic salts. This is accomplished by use of a balanced salt solution—the basis of all chemically defined media—and buffers. Carbon dioxide-bicarbonate is the most commonly used buffer system; this buffer system best simulates in vivo conditions and bicarbonate is essential for growth of some cells. However, loss of carbon dioxide quickly raises the pH above the tolerable physiologic limits of cells and, therefore, closed culture systems or incubation in 5 percent carbon dioxide atmosphere must be used. In addition, the buffering capacity within the physiologic pH range is small and limited. For these reasons, the use of organic buffers, particularly HEPES and TRICINE, are becoming increasingly popular.

Phenol red is incorporated into media to signal changes in pH that are usually associated with altered metabolic activity of the cells. Many basal salt solutions also contain glucose or lactose as an energy source and are supplemented with amino acids, vitamins, and other nutrients. Many different formulations are used in a vast variety of available media. The medium most commonly used in clinical virology is Eagle minimum essential medium (MEM). It contains, as the name implies, minimal but also optimal concentrations of amino acids and vitamins. The medium is easily modified to suit special needs. For example, prepared without calcium, it is used to grow cells in suspension culture, or, supplemented with nonessential amino acids, it meets special growth requirements.

Essentially all media must be supplemented with serum for cell growth and virus propagation. Serum, especially from fetal sources, provides growth-promoting factors. Fetal bovine serum, the most commonly used serum source in diagnostic virology, is expensive; and, due to ever increasing demands, adequate supplies are decreasing.

Serum may also contain mycoplasmata, animal and bacterial viruses as well as antibodies and components that adversely affect cellular and viral growth. Therefore, serum must be screened and subjected to effective quality control procedures before it may be used in diagnostic virology. The serum component in media also causes problems in standardization and quality control of media. For example, the quality of serum in virus isolation media can significantly affect the isolation rate of some viruses. This problem is usually discovered in retrospective analysis because the effect of serum on this crucial parameter cannot be tested directly prior to use within reasonable practical and economical limits. In spite of extensive research efforts, adequate synthetic substitutes have not yet been found.

For practical purposes, media used in diagnostic virology can be divided into two groups: growth media and maintenance media. *Growth media* are used basically for preparation of cell cultures; they are nutritionally rich and contain a high concentration of serum (10 percent) to promote cell growth. *Maintenance media,* in contrast, are designed to slow down growth and to maintain cells in a steady metabolic state during viral replication. They contain minimal concentrations of nutrients and ideally should be serum-free. However, most cells will not tolerate serum-free medium for the prolonged time period that is necessary for virus isolation. Therefore, the serum concentration is commonly reduced to 1 percent, a good compromise in most situations.

Most virology laboratories also supplement media with antimicrobials to control bacterial contamination. Formidable arguments for and against this practice could be raised, particularly with cell growth media. Antimicrobics are indispensible in viral isolation since most specimens derived from humans are contaminated with bacteria.

Cell Cultures

Cell culture is the term used to denote the growing of cells under conditions where they are no longer organized in tissues or organs. *Tissue or organ culture,* in contrast, refers to the maintenance or growth of tissues, whole organs or part of organs under conditions that allow for differentiation or preservation of architecture and function.

Cells may be cultivated in three differing systems. In *explant culture,* a tissue fragment is used to initiate a culture. This culture is rarely used for viral inoculation, but instead is sometimes used to isolate viruses from infected tissues by cocultivation techniques. *Monolayer cultures* consist of a single layer of cells on a surface. They are prepared by treating a tissue or an established cell line with cell dispersing agents (such as proteolytic enzymes or chelating agents), resuspending the cells in growth medium, and placing the suspension into culture vessels. The cells are allowed to settle and attach to the surface; they will replicate and eventually form a single layer of firmly attached cells. This is the system most commonly used in viral isolation. In *suspension culture,* cells are grown suspended in liquid medium under agitation. Cells can be easily manipulated; they can be kept in logarithmic growth status and large numbers of cells can be grown. These can be used to prepare cell monolayers. They are very useful in the preparation of viral antigens or vaccine production. However, not all types of cells can be propagated in suspension culture.

Cells are also described in terms based on cell characteristics that are most useful in diagnostic virology. *Primary cell cultures* consist of cells derived directly from tissue or organs (e.g., lung, kidney). When a primary cell culture is subcultured, it becomes a *cell line.* Cell lines may be of several types. *Diploid or semicontinuous cell lines* are derived from primary cell lines by subculture. At least 75 percent of the cells have the same karyotype as normal cells of the same species from which they were originally cultured. They will undergo only a finite number of subpassages, generally about 50; many will become insensitive to viral infection after the 30th subculture. *A heteroploid cell line* is one which has fewer than 75 percent of cells with diploid chromosome number.

Established or continuous cell lines can be subcultured indefinitely; they are generally derived from malignant tissue of epithelial nature, and they are fast growing. Since all continuous cell lines are heteroploid, the term heteroploid cells is often used, albeit incorrectly, to denote cells which can be perpetuated indefinitely.

Embryonated Hen's Eggs

Embryonated eggs for viral isolation have been largely replaced by cell culture systems. They are still used in special circumstances and in situations where they may be more sensitive than cell culture; for example, with some influenza virus strains and arenaviruses. Eggs may be inoculated at various stages of embryonic development and by different routes. Three main routes are used for inoculation: the allantoic cavity, amniotic cavity, and chorioallantoic membrane (Fig. 22-1). The amniotic route is preferred for virus isolation. Material injected into the amniotic cavity becomes disseminated in the embryo via the respiratory tract and alimentary canal. Viruses present in the inoculum are thus exposed to a larger variety of cells and tissues, increasing the chance for isolation. The allantoic route is often used to prepare larger quantities of viruses after isolation in the amniotic cavity. The yolk sac is primarily used for the isolation and propagation of rickettsias and chlamydias. Viral replication can be detected by the development of pocks on the chorioallantoic membrane (poxviruses, vaccinia viruses, and herpes simplex viruses), appearance of hemagglutinins in the allantoic and amniotic fluids (influenza viruses), and by death of the embryo.

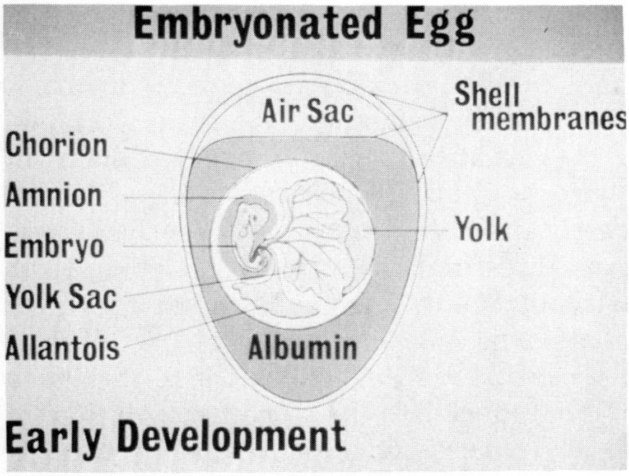

FIGURE 22-1. Schematic representation of viral inoculation sites in embryonated hen's eggs. (Courtesy of Centers for Disease Control, Atlanta, GA 30333.)

Laboratory Animals

Of all the laboratory animals available, only mice (adult or infant) have been used extensively in virus isolation. They too have been largely replaced by cell cultures. However, there remains a number of known viruses that are best isolated in animals and a few that can only be isolated in animals. These viruses are members of Togaviridae and Bunyaviridae, most types of coxsackie A viruses and several members of Arenaviridae. Mice are inoculated intracerebrally or intraperitoneally and are observed for development of specific disease syndromes; for example, paralysis or death. In many cases, hemagglutinins can be extracted from tissues, and viruses can be selected for growth in cell culture where they can be identified or serotyped. In some cases, this is not possible, and serotyping, for example, must be carried out in mice. These procedures are expensive and tedious; they are also often very hazardous and should be carried out only in extremely well-equipped laboratories or reference centers.

Organ Culture

Organ culture is primarily used in research laboratories for isolation and propagation of viruses that will not replicate in cell culture but will sometimes replicate in small tissue fragments or pieces of organs whose normal architecture has been preserved. Organ cultures are established by placing small organ fragments (for example, tracheal or intestinal rings) into cell culture dishes containing enriched growth medium. Specimens are inoculated into the medium and viral replication can be detected by impairment of ciliary movement or cell degeneration, and by electron microscopy. These approaches have been used successfully with coronaviruses and rotaviruses. Organ culture also has been used in research to isolate latent herpes simplex virus from sensory ganglia. Ganglia were placed intact into cell growth medium and incubated. Aliquots of medium were then periodically removed and inoculated into sensitive indicator cell cultures.

Cocultivation and Cell Fusion

Cocultivation and cell fusion are used to detect latent or slow viruses. In cocultivation, either minced tissue suspensions or explants of tissues that are suspected of harboring virus are placed onto sensitive indicator cell monolayers. The indicator cells are then observed and tested for virus.

In the cell fusion technique, membranes of suspected cells and indicator cells are actually being fused by either chemical agents or irradiated viruses with known cell-fusing capabilities (Sendai virus). Cultures are then observed for development of cytopathic effects or are subjected to other virus detection methods. Neither of these techniques is used routinely in diagnostic laboratories, but some have the capability for cocultivation and will perform it on special request; for example, in cases of subacute sclerosing panencephalitis.

CULTIVATION AND IDENTIFICATION OF VIRUSES

All viruses depend on living cells for replication, but some require even the help of other viruses for propagation. Adeno-associated viruses, for instance, must be helped by adenoviruses in order to complete their replicative cycle. Human viruses also exhibit varying degrees of host and cell tropism. Herpes simplex virus can be grown in a variety of mammalian cells; poliovirus, on the other hand, can be propagated only in cells of higher primates; Epstein-Barr virus, the most restricted, will replicate only in transformed human lymphocytes.

The discovery by Enders and co-workers in 1949 that poliovirus could be grown in various human cells of nonneural origin provided one of the two major stimuli for the development of cell culture. The other significant stimulus came from basic research on the nutritional requirements of mammalian cells, by Eagle and coworkers in 1959. Their studies resulted in the formulations of chemically defined media. Continuous progress in these areas, separately or concurrently, has resulted in the development of the science of cell culture. Because of these rapid developments, most cell culture systems in clinical virology have become routine; all components are commercially available, are economical and, most importantly, can be standardized to a large degree. Indeed, many laboratories buy cell cultures ready for use just as bacteriologic laboratories purchase prepared media.

None of the culture systems described here can isolate all viruses of medical importance; in fact, there are several known viruses that cannot be propagated in any of these systems (for example, hepatitis viruses and rotaviruses, or recently discovered intestinal adenoviruses). Cell culture, without doubt, has the broadest spectrum of all virus isolation systems. New cell lines are continuously developed that either have increased sensitivity for the isolation of known viruses or allow propagation of viruses that previously could not be cultivated. The properties that make cells susceptible to viruses are poorly understood: specific receptor sites on cell surfaces as well as intrinsic factors have been implicated.

Consequently, most cell lines have been chosen empirically.

In clinical virology it is not enough that a cell line is simply capable of propagating viruses. To be useful, the cells must also exhibit some signs of viral infection that make the presence of viral growth easily demonstrable. One of the most useful criterion used to identify viruses is the development of characteristic cytopathic effects during viral infection. Thus, infected cells might become stringy and granular; or swollen, rounded, and even aggregated and fused into syncytia. For example, the cytopathic effects of two human coronaviruses in human fetal tonsil cells are shown in Fig. 22-2. The characteristics of the cytopathic effect, its speed of development, and the cell line in which it occurs are employed in differential diagnostic systems as shown in Table 22-4. In some cases the cytopathic effect is so characteristic and pronounced that a presumptive diagnosis can be made, for instance with herpes simplex virus.

Some viruses cause little or no cytopathic effect. However, during the process of infection, the cell membranes become altered in such a way that erythrocytes of some animal species will firmly attach to them when added to the culture. This phenomenon, called hemadsorption, is commonly found with members of Orthomyxoviridae and Paramyxoviridae. Since hemadsorption is blocked by specific antisera, serologic typing can be used to identify influenza, parainfluenza, and mumps virus. In some cases a noncytopathic virus can be detected because infected cells become refractary to superinfection with another cytopathic virus. Such a system is used in the detection of rubella virus in cell culture. Alternately, cell cultures can be tested for presence of virus with specific immunologic reagents before cytopathic effects appear. Thus, specific antibodies coupled to enzymes or fluorescent dyes are being used to detect rubella, varicella-zoster and cytomegaloviruses.

Virus isolation is the focal point of diagnostic virology, it has the broadest spectrum of all detection methods, it is often the first step in virus identification, which makes it possible to assess the association of a particular virus with disease; it allows for drug-sensitivity testing; it is the "gold standard" to which the sensitivities of other diagnostic tests are compared. Most importantly, it is the only method, often in conjunction with electron microscopy, by which new viral agents can be detected. On the

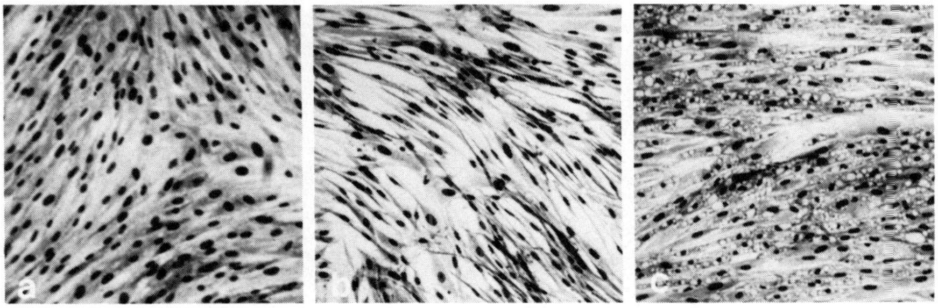

FIGURE 22-2. Cytopathic effect of coronavirus 229E and OC43 in human fetal tonsil cells. *(a)* Uninfected cells. *(b)* 229E-infected cells. *(c)* OC43-infected cells. May-Grunwald-Giemsa stain.

TABLE 22-4. Isolation Scheme for Common Viral Pathogens

Virus	*Primary cell cultures*	*Human diploid cell lines*	*Human heteroploid cell lines*	*Embryonated eggs*	*Mice, 24-72 hours old*
Polio	+ + +	+	+ +	− − − −	− − − −
Echo	+ + +	+	− − − −	− − − −	− − − −
Coxsackie A	− − − −	− − − −	− − − −	− − − −	+ + +
Coxsackie B	+ + +	+	+ + +	− − − −	+ + +
Mumps	+ (HAD, usually no CPE)	− − − −	±	+ + (HA)	− − − −
Arbovirus	− − − −	− − − −	− − − −	− − − −	+ +
Adenovirus	±	+	+ + +	− − − −	− − − −
Influenza	+ (HAD, usually no CPE)	?	− − − −	+ + (HA)	− − − −
Parainfluenza	+ (HAD, usually no CPE)	?	− − − −	− − − −	− − − −
Rhinovirus	±	±	±	− − − −	− − − −
Respiratory syncytial	±	?	± +	− − − −	− − − −
Herpesvirus	±	+ + +	+ +	+ + (Pocks)	+ + +
Varicella-zoster	− − − −	+ + (1-3 wks)	− − − −	− − − −	− − − −
Vaccinia	+ +	+ +	+ +	+ + (Pocks)	±
Cytomegalovirus	− − − −	+ + (1-3 wks)	− − − −	− − − −	− − − −

HA = Hemagglutination; HAD = Hemadsorption; CPE = Cytopathic Effect
From Laboratory Manual Children's Orthopedic Hospital, Seattle: modified and used with permission.

other hand, isolation techniques are costly and time consuming. Consequently, an ever increasing number of diagnostic tests are being developed for specific viruses that are rapid and independent of culture.

Quantitation

Quantitative measures are necessary for the performance of standardization of tests. One of the most commonly used measures is the degree of cytopathic effect rated 1+ to 4+ depending on the area of the cell sheet that is infected. In most instances the cytopathic effect is directly related to the viral concentration and, together with the speed it has developed, a rough estimate can be made in a given system. However, a more exact quantitation is required in most cases. In order to detect the number of infectious particles in a given virus suspension, serial dilutions are prepared and aliquots inoculated into cell culture tubes or laboratory animals. The dilution that produces cytopathic effect or death in 50 percent of the cell cultures or animals, respectively, can be calculated from the experimental data by statistical methods. This dilution is the 50 percent tissue culture infectious dose ($TCID_{50}$) or 50 percent lethal dose (LD_{50}). Viral infectivity can be measured even more precisely by plaque assay. Aliquots are plated onto cell monolayers prepared in small cell culture dishes and overlaid with semisolid medium to prevent viral spread via the medium. Thus, localized foci of virus-infected cells develop, called plaques, which under optimal conditions originate from a single infectious viral particle. Plaques can be visualized by staining with dyes. An example of coronavirus plaques in human fetal tonsil cells is shown in Figure 22-3. Plaque assay is a highly quantitative measure of viral infectivity; consequently, reduction in the number of plaques is a very sensitive technique used to measure virus neutralizing ability of antibody and antiviral substances. In addition to quantitative measurements, plaque purification is very important in the separation of virus mixtures and in the preparation of pure virus strains. Plaque purification is accomplished by repeated plaquing of progeny derived from one plaque.

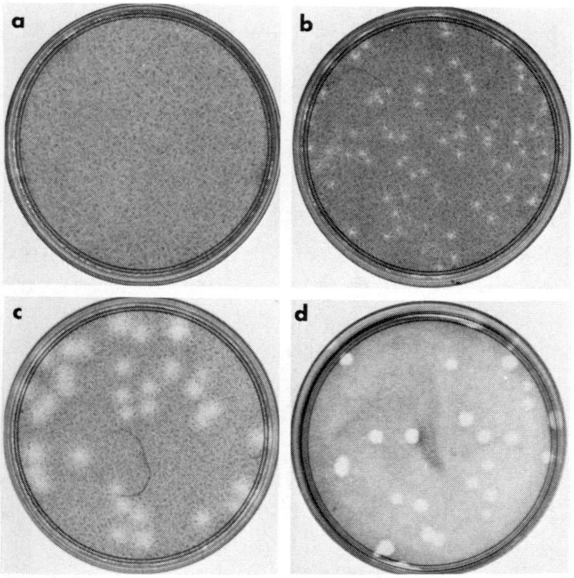

FIGURE 22-3. Coronavirus plaques on two human cell lines at 4 days. *(a)* Uninfected human fetal tonsil cells. *(b)* B814 and *(c)* 229E plaques on fetal tonsil cells. May-Grunwald-Giemsa stain. *(d)* OC43 plaques on human rhabdomyosarcoma cells. Neutral red stain.

Total counts of viral particles may be done by electron microscopy. Together with infectivity assay, the ratio of total number of virions to infectious virions can be calculated. Such measurements are rarely, if ever, taken in diagnostic virology. However, the ratios vary greatly between viruses, ranging from 1:1 to 10,000:1. This information is often useful in deciding the route to pursue in the development of a sensitive detection method for a given virus.

Viral concentrations are also defined in terms of protein or nucleic acid content. These quantitations are important in enzyme-linked immunoassays and nucleic acid hybridization tests. They also best demonstrate the extraordinary sensitivity and specificity of these tests if one realizes that picograms (10^{-12} g) of virus or viral components can be detected and differentiated from similar biologic materials.

Special units of measurement are employed in serologic tests. For instance, in complement-fixation and hemagglutination-inhibition tests, one unit of viral antigen is defined as the dilution of a stock preparation that inhibits complement fixation or agglutinates a defined quantity of erythrocytes. Multiple units (4–8) are used in the actual test to prevent false reactions.

Rapid Viral Detection and Identification Methods

Most of the strategies and methods used in clinical virology laboratories are aimed at detecting and identifying known viruses. Consequently, of the many parameters that are applied to differentiate and classify viruses (physical properties, chemical and antigenic composition, modes of replication, etc.), only a few are used in diagnostic virology. Antigenic composition is by far the most useful property for two reasons: First, immunologic recognition creates an additional dimension in specificity and sensitivity, which is needed to recognize and to differentiate minute quantities of similar biologic materials. Second, many immunologic tests can be arranged to detect either viral antigens or antiviral antibodies. The specificity of any chosen test must be high, and, ideally, so should be its sensitivity. However, the degree of sensitivity needed varies. For example, extraordinary sensitivity might be required to detect a small number of virions present in a clinical specimen, whereas considerably less sensitivity might be needed after the antigenic mass has been greatly amplified during viral replication in cell culture. Most importantly, methods should be used that best result in an etiologic diagnosis at the level required for patient management, in the shortest time and at minimal cost. It is not surprising, then, that rapid diagnosis of viral infections, independent of viral cultivation, has become the fastest developing field in diagnostic virology. There are two main groups of diagnostic tests: those that do require an immunologic component (the vast majority), and those that do not.

Direct microscopic examination of cytologic and histologic preparations does not require an immunologic component; specific stains are used to demonstrate characteristic morphologic changes in cells infected with certain viruses. For example, characteristic cytoplasmic or nuclear inclusion bodies can be seen with herpesviruses, poxviruses, and rabies viruses. Inclusion bodies are materials consisting of virions, viral components, or products of viral synthesis that accumulate during viral replication. Similarly, giant multinucleated cells or syncytia can be demonstrated in measles infections. The specificity of these tests is obviously poor, as is their sensitivity. Therefore, these tests are not used as primary detection methods in virology laboratories.

Electron microscopy is the major rapid diagnostic technique that does not require an essential immunologic component. Its use in diagnostic virology was greatly enhanced by development of the negative-staining technique: addition of heavy metal salts to the specimen in order to provide an electron-dense background against which virus particles become visible in the electron beam. Electron microscopy is most useful in situations where the specimen contains at least 10^6 to 10^7 particles per ml, or where these concentrations can be obtained by enhancement techniques (for example, ultracentrifugation of urine specimens). Besides its use in the direct detection of viruses in clinical specimens, electron microscopy can also be used to substantially speed up the identification of viruses recovered in cell or organ cultures (Fig. 22-4). Sensitivity and specificity, in particular, are markedly increased in immunoelectron microscopy. In this technique, specimens are preincubated with specific antiserum before examination. The subsequent observed clumping of viral particles not only improves visualization and, therefore, sensitivity, but also imposes immunologic specificity on the assay, which allows for serologic typing (Fig. 22-5).

Although electron microscopy is very useful in diagnosis of viral infections—only 15 to 20 minutes are needed for an etiologic diagnosis in some cases—the technique is poorly suited for processing large numbers of specimens. In addition, since the costs of electron microscopes are very large, alter-

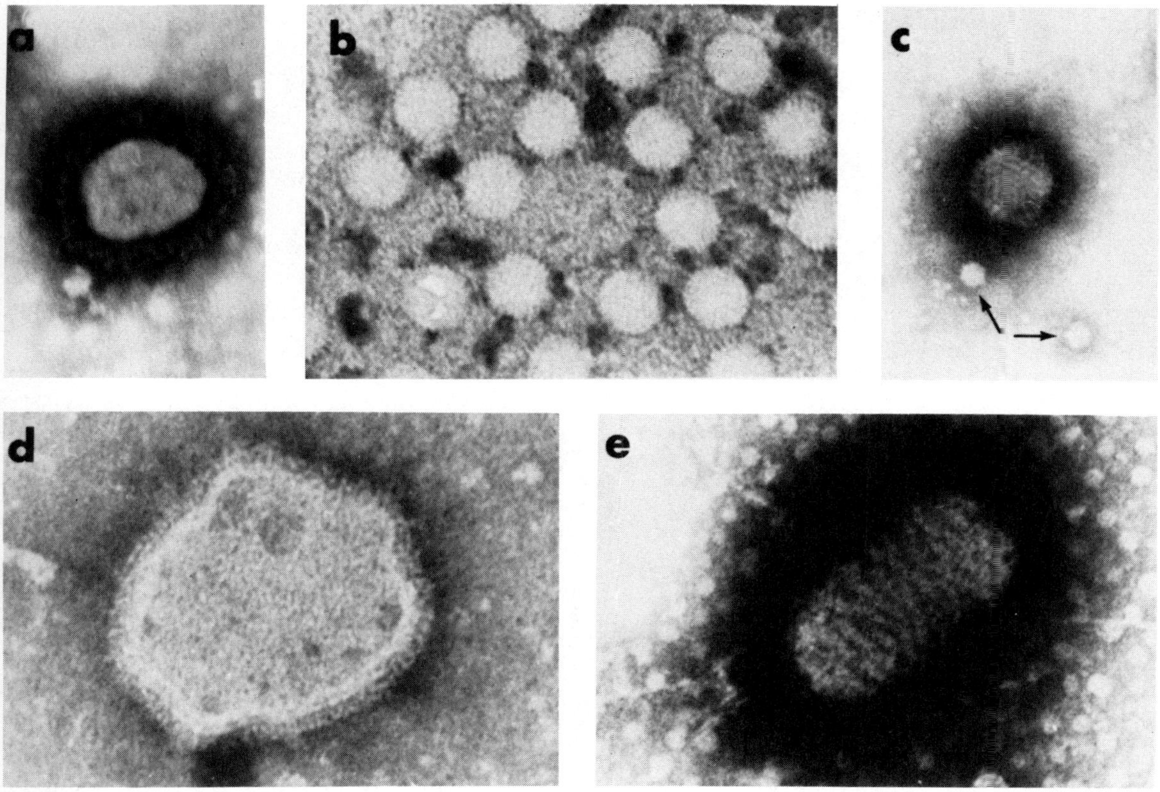

FIGURE 22-4. Human viral pathogens demonstrated by electron microscopy. All specimens are negatively stained with phosphotungstic acid. *(a)* Coronavirus 229E from tracheal organ culture. Magnified 154,000×. *(b)* Adenoviruses (noncultivable) from stool specimen. Magnified 83,000×. *(c)* Adenovirus type 3 and adeno-associated virus type 3 *(arrows)* from eye specimen. Magnified 104,000×. *(d)* Parainfluenza virus from cell culture. Magnified 122,400×. *(e)* Orf virus ("M" form) from finger lesion. Magnified 84,000×. (*d* and *e* courtesy of Joan Dragavon, Children's Orthopedic Hospital, Seattle.)

nate methods are usually substituted for rapid diagnosis. On the other hand, electron microscopy has often been instrumental in the discovery of new viruses, as well as in the development of other diagnostic tests. It is still very useful in the examination of stools from patients with gastroenteritis, because many of the viral agents involved (rotaviruses, Norwalk agents, astroviruses, certain adenoviruses, calciviruses, and coronaviruses) cannot be cultured sufficiently to produce sensitive immunologic reagents.

Many rapid diagnostic techniques with an immunologic basis are used to directly detect viruses or viral antigens in specimens: double immunodiffusion, counter immunoelectrophoresis, indirect (or passive) hemagglutination, mixed hemagglutination, immune adherence (IEA), radioimmunoassay (RIA), and enzyme-linked im-

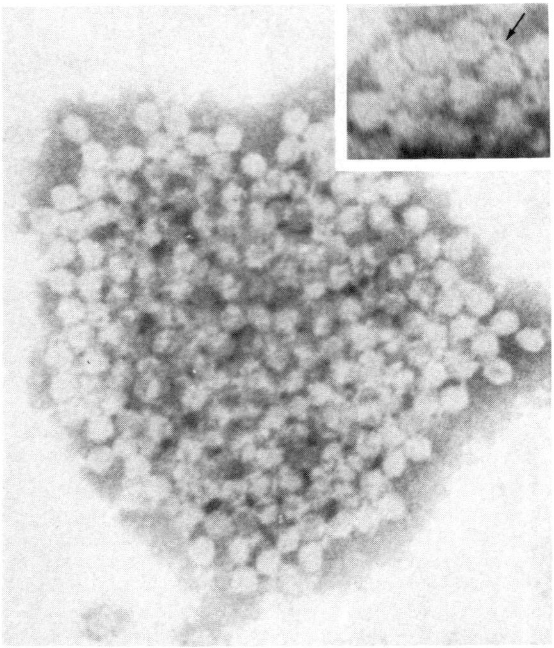

FIGURE 22-5. Adeno-associated virus type 3 immune complex. Magnified 104,000 ×. *(Insert)* immune complex with threadlike structures *(arrow),* presumably antibody. Magnified 258,000 ×.

munosorbent assay (ELISA). The degree of sensitivity needed, and the availability of specific immunologic reagents, determine largely which test should be used to detect a given virus. For example, double immunodiffusion can detect antigens only at microgram levels: it is the least sensitive method of those already discussed. This degree of sensitivity is adequate to detect rotavirus antigens in stools of infected persons. Radioimmunoassay, the most sensitive method, can detect picograms of antigens. This level of sensitivity—a million-fold increase over double immunodiffusion—might be required to demonstrate presence of measles virus antigens in brain tissue of patients afflicted with subacute sclerosing panencephalitis. Such extraordinarily sensitive methods require very potent and specific immunologic reagents. The supply of potent antisera has improved dramatically, in part due to increased demands, but mostly due to rapid developments in monoclonal antibody production.

Regardless of its sensitivity, success or failure of any method is determined by selection of the appropriate antiserum. Proper selection is only possible if the identity of the suspected viral agent is either known or guessed correctly. The decision is largely based on the clinical syndrome and to a lesser degree on the patient's history and the seasonal prevalence of viral pathogens. In order to improve the chances for correct matching, panels or batteries are used that cover the most prevalent viral pathogens associated with certain syndromes. For example, in respiratory syndrome, the antiserum panel would cover influenza A and B, respiratory syncytial, parainfluenza 1, 2, and 3, and adenoviruses. This problem, only one of many, shows why viral isolation in cell culture is still indispensable.

Fluorescent antibody technique (FA), direct or indirect, is the most widely used method where antigens are cell associated and the infected cells are accessible. In the *direct method,* virus-specific antiserum is labelled and reacted with the specimen. In the *indirect method,* virus-specific antiserum is reacted with specimen. Labelled antispecies immunoglobulin, directed against the animal species in which the virus-specific antiserum was prepared, is then added to the complex. Thus, the same conjugated immunoglobulin preparation can be used in many different tests. Since the original antigenic mass is becoming amplified, the indirect FA method is more sensitive than the direct FA method. On the other hand, the direct method is more specific but more individual antisera must be conjugated. Nonspecific reactions can be a problem especially with specimens containing mucus or lymphocytes. Virions are invisible in the light microscope and, therefore, differential criteria such as morphology and size cannot be applied. Instead, characteristic patterns in fluorescent staining are used for interpretation. For example, cytoplasmic and nuclear fluorescence is typical for influenza viruses and herpesviruses; it reflects the mode of viral replication. Similarly, cytoplasmic staining only is expected with parainfluenza and respiratory syncytial viruses.

The advent of monoclonal antibodies has greatly stimulated the use of fluorescent antibody techniques. Nonspecific reactions have been greatly reduced; and the extraordinary specificity of these reagents has made it possible to apply FA techniques to specific purposes previously deemed difficult or unfeasible. For example, the differentiation of herpes simplex virus type 1 from type 2, which is clinically important, presented great difficulties with conventional reagents due to the large number of shared antigens between the two virus types. The application of type-specific monoclonal antibodies in FA techniques allows now for identification and also differentiation of HSV in one test (Fig. 22-6).

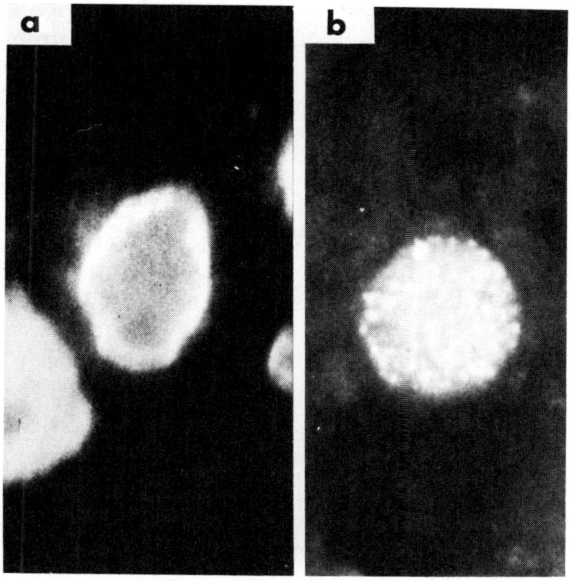

FIGURE 22-6. Identification and differentiation of herpes simplex virus type 1 and type 2 by immunofluorescence assay using monoclonal antibodies. The characteristic fluorescent staining patterns with both anti-HSV-1 and HSV-2 monoclonal antibodies are shown. *(a)* Rimming in HSV-1 infected cells stained with anti-HSV-1 monoclonal antibody G11. *(b)* Cytoplasmic speckling in HSV-2 infected cells with anti-HSV-2 monoclonal antibody E12. (Courtesy of Eric Peterson, Children's Orthopedic Hospital, Seattle.)

FA methods are, of course, also used in rapid identification of viral isolates and for early detection of antigens in cell culture. However, antibodies conjugated with enzymes (for example, horseradish peroxidase) are becoming increasingly popular for this purpose. Enzyme immunoassays are more amenable to mass screening of specimens and do not require special equipment.

For any of these tests, the number as well as the condition of infected cells present in the clinical specimen are crucial. For example, in cases of HSV encephalitis, virus can be isolated in cell culture from brain biopsy material, but FA tests yield false negative results when the antigenic concentration is below the sensitivity threshold of the assay. In addition, only cell-associated antigens can be detected by this technique and the concentration of these antigens can be considerably reduced in a largely necrotic cell population. This is due to spilling of cellular contents or leaching during processing and testing. Although the sensitivity of fluorescent antibody and immunoenzyme assays can be boosted by several techniques, success or failure ultimately depends on adequate cell samples.

Solid Phase Immunosorbent Assays

A new generation of tests, solid phase immunosorbent assays, has emerged in which antigens are used in soluble state and, consequently, dependence on cell-associated antigens has been eliminated. In addition, the sensitivity of these tests is extremely high, they are well suited for automated procedures, and, therefore, have great promise for future developments. Solid phase immunosorbent assays are used in conjunction with radioactive labels (radioimmunoassay, RIA) or enzymes (enzyme-linked immunosorbent assay, ELISA) to detect either antigen or antibody. Both methods are based on similar principles and only the indicator systems are different: measurement of radioactive counts is used in RIA, and colorimetric or fluorimetric measurements are used in ELISA. There are many variations of these tests depending on the level of sensitivity and specificity desired, and what special needs are to be satisfied. In one of the most widely used methods in antigen detection, specific antiviral antibody (capture antibody) is adsorbed onto a solid surface, usually plastic; specimen is added that has been treated to convert antigens into soluble form; and antigens are captured and bound immunospecifically (Fig. 22-7). Another virus-specific antibody, prepared in a different animal species, is added and, finally, labelled antispecies antibody (detection antibody) is added, which binds to the second virus-specific antibody.

In RIA, detection of antigen is accomplished by measuring radioactive counts; in ELISA, a substrate is added and enzymatic cleavage products are measured colorimetrically. The technique described is an indirect "sandwich" technique. The term sandwich is used because the antigen is wedged between two antibodies; it is indirect because the labelled indicator antibody does not react with the antigen that is to be detected. Even more complex systems are used, especially in ELISA, to increase sensitivity and specificity or to detect subclasses of immunoglobulins when ELISA is employed to detect antibody. In some systems, labelled *Staphylococcus aureus* protein A is substituted for indictor antibody. *S. aureus* protein A binds to the Fc domain of IgG antibody molecules. Indeed, stacks of sandwiches can be produced for very specific purposes.

Increases in sensitivity and quantitative capability are also achieved in competition assays. Such assays are set up so that the component one wishes to detect or quantitate—antigen or antibody—competes with an identical known and quantitatively defined component in the assay system for the same attachment site on antigens or antibody. Consequently, even minute deviations from results obtained without the competitor can be detected and precisely quantitated. In spite of their exotic and esoteric appearances, many of these tests are used in large diagnostic virology laboratories and, in fact, a good number have become commercially available. For example, RIA is available for detection of hepatitis A and B antigens and antibodies.

Serologic Diagnosis of Viral Infections

Serologic diagnosis of viral infections is based on the principle that infection of the

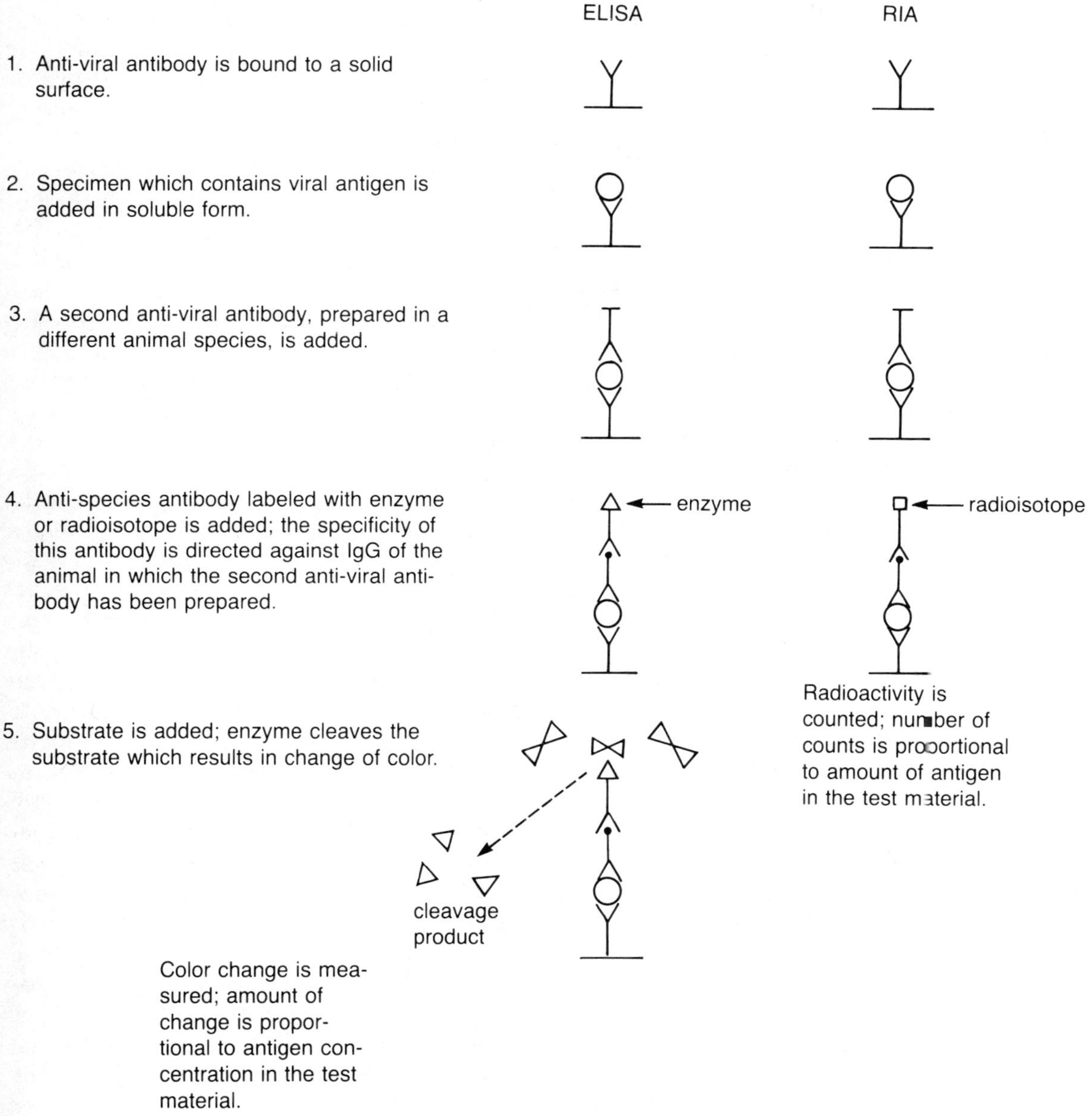

FIGURE 22-7. Schematic diagram for immunosorbent assay for detection of viral antigens.

host by a virus results in a specific antibody response. Consequently, development of specific antibodies during or after illness is diagnostically significant. Antibody response is measured by comparing antibody titers in serum samples collected during acute and convalescent stages of the disease, at least 2 to 3 weeks apart. Four-fold or larger rises in specific antibody titers are judged as being significant, since they occur only in clinical infections.

In the past, this retrospective approach was often the only available means to diagnose viral infections; it usually had little or no impact on patient management. Undoubtedly, the importance and usefulness of serodiagnosis generally lies in epidemiologic studies and in areas concerned with the welfare of public health. However, situations do arise where it is still important for patient management. For example, a positive diagnosis can sometimes be made when virus isolation or detection of viral antigens were negative or impossible. Conversely, it is sometimes used to establish that viral infection has not occurred, for instance, in pregnant females after exposure to rubella virus. Importantly, serodiagnosis can, and in fact, should be employed to establish the clinical relevance of viral isolates that are obtained from noninfected sites.

Competent laboratories will mandate paired serum specimens for serodiagnosis. However, assays performed on a single specimen are justified and useful in specific cases; for example, assessment of a patient's immune status to specific viral diseases, or testing for IgM or high levels of complement-fixing antibodies in a suspected case of a highly infectious disease at a stage when virus isolation is no longer possible. Indeed, the detection and quantitation of specific IgM antibodies and correlation with the acute phase of illness in various viral diseases is one of the prime research targets in clinical virology.

The majority of serologic tests are performed by state health laboratories. The antigenic diversity of human viruses is enormous at any level of their classification. For example, well over 100 serotypes of rhinoviruses alone have been identified that lack a common group antigen. Obviously, a very large number of viral antigens must be available and maintained for broad serodiagnostic testing. This exceeds the capabilities of clinical virology laboratories. However, some of the larger laboratories will perform serologic tests for the most important viral diseases. No matter who does the tests, requests must be confined to specific diseases or viruses.

Many serologic techniques are used. The basis in all of them is, of course, the interaction between antigens and antibodies. However, tests vary greatly in complexity depending on how many components are involved and also what indicator system is employed to demonstrate antigen-antibody interaction. For example, only two components are involved in precipitin reactions that can be visualized as precipitin lines in agarose gels. On the other hand, two competing antigen-antibody systems are used in the complement-fixation test and sheep erythrocytes are required in the indicator system.

Human antibody responses to different viral components are measured in varying degrees by different serologic tests. Complement-fixation test, for instance, adequately measures immune responses to nucleocapsid antigen of influenza virus but is poor in measuring hemagglutinins located on the spikes. In contrast, the hemagglutination-inhibition test precisely measures immune response to hemagglutinins but not at all to nucleocapsid antigens. With the conventional tests—virus neutralization, complement fixation, hemagglutination inhibition—the reactions to many viruses are known. Consequently, assessments can be made that are of clinical importance in many viral diseases. In contrast, correlations have not yet been established with many of the newer, more sensitive methods. This is one of the main reasons why they have been so little used in serologic testing, in spite of very successful employment in virus detection.

Virus neutralization test. This test is a very important test in clinical virology. It is used to identify and serotype virus isolates and also to determine the immune status of a patient. The test is based on the principle that specific antibody binds to viral surface antigens; thus, physical entrapment of virions prevents infection. For virus identification, antibody must be present in sufficient concentration to neutralize. The proper concentration is determined by testing against a standard virus; a 30- to 100-fold excess is used with the unknown virus. Intersection pools of antisera can be used to identify multiserotypic viruses, for example, rhinoviruses. If the serotype is not important, individual specific antisera can be pooled to identify virus groups (i.e., coxsackie type B, or polioviruses. When used to determine neutralizing antibody titers, the virus dose must be quantitated. Usually, serial dilutions of the patient's serum are tested against 100 to 300 fifty percent tissue culture infectious doses.

Hemoagglutination-inhibition test. This test is very important in serologic diagnosis of infections with influenza, parainfluenza, rubella, and arboviruses. The test is based on the principle that the ability of certain viruses to agglutinate erythrocytes of some animal species can be inhibited by specific antiserum. In most tests, serial dilutions of patient serum are tested against four to eight hemagglutinin units of virus.

Passive (indirect) hemagglutination. This test is based on the principle that many viruses can be coupled to the surface of erythrocytes by using a chemical agent such as tannic acid. When mixed with virus-specific antibody, the cells are passively agglutinated. Passive hemagglutination tests are very sensitive for antibody detection. The test is used to detect and measure antibody to hepatitis B surface antigen (anti-HBs).

Complement-fixation test. This test can be used with most viruses. However, large amounts of antigen are needed This is the major reason why the test cannot be used in some cases. Significant CF antibody rises or, in some cases, even antibody declines usually indicate recent infection: antibodies detected by CF testing develop later and decline more quickly. Besides the need for large amounts of viral antigen, many anticomplementary substances may be present in sera that sometimes cannot be removed.

Immune adherence hemagglutination. Both of the problems with the complement-fixation test are eliminated in immune adherence hemagglutination. This test is based on the principle that antigen-antibody complement complexes adhere to and hemagglutinate human type O erythrocytes from some individuals. The test is extremely sensitive for both antigen and antibody detection. Although not yet widely used, it has proven useful in hepatitis detection.

New Developments in Diagnostic Virology

The impact of dramatic developments in molecular biology on diagnostic virology has been felt with the arrival of monoclonal antibodies. On the other hand, only a faint ripple has emanated from gene cloning technology. There are indications that this is about to change. Nucleic acid hybridization techniques have been tried for virus detection for quite some time. Their application has been limited by the availability of reagents. However, with gene cloning tech-

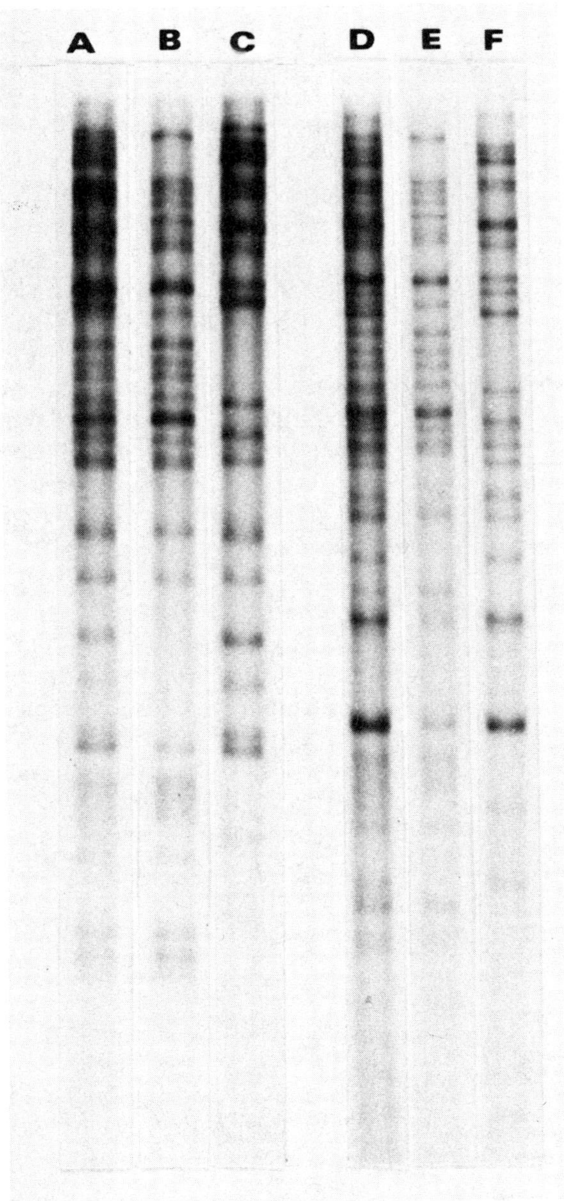

FIGURE 22-8. Demonstration of simultaneous infection with herpes simplex virus types 1 and 2 by restriction enzyme (endonuclease) analysis. The Bam H1 restriction enzyme digests of primary and plaque-purified isolates from two patients with genital lesions are shown. Patient 1, lanes A-C. Lane A is the digest from the primary isolate showing fragments characteristic of both HSV-1 and HSV-2 DNA. Plaque-purified isolates from this specimen are shown in lane B (HSV-2) and lane C (HSV-1). Similar findings are shown in lanes D-F for patient two. Lane D is a primary isolate showing dual infection. E is a plaque-purified HSV-2. F is a plaque-purified HSV-1 isolate. The HSV-1 and HSV-2 isolates between the two patients differ in their restriction enzyme patterns and, therefore, are not epidemiologically related.

niques, specific segments of viral genomes can be mass produced in pure form and used as very specific probes for virus detection and identification. This approach is currently being tried with intestinal adenoviruses. Others are soon to follow.

Another interesting development is the use of restriction endonucleases in viral diagnosis. These restriction enzymes cut viral DNA at specific and unique cleavage sites. Fragments can be separated from each other in agarose gels by electrophoresis according to their molecular weights. They can then be visualized by autoradiography on x-ray film if the viral nucleic acid is radioactively labelled before enzymatic digestion. This technique has proven extremely useful for detection of herpes simplex virus, allowing for unequivocal differentiation for HSV type 1 from type 2. In addition, individual strains could be identified in both types. With the use of restriction endonucleases it was discovered that persons infected with HSV carry their own strain, unless they are epidemiologically related. Consequently, exact studies in the transmission and epidemiology of HSV are now possible. Examples are shown in Figures 22-8, 22-9, and 22-10. The technique is now used routinely in some larger health care institutions.

CLINICAL IMPLICATIONS

As with bacterial diseases, proper specimen collection is essential for the accurate diagnosis of viral disease. Unfortunately, the duration between symptoms and diagnosis is often considerably longer with viral as compared with bacterial disease. It is frequently necessary to initiate therapy on the basis of a presumptive diagnosis. Indeed, the diagnosis may be confirmed only after the patient is well. In other cases facilities may be inadequate to make a laboratory diagnosis at all. Though rapid strides are being made in diagnostic viral techniques, there is still much progress to be made. In the meantime, supportive care for patients with viral illnesses is sometimes the most important treatment available.

The advent of acyclovir in the treatment of herpes simplex virus infections appears to herald the beginning of a new and exciting era in clinical virology. For the first time, a viral disease can be treated with a specific antiviral drug whose concentration and ef-

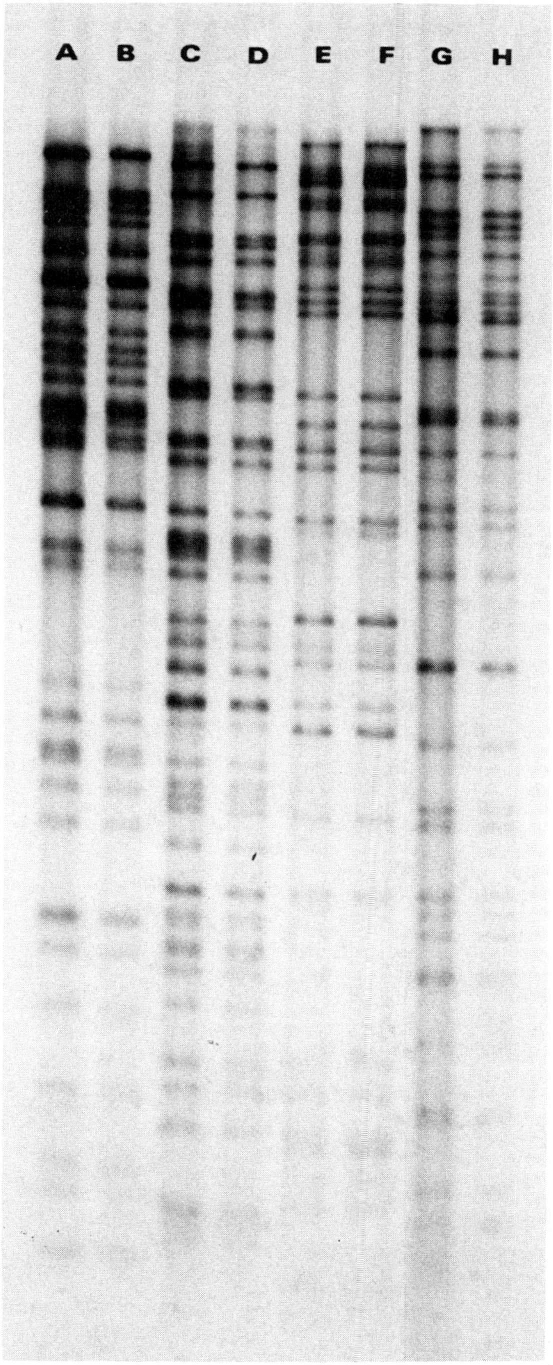

FIGURE 22-9. Application of restriction enzyme "fingerprinting" to identify herpes simplex virus isolates obtained from two different anatomic sites of the same patient. Endonuclease digests from genital and throat isolates of one person are shown in lanes A-D. Lane A represents the genital and lane B the throat isolate digested with Bam H1; digests of the same specimens with Sal 1 are presented in lanes C and D. Findings for another person are shown in lanes E-H. Lanes E and F represent the Bam H1 digests of the genital and throat isolates, respectively; results with Sal 1 are shown in lanes G and H. The endonuclease "fingerprints" indicate that the HSV-2 isolates from the first patient are identical; similarly, the HSV-1 isolates from the second patient are also identical.

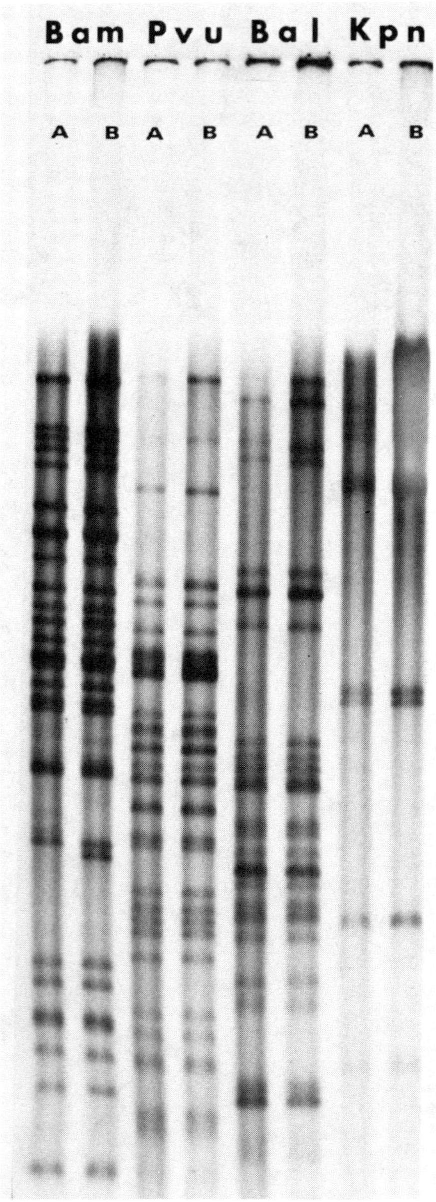

FIGURE 22-10. Use of restriction enzyme analysis to determine sexual transmission of herpes simplex virus. Endonuclease "fingerprints" from the genital isolates of two sexual partners, A and B, are shown. The patterns with each of the four enzymes are the same; the HSV-2 isolates are identical and, therefore, epidemiologically related.

fectiveness in the patient are monitored and tested in a manner similar to methods used in antimicrobial sensitivity testing. With more antiviral drugs on the horizon, a significant impact should undoubtedly be felt in the virology laboratories. However, the greatest benefit should be expected in the management and treatment of patients afflicted with viral diseases.

SUGGESTED READINGS

Bidwell DE, Bartlett A, Voller A. Enzyme immunoassays for viral diseases. J Infect Dis 1977;136(suppl.):S274-278.

Buchman TG, Roizman B, Adam G, Stover H. Restriction endonuclease fingerprinting of herpes simplex virus DNA: a novel epidemiological tool applied to a nosocomial outbreak. J Infect Dis 1978;138:488-498.

Chernesky MA, Ray CG, Smith TF, Drew WL, eds. Cumulative techniques and procedures in clinical microbiology: laboratory diagnosis of viral infections. American Society for Microbiology, 1982.

Cooney MK. Relative efficiency of cell cultures for detection of viruses. Health Lab Sci 1973;10:291-302.

Daugherty H, Ziegler DW. Radioimmunoassay in viral diagnosis. In: Kurstak E, Hurstak C. Comparative diagnosis of viral diseases, volume 2. New York: Academic Press, Inc., 1977, pp 459-488.

Drew WL, Stevens GR. How your laboratory should perform viral studies: laboratory equipment, specimen types, cell culture techniques. Lab Med 1979;10:741-746.

Drew WL, Stevens GR. How your laboratory should perform viral studies (continued): isolation and identification of commonly encountered viruses. Lab Med 1980;11:14-23.

Emmons RW, Riggs JL. Application of immunofluorescence to diagnosis of viral infections. Methods Virol 1977;6:1-28.

Herrmann EC Jr. New concepts and developments in applied diagnostic virology. Prog Med Virol 1974;17:221-289.

Hsiung GD, Fong CKY, August MJ. The use of

electron microscopy for diagnosis of virus infections: an overview. Progr Med Virol 1979;25:133-159.

Lennette EH, Schmidt NJ, eds. Diagnostic procedures for viral, rickettsial, and chlamydial infections, 5th edition. Washington, D.C.: American Public Health Association, 1979.

Lennette DA, Specter S, Thompson KD, eds. Diagnosis of viral infections: the role of the clinical laboratory. Baltimore: University Park Press, 1979.

Yolken RH. Enzyme immunoassays for detection of infectious antigens in body fluids: current limitations and future prospects. Rev Infect Dis 1982;4:35-68.

23 NORMAL BODY FLORA

KEY POINTS

- The human body is inhabited by hundreds of species of bacteria.
- In general, members of the normal flora cause no harm, but "opportunistic infections" are often caused by organisms that are long-term or permanent inhabitants of the host.
- Normal flora is altered by many factors, but especially by changes in general health status and when antimicrobial or immunosuppressive therapy is administered.
- The effects of normal flora are primarily positive: they prevent the growth of more pathogenic organisms, help to keep the immune response primed, and certain species synthesize nutrients.

Before birth the human fetus exists in a sterile environment and is free of indigenous microbial flora. In a matter of minutes, even as the neonate travels through the birth canal, however, microorganisms are contacted and begin to take residence on the skin, mucous membranes, and in the GI tract. Under normal circumstances the internal organs and the circulatory system remain sterile throughout life (see Table 23-1).

TABLE 23-1. Body Sites that are Normally Sterile

Site	*Mechanism*
Lower respiratory tract	Mucociliary blanket
Stomach (usually)	High acidity
Urinary bladder	Mechanical flushing of urine
Blood and internal organs	Physical separation from external contamination, phagocytosis

Since it is inevitable that we are colonized with hundreds of thousands of microbes, it is comforting to know that, on balance, the effects of the body's flora are primarily positive. One important function of the microbial flora is to prevent the growth of more pathogenic organisms. Many skin bacteria produce antibacterial substances such as bacteriocins. Such inhibitors produced by gram-positive bacteria such as

Staphylococcus epidermidis have a wide range of activity, while those produced by gram-negative bacteria have a more narrow spectrum. Bacteriocins adsorb into the cell walls of susceptible bacteria and cause lysis. These inhibitors produced by normal flora are a major protective mechanism to the body.

Normal flora may also induce low antibody levels that cross react with pathogens so that the immune response is primed to respond quickly to pathogenic invaders. In the gut, normal bacterial inhabitants are essential for the synthesis of certain nutrients such as vitamin K and riboflavin.

Of course, microbial inhabitants of the body can also be harmful, primarily as opportunistic pathogens when the host defenses are not intact. It is also possible that low-grade toxicity or hypersensitivity to endotoxin produced by the GI flora may develop. In some tissues, normal commensals may even enhance the growth of pathogens. For example, penicillinase-producing staphylococci in the genitourinary tract can allow the growth of the gonococcus even though the gonococcus may be susceptible to penicillin. Helpful and harmful effects of normal flora are listed on Table 23-2.

Much is known about the bacterial populations inhabiting the human body, less is known about viral populations. The flora of the body is characterized by marked site variation and variation between individuals. The body flora is influenced by environmental factors such as diet, climate, and bathing. Fungal skin infections, for example, are more common in the tropics.

Though numbers and types of organisms inhabiting a particular individual may vary significantly from those inhabiting other persons, the numbers and types remain amazingly stable within that individual despite external influences. It has been shown that an individual's skin flora remains stable for months and years, and those who harbor large numbers of certain species continue to do so, whereas the sparse bacterial populations of others continue, in general, to be sparse. Types of bacteria on the skin remain the same even if one refrains from bathing for 5 to 7 days.

Factors such as those listed on Table 23-3, however, can cause marked changes in normal flora. Three of the four factors listed—modified immunologic responses, loss of intact physical barriers, and antimicrobial therapy—are often imposed on the

TABLE 23-2. Helpful and Harmful Effects of the Body's Normal Flora

Helpful	*Harmful*
Competition or antagonism to prevent growth of other microbes	May enhance or synergize growth of pathogens in certain tissues
Allows maturation of the immune responses	Low grade toxemia due to endotoxins of GI flora is possible
Essential nutrients such as vitamin K and riboflavin are synthesized	Hypersensitivity to endotoxins may develop
May induce low antibody levels that cross react with pathogens	Flora may become opportunistic pathogens
	Some bowel flora, selected out by certain diets, may form carcinogens

TABLE 23-3. Factors that Cause Major Changes in the Body's Normal Flora

Modified immunologic responses (neoplasms, corticosteroid therapy)
Extremes of age
Loss of intact physical barriers such as skin and mucous membranes (trauma, surgery, invasive procedures, mechanical ventilation)
Antimicrobial therapy

patient as part of a required or recommended medical therapy regimen. Thus, it behooves us as health care personnel to determine whether such modifications in normal flora are having harmful effects on the patient. For example, patients frequently develop diarrhea while taking antimicrobial therapy. We have traditionally considered that to be a minimal side effect of such drugs. However, diarrhea may also be a symptom of pseudomembranous colitis, a serious disease that is associated with the overgrowth of a normal inhabitant of the GI tract *(Clostridium difficile)* and is related to use of antimicrobial drugs.

Other more transient organisms may come and go on the body. Conditions, for many reasons, may not be suitable for such microbes to establish residence. In some cases, the immunologic system may be activated to rid the body of certain species. The adherence phenomenon is also a major factor. Cell receptors have an affinity for certain organisms, which causes them to "stick." Such site attachment is highly specific, in regions of the mouth or GI tract, for example.

Nutrients available in certain microenvironments select for organisms with specific metabolic requirements. The difference between bowel flora of babies fed human breast milk and those fed cow's milk is an example. The bifidobacterium of breast-fed babies thrives in a high-carbohydrate, low-protein, high-acid environment, whereas the more mixed flora of infants fed with cow's milk result from lower carbohydrate and higher protein content. Other environmental factors such as the pH, presence or absence of fatty acids, proteolytic enzymes, and moisture help determine which organisms survive well. These factors are listed on Table 23-4. Of course some organisms produce substances (propionibacteria produce fatty acids) that make the environment unsuitable for the growth of others. Specific flora of various body sites are the subjects of the discussion that follows.

SKIN

There are about 10^{12} (1,000,000,000,000) organisms on the skin. The numbers of organisms recovered from any site vary tremendously according to the sampling technique used, but Table 23-5 provides representative bacterial counts obtained with a scrub technique. Types of bacteria on the skin (Table 23-6) are largely determined by moisture content and by numbers of pilosebaceous units at each site. Also important are the presence of skin-surface lipids, lysozyme, and bacteriocins. The largest numbers of organisms are on the moist body surfaces such as the hands and feet, the groin and axillae, and the head. Most organisms

TABLE 23-4. Factors Determining which Organisms Become "Resident" Flora

Response of the immune system
Adherence, affinity for organisms to specific cell surfaces
Nutrients available
Other environmental conditions (moisture, pH, enzymes)

TABLE 23-5. Representative Aerobic Bacterial Counts of Skin Obtained Using Scrub Technique

Area	*Number of samples*	*Counts/cm²*	*Range*
Scalp	24	1.46×10^5	0.22×10^6-14.3×10^6
Axilla	31	2.41×10^6	6300-16.7×10^6
Forearm*	17	4500	400-19,000
	15	105	25-300
Back	19	314	50-1450
Forehead	19	0.20×10^6	0.03×10^6-0.55×10^6

*Individual forearm results could be conveniently grouped into high and low counts.

live in superficial layers of the stratum corneum, in the sebaceous glands, and in the upper parts of hair follicles (Fig. 23-1). About 20 percent of the flora, however, resides deep in follicular canals and cannot be removed. Propionibacteria multiply in hair follicles and are highly anaerobic. It is not possible to sterilize the skin; only surface contamination is removed with disinfection. The fact that more organisms are usually shed immediately after a bath or shower (see Table 23-7) is probably due to movement of organisms from follicles and sebaceous glands to the skin surface. All persons shed epidermis, but some individuals called dispersers shed unusually large numbers of organisms in skin squames. These people may be normal, or may have skin disease. Antiseptic soaps help to reduce the number of viable bacteria that are shed with the epithelium during washing.

The predominant aerobic organisms on all areas of the skin are *Staphylococcus epidermidis*. *S. aureus* is present in small amounts on the skin, and is an inhabitant of the anterior nares in about 25 percent of normal individuals. Streptococci are unusual on the skin.

The anaerobic *Peptococcus saccharolyticus* and *Propionibacterium acnes* are also common (Fig. 23-2). Anaerobes, especially *P. acnes*, outnumber aerobes by 10- to 100-fold. *P. acnes* is often isolated from comedomes (acne lesions) and may have some

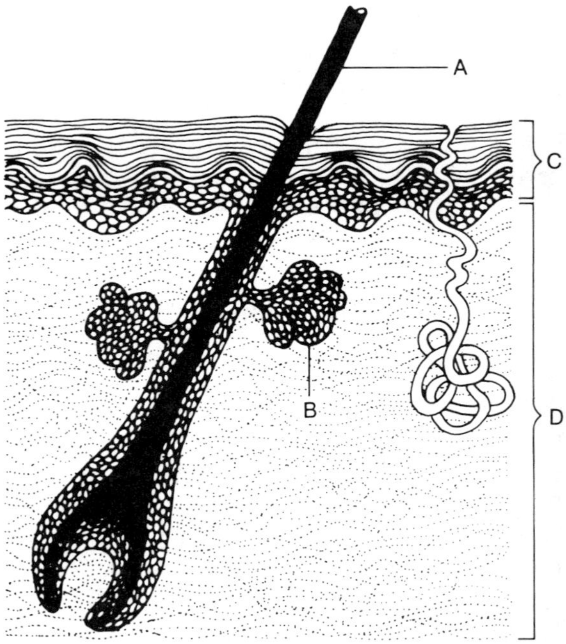

FIGURE 23-1. Microscopic section of skin. *A*, hair in hair follicle. *B*, sebaceous gland. *C*, epidermis. *D*, dermis. Bacteria live mostly in sebaceous glands and hair follicles.

TABLE 23-6. Percentage Frequency of Bacteria and Fungi on the Skin of Normal Adults*

Organism	*Face*	*Hands and forehead*	*Axilla*	*Perineum and groin*	*Toe webs*	*Trunk*
Staphylococcus aureus		5 (100) 8 (50) 18 (479) 25 (663) 40 (50)	8(50) 8(50) 12(50)	16 (60) 18 (50) 26 (50)	16 (125)	12-16(50)
Staphylococcus epidermidis	76-78(134)		98(50)			
Streptococcus pyogenes		0.4(248) 3.1(181) 0.5(1238)		0 (160)	1 (125)	
Streptococcus sp. (not group A)		5.5(181)		6.9(160)	3 (125	
Escherichia coli			6(50)			
Klebsiella-Enterobacter			44(50)			
Proteus			10(50)			
Acinetobacter			11(200)		8 (100) 20 (100)	
Coliforms	6-7(184)	3.6(500) 65 (34)				
Propionibacterium acnes			0(50)			
Candida albicans		0.5(407) 0.5(200)			1 (200)	
Dermatophyte fungi					18.6(124) 1.5(192)	

*Numbers in parenthesis equals numbers of subjects.

Adapted with permission of the publisher from Maibach HI, Hildick-Smith G. Skin bacteria and their role in infection. New York: McGraw-Hill, 1965.

relationship with acne, though this is still being debated.

Most gram-negative bacteria are very sensitive to desiccation and hence cannot survive on skin surfaces. This is especially true of *E. coli*. However, *Klebsiella* and *Enterobacter* species, as well as *Acinetobacter,* seem to survive well and sometimes become established for long periods of time. *Acinetobacter* species have been demonstrated on the skin of about one-fourth of the normal population. Nonlipophilic yeasts such as *Torulopsis glabrata* and *Candida* species are present. Little is known about viral residents of skin, though herpes as a latent virus may survive for long periods on the skin or mucous membranes.

Skin areas such as scalp, hands, and feet,

TABLE 23-7. Changes in Bacterial Skin Populations Following Showering (Contact Plate Method)

Site	Contact plate counts		
	Before	After	10 hours after
Angle of jaw	TNTC*	TNTC*	TNTC*
Breast	31	480	8
Axilla	28	473	95
Arm (above elbow)	6	327	2
Abdomen	17	484	22
Groin	155	533	271
Leg (above knee)	46	502	34
Ankle	169	TNTC*	145
Neck (back)	145	369	78
Back (mid-line waist)	63	134	45
Buttock	TNTC*	TNTC*	TNTC*
Popliteal space	63	339	60
Foot (sole)	TNTC*	TNTC*	TNTC*

*Too numerous to count

Reprinted with permission of the publisher from Maibach HI, Hildrick-Smith G. Skin bacteria and their role in infection. New York: McGraw-Hill, 1965.

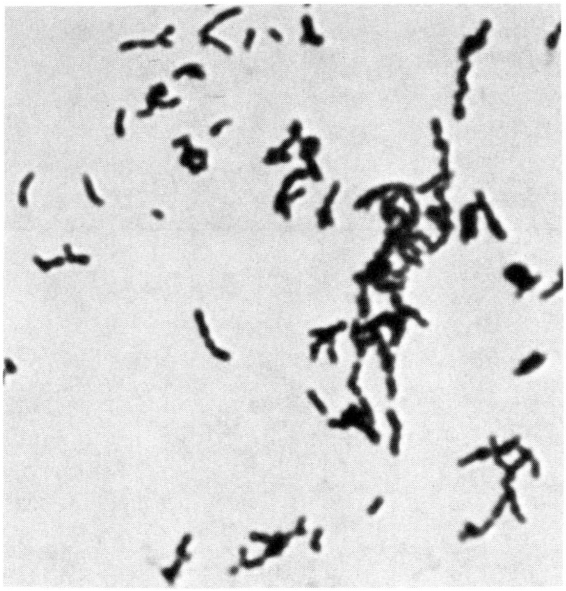

FIGURE 23-2. *Propionibacterium acne,* normal anaerobic flora of the skin. Gram stain. (Courtesy of Dr. James Plorde.)

and especially interdigital spaces, axillae, and groin, are moister, higher in temperature, and often have a larger concentration of sebaceous glands. These attributes provide an excellent residence for numerous molds and lipophilic yeasts such as *Pityrosporum ovale* and *orbiculare*. Gram-negative bacteria are also more common in these regions.

Hands

Since hands are of primary concern to clinicians in the spread of infection, their flora will be considered in more detail. Of course, any organisms can be acquired in the course of patient contact and briefly carried to other patients before the organisms die. Indeed it has been demonstrated that even the briefest and most superficial of contacts (touching an arm, straightening covers) can result in the transmission of adequate numbers of organisms to cause infection in a susceptible host. In addition to these transient inhabitants, the normal flora of hands consist of organisms that can serve as opportunistic pathogens. The skin of the hands has bacterial flora similar to flora in other skin sites, with coagulase-negative staphylococci and coryneform bacteria being predominant. However, various parts of the hand, such as the nail folds and the interdigital spaces, offer unique microenvironments that could support organisms with varying growth requirements.

Numbers of organisms on the hands show tremendous variation. Reports range from 1 to 1 million bacteria per square centimeter (a result of differing sampling techniques as well as individual variation). Probably the most accurate estimate of total numbers of aerobic bacteria on hands and forearms was made by Price in 1938. Using a series of 14 sterile basins for sequential handwashing, he was able to determine that there were from 3 to 5 million aerobic colony-forming units on the area washed. A major portion of resident hand flora is contributed by the area around the nail folds.

In general, gram-negative bacteria have not been considered to be normal inhabitants of the skin. However, increasing evidence suggests that certain species of *Moraxella, Acinetobacter,* and members of the tribe Klebsiellae may reside and proliferate on skin (Fig. 23-3). Price reported five decades ago that

> Colon bacilli placed on the writer's hands did not disappear . . . Certain of these transient bacteria are able to change their status and become permanent residents. This process seems to be a slow one, and just how it takes place is not altogether clear. But the fact that it can occur is important. (Price 1938, p. 309).

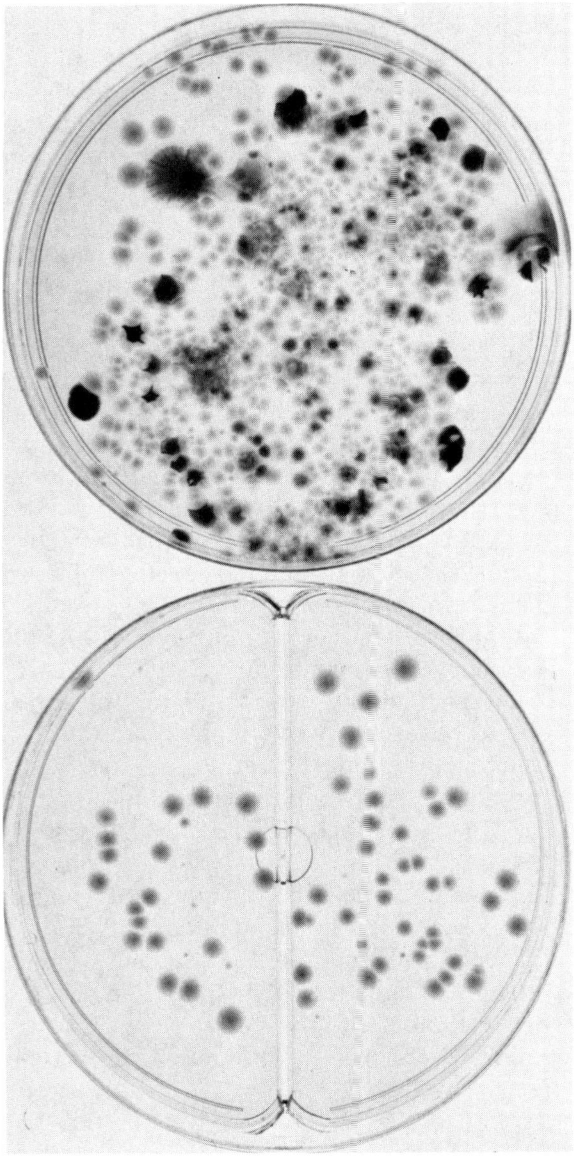

FIGURE 23-3. Typical morphologic appearance of gram-negative bacteria isolated from hands in samples taken immediately after handwashing. *(Top)* An extreme example containing four different species. *(Bottom)* A more typical example containing *Acinetobacter anitratus* and *lwoffi.* Both plates contain MacConkey agar.

Increased humidity, which occurs for example when wearing gloves during surgery, facilitates the establishment of gram-negative bacteria on the skin. Persons exposed to heavy environmental contamination can apparently develop a carrier state of gram-negative bacteria on the hands, though it is unclear whether persistent colonization occurs with organisms endogenously acquired from the individual's own intestinal flora. Persons with malignancy or prolonged hospitalization have an increased prevalence of gram-negative bacteria on their hands and other areas of skin. The prevalence on hands of normal individuals has been estimated to be from 10 to 80 percent. Maki (1978) found that though essentially all hospital personnel had gram-negative bacteria on their hands at times, prolonged carriage of a single strain was rare, with only 16 percent of samples taken every other day yielding the same organism consecutively.

Despite what is known about the physiologic bases for variations in hand flora, much remains unexplained. Persistent differences in individuals cannot be fully accounted for by factors currently understood. Persons with continuously low or high bacterial counts, despite various handwashing regimens, have been identified. Though it has been well documented that the skin and mucous membranes of hospitalized patients become increasingly colonized with gram-negative bacteria, the mechanisms are poorly understood.

MOUTH AND RESPIRATORY TRACT

There are about 10^{10} organisms in the mouth. The saliva averages 10^{8}–10^{9} aerobic and 5 to 500 $\times$ 10^{6} anaerobic bacteria per milliliter. During normal activities such as talking, breathing, and laughing, oral secretions full of bacteria and viruses are expelled into the air (Fig. 23-4). The upper respiratory tract is one of the most common routes to spread infection. Under normal circumstances, the lower respiratory tract is sterile.

Organisms that have frequently been isolated from the mouths of neonates include streptococci, staphylococci, *Veillonella*, *Neisseria*, and lactobacilli. Streptococci, primarily *Strep. salivarius*, isolated by the second day of life, represents about 98 percent of the organisms that can be cultivated in the early neonate. Increasing numbers of other bacteria take residence with the eruption of teeth, and *Strep. salivarius* then comprises about 70 percent of the oral flora.

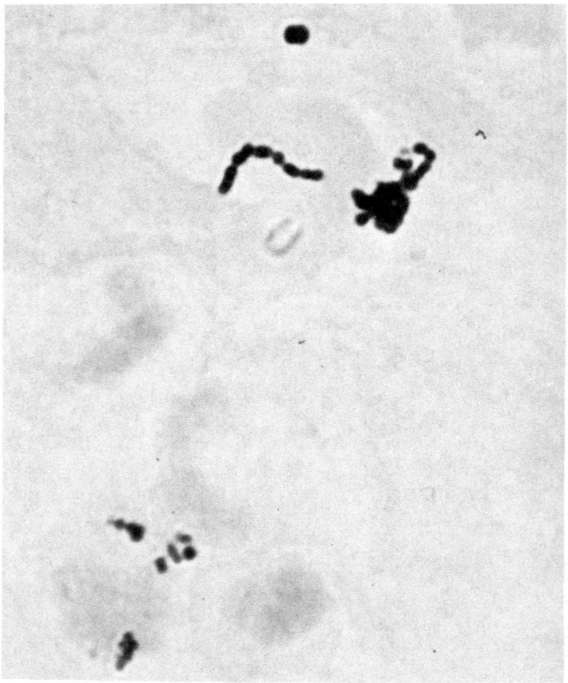

FIGURE 23-4. Normal oral flora isolated from a poorly collected sputum specimen. Note chaining cocci. (Courtesy of Dr. James Plorde.)

Strep. sanguis becomes established only after the emergence of teeth, and preferentially colonizes the tooth surfaces. *Bacteroides melaninogenicus* appear in increasing numbers and find congenial growing conditions in the gingival crevices between the teeth. *Strep. mutans* is found mainly on the tooth surface, and most commonly if there are dental caries. *Strep. sanguis* and *Strep. mutans* are lost when adults lose their teeth, are restored after the fitting and wearing of dentures, but can be lost again if the dentures are not worn for several days.

Strep. sanguis is one of the organisms that preferentially adheres to teeth. *Strep. salivarius,* however, seems to have a marked affinity for epithelial surfaces, and little adherence for the tooth. The ability of predominant bacteria to adhere to oral surfaces is outlined in Table 23-8. Bacterial adherence is influenced by specificity of both the bacteria and their colonizing surfaces. Gram-negative cocci from the genus *Veillonella* are also common in the mouth and are prevalent on the tongue surface and in the saliva. *Neisseria* is one of the first organisms to colonize clean tooth enamel.

Gram-positive rods present in the oral cavity are the *Actinomyces* species and are most common on the tooth surfaces, gingiva, and periodontal areas. Lactobacilli are usually seen in low numbers, but their counts rise considerably in the presence of large, untreated dental caries. Gram-negative rods have been isolated infrequently, and appear to be transient rather than regular inhabitants of the mouth. These are often associated with respiratory tract infections and debilitated hosts. *Bacteroides melaninogenicus,* however, is seen in the gingival crevice. Other microorganisms, such as *Candida albicans* and herpes simplex virus type 1, occur regularly in the oral cavity; though they are regarded as transient, they may become resident and sometimes cause local inflammatory conditions. Table 23-9 presents the distribution of some of the major organisms in the adult mouth. Dental plaque is 60 to 70 percent bacteria, about 10^8/ml (Fig. 23-5).

The organisms in the pharynx are a continuation of those on the oral epithelial surfaces and in the saliva. The microbial population of the saliva is considered to be

TABLE 23-8. Variations in Adherence of Bacteria to Oral Surfaces as a Determinant of Selective Colonization

Bacteria	*Teeth*	*Tongue*	*Cheek mucosa*
Neisseria sp.	+	+	+
Strep. mutans	+ to +++	+	+
Strep. salivarius	+	+++	++
Strep. sanguis	+++	++	++
Veillonella sp.	+	+++	+

+ = low
++ = moderate
+++ = high

Adapted from Gibbons RJ, VanHoute JV, Lilyemark WF. Parameters that effect the adherence of *Streptococcus salivarius* to oral epithelial surfaces. J Dent Res 1971;51:424-435.

TABLE 23-9. Common Organisms in Various Regions of the Upper Respiratory Tract

Teeth	*Tongue and saliva*	*Nasopharynx and oropharynx*
Strep. mutans	*Strep. salivarius*	*Strep. viridans*
Strep. sanguis	*Veillonella*	*Neisseria catarrhalis*
Anaerobic *Bacteroides*	*Neisseria*	*H. influenzae*
Fusobacterium		*S. aureus*
Lactobacillus		*S. epidermidis*
		Corynebacteria
		Adenovirus
		Actinomyces
		Candida sp.
		Mycoplasma
		Spirochetes

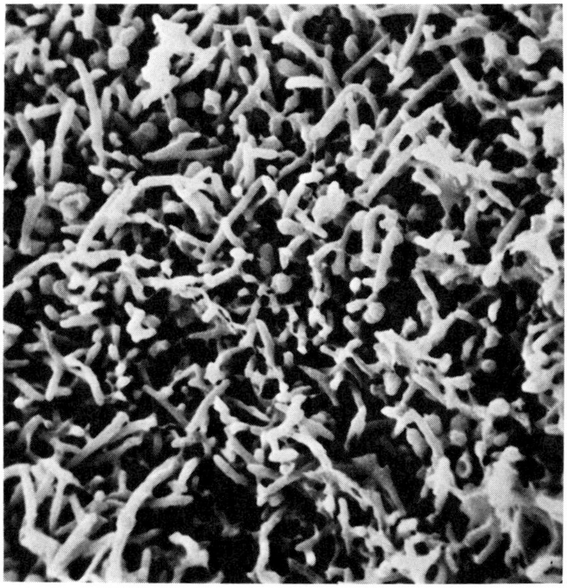

FIGURE 23-5. Scanning electron micrograph. Mixture of filaments, coccal, and bacillary forms in the superficial plaque. (Reprinted with permission of the publisher from Skinner FA, Carr JG, eds. The normal microbial flora of man. New York: Academic Press, 1974.)

transient, derived largely from the tongue and dental plaque. The environment slowly becomes sterile further down the respiratory tree.

GASTROINTESTINAL TRACT

At birth the GI tract is sterile. However, organisms are soon acquired through food. The esophagus contains bacteria from swallowed food. The stomach, because of its high acidity, generally remains sterile except perhaps for a few lactobacilli. The composition of the gut flora is influenced greatly by diet (see Table 23-10) and in some parts of the intestine by interactions between bacterial species. The pH of the small intestine is more alkaline than that of the stomach, allowing the development of a resident flora of about 10^5 to 10^7/ml. Organisms are predominantly gram-positive, streptococci and lactobacilli.

Numbers of bacteria increase as one moves down the intestinal tract, and the

TABLE 23-10a. Fecal Microflora of Different Human Populations

		Mean log_{10} number of bacteria/g of feces					
Diet	*Country*	*Enterobacteria*	*Enterococci*	*Lactobacilli*	*Clostridia*	*Bacteroides*	*Bifidobacteria*
Largely carbohydrate	India	7.9	7.3	7.6	5.7	9.2	9.6
	Japan	9.4	8.1	7.4	5.6	9.4	9.7
	Uganda	8.0	7.0	7.2	5.1	8.2	9.4
Mixed Western	England	7.9	5.8	6.5	5.7	9.8	9.9
	Scotland	7.6	5.3	7.7	5.6	9.8	9.9
	U.S.	7.4	5.9	6.5	5.4	9.7	9.9

TABLE 23-10b. Fecal Microflora of Americans

	Mean and (range) log_{10} number of bacteria/g of feces					
Subjects investigated	*Enterobacteria*	*Enterococci*	*Lactobacilli*	*Clostridia*	*Bacteroides*	*Gram-positive, non-sporing anaerobes*
African extraction (12 subjects)	7.3 (6.2-9.0)	5.0 (4.1-8.9)	8.0 (4.5-9.0)	5.1 (2.6-6.2)	9.8 (9.1-10.0)	10.0 (8.4-10.3)
European extraction (22 subjects)*	7.4 (4.9-9.7)	5.9 (3.0-8.7)	6.5 (3.6-9.3)	5.4 (2.3-6.0)	9.7 (8.2-10.8)	10.0 (8.1-10.6)

*One sample from each subject examined

Reprinted with permission of the publisher from Skinner FA, Carr JG eds. The normal microbial flora of man. New York: Academic Press, 1974.

TABLE 23-11. Flora of the GI Tract

	Colony count/g	*Predominant organisms*
Stomach	Less than 10^3	Rare lactobacilli
Small intestine, duodenum	10^3-10^6	Streptococci Lactobacilli
Ileum, jejunum	10^5-10^8	Coliforms Clostridia Enterococci Anaerobes Over 100 types
Cecum	10^8-10^{10}	99 percent anaerobes: Bifidobacteria (anaerobic lactobacilli) *Bacteroides*

flora becomes progressively more diverse and anaerobic. More than 100 species of bacteria occur in normal feces (Table 23-11). In adults, anaerobes outnumber *E. coli* by about 1000 to 10,000 to 1. Bacterial fermentation also accounts for much of the 500 to 700 ml of gas expelled from the GI tract daily by normal humans. Viruses found commonly in the stool of children and some adults include echoviruses, coxsackie viruses, and adenoviruses. *Candida* species are also present.

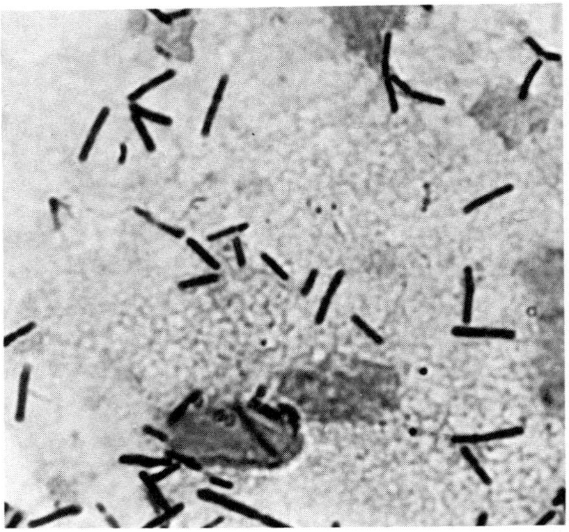

FIGURE 23-6. Endocervical specimen containing multiple lactobacilli, which are normal flora. (Courtesy of Dr. James Plorde.)

GENITOURINARY TRACT

Flora of the vagina change with age. For a few weeks after birth and during the childbearing years, vaginal pH is acid, between 4.8 and 5.7. During these phases, eight to ten species of lactobacilli predominate in the vagina (Fig. 23-6). These organisms convert glycogen to lactic acid and thrive in acid pH. In childhood and after menopause, when pH is neutral, vaginal flora is more mixed than at other times in life. Members of the normal flora include *S. epidermidis, Gardnerella vaginalis* (previously called *Hemophilus vaginale*), coliforms (except for *E. coli*), beta-hemolytic streptococcus, group B, and *Peptostreptococcus*. Beta streptococcus, group B, has become recognized during the 1970s as a major cause of postpartum endometritis and neonatal infections. Saprophytic yeasts as listed on Table 23-12 are also common. The urinary bladder in both sexes is usually sterile, though the anterior urethra may contain a few skin or bowel organisms.

CONJUNCTIVA

Microbial populations in the conjunctiva of the eye are minimized by the mechanical flow of tears and by the presence of lysozyme. Small numbers of *Propionibacterium acnes, Corynebacteria xerosis, Staphylococcus epidermidis,* and *Moraxella* species are present. *Moraxella* species have been known to cause conjunctivitis (Table 23-13).

CLINICAL IMPLICATIONS

Knowledge of the body's normal flora assists the clinician in interpreting results of microbiologic cultures and understanding why the sterility of certain areas must be protected. Of concern to the health care practitioner is the increase in clinical disease resulting from *endogenous* sources of opportunistic pathogens among patients with lowered host resistance. Table 23-14 describes the role of normal flora as potential pathogens; factors that lower resistance have been discussed in Chapter 4.

TABLE 23-12. Yeasts Indigenous to the Vagina

Candida species
albicans
guilliermondii
krusei
parapsilosis
pseudotropicalis
stellatoidea
tropicalis
Cryptococcus species
Hansenuls species
Saccharomyces
Pichia species
Torulopsis

TABLE 23-13. Normal Flora of the Eye

Organism	*Proportion (percent)*
S. epidermidis	65-80
Coryneform bacteria	65-70
S. aureus	10-25
Strep. pneumoniae	2-4
Gram-negative bacteria	2-6
No growth	Less than 1

There have been documented changes in the normal flora in the severely ill. A study of 266 acutely ill patients revealed a clear change in the superficial flora of the nose, throat, axilla, and groin to include a significant number of gram-negative bacteria—even though the normal flora of these areas are predominatly gram-positive cocci. The same study recorded almost complete replacement of the normal flora by *E. coli* within several hours following cardiac arrest (Stratford et al. 1968). Work by Lefrock has revealed a significant shift in fecal microflora within 21 days of hospitalization to include a population of coliforms not normally seen. This includes an increased number of *Klebsiella, Enterobacter,* and *Proteus,* and a decrease in numbers of *E. coli.* It is perhaps not surprising that gram-negative bacteria appearing in the oropharynx of hospitalized patients are usually the same species as the newly appearing fecal flora (LeFrock, Ellis, and Neistein 1979). The fact that many infections are caused by normal flora has several implications for clinicians. We are constrained to help keep patients' flora in their proper niches, avoiding the

(text continues on page 628)

TABLE 23-14a. Normal Flora in Commonly Cultured Sites in Healthy Individuals

Location	*Name*	
Anterior nares	*Staphylococcus epidermidis* *Streptococcus pneumoniae (Pneumococcus, Diplococcus)* *Neisseria* sp., including *meningitidis* (except gonorrhoeae)	Nonhemolytic streptococcus Diphtheroids *Staphylococcus aureus*
Mouth and nasopharynx	*Neisseria catarrhalis (Branhamella catarrhalis)* *Streptococcus viridans* (alpha-hemolytic streptococcus) *Staphylococcus epidermidis* or *aureus* *Streptococcus,* nonhemolytic and anaerobic Various fungi, viruses, and mycoplasma	*Fusobacterium* *Bacteroides* sp. *Borrelia* sp. Diphtheroids *Hemophilus* species
Trachea, bronchi and lungs, sinuses, inner or middle ear, stomach	Usually sterile This group may often become contaminated with many nasopharyngeal and oropharyngeal organisms Often sterile but varies with digestive activity	
Small bowel	Few aerobic organisms	
Large bowel	*Bacteroides* and Fusobacteria (Bifidobacteria in breast-fed infants) Clostridia (*perfringens* and *tetanus* among others) *Enterococcus* (group D streptococcus) Viruses and fungi, protozoa	Lactobacilli *Escherichia coli* *Klebsiella-Enterobacter* group *Proteus* species *Pseudomonas* sp. *Serratia* sp.
Vagina: Premonopausal	Lactobacilli (small numbers of oral-nasopharyngeal flora) Anaerobes including *Bacteroides, Clostridium, Peptostreptococcus, Propionibacterium, Selenomonas* sp.	
Prepubertal and Postmenopausal	Enterobacteriaceae family Streptococci, yeasts	
Skin	Diphtheroids (propionibacteria and aerobic corynebacteria) *Staphylococcus epidermidis* (coagulase negative) *Staphylococcus aureus* (coagulase Positive) *Acinetobacter (Herellea),* yeasts Anaerobic streptococci	
Urinary bladder, blood, or spinal fluid	Usually sterile or contaminated with low numbers of organisms in distal urethra Sterile	

TABLE 23-14b. Normal Flora as Potential Pathogens

Organism	*Normal location*	*Frequency*	*Usual mode of spread*
I. Aerobic or facultative			
A. Gram positive			
1. *Staph. aureus* (coagulase positive)	Skin, hair, naso-oro pharynx	Common	1,2,3*
2. *Staph. epidermidis* (coagulase negative)	Skin, hair, naso-oro-pharynx	Very common	1,3
3. *Streptococcus* group A	Oropharynx, perianal, anal	Uncommon	1,3
4. *Streptococcus* group B	Adult vagina, genitalia, colon	Uncommon (8 to 20 percent)	1
5. *Streptococcus* group D	Colon	Common	1,2
6. *Strep. pneumoniae* (Diplococcus, Pneumococcus)	Anterior nares, oropharynx	Uncommon	1,3
7. *Strep. viridans* (alpha Strep)	Mouth, skin	Very common	1,3
8. Yeasts (i.e., *Candida albicans*)	Mouth, skin, vagina, colon	Common	1,2,3
B. Gram negatives			
1. Enterobacteriaceae (coliforms)	Intestinal tract		
a. *E. coli*	Colon, perineum	Very common	1,3
b. *Klebsiella* sp.	Colon, perineum	Common	1,2,3
c. *Proteus* sp.	Colon, perineum	Common	1,2
d. *Serratia* sp.	Colon, perineum, environment	Not common	1,2
2. *Hemophilus influenzae*	Naso-oropharynx	Rather common	1,3
3. *Neisseria meningitidis*	Oropharynx	Uncommon, epidemic	1,3
4. *Pseudomonas aeruginosa*	Colon (5-10 percent out-patients, up to 30 percent long-term hospital patients)	Not common, but varies	1,2,3
II. Anaerobic bacteria			
A. *Bacteroides* species	Mouth, colon, vagina	Very common	1
B. *Clostridium* species	Colon, dirt	Common	1,2
C. Streptococci, anaerobic	Mouth, colon, vagina	Very common	1
D. Fusiform bacteria	Mouth, colon	Common	1
III. Fungi: Ubiquitous. Disease is due to compromising host factors			
IV. Viruses: More than 400 are currently classified and the number grows. Role as normal flora undetermined			

*Spread by

1. *Contact:* Direct—between people in close proximity, with hands, body, clothing. Indirect—personal articles contaminated by infected person or environment
2. *Vehicle route:* Through food, water, drugs, blood contaminated from infected persons or animals
3. *Airborne:* Respiratory tract droplet nuclei from infected person or carrier, contaminated dust inhaled by or deposited on susceptible host, or from organisms shed into the environment from skin, hair, wounds, or perineal area
4. *Vector-borne:* Via contaminated insects such as flies, mosquitoes, turtles, ticks, etc.

Reprinted with permission of the publisher from Infection control manual. Seattle: University of Washington Hospital, 1983.

introduction of organisms into sterile areas via suction, urinary, or intravascular catheters. Such basic nursing tasks as bathing the patient, providing a wash cloth after elimination, and eliciting from visitors information regarding their own health status to prevent the spread of infections prevalent in the community take on importance for infection control as well as for patient comfort.

Neonates, because of their lack of a stable population of skin flora, are highly prone to skin infections with such organisms as *S. aureus* and *C. albicans*. Skin colonization with *S. aureus* in newborns may be greater than 50 percent in some newborn nurseries. Meticulous hand-scrub techniques and dress codes are essential to minimize the colonization and infection rates among newborns in the hospital.

In addition, the fact that the skin on the hands of health care personnel carries gram-negative bacteria and other organisms known to be opportunistic pathogens adds two reasons for handwashing: (1) to prevent cross infection of one patient's organisms to another and (2) to avoid the spread of indigenous hand flora. Washing with antiseptic soaps has been shown to decrease numbers of bacteria on hands without changing the ecologic balance of the flora. Frequency of handwashing is directly correlated with prevalence of carriage of gram-negative bacteria; those who wash more are less likely to have gram-negative organisms on their hands. The relative effectiveness of various handwashing agents is compared on Table 23-15.

Normal flora can contaminate specimens obtained for microbiologic culture unless appropriate collection techniques are used. For example, flora from the oral cavity can contaminate sputum specimens, perineal flora can contaminate urine specimens, and skin flora can contaminate blood samples to be used for blood culture. It is vital that such contamination be avoided so that an accurate infectious diagnosis can be made; improper specimen collection will result in lost time and money.

SUGGESTED READINGS

Bloomfield AL. Dissemination of bacteria in the upper air passages: the circulation of foreign particles in the mouth. Am Rev Tb and Pulm Dis 1921;5:903-914.

Brandtzaez P, Fjellenger I, Gjeruldsen ST. Adsorption of immunoglobulin A onto oral bacteria in vivo. J Bacteriol 1968;96:242.

Carlson J. Nutritional requirements of *Streptococcus mutans*. Caries Res 1970;4:305.

Soderholm G, Almfeldt I. Prevalence of *Streptococcus sanguis* and *Streptococcus mutans* in the mouth of persons wearing full denture. Arch Oral Biol 1969;14:243-246.

Casewell M, Phillips I. Hands as route of transmission for *Klebsiella* species. Br Med J 1977;2:1315-1317.

Christie RW. Bacterial variations in the nasopharyngeal and skin flora of isolated arctic scientists. N Engl J Med 1958;258:531-533.

Davies RM. General ecology of the commensal microflora of the mouth. In: MacPhee T, ed. Host resistance to commensal bacteria. London: Churchill, Livingstone, 1972.

Dewalt EM. Effects of timed hygienic measures on oral mucosa in a group of elderly subjects. Nurs Res 1975;248:104-108.

Dewalt EM, Haines AK. A comparative study of the effect of two oral hygiene practices at varying time intervals. (Unpublished master's thesis) Columbus: Ohio State University, 1968.

Dewalt EM, Haines AK. The effects of specified stressors on healthy oral mucosa. Nurs Res 1969;18:22-27.

Ellen RP, Gibbons RJ. Parameters affecting the adherence and tissue tropism of *Streptococcus pyogenes*. Infec Immun 1974;9:85-88.

TABLE 23-15. Characteristics of Common Handwashing Agents

Active agent	Preparation	Activity	Substantive effect?*	Comments
Hand soap	Bars, leaflets, powder, liquid	Little chemical effect; main action is mechanical removal of organisms	No	May become contaminated unless small bars are used; granules and leaflets are more expensive but useful since more contact time is required for soap removal
Alcohol	70% ethyl or isopropyl	About 90% reduction in bacterial counts	No	Very effective though inappropriate for routine use due to drying; ethyl alcohol may be better viricide
Iodophors	Water-soluble iodine complexes (Betadine and others)	60-90% reduction in bacterial counts	Yes†	Can be drying to skin; stability and rate of iodine release variable; broad spectrum antisepsis
Hexachlorophene	pHisoHex and other brands	60% reduction in bacterial counts after one wash; 60-70% after repeated applications	Yes†	"Build-up" time of daily use for 3-5 days required for maximum effectiveness; may rarely cause neurologic damage; active primarily against gram-positive organisms
Chlorhexidine	4% with 4% isopropyl alcohol (Hibiclens)	About 90% reduction in bacterial counts	Yes†	Broad spectrum of antiseptic effect

*A residual, continued reduction in bacteria after rinsing off the product
†Only if agent is not removed by alcohol or subsequent washing with another agent

Evans CA, et al. Bacterial flora of the normal human skin. J Invest Dermatol 1950;15:305-324.
Evans CA. Persistent individual differences in the bacterial flora of the skin of the forehead. J Invest Dermatol 1976;64:42-46.
Gibbons RJ, VanHoute JV, Lilyemark WF. Parameters that effect the adherence of *Streptococcus salivarius* to oral epithelial surfaces. J Dent Res 1971;51:424-435.
Ginsberg MK. A study of oral hygiene nursing care. Am J Nurs 1961;61:67-69.
Gross PA, et al. Deaths from nosocomial infection, experiences in a university hospital and community hospital. Am J Med 1980;68:219-223.
Hann J. The source of the 'resident' flora. Hand 1972;5:247-252.
Harbour J, et al. Excretion of faecal viruses during the first year of life J Clin Pathol 1980;33:774-778.
Higuchi JH, Chaudhuri T, Johanson WG. Determinants of postoperative respiratory tract colonization with gram negative bacilli. Am Rev Respir Dis 1978;117:274.
Johanson WG, et al. Nosocomial respiratory in-

fection with gram negative bacilli, the significance of colonization of the respiratory tract. Ann Intern Med 1972;77:701-706.

Johanson WG, Woods DE, Chaundhuri T. Association of respiratory tract colonization with adherence of gram negative bacilli to epithelial cells. J Infec Dis 1979;139:667-673.

Larson EL. Persistent carriage of gram-negative bacteria on hands. Am J Infec Cont 1981;9:112-119.

Larson HE, Price AB, Borriello SP. Epidemiology of experimental enterocecitis due to *Clostridium difficile*. J Infec Dis 1980;142:408-413.

LeFrock J, Ellis CA, Neistein L. The relation between aerobic fecal flora and oropharyngeal microflora in hospitalized patients. Am J Med 1979;277:275-281.

MacFarlane TW, Mason DK. Local environmental factors in the host resistance to the commensal microflora of the mouth. In: MacPhee T, ed. Host resistance to commensal bacteria. London: Churchill, Livingstone, 1972.

Maibach HI, Hildick-Smith G. Skin bacteria and their role in infection. New York: McGraw-Hill, 1965.

Maki DG. Control of colonization and transmission of pathogenic bacteria in the hospital. Ann Intern Med 1978;89 (suppl, part 2):777-780.

Marples M. The ecology of the human skin. Springfield: Charles C Thomas, 1965.

Maurer J. Providing optimal oral health. Nurs Clin N Am 1977;12:671-685.

McBride ME, et al. Hand flora of nurses. II. Qualitative differences in occupational groups. Intern J Dermatol 1972;11:49-53.

McBride ME, Duncan WC, Knox JM. Physiological and environmental control of gram negative bacteria on skin. Br J Dermatol 1975;93:191-199.

McBride ME, et al. Microbial skin flora of selected cancer patients and hospital personnel. J Clin Microbiol 1976;3:14-20.

McCarthy C, Snyder ML, Parker PB. The indigenous oral flora of man, the newborn to the one year old infant. Arch Oral Biol 1965;10:61-69.

Nolte WA. Oral microbiology. St. Louis: CV Mosby, 1968.

Passos JY, Brand LM. Effects of agents used for oral hygiene. Nurs Res 1961;15:196-202.

Penn RG, Sanders WE, Sanders CC. Gram negative bacilli in the oropharynx and pneumonia. Am J Infect Cont 1981;9:2.

Price PB. The bacteriology of normal skin. J Infec Dis 1938;63:301-318.

Rebell G, et al. Factors affecting the rapid disappearance of bacteria placed on the normal skin. J Invest Dermatol 1950;14:247-264.

Ritz HL. The role of aerobic *Neisseria* in the initial formation of dental plaque. In: McHugh WD, ed. Dental plaque. Edinburgh: Livingstone, 1970.

Rosenthal S, Tager IB. Prevalence of gram negative rods in the normal pharyngeal flora. Ann Intern Med 1975;83:355-357.

Russell C, Melville TH. A review of bacteria in the human mouth. J Appl Bacteriol 1978;44:163-181.

Sanford JP, Pierce AK. Lower respiratory tract infections. In: Bennett JV, Brachman PS, eds. Hospital infections. Boston: Little, Brown and Company, 1979.

Schimpff SC, et al. Origin of infection in acute nonlymphocytic leukemia: significance of hospital acquisition of potential pathogens. Ann Intern Med 1972;77:707-714.

Skinner FA, Carr JG, eds. The normal microbial flora of man. New York: Academic Press, 1974.

Socransky SS, Manganiell SD. The oral microbiota of man from birth to senility. J Periodon 1971;42:485-490.

Stevens RM, et al. Pneumonia in an ICU. Arch Intern Med 1974;134:106-111.

Stratford B, et al. Alteration of superficial bacterial flora in severely ill patients. Lancet 1968;1:68.

Taplin D, Rebell G, Zaiss N. The human skin as a source of *Mimae herellea* infections. JAMA 1963;186:952-954.

Valenti WM, Trudell RG, Bentley DW. Factors predisposing to oropharyngeal colonization with gram negative bacteria in the aged. N Engl J Med 1978;298:1108-1111.

VanDrimmelen J, Rollin HF. Evaluation of a commonly used oral hygiene agent. Nurs Res 1969;18:327-333.

VanHoute J, Gibbons RJ, Banghart SB. Adherence as a determinant of the presence of *Streptococcus salivarius* and *Streptococcus sanguis* on the tooth surface. Arch Oral Biol 1972; 15:1025-1029.

Venkataramani TK, et al. Skin mucous membranes as reservoirs of *Bacterium anitratum* and *Mima polymorpha* in patients. J Ind Med Assoc 1972;59:425-428.

24 NOSOCOMIAL INFECTIONS

KEY POINTS

- Nosocomial infections are usually transmitted by hands of health care personnel or by contaminated equipment; direct patient-to-patient transmission is uncommon.
- Nosocomial infections are a major health problem in the United States.
- The most common causative agents of nosocomial infections are gram-negative bacteria, the most common sites are the urinary tract and the respiratory tract.

Nosocomial infections (NI) are a costly national problem; about 5 percent of hospitalized patients are affected, with probably a higher percentage of infection in residents in long-term care facilities. It has been estimated that an additional 6 million patient days can be attributed to prolonged stay due to hospital-acquired infections, and nearly $2 billion per year additional cost. For an individual who acquires a NI, the average hospital stay is prolonged for 1 to 8 days and costs are increased by $200 to $1500.

Though infectious diseases such as tuberculosis have been known to spread in institutional settings, the diseases recognized and treated as infectious are of much less concern and danger than those caused by opportunistic pathogens such as gram-negative bacteria and some viruses. The role of opportunistic pathogens may be unrecognized and they may be spread from patient to patient by personnel who are unaware of their role in disease transmission. Changes in the past decade in NI include increasing gram-negative and urinary-catheter-associated bacteremias, higher rates of septic shock, and increasing costs of infection. Most NI are caused by aerobic bacteria and are endemic; point source or common pathogen epidemics are unusual. (Fig. 24-1)

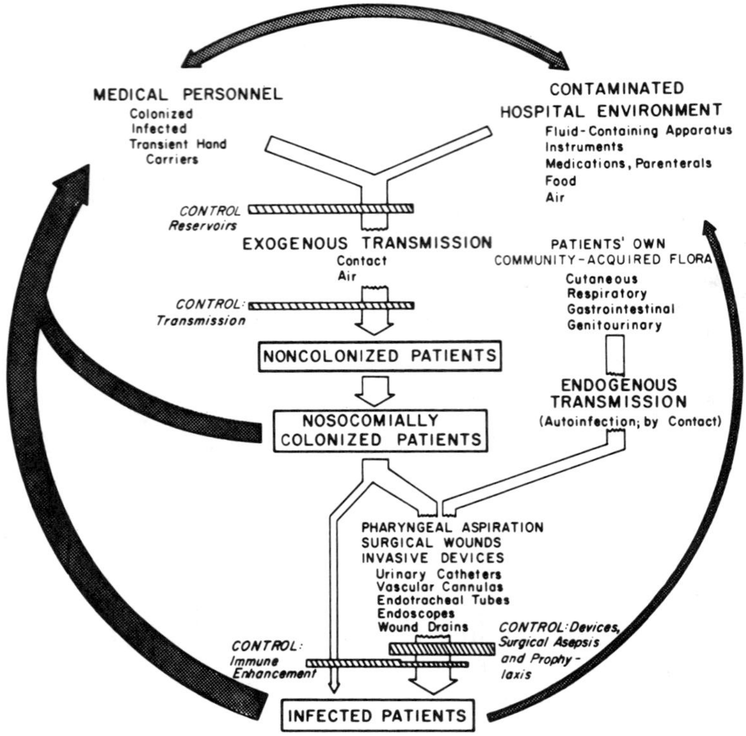

FIGURE 24-1. The epidemiology of nosocomial infection. Transmission occurs mainly by contact spread, and to a much lesser extent by the airborne route. Aspiration, surgical wounds, and exposure to invasive medical devices amplify transmission, colonization, and susceptibility to infection. (Reprinted with permission of the publisher from Maki DG. Control of colonization and transmission of pathogenic bacteria in the hospital. Ann Intern Med 1978; 89 (Part 2):777.)

RISK FACTORS

Major risk factors for developing NI include poor general health status, extremes of age, presence of other infections, diabetes mellitus or hormonal disorders, cancer, immunosuppressive therapy, extensive invasive procedures (surgery longer than 4 hours, multiple blood transfusions, prolonged catheterizations), and antimicrobial therapy. The effect of antimicrobial therapy on the emergence of resistant strains of bacteria is illustrated in Figure 24-2. Individuals with a history of smoking, and thoracic or upper abdominal surgery are especially prone to nosocomial pneumonia.

Forty-five percent of NI are related to invasive devices, especially urinary catheters (26 percent of device-related NI), intravascular equipment (25 percent), hemodialysis

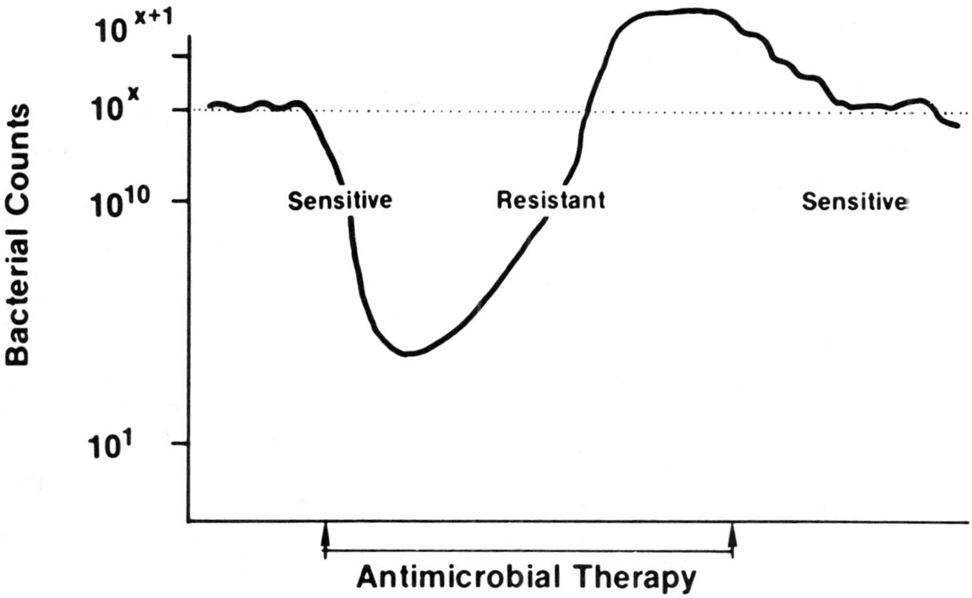

FIGURE 24-2. During antimicrobial therapy there is an initial drop in counts of bacterial flora, often followed by a rise in resistant organisms (which may grow to numbers greater than pretreatment levels as therapy is continued). (Reprinted from DuPont HL. Rational antibiotic therapy. Top Clin Nurs 1979;1:45-52 by permission of Aspen Systems Corporations, © 1979.)

(19 percent), and respiratory therapy (11 percent). These devices generally become contaminated during use. Other devices associated with infections include cerebrospinal fluid shunts, prosthetic devices such as limbs, vascular prostheses, heart valves, and eye lenses. Scalp abscesses in babies after fetal monitoring may be as high as 5 percent (Stamm 1978). Following renal transplantation, infection is the major cause of mortality. Many NI, however, occur in patients who are immune competent. Hospital food has been shown to transmit gram-negative bacteria, but the most important reservoir of drug-resistant gram-negative bacteria is the colonized patient rather than the environment. The organisms are then transmitted by personnel hands to other patients.

TYPES OF INFECTIONS

Urinary tract infections (UTI) comprise 30 to 40 percent of all NI. Following bladder catheterization, the cumulative risk of infection has been estimated to be 5 to 10 percent per day of catheterization. In a nursing home population, 85 percent of persons with catheters had significant bacteriuria, many with antimicrobial-resistant bacterial strains (Garibaldi, Brodine, and Matsumiya 1981).

Only about 10 to 15 percent of UTI are caused by cross infection between patients via personnel hands. The rest are due to endogenous infection with bowel or perineal flora. Meatal care with antibiotic ointments has been shown to be useless in preventing

UTI (Burke et al. 1981). Urine-measuring containers and specific-gravity instruments have also been implicated in epidemics of UTI with multiply drug-resistant gram-negative species. About 2 percent of patients with UTI develop secondary bacteremia.

Surgical wound infections are the second most common type of NI. The probability of infection can be predicted by classification of surgical procedures according to the American College of Surgeons criteria into clean, clean-contaminated, contaminated, and dirty cases. With clean surgical procedures such as craniotomy or corneal transplantation, the probability of developing a NI is about 2 percent. For clean surgical cases, likely sources of contamination are environmental (air or surgical personnel). Frequently changed masks (to prevent saturation with moisture) are essential in the operating room.

There is little evidence that special methods of handling air, such as laminar air flow, contribute significantly to decreasing the incidence of NI. Architectural changes (segregation of patients in intensive care units) are also not instrumental in greatly modifying NI rates. The doubling in rates of cesarean section since 1967 has resulted in increased rates of maternal postpartum infections. Even with clean surgery, but especially with contaminated surgical cases such as perforated appendix with peritonitis where NI rates can be as high as 50 percent, endogenous sources of infections are much more common than exogenous. The intestine, like the skin, cannot be sterilized, even with long-term antimicrobial therapy.

Lower respiratory tract infections are the third most common type of NI. Contaminated respiratory therapy equipment has been associated in the past with epidemics of hospital-acquired pneumonias. Nebulizers that form droplets that are delivered to the lower respiratory tract, suction catheters, solutions for rinsing catheters, respirators, and resuscitation bags have all been demonstrated to transmit gram-negative bacteria. Currently, however, contaminated equipment is rarely a cause of hospital-associated respiratory tract infection.

The ability of gram-negative bacteria to adhere to the damaged epithelium of the respiratory tract in patients who have had prolonged anesthesia and respiratory assistance is the major factor that leads to colonization and disease from these opportunistic pathogens. The removal of normal oropharyngeal flora by antimicrobial drugs interferes with the natural inhibitory action of these organisms against invaders. Unfortunately, attempts to prevent oropharyngeal colonization with gram-negative rods by using local antimicrobial therapy (in the form of mouth sprays or rinses) result in the emergence of resistant strains. Most hospital-acquired respiratory infections in children are seasonal and viral in origin.

Bacteremias occur in about 5 per 1000 hospital patients and the incidence is increasing. A nation-wide epidemic of *Enterobacter agglomerans* (formerly called *Erwinia*) bacteremias in 1971 was due to contaminated commercial intravenous solutions. Such epidemics are extremely rare, however, and most cases are endemic. The majority of bacteremias are secondary to wound, urinary tract, or respiratory tract infections. As shown in Figure 24-3, primary bacteremias, usually due to intravascular devices such as intravenous catheters, total parenteral nutrition apparati, and central monitoring devices, also occur.

In a survey of nursing homes, the prev-

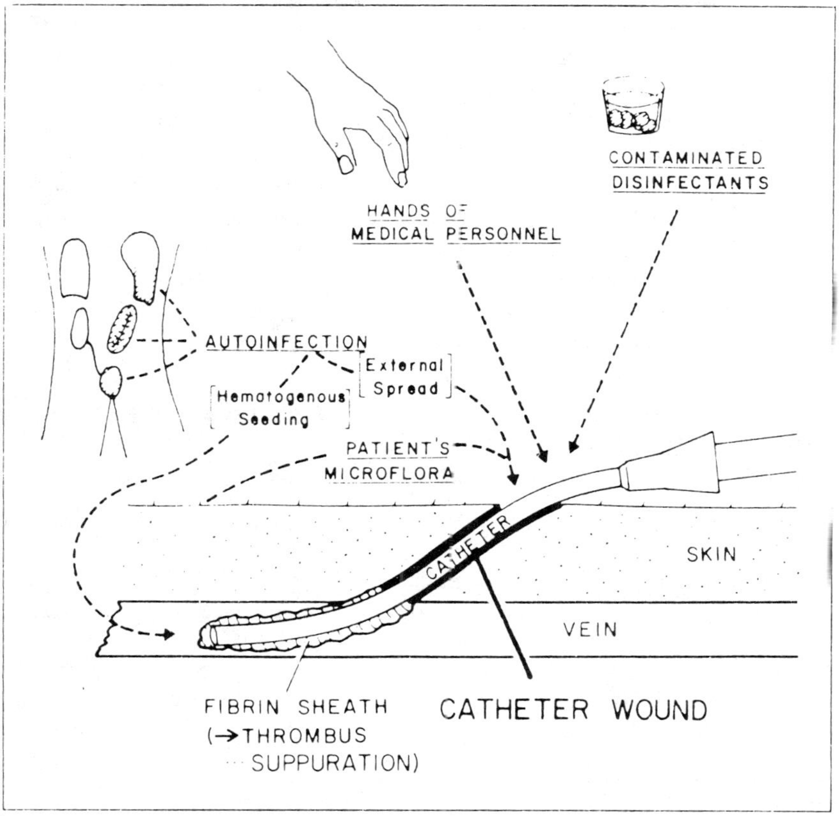

Sources of vascular-cannula-related infection.

FIGURE 24-3. Sources of vascular-cannula-related infection. (Reprinted with permission of the publisher from Maki DG. Nosocomial bacteremia. Am J Med 1981;70:726.)

alence of NI was 16.2 percent. Infected decubiti and conjunctivitis, as well as urinary tract and respiratory tract infections, were common.

CAUSES

The three single organisms most frequently causing NI in 81 hospitals participating in the Centers for Disease Control's (CDC) National Nosocomial Infections Study (NNIS), 1970 to 1979, were *E. coli, S. aureus,* and group D streptococcus. Gram-negative bacteria in both endemic and epidemic NI are the organisms of major concern. In the NNIS group, gram-negative bacterial infections (except those caused by *E. coli*) peaked in summer. Of 53 device-related epidemics investigated by CDC, 1965 to 1975, one was due to *S. aureus,* 10 to hepatitis, one to Epstein-Barr virus, from two no pathogens were isolated, two due to *Plasmodium,* and 37 (about 70 percent) were due to gram-negative bacteria. Types of device-related infections are summarized in Table 24-1. In 223 epidemic investigations of infections by CDC from 1956 to 1979, which were not necessarily device-related, 36 percent were caused by gram-negative bacteria.

In the first month postrenal transplantation, aspiration pneumonia, and UTI with gram-negative bacteria are the most common NI, but after 1 to 4 months organisms such as *Aspergillus, Cryptococcus neoformans,* and *Listeria monocytogenes* become more prevalent. These opportunistic pathogens are also common among patients with neoplasms or those receiving long-term immunosuppressive therapy. Note the high relative frequency of yeast and fungal infections among patients with leukemia and lymphoma shown on Table 24-2. Septicemias due to such organisms as *Enterobacter agglomerans, Pseudomonas cepacia* and *maltophila,* and *Flavobacter* can often be traced to a common source of contamination. These organisms are not normally inhabitants of humans and may be environmental contaminants of intravenous solutions or other equipment, especially if they contain water or other solutions (Table 24-3).

In addition, other bacterial pathogens are being newly recognized. Because they are slow growing, fastidious, or resemble commensals, they have been less readily rec-

TABLE 24-1. Major Types of Device-Related Infections

Device	*Hospitalized patients with device (percent)*	*Related infections*	*Those with device developing infection (percent)*	*Estimated number of cases annually**	*Case-fatality ratio (percent)*
Infections related to commonly used devices					
Urinary catheter	10	Bacteriuria	20	760,000	<1
		Epididymitis	—	—	—
		Septicemia	0.5 to 2.5	28,500	20 to 40
Intravascular infusion†					
Scalp vein needle	12	Suppurative phlebitis	<0.01	—	
		Septicemia	<0.1	2280	20 to 40
Plastic catheter	12	Suppurative phlebitis	—	—	
		Septicemia	0 to 8	22,800	20 to 40
Subclavian catheter	0.5	Suppurative phlebitis	—	—	
		Septicemia	3 to 7	7600	20 to 40
Total parenteral nutrition	0.1	Suppurative phlebitis	—	—	
		Septicemia	0 to 27	2700	20 to 40
Arterial catheter	0.2	Arteritis (?)	—	—	
		Septicemia	0.2	150	20 to 40
Respiratory therapy					

TABLE TABLE 24-1. *(continued)*

Device	Hospitalized patients with device (percent)	Related infections	Those with device developing infection (percent)	Estimated number of cases annually*	Case-fatality ratio (percent)
Infections related to commonly used devices					
Continuous ventilator	2	Tracheitis	—	—	5
		Pneumonia	—	51,650	40
Intermittent positive-pressure breathing	7	Pneumonia	—	12,120	40
Infections related to special devices					
Hemodialysis	—	Hepatitis	3 to 35	—	5 to 10
		Septicemia	20 to 25	—	10 to 20
		Shunt infection	15 to 20	—	<5
Cerebrospinal fluid shunt	—	Ventriculitis	6 to 23	—	30 to 35
		Septicemia	Variable	—	30 to 35
Prosthetic heart valve	—	Endocarditis	1 to 2	—	50 to 60
Vascular graft	—	Graft infection	1.3 to 1.5	—	20 to 45
Artificial hip	—	Deep wound	0 to 10	—	5 to 45
Prosthetic lens	—	Endophthalmitis	—	—	—

*Calculations based on 38 million acute-care hospital admissions annually times percent of hospitalized patients with each device times estimated rate of device-associated infection. With pneumonia, rates of device-associated infections are unavailable, and estimates were derived as described in the text. Since use of the specialized devices noted above varies widely from hospital to hospital, neither the percent of hospitalized patients with these devices nor the number of infections occurring annually could be accurately estimated.

†Catheter phlebitis was not included among device-related infections.

Reprinted with permission of the publisher from Stamm WE. Infections related to medical devices. Ann Intern Med 1978;89(Part 2):764-769.

ognized. *Mycobacterium chelonei* and *fortuitum* cause nosocomial endocarditis, pericarditis, and osteomyelitis. *Legionella pneumophila* and other species can cause pneumonia and lung abscesses. There were 83 nosocomial cases of legionellosis reported to CDC by May, 1980. Most cases occurred in immunocompromised hosts or in those who had diseases such as emphysema, diabetes mellitus, chronic bronchitis, or cancer. Methods to control these infections are not fully understood, but are probably similar to methods for other gram-negative organisms such as *Pseudomonas* and *Serratia,* which are associated with water sources. Postpartum NI are often due to group B

TABLE 24-2. Common Isolates from Nosocomial Infections in Patients with Leukemia and Lymphoma in an Acute-Care Hospital, Showing Preponderance of Gram-Negative Bacteria and Yeasts

	Percent of total isolates from all sites			
Isolate	*2/74-2/77*	*3/77-2/79*	*3/79-9/79*	*10/79-4/81*
	n=150	n=79	n=39	n=276
Gram positive				
S. aureus	15.3	17.7	12.8	2.2
Other	8.6	11.3	2.5	18.8
Gram negative				
E. coli	6.7	3.8	5.1	9.0
Klebsiella	7.3	13.9	12.8	9.8
P. aeruginosa	7.3	10.1	0	3.3
Other	1.3	2.5	7.7	4.0
Yeasts				
C. albicans	27.3	19.0	18.0	23.5
Other yeasts	1.3	5.0	10.2	5.8
Viruses				
Herpes sp.	1.3	2.5	0	4.6
Other	0	0	0	1.1

Data compiled by C. Hargiss, RN, MS, nurse epidemiologist, University of Washington Hospital, Seattle, 10/81.

streptococcus and the anaerobic *Bacteroides fragilis,* which are part of vaginal or bowel flora. A highly antibiotic-resistant *Corynebacterium,* first recoverd in 1977 as a nosocomial cause of sepsis, has been isolated from a number of specimens. Common nosocomial bacteria are listed on Table 24-4.

NOSOCOMIAL VIRAL INFECTIONS

What we know about viral NI is probably very limited since viruses have not often been sought in clinical specimens until recently. Even now, lack of simple diagnostic techniques makes viral identification costly and difficult. Viruses with documented nosocomial transmission include hepatitis A, B, and non-A, non-B, cytomegalovirus, respiratory syncytial virus, influenza, adenovirus, parainfluenza, and Epstein-Barr virus (EBV). In one study, the average hospital stay of patients who developed NI due to viruses was about 9 days longer than in others. Hepatitis A occurs in long-term care facilities (nursing homes, residences for mentally retarded), in food-borne epidemics, or in cases of fecal contamination in conditions of crowding or poor hygiene. Hepatitis B is of much more concern in acute care facilities because of the congregation of high-risk persons (those receiving multiple blood transfusions or hemodialysis, and drug abusers). Non-A, non-B hepatitis now causes 85 to 90 percent of posttransfusion hepatitis. Despite careful monitoring of blood donors, such hepatitis still occurs in about 15 percent of patients who receive multiple transfusions.

Immunocompromised individuals are highly prone to varicella-zoster virus (VZV) infections, which are transmitted either by

TABLE 24-3. Microbial Pathogens Associated with Infusion-Related Septicemia

Source of septicemia	*Major pathogens*
Conventional intravenous therapy	
Cannula	*Staphylococcus aureus* (>50 percent) *Staphylococcus epidermidis* *Klebsiella-Enterobacter* *Serratia marcescens* *Enterococcus* *Acinetobacter* sp. *Pseudomonas aeruginosa* *Pseudomonas cepacia* *Corynebacterium* sp. (diphtheroids)
Contaminated infusate	Tribe Klebsielleae (90 percent) *Klebsiella* *Enterobacter** *Serratia* *Ps. cepacia** *Citrobacter freundii* *Flavobacterium*
Hyperalimentation (most catheter related)	*Candida* sp. (30 percent-50 percent) *Torulopsis glabrata* *Staph. aureus* *Klebsiella-Enterobacter* *Enterococcus*
Contaminated blood products (most catheter related)	Pseudomonads other than *Ps. aeruginosa* (>50 percent)* *S. marcescens* *Achromobacter* *Salmonella choleraesuis* *Citrobacter* *Flavobacterium*
Arterial pressure monitoring (contaminated infusate)	*Ps. cepacia** *Ps. acidovorans* *Ps. aeruginosa* *Serratia* *Enterobacter* *Flavobacterium* *Candida* (rare)
Regional intra-arterial cancer, chemotherapy	*Staph. aureus*

*Septicemia caused by *E. cloacae*, *E. agglomerans*, or *Ps. cepacia* should especially prompt investigations for contaminated infusate.

Reprinted with permission of the publisher from Maki DG. Nosocomial bacteremia. Am J Med 1981;70:719-732.

direct contact or aerosol. However, since most adults have contacted the virus, NI is unusual. NI with VZV is also difficult to trace because of a long incubation period. Reactivation of latent infection may occur more frequently in the hospitalized patient who is physiologically and psychologically stressed. Since varicella (chickenpox) is highly contagious and can be lethal among the immunosuppressed, hospitalization of persons with chickenpox should be avoided or terminated whenever possible. It is possible but rare to develop varicella after exposure to zoster (shingles), so patients with herpes zoster should be placed in wound and skin precautions.

Herpes simplex virus, type 1 (HSV-1) can cause nosocomially acquired stomatitis, pharyngitis, and, especially among hospital personnel, whitlow (infectious swelling of fingers). Finger infections acquired by nurses from patients have been documented from exposure both to HSV-1 and HSV-2 (genital herpes). Both mothers with HSV-2

TABLE 24-4. Common Nosocomial Bacteria (Schizomycetes)

Aerobic		*Anaerobic*	
Gram positive	*Gram negative*	*Gram positive*	*Gram negative*
Staphylococcus *S. aureus* *S. epidermidis* Beta-hemolytic streptococcus *S. pyogenes* (group A) *S. agalactia* (group B) *S. equi* (group C) *S. faecalis* (*Enterococcus* or group D) Alpha-hemolytic streptococcus *S. viridans* (*viridans* group) *S. pneumoniae (Pneumococcus)* Gram-positive rods Lactobacilli Corynebacteria	Enterobacteriaceae (facultative anaerobes) *Enterobacter* species *Escherichia coli* *Klebsiella* species *Proteus* species *Providencia* species *Salmonella* species *Shigella* species *Yersinia* species *Serratia* species Other gram-negative rods *Pseudomonas* species *Legionella* species *Hemophilus* species *Acinetobacter* species Gram-negative cocci *Neisseria* species *N. meningitidis* *N. gonorrhoeae*	*Peptococcus* *Peptostreptococcus* Gram-positive rods, *Clostridium* species *Propionibacterium* species	*Bacteroides* species

Courtesy of Centers for Disease Control. Management Skills for Infection Control Nurses (training manual). 1982.

infections and their babies can transmit the virus.

Cytomegalovirus (CMV), a type of herpes, is a very rare nosocomial pathogen and requires close contact for spread. It can be transmitted via blood transfusion or from infection in a transplanted kidney. Horizontal transmission from mother to fetus can result in congenital defects and affected babies should be segregated from other babies in the nursery. Pregnant personnel should avoid caring for patients with documented CMV infection, though nosocomial transmission by direct contact has not been unequivocally demonstrated. Epstein-Barr virus, which causes infectious mononucleosis, has been transmitted by exposure to infectious blood, but not to respiratory secretions in the institutional setting. Hence, respiratory isolation is not indicated.

Respiratory syncytial virus (RSV) is a common nosocomial pathogen in infants and adults. As many as half of children admitted to hospitals with other problems during periods of high RSV prevalence in the community may acquire infection in the hospital. Large amounts of virus are shed from the respiratory tract for up to 21 days (averaging 5 to 7 days after infection). Infection rates are independent of age or underlying disease. Experimentally, RSV infection can be transmitted by inoculation of infectious secretions into eyes or nose, but not into mouth. In one series, transmission was not demonstrated from infant-to-infant, but over 50 percent of adult personnel were infected. The virus was probably transmitted between babies by the staff.

Influenza is one of the most highly infectious human agents and nosocomial epidemics do occur. There is also documented nosocomial transmission of parainfluenza types 1, 3, 4A, and rhinoviruses, which cause about 30 percent of common colds and 40 percent of exacerbations of chronic bronchitis. Transmission of rhinoviruses is primarily hand-to-hand, on clothing or equipment with subsequent inoculation onto nose or eyes. Handwashing is the most effective control measure.

CLINICAL IMPLICATIONS

Accepted standards for control of NI (Joint Commission for Accreditation of Hospitals) include the requirement for a multidisciplinary infection-control committee, written infection-control policies, and an active surveillance program.

A major breakthrough in the surveillance and control of NI came in the 1960s with the advent of the infection-surveillance practitioner in hospitals and other acute-care institutions. This individual is often a nurse or medical technologist and is variously called nurse epidemiologist, infection control practitioner (ICP), or infection surveillance officer. The efficacy of using such individuals in an infection control program has been studied by personnel at Centers for Disease Control in a project entitled "Study on the Efficacy of Nosocomial Infection Control" (SENIC). Initial reports from that study are quite positive. Currently, essentially every major health care institution in the United States has at least one infection control practitioner. Standards have been developed for their practice (Soule 1983).

When evaluating any infection control measures, issues to be considered include cost benefits of prevention, and the impact of such measures on survival (e.g., does laminar air flow or protective isolation chambers, which are quite expensive, reduce infectious mortality in patients with neoplasms?). Much use of antimicrobial agents is inappropriate and could be eliminated. Attempts to enhance host defenses would probably meet with more success in controlling NI than other, less direct, approaches. Maximization of nutritional status, tissue perfusion, oxygenation, and immune stimulation are areas in which clinicians can directly intervene. In addition, it should be possible to enhance natural interactions among bacteria that help prevent adherence and colonization of gram-negative bacteria to epithelial surfaces. This can be done by avoiding trauma (for example, during tracheal suctioning or catheter insertion), by maintaining hydration of mouth and skin, by providing as much natural stimulation of the GI tract as possible, and by minimizing use of antibacterial agents. Some investigators have implanted alpha-hemolytic streptococci or *S. epidermidis* in infants to prevent colonization with more pathogenic strains. Though this has been somewhat successful, clinical disease caused by the "harmless" organisms have occurred and the practice is not generally accepted.

There are also more specific control measures to be used. Careful management of wounds, delayed primary wound closure, improved design of invasive devices (intravenous catheters, for example) to enhance safety, and the development of practice standards for all personnel can help to minimize surgical wound infections and bacteremias. To prevent epidemic spread of viral febrile respiratory illness, isolation should be maintained as shown on Table 24-5. For influenza, virus shedding continues for 5 to 7 days after onset of symptoms and respiratory precautions should be maintained throughout that time. Patient cohorting, staff gown changes, and handwashing between contacts have been shown to decrease transmission of RSV between patients even when staff infections did not decrease. Personnel caring for a group of patients with respiratory infections should avoid simultaneously caring for other patients whenever possible. Family visits during peak epidemic times in the community should be curtailed.

Personnel with no history of chickenpox should avoid patients with the disease. Passive immunization of individuals at high risk for VZV (those with postrenal transplantation, neoplasm) can be considered. Routine screening of personnel for varicella and rubella titers at time of employment has been recommended. Screening on admission of patients at high risk of selected infectious disease such as hepatitis and resistant gram-negative bacteria might serve to prevent spread of infection from unrecognized reservoirs.

Careful attention should be paid to placement of patients in the hospital (see Table 24-6) to optimize their environmental conditions. Babies with herpes require strict isolation, and mothers require wound and skin precautions. Gloves should be worn when caring for patients with HSV lesions. Personnel with HSV lesions should avoid working with newborns, burn patients, or compromised hosts until lesions have crusted and dried. Preventing the spread of hepatitis A does not require a private room unless the patient has fecal incontinence, though careful handwashing, of course, is essential. Hepatitis B immune globulin can be administered to contacts within 2 weeks of exposure to minimize or prevent infection. Blood precautions are necessary for patients with hepatitis B and patients must be segregated if massive bleeding is present.

Catheter-Associated Urinary Tract Infections

About 4 million patients per year undergo bladder catheterization. A single catheterization has a 1 to 5 percent risk of infection and catheters indwelling for more than 4 days with open drainage will cause infection in virtually every case. Even with the closed drainage systems used currently, 20 percent of catheterized patients will develop urinary tract infections. Most catheter-associated urinary tract infections are caused by coliforms. Control measures include the use of alternative methods of urinary drainage (condom drainage in males or intermittent catheterization) whenever possible, aseptic insertion technique, and strict maintenance of a closed drainage system. Efforts such as daily meatal cleansing, application of antimicrobial ointments, and bladder irrigation with antimicrobial solutions are of unproven value (Wong 1983).

SUGGESTED READINGS

Allen JR, et al. Secular trends in nosocomial infections: 1970-1979. Am J Med 1981;70:389-392.

(text continues on page 647)

TABLE [illegible]. Guidelines for Isolation of Patients with Respiratory, Herpes, and Hepatitis Viruses

Category of illness or virus	Hand-washing	Private room	Mask	Gown	Gloves	Max. rm cleaning	Precautions with: Dishes	Excreta/excreta-soiled articles	Blood/needles	Secreta/secreta-soiled articles	Linen	Duration of isolation/comments
Respiratory viruses												
Influenzavirus A, B	X	X*	X									Duration of illness
Respiratory syncytial virus	X	X*		X								Duration of illness
Parainfluenza virus	X	X*		X								Duration of illness
Rhinovirus	X											
Herpesviruses or herpes zoster (disseminated)												
Varicella	X	X*	X	X	X			X		X	X	Until lesions are dried and crusted. Any personnel with negative or unknown history of chickenpox should avoid contact with these patients
Herpes zoster (localized)	X	X	X**	X**	X**			X		X	X	
Herpes simplex type 1	X											Duration of vesicles
Herpes simplex type 2	X											See text
Epstein-Barr	X											
Cytomegalovirus												
(acquired)	X											
(congenital)	X	X††		X**	X**			X			X	Pregnant women should avoid contact with patients with CMV infection

TABLE TABLE 24-5. *(continued)*

Category of illness or virus	Hand-washing	Private room	Mask	Gown	Gloves	Max. rm cleaning	Precautions with: Dishes	Excreta/excreta-soiled articles	Blood/needles	Secreta/secreta-soiled articles	Linen	Duration of isolation/comments
Hepatitis												
Hepatitis A	X	X†	X‡	X‡	X‡			X	X	X	X	Maintain isolation until liver enzymes (SGOT, SGPT) return to normal levels
Hepatitis B	X	X†	X‡‡	X‡‡	X‡‡			X	X	X	X	Until patient is negative for HBsAg
Hepatitis non-A, non-B	X	X†	X‡‡	X‡‡	X‡‡			X	X	X	X	Epidemiology not completely known

* = or cohort with patient with same illness
† = if incontinent of feces or bleeding
‡ = for direct contact with blood or feces
** = direct patient contact
†† = desirable
‡‡ = for direct contact with blood, feces, or body fluids

Reprinted with permission of the publisher from Valenti VM, et al. Nosocomial viral infections: II. Guidelines for prevention and control of respiratory viruses, herpesviruses, and hepatitis viruses. Infect Cont 1980;1:165-178.

TABLE 24-6. Assignment of Patients in the Hospital Environment to Prevent Nosocomial Infections

1. Does patient have communicable disease or condition hazardous to other patients or personnel?	______ No;	______ Yes
If so, kind of infection: ______		
System involved: ______		
Precautions required: ______		
2. Does patient have seriously compromised		
Respiratory system?	______ No;	______ Yes
Cardiac system?	______ No;	______ Yes
Renal system?	______ No;	______ Yes
Immune system?	______ No;	______ Yes
If so, degree of protection necessary ______		
3. Is patient able to understand and carry out instructions?	______ No;	______ Yes
4. Can patient share a room?	______ No;	______ Yes
If so, kind of unit: ______		
Kind of roommate: ______		

Adapted with permission of the publisher from Hargiss CO. The patient's environment: haven or hazard. Nurs Clin North Am 1980;15:671-688.

Bennett JV. Human infections: economic implications and prevention. Ann Intern Med 1978;89 (Part 2):761-763.

Bennett JV, Brachman PS, eds. Hospital infections. Boston: Little, Brown and Company, 1979.

Burke JP, et al. Prevention of urinary catheter-associated infections: efficacy of daily meatal care regimens. Am J Med 1981;70:655-658.

Cross AS, Roup B. Role of respiratory assist devices in endemic nosocomial pneumonia. Am J Med 1981;70:681-685.

Dixon RE. Effect of infections on hospital care. Ann Intern Med 1978;89 (Part 2):749-753.

Eickhoff TC. General comments on the study on the efficacy of nosocomial infection control. Am J Epidemiol 1980;111:465-469.

England AC, Fraser DW. Sporadic and epidemic nosocomial legionellosis in the US. Am J Med 1981;70:707-711.

Gardner P, et al. Nosocomial management of resistant gram negative bacilli. J Infect Dis 1980;141:415-417.

Garibaldi RA, Brodine S, Matsumiya S. Infections among patients in nursing homes. N Engl J Med 1981;305:731-735.

Garibaldi RA, et al. Risk factors for postoperative pneumonia. Am J Med 1981;70:677-680.

Greaves WL, et al. The problem of herpetic whitlow among hospital personnel. Infect Cont 1980;1:381-385.

Hall CB. Nosocomial viral respiratory infections: perennial weeds on pediatric wards. Am J Med 1981;70:670-675.

Hargiss CO. The patient's environment: haven or hazard. Nurs Clin N Am 1980;15:671-688.

Hunt TK. Horizons of infection control in surgery—or, has anyone here seen an infected pile? Am J Med 1981;70:712-718.

Jackson MM. Viral hepatitis. Nurs Clin N Am 1980;15:729-746.

Kibrick S. Herpes simplex infections at term—what to do with mother, newborn and nursery personnel. JAMA 1980;243:157-160.

LaForce FM. Hospital-acquired gram negative rod pneumonias: an overview. Am J Med 1981;70:664-669.

Ledger WJ. Hospital infections gynecologic, obstetric, and perinatal infections. Ann Intern Med 1978;89 (Part 2):774-776.

Maki DG. Control of colonization and transmission of pathogenic bacteria in the hospital. Ann Intern Med 1978;89 (Part 2):777-780.

Maki DG. Nosocomial bacteremia. Am J Med 1981;70:719-732.

Maynard JE. Nosocomial viral hepatitis. Am J Med 1981;70:439-444.

Noble RC. Infectiousness of pulmonary tuberculosis after starting chemotherapy. Am J Infec Cont 1981;9:6-10.

Penn RG, Sanders WE, Sanders CC. Colonization of the oropharynx with gram negative bacilli: a major antecedent to nosocomial pneumonia. Am J Infec Cont 1981;9:25-34.

Polk HC. Prevention of surgical wound infections. Ann Intern Med 1978;89 (Part 2):770-773.

Preston GA, Larson, EL, Stamm WE. The effect of private isolation rooms on patient care practices, colonization, and infection in an intensive care unit. Am J Med 1981;70:641-645.

Rubin RH. Infection in renal transplant patients. Am J Med 1981;70:405-411.

Rutala WA, et al. *Serratia marcescens* nosocomial infections of the urinary tract associated with urine measurement containers and specific gravity instruments. Am J Med 1981;70:659-663.

Scheckler WE. Septicemia and nosocomial infections in a community hospital. Ann Intern Med 1978;89 (Part 2):754-756.

Soule B. The APIC curriculum for infection control practice. Dubuque, Iowa, Kendall/Hunt Publishing Co, 1983.

Stamm WE. Comparison of endemic and epidemic nosocomial infections. Am J Med 1981;70:393-397.

Stamm WE. Infections related to medical devices. Ann Intern Med 1978;89 (Part 2):764-769.

Turck M, Stamm WE. Infections of the urinary tract. Am J Med 1981;70:651-654.

Valenti WM, et al. Nosocomial viral infections: I. Epidemiology and significance. Infec Cont 1980;1:33-37.

Valenti WM, et al. Nosocomial viral infections: II. Guidelines for prevention and control of respiratory viruses, herpesviruses, and hepatitis viruses. Infec Cont 1980;1:165-178, 1980.

Weinstein RA. Strategies for control of multiply drug resistant nosocomial infections. Am J Med 1981;70:449-453.

Wong ES. Guidelines for prevention of catheter-associated urinary tract infections. Am J Infec Cont 1983;11:28-33.

Young VM, et al. The emergence of coryneform bacteria as a cause of nosocomial infection in compromised hosts. Am J Med 1981;70:646-650.

25 STERILIZATION AND DISINFECTION

KEY POINTS

- Sterilization is the complete destruction of all forms of microbial life. Disinfection is the process of destroying pathogens on inanimate objects (one does not disinfect a human). Antisepsis renders microoganisms harmless on the skin by killing them or by inhibiting their growth.
- A variety of environmental factors can influence the effectiveness of sterilization/disinfection regimens, especially temperature, pH, and presence of organic matter.
- All chemical forms of sterilization/disinfection have potential side effects or toxic effects.
- The major forms of sterilization used in health care institutions are the autoclave (moist heat) and the ethylene oxide gas sterilizer.

Methods of disinfection have been practiced for centuries. Many of the very early measures used were primitive but they have served as precursors for our modern day sterilization and disinfection procedures. As the science of microbiology developed so has the practice of disinfection and sterilization. In the early nineteenth century methods of sterilization were introduced into surgical procedures as a means of preventing sepsis. Some of the basic techniques and agents first employed in the early nineteenth century are still used today.

There have been numerous refinements made in the methods and agents used for sterilization and disinfection as our knowledge of the dynamics of microbial life and death have increased. Many new antibacterial and chemotherapeutic agents have replaced older chemical agents once used in medical therapy. However, some of these chemical agents still have use as disinfectants for inanimate objects. Sterilization by physical methods has improved immensely with the technological advances made in equipment. However, there are limitations to success for all sterilization and disinfection methods and procedures. Therefore, it is essential that nurses, physicians, dentists, and other health professionals increase their knowledge and understanding of the mechanisms and factors affecting microbial life and destruction, as well as the agents and methods by which potential pathogens are destroyed or inhibited.

The methods of sterilization and disinfection that will be emphasized in this chapter are those most frequently used in health care settings. Other recognized methods will be discussed in less detail.

DEFINITIONS

The terms that are most frequently used when discussing the damaging effects that certain physical and chemical agents have on various microorganisms need to be succinctly defined.

Sterilization refers to complete destruction of all forms of microbial life including viruses. It must be remembered that an article is either sterile or nonsterile.

Disinfection is the destruction of pathogenic microorganisms on inanimate objects by chemical or physical agents.

Antisepsis refers to the chemical disinfection of the skin on living tissue. Substances known as antiseptics when applied to microorganisms render them innocuous by either killing them or by inhibiting their growth in accordance to the characteristics of the preparation or the method of application.

Other terms often seen in reference to chemical agents are germicide, bactericide, fungicide, sporocide, and virucide. Addition of the suffix *cide* to the above words indicates that these agents kill the microorganisms with which they come in contact. Another suffix added to bacteria, spores, fungi and viruses is *stasis*. Chemical agents with this suffix indicate that the microorganisms will be inhibited by the presence of the agent, but not destroyed.

It is important that this terminology be understood in order that the appropriate procedure or agent be chosen as dictated by the specific situation. In surgical procedures, the aim is to have destruction of all microorganisms in or on materials used. The same would be true in microbiology for glassware, media, and other special laboratory equipment. The method of choice for sterilization about 95 percent of the time is the autoclave, or steam under pressure. For materials that would be destroyed by heat, chemical sterilization is most often applied. Where the intent is primarily to prevent cross infection, as in communicable disease, a disinfectant is usually employed, especially as a cleaning procedure.

FACTORS INFLUENCING STERILIZATION/DISINFECTION

A variety of factors impact on the effectiveness of a sterilization or disinfection process. One of the foremost considerations should be the population of microorganisms to be treated. To choose an adequate agent, one must assume that the most resistant microbial forms or sporulating bacteria are present. Therefore, in exposing these microbial forms to the sterilization or disinfection process, recognition must be given to the differences in the metabolic states, the differences in the microenvironment, and the differences in the phases of the growth cycle of the microbial population likely to be present. Thus, it would be expected that 90 percent of the population would be killed in 1 minute, 90 percent of the remaining 10 percent in the next minute, with 90 percent of the 1 percent in the next minute, and so on. Because all microorganisms in a population are not killed at the exact same time, sufficient time must be allotted to the sterilizing or disinfecting process to ensure that the treatment has affected the entire population.

Certain microorganisms or forms of microorganisms tend to react differently. For example, spore-forming organisms are more resistant to heat and chemical agents. *Clostridium botulinum* can resist a temperature of 212°F (100°C) for 3 to 5 hours. Vegetative bacteria, on the other hand, are usually more susceptible to heat and most are killed at temperatures of 212°F (100°C) in 30 minutes. A few bacteria such as *Mycobacterium tuberculosis* and *Pseudomonas* species are particularly resistant to chemical disinfectants. The resistance of the tubercle bacillus is attributed to the waxlike composition of its cell wall.

The hepatitis virus, types A and B, and the more recently identified hepatitis viruses, non-A, non-B, are transmissible in blood and serum, as well as urine, feces, saliva, and semen. Nurses, physicians, laboratory workers, and dentists are all at high risk because of frequent contact with organic substances contaminated with these viruses. Hepatitis B and non-A, non-B viruses are killed by autoclaving, high temperatures of dry heat, and ethylene oxide. Hepatitis B virus can also be killed in 10 minutes in boiling water at 212°F (100°C). Disinfectants kill the virus but the exposure time is much longer (e.g., 2 percent glutaraldehyde requires 10 hours and 2.5 percent sodium hypochlorite [household bleach] requires 30 minutes).

Additional conditions that have a bearing upon the rate and level of kill of microorganisms are organic materials, pH, time-temperature relationships, and moisture (Table 25-1). Changing the pH (hydrogen ion concentration) of a solution can alter the concentration of a chemical agent at the surface of the bacterial cell. As bacteria are exposed to increased temperature, the time necessary to kill them decreases (e.g., a standard bacterial inoculum exposed to moist heat that may be destroyed in 2 1/2 hours at 230°F, (110°C), can be killed in 15 minutes at 250°F, (121°C). For those articles or materials that are heat sensitive and can withstand the process, autoclaving is the method of choice.

In the health related fields, materials to be sterilized should be free of all soil and organic matter such as blood, serum, saliva, urine, etc. Microorganisms are protected by these substances, which especially impede the action of disinfectants and heat. Therefore, cleanliness of syringes, instruments, hypodermic needles, and other materials for sterilization is necessary to ensure effective sterilization. Likewise, if an article is properly sterilized but carelessly handled upon completion of the process and becomes contaminated, a new population of microorganisms can quickly multiply. Personnel responsible for the sterilization process must be always alert to possible sources of contamination, as well as to the initial preparation of supplies to be sterilized and the

TABLE 25-1. Factors that Influence the Effectiveness of Disinfectants and Antiseptics

Exposure time
Concentration
Presence of organic materials (body fluids, tissues)
pH
Moisture
Time-temperature relationships

appropriate method by which this might be achieved.

PHYSICAL METHODS OF ANTIMICROBIAL ACTIVITY

Sterilization and disinfection procedures are customarily divided into physical and chemical methods. The physical methods include moist heat, dry heat, cold, ultrasonics, and irradiation.

Moist Heat

Moist heat, because it is reliable, is the most universally used method of sterilization. It is preferred over dry heat because of its rapid killing action. Mesophilic bacteria in the vegetative form are susceptible to moist heat and are killed at 140°F (60°C) for 30 minutes. In order to destroy a few of the more resistant species of vegetative bacteria such as *Staphylococcus aureus* and *Streptococcus faecalis,* it is necessary to expose them to a temperature of 140°F (60°C) for 60 minutes. At temperatures of 212°F (100°C) most forms of vegetative bacteria, yeast, and fungi are destroyed in 10 to 15 minutes. However, one of the most resistant spore-forming organisms, *Clostridum botulinum,* is killed only after 5 1/2 hours of exposure to a temperature of 212°F (100°C), but it is destroyed in 4 minutes at 250°F (121°C).

Saturated steam under pressure. Saturated steam under pressure is the most effective and efficient method of heat sterilization. The autoclave makes it possible to increase the temperature of moist heat above the standard 212°F (100°C) by increasing the pressure. The temperature of moist heat can be raised to 250°F (121°C) if the pressure in the autoclave in increased to 15 pounds per square inch (psi). Most microorganisms will be destroyed when subjected to this temperature for 15 minutes. If the pressure is increased to 20 psi, the temperature is increased to 270°F (126°C). At this temperature microorganisms are killed in 3 minutes.

It is this combination of steam under pressure, temperature, and exposure time that is instrumental in the destruction of the microorganisms. Steam gives up its heat as it condenses upon an object that is cooler than itself. This condensation process is very rapid. As the steam contacts the outer layer of fabric, wrapped packs, or other porous substances, this cooler outer layer causes the steam to condense, penetrating the fabric of this layer and leaving it moist. This process continues with the steam condensing on each successive fabric layer until the interior of all the packages or bundles have been permeated. All layers of the packages will attain equivalent amounts of moisture and heat. The process is extended to other objects in the surrounding area that absorb the latent surface heat produced by the condensation process, thereby exposing all materials to the constant temperature of the surrounding steam.

A similar mechanism of condensation and heating occurs with instruments and other utensils undergoing surface sterilization. There is no permeation of steam through the metal, but sterilization is accomplished by steam condensation on the cold surfaces of the instruments. Less time is required to sterilize unwrapped instruments alone, because permeation of layers of porous materials is obviated.

It should be emphasized that the autoclave pressure, except to increase the temperature, does not contribute to the destruction of the microorganisms. It is the maintenance of a sustained high degree of temperature for a specified period of time to various articles that kills the microorganisms and results in sterilization.

An extremely important element in the effectiveness of sterilization by autoclaving is the presence of air within the chamber. The air, because it is denser than steam, gravitates toward the bottom of the sterilizer. The temperature of this air is cooler, causing interference with the penetration of steam, thereby reducing the pressure at a given temperature. Therefore it is necessary to evacuate all air prior to autoclaving. The most frequently used temperatures for sterilizing supplies in the autoclave are shown in Table 25-2.

The autoclaves today are equipped with controls that allow them to operate automatically. The discharge of air from the autoclave along with the pressure and timing are automatically regulated, reducing the margin for human error in the process. The sterilization process can also be manually controlled if some phase of the cycle needs to be changed. Because some autoclaves are designed to achieve high temperatures as the pressure is increased, it is possible to sterilize supplies in a much shorter time period. In the autoclave, which reaches a temperature of 274°C (134°C) at a pressure of 30 psi, the sterilization time is reduced (e.g., 15 minutes for wrapped instruments, needles, syringes, and catheters). These same supplies if sterilized at 250°F (121°C) at a pressure of 15 psi would require a 30-minute exposure period. The exposure time refers to length of time materials must remain at a given temperature once it has been attained. Therefore, the completed sterilization cycle for supplies is longer than the recommended exposure time.

TABLE 25-2. Autoclave Steam Pressures and Temperatures

Steam pressure (lbs/sq in)	Temperature C°	Temperature F°
15	121.5	250
20	126.5	260
30	134.5	274

Timing of the autoclave load also depends on the size of the load, the characteristics of the material, and the item placement within the autoclave. If the autoclave is completely filled, timing must be sufficient to allow for all items to be exposed to the optimum temperature for the recommended period. The autoclave must be loaded so that the saturated steam can freely circulate around and reach all items. All sides of an article should be easily accessible for steam penetration. Canisters and deep trays should be placed on their sides so that air is not trapped in the bottom. Trapped air prevents the steam from penetrating articles. Instruments, whether wrapped or unwrapped, should be packed in perforated trays to allow the steam to circulate freely around them. Hollow objects should be flushed with water prior to sterilization so that moisture in the lumen can generate steam. All materials should be clean and free of any organic substance which might protect resistant microorganisms and prevent sterilization. After wrapped supplies have been sterilized, they need to be carefully handled so that they are not contaminated. Wrapped supplies should be dry before they are removed from the autoclave. Storage should be in a closed cabinet if possible.

Boiling water. Boiling water effectively kills vegetative bacteria and some spore-forming organisms when they are exposed for 30 minutes. The temperature of boiling water at sea level is 212°F (100°C). At higher elevations water boils at a lower temperature, therefore it is recommended that the 30-minute boiling time be extended by 5 minutes for every 1000 feet of altitude. If boiling water is used for sterilization it is important to remember that the articles must be completely covered by the water, hinged instruments should be opened, and syringes should be separated. Timing should start when the water is boiling vigorously. Because boiling water does not kill all bacterial spore-forming organisms (such as *Clostridium tetani* and *Clostridium botulinium*) in the 30 minutes recommended for vegetative forms of bacteria, and, because more sophisticated methods of sterilization are usually available, this method of sterilization should be discouraged in a modern hospital.

Dry Heat

Dry heat is a method of sterilization that requires high temperatures and long periods of time. It is used more frequently in the laboratory than in the hospital. However, materials or articles that cannot withstand exposure to saturated steam, or are impervious to it, can be sterilized by the dry heat method. Materials such as glassware, syringes, sharp instruments, jellies, and powders are more effectively sterilized in this manner.

For dry heat sterilization, a special sterilizer is required in which the hot air is controlled within narrowly defined limits. The most widely used device is the hot air oven.

A dry heat or hot air sterilizer acts by conduction. The heat from the hot air is first absorbed by the exterior surface of materials or articles in the sterilizer. Eventually this heat is transferred to the interior of the articles. Therefore, as the heat penetrates these materials, the organisms in contact with them are killed. It is extremely important that the entire material or item to be sterilized is heated uniformly, because success of the process is dependent on the heat penetration and not just the hot air surrounding it.

The minimum exposure times for dry heat sterilization are
340°F (172°C) for 1 hour
320°F (160°C) for 2 hours
250°F (121°C) for 6 hours or longer
These times refer to the actual temperatures necessary for the sterilizer load and do not allow for heat-up time. Most of these time-temperature ratios allow for a margin of safety in the exposure time.

Dry-heat exposure time required for bacterial destruction of vegetative and spore-forming organisms varies considerably with different species. *Bacillus stearothermophilis* is maximally resistant to destruction by moist and dry heat and requires 60-minutes exposure at a temperature of 320°F (160°C). *Clostridium botulinium* requires only 25-minutes exposure at the same temperature. Exposure time should be based on the nature and properties of materials to be sterilized, the sterilizer load, and the preparation of supplies (e.g., size, container, and wrappings). If these factors are ignored, the exposure time selected may be inadequate for the destruction of the most resistant microorganisms.

Flaming and incineration. While the flaming of transfer needles by the use of a bunsen burner in the laboratory is considered an effective sterilization method, the flaming of needles or sharp instruments by using an alcohol lamp in the hospital is to be discouraged. The heat created by the alcohol lamp flame is relatively cool and does not allow for quick incineration of microorganisms, therefore, the organisms may splatter into the environment and may not be killed. Incineration is used for destroying materials.

Cold

Freezing temperatures slow the metabolic activities of microorganisms and inhibit their growth. Freezing is used in microbiology laboratories to preserve cultures of microorganisms for prolonged periods of time. Freezing and drying under vacuum stops the growth or reproduction of bacteria although under these conditions some microorganisms survive for years. At normal refrigeration temperatures (0 to 8°C), the growth of microorganisms is slowed; at freezer temperatures (−5°C to −25°C) the quality and flavor of foods can be maintained and microbial growth inhibited.

In microbiology laboratories cold or freezing temperatures are used mainly for the preservation rather than the sterilization of cultures and specimens, since cold does not kill all the microorganisms. Food services use cold to preserve and to prevent the multiplication of microorganisms in food.

Ultrasonics

Ultrasonics have been used in the laboratory for cell disintegration and in other areas of the hospital for cleaning purposes. Ultrasonic cleaners contain liquids through which sonic waves pass at frequencies of 100,000 per second or higher. These waves have profound effects on materials suspended in solutions. The high intensity sound waves traverse the liquid media rapidly and generate the formation of submicroscopic bubbles that subsequently collapse. This process creates a suction type action or a negative pressure on the particles in the suspension. Bacteria are disintegrated and protein materials are coagulated by this action. Soil on instruments is removed in this manner.

While these untrasonic vibrations may be lethal to some microorganisms (gram-negative rods are the most susceptible, and staphylococci are the most resistant), many survivors remain in the bacterial population and this prevents the use of ultrasonics as a sterilizing method. In the hospital ultrasonic cleaners are used for cleaning glassware and instruments, especially those with serrated and hinged areas. Although ultrasonic equipment is costly it does provide a sophisticated method for cleaning and decontaminating instruments. However, items emerging from this cleaner must be washed free of the solvent cleaner and resterilized prior to use.

Radiation

Ultraviolet (UV) light and ionizing radiation are recognized measures used to destroy microorganisms. But, the application of UV light and ionizing radiation as agents for sterilization or disinfection is limited to specific products and situations.

UV rays of the light spectrum are in the range from 4000 Ångströms (Å) downward to 2000 Å. At the lower end of this scale (2800 to 2500 Å) the rays have a highly destructive effect on microorganisms, particularly bacteria. It is thought the UV rays are absorbed by the nuclear proteins of the bacterial cell thus inducing cell mutations or death based on chemical changes that effect the genetic mechanisms. The optimum bactericidal effect is around 2600 Å, which coincides with the absorption peak of DNA for UV rays. The utilization the UV component of sunlight as a sterilizing or disinfecting agent is not reliable because only those organisms directly exposed to the rays are killed.

Practical application of UV radiation for the control of microorganisms has been primarily for reducing the number of microorganisms in the air. This is usually done by strategically placing low-pressure mercury lamps that generate UV rays of a 2537 Å wavelength in the air ducts of buildings, hospital rooms, operating rooms, entry ways, and nurseries. These germicidal lights are installed in such a manner that the occupants of the rooms are not affected by the light.

Commercially, UV radiation is used for the sterilization of vaccines, sera, and toxins. It is also used for sanitizing packaged foods, paper products, and the contents of storage cabinets.

A disadvantage associated with the use of UV light in controlling microorganisms is its low penetrability. The organisms must be in direct contact with the UV rays and not shielded by glass or materials such as fabrics. The effectiveness of the light is limited by the life of the radiation source. It is necessary to measure the output of the source every 1 thousand hours of use. Because these lights continue to glow visibly after their short-ray output is ineffective, they are dangerous in that they create a false sense of security. Personnel whose work requires them to be directly in contact with these lights must protect their eyes and skin from exposure to the radiation to prevent skin burn and eye irritation.

Ionizing radiation has a shorter wavelength and greater energy than UV radiation. This type of radiation has a lethal effect on many tissues, cells, nonspore-forming bacteria, and viruses. The spore-forming bacteria are the most resistant and the gram-positive, nonspore-forming organisms are more resistant than gram-negative organisms. Gram-negative pseudomonads are most readily killed by this method. The shorter rays cause ionization of molecules that lies in their path. A dose of 2.5 millirads of ionizing radiation has been accepted as a sterilizing dose for most resistant microorganisms.

Ionizing radiation has the advantages of not requiring heat or moisture, having great penetrability, and permitting prepackaged and sealed items to be sterilized. This process has been found particularly useful for sterilizing pharmaceuticals, plastic syringes, catheters, and suture materials. It has been used to preserve foods. The disadvantages of this method are that expensive equipment is required and rigid safety measures must be enforced. In addition, materials sterilized in this manner may be adversely affected. Problems include loss in tensile strength of fabrics, discolorations of the product, chemical changes in pharmaceuticals, appearance change, and loss of nutritional value and palatability in food.

MONITORING AND STERILITY INDICATORS

Physical, chemical, and biologic indicators are used to determine the effectiveness of the sterilization process. Physical indicators such as thermometers, pressure gauges for temperature, humidity and load sensors are usually built into the sterilizers. The temperature and pressure gauges of the autoclave cannot detect air pockets and air leaks, which cause a lowering of temperature and underprocessing of supplies in those selected areas.

The chemical indicators, regardless of the company marketing them, consist of a sensitive ink dye that changes color under designated conditions. The biologic indicators have a known population of microorganisms in or on a carrier that are highly resistant to the method of sterilization being tested. These indicators, when used in a sterilizer, demonstrate whether the conditions necessary for sterilization were met during the cycle being tested. The organisms used in these commercially prepared biologic indicators are spore-forming bacteria. One of the most resistant spore formers is *Bacillus stearothermophilus*. Its resistance is considered greater than that of the pathogenic clostridia and bacilli. It has been recommended for testing the efficiency of the autoclave.

The ethylene oxide sterilizer and the hot-air sterilizer are tested using an indicator containing the spore-forming organism *Bacillus subtilis*. There has been some question regarding the reliability of the results of this indicator in testing the ethylene-oxide sterilizer because of the time between when the sterilizer door is opened and when the indicator can be removed for testing (exposure of personnel to ethylene oxide prohibits immediate removal of the indicator).

The Operating Room Association, the Joint Commission on Accreditation of Hospitals, the Centers for Disease Control, the Association for the Advancement of Medical Instrumentation, and the American Hospital Association require that all materials sterilized have internal and external indicators. Indicators should be evaluated and used only in situations for which they are designed. The manufacturer's instructions must be strictly followed. It should be noted that although chemical indicators are not absolutely sensitive, they do demonstrate that an article has been submitted to the sterilizing process.

Chemical indicators should be attached internally and externally to all wrapped supplies. Autoclaves should be checked weekly with a biologic indicator. For the high vacuum autoclave, a Bowie Dick Type test should be used daily on the first run of the day. If the steam supply has been turned off, a minimum of 30 minutes is needed between turning on the sterilizer and running the test cycle. The test evaluates the ability of the prevacuum sterilizer to reduce air residuals effectively from the chamber, to prevent air re-entrainment into the load, and to identify air leaks.

In the ethylene oxide sterilizer, a biologic indicator (e.g., an indicator containing *Bacillus subtilis*) should be run weekly. When implantables are being sterilized a biologic indicator containing spores should be included each time. Where it is feasible, the results of the spore test should be reported prior to use of the sterilized article or articles.

If positive cultures are produced from the biologic indicator, the sterilizer should not be used until it has been checked by an engineer. The biologic indicators should also be examined for accuracy. Inappropriately stored indicators may give false positive results. Therefore, a positive culture needs to be taken seriously but should be carefully reviewed by having both the sterilizer and the indicators checked.

ANTIMICROBIAL ACTIVITY OF CHEMICAL AGENTS

Chemical agents used to control microorganisms are referred to as disinfectants and antiseptics. The chemical agents used for inhibiting or destroying organisms on inanimate objects are disinfectants, while agents applied to body surfaces or tissue are called antiseptics.

Substances such as ethylene oxide, formaldehyde, glutaraldehyde, betapropialatone, phenolics, halogens, and heavy metals are called chemosterilizers. They may be used for either sterilization or disinfection, depending upon the response of the organism and environmental conditions such as pH, temperature, and concentration of the chemical agent.

Evaluation of Disinfectants

In evaluating a disinfectant, the official quantitative tests used by the Food and Drug Administration of the United States Department of Agriculture has been the phenol coefficient test. It tests, under defined conditions, disinfectants against strains of *Staphylococcus aureus, Salmonella typhi,* and *Pseudomonas aeruginosa* whose susceptibility to phenol is known. A comparison is made between the effects of the test disinfectant with the effects of phenol, and a phenol coefficient is calculated. For example, if the phenol coefficient is 5 it means that the agent tested was 5 times more effective as a germicide than phenol.

The phenol coefficient test has been criticized because it measures activity against only three organisms under specified conditions. It does not provide information about toxicity and diffusibility of the substance in tissue or the effects that organic matter has on its activity. It has been suggested that a series of in-use supplemental tests need to be designed to determine additional properties of the disinfectants or germicides. The variables that affect the bactericidal power of a disinfectant or germicide are time, temperature, and concentration. To obtain an accurate estimate of the effectiveness of a disinfectant, these variables need to be tested against a variety of organisms including spores and viruses both in vitro and in vivo.

In practical terms, the selection of a disinfectant or germicide is based on several parameters. Consideration should be given to the situation for which it is to be used, the type of microorganism present, the possible population number, the time frame for use, and the past success the personnel have had in using the agent. Each disinfectant or germicide must be selected with care, and should be chosen to fit the specific situation.

Alkylating Agents

Ethylene oxide. Ethylene oxide has been used for many years to sterilize temperature-sensitive medical equipment. It is a colorless gas that is highly penetrative in the absence of heat or moisture. It is potentially explosive in concentrations of 3 percent if mixed with air. For this reason, when used as a medical sterilant, it is mixed with an inert gas such as freon or carbon dioxide. The gas is very toxic if too much is inhaled and it is a vesicant (induces blistering).

Ethylene oxide not only dissolves in ordinary solvents including water, but it dissolves in rubber and plastics. It is especially useful in sterilizing catheters, dental and medical laboratory items, textiles, sutures, vascular transplants, and heart and lung machines.

Ethylene oxide sterilization should be carried out in an automated and programmed ethylene oxide sterilizer (Fig. 25-1). The manufacturer's instructions should be followed. It is of the utmost importance that the correct concentration of the sterilant be used. An effective margin is between 450 mg to 1000 mg/liter of chamber space, regulated relative humidity of 40 to 60 percent, and temperature of 70°F (22°C) to 140°F

FIGURE 25-1. Eagle 2000 ethylene oxide sterilizer, manufactured by AMSCO/American Sterilizer Company, is microcomputer controlled, and includes an alphanumeric printer and controller. (Courtesy of American Sterilizer Company, Erie PA.)

(60°C). The exposure time during the cycle of sterilization depends on the material and the size of the load. A standard time of 4 hours of exposure is given to ensure that all materials are sterilized, with 99 percent being sterilized in 2 hours, and 90 percent in 1 hour.

Articles to be sterilized by ethylene oxide should be packaged properly in material such as paper, cloth, and certain plastics that are permeable to the gas. The article must be dry prior to packaging and should be separated. Caps, plugs, and valves should be removed from instruments or equipment. The items, when placed in the sterilizer, must be positioned to allow for free circulation and penetration of the ethylene oxide gas. Immediately on completion of the sterilization cycle the door should be opened approximately 6 inches to permit the unexhausted gas to escape. The sterilizer should remain with the door open for 15 minutes before any materials are removed. Persons handling the sterilized materials are in danger of receiving skin burns if they remove items from the sterilizer before they aerate. Gloves impervious to the gas should be used for removing these supplies from the sterilizer.

Immediately after sterilization still further aeration of supplies is crucial (Fig. 25-2). The length of aeration depends on the composition and porosity of the sterilized items, the wrap of the items, the concentration of the ethylene oxide used for the process, plus the air flow rate and temperature during aeration. For example, it is very difficult to eliminate ethylene oxide from polyvinyl chloride (PVC). If the composition of the article is not known it should be treated as PVC. With a mechanical aerator the minimum of four air changes per minute is required for 8 hours at 140°F (60°C), or for 12 hours at 120°F (50°C). If materials must be aerated by ambient air at room temperature, they need 7 days in a room that is well ventilated and vented to the outside and is away from heavily trafficked areas. The room temperature should be well controlled between 65°F (18°C) and 72°F (22°C). Access to the room should be restricted since ex-

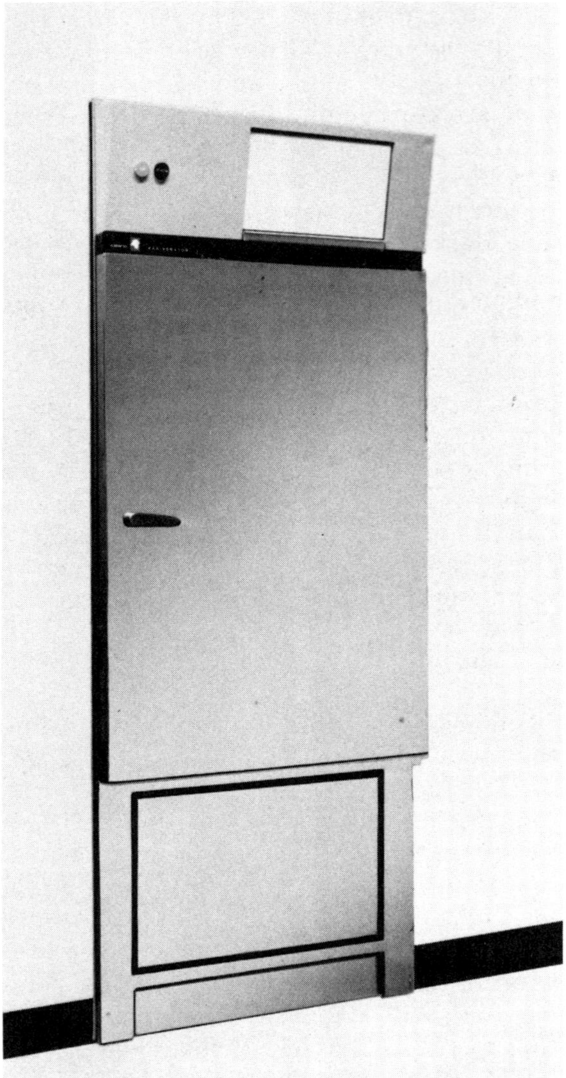

FIGURE 25-2. AMSCO ethylene oxide aerator (Courtesy of American Sterilizer Company, Erie PA.)

cessive exposure to ethylene oxide gas does present a health hazard.

Inadequate aeration of equipment can result in skin burns. The gas has been recovered from the blood of patients undergoing surgery with cardiopulmonary bypass in cases where the bypass machine and the polyvinyl tubing have been aired for less than 3 days. In addition, ethylene oxide has been found to be carcinogenic and teratogenic in animals. Ethylene oxide sterilization has gained in popularity over the last two decades but it should not be used if the materials can be sterilized by the autoclave.

Formaldehyde. Formaldehyde is used as a solution of 37 percent formaldehyde with methanol (formalin). It is colorless and has a pungent odor. It has a powerful effect on all tissue. It hardens the skin, and kills or inhibits bacterial growth. Formalin in 0.2 to 0.4 percent solution has been used to inactivate viruses in the preparation of vaccines. As a gas it is used to disinfect rooms, clothing, bedding, and excreta.

Glutaraldehyde (Cidex). This chemical agent is categorized as a dialdehyde. It is more active than formaldehyde and lacks its toxic properties. It is irritating to the skin, mucous membranes, and eyes. Used in a 2 percent solution at a pH of 7.5 to 8.5, it is a reliable sterilizing agent that possesses speed of action and destroys a broad spectrum of microorganisms. The exposure times required to kill various microorganisms are 10 hours for spore-forming bacteria, 10 minutes for viruses. It kills vegetative bacterial forms almost instantly. In actual use, times of exposure are usually extended to provide a margin of safety (e.g., respiratory equipment and endoscopes are soaked for a minimum of 30 minutes). As a sterilant, glutaraldehyde is one of the most effective of chemical agents used today. It retains its activity for 2 weeks, provided it does not become accidentally diluted.

Betapropiolactone. This chemical disinfectant is used in its gaseous state to fumigate closed rooms or chambers. It is very toxic and will kill most forms of microbial life in several hours. The vapor can cause blisters or burns to tissue and it is irritating to the eyes. It is also carcinogenic. Because use of this disinfectant must be supervised by a professional, betapropiolactone has limited application for general hospital disinfection.

Phenol and Phenolic Compounds

Phenol is a disinfectant that has been used for many years. It is often called carbolic acid. When used in high concentrations it is bactericidal. It is colorless to light pink, is caustic, and has a slight anesthetic action. High concentrations precipitate proteins, low concentrations denature proteins without coagulating them. Phenol penetrates the tissues, is very toxic, and in a pure state it is corrosive to the skin. While it is not used as widely today as previously, it still serves as the basic compound for a large number of disinfectants referred to as phenolics, phenol derivatives, and phenol-related compounds. These are suitable agents for disinfecting saliva, feces, and similar materials because phenolics are active in the presence of organic matter. Most are also tuberculocidal. Phenolics are quite effective for environmental sanitation.

Hexachlorophene. The bisphenolic compounds, of which hexachlorophene is one, are more effective than the monomers and their antiseptic activity is increased by their chlorine content. Hexachlorophene is deposited on the skin for prolonged periods of time in low concentrations and has been found to be an excellent skin antibacterial agent. It is very effective against gram-positive bacteria, however, gram-negative bacteria tend to be more resistant, as do spores, to its action. It has been added to soaps, detergent creams, and oils in a 2 and 3 percent concentration.

In 1972, the Food and Drug Administration restricted the use of this antiseptic to physician's prescription after studies demonstrated the neurotoxicity of hexachlorophene, especially to newborns. In Sweden, among infants of 460 nurses who washed their hands 60 times per day during their first trimester of pregnancy there were 25 severe malformations and 46 minor malformations. This incidence for malformation was 12 percent higher than the normal for the Swedish population. The teratogenic effect of this antibacterial agent is being further studied, and its use limited to selected populations.

Alcohol

Ethyl alcohol is most bactericidal at a concentration of 70 percent. It is widely employed as a skin disinfectant. It does not kill spores, which limits its use for the sterilization of instruments. It is active against gram-positive, gram-negative, and acid-fast organisms in concentrations of 50 to 70 percent.

Isopropyl alcohol is as good a skin antiseptic as ethyl alcohol and its bactericidal activity is greater. Its toxic effects are greater and longer lasting than those produced by ethanol. Alcohols in 50 and 70 percent concentrations denature proteins and hence are more effective than absolute alcohol for skin cleaning.

Halogens (Chlorine, Iodine, Iodophors)

The halogens most often used as disinfectants are iodine and chlorine. Chlorine is an inexpensive bactericidal agent most often used for disinfection of water and sewage. The sodium or potassium hypochlorites can be used to disinfect small objects or surfaces in a sick room. All the hypochlorites deteriorate with age and their germicidal effectiveness is reduced in the presence of organic matter.

Tincture of iodine contains 2 percent iodine and 2 percent sodium iodide. This is an excellent skin disinfectant. Some tinctures contain as much as 7 percent iodine. The tinctures leave stains, and those prepared with iodides tend to cause allergies in some individuals.

By using surfactants as carriers for iodine, complexes are formed that possess antibacterial properties. In iodophors, 80 percent of iodine is available as bacteriologically active. Iodophors are fungicidal and active against tubercle bacilli. They are effective in moderate concentration against *Bacillus subtilis*. Iodophors are nonallergenic, nonstaining, and nonirritating, thus they have advantages over iodine. One iodophor frequently used for surgical scrubs is povidone-iodine (Betadine).

Quaternary Ammonium Compounds

The quaternary ammonium compounds have enjoyed wide usage as disinfectants and antiseptics, but are relatively bland. They are cationic surface active antiseptics and are more potent at a higher pH and temperature. They are neutralized by anionic soaps, which limits their value as antiseptics. They have been found to have certain limitations such as being selectively absorbed by fabrics so that a 1:1000 solution is diluted to 1:1500 or 1:2000 if gauze is immersed in solution. The efficiency of the active concentration of benzalkonium chloride (Zephiran) is reduced by diluting it with hard water. The "quats" are not tuberculocidal and have little activity against *Pseudomonas*. Personnel responsible for applying these chemical agents must be certain that all soap has been removed from the epidermal surface since any trace of soap decreases the power of a "quat" such as benzalkonium chloride. In addition, the effectiveness of these agents is influenced by organic soil, which, like the soaps, inhibit their activity. Other common representatives of the quaternary ammonium compounds are cetyl pyridium chloride (Ceepryn), benzethonium chloride (Phemerol), cetyltrimethylammonium bromide (Cetrimide and Cetavlon) and alkyldimethylbenzylammonium chloride (Roccal).

Heavy Metals

Most of the heavy metal disinfectants contain mercury or silver. While these preparations were widely used at one time they have been replaced by more effective compounds. For the most part they were slowly bactericidal and primarily acted as bacteriostatic agents. Silver nitrate is one compound that is still used in a 0.5 to 1 percent solution, sometimes for instillation into the eyes of newborns to prevent gonococcal conjunctivitis and in the treatment of some burns. However, other antibacterial agents and treatment modalities are now available that may be substituted.

At one time mercurials were enthusiastically accepted as skin antiseptics, but they are far less popular today. The aqueous preparations of such organic mercurials as Merthiolate, Mercresin, Metaphen, and Mercurochrome were not bacteriostatic, and the tinctures were found to be less effective than simple 70 percent ethyl alcohol. Therefore, these agents have no place in the treatment of microbial contamination.

CLINICAL IMPLICATIONS

Sterilization and disinfection are the responsibility of the health care practitioner. Health care practitioners must participate in the development of policies and procedures related to the preparation and use of patient supplies, the care of patient contact areas, and the prevention of patient-to-patient, patient-to-personnel, or patient-to-environment infection. Guidelines for preparing items for sterilization/disinfection are listed in Table 25-3; Table 25-4 summarizes sterilization/disinfection methods commonly used.

Before a surgical procedure, the patient's skin is scrubbed in order to minimize the numbers of organisms present. The choice of antiseptic for presurgical scrub should be based on knowledge of skin flora, microorganisms sensitive to the antiseptic, and the operative procedure. The resident flora varies quantitatively and qualitatively depending on the body site and the individual host. For example, patients who undergo orthopedic procedures are more likely to develop infections with gram-positive organisms such as *S. aureus*. For that reason, a hexachlorophene-containing antiseptic might be the best skin preparation to use. On the other hand, patients following abdominal surgery more often develop infections with gram-negative bacteria and an iodophor would be the agent of choice.

The physical composition of supplies is a major determinant of the type of sterilization process to be used. If the supplies can withstand heat and moisture, the most efficient and effective method of sterilization is the autoclave. However, bottles of liquid

(text continues on page 667)

TABLE 25-3. Preparing Items for Sterilization/Disinfection

General guidelines
- Rinse item thoroughly, removing organic material or dirt
- Pack items loosely, do not stack
- Leave instruments slightly open
- Separate items that have fitting parts (e.g., syringes)
- Sort equipment into like items (do not combine plastic and metal, for example)

	Steam sterilization	*Gas sterilization*	*Other*
Specific guidelines			
Description	Steam under pressure; 212°F for 10 min	Ethylene oxide mixed with steam, but no pressure; 130°F for 2 hours	Various chemical disinfectants, dry heat, flaming, cold, radiation ultrasonics
Items to be processed	Anything that can withstand moist high heat; glass, metal, rubber	Plastic, polyethylene; items with gauges, electrical or motorized parts; fabric	Delicate items; procedure trays, delicate instruments, thermometers, respiratory therapy equipment
Toxicity	None, sterilization method of choice whenever possible	Extremely toxic; aeration necessary	Varies with method

TABLE 25-4. Summary of Sterilizing/Disinfecting Agents

Agent	Application of sterilization or disinfectant	Advantages	Disadvantages
PHYSICAL AGENTS			
Moist heat			
Autoclave	Metal instruments, metal basins, needles, gowns, towels, linen packs, rubber items (catheters, tubing, gloves)	Rapid killing action for vegetative or spore-forming bacteria, viruses, and fungi. Most efficient method for heat- and moisture-stable items	Not recommended for heat-sensitive items, or materials impermeable to moisture
Boiling water	Metal instruments, glassware, rubber tubing, catheters	Convenient if other methods are not available especially in emergency situation	Spore-forming bacteria not killed; corrosive to metal instruments and glassware
Dry heat			
Hot air oven	Glassware, syringes, needles, oils, powders	Sterilizes moisture-impervious materials; noncorrosive to glassware	High temperature required; long exposure time. Destructive to cotton or linen materials
Incineration			
Flaming	Laboratory loops and needles, forceps	Destruction of microorganisms quickly by incineration	Destruction of microorganisms dependent on heat of flame
Alcohol lamp	Needles, forceps	Emergency method if others not available	Flame not sufficient to destroy all microorganisms; contaminated materials may splatter
Cold	Refrigeration of cultures, specimens, foods	Preservation of cultures; decreases multiplication of microorganisms in food	Does not sterilize but preserves perishable items
Ultrasonics	Cleaning of instruments; cell disintegration in the laboratory	Disintegrates bacteria and protein materials; removes soil from instruments	Materials must be sterilized after processing
Radiation			
Ionizing	Pharmaceutical sterilization. Vaccines, sera, and toxins. Disposable medical and dental supplies—plastic syringes, surgical gloves, etc. Preservation of foods	High penetrability; prepackaging of supplies and sealing; does not require heat or moisture for sterilization	Expensive equipment needed; adherence to stringent safety measures. Effect on materials exposed (e.g., loss of tensile strength with fabric, discoloration, and chemical

TABLE 25-4. *(continued)*

Agent	Application of sterilization or disinfectant	Advantages	Disadvantages
			changes in pharmaceuticals). Nutritional value and palatibility of foods may change
Ultraviolet light	Germicidal lights for laboratory, patient areas, and operating rooms	Reduces number of microorganisms in the air	Low penetrability of rays; destruction of microorganisms only in direct contact with ray. Effectiveness of ultraviolet determined by source output; irritating to eyes and skin.
CHEMICAL AGENTS			
Alkylating agents			
Ethylene oxide	Plastic materials—syringes, tubing, and catheters. Sutures, vascular transplants, valve prostheses. Anesthesia equipment—tubing, face mask, endotracheal tubes. Dental equipment, rubber materials	Highly penetrative gas; items may be wrapped and sealed for sterilization; effective sterilization for heat-sensitive materials	Explosive in pure form; long exposure time. Time required poststerilization for elution of gas to prevent skin burns. Gas may be carcinogenic to exposed personnel; equipment is expensive
Formaldehyde	Sickrooms, clothing, bedding in gaseous state. Small items (combs, brushes in gas sanitizer)	Sporicidal in high concentrations	Pungent odor; highly damaging to tissue. Poorly penetrating and slow acting
Glutaraldehyde	Instrument sterilization (endosopes or other lensed instruments). Respiratory equipment	2 percent solution bactericidal, tuberculocidal, and virucidal in 10 minutes. No deleterious effects on lensed instruments, noncorrosive. Not inactivated by organic material	Irritating to the skin, mucous membranes, and eyes. Sporicidal in 10 hours. Stability of solution limited
Betapropiolactone	Gaseous state fumigant for closed room	Odor dissipates in 48 hours. Liquid form decomposes in few hours to inactive form. Destroys most microbial forms in several hours	May be carcinogenic; vapor is a vesicant and is irritating to eyes

TABLE 25-4. *(continued)*

Agent	*Application of sterilization or disinfectant*	*Advantages*	*Disadvantages*
Phenol and phenolic compounds			
Phenol	Base for disinfecting compounds. Compounds used for disinfecting saliva, faces, environmental surfaces	Bactericidal in high concentrations. Residuals of compounds bactericidal and bacteriostatic for floors, furniture, and walls	Corrosive to the skin; not sporocidal
Hexachlorophone	Skin antibacterial agent; used in creams, soaps, and oils	Effective against gram-positive organisms. Bland skin antiseptic	Not as effective against gram-negative or spore-forming organisms. A possible teratogen that also may be neurotoxic for newborns
Alcohol	Skin antibacterial at 70 percent; small instruments disinfectant	Easily accessible as a skin antiseptic. Active against gram-positive, gram-negative, and acid-fast bacteria	Not sporicidal; drying to skin
Halogens			
Chlorine	Disinfectant for water and sewage; food sanitization compounds of chlorine used for disinfecting sick rooms, dishes, laundry, and dairy equipment	Inexpensive bactericidal agent	Irritant to skin; inactivated by organic compounds.
Iodine	Skin antiseptic	Excellent disinfectant; active against bacteria, tubercle bacillus, fungi, and viruses	Causes skin allergies, stains
Iodophors	Skin antiseptics; disinfectants for instruments used for environmental cleaning	Toxic effect on skin less than iodine; wide range of antimicrobial activity	Instability of solution, temporarily staining. Combined with organic matter it loses effect as disinfectant
Quaternary			
Ammonium compounds	Skin antiseptic; disinfectant for instruments	Nontoxic, effective surface tension reductants	Concentration instability caused by soaps, water incapatibility and plant proteins

TABLE 25-4. *(continued)*

Agent	*Application of sterilization or disinfectant*	*Advantages*	*Disadvantages*
			(e.g., in fabrics and gauze) Microbial effectiveness limited; nonsporicidal.
Heavy metals			
Mercury	Skin antiseptic in compounds (Merthiolate, Mercurochrome); preservative in microbial vaccine ointments	Bland when applied to skin in a gaseous solution	Slow-acting bactericide; ineffective in reducing cutaneous flora
Silver Nitrate	Aqueous solution—antiseptic as prevention for gonorrheal infection in eyes of newborns; treatment of burns	Bactericidal for gonorrhea	Practical value restricted for inorganic Silver salts—they are irritating and caustic

must have tops loosened to allow penetration of steam into the bottle. For materials such as powders, oils, and glassware, which are impervious to saturated steam, dry heat is the method of choice. The temperature for dry heat is higher, exposure time is longer, and the high degree of heat is destructive to some materials.

Personnel responsible for preparing supplies prior to sterilization should make sure that—regardless of the method used—linens are freshly laundered, and instruments and glassware are free of organic soil. In carrying through the process, the packing or loading of the sterilizer is extremely important as well as time exposure. Supplies should be dry prior to removal from sterilizer to prevent transmission of bacteria through moist material. If the specific microorganisms to be destroyed are unknown, personnel should always allow for a margin of safety by using the maximum time exposure for the most resistant spore-forming bacteria (e.g., *Clostridum botulinum*).

The introduction of ethylene oxide as a sterilant has provided a very efficient method for sterilizing heat-sensitive materials. However, this method has its limitations: it requires a longer exposure time to effect sterilization, an extended aeration period, and it is toxic to personnel if inhaled. Careless handling of supplies by not allowing enough time for aeration can result in severe chemical burns to personnel and to patients when the materials come in contact with the skin. Strict adherence to aeration times is essential. Periods of aeration, depending on the item's composition and the method used, can range from 1 to 7 days. It is crucial that a sufficient number of supplies be available for use to prevent personnel from hastily choosing supplies from which the gas has not eluted.

All sterile supplies should have an expiration date plainly marked on them. Prepackaged supplies should have stamped on them, "sterile until package is opened or torn." Personnel must monitor the sterile supplies to prevent those that are outdated or contaminated from being used.

Regulations have been set up for personnel exposure to ethylene oxide by the Occupational Safety and Health Administration (OSHA). They have established that health workers may not be exposed to more than 50 parts per million (ppm) of ethylene oxide gas averaged over an 8 hour period. However, OSHA has recently stated that they have no jurisdiction over hospital employees who work with this sterilant. Regulations for hospital employees come under the Environmental Protection Agency (EPA). In this case, it means the EPA regulates the use of ethylene oxide as a sterilant. EPA has not established an exposure limit; it only requires that hazardous substances be properly labelled and that the suppliers are registered.

Since 1981, the American Congress of Industrial Hygienists have recommended a change in the maximum exposure level of ethylene oxide to 10 ppm as a time-weighted average with no short-term exposure limit. This group plans to lower this limit further to 5 ppm. In the meantime, hospitals have been concerned about the exposure of employees in the central supply and operating room areas. Monitoring devices have been suggested for employees who are exposed to the sterilant on a daily basis. In addition, equipment is available that will monitor the sterilizer area, and the exhaust hoses from the sterilizer and the aerator equipment. These monitors would allow personnel to recognize when air exposures were too high.

The choice of antiseptic or germicide differs if it is to be used for the skin, for instruments, or for cleaning inanimate objects. It should be determined if the agent selected will destroy the microbial population at which it is aimed. If all microorganisms are to be killed then the manufacturer's label must be reviewed to see what organisms are destroyed and at what concentration. It is imperative that the instructions for preparation be followed. Items to be washed or soaked in a germicidal solution must be completely covered by the solution and be free of organic material. Many chemical agents are inactivated by organic material; in other cases the organic material prevents contact of the solution with the organism.

Antiseptics and disinfectants do inhibit or kill microorganisms but some preparations can actually support the growth of microorganisms. Hexachlorophene soap and detergent hand creams have been found to be contaminated with gram-negative bacilli, primarily *Klebsiella* species, *Pseudomonas aeruginosa, Escherichia coli,* and *Candida albicans.* Benzalkonium chloride has supported the growth of such microorganisms as *Serratia marcescens, Enterobacter cloacae, Enterobacter agglomerans,* and *Pseudomonas cepacia.* The latter has also been found in povidone-iodine.

Knowledge that the agent for sterilizing or disinfecting materials may be contaminated points out the need for clinicians to monitor closely these materials and their use. It is extremely important that the decisions concerned with the methods and agents used for sterilization be made by personnel who have an excellent understanding of microbial life and death, and the relationship of microorganisms to the patient and the environment.

SUGGESTED READINGS

An evaluation of a new and novel ETO monitor. (editorial) Hosp Topics 1980;58:37-41.

AORN recommended practices for hospital sterilization. (editorial) SCORE 1981;6:12-20.
Baker BM. Expiration dates sterile or unsterile? That is the question. Hosp Topics 1980;58:34-35.
Benarde MA ed. Disinfection. New York: Marcel Dekker, 1970.
Boyd RB, Hoerl B. Basic medical microbiology, 2nd edition. Boston: Little, Brown and Company, 1981.
Carson LA, et al. Morphological, biochemical and growth characteristics of *Pseudomonas cepacia* from distilled water. Appl Microbiol 1973;25:476-483.
Check W. New study shows hexachlorophene is teratogenic in humans. JAMA 1978;240:513-514.
Danielson NE. Ethylene oxide update. SCORE 1979;4:12-16.
Dineen P. Penetration of surgical draping materials by bacteria. Hosps 1969;43:82-84.
Done AK. The occasional hazards of antiseptics. Emerg Med 1980;12:89-91.
Frank MJ, Shaffner W. Contaminated aqueous benzalkonium chloride: an unnecessary hospital infection hazard. JAMA 1976;236:2418.
Fuller JR. Principles of sterilization and disinfection. In: Fuller JR. Surgical technology principles and practices. Philadelphia: WB Saunders, 1981, pp. 28-37.
Infection control rounds. St. Paul, Minn., 3M Company, Medical Products Division, 1982;7:1-7.
Joklik WK, Willett HP, Amos DB. Zinsser microbiology, 17th edition. New York: Appleton-Century-Crofts, 1980.
Kaul AF, Jewett JF. Agents and techniques for disinfection of the skin. J Enteros Therapy 1981;8:19-22.
Lee CH, Montville TJ, Sinskey AJ. Comparison of the efficacy of steam sterilization indicators. Appl Environ Microbiol 1979;37:113-117.
Lilly HA, Lowbury EJL. Transient skin flora: their removal by cleansing or disinfection in relation to their mode of disposition. J Clin Pathol 1978;31:919-922.
Lindop CR, et al. Exposure of patients to ethylene oxide during cardiopulmonary by-pass using gas sterilized pump components. J Thorac Cardiovasc Surg 1980;79:845-860.
Lowbury EJL, et al, eds. Control of hospital infection, 2nd edition. Philadelphia: JB Lippincott, 1975.
Mallison GF. Monitoring ethylene oxide. Hosp Infec Contr 1979;6:158.
Perkins JJ. Principles and methods of sterilization in health sciences, 3rd edition. Springfield, Ill: Charles C Thomas, 1969.
Perkins RE, et al. Monitoring steam sterilization of surgical instruments: a dilemma. Appl Environ Microbiol 1981;42:383-384.
Phillips GB. Expiration dating not needed on commercially sterilized items. Hosps 1981; 55:85-86.
Phillips GB. Prevention and control of infection. In: Infection control in the hospital, 4th edition. Chicago: American Hospital Association, 1979, pp. 77-129.
Reich RR. Storage stability of *Bacillus subtilis* ethylene oxide biologic indicator. Appl Environ Microbiol 1980;39:277-279.
Roberts RB. Infections and sterilization problems. Boston: Little, Brown and Company, 1972.
Samuels TM. Ethylene oxide environmental monitoring. Hosp Topics 1980;58:37-41.
Will government write a new standard for ethylene oxide? AORN J 1982 35:965-966.
Wilson CO, et al. Textbook of organic medicinal and pharmaceutical chemistry, 5th edition. Philadelphia: JB Lippincott, 1966.
Wilson M, Mizer HE. Prevention and control of infectious disease. In Wilson M, Mizer HE. Microbiology in patient care, 2nd edition. New York: MacMillan, 1974, pp. 192-235.

26 ANTIMICROBIAL AGENTS

KEY POINTS

- Effective antimicrobial agents are available against most bacteria.
- Agents used against fungal diseases are more toxic and less effective; there are very few chemotherapeutic agents of proven efficacy against viral diseases.
- Antimicrobial agents must be used only when there are clear indications; inappropriate use increases the risk of emergence of resistant bacterial strains.
- The antimicrobial effects of drugs can occur by five major mechanisms: interference with nucelic acid metabolism, damage to cell membranes, inhibition of protein synthesis, prevention of cell wall synthesis, or antimetabolic activity.
- Major groups of antimicrobial agents in current use are the aminoglycosides, the cephalosporins, and the penicillins.
- Common adverse effects of antimicrobial therapy include direct toxicity (e.g., kidney or eighth cranial nerve damage with aminoglycosides, or bone marrow toxicity with chloramphenicol) or hypersensitivities (penicillin is the most common example).

The introduction of chemotherapy into clinical practice during this century has probably resulted in a greater reduction in human morbidity and mortality than has any other factor with the exception of the introduction of public health measures and sanitation. Many infectious diseases have evolved, as a result of antimicrobials, from dreaded, life-threatening or maiming events to mere annoyances. The value of chemical antimicrobial therapy cannot be overestimated, but with the introduction of each new agent also comes problems sometimes adverse reactions worse than the disease itself, and some untoward effects for populations in general, beyond the individual patient being treated. The emergence of resistant strains is one such untoward effect.

In 1910, Ehrlich introduced the first experimentally developed antimicrobial, a compound called salvarsan that is active against *Treponema pallidum*.

In 1928, while Fleming was studying lysozyme, he accidentally discovered the antibacterial properties of a mold, *Penicillium notatum*. The penicillins, however, were not in clinical use until the early 1940s. In 1935 Domagk discovered prontosil, the forerunner of the sulfonamides. The development of a series of antimicrobials derived from streptomyces molds began with the discovery of streptomycin in 1944 by Waksman. Many of the antimicrobial agents used today are developed from streptomyces molds (some aminoglycosides, chloramphenicol, erythromycin, tetracyclines). Major antimicrobials available currently are listed on Table 26-1. Developments in antimicrobial therapy since the 1960s have not been so much discoveries of new agents but rather modifications in the molecular structure of the same basic compounds to enhance their therapeutic effectiveness. There are about 100 antimicrobial drugs under investigation in the United States at any given time.

Overall use of antimicrobials steadily increased until the early 1970s, when it plateaued; it has been relatively constant since. Use of such agents as the tetracyclines and sulfonamides has decreased, whereas use of some newer agents such as cephalosporins and aminoglycosides is still increasing. The total cost of antimicrobials in 1976 accounted for 16 to 41 percent of pharmacy costs in 19 hospitals surveyed. Cephalosporins and aminoglycosides accounted for 66 percent of those costs. In a review of the clinical records of over 5000 patients, it was determined that about 28 percent received at least one antimicrobial during their hospital stay. There is evidence that as much as half of such drug use is inappropriate, for drugs are given to patients without adequate evidence of infection (Craig et al. 1978).

MODES OF ACTION

Drugs can cause their antimicrobial activity in one of five major ways (Fig. 26-1): they can interfere with nucleic acid metabolism, damage the cell membrane, inhibit protein synthesis, prevent cell wall synthesis, or serve as antimetabolites (sulfonamides, para-aminosalicylic acid, trimethoprim). Agents such as penicillins and cephalosporins, which affect cell wall synthesis, are effective only when cells are multiplying, and because of the structural differences in cell walls, are more active against gram-positive than against gram-negative bacteria. Use of a bacteriostatic agent like tetracycline or chloramphenicol with penicillins or cephalosporins can decrease their effectiveness by slowing cell division.

Drugs that affect cell wall synthesis have different sites of action. If this were not the case, all such antimicrobials would have cross resistance. Penicillins and cephalosporins, which both act during the terminal cross linking of peptidoglycan, are often cross resistant, but vancomycin and bacitracin work during the formation of the linear peptidoglycan and organisms resistant to penicillin or cephalosporins are susceptible to vancomycin.

Antimicrobials that affect bacterial cell wall synthesis have relatively few toxic effects in humans because mammalian cells, being eukaryotic, do not have the distinct cell wall of prokaryotes. Likewise, many of the antimicrobials that affect protein synthesis act on cell components (ribosomes at various subunits such as 50S, 30S, 70S) that are chemically dissimilar between micro-

(text continues on page 676)

TABLE 26-1. Major Antimicrobial Agents

				Primary activity against				
Class	*Agent*	*Static*	*Cidal*	*Gram positive*	*Gram negative*	*Ricketsias/ Chlamydias*	*Fungi*	*Viruses*
Aminoglycosides	Amikacin		X	X	X			
	Gentamicin							
	Kanamycin							
	Neomycin							
	Paromomycin							
	Streptomycin							
	Tobramycin							
Antituberculosis	Ethambutol	X		X				
	Isoniazid		X					
	Para-aminosalicylic acid (PAS)	X						
	Rifampin		X					
Cephalosporins	Cefamandole							
	Cefazolin		X	X	X			
	Cefoperazone							
	Cefotaxime							
	Cephalexin							
	Cephaloglycin							
	Cephalothin							
	Cephapirin							
	Cephradine							
	Moxalactam							
Chloramphenicol		X		X	X	X		
Erythromycins		X		X		X		
Penicillins	Amoxicillin		X	X	X			
	Ampicillin							
	Azlocillin, Bacampicillin							
	Carbenicillin							
	Cloxacillin							
	Dicloxacillin*							
	Mazlocillin							
	Methicillin*							

TABLE 26-1. *(continued)*

				Primary activity against				
Class	*Agent*	*Static*	*Cidal*	*Gram positive*	*Gram negative*	*Ricketsias/ Chlamydias*	*Fungi*	*Viruses*
	Nafcillin*							
	Oxacillin*							
	Penicillin G,V							
	Piperacillin							
	Ticarcillin							
Sulfonamides	Co-trimoxazole	X		X	X			
	Silver sulfadiazine							
	Sulfacytine							
	Sulfadiazine							
	Trisulfapyrimidines							
	Sulfamethizole							
	Sulfamethoxazole							
	Sulfisoxazole							
Sulfones	Dapsone	X		X				
	Sulfoxone							
Tetracyclines	Doxycycline, minocycline	X		X	X	X		
Urinary germicides	Cinoxacin		X		X			
	Methenamine							
	Nalidixic Acid							
	Nitrofurantoin							
	Phenazopyridine							
Others	Bacitracin		X	X				
	Clindamycin		X	X				
	Metronidazole		X	X				
	Polymyxin B/ colistin		X	(Anaerobes)	X			
	Spectinomycin		X		X			
	Vancomycin		X	X				

Antifungal	Amphotericin B	X	X	
	Candicidin, clotrimazole			
	Flucytosine			
	Griseofulvin			
	Ketoconazole			
	Miconazole			
	Nystatin, tolnaftate			
Antiviral	Acyclovir			X
	Amantadine			
	Cytarabine			
	Idoxuridine			
	Methisazone			
	Trifluridine			
	Vidarabine			

*Beta-lactamase resistant

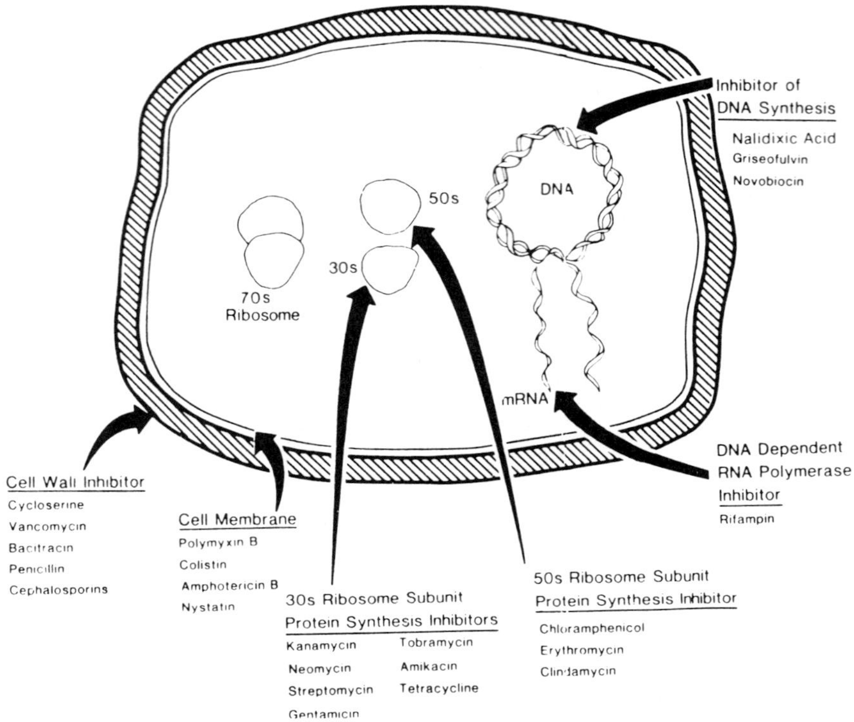

FIGURE 26-1. The relative locations of bacterial structures in a typical bacterial cell. Antimicrobial agents are arranged by site of action in the cell. (Reprinted with permission of the publisher from Arking L, Saravolatz L. Antimicrobial treatment: a coordination challenge for nursing. Nurs Clin North Amer 1980;15:689-702.)

organisms and humans. This capacity to selectively act against microbial invaders without causing damage to human cells in called *selective toxicity,* and is a very important characteristic of any acceptable antimicrobial.

Some drugs act on the cell membrane, which is similar in host and parasite cells. This relative lack of selective toxicity is one reason why such drugs cause more severe side effects. The sulfonamides act by inhibiting an enzyme system necessary for the synthesis of folic acid. Human cells do not synthesize folic acid; thus, sulfonamides are selectively toxic by inhibiting metabolites essential to bacterial but not mammalian cells. Disinfectants, unlike antimicrobials, have no selective toxicity.

FACTORS INFLUENCING EFFECTIVENESS

Antimicrobial agents are often characterized by whether they are bacteriostatic (resulting in slowed growth), or bactericidal (resulting in immediate cell death). In general, drugs that affect the cell membrane or cell wall tend to be bactericidal, those that affect protein synthesis may be bacteriostatic or bactericidal. In some cases the dosage and route of administration determine whether the same agent is bacteriostatic or bactericidal. The difference is usually of little therapeutic concern since organisms that are static and not growing or reproducing will eventually die. However, bacteriostatic agents do require the assistance of the host's immunologic defenses to phagocytize and eventually rid the body of the invaders. That is, bacteriostatic agents hold the bacterial population down while the natural immune systems can work. Bactericidal agents are therefore necessary when the body's defenses cannot work.

The effectiveness of antimicrobial therapy is influenced by the ability of the drug to penetrate tissues (many drugs do not cross the blood-brain barrier and are useless in the treatment of central nervous system infections unless given intrathecally), the degree to which the drug is bound to protein, and the dose and route of administration. Penicillin G, for example, is only 20 percent bound to protein, methicillin is 30 percent, but oxacillin is 97 percent bound so that only 3 percent of the drug is in the circulation available to react with infecting microorganisms. The result of having a drug that is highly bound to proteins is that the serum drug levels will be lower and more prolonged. The location and type of infectious lesion also affects drug activity. Organisms in abscesses or bone are often inaccessible to antimicrobials.

For each species of microorganism there are some antimicrobial agents to which they are naturally resistant and some to which they are susceptible (Table 26-2a) at the drug concentrations normally attained in the serum (Table 26-2b). Organisms can also acquire resistance to chemotherapeutics while they are used in vivo. A major factor in the effectiveness of an antimicrobial is the degree to which such resistant organisms emerge. There are many mechanisms by which acquired antimicrobial resistance can develop (Table 26-3). Essentially, no antimicrobial drug exists to which resistant strains have not emerged. The spread of resistant organisms is enhanced under the selective pressure of antimicrobial therapy during which susceptible strains are killed and resistant ones can grow in abundance.

There are ways to minimize the emergence of such resistant strains: the sparing use of antimicrobials, the prevention of cross infection by barrier techniques such as isolation and disinfection, and the use of combined chemotherapy. Prophylactic antimicrobials for surgery should be given for no longer than 24 hours before and 72 hours after the procedure. Some drugs like streptomycin are almost always used in combination with other agents because resistant organisms tend to emerge with such rapidity. With the long-term treatment necessary for such chronic diseases as tuberculosis, resistant organisms invariably appear, so combination therapy is necessary.

Combining chemotherapeutic agents may not always be desirable. The resultant activity can be synergistic, in which the combined effect is greater than the sum of the agents; additive, in which the combined treatment is the same as adding its components; or antagonistic, in which the combined effect is less than its components. In general, bactericidal agents act synergistically if their modes of action are different, bacteriostatic agents are additive (exception: combinations of sulfonamide and trimethoprim are synergistic), and combined bactericidal and bacteriostatic agents may be antagonistic. Frequently used chemotherapeutic combinations are listed on Table 26-4.

From this discussion we can develop cri-

TABLE 26-2a. Usual Susceptibility Patterns (Disc Diffusion Method) of Selected Bacteria to Commonly Used Antimicrobials

	PEN	*AMP*	*OXA*	*CEPH*	*TET*	*CHLO*	*MAND*	*KANA*	*POLY*	*THIO*	*FURA*	*ERY*	*CLIN*	*GENT*	*TOBRA*	*CARB*	*CEFOXITIN*	*AMIKACIN*	*VANCO*	*SMX-TMP*
Acinetobacter	-	R	-	R	S	R	R	S	S	S	R	-	-	S	S	I	R	S	-	S
B. fragilis	R	R	-	I	R	S	I	-	-	-	-	I	S	-	-	S	S	-	-	-
Citrobacter freundi	-	R	-	R	S	S	S	S	S	S	S	-	-	S	S	S	R	S	-	S
Enterobacter	-	R	-	R	S	S	S	S	S	S	S	-	-	S	S	S	R	S	-	S
Enterococcus	I	S	-	I	R	S	R	-	-	-	S	S	R	-	-	-	R	-	S	-
E. coli	-	S	-	S	S	S	S	S	S	S	S	-	-	S	S	S	S	S	-	S
H. influenzae	S	S	-	S	S	S	S	-	-	-	-	S	R	-	-	S	S	-	R	-
K. pneumoniae	-	R	-	S	S	S	S	S	S	S	S	-	-	S	S	R	S	S	-	S
P. mirabilis	-	S	-	S	R	S	S	S	R	S	R	-	-	S	S	S	S	S	-	S
P. aeruginosa	-	R	-	R	R	R	R	R	S	R	R	-	-	S	S	S	R	S	-	R
Serratia	-	R	-	R	R	S	R	S	R	S	R	-	-	S	S	S	I	S	-	S
S. aureus	R	-	S	S	S	S	S	-	S	S	S	S	S	-	-	S	-	S	S	-
S. epidermidis	R	-	S	S	R	S	S	I	-	S	S	S	S	S	-	-	S	-	S	S

- = not tested
R = resistant
I = intermediate
S = sensitive

Data derived from University of Washington Hospital Clinical Microbiology Laboratory, 1980.

teria for an ideal antimicrobial agent (Table 26-5). Of course, no drug has all or even most of the desired characteristics, but its advantages and disadvantages must be weighed against such ideal standards and the most acceptable agent for the patient's needs can then be chosen.

ADVERSE EFFECTS OF ANTIMICROBIAL THERAPY

All antimicrobial therapy results in some untoward effects, either due to direct toxicity or to hypersensitivity. Some of the more serious toxic reactions occur with aminoglycoside use (kidney and eighth cranial nerve damage) and with chloramphenicol (bone marrow toxicity). A more complete list of possible reactions is found in Table 26-6. Toxic reactions are much more likely to occur in patients with impaired ability to excrete or metabolize drugs (renal or liver disease), especially if drug blood levels are not carefully monitored. Some important general rules are that if a drug is excreted by the kidneys or metabolized by the liver, and the patient has renal or liver dysfunction, (1) dosages should be decreased, (2) blood levels of the drug should be monitored at peak and low (trough) levels, (3) the patient's renal or liver function should be monitored at regular intervals, and (4) at

TABLE 26-2b. Peak Serum Concentrations Frequently Obtained with Some Antimicrobics

Antimicrobic	*Serum concn (μg/ml)*	*Antimicrobic*	*Serum concn (μg/ml)*
Penicillins*		Amphotericin B	0.2-2.0
Depot (benzathine penicillin G)	0.01-0.06	Clindamycin	5-26
Depot (procaine penicillin G)	0.1-18	Chloramphenicol	
Oral	3-20	Oral	3-12
Intramuscular or intravenous	2-200†	Intravenous	20-40
Cephalosporins		Colistin	6
Cefamandole	18-123	Erythromycin	
Cefazolin	50-150	Intravenous	5-30
Cefoxitin	15-25	Oral	<1-10
Cephalexin	5-35	Ethambutol	5-10
Cephaloridine	14-28	Flucytosine	10-100
Cephalothin, cephapirin	15-150	Isoniazid	2-3
Cephadroxil	16-28	Lincomycin	2-20
Cephradine (intravenous; oral as for cephalexin)	17-49	Metronidazole	4-10
Cefaclor (oral)	5-7	Nalidixic acid	20
Aminocyclitols		Nitrofurantoin	ND‡
Amikacin	12-30	Oxolinic acid	1-2
Gentamicin	1-12	*p*-Aminosalicylic acid	60-100
Kanamycin	10-25	Polymyxin B	2-8
Spectinomycin	100	Rifampin	10
Streptomycin	10-25	Sulfonamide	30-150
Tobramycin	1-12	Vancomycin	20-50
Tetracyclines			
Tetracycline			
Oral	1-5		
Intravenous	5-30		
Demethylchlortetracycline (demeclocycline)	2-5		
Doxycycline (oral)	1-6		
Minocycline	0.7-4.5		

*Penicillins include benzylpenicillin, phenoxymethylpenicillin, ampicillin, amoxicillin, carbenicillin, ticarcillin, methicillin, oxacillin, cloxacillin, dicloxacillin, and nafcillin.

†Wide ranges represent the effect of different doses, routes of administration, or preparations.

‡Not detectable in serum after usual oral dose.

Reprinted with permission of the publisher from Lennette EH, et al. Manual of clinical microbiology, 3rd edition. Washington DC: American Society of Microbiology, 1980.

any sign of toxicity or overdose the drug should be discontinued.

Some drugs that are not very toxic may still be associated with a number of adverse reactions by their ability to induce an allergic response in the host Penicillins, among the least toxic antimicrobials, frequently result in hypersensitivity. Hyper-

TABLE 26-3. Mechanisms by Which Resistance to Antimicrobial Agents Can Be Acquired

Mechanism	*Example*
Change in the target site (e.g., receptor sites, L forms without cell walls)	Erythromycin, penicillin
Chemical modification of antimicrobial (e.g., acetylation, phosphorylation)*	Aminoglycosides, chloramphenicol
Destruction of the antimicrobial (e.g., beta-lactamase production)*	Penicillins
Loss of permeability	Tetracyclines

*Usually mediated by plasmids

TABLE 26-5. Criteria for an Ideal Antimicrobial Agent

1. Possessing widest possible antimicrobial spectrum
2. Bactericidal
3. Nontoxic to human cells
4. Suitable for administration by any route, well absorbed
5. Penetrates across tissue barriers and into cells
6. Unbound by protein or in any way inactivated in the body
7. Synergistic or additive (not antagonistic) with other antimicrobials
8. Highly acceptable to patients (taste and odor unoffensive, no side effects)
9. Resistant organisms do not emerge

sensitivity can occur to some degree with all chemotherapeutic agents. Other very common adverse effects of antimicrobials are drug fever and gastrointestinal disturbances (when the agent is administered orally). In addition, broad-spectrum antibiotics are associated with changes in normal body flora and may cause superinfections (Table 26-7).

TABLE 26-4. Frequently Used Combinations of Antimicrobials

Drugs	*Effective against*
Gentamicin/chloramphenicol	*Klebsiella pneumoniae* infections
Carbenicillin/gentamicin	*Pseudomonas* infections
Penicillin G/cloxacillin	*Staphylococcus aureus* infections
Penicillin/streptomycin	*Streptococcus faecalis* endocarditis
Tetracycline/streptomycin	Brucellosis
Streptomycin/PAS, INH	Tuberculosis
Ethambutol/INH	Tuberculosis
Rifampin/INH	Tuberculosis

AMINOGLYCOSIDES

Pharmacologic characteristics. As their name implies, aminoglycosides are composed of one or more sugars attached to at least one amino group. They act by binding to receptor proteins on the small subunit (30S) of the bacterial ribosomes. This interferes with the functioning of mRNA, thus blocking protein synthesis and causing cell death. Since all aminoglycosides have a similar mode of action, cross resistance between agents is the rule.

The first aminoglycoside, streptomycin, was marketed in 1944. It was isolated from a fungus, *Streptomyces griseus*. Kanamycin, neomycin, and tobramycin were also developed from *Streptomyces* species. Amikacin is a synthetic analogue of kanamycin. Gentamicin was derived from a different fun-

TABLE 26-6. Adverse Reactions to Antimicrobial Agents

Reaction	*Common causes*
Gastrointestinal	
Diarrhea, pseudomembranous colitis	Clindamycin, ampicillin
Hematologic	
Aplastic anemia	Chloramphenicol, sulfonamides
Bone marrow suppression	Chloramphenicol, sulfonamides
Agranulocytosis, leukopenia	Sulfonamides, semisynthetic penicillins
Immune hemolytic anemia	Sulfonamides, INH, rifampin, penicillins, cephalosporins
Thrombocytopenia	Penicillins, chloramphenicol, sulfonamides, INH, rifampin
Bleeding disorders	Carbenicillin, ticarcillin
Kidney	
Allergic interstitial nephritis	Methicillin
Glomerulonephritis	Penicillins, sulfonamides
Acute tubular necrosis	Aminoglycosides
Obstructive uropathy	Sulfonamides
Liver	INH, rifampin, oxacillin, tetracyclines, nitrofurantoin, erythromycin estolate, sulfonamides
Neurologic	
Allergic meningitis	Sulfonamides
Encephalopathies	Penicillins
Neuromuscular blockade	Polymyxin, oxytetracycline, clindamycin
Optic neuritis	Ethambutol, chloramphenicol
Ototoxicity	Aminoglycosides, vancomycin
Pseudotumor cerebri	Tetracycline
Seizures	Penicillins
Pulmonary	
Pulmonary infiltrate with eosinophilia (PIE)	Amphotericin B, nitrofurantoin
Pulmonary fibrosis	Nitrofurantoin
Skin	
Erythema multiforme	Sulfonamides, penicillins, tetracyclines
Exfoliative dermatitis	Sulfonamides, penicillins, INH
Hypersensitivity angitis	Sulfonamides, penicillins
Photosensitivity	Tetracyclines
Systemic	
"Gray syndrome"	Chloramphenicol
Lupus syndrome	INH, penicillins, sulfonamides, streptomycin, tetracyclines
Anaphylaxis	Penicillins

Note: False positive test for glucosuria using Clinitest occurs with use of many antimicrobials; another reagent such as Testape should be substituted.

False positive tests for urine protein are common if measured with sulfasalicylic acid; use reagent strip bromphenol

TABLE 26-7. Problems that Occur When Normal Flora is Altered by Antimicrobial Therapy

Problem	*Frequently associated with*
Mucocutaneous candidiasis	Ampicillin, tetracyclines, other broad-spectrum antimicrobials
Prolonged carriage of *Salmonella*	Any broad-spectrum antimicrobials
Pseudomembranous colitis	Clindamycin, ampicillin, any broad-spectrum antimicrobials
Superinfection with gram-negative bacteria carrying plasmid-mediated resistance factors	Any broad-spectrum antimicrobials

gus, *Micromonospora purpura*. It is recommended that gentamicin be spelled with an "i" (rather than gentamycin) to differentiate it from those aminoglycosides that derive from *Streptomyces*.

Penicillins are often given in conjunction with aminoglycosides because the penicillins work on the cell wall, facilitating the aminoglycoside's inhibition of protein synthesis. A new group of drugs, beta-lactam antimicrobials, may replace aminoglycosides in the next decade. They have the basic structure of cephalosporins and penicillins linked together in a new molecule. Such a drug is *moxalactam* (see discussion under "Cephalosporins").

Uses. Aminoglycosides are primarily used for infections with aerobic gram-negative bacteria. The six most common types are listed on Table 26-8. None are active against anaerobes. Because resistance to streptomycin occurs so readily, it is usually given with other antimicrobials. Its use currently is primarily as an antituberculosis drug. Kanamycin, active against many gram-negative bacteria except *Pseudomonas*, is not used frequently now because bacterial resistance is more common than with the other aminoglycosides.

Neomycin has a limited but important use as a bowel antiseptic. In hepatic encephalopathy neomycin can be useful by decreasing numbers of bacteria in the intestine and the subsequent ammonia that they produce. Neomycin is also used for intestinal "sterilization" prior to bowel surgery.

Gentamicin is useful in systematic infections caused by gram-negative bacteria. Tobramycin has similar activity, though it may be more active against *Pseudomonas* and less nephrotoxic (see "Adverse Effects"). Amikacin, a structural analogue to kanamycin, should be reserved for use against organisms that are resistant to gentamicin and tobramycin. There is, as yet, no cross resistance between gentamicin or tobramycin and amikacin. Because aminoglycosides are excreted almost unchanged in the urine, they are efficacious in the treatment of urinary tract infections. There are several aminoglycosides undergoing clinical trials including sisomicin and netilmicin.

Dosage and administration. Usual adult dosages are listed on Table 26-8. Aminoglycosides are very stable, therefore, they can be premixed. However, they precipitate with heparin; if given through a heparin lock, the heparin should be flushed out first. Though penicillins and cephalosporins potentiate the effect of aminoglycosides, they should not be mixed or given in the same solution or they will neutralize each other. The two types of antimicrobials can either be given through a different line or given on a different time schedule.

Because of poor gastrointestinal absorp-

TABLE 26-8. Uses of Aminoglycosides

Drug	*Usual adult dosages*	*Indications*	*Precautions*
Amikacin	15 mg/kg/day I.V., I.M. in 3-5 doses	Gram-negative bacteria resistant to gentamicin and tobramycin	One of the more toxic aminoglycosides; restrict use to severe infections and for < 2 weeks duration
Gentamicin	80 mg q8h I.M., I.V.	Gram-negative bacterial infection	Nephrotoxicity most common with this drug, as well as hearing loss*
Kanamycin	250 mg q6h I.M. 4-6 g/day po	Neonatal meningitis, sepsis due to susceptible gram-negative organisms, suppression of normal bowel flora for treatment of hepatic coma	Not active against *Pseudomonas*
Neomycin	4-12 g/day po for hepatic coma, prebowel surgery	Topical, preoperative bowel disinfection, hepatic encephalopathy (decreases bacterial production of ammonia), GU irrigant, UTI	Avoid I.M. use due to toxicity
Paromomycin	25-35 mg/kg/day po in 3 doses	Intestinal amoebiasis or other intestinal parasites	Watch for superinfection with candidiasis
Streptomycin	0.75-1.0 g/day I.M., I.V., S.Q.	Tuberculosis, *S. faecalis* endocarditis (with penicillin), brucellosis, plague, tularemia	Used in combination with other agents due to rapid emergence of resistant strains
Tobramycin	3-4 mg/kg/day I.M. or I.V. in 3 divided doses	Gram-negative bacteria, especially *Pseudomonas*	Mix I.V. dose in 50-100 ml solution, infuse over 20-60 min; do not mix with other drugs

*All aminoglycosides are toxic to eighth cranial nerve to some degree

tion all aminoglycosides are given parenterally. Neomycin is the exception because it is used for its local effect on gut flora. There is rapid absorption of aminoglycosides intramuscularly and subcutaneously, as well as intravenously, but the blood-brain barrier is not passed. The intravenous route is preferred for gentamicin because intramuscular injection does not yield even or as rapid distribution of the drug. This can cause unpredictable changes in blood levels. Peak serum levels of the drug are reached in 30 minutes after intravenous injection, 60 minutes after intramuscular injection.

Adverse effects. Aminoglycosides have the narrowest therapeutic:toxic ratio of any other commonly used antimicrobial agents, so their effects must be monitored carefully. About 5 to 10 percent of patients have toxic reactions, either decreased kidney function or hearing impairment. Half of these develop permanent ear damage as a result of eighth cranial nerve damage. This damage may cause vestibular dysfunction, as well as hearing loss. Streptomycin is especially prone to affect the vestibular branch of the eighth cranial nerve, resulting in dizziness, and nausea and vomiting. Hearing loss due to toxic effects of the auditory branch is rare after streptomycin, most common after gentamicin. The ototoxic effects are directly related to mean *peak* levels, which should never exceed 10 to 12 μg/ml for tobramycin and gentamicin, 25 μg/ml for kanamycin and amikacin. Use of diuretics such as ethacrynic acid and furosemide can increase ototoxicity associated with aminoglycosides. Ototoxicity can develop even after the drug is discontinued.

Nephrotoxicity also occurs and can be avoided by keeping the *trough* levels less than 2 μg/ml (5 μg/ml for amikacin). A gentamicin-cephalothin mix increases chances of nephrotoxicity. Gentamicin has been shown to cause decreased glomerular filtration rate—even with accepted dosage levels in persons with normal renal function. Gentamicin, but not tobramycin, also causes significant dose-related decreases in creatinine clearance. Renal damage is most common after amikacin, least common after tobramycin. See "Clinical Implications" for further discussion of these serious side effects. Streptomycin can also cause respiratory paralysis due to neuromuscular blockade, especially after anesthesia or use of muscle relaxants.

As with other antimicrobials, a major problem of aminoglycoside therapy is the development or emergence of resistant organisms. Aminoglycoside resistance is due to (1) altered ribosomal binding site, (2) modified enzymes mediated by plasmids, or (3) decreased permeability through the cell envelope. A reduction in aminoglycoside use has been documented to result in decreased resistance of the bacterial strains prevalent in the environment. For example, during therapy with amikacin, a *Serratia marcescens* carrying a plasmid coding for resistance to gentamicin and tobramycin lost the plasmid and again became sensitive. It was postulated that the sensitive variant resulted from loss of the plasmid by removal of selection pressure of using gentamicin and tobramycin (Rubens, McGee, and Farrar 1980).

ANTITUBERCULOSIS DRUGS (Table 26-9)

Ethambutol

Ethambutol is a synthetic compound effective only against *Mycobacterium tuberculosis*. In actively dividing cells it enters and interferes with metabolic processes (the exact mechanism is as yet unexplained). It is rapidly absorbed after oral administration. It is excreted by the kidneys and in at least half of individuals causes increased serum uric acid, which can precipitate gout. Like the other antitubercular drugs, resistance occurs rapidly and therefore combination therapy is the rule.

Before an individual begins therapy with ethambutol, baseline measurements of visual acuity should be obtained because a significant side effect in about 1 percent of individuals is optic neuritis, affecting one or both eyes. Loss of peripheral vision as well as acuity can occur. Symptoms of blur-

TABLE 26-9. Antituberculosis Drugs

Drug	*Usual adult dosages*	*Major side effects*
Ethambutol	25 mg/kg/day po	Optic neuritis, increased serum uric acid
Isoniazid	For prophylaxis: 300 mg/day po For treatment: 5mg/kg/day (maximum 300 mg)	Liver damage, vitamin B_6 deficiency, rheumatic or lupus syndromes
Rifampin	600 mg/day po	Liver or kidney damage, GI disturbances
Para-aminosalicylic acid (PAS)	5 g tid po	GI disturbances, liver damage
Streptomycin	1 g tid po	Vestibular imbalance, paresthesias, hearing loss

ring or inability to discriminate colors should alert one to optic toxicity. Eye problems only arise after several months of therapy and disappear on discontinuation of the drug. Allergic reactions and gastric upset can result, but ethambutol is in general well tolerated.

Isoniazid

Isoniazid (isonicotinic acid hydrazide or INH) probably works by inhibiting the synthesis of mycolic acids, which are necessary components in mycobacterial cell walls. INH is effective only in the treatment of tuberculosis and some atypical mycobacterial infections. It is most effective when cells are actively dividing. The drug can be given orally or parenterally, being well absorbed by either route, but is usually given in tablet form. It is excreted by the kidneys. INH is also used prophylactically for intimate contacts with patients with tuberculosis and to prevent active disease in individuals whose tuberculin skin tests convert from negative to positive (indicating infection). It is usually given for 1 year prophylactically. Like the other antituberculosis agents, it must be given in combination with other drugs.

There are many possible adverse effects of INH. Neurologic toxicity can be manifested by numbness, paresthesias, peripheral neuritis, seizures, depression, and psychosis. Persons receiving INH excrete excessive amounts of pyridoxine (vitamin B_6), which is the structural analog to INH. Individuals who are malnourished may become deficient in pyridoxine and have symptoms of pellagra (anemia, skin lesions, peripheral neuritis). This can be prevented by concomitant administration of vitamin B_6. As many as 20 percent of individuals taking INH sustain some degree of liver injury. Some develop progressive liver damage and hepatitislike symptoms that can be fatal. Liver function must be monitored throughout treatment. Other adverse effects include gastrointestinal disturbances, blood dyscrasias, hyperglycemia, metabolic acidosis, gynecomastia, and a lupuslike syndrome.

Rifampin

Rifampin is the only one of several derivatives of rifamycin available in the United States. It is obtained from *Streptomyces mediterranei* and inhibits RNA synthesis by interacting with RNA polymerase in bacterial but not human cells. Its use is primarily as an antituberculosis drug, but is also used to eradicate the carrier state of the meningococcus and to treat leprosy and some infections due to atypical mycobacteria. It is well absorbed through the gastrointestinal tract and excreted in the bile. Though *M. tuberculosis* develops resistance to rifampin after long-term therapy, there is no cross resistance with other antituberculosis drugs.

Rifampin decreases the effectiveness of coumarin anticoagulants and oral contraceptives; dosages of these agents may need modification when an individual is taking rifampin. Disturbances in liver function can occur with increased bilirubin and transaminase, especially when rifampin is combined with INH. Rarely, hepatitislike disease can be very serious, especially in alcoholics or others with preexisting liver damage. Rifampin should be administered daily; an increased incidence of both hypersensitivity reactions and renal failure have been noted if the drug is given intermittently. The reason for the additional risk with intermittent therapy is not known. The most common side effects are minor gastric disturbance. Body secretions (urine, sweat, tears, etc.) are turned brownish-orange during rifampin administration, but this is harmless. Nervous system symptoms such as headache, drowsiness, or visual disturbances have been reported.

Streptomycin and PAS are discussed under aminoglycosides and sulfonamides, respectively. Other minor antituberculosis drugs, used primarily for treatment of persistent disease or recurrences include capreomycin, cycloserine, ethionamide, kanamycin, pyrazinamide, and viomycin.

BACITRACIN

Bacitracin is derived from *Bacillus subtilis*. It blocks cell wall synthesis by preventing the regeneration of the lipid carrier, a step necessary for peptidoglycan production. It is excreted by the kidneys. Its primary use is as a topical ointment for skin and superficial infections caused by gram-positive organisms like staphylococci and streptococci. It has no effect on gram-negative bacteria. There is an injectable form for intramuscular use, but when given systemically it is extremely nephrotoxic. It is used very rarely, primarily with infants who have staphylococcal pneumonia or empyema. The intramuscular dose for infants is 1000 U (50 U = about 1 mg)/kg/day in 2 to 3 divided doses. Bacitracin is available in ointment either alone or in combination with neomycin and polymyxin. To reduce the chance of nephrotoxicity, the patient's renal function must be monitored very carefully and an adequate fluid intake must be maintained throughout the duration of therapy. Nausea, vomiting, rash, and pain at injection site also occur.

CEPHALOSPORINS

Pharmacologic characteristics. Cephalosporins, originally developed from the fungus, *Cephaloporium acremonium,* were introduced in about 1962, and there are now at least 10 derivatives on the market (Table 26-10). They share a basic structure, 7-amino-cephalosporanic acid, which has a beta-lactam and dihydrothiazine ring structure. They are generally bactericidal as a

TABLE 26-10. Cephalosporins

Basic structure:

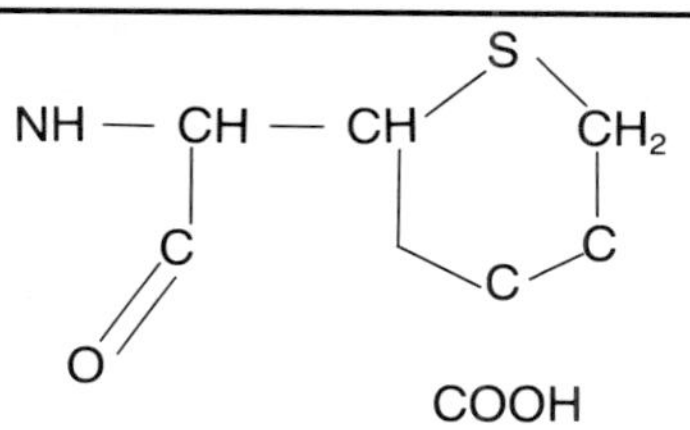

Drug	Comments
Cefaclor	Administer orally; used primarily in pediatrics
Cefadroxil	For UTI
Cefamandole	For UTI
Cefazolin	Less pain and phlebitis on injection; high concentration in bile
Cefoperazone	
Cefotaxime	
Cefoxitin (cephamycin)	Severe thrombophlebitis; use central vein for long-term administration
Cephalexin	Oral only; *H. influenzae* is resistant
Cephaloglycin	Oral only
Cephaloridine	Nephrotoxicity; not used much currently
Cephalothin	Painful when given I.M.
Cephipirin	
Cephradine	*H. influenzae* is resistant
Moxalactam	

result of inhibiting mucopeptide synthesis in the bacterial cell wall. Their action is very similar to that of the penicillins, except that the cephalosporins are resistant to beta-lactamase. The drugs may be bacteriostatic at low concentrations. The half-life of cephalosporins in blood is generally about 2 1/2 times that of the penicillins, they are excreted by the kidneys, and they enter the CSF poorly, though more easily in the presence of meningeal irritation. The cephamycins (cefoxitin, for example) are like cephalosporins but differ by containing a methoxy group on the beta-lactam ring.

Uses. Cephalosporins are active against gram-positive bacteria, and especially useful for penicillinase (beta-lactamase) producers. Some gram-negative bacteria are also susceptible, but the drugs are generally not active against anaerobes and not recommended for central nervous infections since their entry into the spinal fluid is unreliable. They are probably overprescribed currently, since a drug with a narrower spectrum could often be substituted.

Cephalothin is a first generation derivative active against gram-positive cocci such as *Strep. pneumoniae* and *S. aureus*. Cefazolin gives the highest blood levels for any given dose, with less pain and phlebitis on injection. However, it should not be used to treat staphylococcal infections because the organism produces enzymes that inactivate cefazolin. Cephalexin is an oral preparation that can be used when parenteral therapy is no longer required. Cefadroxil is specifically for urinary tract infections, but it has no real advantages over other cephalosporins.

Cefamandole and cefoxitin, the more recent cephalosporins, are active against many cephalothin-resistant strains of bacteria. Cefamandole has a broader spectrum against gram-negative bacteria, both aerobes and anaerobes. Cefoxitin is also used against gram-negative rods and anaerobes. Cefaclor, given orally, is effective against *Hemophilus*, pneumococci, alpha- and beta-hemolytic streptococci, and is used primarily in pediatrics. Cefoperazone and cefotaxime are third generation cephalosporins, the first to have some activity against *Pseudomonas aeruginosa*. Cefotaxime is also active against penicillinase-producing *N. gonorrhoeae*.

Moxalactam, a beta-lactamase-resistant drug available since 1982, is structurally related to the third generation cephalosporins. It is active against gram-negative bacteria, both aerobes and anaerobes, and penetrates the blood-brain barrier when there is meningeal inflammation. It is expensive, and resistance and superinfections can occur. Because it can be used against organisms resistant to other agents such as amoxicillin, aminoglycosides, cephalosporins, and amikacin, it is a valuable new drug. Its use should be limited to serious systemic infections with Enterobacteriaceae, not for infections with *Pseudomonas*, meningococci, or gram-positive cocci.

Dosage and administration. Cephalosporins may be given orally, intramuscularly, or intravenously, depending upon absorption qualities of the particular compound. The usual adult dosage is 500 mg to 1 g every 4 to 12 hours. Most (60 to 80 percent) of the drug is excreted by the kidneys. Like all antimicrobials excreted in the urine, dosages must be decreased for patients with renal dysfunction.

Adverse effects. Cephalosporins have been shown to stimulate the production of lymphokines, which might be the mechanism for some of the immunologic reactions to them (urticaria, neutropenia, a positive Coomb's test). Penicillins do not produce these effects on lymphokines. Cephaloridine (Loridine) causes renal toxicity and for that reason is rarely used today, but the other preparations do not have this effect. Other toxic effects are rare. Patients who have had immediate hypersensitivity reactions to penicillins should not be given cephalosporins because of a high incidence of cross-reaction between the two. I.V. administration is frequently associated with phlebitis. Consumption of alcohol within 72 hours of receiving cefamandole, cefoperazone, or moxalactam should be avoided. Alcohol intolerance with flushing, trembling, dyspnea, and hypotension can occur.

CHLORAMPHENICOL

Pharmacologic characteristics. Chloramphenicol (Chloromycetin) was first discovered in 1947 from a soil *Streptomyces* mold in Venezuela, but is now produced synthetically. It was the first truly broad-spectrum antimicrobial agent and is chemically unrelated to any other agent. Its mode of action is similar to that of erythromycin and clindamycin; protein synthesis is inhibited because the ribosomes attach to mRNA but no peptide bonds are produced. In therapeutic doses, its action is bacteriostatic rather than bactericidal. It is widely distributed in tissues, including the eye, central nervous system, and subarachnoid space. Even without meningeal inflammation, 50 percent of serum levels are attained in the spinal fluid. Chloramphenicol is excreted in the kidneys, but the glucuronyl-transferase system in the liver must be functioning to inactivate the drug.

Uses. Chloramphenicol has an antibacterial spectrum similar to the tetracyclines and whenever possible (e.g., in rickettsial and chlamydial infections) tetracycline should be used because of the toxic effects of chloramphenicol. Chloramphenicol is the drug of choice, however, in typhoid, and it is also useful in treating other *Salmonella* diseases. Because of its high concentration in the spinal fluid, chloramphenicol is useful in the treatment of *H. influenzae* meningitis as an alternative to ampicillin. Chloramphenicol is also active against many anaerobes and gram-negative bacteria but its use should be limited to serious systemic diseases. Ointments and drops are also available for skin, eye, or ear infections due to susceptible organisms.

Dosage and administration. Because of its excellent gastrointestinal absorption, the drug is often given orally; peak blood levels are reached in 2 hours. It is not given intramuscularly because adequate blood levels are not maintained with this route. A sodium succinate preparation is available for intravenous administration, though absorption is no better than by the oral route.

Adverse effects. The major adverse effect is blood dyscrasias. About 10 percent of persons receiving chloramphenicol suffer from some depression in bone marrow function and in about 1 in 25,000 to 40,000 persons severe or fatal complications such as thrombocytopenia, granulocytopenia, aplastic or hypoplastic anemia occur. These serious complications can occur even after only one dose of chloramphenicol or after treatment has been discontinued for months; the toxic hematopoietic effects are not dose related. Bone marrow depression is due to the inhibition of mitochondrial protein synthesis in human cells. During therapy with chloramphenicol, the leukocyte count should be monitored several times each week. If the WBC count becomes less than 4000 the drug should be discontinued.

Chloramphenicol is contraindicated in pregnancy, and in infancy due to immature liver function, which makes it impossible for the infant to metabolize the drug. The "gray syndrome," characterized by cyanosis, hypothermia, abdominal distention, and sometimes fatal vasomotor collapse occurs in babies or in others with immature metabolic functions who receive chloramphenicol.

CLINDAMYCIN

Pharmacologic characteristics. Lincomycin (Lincocin) was discovered in 1962 in a *Streptomyces* mold in soil near Lincoln, Nebraska. Clindamycin (Cleocin) is a derivative. This drug suppresses protein synthesis, binding to the same site as erythromycin and chloramphenicol, so they inhibit each other. It is primarily metabolized in the liver.

Uses. This is the drug of choice for anaerobic bacteroidal infections. Clindamycin is a good alternative to penicillins against many gram-positive cocci and anaerobic gram-negative bacteria. Since it yields high concentrations in bone, clindamycin can be used to treat osteomyelitis. It is not useful against *H. influenzae* or *Pseudomonas*.

Dosage and administration. Clindamycin is well absorbed from the intestinal tract; the presence of food does not decrease its absorption. Its half-life is about 2 1/2 hours. It is distributed well in the body but not the spinal fluid, and is given orally and parenterally (Table 26-11).

Adverse effects. The major complications of clindamycin are gastrointestinal: diarrhea and gastroenteritis are common. Occasionally pseudomembranous colitis, due to an overgrowth of *Clostridium difficile* in the bowel, occurs, though this syndrome has also been reported in conjunction with tetracyclines, ampicillin, cephalosporins, and trimethoprim-sulfamethoxazole. The frequency of stools should be carefully monitored in patients receiving clindamycin.

METRONIDAZOLE

Pharmacologic characteristics. Metronidazole (Flagyl) is a synthetic compound whose actions are not completely described. It produces a metabolite that interferes with DNA transcription and replication in certain organisms. It is excreted primarily in the urine. Metronidazole is carcinogenic to animals, but there is no evidence that the risk of cancer in humans is increased with short-term use of the drug.

Uses. Metronidazole can be used to treat infections caused by anaerobes such as *Bacteroides, Clostridium, Fusobacterium,* and *Peptostreptococcus.* Its primary use, however, is to treat amoebic dysentery, amoebic liver abscesses, giardiasis, and trichomoniasis.

Dosage and administration. The drug may be give intravenously for serious infections, but the oral route should be used whenever possible to prevent serious side effects. Oral therapy for trichomoniasis, giardiasis, and amoebiasis as well as anaerobic infections usually continues for at least 7 to 10 days. Intravenous doses should never exceed 4 g/

TABLE 26-11. Chloramphenicol, Erythromycin, Clindamycin

Drug	*Uses*	*Usual adult dosage*	*Precautions*
Chloramphenicol	Typhoid, other salmonella, rickettsias, chlamydias, serious gram-negative infections, *H. influenzae* meningitis	50 mg/kg/day po or I.V. (sodium succinate), topical	Can cause serious or fatal bone marrow depression and blood dyscrasias
Clindamycin	*B. fragilis,* many gram-positive cocci, bone infections	500 mg 3×/day po; 600 mg q12h I.M.; 0.6-1.0 g q8-12h I.V.	Can cause pseudomembranous colitis, gastroenteritis
Erythromycin	*M. Pneumoniae, Listeria,* alpha- and beta-strep., amoebiasis, legionnaires' disease, *C. diphtheriae,* substitute for penicillin in sensitive individuals	250 mg po q6h; 15-20 mg/kg/day I.V.	GI distress in 10 to 15 percent of cases; estolate preparation can cause reversible jaundice

24 hours; oral doses range from 250 to 750 mg, three times a day.

Adverse effects. Neurologic symptoms such as seizures, numbness, and paresthesias of extremities can occur, and they resolve slowly. A metallic taste is common whether the drug is administered orally or parenterally. The urine may also have a reddish-brown color. Alcoholic beverages should be avoided during treatment because a disulfiram (Antabuse)-like response with vomiting, cramping, and flushing can occur.

ERYTHROMYCIN (Table 26-11)

Pharmacologic characteristics. Erythromycin was isolated in 1952 from a soil *Streptomyces* fungus in the Philippines. It is a member of the macrolide group of antimicrobials, which include oleandromycin, carbomycin, and spiramycin, the second two being unavailable in the United States. Spiromycin is used for toxoplasmosis in Europe. These drugs attach to the 50S portion of the ribosomal subunit, inhibiting protein synthesis by preventing incorporation of some amino acids into polypeptide linkages. Oleandomycin has cross resistance with erythromycin and is less effective. Erythromycin is primarily a bacteriostatic agent and interferes with the action of chloramphenicol and clindamycin because it binds on the same ribosomal subunit. The drug is excreted in bile and passes the placental barrier.

Uses. The antimicrobial spectrum of erythromycin resembles that of penicillin G. It is active against gram-positive bacteria, but is not clinically effective against most gram-negative organisms. It is the best alternative to penicillin for prophylaxis against carditis in persons with rheumatic fever, and it can be used as a substitute for penicillin-sensitive individuals with staphylococcal, streptococcal, or pneumococcal infections. It is effective against *M. pneumoniae, T. pallidum, C. diphtheriae,* the *Bordetella* and tetanus bacteria, some *H. influenzae* strains (when combined with sulfonamides), atypical mycobacteria, some *Rickettsia,* and systemic chlamydial infections. It is currently the treatment of choice for legionnaires' disease. Though it has been suggested for use as an ophthalmic ointment to prevent chlamydial conjunctivitis, its efficacy is not proven. Strains of pneumonococci resistant to erythromycin have recently emerged.

Dosage and administration. Erythromycin is supplied in four forms: base erythromycin, erythromycin estolate (Ilosone), erythromycin ethylsuccinate, and erythromycin stearate. Though the base and stearate are absorbed poorly from the gastrointestinal tract, the estolate preparation has better absorption and is unaffected by food in the stomach. The base erythromycin and erythromycin stearate should be taken between meals when the stomach is empty, but the absorption of erythromycin ethylsuccinate is actually improved with a full stomach. Enteric-coated tablets are necessary since gastric acid destroys some preparations. Serum levels of the drug reach their peak about 2 hours after oral administration. Erythromycin diffuses to all body tissues, and to the spinal fluid when there is meningeal irritation. It is given topically, orally, and intravenously, but intramuscular injection is too painful. Use is contraindicated in pregnancy.

Adverse effects. Erythromycin is a safe drug in comparison with many other antimicrobials. It does cause gastrointestinal distress in 10 to 15 percent of individuals, but this is minimized if the drug is given in the preparation that can be taken with meals. The only major side effect of the estolate preparation is a jaundice due to cholestatic hepatitis that occurs in 2 to 4 percent of individuals. It is completely reversible upon discontinuation of the drug and is more of an allergic rather than a toxic reaction. Intravenous use of erythromycin is quite irritating to veins and thrombophlebitis is common. Unfortunately, resistance to erythromycin emerges rapidly and for this reason it is not a recommended drug for serious infections due to *S. aureus*. Resistant staphylococci harbor a plasmid that alters the bacterial RNA, coding for the production of an essential enzyme, so that bacteriostasis no longer occurs.

PENICILLINS

Pharmacologic characteristics. Though the antibacterial properties of penicillins, extracted from the mold *Penicillium notatum,* were first noted and described by Alexander Fleming in 1928, penicillins did not become commercially available until the 1940s. Today there are at least 15 penicillins available, which differ in their molecular side chains, modified to enhance their spectrum of activity or resistance to beta-lactamase (Table 26-12). The penicillins all share the same basic structure, fused thiazolidine and beta-lactam rings with the side chains attached by peptide linkage to the beta-lactam ring. They exert their antibacterial effect by preventing the cross linking of peptidoglycans in bacterial cell walls. They do this by inhibiting several essential enzymes such as transpeptidase, carboxypeptidase, and endopeptidase. Penicillins also activate murein hydrolase (autolysin) to result in cell lysis. Because of their action on cell wall production, they are most effective against bacteria that are actively multiplying.

The enzyme beta-lactamase was discovered in the 1950s. It is produced by some staphylococci and other organisms and can split the penicillin molecule, rendering it useless. The second, third, fourth, and now fifth generation penicillins have been developed in part to increase the spectrum of antibacterial activity of the agents and in part to produce penicillins resistant to this cleavage of beta-lactamase (sometimes called penicillinase).

The second generation penicillins were produced throughout the 1950s and 1960s. Even now a number of improved compounds are becoming available, including azlocillin (similar to ticarcillin but with 4 times the activity against *Pseudomonas*), mecillinam (for use against *E. coli* and *Klebsiella*), mezlocillin and piperacillin (for gram-negative and anaerobic bacteria), and pivmecillinam, an oral form of mecillinam.

Uses. Even after four decades, aqueous penicillin G remains the drug of choice for clinical infections due to pneumococci, meningococci, gonococci, streptococci, clostridia, and respiratory tract anaerobes. It is also useful for treating syphilis and diphtheria. Unfortunately, penicillin-resistant gonococci and pneumococci have recently emerged, though the *S. pneumoniae* strains isolated in the United States and Great Britain are still almost consistently susceptible. Group A streptococci are universally sensitive. The majority of strains of *S. aureus* in the 1940s were sensitive to penicillin G, but the plasmid mediated transfer of beta-lactamase is very prevalent; about 90 per-

(text continues on page 695)

TABLE 26-12. Commonly Used Penicillins

Basic structure:

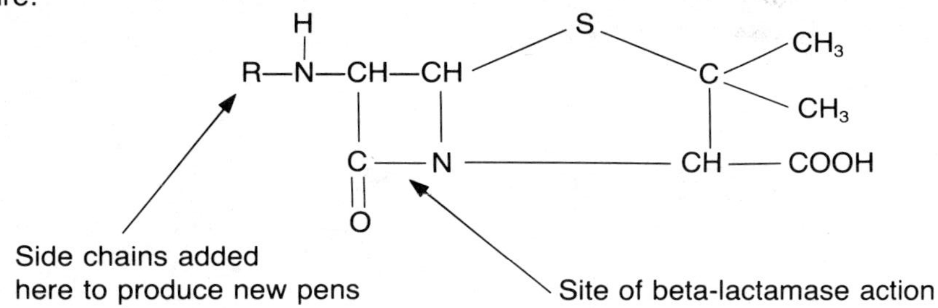

Generic name	Spectrum: Narrow	Spectrum: Broad	Penicillinase: Sens	Penicillinase: Resis	Side chain	Dosage	Uses
Penicillin G	X		X		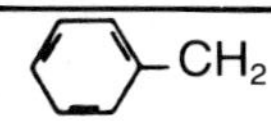	0.5-2 g/day po; 5-30 mill U/day I.M., I.V.	Least expensive; *Streptococcus, Neisseria,* some anaerobes, spirochetes
Penicillin V	X		X		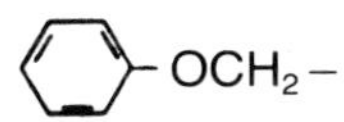	0.5-2g/day po	As above, but better oral absorption
Ampicillin		X	X		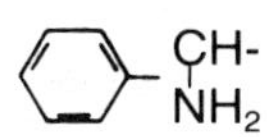	1-2g/day po; 8-16g/day I.M., I.V.	As above, but also *S. faecalis,* many gram-negative bacteria, *H. influenzae*
Amoxicillin		X	X		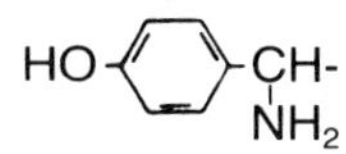	0.75-1.5 g/day po	As ampicillin except better oral absorption
Carbenicillin		X	X		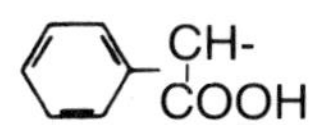	8-30 g/day I.V., I.M.	Gram-negative bacteria, especially pseudomonas
Cloxacillin	X			X	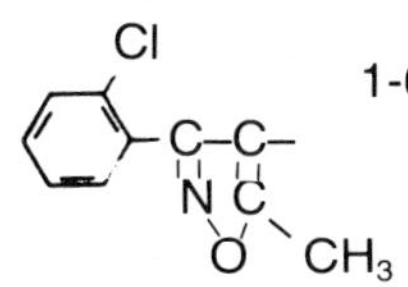	1-6 g/day po	Penicillinase-producing gram-positive cocci, especially *S. aureus;* not gram-negatives

TABLE 26-12. *(continued)*

Generic name	Spectrum: Narrow	Spectrum: Broad	Penicillinase: Sens	Penicillinase: Resis	Side chain	Dosage	Uses
Dicloxacillin	X			X	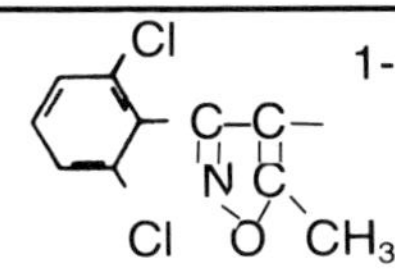	1-4 g/day po	As above
Methicillin	X			X	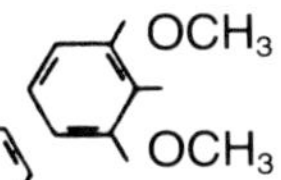	8-16 g/day I.M., I.V.	As above
Nafcillin	X			X	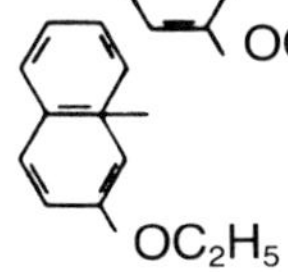	2-6 g/day po; 2-12 g/day I.M., I.V.	As above
Oxacillin	X			X	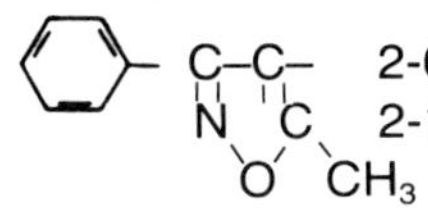	2-6 g/day po; 2-12 g/day I.M., I.V.	As above
Ticarcillin		X	X			3-20 g/day I.M., I.V.	As carbenicillin, especially *Pseudomonas*

cent of strains are now resistant. The resistance of pneumococci to penicillin G is not mediated by beta-lactamase but by some other mechanism.

Penicillin G is often combined with an ammonium or procaine base to slow absorption by the intramuscular route (procaine or benzathine penicillin G). Penicillin V has the same spectrum as penicillin G with the advantage that it is not destroyed by gastric acid and is the preferred preparation for oral use (e.g., in streptococcal pharyngitis).

All the other penicillins have spectra that are expanded from those of penicillin G and V. Ampicillin, for example, is active against *Streptococcus faecalis* and is used to treat enterococcal endocarditis. It is also used to treat clinical infections caused by *H. influenzae, E. coli, Neisseria,* and *P. mirabilis.* Ampicillin-resistant strains of *H. influenzae* were first reported in the United States in 1973. The prevalence is now about 20 percent. Amoxicillin has identical action to ampicillin, but its oral absorption is not diminished if there is food in the stomach. Bacampicillin is hydrolyzed to ampicillin in the GI tract. It may be substituted for ampicillin because it is more completely absorbed by the oral route. Molecules of methicillin, cloxacillin, dicloxacillin, and oxacillin are not split by the beta-lactamase enzyme, so they are especially useful for gram-positive cocci (primarily *S. aureus*), which produce beta-lactamase. Methicillin-resistant *S. aureus* must be treated with vancomycin since they will probably be resistant to cephalosporins as well as the penicillins.

Carbenicillin and ticarcillin are broad-spectrum penicillins with similar activity against many gram-negative organisms including *P. aeruginosa.* With ticarcillin, higher blood levels can be attained with lower doses. It is also less expensive and less toxic than carbenicillin. Azlocillin is even more active against *P. aeruginosa,* but should be used in combination with an aminoglycoside for patients with life-threatening infections. Indanylcarbenicillin is useful only as a urinary antibiotic in urinary tract infections caused by organisms sensitive only to carbenicillin and ticarcillin.

Penicillin in some places is given to newborns as prophylaxis against ophthalmia neonatorum. It is argued that sensitization is not a problem because of the immunologic immaturity of neonates. Results of trials to give prophylactic penicillin to prevent neonatal streptococcal disease have varied, but during a large trial (15,010 subjects) the flora isolated from babies did not change as a result of the penicillin administration (Hodes 1980).

Dosage and administration. Because early penicillin preparations contained impurities, it was helpful to have a measurement system that insured standardization in dosages. Though penicillins are now available in the pure forms, units are still used to measure penicillin G. One unit of penicillin G equals 0.5 μg. The usual dose for penicillin G is 125 to 500 mg (20,000 to 80,000 U) 2 to 3 times per day. It is necessary to administer larger doses (about 2 1/2 times normal) for meningococcal meningitis, as absorption in the spinal fluid is poor. Since piperacillin inactivates aminoglycosides when they are mixed together, they should be given by different routes or at different times.

Adverse effects. The penicillins are among the least toxic of any antimicrobials. Rarely, neuropathy or hemolytic anemia can follow prolonged intravenous administration of penicillin. Penicillins are excreted by the kidneys. Because 1 million units of penicillin G contain 1.5 mEq of potassium, individuals with impaired renal function can suffer from potassium overload. Signs and symptoms of hyperkalemia should be monitored in such patients. On the other hand, about one-third of patients on broad-spectrum penicillins (carbenicillin and ticarcillin) get hypokalemic. High doses of penicillin cause altered platelet aggregation and increased bleeding, especially if aspirin is also ingested. Patients receiving penicillins should avoid aspirin. Interstitial nephritis occurs rarely, most commonly with methicillin or oral ampicillin.

The major problem with penicillin administration is the high incidence of hypersensitivity phenomena, ranging from slight rash or urticaria to rare immediate anaphylactic shock and death. About 20 percent of hospitalized patients claim to be allergic to penicillin, but this is probably greatly overestimated. A more realistic estimate of the proportion of the general population with penicillin allergy is from 1 to 5 percent. Reactions may occur immediately or may be delayed for more than 3 days (primarily skin eruptions), and may be seen after any route of administration. Anaphylaxis occurs in about 0.4 to 1.3 per 1000 patients receiving penicillins and death in about 1 in 100,000. Once an individual has developed hypersensitivity to one penicillin, they are allergic to all, as cross reaction exists among all compounds. There is also some cross sensitization with the related compounds, cephalosporins. It is possible to desensitize allergic persons who require treatment by giving them small oral or parenteral doses of penicillin. Skin tests are available that predict with high probability those individuals who will develop subsequent hypersensitivity reactions. Though they are not completely accurate, skin-testing methods are being continually improved.

POLYMYXINS

Pharmacologic characteristics. Polymyxin B and colistin (polymyxin E), the two polymyxins used clinically, act as cationic surface active detergents at physiologic pH, which destroys the osmotic integrity of bacterial cell membranes. They are bactericidal. For unknown reasons, certain gram-negative bacteria are more susceptible than gram-positive bacteria. Polymyxins are strongly bound to phospholipids in human cells, are inhibited by pus, and do not enter the spinal fluid. Concentrations of the drug in the liver and kidneys are high.

Uses. Polymyxins are used to treat gram-negative bacterial infections caused by organisms such as *P. aeruginosa,* which are resistant to other antimicrobial agents. Because of their low concentration in tissues, polymyxins are most useful in treating *Pseudomonas* bacteremias and urinary tract infections rather than tissue infections. *Proteus* and *Neisseria* are not susceptible. Polymyxin B is also available as a cream for topical use or as an ophthalmic solution.

Dosage and administration. Oral absorption is very poor, so the drugs are usually given intramuscularly or intravenously. *Pseudomonas* meningitis can be treated with intrathecal administration. Usual dosages are

	I.V.	*I.M.*
Polymyxin B	15,000-25,000 U (100,000 U = 1 mg)/kg/day q12 hr	25,000-30,000 U/kg/day
Colistin	2.5-5 mg/kg/day q6-8 hr	Same as I.V.

The antibacterial effect of polymyxins is decreased by soap (which is important with topical use of polymyxin B).

Adverse effects. Side effects of the two drugs are the same, but colistin causes fewer than polymyxin B when given parenterally. Neither should be given unless the patient is hospitalized so that renal function and neurologic status can be carefully monitored. Signs of nephrotoxicity are indications that the drug should be discontinued. Polymyxin B causes albuminuria and cellular casts in the urine. Patients with impaired renal function or with nephrotoxicity developing during treatment are at increased risk of respiratory arrest due to neuromuscular blockade. Facial flushing, weakness, blurred vision, numb extremities, drowsiness, ataxia, irritability are some signs of neurotoxicity. Fever, rash, and thrombophlebitis after intravenous injection also occur.

SPECTINOMYCIN (TROBICIN)

Spectinomycin is closely related to the aminoglycosides; it has a slightly different structure and mechanism of action. It is derived from *Streptomyces spectabilis,* and inhibits protein synthesis at the 30S ribosomal subunit. Spectinomycin is used to treat genital gonorrheal disease: cervicitis, proctitis, or urethritis. It is the drug recommended by Centers for Disease Control for gonococcal disease in individuals sensitive to penicillins or gonococcal strains resistant to penicillin. A single, deep intramuscular injection of 2 g is the usual dose. Resistant strains occur. The treated individual may experience nausea, fever, dizziness, or urticaria.

SULFONAMIDES AND SULFONES

Pharmacologic characteristics. The sulfonamides were the major chemotherapeutic agents available in the 1940s before penicillins. Because they are analogs of para-aminobenzoic acid (PABA) they compete with PABA in bacterial cells in forming folic acid. The bacteria continue to grow and multiply in the presence of sulfonamides, however, until all the folic acid is depleted; hence they are usually bacteriostatic. Since human cells require preformed folic acid from the diet, sulfonamides are selectively toxic for bacterial cells. Para-aminosalicylic acid (PAS) and the sulfones act in a similar manner with a more limited spectrum of antibacterial activity.

Trimethoprim is a synthetic drug that also works by inhibiting purine synthesis at a point subsequent to the site of sulfonamide activity. Hence trimethoprim more rapidly stops bacterial growth and a combination of trimethoprim and sulfonamides results in synergistic effects. Trimethoprim is usually bacteriostatic but can be bactericidal if the growth medium lacks thymine. Sulfonamides are excreted through the kidneys; PAS and sulfones are excreted primarily by the liver.

Uses. Since sulfonamides are less potent than other antimicrobials, and since many organisms have natural or acquired resistance, their use is limited primarily to treatment of urinary tract infections. Sulfisoxazole (gantrisin) and co-trimoxazole (a combined trimethoprim-sulfamethoxazole with trade names of Bactrim or Septra), are most commonly used, though other preparations are available (Table 26-13). Their other major therapeutic usefulness is in treatment of nocardiosis. In conjunction with other drugs they are also used to treat fal-

TABLE 26-13. Common Sulfonamides

Generic name	*Trade name(s)*	*Usual adult dosages*	*Uses/characteristics*
Sulfachloropyridazine	Sonilyn	2-4 g/day in 3-6 doses	Acute, chronic UTI, malaria, or toxoplasmosis (as adjuncts to other therapy)
Sulfacytine	Renoquid	250 mg 4×/day po	Short-acting, UTI
Sulfadiazine	Suladyne (125 mg sulfadiazine and sulfamethizole)	2 Tabs 4×/day po	Acute UTI
Sulfamethizole	Azotrex Suladyne Thiosulfil	0.5-1.0 g 3-4×/day po	UTI
Sulfamethoxazole	Gantanol	1 Tab q12h (160 mg trimethoprim, 800 mg sulfamethoxazole); 8-20 mg/kg/day I.V.	UTI
Co-trimoxazole	Septra Bactrim		UTI; *Pneumocystis carinii* pneumonia; otitis media due to *H. influenzae* or *S. pneumoniae;* shigellosis
Sulfisoxazole	Gantrisin	1-1.5 g q4-6h po; 100 mg/kg/day I.M., SQ, I.V.	Short-acting, UTI
Silver sulfadiazine	Silvadene Cream	Applied topically 1-2×/day	To prevent sepsis in burns
Trisulfapyrimidines (combines sulfadiazine, sulfameruzine, sulfamethazine)	Terfonyl	2-4 g initially, 2-4 g/day po	Recurrent, chronic UTI with susceptible *E. coli, Klebsiella, P. mirabilis;* otitis media with *H. influenzae* (in combination with other antimicrobials)

ciparum malaria and toxoplasmosis, but are unfortunately no longer an acceptable therapy for meningococcal disease because of the prevalence of resistant *N. meningitidis*. Silver sulfadiazine is applied topically to minimize bacterial colonization of burns. PAS was used commonly in the treatment of tuberculosis in the past, but it has been replaced by more effective drugs (rifampin, ethambutol). Use of sulfones (especially Dapsone, but also acetosulfone, glucosulfone, and sulfoxone sodium) is limited to treatment of leprosy and occasionally malaria. Sulfisoxazole can be given as an ophthalmic solution for eye infections.

Dosage and administration. All of these drugs are generally given orally as they are well absorbed by that route. Sulfones penetrate into all body fluids including the spinal fluid. Unfortunately, organisms causing central nervous system infections are usually resistant to sulfonamides. Some preparations (sulfisoxazole) can be given parenterally as well.

Adverse effects. Direct toxic effects of sulfonamides are very rare, but hypersensitivity reactions can occur in up to 3 percent of individuals. Common manifestations are rash, fever, eosinophilia, and rarely, Stevens-Johnson syndrome characterized by fever, and vesicular and hemorrhagic rash, which can be fatal. Liver damage, agranulocytosis, and thrombocytopenia are other rare allergic reactions. Long-acting sulfonamides are not available in the United States because of the higher incidence of these hypersensitivity reactions, but they also occur after use of short-acting compounds.

Dark-skinned persons (about 10 percent of black males, but also Asians and those from Mediterranean countries) may have a genetic deficiency of glucose-6-phosphate dehydrogenase activity, an enzyme necessary for the integrity of the red blood cells. In such individuals, sulfonamide therapy results in hemolytic anemia. Patients with extensive burns being treated with silver sulfadiazine cream can develop metabolic acidosis.

With older preparations of sulfonamides, insoluble precipitates collected as crystals in the urine and caused obstruction. However, newer preparations like sulfisoxazole are much more soluble, therefore, this problem is minimal. Nevertheless, patients receiving sulfonamides should be encouraged to maintain a high fluid intake. Solubility is also increased at an alkaline pH, so the patient can take bicarbonate to increase the pH of the urine. Trisulfapyrimidines, by combining three sulfones, allow full therapeutic benefits while minimizing the risk of renal damage.

PAS is very irritating to the gastrointestinal tract and also occasionally causes allergic reactions, liver damage, and thyroid enlargement. Sulfones are quite toxic, often causing anemia, liver damage, erythema nodosum, and agranulocytosis. They sometimes cause a neuritis that can lead to muscle paralysis unless the drug is discontinued.

TETRACYCLINES

Pharmacologic characteristics. Tetracycline was found in 1948 during a systematic screening of soil throughout the world in search of microorganisms that might be used for antimicrobials. There are about 3000 tetracycline compounds available, but only a few are useful for clinical therapy. Cross resistance generally occurs between tetracyclines. Tetracycline is the least expensive antimicrobial available. The mode of action is the inhibition of protein synthesis in a manner similar to aminoglycosides, primarily binding to the 30S ribosomal subunit as well as to mRNA. Tetracyclines are selectively toxic to bacterial cells (as opposed to human cells) in part because the drug differentially enters bacterial cells. In therapeutic doses tetracycline is bacteriostatic. Tetracyclines are excreted in the kidneys and in the feces after passing through the bile. They are well distributed in the body, but the concentrations in spinal fluid are much lower than those in blood.

Uses. Various preparations of tetracyclines are available (Table 26-14), all of which have the same antimicrobial spectrum, but slightly different pharmacologic characteristics. Tetracycline is a drug of choice for *M. pneumoniae* infections, some rickettsial and chlamydial diseases, and is a good alternative to penicillin for susceptible organisms, especially the gonococcus and *H. influenzae*. Some gram-negative bacteria are also susceptible but at the serum levels obtained most enteric organisms are

TABLE 26-14. Tetracyclines*

Generic	*Brand*	*Characteristics*
Chlortetracycline	Aureomycin	Least stable
Demeclocycline	Declomycin	Higher incidence of photosensitivity
Doxycycline	Vibramycin	Long acting, half-life about 15 hours; can be used in patients with renal insufficiency
Methacycline	Rondomycin	
Minocycline	Minocin Vectrin	Causes vertigo, long acting
Oxytetracycline	Terramycin	
Tetracycline hydrochloride	Achromycin Panmycin Sumycin Tetracyn Mysteclin	Least expensive

*Usual adult dosage: 250 mg q6h po; 250 mg I.M. qd; 250-500 mg I.V. q12h

resistant. Tetracylines may be used for urinary tract infections. Many strains of *Bacteroides* are now resistant, but tetracycline combined with chloramphenicol can sometimes eradicate infections with anaerobes. Thirty-five to 75 percent of streptococci, as well as many *S. aureus,* are resistant to tetracyclines.

Minocycline can be a useful alternative to sulfonamides to eradicate the carrier state of meningococci. Tetracycline is sometimes used for long-term treatment of acne vulgaris. Because such low doses are used there do not seem to be any adverse drug effects, though occasionally superinfections result. Its effectiveness may be the result of its inhibition of the purified lipase enzyme of *Propionibacterium acnes*. Doxycycline has been used as prophylaxis against "traveler's diarrhea" but its efficacy is unproven. Because doxycycline is excreted primarily by the liver, it may be a better tetracycline to use for patients with renal dysfunction.

Dosage and administration. Since the half-lives of chlortetracycline, oxytetracycline, and tetracycline hydrochloride are short, oral doses must be taken every 6 hours. Demeclocycline and methacycline are intermediate, given 2 to 4 times per day, and minocycline and doxycycline are long acting and can be administered every 12 hours. Oral absorption is good, and all body tissues are contacted except the central nervous system (where the drug penetrates poorly). The usual adult oral dosage is 1 to 2 g divided into 2 to 4 doses. Dosage intramuscularly is 250 mg per day and intravenously is 250 to 500 mg every 12 hours. Antacids and milk products interfere with gastrointestinal absorption; tetracyclines should not be given with them.

Adverse effects. An important side effect is brown tooth discoloration if the drug is given during the time of enamel formation. Tetracyclines are hence contraindicted in the last half of pregnancy and in the first years of life. This effect is dose-related. Tetracyclines are irritating both intravenously, causing a high incidence of thrombophlebitis, and orally, causing gastric distress. Gastrointestinal symptoms can be minimized if the drug is taken immediately after meals. Minocycline therapy causes a rather high incidence of dizziness. It can be so disturbing that the drug must be discontinued. After high intravenous doses of tetracyclines, hepatic toxicity can result. Though this serious problem is rare, mortality is high. Tetracyclines can also cause photosensitivity with increased susceptibility to sunburn; patients should be cautioned to avoid direct exposure to sunlight.

URINARY TRACT ANTISEPTICS

Pharmacologic characteristics. There is a group of antimicrobial agents that are distributed poorly or in inactive form in body tissues, but are concentrated in high therapeutic doses in the urine. Thus, their major clinical usefulness is in the treatment of urinary tract infections (Table 26-15). Such drugs include nalidixic and oxolinic acids, nitrofurantoin, methenamine mandelate and hippurate. Nalidixic and oxolinic acid inhibit semiconservative bacterial DNA synthesis. Cinoxacin is a synthetic compound structurally similar to nalidixic acid. The exact mechanism of action for nitrofurantoin is not understood, but it does inter-

TABLE 26-15. Urinary Antiseptics

Generic	*Brand*	*Effective against*	*Usual adult dosages*	*Side effects*
Methenamine hippurate	Hiprex Urex	Most gram-negative bacteria, but those that split urea (*Proteus*, some *Pseudomonas*, *Klebsiella*)	2 g/day po	Minor; occasional GI upset or rash
Methenamine mandelate	Donnasep Mandelamine Uroqid-acid	As above	4 g/day po	As above
Nalidixic and oxolinic acid	NegGram	Most gram-negative bacteria except *Pseudomonas*	1 g 4×/day po	CNS reactions: increased intracranial pressure, seizures, psychosis, headache, vertigo; visual disturbances; photosensitivity; GI upset
Nitrofurantoin	Cyantin Furadantin Macrodantin	*E. coli* and some other gram-negative bacteria, enterococci, *S. aureus*	50-100 mg 4×/day po	Nausea, vomiting; peripheral neuritis; pulmonary reactions: pneumonitis, fibrosis; rash; hemolytic anemia

fere with several bacterial enzyme systems. Methenamine does not act directly to kill bacteria, but if the urine has an acid pH, it breaks down into ammonia and formaldehyde. The formaldehyde denatures bacterial proteins and is bactericidal.

Uses. Nalidixic acid and cinoxacin are active against most gram-negative bacteria except *Pseudomonas* and can be safely used in patients with renal disease. However, bacterial resistance emerges readily and cross resistance between the two agents occurs. Nitrofurantoin is active primarily against *E. coli* and enterococci with other gram-negative bacteria often being resistant. Fortunately, the development of bacterial resistance is not a problem with nitrofurantoin. Methenamine mandelate and hippurate have the same bacterial spectrum; they are effective against most gram-negative bacteria when the urine has a pH less than 5.5 and formaldehyde is formed. Many *Proteus* species, some *P. aeruginosa,* and *Klebsiella* split urea, causing an alkaline urine that results in resistance to these antimicrobials. Methenamine hippurate contains hippuric acid, which helps maintain an acid pH in the urine. Because of its mechanism of action, no resistance to methenamine developes.

Dosage and administration. All of these agents are well absorbed from the gastrointestinal tract and concentrate in the urine. Drinking cranberry juice will help to maintain a urinary acidic pH in individuals taking methenamines. It is important for individuals taking urinary antiseptics to maintain a good fluid intake, as bacteria are then more rapidly flushed out of the urinary tract. Oxolinic acid has a longer half-life than nalidixic acid and can be administered only twice instead of 4 times per day. Nitrofurantoin can be given intravenously.

Adverse effects. Side effects of nalidixic acid and oxolinic acid are varied. Central nervous system complaints, only reported with nalidixic acid, can be mild (headache, drowsiness, or dizziness) or severe (seizures, toxic psychosis, and increased intracranial pressure). These reactions are more common if the recommended dose is exceeded (overdosage) and are reversible within hours after the drug is discontinued. The usual gastrointestinal symptoms of nausea, vomiting, and diarrhea can occur, as can dermatologic reactions. Photosensitivity can be a problem; individuals should be warned to limit sun exposure. Bulbous lesions can continue to erupt for several months after discontinuation of the drug. Rarely metabolic acidosis or blood dyscrasias such as thrombocytopenia occur. Hemolytic anemia after treatment with nalidixic acid or nitrofurantoin may be associated with the glucose-6-phosphate dehydrogenase deficiency discussed in conjunction with sulfonamides. Cinoxacin has similar, though generally milder, side effects.

Complications accompanying use of nitrofurantoin include nausea, vomiting, pulmonary infiltration and pneumonitis, and peripheral neuritis. If paresthesias occur during use of nitrofurantoin, the drug should be stopped since neurologic damage can be serious. Nitrofurantoin may interfere with oral contraceptive therapy. The very few side effects of methenamines include an occasional gastric upset or rash.

VANCOMYCIN

Pharmacologic characteristics. Vancomycin is also, like many antimicrobials, derived from a *Streptomyces* mold. It works by inhibiting the formation of peptidoglycan by interfering with the enzyme peptidoglycan synthetase. It is excreted primarily by the kidneys and can diffuse into the spinal fluid when the meninges are inflamed.

Uses. Vancomycin has a limited but invaluable use as a potent antistaphylococcal agent, and should be reserved solely for use in patients with disease caused by staphylococci resistant to other forms of therapy (e.g., methicillin-resistant staphylococci; MRSA). The combined effect of vancomycin and cephalothin or cefamandole is synergistic against MRSA. Vancomycin is effective for staphylococcal endocarditis, osteomyelitis, pneumonia, septicemia, or other infections. It can also be used for enterococcal or streptococcal endocarditis if an alternative to penicillin is required. It is the drug of choice for pseudomenbranous colitis.

Dosage and administration. For systemic disease, vancomycin must be administered intravenously since gastrointestinal absorption is very poor and since tissue necrosis may occur if the drug is given intramuscularly. Dosage is 0.5 to 1 g every 6 to 12 hours. The same dose of drug is administered orally for staphylococcal enterocolitis. Rapid intravenous injection can cause release of histamine and potential cardiovascular collapse. If the patient's skin becomes flushed during treatment, it is probably due to histamine release. In patients in renal dialysis, only 1 dose every 10 days is needed.

Adverse effects. Vancomycin is a highly toxic drug, sometimes causing damage to both the eighth cranial nerve and the kidneys. The same precautions described under "Clinical Implications" for aminoglycosides also apply to vancomycin. It is also very irritating to the veins, with thrombophlebitis, drug fever, occasionally hypotension occurring. Vein irritation can be lessened by adding a small amount of corticosteroid (hydrocortisone) to the infusion. When the drug is administered orally for enterocolitis, vomiting is common. It may be necessary to insert the drug through a nasogastric tube that extends beyond the pylorus in order to ensure that the drug is retained.

ANTIFUNGAL AGENTS

The major agents that act specifically against fungal cells are the polyenes (amphotericin B and nystatin), flucytosine, griseofulvin, candicidin, clotrimazole, ketoconazole, miconazole, and tolnaftate. Amphotericin B and nystatin exhibit selective toxicity to fungi because they act by binding to sterols on the fungal cell membrane, changing cell permeability and causing cell death. Since human and most bacterial cells do not contain sterols, they are resistant to polyenes. Flucytosine results in fungal killing because it is metabolized in the fungal cell (but not in human cells) into 5-fluorouracil, which probably results in the production of faulty RNA. It is fungistatic, used for generalized cryptococcal and candidal infections such as meningitis, pneumonia, endocarditis, or septicemia.

Griseofulvin, derived from *Penicillium* mold, inhibits fungal division, but is useful only in the treatment of dermatophytes. Candicidin is used to treat vaginal infections caused by *Candida* species. Miconazole and clotrimazole, which alter cell membrane permeability, are also used for athlete's foot and other dermatophyte infections as well as for vaginal candidiasis. Miconazole is administered parenterally to treat systemic mycoses. Ketoconazole causes damage to fungal cell walls by impairing synthesis of an essential chemical ergosterol (action similar to that of amphotericin). It is used orally for systemic infections such as candidiasis, cocidioidomycosis, histoplasmosis, or paracoccidioidomycosis. These agents are discussed in Chapter 13.

ANTIVIRAL AGENTS

Viruses do not respond to the antimicrobial effects of conventional therapy because their mode of replication and metabolic functioning are unaffected. There are currently three types of agents used in antiviral therapy: amantadine, thiosemicarbazones, and certain thymidine inhibitors such as idoxuridine and cytarabine. Even these agents, however, are very limited in usefulness and effectiveness.

Amantadine. Amantadine is a synthetic amine that inhibits the replication (probably by preventing penetration or uncoating) of certain myxoviruses such as influenza A (but not B), parainfluenza, and rubella. Its only clinical application is in prophylaxis against influenza A. There is some evidence, though it is conflicting, that amantadine is useful in minimizing the duration and severity of symptoms in clinical influenza. Thus, its use at the current time is of minimal clinical importance. It is given orally and excreted by the kidneys. Usual dose is 200 mg per day for 10 days after known exposure, or up to 30 days if there are repeated exposures to influenza A. Toxic effects include confusion, depression, dizziness, psychosis, hallucinations, hypotension, and urinary retention. Two analogues of amantadine, cyclooctylamine and rimantadine, are also being tested. Amantadine has been abused for its hallucinogenic characteristics.

Thymidine inhibitors. Several compounds, idoxuridine, cytarabine, vidarabine, acyclovir, deoxy-O-glucose, ribavirin, and trifluridine are used as antiviral agents because they inhibit the replication of some DNA viruses by replacing thymidine in the nucleotide sequences. Potentially susceptible viruses are herpes simplex, SV-40, polyoma, pox, vaccinia, and varicella-zoster.

Idoxuridine (Stoxil) is used as an ophthalmic solution or ointment to treat eye infections due to herpes simplex. Cytarabine (cytosine arabinoside) has been given systemically for treatment of varicella-zoster infections in those with reduced immunologic functioning. It is also used in the treatment of leukemias. It is inactivated by the intestinal tract and must be given intravenously. Vidarabine (adenine arabinoside or Vira-A) has also been used both as an anticancer and antiherpes agent. It is given I.V. for encephalitis due to herpes. It is effective in ophthalmic solutions for herpes keratitis and is much less toxic than the other agents when given systemically, although intravenous administration frequently causes vomiting.

Acyclovir is administered topically for herpes genitalis and mild mucocutaneous infections in the immunocompromised host, and I.V. for more severe herpes infections. Trifluridine inhibits viral replication and is used as an ophthalmic solution to treat infections due to herpesviruses and vaccinia. The major side effect of the thymidine inhibitors is bone marrow suppression. Thymidine inhibitors can also cause leukopenia, loss of hair, stomatitis, liver injury, and possibly carcinogenesis.

Thiosemicarbazones. Methisazone was first found to be effective against the poxvirus in mice in 1950. Some early clinical trials found the drug to be efficacious in protecting against smallpox among those exposed, while more recent trials have failed to show any effect. The drug has no usefulness once the disease is active. Since smallpox has been declared to be eradicated from the world, the usefulness of methisazone is limited indeed. It is taken by mouth and its major side effect is vomiting.

CLINICAL IMPLICATIONS

There are three basic responsibilities of the clinician monitoring antimicrobial therapy:

1. To administer the drugs correctly: by appropriate route, at appropriate times, prepared according to directions
2. To assess the effectiveness of the drug in treating the infectious disease for which it was intended
3. To evaluate the patient for signs and symptoms of untoward or toxic side effects of the agent used

Before administration of any agent, however, the clinician should ascertain whether the patient has a history of allergy to the drug ordered.

Correct Administration

The goal of therapy is to maintain adequate blood and tissue levels of the antimicrobial so that organisms will be killed with minimal toxic effects to the host. Because of varying absorption times, drugs must be given in different ways; some must be infused slowly over 30 to 60 minutes so that toxic peak levels are avoided (aminoglycosides), others must be given between meals (erythromycin), and still others are better taken with meals to avoid gastrointestinal disturbance and enhance absorption.

Certain agents have special requirements for administration. For example, aminoglycosides are most effective in an alkaline medium. Hence, patients with urinary tract infections should not drink cranberry or prune juice or eat foods that produce an acid urine if they are being treated with aminoglycosides. On the other hand, methenamine mandelate is active only when the urine is acidic. It has been recommended that, whenever possible, bacteriostatic water rather than sterile water be used for injection as the 0.9 percent benzyl alcohol in bacteriostatic water decreases the chances of contamination of solutions.

To ensure that adequate drug levels are maintained, blood is often drawn at the peak of drug activity and again at the lowest point (trough). If peak levels are ordered, they must be drawn at a time coordinated with the last drug dose. Peak blood levels are usually reached within minutes after intravenous administration, within 2 to 4 hours after oral administration, and 1 to 2 hours after intramuscular or subcutaneous administration. These times differ, however, and accurate information must be obtained from the drug package inserts or from a pharmacist. Trough levels are usually drawn about 30 minutes before the next dose of drug is due.

Effectiveness of the Drug

If the patient does not show beginning resolution of infection within 5 days of the initiation of therapy, there is something wrong with the treatment or with the patient's ability to respond. Possible causes of such nonresponse are incorrect diagnosis (may be a viral disease), inappropriate drug (organism not susceptible), inadequate dose, resistance developed during treatment, infection caused by several organisms, superinfection with another organism, fever due to drug, poor host defenses, abscess formation requiring drainage, or noncompliance with therapy. The clinician can help to assess the status of the infection and the patient's clinical response. The patient's temperature should be taken at least 4 times each day at regular intervals so that daily patterns and fluctuations can be observed. It is also imperative that cultures taken be well collected. *E. coli,* for example, can multiply rapidly in a contamined urine specimen. Urine specimens should be sent to the laboratory as soon as possible after collection, at least within an hour, and in the meantime they should be refrigerated. To calculate dosages, it is necessary that the patient be accurately weighed just prior to initiation of drug therapy.

Evaluate Side Effects

Any time a drug that is excreted by the kidneys is administered, the patient's renal function must be monitored to prevent both drug overdosage and renal toxicity. The following are guidelines for prevention of eighth cranial nerve damage and renal toxicity, both of which are especially common with aminoglycoside therapy:

1. Assess patient's hearing and balance (eighth cranial nerve function) before and during treatment.
   a. Remember that 30 percent of people over age 65 already have some hearing deficit.
   b. Use an audiometer or talk in a normal voice as you enter the room to determine whether you can be heard by the patient.
   c. Signs of vestibular dysfunction include vertigo, tinnitus ("roaring noise" in ears), nystagmus, loss of balance with unsteady gait.
   d. Monitor BUN and creatinine, urine output, and fluid retention at the onset of treatment and at least 2 to 3 times per week during treatment.
2. Maintain appropriate peak and trough blood levels.
   a.

	Peak (μg/ml)	*Trough (μg/ml)*
Amikacin	10-30	Less than 10
Gentamicin	5-10	Less than 2
Tobramycin	5-10	Less than 2

   b. Troughs greater than those listed can increase the probability of kidney damage; peaks greater than those listed can increase the probability of both kidney and eighth cranial nerve damage.
3. Help the patient to maintain adequate hydration to prevent the accumulation of the drug in tissues and to enhance renal function.
4. Be aware of factors that increase the risk of toxic effects:
   a. Impaired renal function
   b. Prior administration of aminoglycosides (cumulative toxicity occurs)
   c. Simultaneous administration of other ototoxic drugs (furosemide, ethacrynic acid) or nephrotoxic drugs (penicillins, cephalosporins)
5. Do not mix animoglycosides in the same intravascular line with other antimicrobials or incompatible drugs: cephalothin, chloramphenicol, ticarcillin, amphotericin B, ampicillin, heparin, vitamin B, vitamin C, or dopamine.

Most antimicrobials given intravenously are irritating to veins. Bolus administration is almost universally to be avoided, both to minimize vein damage and to ensure that blood levels do not rise too rapidly. Other guidelines for use of intravenous antimicrobials are found in Table 26-16 and the interactions of antimicrobials with other drugs are listed in Table 26-17.

Symptoms such as rash, flushing, joint swelling, pruritus, hypotension, or wheezing while the patient is receiving or has recently received antimicrobials, especially penicillins, should alert the clinician to a possible allergic reaction. A careful description of signs and symptoms should be reported to the physician immediately. When a hypersensitivity reaction is confirmed, the patient needs to be warned about future use of the same and related compounds. As a rule, any new or unusual symptom or changes in hematologic status that occur while the patient is receiving antimicrobials must be evaluated for possible relationship to the therapy. Symptoms such as nausea, vomiting, diarrhea, and oral or genital candidiasis may accompany administration of most antimicrobials. Finally, the clinician needs to be aware of the possibility of

(text continues on page 709)

TABLE 26-16. I.V. Antibiotics: Reducing the Problems

When giving	*Reconstitute with*	*Dilute with dextrose or saline solution to get*	*Infuse over*	*Comments*
Ampicillin sodium (Amcill-S, Omnipen-N, Penbritin-S, Polycillin-N, Totacillin-N)	BWFI*	50 to 100 ml, preferably with saline	10 to 30 minutes	Use solution within 1 hour after reconstitution. Drug degrades much more slowly in saline than in dextrose. If diluent is saline, give cautiously to cardiac patients. Usual dose is 250 to 500 mg q6h (maximum, 2 g q6h)
Carbenicillin disodium (Geopen, Pyopen)	BWFI	At least 1 g/ 20 ml	15 to 30 minutes. Infuse high single doses (4 g or more over 1 hour)	Watch fluid status in patients getting high doses; each gram contains 4.7 mEq of sodium. Usual dose is 300-500 mg/kg per day (20 gm to a maximum of 40 gm)
Cefazolin sodium (Ancef, Kefzol)	BWFI or normal saline	50 to 100 ml (for 500 mg to 1 g)	15 to 30 minutes	Usual dose is 500 mg to 1 g q6h to q8h. Maximum dose is 6 g/day
Cephalothin sodium (Keflin)	BWFI	At least 1 g/ 50 ml	15 minutes (1 gram)	Small-vein needles are less likely to produce phlebitis. Usual dose is 4 to 12 g per day in divided doses

TABLE 26-16. *(continued)*

When giving	*Reconstitute with*	*Dilute with dextrose or saline solution to get*	*Infuse over*	*Comments*
Chloramphenicol sodium succinate (Chloromycetin, Mychel-S)	BWFI or dextrose, 10 ml to make 100 mg/ml	50 ml	10 to 15 minutes	Each vial contains 1 g. Usual dose is 50 mg/kg per day, q6h
Clindamycin phosphate (Cleocin)	(Already in solution)	50 ml (for 300 mg)	10 minutes (for 300 mg)	Usual dose is 600 mg to 1.2 g per day in 2, 3, or 4 equally divided doses
Erythromycin lactobionate (Erythrocin)	SWFI† (10 ml for 1 g vial)	100 ml (for 250 to 500 mg)	20 to 60 min (for 250 to 500 mg)	Avoid reconstituting with BWFI or saline; precipitation may result. Doses larger than 100 ml cannot be given by volume-control sets. One-gram doses must be given in 250 ml of saline or dextrose solution, phlebitis is common. Usual dose is 15-20 mg/kg/day
Gentamicin sulfate (Garamycin)	(Already in solution)	1 mg/ml	30 to 60 minutes	Reduce dose for patients with renal impairment. Usual dose is 3-5 mg/kg per day in 3 equally divided doses
Kanamycin sulfate (Kantrex)	(Already in solution)	At least 200 ml (for 500 mg)	2 to 4 ml of listed dilution per minute	Use I.V. route only if I.M. injection is not possible. Usual dose requires dilution in a bottle. Doses larger than 100 ml cannot be given by volume-control sets. Usual dose is 15 mg/kg per day in 2 equally divided doses
Nafcillin sodium (Unipen, Nafcil)	BWFI	100 ml	30 minutes	Usual dose is 500 mg to 1 g q4h
Oxacillin sodium (Bactocill, Prostaphlin)	BWFI or saline	25 ml (for 1 gram)	15 to 30 minutes	Usual dose is 250 mg to 1 g q4-6hr (maximum, 8 g per day)

TABLE 26-16. *(continued)*

When giving	*Reconstitute with*	*Dilute with dextrose or saline solution to get*	*Infuse over*	*Comments*
Penicillin G, potassium (multisource)	BWFI, saline, or dextrose solution	50 to 100 ml	15 to 30 minutes	Monitor electrolytes; each 1,000,000 units contains 1.7 mEq of potassium. Use with caution for patients with renal impairment. Usual dose is 300,000 to 1.2 million units per day in divided doses
Tetracycline hydrochloride (Achromycin intravenous)	BWFI, saline, or dextrose solution	At least 10 mg/ml	5 minutes maximum for 100 mg	Maximum dose is 2 g per 24 hours. Usual dose is 250 to 500 mg per day q12h
Vancomycin hydrochloride (Vancocin)	BWFI	100 to 200 ml	20 to 30 minutes	To minimize phlebitis (a very common problem), dilute 200 ml and rotate infusion sites. Usual dose is 2 g per day in 2 to 4 divided doses

*BWFI = bacteriostatic water for injection.
†SWFI = Sterile water for injection.
Reprinted with permission of the publisher from Stricklorius C: A safe approach to intravenous antibiotics. RN 1981; 44:37–39; as adapted from a guide developed at the Hospital of the University of Pennsylvania, Philadelphia.

superinfection, especially with fungi or normal flora, when the patient has been on a regimen that includes use of broad-spectrum antimicrobial agents. The health care practitioners can help to prevent the dissemination of resistant organisms by careful prevention of cross infection between patients. Inappropriate antibiotic usage has been decreased by intervention programs and utilization review mechanisms. Infection surveillance personnel may become involved in such programs.

Prophylactic Antimicrobial Therapy

Antimicrobial agents are sometimes used before infection develops, especially prior to major surgical procedures. Prophylaxis is not recommended in clean surgical cases because the costs and risks outweigh the benefits. Prophylactic antimicrobial therapy is indicated in cases when

1. The incidence and consequences of infection are great

TABLE 26-17. Adverse Interactions Between Antimicrobial and Other Drugs

Antimicrobial +	*Interacting drug →*	*Effect*
Aminoglycosides	1. Cephalothin, cephaloridine	Nephrotoxicity
	2. Succinyl choline	Neuromuscular blockade
	3. Carbenicillin	Inactivation of gentamicin in patients with renal failure
Cephalosporins	1. Probenecid	↑ blood levels of cephalosporins
Chloramphenicol	1. Alcohol	Antabuse-like symptoms
	2. Coumarin	↑ anticoagulant effect
	3. Dilantin	Dilantin toxicity
Griseofulvin	1. Anticoagulants	↓ anticoagulant effect
	2. Phenobarbital	↓ in absorption of griseofulvin
INH	1. Alcohol	↑ Incidence of hepatitis
	2. Antacids (aluminum)	↓ INH effect
	3. Antabuse	Psychotic episodes
	4. Dilantin	Dilantin toxicity
PAS	1. Aspirin (ASA)	↑ blood levels of ASA
	2. Dilantin	Dilantin toxicity
	3. Probenecid	PAS toxicity
Penicillin	1. Aspirin	↑ half-life of penicilliln
	2. Indomethacin	
	3. Phenylbutazone	
	4. Probenecid	
Rifampin	1. Anticoagulants	↓ anticoagulant effect
	2. Contraceptives	↓ contraceptive effect
Sulfonamides	1. Anticoagulants	↑ anticoagulant effect
	2. Dilantin	Dilantin toxicity
	3. Tolbutamide	↑ hypoglycemia
Tetracycline	1. Antacids	↓ in absorption of tetracyclines
	2. Barbiturates	↓ effect of tetracycline (doxycycline)
	3. Iron (oral)	↓ effect of tetracycline

References
Kabins SA. Interactions Among Antibiotics and Other Drugs. *JAMA* 1972;219:206-212.
Adverse Interactions of Drugs. *Medical Letter* 1975;17:17-24.
Reprinted with permission of the publisher from Eisenberg M. Manual of antimicrobial therapy and infectious diseases. Seattle: University of Washington School of Medicine, 1979.

2. The offending organisms are predictable and susceptible to antimicrobial therapy
3. The period of exposure to possible infection is brief

Thus, antimicrobial prophylaxis is recommended in the following types of surgery: coronary artery bypass graft, heart valve replacement, orthopedic surgery such as joint replacement, colorectal surgery, head and neck surgery when the oral cavity or pharynx is entered, vaginal or abdominal hysterectomy, ruptured viscus, traumatic wounds, and, in high-risk patients, gastroduodenal, biliary tract, surgery or cesarean sections (Fernsebner 1982).

The timing of such therapy is crucial, since the period when drugs are at their peak effectiveness is short. Sometimes antimicrobial agents are given intravenously with anesthesia to ensure that correct blood and tissue levels are achieved during the period of greatest risk for contamination. It is vital in that case that operating room personnel observe for adverse reactions, especially since the patient will be anesthetized and unable to communicate.

SUGGESTED READINGS

Antimicrobial prophylaxis. Med Letter on Drugs Therapeutics 1981;23:80.

Arking L, Saravolatz L. Antimicrobial treatment: a coordination challenge for nursing. Nurs Clin N Am 1980;15:689-702.

Barr NL. The penicillins. Ear Nose Throat J 1981;60:215-218.

Calbreath DF. Assessing gentamycin toxicity. J Nurs Care 1979;12:7.

Calderwood SB, Moellering RC. Common adverse effects of antibacterial agents on major organ systems. Surg Clin N Am 1980;60:65-81.

Cortez LM, et al. Penicillins: which, when, how, why? Pat Care 1980;14:14-77.

Craig WA, et al. Hospital use of antimicrobial drugs. Ann Intern Med 1978;89 (Part 2):793-795.

Done AK. More anti-infectives: fungicides and antivirals. Emerg Med 1981:13:51-53+.

DuPont HL. Rational antibiotic therapy. Top Clin Nurs 1979;1:45-51.

Eisenberg M. Manual of antimicrobial therapy and infectious diseases. Seattle: University of Washington School of Medicine, 1979.

Fernsebner B. Antimicrobial therapy for surgical patients. AORN J 1982;36:479-486.

Finkel MJ. Magnitude of antibiotic use. Ann Intern Med 1978;89 (Part 2):791-792.

Finland M. New developments in antibiotics. Ann Intern Med 1978;89 (Part 2):849-853.

Harding GM, et al. Prospective, random comparative study of clindamycin, chloramphenicol, ticarcillin, each in combination with gentamicin, in therapy for intraabdominal and female genital tract sepsis. J Infect Dis 1980;142:384-393.

Hayden FG, Hall WJ, Douglas RG. Therapeutic effects of aerosolized amantadine in naturally acquired infection due to influenza A virus. J Infect Dis 1980;141:535-542.

Hodes HL. Penicillin prophylaxis and neonatal streptococcal disease. Hosp Pract 1980;15:115-118+.

Hofstra W, deVries-Hospers HG Van der Waaif D. Concentrations of nystatin in faeces after oral administration of various doses of nystatin. Infec 1979;7:166-170.

Kass EH. Antimicrobial drug usage in general hospitals in Pennsylvania. Ann Intern Med 1978;89 (Part 2):800-801.

Langslet J, Habel ML. The aminoglycoside antibiotics. Am J Nurs 1981;81:1144-1146.

Lorian V, ed. Antibiotics in laboratory medicine. Baltimore: Williams and Wilkins, 1980.

Luft FC, Evans AP. Comparative effects of tobramycin and gentamycin on glomerular ultrastructure. J Infect Dis 1980;142:910-914.

Magnussen MH. Aminoglycoside therapy. Nurs 80 1980;10:82-88.

Mar DD. Drug-induced hepatotoxicity. Am J Nurs 1982;81:124-126.

Mar DD. Penicillins—the fifth generation. Am J Nurs 1981;81:2072-2073.

McFarland JJ. The tetracyclines. Ear Nose Throat J 1981;60:226-227.

Meade RH. Antimicrobial spectrum, pharmacology and therapeutic use of erythromycin and its derivatives. Am J Hosp Pharm 1979;36:1185-1189.

Mirtallo JM. Antimicrobial therapy of immunosuppressed patients with malignant diseases. Hosp Form 1979;14:304-311+.

Nichols RL. Use of prophylactic antibiotics in surgical practice. Am J Med 1981;70:686-692.

Plotkin GR. Salient features of the newer antimicrobial agents. Hosp Form 1979;14:279+, 433+.

Polk R. Moxalactam (Moxam). Drug Intell Clin Pharm 1982;16:104-112.

Pratt WB. Chemotherapy of infections. New York: Oxford University Press, 1977.

Pumphrey RE. The cephalosporins. Ear Nose Throat J 1981;60:219-221.

Pumphrey RE. The erythromycins. Ear Nose Throat J 1981;60:222-223.

Rubens CE, McGee ZA, Farrar WE. Loss of an aminoglycoside resistance plasmid by *Serratia marcescens* during treatment of meningitis with amikacin. J Infect Dis 1980;141:346-350.

Ryan SA, Clayton BD. Handbook of practical pharmacology, 3rd edition. St. Louis: CV Mosby, 1983.

Sarson S, et al. Direct stimulation of lymphokine production by cephalothin. J Infect Dis 1980;142:265-272.

Schwartz RH. The sulfonamides. Ear Nose Throat J 1981;60:230-232.

Shapiro M, et al. Antibiotic use on the surgical services of two Jerusalem hospitals, as determined by surveillance and influenced by an intervention program. Rev Infect Dis 1981;3:754-759.

Shapiro M, et al. Use of antimicrobial drugs in general hospitals: II. Analysis of patterns of use. J Infect Dis 1979;139:698-706.

Smith H. Antibiotics in clinical practice, 3rd edition. Baltimore: University Park Press, 1977.

Stiklorius C. A safe approach to intravenous antibiotics. RN 1981;44:37–39.

Sullivan T, et al. Desensitization of patients allergic to penicillin using orally administered B-lactam antibiotics. J Allergy Clin Immunol 1982;69:275-282.

Trollfors B, et al. Quantitative nephrotoxicity of gentamicin in nontoxic doses. J Infect Dis 141:306-309, 1980.

Tuazon C. The aminoglycosides. Ear Nose Throat J 1981;60:233-238.

Vincent PC, Fabri S, Jennis F. Management of bacterial infections in severely neutropenic patients. Med J Aust 1977;(suppl)2:30-33.

Webb BM. Clindamycin-lincomycin. Ear Nose Throat J 1981;60:224-225.

Weinstein RA, et al. Endemic aminoglycoside resistance in gram-negative bacilli: epidemiology and mechanisms. J Infect Dis 1980;141:338-345.

William TW, Jackson D. New antibiotics—uses and dangers. Heart Lung 1980;9:1089-1095.

27 MICROORGANISMS IN PUBLIC HEALTH

KEY POINTS

- The major public health measures to control disease spread are food and water sanitation and immunization.
- Currently, the recommended schedule for immunizations in developed countries includes protection against seven organisms: the agents of diphtheria, pertussis, tetanus, polio, mumps, measles, and rubella.
- Active immunization against influenza, pneumococci, *Hemophilus*, the meningococci, and hepatitis B is recommended in developing countries only for special populations of individuals at high risk of infection.
- Passive immunization is practiced after exposure to botulism, diphtheria, hepatitis, rabies, tetanus, vaccinia, and varicella (the last only for immunocompromised).
- Food-borne outbreaks are preventable by proper preparation and handling of food.
- Water is purified with rapid sand filtration and chlorine, though many viruses are probably not killed by current purification processes.
- Adequate treatment and disposal of wastes are essential to destroy disease agents, convert wastes into reusable materials, and prevent pollution of large bodies of water.

Most of the text to this point has been concerned with an understanding of the agents of infectious diseases and their treatment. Little emphasis has been placed on preventive measures, and yet many of the infectious diseases that affected large population masses in the past are now rare because of public health measures such as milk pasteurizaton, water purification, and the regulation and inspection of commercial food producers and packagers. As is demonstrated in Table 27-1, such primary and sec-

TABLE 27-1. Methods of Controlling Infectious Diseases

Levels of prevention	*Definition*	*Examples*	*Responsible groups: Government agencies*	*Responsible groups: Individual health care practitioner*	*Responsible groups: Private citizen*
Primary	Prevention of disease occurrence	Monitoring commercial food production	X		
		Purification of public water supply	X		
		Waste disposal	X		
		Milk pasteurization	X		
		Immunizations	X		
		Control of vectors	X		
		Hygienic food handling			X
		Personal hygiene			X
		Prophylactic use of antimicrobials		X	
Secondary	Early detection and rapid treatment of disease	Screening programs (e.g., TB, syphilis)	X	X	
		Outbreak investigation	X	X	
Tertiary	Limitation of disease or infectivity	Use of antimicrobial agents		X	
		Barrier, isolation techniques		X	X
		Enhancing host defenses		X	X

ondary prevention is in the realm of the public domain and is administered predominantly through local, state, or federal agencies and departments, whereas tertiary prevention becomes the responsibility primarily of the private health care sector and the individual citizen.

This chapter is concerned with the control of microorganisms by public health measures (primary and secondary prevention). The role of immunizations and food and water sanitation in the prevention of infections will be discussed. Municipal and state health departments on the local level, and the Department of Health and Human Services (especially the Centers for Disease Control [CDC]) on the national level, are the agencies through which the planning and administration of activities to prevent infectious diseases originate in the United States.

IMMUNIZATION

Immunity, from the Latin word *immunitus* (meaning exemption from taxation or jurisdiction) is acquired naturally when an individual encounters an infectious agent and develops specific antibody, thereby gaining freedom from further attacks by the same agent. However, some infectious diseases are fatal, have serious sequelae, or at least result in loss of health for variable lengths of time. It is desirable therefore to incur immunity to infectious diseases without the necessity of acquiring and successfully combating a disease agent. Artificial immunization accomplishes that task in one of two ways: either by administering to the individual a very small dose of the infectious agent in a modified (attenuated) or harmless (killed) form so that the individual can produce an immune response without danger of becoming clinically ill (active immunity) or by administering to the individual preformed antibodies to the disease (passive immunity). Types of immunization and vaccination are described in more detail in Chapter 4, and Table 4-8 lists disease agents for which there are currently available immunizing preparations.

The preparation of immunizing products has improved tremendously since Jenner first experimented in 1796 with scratching the skin and placing a small dose of cowpox virus on the abraded area (variolation) in order to incur resistance to the closely related and virulent smallpox virus. There were no quality control measures, only a guess about the dose given, and no assurance that other microorganisms were not at the same time transferred. There have been episodes of production of faulty vaccines or ones containing contaminants even as late as the 1950s and 1960s that resulted in sickness and death. For this reason, there are strong federal regulatory mechanisms for vaccine manufacturers. When a vaccine is developed its efficacy must be established; in addition, tests for extraneous agents and production methods must be performed.

Immunizations for the General Population

A century ago there were only two immunizing agents: smallpox and rabies vaccines. Currently there are dozens, with many new ones being investigated, but only a few are recommended for routine general use. The Advisory Committee on Immunization Practices (ACIP) of the Public Health Service updates recommendations for the use of immunizing agents every 2 years. The currently recommended schedule for developed countries, Table 27-2, includes immunizing against diphtheria, pertussis (whooping cough), tetanus, polio, measles, mumps, and rubella.

Immunization against diphtheria, pertussis, and tetanus has been practiced for many decades. Diphtheria and tetanus toxoids are combined with pertussis vaccine in one injection (called DPT or DTP). Though the pertussis vaccine has been used since the 1930s, its exact efficacy is unknown. Rates of pertussis were declining before vaccine use was prevalent, but, after the implementation of the vaccine, they did fall faster than would have been expected if the vaccine were not successful (Fig. 27-1).

An additional advantage to using pertussis vaccine is that *B. pertussis* acts as an adjuvant to the diphtheria and tetanus agents. It does have some rare adverse effects: prolonged crying several hours after injection, convulsions, and encephalopathy. Because the incidence of whooping cough in developed countries is so low, and because the actual protective effect of the vaccine is unknown, there has been some controversy about whether the pertussis vaccine should be removed from the DPT series, but evidence to date suggests that the benefits probably still continue to outweigh the risks.

Smallpox was officially declared by the World Health Organization (WHO) to be

TABLE 27-2. Recommended Immunization Schedule in Developed Countries

Vaccine	*Age*	*Comments*
Diphtheria-pertussis-tetanus (DPT) and trivalent oral polio vaccine (OPV)	2 months and 4 months and 6 months and 18 months and 4-6 years	Persons receiving OPV can develop clinical polio (1 per 5-15 million) or shed the virus and infect close contacts (number of OPV-induced polio cases is 3 times as great among unimmunized parents of children receiving the vaccine)
Measles-mumps-rubella (MMR)	15-18 months	Immunologic response is adequate only when given after age 12 months
Diphtheria-tetanus (Td) booster	14-16 years	Often missed. Patient reminders are usually necessary. Tetanus boosters every 10 years are quite adequate

eradicated from the world in May, 1980. Up to a decade prior to that time, vaccination against smallpox was also required, but the disease in the 1970s became less common than the serious side effects (such as encephalitis) of the vaccine, so use of vaccine is now recommended only for laboratory personnel working closely with poxviruses. Only time will tell if smallpox becomes completely a disease of the past.

There are a few other diseases for which eradication is also a possibility. Such diseases usually share certain characteristics: there are no animal reservoirs; the disease (or artificial immunization against the disease) results in permanent, life long immunity; there is no relapse or carrier state; the mode of transmission is not complicated; and the disease is relatively easy to recognize and diagnose.

Measles (rubeola) has many of these characteristics and the eradication of measles in the United States has been set as a goal for the 1980s. To date, however, the goal has not been reached. A large proportion of the general population remains unimmunized, and there has been a marked change in the age-specific attack rate (Fig. 27-2). Whereas the disease was primarily one of young children, by 1977 60 percent of cases occurred in individuals past age 10 years, and 26 percent in those past age 15. This was due to the large numbers of nonimmunized susceptibles. About 60 to 70 percent of the United States population is immunized. In the late 1970s a measles outbreak occurred in a university student population that consisted of 91 percent immune individuals. Thus measles can spread despite high levels of herd immunity. Initially, measles vaccine was administered to children less than 12 months of age.

Ample evidence has since accumulated that though the vaccine is about 90 percent efficacious, when given prior to 1 year of age maternal antibody can interfere with seroconversion. Those vaccinated prior to age 12 months were found to have 7 times the attack rate of those vaccinated after age 12 months. Thus, children who were among

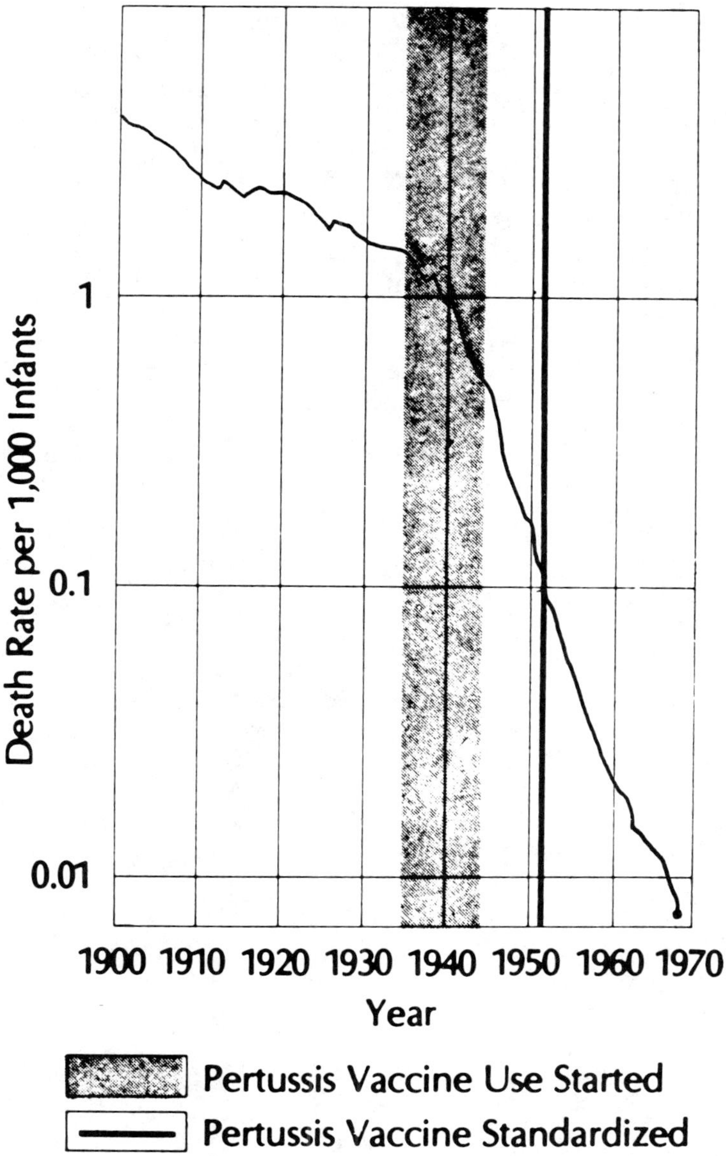

FIGURE 27-1. Before vaccine was available in the United States, pertussis mortality was already declining; widespread vaccination accelerated the trend. (Reprinted with permission of the publisher from Mortimer EA. Pertussis immunization: problems, perspectives, and prospects. Hosp Pract 1980; 15(10):106.)

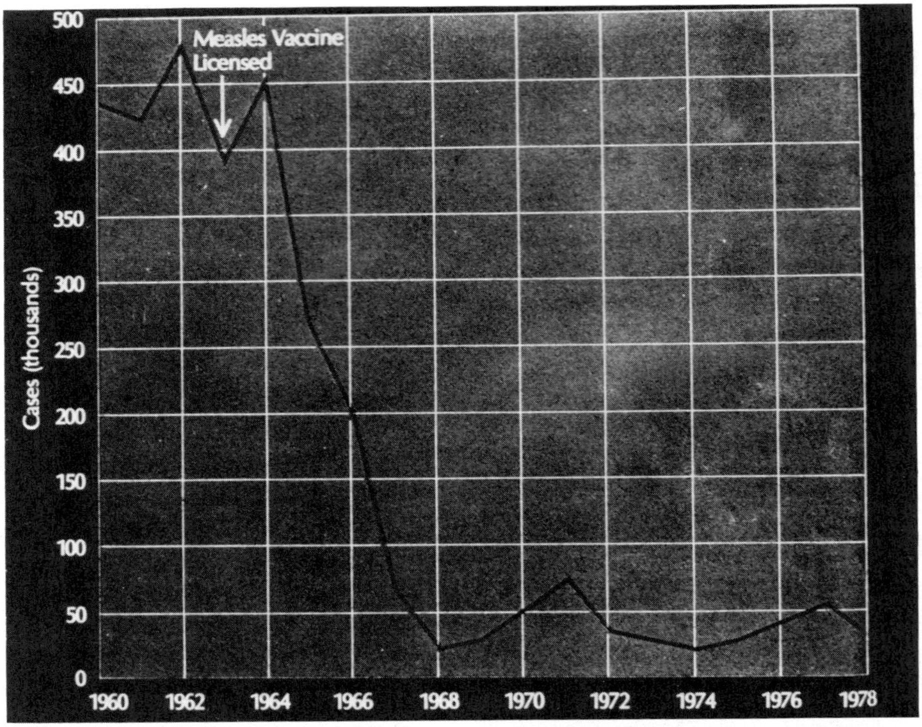
Measles Vaccine Licensed
Cases (thousands)
500
450
400
350
300
250
200
150
100
50
0
1960
1962
1964
1966
1968
1970
1972
1974
1976
1978

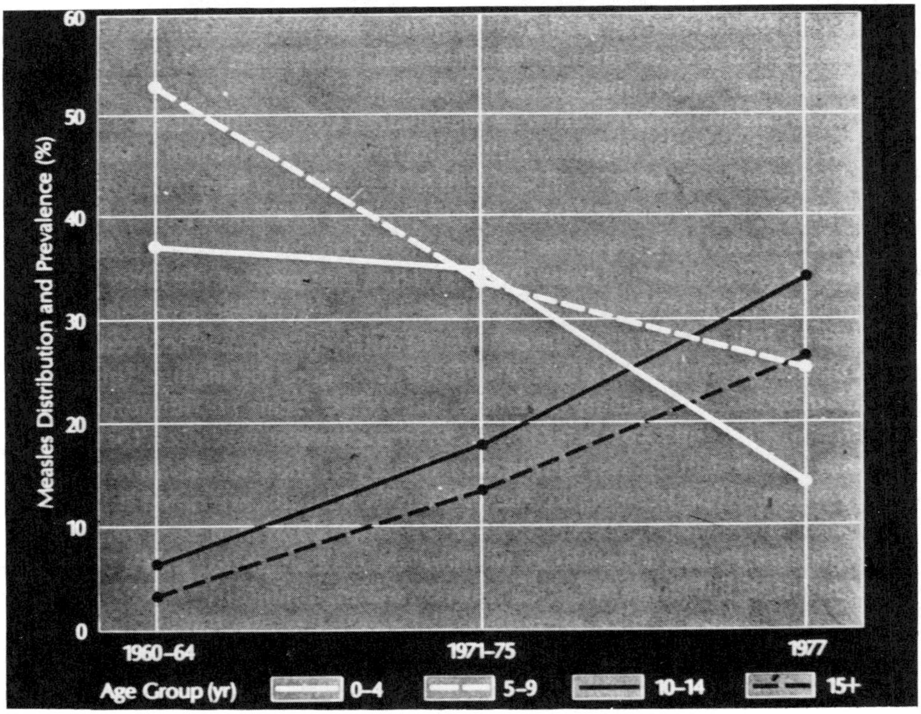
Measles Distribution and Prevalence (%)
60
50
40
30
20
10
0
1960–64
1971–75
1977
Age Group (yr)
0–4
5–9
10–14
15+

FIGURE 27-2. Since the introduction of measles vaccines in 1963, the overall prevalence of the disease in the United States has dropped by more than 90 percent, from about 400,000 cases per year in the early 1960s to about 30,000 in recent years *(top)*. At the same time, however, there has been a dramatic shift in age-specific attack rates, with relative measles prevalence in older age groups rising: in the early 1960s, less than 10 percent of cases involved patients aged 10 or older, compared with over 60 percent in 1977 *(bottom)*. (Reprinted with permission of the publisher from Cherry JD. The "new" epidemiology of measles and rubella. Hosp Pract 1980; 15(7):50-51.)

the first to receive the vaccine may be unprotected. This same phenomenon is true for mumps and rubella vaccine; all three are now given at age 15 to 18 months. The effectiveness of measles, mumps, rubella (MMR) vaccine is the same whether the agents are given together or separately. They can also be administered in conjunction with DPT or polio vaccines without loss of effectiveness.

Rubella vaccine has been available since 1965. The purpose of this vaccine is to prevent the fetal malformations of congenital rubella rather than to prevent the clinical disease, which is relatively benign except for its teratogenic effects. The vaccine is prepared as a live vaccine in human diploid cell culture and is available alone, as measles-rubella, or as MMR. Life long protection results in 95 percent of vaccinees. Its effectiveness is evidenced by the fact that in the United States there have been no epidemics among children since 1969. Side effects are mild; 40 percent of individuals have joint pain and some develop peripheral neuritis for several weeks after immunization. A potential hazard exists in that small amounts of virus are shed from the pharynx for 1 to 3 weeks after immunization, but no transmission to susceptibles has been demonstrated.

All children should be immunized against rubella, as well as any nonimmunized adults working in educational, health, or training institutions where they are likely to contact or spread rubella. The vaccine should not be given within 3 months of pregnancy; therefore, females of child-bearing age must be carefully screened before immunization. About 20 percent of nonimmunized women of child-bearing age are susceptible to rubella. Vaccination is especially important because immune serum globulin will neither prevent rubella infection nor congenital rubella even though symptoms may be lessened.

There are two types of polio vaccine available: the Salk inactivated polio virus vaccine (IPV) and the Sabin oral polio vaccine (OPV) (compared in Table 27-3). Both are trivalent; that is, they are active against all three serotypes of polio. OPV is currently the vaccine of choice for routine immunization because it incurs long-term immunity, making boosters unnecessary. It results in intestinal as well as serum immunity so that the chance of gastrointestinal infection with a "wild" strain of polio is less, and

TABLE 27-3. Comparison of Available Polio Vaccines

	Inactivated polio virus (IPV)	*Oral polio vaccine (OPV)*
Developer	Salk	Sabin
Trivalent	Yes	Yes
Long-term immunity	No	Yes
Periodic booster required	Yes	No
Prevents subsequent asymptomatic polio infection	No	Yes
Prompt immune response	No	Yes
Enteric viruses may interfere with action	No	Yes
Virus shed/transmitted	No	Yes
Recommended use	Unimmunized adults; immunocompromised individuals	All children, adults who need boosters, vaccine of choice during outbreaks (faster immunologic response)
Method of administration	Parenteral	Oral
Side effects	Guillain-Barré	Risk of developing clinical polio

it is given orally. It is also the vaccine of choice during outbreaks because it induces a more rapid immunologic response. The correlation between immunization history and presence of serum antibody is high; in one study, 80 percent of those who stated that they were immunized had neutralizing antibody to all three polio serotypes. Because it is a live attenuated viral preparation, however, it carries with it the possible risk of causing clinical disease either to the recipient or to close contacts of the recipient. Nonimmune parents of recently immunized children, for example, have developed paralytic polio on rare occasions. Therefore, OPV is contraindicated in adults, who are more prone to developing clinical disease than children, and in individuals who lack immunocompetence. Such individuals should receive the IPV.

Immunizations for Special Populations

In geographic areas where certain diseases are endemic, or in groups of high-risk or exposed individuals, other immunizing agents are also used (Tables 4-8 and 27-4). Agents used under special circumstances include influenza, pneumococcal, hemophilus, meningococcal, and hepatitis B vaccines, as well as a number of passive immunizing products. Influenza vaccine is usually bivalent (types A and B) and is adjusted each year to the most current strains anticipated. Influenza A is classified by two antigens: hemagglutination (H) and neuraminidase (N). There are four subtypes of H (H0-H3) and two of N (N-1, N-2). Immunity to these antigens decreases the chance of developing disease or decreases the severity of symptoms. However, anti-

TABLE 27-4. Uses of Vaccines

Everyone	*Only travellers to endemic areas*	*Only select high-risk groups*	*Only after exposure to disease (passive immunization)*
Diphtheria	Cholera	*H. influenzae*, B*	Antitoxins
Pertussis	Plague	Influenza	Botulism
Tetanus	Typhoid fever	Meningococci, A, C	Diphtheria
Polio	Typhus	Pneumococci	Tetanus
Measles	Yellow fever	Rabies	Immune globulins
Mumps		Rocky Mountain spotted fever	Immune serum globulin
Rubella		Smallpox	Hepatitis B immune globulin
		Varicella*	Vaccinia
			Varicella-zoster

*Still undergoing clinical trials

genic variation, called antigenic drift, occurs in the same subtype over time, so that immunity to a previously prevalent strain may not result in immunity to newer "drifted" strains. The same phenomenon occurs to a lesser extent in influenza B. When researchers develop a vaccine, they use the antigenic composition of the most current strain.

Only individuals at particular risk for serious sequelae from influenza should be immunized. Those for whom immunization is recommended include people with acquired or congenital heart disease, compromised pulmonary function, chronic renal disease, diabetes mellitus or other metabolic disease, anemia, immunologic incompetence, and those over age 65. Side effects of the vaccine include fever, malaise, allergic responses, and Guillain-Barré syndrome (ascending paralysis) in about 1 per 100,000 people. Its efficacy is 70 to 90 percent for 1 to 1 1/2 years. Whole virus vaccine (attenuated) is more immunogenic but causes more side effects. A split virus vaccine (killed) has fewer side effects, though Guillain-Barré can occur after either type.

There are three other organisms that cause extensive morbidity and mortality among the very young and the very old: *H. influenzae*, the meningococcus, and pneumococcus. Vaccines of varying efficacy are available for each. *H. influenzae* is the commonest cause of bacterial meningitis in preschoolers (1 in 1000). Unfortunately, the vaccine for *H. influenzae* type B does not stimulate protective antibody levels in children less than age 2 years. The vaccine has been used prophylactically among close contacts of cases, but at this point, use of rifampin is the preferred treatment.

Meningococcal vaccines for types A and C are 90 percent effective; type B vaccine is being developed. These vaccines are used for small, high-risk groups during outbreaks. Type A vaccine is effective from age 3 months and side effects are minimal. Type C vaccine does not produce an adequate antibody response until age 18 to 24 months. The duration of immunity is at least 2 years.

There is a pneumococcal polysaccharide vaccine effective against the 14 most common types of *S. pneumoniae* that cause clinical disease. Duration of immunity is at least 2 years and immunity is 80 percent effective. Reactions are minor. Widespread use of the vaccine is not practical but it is useful for high-risk groups such as those who are debilitated, bedridden, or suffering from chronic lung disease.

In Japan a live, attenuated varicella (chickenpox) vaccine has been developed. Even though the vaccine is efficacious, it is not recommended for use with normal children until more is known about characteristics of the varicella-zoster virus, especially in terms of latency and reactivation. It is possible, too, that the vaccine could be hazardous to immunocompromised individuals, for whom passive immunization would be preferred.

The hepatitis B vaccine was approved in late 1981. It is a killed virus vaccine made from hepatitis B surface antigen (HBsAG) from plasma of human carriers. It is recommended for immunization prior to exposure of high-risk groups such as health care personnel working in dialysis units or blood centers, drug abusers, or male homosexuals. The immune response induced lasts for at least a year. It is not recommended for use after exposure to hepatitis B since antibody formation takes 7 to 10 days; rather immune serum globulin (ISG) or hepatitis B immune globulin (HBIG) should be used (see below). The only side effects noted to date are soreness and redness at the injection site.

Passive Immunization

It is possible to minimize the clinical manifestations of certain diseases by administration to an individual, as soon as possible after exposure, serum globulin that contains preformed antibody to the infectious agent. This is commonly practiced for botulism, diphtheria, hepatitis, rabies, tetanus, vaccinia, and varicella. Use of antitoxins that are made from horse serum can result in allergic reactions to the protein material contained in the serum. Reactions may be of the immediate anaphylactic type or slower developing serum sickness, which causes rash, fever, and itching and occurs 5 to 14 days after administration. Treatment for serum sickness is symptomatic, usually with use of antihistamines. Precautions to be taken to prevent such reactions are described under "Clinical Implications."

Botulism antitoxin is bivalent for types A and B. If *C. botulinum* type E is the cause of disease, a monovalent (type E) or trivalent (types A, B, E) antitoxin is available from CDC. Human immune serum globulin (ISG) is derived from the pooled plasma of many donors that has been processed to contain a known quantity of gamma globulin per milliliter. It is used to provide antibodies that can modify certain diseases such as measles (rubeola), rubella, chickenpox, and hepatitis. It is given intramuscularly and only rarely causes hypersensitivity reactions (much less so than antitoxin that is developed from horse serum). ISG should always be given as soon as possible after exposure; time delays decrease effectiveness of the protection. Hepatitis B immune globulin (HBIG) is also available. Reports of its efficacy are variable, but it is the preferred agent after exposure to hepatitis B.

For rabies, both the vaccine that induces an active immune response in 7 to 10 days, and globulin that provides short-term (half-life about 3 weeks) passive protection should be used after exposure (Table 27-5). Rabies has developed after antirabies vaccination, but the vaccine does seem to have some efficacy. It is essential with rabies immunization that the series of injections be completed and that antibody titers be obtained, especially after use of the duck embryo vaccine (DEV), to ensure effectiveness.

TABLE 27-5. Rabies Immunizing Products

Type	Product	How supplied	Comments
Vaccine	Duck embryo vaccine (DEV)	1-ml single dose vials with diluent	Not evaluated in clinical trials. Appears to be efficacious, though rabies has occurred after its use. 85-90% of those vaccinated develop adequate antibody titer
	Human diploid cell rabies vaccine (HDCV)	1-ml single dose vials with diluent	Probably more efficacious with fewer side effects than DEV. Average peak antibody titer is 10 times higher than those seen with DEV
Globulin	Antirabies serum, equine (ARS)	1000-IU vial (about 5 ml)	From hyperimmunized horses. Serum sickness occurs in about 50% of adult recipients
	Rabies immune globulin (RIG)	Pediatrics: 2-ml (300-IU) vials Adults: 10-ml (1500-IU) vials	From hyperimmunized humans. Adverse reactions are very rare, much preferred over ARS if available

Varicella-zoster immune globulin (VZIG), licensed in January 1981, is useful for both shingles and varicella. Its use is limited to immunocompromised children (including newborns with mothers who had varicella 4 days before to 2 days after delivery), with intimate exposure in the previous 3 days and no known history of varicella (Table 27-6). Vaccinia immune globulin (VIG) is used for adverse effects of smallpox vaccine such as accidental inoculation of the eye or eczema vaccinatum. The potential for new immunizing agents is great and much research is occurring in that field (Table 27-7).

FOOD SANITATION

Food-borne infectious diseases are unique in that they are almost completely preventable with adequate hygiene. A food-borne outbreak occurs when two or more people experience a similar illness after eating a common food and evidence implicates the food as the source (exceptions are botulism and chemical intoxication, cases in which one victim constitutes an outbreak). About 400 outbreaks involving about 13,500 cases are reported each year to CDC and this is certainly just a minority of cases, since many such outbreaks go unidentified or unreported.

The etiologic agent is identified in only a little more than one-third of food-borne outbreaks. During the 1970s 67 percent of the outbreaks were from unknown cause; of those with known cause, 66 percent were bacterial, 24 percent were chemical, 8 percent were parasitic, and 3 percent were viral. Many of the "unknown cause" epidemics could well have been due to viruses, for which isolation techniques are still cumbersome, slow, or inadequate.

Obviously, food can be contaminated (and

TABLE 27-6. Indications and Guidelines for the Use of Varicella-Zoster Immune Globulin (VZIG) for the Prophylaxis of Chickenpox (Varicella)

1. One of the following underlying illnesses or conditions
   a. Leukemia or lymphoma
   b. Congenital or acquired immunodeficiency
   c. Under immunosuppressive treatment
   d. Newborn of mother who had onset of chickenpox < 5 days before delivery or within 48 hours after delivery
2. One of the following types of exposure to chickenpox or zoster patients(s)
   a. Household contact
   b. Playmate contact (>1 hour play indoors)
   c. Hospital contact (in same 2- to 4-bed room or adjacent beds in a large ward)
   d. Newborn contact (newborn of mother who had onset of chickenpox <5 days before delivery or within 48 hours after delivery)
3. Negative or unknown prior history of chickenpox (see text)
4. Age of <15 years, with administration to older patients on an *individual* basis (see text)
5. Time elapsed after exposure is such that VZIG can be administered within 96 hours

Reprinted from Centers for Disease Control. Varicella-zoster immune globulin. Morb Mort Week Rep 1981;30:22.

TABLE 27-7. Potential New Immunizing Agents

Bacterial	*Viral*
Cholera toxoid	Adenovirus
E. coli enterotoxoids	Cytomegalovirus
Gonococci	Hepatitis B
Pseudomonas and other gram-negative bacteria	Herpes simplex virus, types 1 and 2
S. mutans (for caries)	Japanese B encephalomyelitis
Syphilis	Respiratory syncytial virus
Trachoma *(Chlamydia trachomatis)*	Russian spring-summer encephalitis
	Varicella-zoster
	Venezuelan equine encephalomyelitis
	Western equine encephalomyelitis

usually is) without causing clinical disease. The outcome (whether or not clinical disease results) of ingesting contaminated food depends upon a number of host and disease agent factors that are summarized in Table 27-8. Some organisms, being more pathogenic, are frequently causes of food-borne disease. The three most common are *Salmonella* (which causes about one-third of reported outbreaks), *S. aureus,* and *C. perfringens*. Other agents include *Shigella,* enteropathic *E. coli, B. cereus, V. parahaemolyticus,* and hepatitis A. Their vehicles are listed in Table 27-9.

Bacterial outbreaks occur more often in the summer. However, botulism is more common in the fall, presumably because more canned foods are eaten then. Trichinosis is most prevalent in the winter. Most outbreaks of food poisoning (about 70 percent) are due to improper food handling (Table 27-10), especially improper holding temperatures or inadequate cooking. Of reported outbreaks, about 70 percent occur in restaurants, 20 percent in homes, and 3 percent are traced to commercial food processing companies. Meat and poultry are implicated in almost half of outbreaks; seafood, sandwiches, salad, and dairy products are also common vehicles.

Salmonellosis is usually due to contaminated meat or poultry and can be controlled by keeping animals and feed disease-free, by minimizing animals' stay in slaughterhouses to no more than 2 to 3 days, by disinfecting farm water, and by proper handling and preparation of meat. *Shigella* is

(text continues on page 727)

TABLE 27-8. Factors Affecting Outcome After Ingesting Contaminated Food

Host factors	*Environmental factors*	*Disease agent factors*
Immunity to agent Degree of exposure to agent (inoculum) Consumer's threshold of tolerance and health status Cultural factors (diet, eating patterns)	Climate Season Geography Presence of vector or vehicles	Invasiveness Ability to multiply in vehicle Toxic substances produced

TABLE 27-9. Representative Food-Borne Pathogens or Toxins*

Organism	*Frequency*	*Distribution*	*Vehicles*	*Secondary factors that influence stringency of sampling*
		1. Severe hazards		
Clostridium botulinum (botulism)	Rare in areas with effective food control	Widespread	Faultily processed (canned or preserved) foods; meat products; raw and smoked fish	High mortality; rapid recognition and specific treatments may be essential for patients' survival
Salmonella typhi, S. paratyphi, and *S. cholerae suis* (typhoid and paratyphoid fevers)	Endemic in many parts of world; occasionally epidemic	Worldwide	Water; raw milk and milk products; meat products and vegetables	Low infective dose; require prolonged medical care; liable to result in carrier state (especially *S. typhi*)
Shigella dysenteriae I (Shigae) (shigellosis, shiga dysentery)	Sporadic or epidemic	Central America and Mexico, north and central Africa, Japan and Southeast Asia	Water, vegetable and salad foods	High mortality rate; frequently misdiagnosed
Vibrio cholerae (cholera)	Sporadic, endemic, and occasionally epidemic	Widely spread in Asia and Africa	Water, various foods	
Brucella melitensis (brucellosis)	Moderately rare, but occasionally endemic	Mediterranean countries	Goat milk and cheese	Convalescence often prolonged

TABLE 27-9. *(continued)*

Organism	*Frequency*	*Distribution*	*Vehicles*	*Secondary factors that influence stringency of sampling*
Clostridium perfringens, type C (enteritis necroticans)	Rare	Sporadic in Europe; Island of New Guinea	Cooked meats	
Infectious hepatitis virus†	Common	Worldwide	Water, milk, and milk products; salad foods; vegetables and shellfish	Very severe for patient with liver disease; long duration
2. Moderate hazards: potentially extensive spread				
Salmonella typhimurium and other *Salmonella* species (salmonellosis)	Common	Worldwide	Poultry and eggs; meats, wide range of other foods	Serious for the young and the old
Shigella (see above) (shigellosis, Flexner and Sonne dysentery)	Common: endemic in certain areas	Worldwide	Water, salads, fruits	Serious for the young and old: difficult to isolate organism from foods
Vibrio parahaemolyticus	Common in Japan; increasingly reported from elsewhere	Reported mainly from Japan and Southeast Asia but probably more widespread	Marine fish; crustacea	Increased risk from raw or inadequately cooked fish
Escherichia coli (enteropathogenic)	Increasingly reported from several areas	Probably worldwide	Meats, raw milk, and milk products	Serious for the young
Beta-haemolytic streptococcus	Infrequently food-borne	Widespread	Raw milk and milk products; egg salads	Certain group A types are liable to cause acute pharyngitis, nephritis, arthritis, and cardiovascular complications

TABLE 27-9. *(continued)*

Organism	*Frequency*	*Distribution*	*Vehicles*	*Secondary factors that influence stringency of sampling*
		3. Moderate hazards: limited spread		
Bacillus cereus	Increasingly reported from several areas	Probably worldwide	Reconstituted cereal products; milk, puddings, and custards; rice	
Brucella abortus	Sporadic in several parts of world	Widespread	Raw milk and cream; fresh cheese	
Clostridium perfringens	Common	Probably worldwide	Cooked meats and poultry	
Staphylococcus (enterotoxigenic) (staphylococcal enterotoxicosis or food poisoning)	Common	Worldwide	Ham, meat, and poultry products; cream-filled pastries and exposed cooked foods; sauces and dressings; milk, cheese, and egg dishes; crustacea	

*Severity of hazard reflects danger to the life of the patient; it does not refer to the overall public health problem for a population. For example, botulism exhibits a high mortality rate but affects few individuals. Salmonellosis, on the other hand, produces a much higher rate of morbidity and, though relatively less severe than botulism, results in many more deaths. Adapted from International Commission on Microbiological Specifications for Foods: *Microorganisms in food. 2. Sampling for microbiological analysis: principles and specific applications.* University of Toronto Press, 1974; reproduced by kind permission of the publisher.

†Virus not yet isolated from food.

Reprinted with permission of the publisher from World Health Organization. Microbiologic aspects of food hygiene. Geneva, Technical report series 339, 1976.

often transmitted in potato, chicken, or tuna salads that are contaminated during preparation. Another problem is the presence of R factors in the gram-negative bacteria that contaminate feed and animals, resulting from the indiscriminate use of antibacterial agents among livestock. The toxin of *C. botulinum* grows in low-acid foods such as home-canned vegetables, cheese, or fish. *S. aureus,* which only causes disease after large doses are ingested (10^6-10^9 organisms/g of food), commonly occurs in cold-cured meats

TABLE 27-10. Major Causes of Food-borne Illness

Handling and processing	Storage
Inadequate refrigeration	Food cooled too slowly before storage
Food prepared in advance or "incubating" at too low a temperature	Stored at improper temperature (should be at less than 4°C or more than 60° C)
Infected food handlers with poor hygiene	
Inadequate cooking	
Contaminated raw materials	
Cross contamination	
Inadequate cleaning of equipment	

like ham or in dairy products. It results when an infected food handler uses poor hygiene. Individuals with open skin lesions should not handle food. Brucellosis can result from ingesting raw milk.

In the past, streptococcal food poisoning frequently resulted from ingestion of raw milk, but it is now quite rare. However, an outbreak occurred in 1981 among several hundred people who were attending a meeting. Half of the food handlers were found to have either throat or hand lesions infected with beta-hemolytic streptococci, group A. At this meeting, the attack rate was about 70 percent for individuals eating one implicated meal (Centers for Disease Control 1982a). The usual pattern of streptococcal outbreaks is a high attack rate and short (1 to 4 days) incubation period. This outbreak illustrated the fact that even "rare" diseases can occur if the vigilance of food handlers wanes. Viruses such as hepatitis A and polio are spread by the fecal-oral route, but they cannot multiply in food because they require the presence of living cells. They can, however, survive for varying periods of time and be transmitted in food. Food-borne diseases are especially common with refugees and immigrants, but there is no evidence to suggest that such persons are a hazard to the general population.

Food can be intrinsically infected (e.g., trichinosis in pork, wild boar, bear, or walrus meat, *Salmonella* in poultry), or can become contaminated in the process of preparation for consumption (*S. aureus*, *Clostridium*). In either case, however, proper handling, processing, and storage can prevent human disease. The major causes of food-borne diseases are listed on Table 27-10).

Methods to Ensure Hygienic Food

The following factors are necessary to break fecal-oral transmission:

1. Hygienic food handling
2. Proper disposal of wastes
3. Fly and other vector control
4. Water purification
5. Identification and treatment of carriers
6. Proper cooking and storage of foods

Foods may be treated in several ways to eliminate or minimize microbial content: heat, radiation, refrigeration, freezing, drying or salting, and treatment with acids. These methods are often used together (extreme temperature, low pH, drying, irradiation, or curing) to be additive or synergistic in effect.

Viruses and bacteria contaminating milk are destroyed by pasteurization (68.3°C for 30 minutes or 79.4°C for 25 seconds) but boiling is required to kill some viruses as well as the spores of *C. perfringens* and *botulinum* and *B. cereus*. Milk fermentation, because of the high concentration of lactic acid, is effective in preventing growth of most harmful contaminants. Staphylococcal enterotoxins require heat for about 10 minutes at 121.1°C for destruction. They are more resistant than botulism toxin, which is readily destroyed by boiling. Fungal toxins are not destroyed by boiling or autoclaving. Microwave heat kills vegetative cells, but not spores unless the temperature exceeds 100° C.

The Food and Drug Administration (FDA) requires that all commercial producers of canned low-acid foods register and submit reports of their heat processes and that procedures be established to recall all products if necessary. Such firms are also routinely inspected. The minimum heat standard for preparation of canned goods is 2.8 minutes at 121.1°C, the highest heat resistance identified for spores of *C. botulinum* type A. Ionizing radiation at about 4.8 megarads can be used to sterilize food, but toxins cannot be inactivated by radiation.

Bacteria, viruses, toxins, and parasites can remain viable (though not active or multiplying) for long periods, sometimes indefinitely, under refrigeration. Many bacteria and parasites are destroyed by freezing, but not spores, viruses, fungi, or toxins. Drying and increased concentrations of solutes can cause bacterial cell death. Glucose, gylcerol, propylene glycol, sodium chloride, and sucrose are used for this purpose. Staphylococcal enterotoxins are not produced at a pH less than 4.5. A pH of less than 4 is lethal to most vegetative cells, and spores also die, but slowly. However, toxins are resistant to low pH.

Cured meat is required to have only one-fifth to one-tenth of the heat treatment applied to other low-acid foods because of the added protection of the salt or nitrite that is present. One problem with the curing of meat using nitrites is the formation of nitrosamines, which have been shown to be carcinogenic. Bacon is a major food in which this is of concern. The actual danger is not known; nitrates in much higher amounts than those ingested occur naturally in water, vegetables, and saliva. Nevertheless, nitrate concentrations in cured meats have recently been decreased, though the minimal level at which *C. botulinum* spores are still killed is not certain.

Adequately cooked food is always safe if eaten immediately unless preformed heat-stable toxin is present. If storage of food is necessary, temperature control is essential. Most organisms will not survive temperatures less than 4°C or greater than 60°C. Cooling, however, must be rapid rather than slow to avoid microbial growth before the temperature reaches an acceptable level.

There are certain situations in which food-borne outbreaks are a special hazard. One such situation occurs on airplanes and ships where *S. aureus, C. perfringens, Salmonella, V. parahaemolyticus, Shigella* have all caused epidemics. It is recommended that crew members, especially pilots, be served several different meals to ensure that not all get sick at the same time.

Institutions with communal or close living quarters are frequent sites of food-borne outbreaks. Day care centers, for example, are the sites of 9 to 12 percent of hepatitis A outbreaks. These outbreaks have three common characteristics:

1. Affected children are often asymptomatic.
2. Household contacts of these children are usually the first recognized cases.
3. Diapered children under the age of 2 years are usually the ones who transmit the disease.

After an index case, 70 percent of day care centers with children under 2 years had outbreaks whereas only 10 percent of day care centers admitting only children over 2 years had such outbreaks. Thus, staff in centers in which young children are housed must be especially cognizant of the risk of hepatitis.

Food handlers can transmit disease in several ways:

1. Inadequate handwashing, especially after contact with feces
2. Touching raw foods or equipment that previously contacted raw foods, and then handling other food
3. Coughing, sneezing, or touching nose or mouth before handling food
4. Working with exposed infected lesions (most common cause of *S. aureus* food poisoning)

Food workers must be instructed and monitored to insure that safe food handling occurs.

TABLE 27-11. Investigation of Food-borne Outbreaks

1. Verify diagnosis. If possible, obtain food samples for culture as well as patient samples (blood, feces)
2. Establish definition and criteria for a "case"
3. Verify existence of an epidemic. Compare current incidence with past levels
4. Describe time, place, person. Plot an epidemic curve or make a spot map.
   Tabulate cases by age, sex, or other relevant attributes (knowledge of disease agent will help determine "relevant")
5. Formulate and test hypotheses. Identify type of epidemic; propagated, common source, or point source.
   Compare ill and well populations regarding exposure
   Calculate food-specific attack rates:

$$\frac{\text{\# ill who ate food}}{\text{Total who ate food}} \qquad \frac{\text{\# ill who did not eat food}}{\text{Total who did not eat food}}$$

Investigation of Food-borne Outbreaks

Food-borne illnesses are diagnosed in several ways. Indicator organisms are used to detect fecal contamination in water and food; usually *E. coli* or other coliforms are used because they grow rapidly and well and warn that the food or water is unhygienic. Unfortunately, the correlation between the presence of coliforms and viruses is poor. In addition, the coliforms used as indicators are not necessarily just fecal contaminants, but also include gram-negative bacteria found in nature. Gas chromatography and spectrometry are used to identify certain bacteria within minutes. Bacterial metabolites can be detected in milk, and fluorescent antibody techniques are also used. To detect parasites in tissues, blood, or food animals, enzyme-linked immunosorbent assay (ELISA), hemagglutination, and latex-agglutination tests are available.

An outline of how to investigate a food-borne outbreak is given in Table 27-11. Such investigations are usually conducted by local or state epidemiologists. CDC also maintains an epidemic intelligence service that provides assistance with investigations. Most food-borne outbreaks occur at one point in time and are due to a common source of infection. *Common source* epidemics can occur over a period of time (e.g., a water supply intermittently contaminated); therefore, they may be more difficult to recognize. Most food-borne illness, however, is due to exposure restricted in time, causing *point source* epidemics that are more readily recognized when epidemic curves are drawn (Fig. 27-3), because all cases usually occur within one incubation period. *Propagated* epidemics can start from a common point of exposure, but then the infectious agent is spread person-to-person. Hepatitis A and influenza are examples of propagated epidemics.

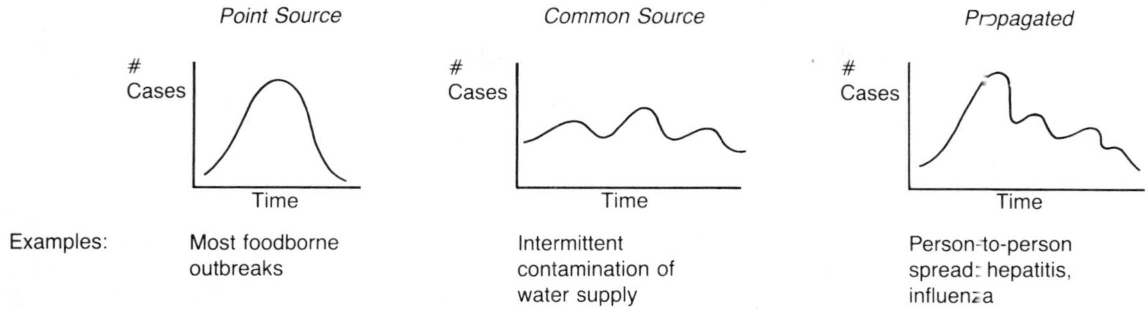

FIGURE 27-3. Types of epidemic curves.

WATER SANITATION

In 1980, 66 water-related outbreaks were reported to CDC, the highest number since surveillance began in 1971. In 44 percent of the outbreaks occurring from water meant for human consumption (50 of the 66 outbreaks) an agent was identified: *G. lamblia* (7), chemical (7), Norwalk agent (7), *Shigella* (1), *C. fetus,* subspecies *jejuni* (1), and hepatitis A (1). Most outbreaks were from community (public) systems (46 percent), with an average number of cases per outbreak (818) being 16 times that of noncommunity system outbreaks (51). Individual systems were implicated in 10 percent of outbreaks. Community systems serve 180 million people, noncommunity systems serve 20 million, and individual systems serve 30 million. The largest single cause of waterborne outbreaks was water treatment deficiencies. Water-borne gastroenteritis occurs most frequently in the summer. Four outbreaks involved hikers and road crews drinking untreated water *(Giardia* and *Campylobacter).*

Recreational areas that have heavy use during certain seasons have a particular problem supplying potable water. At Crater Lake and Yellowstone National Parks, for example, there were large water-borne outbreaks of disease in the 1970s. Thirteen outbreaks were due to use of recreational water; that is, swimming in lakes or pools.

At least 30 infectious diseases are known to be transmitted via water (a partial list is on Table 27-12). Viruses transmitted in water include hepatitis A and enteroviruses (polio, coxsackie viruses, echoviruses). Adenoviruses and reoviruses have been found to survive for 9 weeks in 20°C pond water. Vibrios are quite prevalent in waters of India. Amoebiasis, a common waterborne parasitic infection, affects about 10 percent of the world's population; in some areas the prevalence is over 80 percent. Certain parasitic worms—*Taenia saginata* (beef tapeworm) and *Ascaris lumbricoides* (roundworm)—affect over half of the population in some areas. They are often transmitted by agricultural irrigation with fecally contaminated water.

In all regions of the world it is estimated by the World Health Organizaton (WHO) that only about 15 to 20 percent of people have reasonable access to safe water (either treated or uncontaminated). In cities and towns with modern water facilities about 300 liters per day per person are used as compared with about 25 liters per day per person in developing countries. The International Drinking Water Supply and Sanitation Decade (1981 to 1990) was initiated by the United Nations General Assembly in

TABLE 27-12. Infectious Diseases that are Frequently Water-borne

Disease	*Severity*	*Ingested (fecal-oral)*	*Penetrates skin*
Amoebic dysentery	+ + +	X	
Cholera	+ + +	X	
Gastroenteritis, diarrhea	+	X	
Hepatitis A	+ +	X	
Leptospirosis	+ +	X	
Maduromycosis	+ +		X
Malaria, yellow fever (water is breeding ground for vectors)	+ + +		X
Paratyphoid	+ +	X	
Schistosomiasis	+ +		X
Sporotrichosis	+ +		X
Typhoid	+ + +	X	

+ = usually self-limited
+ + = often chronic or resulting in severe sequelae
+ + + = very serious, may be fatal

November, 1980. Two billion people currently lack adequate water supplies and waste disposal. We are extremely fortunate in developed countries to have ready access to clean drinking water.

In 1905 the first edition of Standard Methods for Examination of Water and Wastewater was published by the American Public Health Association (APHA). It is now in its fourteenth edition. Since 1914, fecal coliform counts have been used to assess the microbiologic safety of potable (drinking) water. The recommended fecal coliform count is a maximum of 1 colony per 100 ml of water. Current disinfection techniques are not designed to inactivate viruses, which are more resistant to chlorine disinfection than bacteria (Fig. 27-4). Viruses are probably much less common in water, however, since there is an estimated virus to coliform ratio of 1:65000 in feces and 1:92000 in urine. Unfortunately, the presence of viruses, *Giardia,* and parasites in water do not correlate well with coliform counts. Since animals excrete much greater numbers of fecal streptococci than humans, a fecal streptococci:coliform ratio can help determine whether contamination is human or animal. The level of turbidity, caused by suspended matter such as clay, silt, organic materials, and microorganisms is also used as a standard for potable water.

The ideal water treatment is rapid sand filtration and then chlorination. Sand filtration removes about 99 percent of microorganisms without the use of chemicals. Chlorination destroys enzymes necessary for survival of most microorganisms, controls algae and slime, and oxidizes some minerals. A certain amount of "residual chlorine" must remain, and this is often measured as a standard of purity. Ozone is used for water purification but requires much technical supervision, as does the use of heat and membrane filtration. Bottled water may also be unsterile and its quality variable.

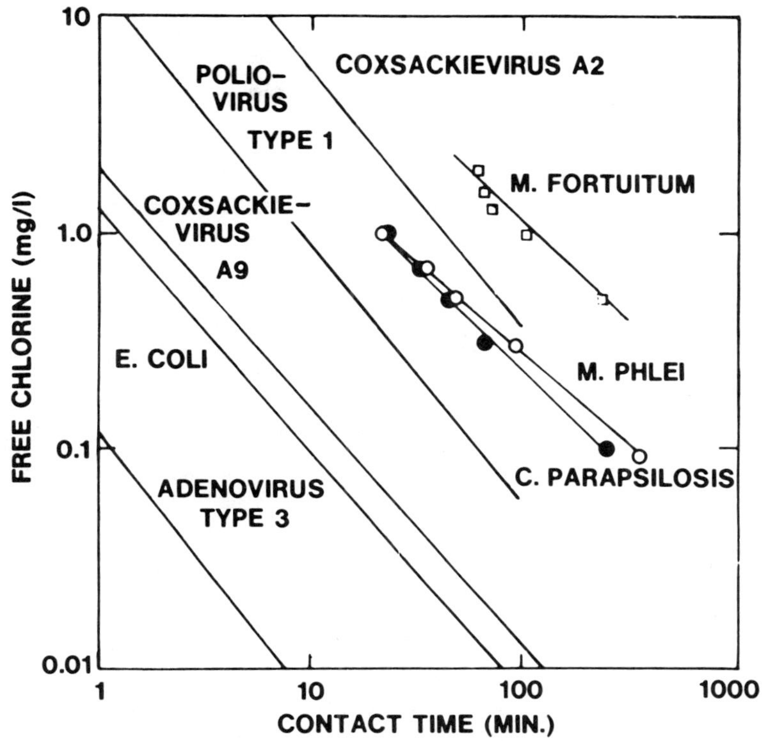

FIGURE 27-4. Free-chlorine residuals and contact times necessary for 99.9 percent inactivation for selected viruses, bacteria, and yeast. As can be seen, most viruses require longer times and higher concentrations of free chlorine for inactivation. (Reprinted with permission of the publisher from Hendricks CW, ed. Evaluation of the microbiological standards for drinking water. Washington DC: U.S. Environmental Protection Agency, 1979.)

Bacterial counts in bottled water often increase with storage.

Bacterial growth can occur in water after treatment is finished, while it is stored in reservoirs. Common organisms found in "clean," drinking water are *Klebsiella, Serratia, Proteus, Bacillus, Corynebacterium, Pseudomonas, Flavobacter,* and *Achromobacter* (Fig. 27-5). Many *Flavobacter* are resistant to chlorine. Ships have a special problem of replenishing potable water supplies and preventing cross contamination with waste water. Epidemics of water-borne disease have been reported with some frequency on cruises.

Diseases are transmitted not only in drinking water but also in water used for swimming and recreation. For example, in a recent study examining swimming pool water for presence of enteroviruses, 10 of 14

UNIVERSITY HOSPITALS • UNIVERSITY OF WASHINGTON
MICROBIOLOGY REQUEST & REPORT • GENERAL & BLOOD

Clinical Problem or Dx ______________ Transplant recipient YES

______________ Immunosuppressive YES

Antibiotics ______________

Specimen Source TAP WATER BEHIND AERATOR ON 4 East

Other Chemotherapy ______________

Other relevant info

EXAM REQUESTED C and S | REQUESTED BY NAME NUMBER | DATE & TIME OF REQUEST MO DAY YR | TIME a.m. p.m.

	PENICILLIN	AMPICILLIN	OXACILLIN METHICILLIN NAFCILLIN	CEPHALOTHIN	TETRACYCLINE	CHLORAMPHENICOL	CEFAMANDOL	KANAMYCIN	POLYMYXIN	THIOSULFIL	NITRO FURANTOIN	ERYTHROMYCIN	CLINDAMYCIN	GENTAMICIN	TOBRAMYCIN	CARBENICILLIN	TRIMETHOPRIM SULFAMETHOXAZOLE	CEFOXITIN	AMIKACIN
Acinetobacter (Herellea)		R		R	R	R		R	S		R			R	R	R	R		
Pseudomonas maltophilia		R		R	I	S		R	S		R			R	R	S	R		
Flavobacter sp.		R		R	R	R		R	R		R			I	R	R	R		

CHART COPY TEST PERFORMED AT UH ☐ HMC ☐ COH ☐ TECHNOLOGIST ______________ DATE ______________

UH A-367 REV NOV 75 UNIVERSITY HOSPITAL USE ONLY Code + + + + = Profuse + + = Scanty S = Sensitive R = Resistant
HMC 0224 NOV 75 HARBORVIEW USE ONLY + + + = Moderate + = Very Scanty I = Intermediate Sensitivity SS = Sensitive w Synergy

FIGURE 27-5. Common organisms isolated from the municipal water supply are listed in this culture taken from a faucet aerator on a hospital unit. The accumulation of such "water bugs" can be minimized by routine cleaning of such aerators. (Courtesy of Clarice Hargiss, RN, MS.)

samples were positive even in two pools in which the water exceeded the free residual chlorine standard. Viruses found included coxsackie, polio, and echo. Coliform counts were not appropriate indicators of virus, as six samples were positive for virus and negative for coliforms (Keswich, Gerba, and Goyal 1981). These pools could certainly serve as modes of viral transmission, and prior to polio immunization, swimming pools were indeed a common mode of transmitting the virus. Swimming pool safety standards, unrevised for 30 years, are probably not adequate to prevent outbreaks.

WASTE DISPOSAL

The provision of clean food and water without sanitary waste disposal would be useless, as the cycle of fecal-oral spread can begin at any point. Inadequately treated wastes, then, would result in continued contamination of clean food and water. There are several reasons why human wastes must be treated:

1. To destroy agents of disease
2. To convert wastes into reusable material (water, nutrients)
3. To prevent pollution of any large bodies of water

The simplest form of sanitary waste disposal is the pit toilet, which must be replaced every 5 to 15 years. Septic tanks are also used in many rural areas. The tanks are located beneath houses and collect household wastes from sinks, toilets, and tubs. The effluent is dispersed by underground pipes into the soil and the solids settle in the tank. Septic tanks have several disadvantages. They are relatively expensive and they require large area of permeable soil that can accept the effluent. If the liquid wastes cannot penetrate the soil they flow across the ground, spreading microorganisms. In a large urban setting septic tanks cannot adequately handle all the wastes.

Sewer systems are expensive but are very hygienic and prevalent in Western countries. A toilet requires 5 to 10 liters to flush, meaning that sewage is about 99 percent water. Hence, sewer systems are not practical where water supplies are scarce. Raw sewage is treated in waste stabilization ponds by natural processes that degrade bacteria and algae. About one-third of sewage treatment in the United States is with ponds that are low cost, simple, and provide excellent removal of bacteria. Some are mechanically aerated. Other treatment techniques are more complex, expensive, and require skilled operators. They consist of some form of initial screening, primary sedimentation, secondary treatment, treatment of sludge that is produced by anaerobic digestion, and dewatering on drying beds (Fig. 27-6).

CLINICAL IMPLICATIONS

Immunization

Though most states have immunization laws, they are often not strictly enforced except when an epidemic occurs. Such laws are of major importance in preventing the spread of disease and they also serve to educate the public. For example in a comparison of 13 low and 10 high measles prevalence areas, there were no differences in demographic characteristics of the population, vaccine utilization, or surveillance systems; however, in the low prevalence areas school immunization laws were found to be more comprehensive and more strictly enforced, with exclusion from school of noncompliant students. Periodic surveys of serum antibody levels serve to identify compliance rates, levels of herd immunity, and/or vaccine failures. Clinicians are often responsible for enhancing compliance, providing education to the public regarding immunizations, and implementing such preventive measures.

In 1977 it was estimated that 40 percent of the 50 million young people in the United States who were under 15 years of age were inadequately immunized. For that reason, the Department of Health, Education and Welfare (now called the Department of Health and Human Services) announced the Childhood Immunization Initiative to attain and maintain high levels of immunization. As a result of the efforts that followed this initiative, well over 90 percent of children newly entering school in 1980 were fully immunized. In addition, there were record decreases in the incidences of measles, mumps, polio, rubella, and tetanus by 1981. Clearly, a concerted effort to increase public compliance with immunization can be successful.

FIGURE 27-6. Sophisticated water treatment plant in Kansas City, Missouri, with aerated ponds. (Reprinted with permission of the publisher from Hendricks CW, ed. Evaluation of the microbiologic standards for drinking water. Washington DC: U.S. Environmental Protection Agency, 1979.)

Health care practitioners are also responsible for the actual administration of immunizing agents and should be aware of potential side effects. Of particular concern are the antitoxins. Antitoxins are usually produced from plasma of hyperimmunized horses and hypersensitivity to the foreign protein is common. Before administration of antitoxin, the following measures can be taken to minimize the risk of serious reactions:

1. Obtain a careful history of allergies, and asthma. Note if there was injection of serum more than 7 days previously (it takes about 7 days for antibody against the serum to develop), or previous sensitivity to horse serum or antitoxin.
2. Perform a skin test. Inject 0.1 ml intradermally of normal saline as a control and in another area inject intradermally 0.1 ml of 1:10 saline dilution of antitoxin. A positive test occurs in 10 to 30 minutes with wheals surrounded by redness. This is a contraindication to injection of horse serum, although it is still possible to have a reaction even after a negative skin test.
3. In the event of history of previous sensitivity to horse serum or antitoxin, the usual intradermal skin test may be dangerous; it may itself result in systemic reactions. In such cases
   a. Use a 1:100 dilution for intradermal injection, or

b. Do a conjunctival test. Drop 0.1 ml of 1:10 dilution of antitoxin into the lower conjunctival sac. In 10 to 15 minutes there will be redness, itching, tearing, and swollen lids if the test is positive. The eyes should be flushed with 1:1000 epinephrine. Such a reaction is also a sign that the antitoxin is contraindicated.
c. If use of antitoxin is essential and possibly life saving it will be necessary to desensitize the individual. It must be done with extreme caution. Small amounts of dilute antitoxin are given subcutaneously at short intervals (no less than 15 minutes), beginning with 1:500 to 1:100 dilutions and gradually increased as tolerated. Then the intravenous route can be attempted.

4. During use of antitoxin, epinephrine should be readily available in case of immediate anaphylactic reaction.

Sanitation

In general, public sanitation is provided for us and we are only aware of it when failures occur. It behooves us to understand something of the public health systems and how they can be accessed.

In addition, clinicians can play a major and identifiable role in the enhancement of public health practices and sanitation. Clinicians plan and implement immunization programs, collect data when outbreaks occur, and provide public education. State and local health departments and acute care institutions employ epidemiologists or infection control practitioners (ICPs) whose roles include the investigation of unusual instances of infection and determination of their cause. Most often, such institutional outbreaks are not food or water related, but sometimes they are, and such vehicles must always be ruled out. CDC provides training for clinicians in outbreak investigation.

In health care institutions policies and procedures in various departments must be developed to prevent infection. Of particular concern (in addition to the patient care units) are the kitchen or dietary department, housekeeping, laundry, and sterile supply and packaging areas. It is also important to have some mechanism to screen visitors to hospitalized patients, especially children, during peak seasons for infectious diseases such as influenza and chickenpox, and to have clear policies for isolating infected patients. Aids for differentiating isolation practices can be helpful (Table 27-13).

SUGGESTED READINGS

American Public Health Association. Standard methods for the examination of water and wastewater, 14th edition. Washington DC: American Public Health Association, 1978.

Bader M. Infection control: immunization. Top Clin Nurs 1979;1:7-21.

Bacley J. Guide to hygiene and sanitation in aviation. Geneva: World Health Organization, 1977.

Benenson AS, ed. Control of communicable diseases in man, 12th edition. Washington DC: American Public Health Association, 1975.

Brenniman GP, Rosenberg SH, Northrop RL. Microbiologic sampling variables and recreational water quality standards. Am J Publ Health 1981;71:283-289.

Brunell PA. Varicella vaccination: where do we go from here? Hosp Pract 1980;15:91-97+.

Centers for Disease Control. Hepatitis A outbreaks in a day care center. Morb Mort Week Rep 1980;29:565-567.

Centers for Disease Control. Influenza vaccine 1980-1981. Morb Mort Week Rep 1980;29:225-226.

Centers for Disease Control. Measles vaccine efficacy. Morb Mort Week Rep 1980;29:470-472.

TABLE 27-13. Variations of Isolation

Type of isolation	*Mask*	*Gown*	*Gloves*	*Thermometer*	*Visitors*	*Env. services*	*What is contaminated*
Strict (containment)	Always	Always	Always	Disposable	Limit gown, mask	Gloves gown, mask	Everything in room. Wipe stethoscope, B/P tubing, and cuff with percent phenolic. Double bag everything including B/P sleeve
Respiratory (containment)	If personnel are susceptible	Not needed	Not needed	Disposable	Visitors limited, patient wears mask	Mask	Any items soiled with patient secretions (including linen). Place mask over mouth and nose of patient for transportation
Skin and wound (containment)	During dressing changes or direct wound contact	When having direct contact with patient	When in direct contact with infected area	Disposable BTI or, if glass used, double bag and send to C.S.	Limit	Gown gloves	Patient, bed, linen, instruments used for dressing changes. Double bag dressing as contaminated waste
Enteric (containment)	Not needed	During direct contact with patient	During direct contact and handling of items contaminated with fecal material or urine	Disposable	Limit	Gown, gloves	Patient to use bedpan/urinal; use disposable dishes. Chlorine bleach used for cleaning for viral hepatitis (A or B or unknown)

Blood (containment)	If exposure expected	If exposure expected	For handling blood contaminated items			Gown, gloves	Anything contaminated with blood or serum—use gloves and disposable paper to clean gross blood; follow by using 1 percent bleach solution over area for 10 to 30 minutes
Multiresistant organisms (containment)	Not needed	For direct contact	When in contact with infected area	Disposable	Set limits prn	Gown and gloves	Patient, bed, linen, instruments, dressings, used equipment, and furniture
Neutropenic (protective; low white count) or immunosuppressed patient	Always	None	None	Disposable	Limit mask only	Mask only if cleaning when patient is in room	Wipe with phenolic and allow to dry any clean items before taking into room, i.e., wheel chairs, stretchers; patient transported with mask on

Reprinted with permission of the publisher from Infection control manual. Seattle: University of Washington Hospital, 1983.

Centers for Disease Control. Measles in high school students. Morb Mort Week Rep 1980;29:261-262.

Centers for Disease Control. Rabies prevention. Morb Mort Week Rep 1980;29:265-272+.

Centers for Disease Control. Smallpox vaccine. Morb Mort Week Rep 1980;29:417-420.

Centers for Disease Control. Waterborne illness. Morb Mort Week Rep 1980;29:585-586.

Centers for Disease Control. Multiple outbreaks of *Salmonella* associated with precooked roast beef. Morb Mort Week Rep 1981;30:569-570.

Centers for Disease Control. Outbreak of illness due to *Clostridium perfringens*. Morb Mort Week Rep 1981;30:171-172.

Centers for Disease Control. Raw milk associated illness. Morb Mort Week Rep 1981;30:90-92+.

Centers for Disease Control. Salmonellosis traced to marijuana. Morb Mort Week Rep 1981;30:77-78.

Centers for Disease Control. School immunization requirements for measles. Morb Mort Week Rep 1981;30:158-160.

Centers for Disease Control. Suspected hepatitis A in a food handler. Morb Mort Week Rep 1981;30:596-597.

Centers for Disease Control. Trichinosis associated with meat from a grizzly bear. Morb Mort Week Rep 1981;30:116-116+.

Centers for Disease Control. A food-borne outbreak of streptococcal pharyngitis. Morb Mort Week Rep 1982a;31:3-5.

Centers for Disease Control. Water-related disease outbreaks in the US—1980. Morb Mort Week Rep 1982b;30:623-634.

Cherry JD. The 'new' epidemiology of measles and rubella. Hosp Pract 1980;15:49-57.

Dudgeon JA. Immunization in times ancient and modern. J Royal Soc Med 1980;73:581-586.

Editorial staff. Immunizations today. Emerg Med 1980;12:72-77.

Editorial staff. The latest rules on rubella. Emerg Med 1981;13:119-120.

Evans AS. The need for serologic evaluation of immunization programs. Am J Epidemiol 1980;112:725-731.

Feachem R, McGarry M, Mara D. Water, wastes and health in hot climates. New York: John Wiley and Sons, 1978.

Fulginiti VA. The problems of poliovirus immunization. Hosp Pract 1980;15:61-67.

Haley CE, et al. Outbreaks of water-borne disease in the U.S., 1978. J Infect Dis 1980;141: 794-797.

Hendricks CW, ed. Evaluation of the microbiologic standards for drinking water. Washington DC: U.S. Environmental Protection Agency, 1979.

Judelsohn RG, Fleissner ML, O'Mara DJ. School-based measles outbreaks: correlation of age at immunization with risk of disease. Am J Publ Health 1980;70:1162-1165.

Keswick BH, Gerba CP, Goyal SM. Occurrence of enteroviruses in community swimming pools. Am J Publ Health 1981;71:1026-1030.

Lopez CE, et al. Water-borne giardiasis: a communitywide outbreak of disease and a high rate of asymptomatic infection. Am J Epidemiol 1980;112:495-507.

Mann JM, et al. The advent of human diploid cell rabies vaccine: impact on vaccine utilization in New Mexico. Am J Publ Health 1981;71:423-424.

Mar DD. New hepatitis B vaccine: a breakthrough in hepatitis prevention. Am J Nurs 1982;82:306-307.

Mortimer EA. Pertussis immunization: problems, perspectives, prospects. Hosp Pract 1980;15:103-107+.

Nelson JD. Infectious disease problems in pediatrics that need investigation. Ann Intern Med 1978;89 (Part 2):854-856.

Plotkin SA. Rabies vaccination in the 1980s. Hosp Pract 1980;15:65-72.

Robbins KB, Brandling-Bennett AD, Hinman AR. Low measles incidence: association with enforcement of school immunization laws. Am J Publ Health 1981;71:270-277.

Sharpe AN. Food microbiology. Springfield, Ill: Charles C Thomas, 1980.

Sour HE, Smith DG. Outbreaks of food-borne disease in the United States, 1972-1978. J Infect Dis 1980;142:122-125.

The water and sanitation decade. J Trop Med Hyg 1981;84:47-48.

World Health Organization. Microbiologic aspects of food hygiene. Geneva: World Health Organization, 1976.

IV GLOSSARY

GLOSSARY

Ancanthosis: thickening and warty growths on the skin.

Acid-fast: not decolorized by 95% ethyl alcohol with 3% hydrochloric acid. Acid-fastness is a unique characteristic of mycobacteria.

Adjuvant: a substance that enhances the antigenicity of an antigen.

Aerobe: an organism that can grow in the presence of atmospheric oxygen.

Agglutination: clumping together.

Alkaline phosphatase: a serum present in many tissues, including bone, intestine, liver, and placenta. In the absence of bone disease and pregnancy, increased levels reflect liver dysfunction. Normal serum contains 1.5 to 4.5 (Bondansky) units. Alkaline phosphatase is increased with hepatitis viral infection.

Alkylation: the replacement of a hydrogen atom in a molecule by an alkyl group is called alkylation. A compound that can effect this replacement of H atom or its addition is known as an alkylating agent.

Anaerobe: an organism that can grow in the absence of atmospheric oxygen.

Anergy: diminished reactivity to specific antigens.

Ångstrom unit: the Ångstrom unit (Å) is used to express the wavelength of light. One Ångstrom unit is equal to 10^{-7} mm (1 ten-millionth of a mm) or 1/250,000,000 of an inch.

Antibiogram: the antimicrobial susceptibility pattern of specific strains of bacteria. For example, a *S. aureus* strain resistant to penicillin, tetracycline, and streptomycin would have an antibiogram of PTS. The antibiogram is used as one epidemiologic screening tool in identifying related organisms.

Antibiotic: a biologic substance developed from a microbe that inhibits or kills some other organisms.

Antibody: a serum protein (globulin) that is formed in response to a substance that is perceived by a vertebrate animal as foreign.

Antigen: an *anti*body *gen*erating substance.

Antigenic drift: gradual and minor alteration in the antigenic constitution of a viral strain. Commonly seen in influenza viruses.

Antigenic shift: a major alteration in the antigenic composition of a viral strain, constituting an essentially new strain.

Antimicrobial: substances, either antibiotics or chemotherapeutics, used to treat microbial infections.

Antiseptic: a substance applied to the skin to kill or inhibit the growth of microorganisms.

Attack rate: number of persons affected in a specified time interval (for example, in an outbreak) divided by the total population.

Attenuated: reduced in virulence.

Bacillus: a rod-shaped bacterium.

Bacteremia: the presence of bacteria in the blood, which may or may not be symptomatic.

Bactericidal: causing bacterial death or lysis.

Bacteriocin: a heterogeneous group of substances that may be proteins, lipopolysaccharides, or phages, which are produced by some strains of gram-negative and gram-positive bacteria to inhibit the growth of other organisms.

Bacteriophage, phage, coliphage: a complex virion that parasitizes bacterial cells.

Bacteriostatic: causing inhibition of a bacteria's growth and multiplication.

Beta-lactamase: an enzyme produced by some bacteria that can split the penicillin molecule, rendering it useless. Also called penicillinase.

Binary fission: dividing into two identical parts.

Bowie-Dick test: a way to monitor the adequacy of air removal from the chamber of an autoclave. A test pack is processed by itself in an otherwise empty chamber. The pack consists of a specified number of towels. Autoclave tape is placed in the center of the pack. The pack is placed horizontal in the front of the sterilizer. After the cycle has run at 273°F (134°C) for 3 to 5 minutes the tape is examined for uniform color change. The test monitors the functioning of the vacuum system of prevacuum steam sterilizer.

Capsid: a protein covering that surrounds the viral genome or nucleic acid of the virus.

Capsomere: a unit of composition of the viral capsid.

Carrier: an individual who, for varying lengths of time, serves as a host for an organism that, though not causing disease in the host, may be transmitted and cause disease in other susceptible persons.

Castañeda stain: a stain with three reagents—phosphate, methyl alcohol and methylene blue, and safranin and acetic acid—in which *Rickettsia* and *Chlamydia* stain blue and host cellular elements stain red.

Cell: a basic unit of structure.

Chemotaxis: attraction of one substance toward another (positive chemotaxis) or away from another (negative chemotaxis).

Chemotherapeutic: a synthetic therapeutic substance that inhibits or kills certain organisms.

Chocolate agar: culture medium containing blood that has been heated to 80°C for 15 minutes to lyse the red blood cells. Useful for isolation of *Neisseria* and *Hemophilus* species.

Chromosome: a threadlike chain in the cell nucleus on which genes are located.

Cleavage: cell division, splitting of fragments.

Coccus: a round-shaped bacterium.

Codons: code words that consist of a sequence of three nucleotides that make up the genetic code. All known organisms use the same genetic series of codons for synthesis of amino acids.

Cohorting: place in groups (e.g., putting all babies born at one time in one room together, putting all individuals with a certain infection together), often for the purpose of preventing cross infections.

Coliform: a general term for facultative gram-negative rods that inhabit the intestinal tract of humans and other animals without usually causing disease.

Colonization: the establishment of an organism as part of the body's flora; does not connote disease.

Colony: a mass of microorganisms on a culture medium that contains sufficient organisms to be seen with the naked eye.

Commensal: an organism that obtains food or benefit from another without causing harm or benefit in return.

Complement: a system of proteins in the blood that helps to mediate antigen-antibody reactions and protects the body against foreign invaders.

Conjugation: a primitive form of bacterial mating that results in the exchange of genetic material.

Cytotoxic: causing damaging effects on a cell (a poison or toxin).

D-value: the decimal reduction time or the time required to destroy 90% of a group of cells. The resistance of an organism many times is expressed in terms of the D_{10} value. This is the exposure time to the lethal agent that is required to reduce a given population to 10% of the original number. It is the time required for the survivor curve to traverse one log cycle.

Darkfield examination: a special microscopic technique that does not throw light directly through a field, but rather across it, making it possible to visualize objects that reflect or refract the light. The negative image appears light; the background is dark and the visualized object is light. It is used for observing objects such as spirochetes, which are very thin and difficult to see under the ordinary light microscope.

Deoxyribonucleic acid (DNA): a polymeric molecule, the substance of which genes are composed. In all living organisms, except for some viruses, it is the basic chemical component responsible for replication and functioning of the cell. It occurs naturally as a double strand and is composed of pyrimidines and purines.

Diarrhea: watery, loose stools with no blood, minimal pus.

Dimorphism: the ability to grow in two forms (e.g., fungi that can be molds or yeasts, depending upon environmental circumstances).

Diploid: containing a double number of basic chromosomes.

Disinfectant: a substance used to rid the environment of microorganisms. It is not used in clinical therapy of infections because it does not exhibit selective toxicity.

Dormant, latent infection: infection that does not cause clinical disease; asymptomatic infection that may result in clinical disease at some future time when host defenses are lowered.

Drug fever: an allergic reaction to a drug characterized by low-grade fever and malaise; sometimes accompanied by a rash.

Dysentery: profuse diarrhea, usually containing blood or mucus, accompanied by pain.

Effluent: material that flows out; eg., sewage after purification treatment.

Endemic: distributed in a community at all times, but not always clinically recognizable.

Endogenous: originating from within the body.

Endotoxin: toxic component of the cell walls of gram-negative bacteria that can cause shock in humans.

Enteric organisms: those organisms that inhabit the gastrointestinal tracts of humans or animals; similar to coliforms, except that pathogens such as *Vibrio, Salmonella,* and *Shigella* are also included.

Envelope: membrane composed of lipid and glycoprotein; encloses some viruses.

Enzootic: disease endemic in animals.

Epidemic: any incidence of disease that is greater than the usual (endemic) and expected level.

Epistaxis: nose bleed.

Epizootic: common to or affecting many animals at the same time; an epidemic among animals.

Eukaryote: a cell with a true nucleus, cytoplasmic organelles, and capable of differentiation. Plants, animals, algae, and fungi are eukaryotic.

Exanthem: a skin eruption or rash.

Exogenous: originating from outside the individual; due to external causes.

Exotoxin: a toxin produced and released by some microorganisms.

Extrinsic incubation period: period of time required for virus to multiply in an insect vector before it is infectious to vertebrate hosts.

Flagella: organs of motility possessed by some microorganisms.

Fomites: articles that can carry microorganisms (e.g., clothing, bedding, personal items).

Gamete: a mature germ cell having a haploid set of chromosomes, able to initiate formation of a new cell by fusion with another gamete.

Gene: a unit of genetic function that determines the structure of a particular protein.

Genome: a set of chromosomes with all of an organism's genes; the nucleic acid of the viral particle.

Germicide: an agent that kills pathogenic microorganisms.

Giant cell formation: polykaryocytosis, fusion of a large number of cells.

Giemsa stain: a commercially available stain useful for identifying spirochetes, *Chlamydia,* and *Rickettsia. Rickettsia* stain purplish-blue; spirochetes stain blue; *Chlamydia* stain red.

Gram stain: the most commonly used stain; differentiates gram-positive (stain purple) and gram-negative (stain pink) bacteria. Staining reagents are crystal violet and safranin.

Guillain-Barré syndrome: a peripheral nerve disorder that causes ascending paralysis, paresthesias, and muscle weakness. May lead to death from respiratory paralysis unless ventilatory support is provided. A rare adverse effect of vaccination against some viruses (influenza) or of viral infection.

Haploid: containing half the number of chromosomes characteristic of the cell; having the gametic number of chromosomes.

Hapten: a substance that cannot serve as an antigenic stimulus unless combined with a protein or other substance, but which can react with antibody once it has been formed.

Helminths: a group of macroscopic worms, many of which parasitize humans.

Hemadsorption: a phenomenon associated with viral infection with hemagglutinating viruses that may alter the infected cell surface. Hemadsorption is detected microscopically by visualizing red blood cells adhering to the cell monolayer. Uninfected cells do not show hemadsorption.

Hemagglutinin: an antigen that causes agglutination of infected red blood cells.

Hemolysins: substances produced by microorganisms that dissolve red blood cells.

Hemolysis: The lysis (killing) of red blood cells.

Alpha: partial lysis of red blood cells. On

blood agar, alpha hemolysis is identified by a greenish pigment surrounding the hemolytic colony.

Beta: complete lysis of red blood cells. On blood agar, beta hemolysis is identified by a clear zone surrounding the hemolytic colony.

Gamma: no lysis of red blood cells; synonymous with nonhemolytic.

Hepatitis B immune globulin (HBIG):: immune globulin prepared from plasma of persons who have recently recovered from hepatitis B infection. It contains greater than 1:100,000 titer of antibody to hepatitis B surface antigen as compared to a titer of 1:1000 in NHIG (see immune globulin). The cost of HBIG is more than 20 times the cost of NHIG.

Herd immunity: protection from an infectious disease provided to susceptibles due to the fact that immune members of the same community do not spread the infectious agent.

Homology: a similarity in structure often due to common origin.

Horizontal transmission: the passing of an organism from animal to animal through the environment.

Hybridization: the process of combining offspring of two heterogeneous varieties, e.g., DNA from two viruses or from a virus and a bacteria.

Hydatid: an encapsulated, watery cyst.

Hyperkeratosis: overgrowth of horny layer of skin.

Icosahedral: bounded by 20 plane faces. One of three shapes of viruses (the others are helical or complex).

Immune globulin: sterile solution of immunoglobulins (antibodies) derived from human plasma that is free of hepatitis B surface antigen. It is also called normal human immune globulin (NHIG), immune serum globulin (ISG), or gamma globulin (gg).

Immunization: the process that results in the development of resistance to a particular infectious agent. A more general term than vaccination.

Impetigo: pyoderma, a contagious skin disease characterized by pustular eruptions; may be caused by *Staphylococcus aureus* or *Streptococcus pyogenes*.

In vitro: in the laboratory.

In vivo: in living beings.

Indigenous: growing or living naturally in a particular environment.

Infection: the establishment of an organism as part of the body's flora; does not necessarily connote clinical disease; synonymous with colonization.

Infectious disease: an infection that causes clinical signs and symptoms of impairment to the host's normal state; common signs include inflammation and fever.

Inoculate: to introduce material containing microorganisms into a human or animal *or* to introduce material suspected of containing microorganisms into a culture medium.

Inoculum: the portion of material inoculated.

L form: a bacterial mutant that is lacking a cell wall.

Lethal dose$_{50}$ (LD_{50}): the number of microorganisms required to kill 50% of susceptible animal host cells. *Or* the amount of an agent (e.g., radiation or an antimicrobial drug) which is sufficient to kill 50% of microorganisms present.

Lysis: disruption of a cell, resulting in death.

Lysogeny: the process, controlled by a phage integration gene, by which phage nucleic acid integrates into host cell chromosome. The host appears normal and is immune to superinfection by the same or closely related phages. Such an integrated phage is termed lysogenic.

Macule: a blotch or spot on the skin.

Meiosis: nuclear changes that result in gametes with half the number of chromosomes present as were in the original cell.

Microaerophilic: ability to grow at subatmospheric levels of oxygen.

Mitogenic: inducing cell division.

Minimum bactericidal concentration (MBC) or minimum lethal concentration (MLC): the concentration of antimicrobial that kills 99.9% of a bacterial inoculum under specified laboratory conditions.

Mold: the form of fungus that produces mycelia and hyphae.

Mononucleosis: the presence in the blood of an abnormal number of leukocytes.

Morbilliform: resembling measles.

Mutation: an alteration in genetic composition.

Mutualism: a relationship between a microorganism and host which is beneficial to both.

Mycetoma: a mass or tumor caused by a fungus, most often in lungs or skin.

Mycosis: an infectious disease caused by a fungus.

Mycotoxin: a poison produced by a fungus which causes disease when ingested (e.g., poisonous mushrooms).

"Negative" strand genome: RNA that has the opposite polarity of messenger RNA, thus requiring an enzyme, transcriptase, to initiate transcription; common to orthomyxoviruses, arenaviruses, bunyaviruses and rhabdoviruses. Negative-strand RNA is noninfectious.

Normal flora: organisms often found living indigenously on or in healthy individuals.

Nosocomial infection: clinical infection manifested during institutional stay, not present or known to be incubating on admission; may not necessarily mean hospital acquired (for example, a newborn who develops chlamydial conjunctivitis has a nosocomial infection even though the organism was acquired from the mother's genitourinary tract).

Nucleocapsid: the combined viral nucleic acid and viral capsid.

Nucleotides: intracellular molecules that function as precursors to DNA and RNA, as energy sources, as regulatory signals, and as components of certain coenzymes. Includes pyrimidines (major bases are cytosine, thymine, and uracil) and purines (major bases are adenine and guanine).

Oncogenic: organism that has the ability to change the host cell into a neoplastic cell. Also called carcinogenic.

Opportunistic pathogen: an organism that under normal conditions does not cause disease, but can do so when host defenses are not intact.

Organotropism: characteristic by which certain organisms select and localize in specific body tissues.

Pandemic: distributed globally, continental.

Papule: a pimplelike skin lesion.

Parasite: an organism that lives at the expense of another.

Pathogen: an organism capable of causing disease.

Pathognomonic: a specific sign, indication, or test of one particular disease.

Penicillinase: beta-lactamase.

Peptidoglycan: a complex substance composed of three parts: a backbone, two identical peptide side chains, and two identical peptide cross bridges. This is the layer of the bacterial cell that is responsible for its strength.

Phenotype: a type or strain of organism distinguishable by some visibly manifested characteristics rather than genetic composition.

Photosynthesis: the conversion of light energy to energy stored in a chemical compound within the cell. Carbohydrates are formed from water and carbon dioxide and oxygen is released.

Phylogenetic: based on natural evolutionary relationships.

Pili: Short, hairlike protrusions from some bacterial cells; fimbriae.

Pinocytosis (viroprexis): similar to phagocytosis, except the process occurs with viruses. The plasma membrane invaginates and virions are engulfed into host cytoplasm. It is the most common mode of penetration among viruses infecting humans.

Plaque: an area on a culture plate in which there is no bacterial growth as a result of destruction of the bacteria by bacteriophage.

Pleomorphic: varying in form or shape.

Plasmid: an extrachromosomal piece of genetic material transmitted by conjugation.

Polymerase: one of at least ten enzymatic proteins involved in DNA replication. There are three types: I, II, III. Other important enzymes include polynucleotide ligases, endonucleases, and helix-stabilizing proteins.

"Positive" strand genome: RNA that serves as messenger RNA to specify information for protein synthesis; common to picornaviruses and togaviruses. Positive-strand RNA is infectious.

Potable: suitable for drinking (refers to water).

Prodrome: a premonitory symptom or sign of impending disease.

Prokaryote: a simple cell with no nuclear membrane, no membrane-bound cytoplasmic organelles, and nondifferentiated. Bacteria are prokaryotic.

Prophage: a temperate viral stage during which the viral genome is joined in the bacterial chromosome without harm to the host cell; it infers immunity to the host from superinfection with the same or related phage; a phage that results from lysogeny.

Pyogenic: causing the formation of pus or suppuration.

Quellung reaction: a precipitation test in which capsule swelling occurs. Used to visualize capsules in organisms such as *Strep. pneumoniae.*

Reiter's syndrome: a disease characterized by arthritis, urethritis, conjunctivitis, and mucocutaneous lesions. The gonococcus and *Chlamydia* have been implicated as precursors.

Replicon: an independent unit of replication that regulates DNA synthesis (a plasmid, for example).

Reticuloendothelial system (RES): more properly called the monocyte-macrophage system of cells, whose primary function is to filter and destroy invaders from the blood stream.

Reverse transcriptase: an enzyme, RNA-dependent DNA polymerase, in retroviruses that causes reversal of the flow of genetic information of synthesizing DNA on an RNA template.

Ribonucleic acid (RNA): a polymer of purine and pyrimidine ribonucleotides: adenine, cytosine, guanine, and uracil (*not* thymine as in DNA). It usually occurs as a single strand. In some viruses it is the basic substance responsible for gene replication and function. There are three main classes.

mRNA: messenger RNA; a single polyribonucleotide strand that serves as a messenger, passing out of the cell nucleus into the cytoplasm to transfer the genetic code from DNA to the protein-synthesizing mechanisms.

tRNA: transfer RNA; molecules consisting of 75 nucleotides. Encounters mRNA on surface of the ribosomes and translates the nucleotide sequence into specific polymerized amino acids.

rRNA: ribosomal RNA; the protein synthesizing material residing in the cytoplasm; gets its "assignments" via mRNA and tRNA.

Salmonella-Shigella (SS) medium: : selective culture medium containing bile salts and brilliant green; inhibits growth of gram-positive bacteria.

Sanitation: promotion of hygiene and prevention of disease by maintenance of healthy, clean conditions.

Sapronosis: arising from vegetative matter.

Saprophyte: an organism that lives on dead organic material.

Schwann cells: cells that surround the myelin sheaths of nerves.

Secondary pathogen: an organism that causes a superinfection following an underlying primary infection.

Septicemia: the presence of microorganisms in the blood with accompanying clinical illness.

Serology: the study of antibody components of the serum.

Serotype: an organism that is identified by its antigenic composition.

SGOT or **serum glutamic-oxaloacetic transaminase:** a serum enzyme present in heart, liver, and skeletal muscle tissue; used as a measure of hepatocellular damage. Normal serum contains less than 40 units. It is increased with hepatitis virus infection.

SGPT or **serum glutamic-pyruvic transaminase:** a serum enzyme present primarily in the liver and to some extent in kidney and skeletal muscle; used as a measure of hepatocellular damage. Normal serum contains less than 40 units. It is increased with hepatitis virus infection.

Sterilization: the complete destruction of all living microorganisms.

Substrate: the base on which an organism lives.

Superinfection: a second infection caused by a different organism that occurs as a complication of a primary infection.

Syncytia formation: multinucleated giant cells caused by a major alteration in the properties of the surface membranes of the infected cells; leads to the fusion of several cells into one cytoplasmic mass. Common with poxviruses.

Synergism: the cumulative effect of two or more drugs that exceeds the sum when their individual effects are added.

Taxonomy: the science of classifying plants, animals, and microorganisms according to presumed natural relationships.

Teratogenic: causing congenital malformations.

Titer: the highest dilution of a serum or antigen that gives a positive reaction. Used to measure levels of antibodies in blood, for example.

Toxigenicity: the ability to produce toxic or poisonous products.

Total magnification: magnification of objective × magnification of eyepiece.

Toxin: poisonous substance generated in living or dead organisms.

Toxoid: a toxin that has been treated so that it is harmless, but can still induce an immune response.

Transcriptase: an enzyme, DNA-dependent RNA polymerase, which mediates transcription.

Transcription: the process by which mRNA, using DNA as a template, copies a nucleotide sequence complementary to one of the DNA strands in the chromosome.

Transduction: the process of conveying genetic material from one bacteria to another via a bacteriophage.

Transformation: process by which a cell changes from a slow quiescent to a rapidly growing state. *Also* the process by which free bacterial DNA is transmitted to another bacteria.

Translation: the process by which tRNA converts the nucleotide codons on mRNA into polypeptides that become a sequence of amino acids.

Transovarial transmission: the transmission of infectious agents through the female eggs to the offspring.

Trophozoite: a parasitic sporozoan at the stage of entering the blood cell of its host; the active, multiplying stage.

Vaccination: inoculation with cowpox virus in order to produce immunity to smallpox. Sometimes more generally used to mean immunization.

Vaccine efficacy:

$$\frac{\text{incidence rate of unprotected} - \text{incidence rate protected}}{\text{incidence rate unprotected}}$$

Vector: animate source of transmission of an infectious agent (e.g., an insect or animal).

Vegetative phage: the stage of a virus that infects bacteria when it is in the host cell (bacterium) replicating.

Vehicle: an inanimate source of transmission of an infectious agent (synonym is fomite).

Vertical transmission: the passing of an organism from mother to fetus through the placenta.

Vesicles: blisters.

Viremia: presence of viruses in the blood.

Virion: a complete virus particle (denotes structural intactness and ability to infect cells).

Waterhouse-Fredrickson syndrome: fulminant septicemia with septic shock frequently associated with meningococcemia.

Wood's lamp: a lamp that emits long-wave ultraviolet light under which certain superficial and cutaneous fungi fluoresce; a very useful tool in diagnosis of mycoses.

Yeast: the form of fungus which reproduces by budding.

Ziehl-Neelsen stain: used to identify acid-fast bacteria (mycobacteria). The specimen is heated 2 to 3 minutes in carbolfuchsin, and washed with 95% alcohol and 3% HCl. Acid-fast organisms retain the stain.

Zoonosis: arising from animals.

INDEX